CHEMISTRY

ACADEMIC PRESS, INC.
111 Fifth Avenue, New York, New York 10003

United Kingdom Edition published by
ACADEMIC PRESS, INC. (LONDON) LTD.
24/28 Oval Road, London NW1 7DD

LIBRARY OF CONGRESS CATALOG CARD NUMBER: 68-26647

PRINTED IN THE UNITED STATES OF AMERICA

THE CARBOHYDRATES

Chemistry and Biochemistry

SECOND EDITION

VOLUME IA

CONTRIBUTORS

Laurens Anderson
E. F. L. J. Anet
S. J. Angyal
J. S. Brimacombe
A. B. Foster
R. D. Guthrie
G. W. Hay
Derek Horton
L. Hough

J. K. N. Jones
Donald L. MacDonald
W. G. Overend
Ward Pigman
A. C. Richardson
Robert Schaffer
Walter A. Szarek
J. M. Webber
Melville L. Wolfrom

THE CARBOHYDRATES

Chemistry and Biochemistry

SECOND EDITION

EDITED BY

Ward Pigman

Department of Biochemistry
New York Medical College
New York, New York

Derek Horton

Department of Chemistry
The Ohio State University
Columbus, Ohio

VOLUME IA

ACADEMIC PRESS   New York and London 1972

We dedicate this work to the persons most responsible for our professional development

HORACE S. ISBELL AND THE LATE MELVILLE L. WOLFROM

CONTENTS

1. Introduction: Structure and Stereochemistry of the Monosaccharides

Ward Pigman and Derek Horton

**2.   Occurrence, Properties, and Preparation of Naturally Occurring
         Monosaccharides (Including 6-Deoxy Sugars)**

Robert Schaffer

3.   Synthesis of Monosaccharides

L. Hough and A. C. Richardson

4. Mutarotations and Actions of Acids and Bases

Ward Pigman and E. F. L. J. Anet

5. Conformations of Sugars

S. J. Angyal

6. Esters

Melville L. Wolfrom and Walter A. Szarek

7. Halogen Derivatives

Melville L. Wolfrom and Walter A. Szarek

8. Phosphates and Other Inorganic Esters

Donald L. MacDonald

9. Glycosides

W. G. Overend

10. Acyclic Derivatives

Melville L. Wolfrom

11. Cyclic Acetal Derivatives of Sugars and Alditols

A. B. Foster

12. Ethers of Sugars

J. K. N. Jones and G. W. Hay

13.   Glycosans and Anhydro Sugars

R. D. Guthrie

14.   Alditols and Derivatives

J. S. Brimacombe and J. M. Webber

15.   The Cyclitols

Laurens Anderson

LIST OF CONTRIBUTORS

Numbers in parentheses indicate the pages on which the authors' contributions begin.

Laurens Anderson (519), Department of Biochemistry, College of Agricultural and Life Sciences, University of Wisconsin, Madison, Wisconsin

E. F. L. J. Anet (165), Division of Food Preservation, C.S.I.R.O., New South Wales, Australia

S. J. Angyal (195), The University of New South Wales, Kensington, New South Wales, Australia

J. S. Brimacombe (479), Department of Chemistry, University of Dundee, Dundee, Scotland

A. B. Foster (391), Chester Beatty Research Institute, Institute of Cancer Research, Royal Cancer Hospital, Fulham Road, London, England

R. D. Guthrie (423), Department of Chemistry, University of Sussex, Falmer, Brighton, Sussex, England

G. W. Hay (403), Department of Chemistry, Queen's University, Kingston, Ontario, Canada

Derek Horton (1), Department of Chemistry, The Ohio State University, Columbus, Ohio

L. Hough (113), Department of Chemistry, Queen Elizabeth College, University of London, England

J. K. N. Jones (403), Department of Chemistry, Queen's University, Kingston, Ontario, Canada

Donald L. Macdonald (253), Department of Biochemistry and Biophysics, Oregon State University, Corvallis, Oregon

W. G. Overend (279), Birkbeck College, University of London, London, England

Ward Pigman (1, 165), Department of Biochemistry, New York Medical College, New York, New York

A. C. Richardson (113), Department of Chemistry, Queen Elizabeth College, University of London, London, England

Robert Schaffer (69), Organic Chemistry Section, Division of Analytical Chemistry, National Bureau of Standards, Washington, D.C.

Walter A. Szarek (217, 239), Department of Chemistry, Queen's University, Kingston, Ontario, Canada

J. M. Webber (479), Department of Chemistry, University of Birmingham, Edgbaston, Birmingham, England

Melville L. Wolfrom (217, 239, 355), Department of Chemistry, The Ohio State University, Columbus, Ohio

PREFACE

This edition of "The Carbohydrates" is a complete revision of the 1957 work which was based on "The Chemistry of the Carbohydrates" (1948). Because of its size, it is divided into two volumes, each in two separate parts. The considerably greater length of this edition is a reflection of the rapid growth of research in the field.

In retrospect, the previous edition has very little that needs correction, but new fields of knowledge have developed. Thus, conformational analysis has made spectacular advances with the development of nuclear magnetic resonance methods. Amino sugars and uronic acids have attained great importance because of their widespread occurrence in important biological substances. Unsaturated sugars and dicarbonyl sugars have become especially important with the development of newer methods of synthesis and structural characterization. A chapter has been added on the effects of ionizing radiations and of autoxidation reactions. Recently developed physical methods and methods of separation are described in additional chapters. The literature on nucleosides and antibiotics has expanded to the extent that these subjects have necessitated full chapters. With the discovery of transglycosylation reactions, the number of known oligosaccharides and enzymes acting on carbohydrates has greatly increased. A new chapter on the biosynthesis of sugars and complex saccharides was required to cover this rapidly growing field.

In the previous edition, the discussion of polysaccharides was reduced to two chapters because of the prior appearance of "The Polysaccharides" by Whistler and Smart. In the present edition, the original practice of having separate chapters for the main types of polysaccharides has been restored. Chapters on the rapidly growing fields of glycolipids and glycoproteins have been introduced.

The two final chapters cover the official nomenclature rules for carbohydrates and for enzymes having carbohydrates as substrates. The latter were extracted from the official report, but the names have been modified to conform as much as possible to official carbohydrate nomenclature. In the other chapters, official carbohydrate nomenclature has been used, but both old and new enzyme names are given.

As in the previous edition, the chapter authors were encouraged to be selective rather than exhaustive in their citation of the literature. The objective has been to achieve a proper balance, in correct historical perspective, between the important early papers and the more recent developments. Even in a work the size of this one, it is possible to cite only a small fraction of the total published literature on the subject, and the material selected reflects the collective judgment of the chapter authors and the editors.

For the chapters in Volume I, selective coverage of articles published through

1971 has been made. For more detailed treatment of individual subject areas, the reader is referred especially to the annual series *Advances in Carbohydrate Chemistry and Biochemistry* for authoritative, in-depth articles. For detailed listing of all articles and patents published in the carbohydrate field, both before and since the date of publication of this book, the reader should consult the Carbohydrates section of *Chemical Abstracts* (currently Section 33) and also the Cellulose and Industrial Carbohydrates sections (currently Sections 43 and 44). For new research articles on the carbohydrates, the international journal *Carbohydrate Research*, inaugurated in 1965, provides a prime source. The series of *Specialist Periodical Reports on Carbohydrate Chemistry* (Chemical Society, London) serves to catalogue, on an annual basis starting in 1967, a major proportion of the work published on the carbohydrates during each year. The *Advances*, *Carbohydrate Research*, the Carbohydrates section of *Chemical Abstracts*, and the *Specialist Periodical Reports on Carbohydrate Chemistry* serve to complement this treatise and with it provide a complete bibliographic core on the subject, including detailed foundations in the past literature, and continuing developments subsequent to the publication of this edition.

For extracting from the literature new articles on carbohydrates as they appear, a most valuable tool is the computer-assisted search of current-awareness journals (such as *Chemical Titles* produced by Chemical Abstracts Service) that are available on magnetic tape. Titles and literature citations are printed out automatically in response to a given search-profile. Search profiles for selecting titles dealing with carbohydrates have been devised [see, for example, G. G. S. Dutton and K. B. Gibney, *Carbohyd. Res.* **19**, 393 (1971)] and can be modified appropriately to reflect individual interests and emphasis.

This book is an international collaborative effort, and sixty-four authors were involved in the writing of the various chapters. They reside in Australia, the British Isles, Canada, France, Germany, Japan, and the United States. Most of the chapters were read by other workers in the field. We thank especially the following for their assistance in this way: Drs. I. Danishefsky, H. El Khadem, J. J. Fox, S. Hanessian, M. J. Harris, R. Hems, M. I. Horowitz, K. L. Loening, R. H. McCluer, D. J. Manners, F. Parrish, N. K. Richtmyer, R. Schaffer, C. Szymanski, R. Stuart Tipson, and the late M. L. Wolfrom.

We owe special appreciation for the help of Drs. Anthony Herp, Hewitt Fletcher, Jr., and Leonard T. Capell. Dr. Herp acted as a co-editor in Volume II and translated or rewrote several chapters, Dr. Fletcher read all of the galley proofs, and Dr. Capell was responsible for the indexes, both important and onerous tasks.

Our own institutions, New York Medical College and The Ohio State University, gave important support and encouragement to us in the preparation of these volumes. Academic Press gave the expected hearty cooperation.

CONTENTS OF OTHER VOLUMES

ERRATA FOR VOLUME IIB

Page 697, *below heading,* *IX. BIOSYNTHESIS.* Insert (This section was written by W. Pigman)

Page 708, *reference* 362. This reference should read: Michael Weiss, "An Investigation of the Primary Structure of the Protein Core of Bovine Submaxillary Mucin," New York Medical College, 1970 (Advisor, Ward Pigman), *Diss. Abstr. Intern.* B (Ann Arbor) **32**, No. 7 72–305 (1972).

1. INTRODUCTION: STRUCTURE AND STEREOCHEMISTRY OF THE MONOSACCHARIDES

WARD PIGMAN AND DEREK HORTON

I. GENERAL RELATIONS

The carbohydrates comprise one of the major groups of naturally occurring organic materials. They are the basis of many important industries or segments of industries, including manufacture of sugar and sugar products, D-glucose and starch products, paper and wood pulp, textile fibers, plastics, foods and food processing, fermentation, and, to a less-developed extent, pharmaceuticals, drugs, vitamins, and specialty chemicals.

1

They are of special significance in plants, the dry substance of which is usually composed of 50 to 80% of carbohydrates. For plants, the structural material is mainly cellulose and the related hemicelluloses, accompanied by smaller proportions of a phenolic polymer (lignin). Lower but important proportions of starch, pectins, and sugars, especially sucrose and D-glucose, are also constituents and are obtained commercially from plants. Many noncarbohydrate organic compounds are found conjugated with sugars in the form of glycosides.

For the higher animals, the principal structural material is protein, not carbohydrate, and frequently the animal carbohydrates are found in loose or firm combination with proteins, as well as with other materials. The amorphous "ground-substance" between cells is composed to a considerable extent of the polysaccharide hyaluronic acid. Other important carbohydrates or carbohydrate–protein complexes are the D-glucose of blood and tissue fluids, particularly the glycogen of liver and muscle, the immuno and blood-group substances, mucopolysaccharides (glycosaminoglycans) of connective tissue, glycolipids, blood glycoproteins, and the mucins. Many foreign substances are removed from the body through the intermediacy of the formation of glycosides or 1-esters of D-glucuronic acid. Of special importance to animals are 2-amino-2-deoxy-D-glucose (glucosamine) and 2-amino-2-deoxy-D-galactose (galactosamine). In the arthropods, a major constituent of the exoskeleton (shell) is chitin, a polymer of 2-acetamido-2-deoxy-D-glucose. Fats, of both animal and plant origin, are fatty acid esters of glycerol, a sugar alcohol.

As far as is known, the carbohydrates in all living cells are the central source for the supply of energy needed for mechanical work and chemical reactions. Phosphoric esters of the sugars are important in these transformations, and such carbohydrate derivatives as adenosine triphosphate and its relatives are key substances in storage and transfer of energy. Related, but polymeric carbohydrate derivatives—namely, the nucleic acids—control the biosynthesis of proteins and the transfer of genetic information. Even more basically, the light energy from the sun is trapped by plants by a mechanism involving chlorophyll, and is rapidly stored by conversion into carbohydrate derivatives, sugars, and hydroxy acids. The man-made chemical classifications of cell components have only formal significance in biological systems, and, through common intermediates such as pyruvate, serine, and acetate, the proteins, fats, and carbohydrates are interconvertible.

II. SOME DEFINITIONS

Although the term "carbohydrate" cannot be defined with exactitude, there is value in an examination of its significance and that of related, commonly used terms. The carbohydrates comprise several homologous series characterized by a plurality of hydroxyl groups and one or more functional groups, particularly aldehyde or ketone groups, usually in the hemiacetal or acetal forms. Natural polymers of these products, having acetal linkages joining the component residues, are a very important portion of the carbohydrate group and are known as oligo- and polysaccharides.

An oversimplified but possibly acceptable definition of the carbohydrates is that they are composed of the polyhydroxy aldehydes, ketones, alcohols, acids, their simple derivatives, and their polymers having polymeric linkages of the acetal type. The full-fledged nonpolymeric carbohydrates are the four-, five-, six-, and higher-carbon members of the several homologous series, having at least one asymmetric carbon atom. With progressively fewer carbon atoms, the carbohydrate characteristics of the compounds degenerate until the atypical one- and two-carbon compounds, like methanol, formaldehyde, ethanol, acetaldehyde, and acetic acid, are reached.

Although they constitute only one representative of the type, the sugars have often been considered to be the typical carbohydrates. The sugars (or saccharides) are the monosaccharides and their lower oligomers (the oligosaccharides). The monosaccharides are polyhydroxy aldehydes (1) and ketones (4), and they usually exist in an inner hemiacetal form (2 or 3). The oligosaccharides (7) contain relatively few (2 to 10) monosaccharide residues connected through acetal (glycosidic) linkages. When the molecules contain

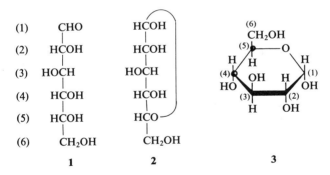

The aldohexose D-glucose in the acyclic Fischer formula (1), the Fischer–Tollens hemiacetal ring formula (2), and the Haworth formula (3).

many bound monosaccharide residues, the compounds belong to the class of polysaccharides (7). For the polysaccharides, the relatively diminished percentage of aldehyde or ketone groups enhances the behavior of the polysaccharides as polyhydric alcohols, except for the acid-labile acetal (glycosidic) linkage (see formula 7).

With the historical, and frequently practical, concept of the monosaccharides

* *References start on p. 65.*

as the fundamental units from which all carbohydrates can be derived, the term glycose has become increasingly used as a basis for class names. A glycose is any monosaccharide, and, by the addition of a suitable ending, various simple derivatives are indicated as classes, such as glycosides (**5, 6**) and aminodeoxyglycoses (**8**). The aldoses afford aldonic (glyconic) acids (**9**), aldaric (glycaric) acids (**11**), and alditols (glycitols) (**12**).

Usually, monosaccharides are further classified according to the number of carbon atoms in the central chain of the molecule and to the type of potential

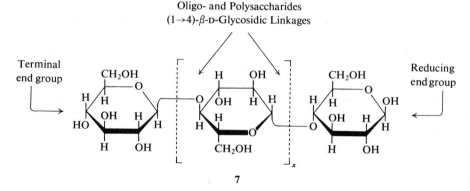

A 2-hexulose or 2-ketohexose, D-fructose (**4**). A glycoside having the common ring formula of a mixed, full acetal (**5**) and the Haworth acetal ring formula (**6**); R is an alkyl or aryl group.

Oligo- and Polysaccharides
$(1 \rightarrow 4)$-β-D-Glycosidic Linkages

7

Polysaccharide: x is large (greater than 8 and usually 100 to 2000).
Oligosaccharides: x is small (0 to 8).
  Disaccharide (biose), $x = 0$
  Trisaccharide (triose), $x = 1$
  Tetrasaccharide (tetraose), $x = 2$
  Pentasaccharide (pentaose), $x = 3$

Acids, Alditols, and Amino Sugars

(1)	CHO	CO$_2$H	CHO	CO$_2$H	CH$_2$OH
(2)	HCNH$_2$	HCOH	HCOH	HCOH	HCOH
(3)	HOCH	HOCH	HOCH	HOCH	HOCH
(4)	HCOH	HCOH	HCOH	HCOH	HCOH
(5)	HCOH	HCOH	HCOH	HCOH	HCOH
(6)	CH$_2$OH	CH$_2$OH	CO$_2$H	CO$_2$H	CH$_2$OH
	8	**9**	**10**	**11**	**12**

An amino sugar, 2-amino-2-deoxy-D-glucose (**8**); the aldonic acid, D-gluconic acid (**9**); the uronic acid, D-glucuronic acid (**10**); the aldaric acid, D-glucaric acid (**11**); the alditol, D-glucitol (**12**).

carbonyl group (aldehyde or ketone) present. This system gives rise to such names as aldotriose, aldotetrose, aldopentose, aldohexose, and aldoheptose. At the one- and two-carbon stage, this series converges into formaldehyde and glycolic aldehyde. The presence of a potential ketone group has been indicated by such names as ketopentose and ketohexose; the present tendency, except for some established trivial names, is to indicate the presence of a potential ketone group by the ending "ulose" in such names as pentulose, hexulose, and heptulose.

An aldehyde or ketone group, usually in the hemiacetal form (**2**), is generally the most reactive of the functional groups present. These groups, called reducing groups, are responsible for the characteristic reactions of reducing sugars. The sugars having such unsubstituted groups are called reducing sugars. Among such reactions are reduction in alkaline solution of the salts of certain heavy metals, changes of optical rotation in solution, formation of such derivatives as osazones and hydrazones, and instability to alkalies. The oligosaccharides that have a reducing group at one end of the molecule are called reducing oligosaccharides (**7**). When no free hemiacetal, aldehyde, or ketone group is present, the compound is a nonreducing oligosaccharide. Such disaccharides as maltose and lactose are reducing, whereas sucrose is nonreducing because the aldehyde and ketone groups of the component D-fructosyl and D-glucosyl groups are combined as full acetals in formation of the ring system and the (disaccharidic) glycosidic linkage. Like the oligosaccharides, the polysaccharides usually have a terminal reducing group (see formula **7**), but the relative proportion of these terminal groups is usually too small to influence the reactions greatly. However, such properties

* *References start on p. 65.*

as alkali instability may be determined by the reducing groups (see formula **7**), even though their proportion is small.

III. NOMENCLATURE

In the early development of carbohydrate chemistry, as in that of many other natural materials, special systems of nomenclature were originally devised that frequently were inconsistent within themselves or with the established nomenclature of organic compounds in general. Organized efforts have been made to systematize carbohydrate nomenclature, and a considerable area of agreement has been reached, as set out in a series of 35 rules[1] (see Vol. IIB, Chapter 46). These rules, originally developed by American and British chemists and adapted for international use, are followed as closely as possible in the present text.

IV. DEVELOPMENT OF CARBOHYDRATE CHEMISTRY[2]

Such carbohydrates as cellulose (cotton) and sucrose (cane sugar) were known to man in very early times in pure or semipure forms. Prehistoric man was acquainted with honey, a fairly pure mixture of the three sugars sucrose, D-fructose, and D-glucose. Starchy pastes were used as adhesives for papyrus sheets in the time of the Pharaohs.

The culture of sugar cane and the use of the juices as a sweetening agent appear to have originated in northeastern India. As early as A.D. 300, crystalline sugar therefrom (cane sugar) was known and used. The culture of sugar cane was extended to China about A.D. 400 and to Egypt about A.D. 640; from Egypt, the culture of the cane and use of cane sugar spread gradually over North Africa to Spain and Sicily. The introduction of sugar into North America is ascribed to Columbus, who took the plant to Santo Domingo on his second voyage. Sugar cane cannot be grown well in Europe, because it requires a tropical or semitropical climate, but cane sugar was known in Europe during the fourteenth and fifteenth centuries and was used as a then-costly sweetening agent. However, by 1600, many sugar refineries had been erected in Europe, and the use of cane sugar had become widespread.

The necessary restriction of the culture of sugar cane to tropical or semi-tropical lands stimulated the search for sweetening materials that could be obtained from plants native to the temperate region. This search led to technical development, on the European continent, of the sugar beet during the latter part of the eighteenth century and, especially, in the early years of the nineteenth, because of the continental blockade during the Napoleonic wars.

This desire to find other sweetening agents stimulated the study of known products and of new sources. Honey, grape juice, and raisins were known to contain material that crystallized under certain conditions. Marggraf in 1747 described a type of sugar occurring in raisins. Lowitz (1792) isolated a sugar from honey and determined that it was different from cane sugar (sucrose). Proust (1802) claimed that grapes contain a sugar that is different from sucrose. The action of acids on starch was shown to produce a sweet syrup from which a crystalline sugar was isolated by Kirchoff in 1811. Later workers established that the sugar contained in grapes is identical with one of those in honey, and with that in diabetic urine and in the acid hydrolyzates of starch and cellulose; it was given the name glucose by Dumas (1838) and of dextrose by Kekulé (1866). Emil Fischer revived the name glucose, and it is now used generally in scientific work.

The presence in honey of a syrupy sugar different from glucose and sucrose was recognized by many early workers; the crystalline material was prepared first by Jungfleisch and Lefranc in 1881. The name levulose seems to have been applied first by Berthelot (1860); Emil Fischer (1890) suggested the name fructose for this sugar.

Because of their ease of isolation and purification, sucrose, lactose (milk sugar), starch, cotton cellulose, D-glucose, and D-fructose were among the first to be studied, and the empirical composition of each was found to correspond approximately to that calculated for the general formula $C_n(H_2O)_x$. Because structural chemistry, and the existence of hydroxyl groups and hydrogen as structural elements, were unknown at the time, the substances were quite naturally looked upon as compounds of carbon with water and were termed carbohydrates (French, *hydrates de carbone*).

It was soon learned that acid hydrolysis converted starch and cellulose, $[C_6(H_2O)_5]_x$, into D-glucose, $C_6(H_2O)_6$, with the uptake of one molecule of water per C_6 unit. Cane sugar, $C_{12}(H_2O)_{11}$, took up one molecule of water to give two $C_6(H_2O)_6$ sugars (hexoses)—namely, D-glucose and D-fructose. Lactose, another $C_{12}(H_2O)_{11}$ compound, gave D-glucose and D-galactose, both $C_6(H_2O)_6$. Hydrolysis of cherry gum yielded L-arabinose, $C_5(H_2O)_5$, a pentose. Another C_6 sugar, L-sorbose, was discovered in an old, fermented sample of sorb-apple juice. Further work showed that L-arabinose, D-glucose, and D-galactose are polyhydroxy aldehydes (aldoses), whereas D-fructose and L-sorbose are polyhydroxy ketones (ketoses). Somewhat later, a third C_6 aldose (aldohexose), D-mannose, was synthesized from D-mannitol and was subsequently found in Nature. The actual structures of the three natural C_6 aldoses were unknown, but after development of the Le Bel–Van't Hoff theory it was evident that they were stereoisomers, as all were nonbranched-chain compounds.

Meanwhile, the series of naturally occurring, homologous, nonbranched-chain polyhydric alcohols—glycol; glycerol; erythritol (C_4); arabinitol (C_5);

mannitol, galactitol ("dulcitol"), D-glucitol ("sorbitol"), iditol (C_6); and perseitol (C_7)—had been discovered and found to have the general formula $C_n(H_2O)_nH_2$ (in modern terms, $HOCH_2(CHOH)_{n-2}CH_2OH$). Erythritol and the higher members were found to be crystalline, sweet, and water-soluble. The four hexitols were known to be isomeric, but their relationship to each other and to the five natural C_6 sugars was not known until Emil Fischer's classical work in the early eighteen-nineties.

Three dibasic acids of the series $HO_2C(CHOH)_{n-2}CO_2H$ were likewise discovered very early, the C_4 L-threaric (tartaric) acid from wine lees, and the isomeric C_6 galactaric ("mucic") and D-glucaric ("saccharic") acids from the nitric acid oxidation of lactose and of cane sugar.

V. STRUCTURES OF D-GLUCOSE AND D-FRUCTOSE[3]

The gross structure of D-glucose is established by the following evidence. Dumas (1843) determined the empirical formula of the sugar to be CH_2O (when water is taken as being H_2O, and not HO, as it appears in the early work). Berthelot established the presence of a number of hydroxyl groups by the preparation of an acetate (indicated by him to be a hexaacetate) and formulated D-glucose as a hexahydric alcohol. However, as a result of additional studies (1862), D-glucose was formulated as an aldehyde-alcohol having five carbon atoms. The six-carbon nature and the various known properties of D-glucose were expressed by Fittig and by Baeyer (1868 to 1870) in the formula:

$$(HO)H_2C–CH(OH)–CH(OH)–CH(OH)–CH(OH)–CHO \text{ (Fittig, Baeyer)}$$

The Baeyer–Fittig formula was confirmed by molecular-weight determinations (B. Tollens and Mayer, 1888), by the formation of pentaacetates and other esters, and by the exhibition of many aldehyde-type reactions. Thus, reduction of the sugar produces a hexahydric alcohol (D-glucitol, "sorbitol"), and oxidation with bromine or nitric acid affords a monobasic acid (D-gluconic acid). These reactions would be anticipated from the presence of an aldehyde group. By reduction (with hydrogen iodide) of the alcohol or acid obtained from D-glucose, 2-iodohexane or hexanoic acid, respectively, is obtained. The formation of 2-iodohexane proves that the sugar has a nonbranched chain. These and many other reactions support the Baeyer–Fittig formulation of D-glucose. However, the formula does not show the stereochemical relationships of the various groups, and many reactions and properties of the sugar cannot therefore be fully expressed by it.

D-Fructose must be somewhat similarly constituted to D-glucose, as it is reduced to a mixture of two hexahydric alcohols (D-mannitol and D-glucitol). The mannitol has a nonbranched-chain structure, as shown by its conversion into 2-iodohexane by the action of hydrogen iodide. Oxidation of the sugar with nitric acid yields erythraric (*meso*-tartaric) acid (HO_2C–CHOH–CHOH–CO_2H), glycolic acid ($HOCH_2$–CO_2H), and oxalic acid, and must cause cleavage of the carbon chain. The formation of the tartaric acid and glycolic acid would be expected were a ketone group present at C-2. The existence of a ketonic group was shown by the formation of a branched-chain acid when D-fructose was treated with hydrogen cyanide; that the acid material was a seven-carbon acid was shown by Kiliani, who reduced it with hydriodic acid to 2-methylhexanoic acid. A 4-hydroxy-2-methylhexanoic acid is also formed.

VI. STEREOCHEMISTRY

A. GENERAL PRINCIPLES[4]

The sugars having the formula $C_6H_{12}O_6$ that were known in 1886 were D-glucose, D-fructose, D-galactose, and L-sorbose. In these hexoses, two types of structure were present, the glucose–galactose type having aldehyde-like structures and the fructose–sorbose type having ketone-like structures.

The occurrence of such sugars as glucose and galactose having identical gross structures presented a challenge to the chemists of the later nineteenth century to provide an explanation for the existence of isomers of a type other than structural isomers. The basis for this explanation was developed almost simultaneously by Le Bel and Van't Hoff and was published in 1874. According to these workers, isomers of a type other than structural isomers should be possible for compounds that contain asymmetrically substituted carbon atoms. This type of isomerism is illustrated for glyceraldehyde ($HOCH_2$–CHOH–CHO). Each of the two isomers is represented by a tetrahedral formula, a wedge-bond formula, and a conventional (projection) formula.

The conventional formulas are derived from the tetrahedral formulas by the use of the convention established by Fischer.[5] The tetrahedra are represented as being held so that the dotted rear edge is in the plane of the paper; the H and OH corners are above the plane of the paper, with the aldehyde group at the top. The formula by the Fischer convention represents the projection of the model onto the plane of the paper.

The two tetrahedra differ only in the configuration of the groups in space, and the compounds are thus stereoisomers. Careful examination of the above figure—or, better, of models—will show that, no matter how the tetrahedra are turned in space, they cannot be made to coincide. However, it should be noted that the two tetrahedra are related as an object and its mirror image. When two of the groups attached to the same carbon atom are identical, isomerism of this type is not possible. The presence of asymmetrically substituted carbon atoms (loosely referred to as "asymmetric carbon atoms") in organic compounds was suggested by Le Bel and Van't Hoff as the cause of the optical activity of the compounds. Compounds that contain such atoms may cause a rotation of the plane of polarization of plane-polarized light when the light is passed through their solutions.

The presence of an asymmetrically substituted carbon atom in a molecule usually gives rise to overall molecular dissymmetry. A molecule that cannot be superposed on its mirror image is termed a dissymmetric molecule, and, to be uniquely described, it requires the use of some terminology to describe the chirality ("handedness") of the molecule, unless a model or projection is used. Although asymmetrically substituted carbon atoms are the commonest cause of molecular dissymmetry, other structural features may give rise to such dissymmetry.[4] Furthermore, it is possible for a molecule to possess asymmetrically substituted carbon atoms but be symmetric overall by virtue of its possessing a plane or center of symmetry.

For each of the two aldotrioses, there are two related aldotetroses. The aldotetroses have two asymmetric carbon atoms; the formulas of the four possible isomers of the aldotetroses are given in both the tetrahedral and the Fischer projection formulas (**13–16**).

The isomeric tetroses differ in their spatial relationships and cannot be brought into coincidence by rotation of the molecules in space, even though free rotation about the bond between the tetrahedra is theoretically possible. Formulas **13** and **16** constitute a mirror-image pair of molecules; **14** and **15** represent another such pair. For the four-carbon aldoses, there are two pairs of mirror images (enantiomorphs) that are related as diastereoisomers, and, altogether, four stereoisomers. In the sugar series, compounds that differ only in the configuration of the carbon atom immediately adjacent to that carrying the carbonyl or carboxyl group are known as epimers. In the above formulas, **13** and **14** are a pair of epimers, and **15** and **16** are another pair. The definition

The structures 13 through 16 are drawn as Fischer projections:

$$
\begin{array}{cccc}
\text{CHO} & \text{CHO} & \text{CHO} & \text{CHO} \\
\text{HO–C–H} & \text{H–C–OH} & \text{HO–C–H} & \text{H–C–OH} \\
\text{H–C–OH} & \text{H–C–OH} & \text{HO–C–H} & \text{HO–C–H} \\
\text{CH}_2\text{OH} & \text{CH}_2\text{OH} & \text{CH}_2\text{OH} & \text{CH}_2\text{OH}
\end{array}
$$

$$
\begin{array}{cccc}
\text{CHO} & \text{CHO} & \text{CHO} & \text{CHO} \\
\text{HOCH} & \text{HCOH} & \text{HOCH} & \text{HCOH} \\
\text{HCOH} & \text{HCOH} & \text{HOCH} & \text{HOCH} \\
\text{CH}_2\text{OH} & \text{CH}_2\text{OH} & \text{CH}_2\text{OH} & \text{CH}_2\text{OH} \\
\textbf{13} & \textbf{14} & \textbf{15} & \textbf{16}
\end{array}
$$

of epimers has now been extended to mean any pair of stereoisomers that differ solely in the configuration of a single asymmetric carbon atom. By this definition, compounds **17** and **18** are 2-epimers, and compounds **17** and **19** are 3-epimers.

$$
\begin{array}{ccc}
\text{CHO} & \text{CHO} & \text{CHO} \\
\text{HCOH} & \text{HOCH} & \text{HCOH} \\
\text{HOCH} & \text{HOCH} & \text{HCOH} \\
\text{HCOH} & \text{HCOH} & \text{HCOH} \\
\text{CH}_2\text{OH} & \text{CH}_2\text{OH} & \text{CH}_2\text{OH} \\
\text{D-Xylose} & \text{D-Lyxose} & \text{D-Ribose} \\
\textbf{17} & \textbf{18} & \textbf{19}
\end{array}
$$

In general, the number of stereoisomers for a structure involving n asymmetric carbon atoms in an unlike-ended chain is given by 2^n. However, when the terminal groups in the molecule are identical, the number of isomers is given by: $2^{(n/2)-1}(2^{n/2}+1)$ when n is an even number, and by 2^{n-1} when n is an odd number. Thus, for the tetraric (tartaric) acids ($HO_2C–CHOH–CHOH–CO_2H$), three isomers are possible; for the pentaric (hydroxyglutaric) acids ($HO_2C–CHOH–CHOH–CHOH–CO_2H$), four isomers are possible. Fewer isomers can exist when the end groups are identical, because of the symmetries that develop, leading in certain cases to molecules having a formal plane of symmetry (*meso* compounds). Thus, in the compounds having an odd number

* *References start on p. 65.*

of asymmetric carbon atoms, the middle carbon atom has attached two groups that may or may not have the same structure. If the two groups are identical, the number of asymmetric centers is really $n - 1$. This relationship may be seen from the formulas given for the pentaric (trihydroxyglutaric) acids.

Pentaric acids

20   21   22

The isomeric tetraric (tartaric) acids

For the tetraric (tartaric) acids, which have an even number of carbon atoms, the number of isomers is decreased to three because of the symmetry of the molecule. The two formulas represented by **22** are identical. This identity may be shown by moving either of formulas **22** through 180°, keeping it in the plane of the paper. It then becomes identical with the other formula. When formula **20** is rotated through 180° in the plane of the paper, it does not become identical with either **21** or **22**. A better test is provided by inspection of the space models; it is possible to construct only three stereoisomers. Note, however, that any monosubstitution of **22** removes the *meso* symmetry, giving rise to enantiomorphs.

Compounds that contain asymmetric carbon atoms are usually dissymmetric and thus rotate the plane of polarization of plane-polarized light. For this reason, they are said to be optically active. A *meso* compound results when the molecule has a formal plane of symmetry; the molecular symmetry is such that the optical activity of one portion of the molecule is cancelled by that of the other portion of the molecule, and the compound is said to be internally compensated. The tetraric acid having formula **22** is such a compound and has been known as *meso*-tartaric acid. The tartaric acids identified as **20** and **21** had been known as *d*- or (+)-tartaric acid and *l*- or (−)-tartaric acid because of the signs of their optical rotations (*dextro* and *levo*, respectively). (The nomenclature of these acids is discussed later in this chapter.) Compounds **20** and **21** are nonsuperposable mirror images, called enantiomorphs. The existence of such dissymmetry is the fundamental basis of optical activity. The dissymmetry may be in either the molecular structure or the crystal structure, but dissymmetric crystals (such as quartz) lose their optical activity when the crystal lattice is broken down, as by fusion. In contrast, dissymmetric molecules exhibit optical activity in any of the physical states, or in solution.

Enantiomorphs (in the same polymorphic modification if in the solid state) are identical in most of their properties, including melting point, infrared and n.m.r. spectra, solubility, and chemical reactivity. However, when compared by an independent chiral reference, as by another dissymmetric molecule or by polarized light, they are markedly different. This behavior is especially pronounced in biological systems, because enzymes are also dissymmetric molecules. Frequently, in biological systems, one enantiomorph is handled quite differently from the other. D-Glucose is readily metabolized by man, whereas its mirror image, L-glucose, is not utilizable.

A mixture of equal amounts of the tartaric acids **20** and **21** is optically inactive and is termed a racemic or DL mixture. Racemic forms are always produced in the chemical synthesis of chiral molecules from optically inactive precursors, reagents, and catalysts, unless dissymmetric substances (such as enzymes) have been used in the synthesis, and subsequent racemization is avoided. Frequently, the two enantiomorphs react to form a racemic compound which has properties (such as solubility and melting point) different from those of the component isomers. In contrast, *meso*-tartaric acid (**22**) is optically inactive because it has a formal plane of symmetry.

Because of the extensive use of isotopes in the study of reaction mechanisms,

TABLE I

NUMBER OF STEREOISOMERS OF THE ALDOSES AND ALDONIC ACIDS CONTAINING TWO TO SEVEN CARBON ATOMS AND OF THE CORRESPONDING ALDITOLS AND ALDARIC ACIDS

		Number of possible forms or isomers	
Parent aldoses	Number of asymmetric carbons (n)	Aldoses (and aldonic acids) CHO(CO₂H) | (CHOH)ₙ | CH₂OH	Alditols (and aldaric acids)[a] CH₂OH(CO₂H) | (CHOH)ₙ | CH₂OH(CO₂H)
(Diose)[b]	0	1	1
Trioses	1	2	1
Tetroses	2	4	3
Pentoses	3	8	4
Hexoses	4	16	10
Heptoses	5	32	16

[a] When *n* is an odd number, the middle carbon atom of the chain may be termed pseudo-asymmetric. [b] Not strictly a sugar.

* *References start on p. 65.*

particularly biological mechanisms, the special case of molecular dissymmetry caused by attachment of two, or three, different isotopes of the same element to the same carbon atom needs to be recognized. Thus, optically active isomers of the type R_1R_2CHD are well known.[6] The existence of such enantiomorphic isomers, which are treated differently in biological systems, is of major significance in studies by the isotope-tracer technique.[7]

On the basis of the foregoing considerations, which are consequences of the Le Bel–Van't Hoff theory, the number of isomers of each of the aldoses having seven or fewer carbon atoms, and of the corresponding aldaric acids and alditols, is given in Table I.

B. Establishment of the Configuration of D-Glucose and Some Other Sugars

The existence of structurally isomeric sugars was a corollary of the Le Bel–Van't Hoff theory. After publication of the theory in the latter part of the nineteenth century, it was soon realized that such sugars as D-glucose and D-galactose are stereoisomers. In a series of brilliant researches, Emil Fischer applied the Le Bel–Van't Hoff theory to the sugar series and established the configurations of many of the individual sugars.

Fischer's proof was published in two papers that appeared[5] in 1891. His proof was expressed in the terminology and conventions of the time. Because expression of the proof in his original fashion would require a detailed explanation of the older concepts of stereochemistry, it seems better in the present discussion to use the data available to Fischer at the time and to introduce the proof in terms of modern concepts and conventions. The present discussion follows the proof of configuration as outlined[8] by C. S. Hudson and, in part, quotes him.

The following facts were available to Fischer at the time of his establishment of the configuration of D-glucose.

1. Three sugars having the formula $C_6H_{12}O_6$ (D-glucose, D-mannose, and D-fructose) react with an excess of phenylhydrazine to give the same product, D-*arabino*-hexulose phenylosazone. The reactions are illustrated in the accompanying formulas.

Carbon No.

1	CHO		HC=N—NHPh		CH$_2$OH	
2	CHOH		C=N—NHPh		CO	
3–5	(CHOH)$_3$	$\xrightarrow{\text{PhNHNH}_2}$	(CHOH)$_3$	$\xleftarrow{\text{PhNHNH}_2}$	(CHOH)$_3$	
6	CH$_2$OH		CH$_2$OH		CH$_2$OH	
	D-Glucose and D-mannose		D-*arabino*-Hexulose phenylosazone		D-Fructose	

These reactions prove that D-mannose and D-glucose are 2-epimers—that is, they differ only in the configuration at C-2; also D-fructose, D-glucose, and D-mannose must have the same configurations for carbon atoms 3, 4, and 5.

2. D-Glucose and D-mannose are oxidized by nitric acid to aldaric acids which are different and which are both optically active. The optical activity observed for the products proves that the configuration of the asymmetric centers (carbon atoms 2 to 5) cannot be of the type that produces internal compensation; that is, neither aldaric acid has a formal plane of symmetry.

$$
\begin{array}{ccccccc}
\text{CHO} & & \text{CO}_2\text{H} & & \text{CHO} & & \text{CO}_2\text{H} \\
| & & | & & | & & | \\
\text{(CHOH)}_4 & \xrightarrow{\text{HNO}_3} & \text{(CHOH)}_4 & & \text{(CHOH)}_4 & \xrightarrow{\text{HNO}_3} & \text{(CHOH)}_4 \\
| & & | & & | & & | \\
\text{CH}_2\text{OH} & & \text{CO}_2\text{H} & & \text{CH}_2\text{OH} & & \text{CO}_2\text{H} \\
\end{array}
$$

D-Glucose   D-Glucaric acid   D-Mannose   D-Mannaric acid

3. L-Arabinose, which had been isolated from beet pulp by Scheibler in 1868 and shown to be an aldopentose by Kiliani in 1887, reacts with hydrogen cyanide with the production of a nitrile which, on hydrolysis, gives a six-carbon, monobasic acid (23). This acid was shown by Fischer to be the mirror image of the acid (24) produced by the mild oxidation of D-mannose.

$$
\begin{array}{ccccccc}
 & & \text{CO}_2\text{H} & & \text{CO}_2\text{H} & & \text{CHO} \\
\text{CHO} & & | & & | & & | \\
| & \xrightarrow{\text{HCN}} & \text{CHOH} & & \text{CHOH} & \xleftarrow{\text{Br}_2} & \text{CHOH} \\
\text{(CHOH)}_3 & & | & & | & & | \\
| & & \text{(CHOH)}_3 & & \text{(CHOH)}_3 & & \text{(CHOH)}_3 \\
\text{CH}_2\text{OH} & & | & & | & & | \\
 & & \text{CH}_2\text{OH} & & \text{CH}_2\text{OH} & & \text{CH}_2\text{OH} \\
\end{array}
$$

L-Arabinose   L-Mannonic acid   D-Mannonic acid   D-Mannose

23   24

In the synthesis of L-mannonic acid (23), a second acid is also formed, which is enantiomorphic with that obtained by the oxidation of D-glucose. The aldaric acid obtained by the nitric acid oxidation of L-arabinose is also optically active.

4. D-Glucaric acid can be obtained not only by the oxidation of D-glucose, as already indicated, but also by the oxidation of another hexose, L-gulose.

* References start on p. 65.

$$
\begin{array}{ccc}
\text{CHO} & \text{CO}_2\text{H} & \text{CO}_2\text{H} \\
\text{HCOH} & \text{HCOH} & \text{HCOH} \\
\text{HOCH} & \text{HOCH} & \text{HOCH} \\
\text{HCOH} & \text{HCOH} & \text{HCOH} \\
\text{HCOH} & \text{HCOH} & \text{HCOH} \\
\text{CH}_2\text{OH} & \text{CH}_2\text{OH} & \text{CO}_2\text{H} \\
\text{D-Glucose} & \text{D-Gluconic acid} & \text{D-Glucaric acid}
\end{array}
$$

with $\xrightarrow{\text{Br}_2}$ from D-Glucose to D-Gluconic acid, and $\xrightarrow{\text{HNO}_3}$ from D-Gluconic acid to D-Glucaric acid.

$$
\begin{array}{cccc}
\text{CHO} & \text{CO}_2\text{H} & \text{CO}_2\text{H} & \text{CO}_2\text{H} \\
\text{HOCH} & \text{HOCH} & \text{HOCH} & \text{HCOH} \\
\text{HOCH} & \text{HOCH} & \text{HOCH} & \text{HOCH} \\
\text{HCOH} & \text{HCOH} & \text{HCOH} & \text{HCOH} \\
\text{HOCH} & \text{HOCH} & \text{HOCH} & \text{HCOH} \\
\text{CH}_2\text{OH} & \text{CH}_2\text{OH} & \text{CO}_2\text{H} & \text{CO}_2\text{H} \\
\text{L-Gulose} & \text{L-Gulonic acid} & & \text{D-Glucaric acid}
\end{array}
$$

with $\xrightarrow{\text{Br}_2}$ from L-Gulose to L-Gulonic acid, $\xrightarrow{\text{HNO}_3}$ from L-Gulonic acid, and $\equiv$ between the two D-Glucaric acid forms.

5. Until 1951, no method was available for the establishment of absolute configurations. Fischer's method of assignment, described next, leads finally to a choice between either of a pair of configurations that have a mirror-image relationship. Fischer's solution of this problem consisted in the arbitrary assignment to D-glucaric acid (derived from D-glucose) of one of two possible formulas. By this action, a convention was established that enabled him to make a choice between the enantiomorphic formulas for other compounds, once their genetic relationships to D-glucaric acid or D-glucose had been established. Fischer's concept, although fundamentally correct, has been somewhat modified and made more precise. (See the discussion of D and L usage later in this chapter.) In conformity with modern concepts, the convention may be expressed by placing the hydroxyl group at C-5 of D-glucose on the right side of the carbon chain (see proof that follows). According to the convention, the sugar known to Fischer as "grape sugar" is termed D-glucose; and, because the mannose and fructose studied by Fischer have the same configuration for C-5, they, too, are regarded as D.

Although of necessity purely fortuitous, Fischer's assignment of the absolute configuration of D-glucose turned out to be correct. By use of a specialized X-ray crystallographic technique, Bijvoet[9] was able to demonstrate that the absolute configuration of a salt of L-threaric [(+)-tartaric] acid was identical with that given in the arbitrary Fischer projection for this acid on the basis of unambiguous correlation with D-glucose. The chirality of all the sugars correlated directly or indirectly with D-glucose is thus established in the absolute sense.

Facts 1 through 4 were known at the time of Fischer. In conjunction with the Le Bel–Van't Hoff theory, they enabled him to select the configuration of D-glucose from those for the eight configurations possible for an aldohexose (when only one of the mirror images of each is considered). The following proof, quoted from a paper by C. S. Hudson, may be regarded as a modernized version of the Fischer proof. Hudson's nomenclature has been modernized in the quotation.

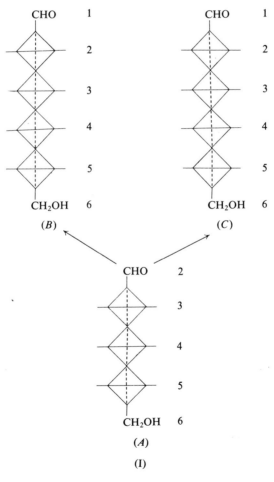

"Write the formulas for a pentose (*A*) and the two hexoses (*B* and *C*) which it yields by the Fischer–Kiliani cyanohydrin synthesis as shown in the accompanying diagram (I), using Fischer's convention that the asymmetric carbon atoms (tetrahedra) have the lower edge in the plane of the paper and the corners which carry the H and OH groups lie above this plane. The arrangement of the H and OH groups is then decided through the following steps, in which the pentose is selected to be D-arabinose and in consequence the hexoses become D-glucose and D-mannose.

"Step 1—By convention for the D-configurational series OH is on the right of C-5 (see II).

"Step 2—(*D*) is optically active; hence OH is on the left of C-3 (see II).

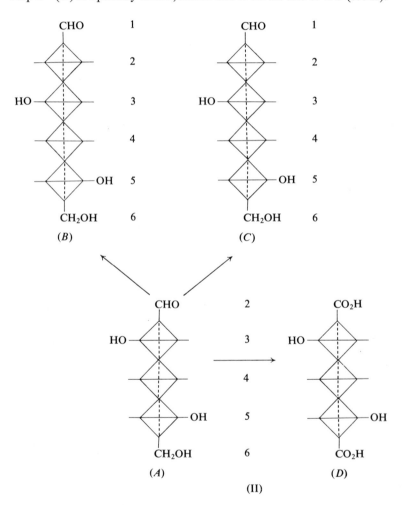

(II)

"Step 3—D-Glucose and D-mannose are epimeric, hence the OH groups on C-2 are opposed. Either (B) or (C) may be selected as having OH on the right, without changing the final result; here the OH is placed to the right of C-2 in (B) and consequently to the left in (C) (see III).

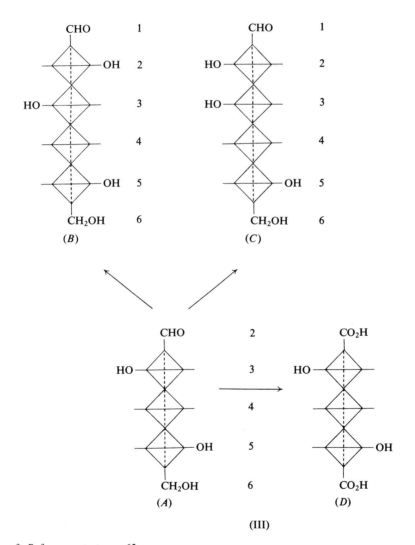

(III)

"Step 4—Since both D-glucaric and D-mannaric acids (*E* and *F*) are optically
active, the configuration of neither of them can possess end-to-end
symmetry; hence the OH on C-4 must be on the right (see IV). (If
it were on the left, (*E*) would have end-to-end symmetry.) At this
stage the configuration of D-arabinose (*A*) and its dibasic acid (*D*)
have become established. D-Glucose and D-mannose have been
limited to the configurations (*B*) and (*C*), but the correlation
within this limit remains to be established. This is done by:

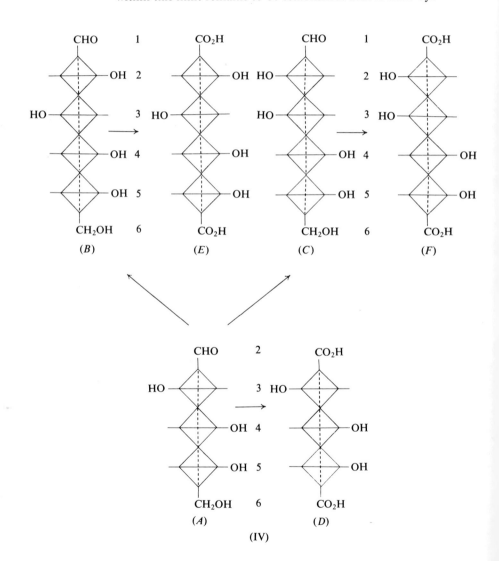

"Step 5—D-Glucaric acid is obtainable from the oxidation of each of two hexoses, namely D-glucose and L-gulose. (E) must therefore refer to D-glucaric acid because (F) cannot result from the oxidation of two hexoses. Hence (B) refers to D-glucose, (C) to D-mannose, and (F) to D-mannaric acid."

The proof is now complete and (V) the formulas become:

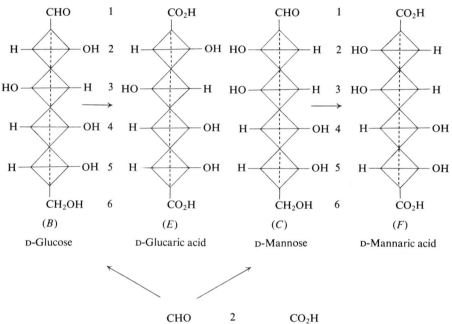

(B)
D-Glucose

(E)
D-Glucaric acid

(C)
D-Mannose

(F)
D-Mannaric acid

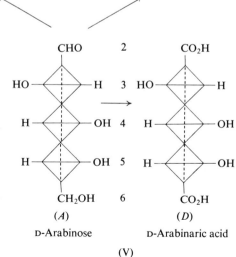

(A)
D-Arabinose

(D)
D-Arabinaric acid

(V)

By means of the Fischer convention, the tetrahedral models for D-glucose, D-mannose, and L-arabinose are equivalent to the planar formulas given. The formula for D-fructose is derived from the fact that D-fructose yields the same phenylosazone as D-glucose when treated with phenylhydrazine (see fact 1, p. 14); it and D-glucose must have identical configurations for carbon atoms 3, 4, and 5.

Carbon Atom No.

1	CHO	CHO	CH$_2$OH	CHO
2	HCOH	HOCH	CO	HOCH
3	HOCH	HOCH	HOCH	HCOH
4	HCOH	HCOH	HCOH	HCOH
5	HCOH	HCOH	HCOH	CH$_2$OH
6	CH$_2$OH	CH$_2$OH	CH$_2$OH	
	D-Glucose	D-Mannose	D-Fructose	D-Arabinose

(Fischer formulas)

C. D AND L NOMENCLATURE

In some types of optically active compound, it was once customary to distinguish between the enantiomorphic modifications by indicating the sign of their optical rotation as "*d*" (dextrorotatory) or "*l*" (levorotatory). Thus, *d*-tartaric acid (the naturally occurring form) is the isomer that has a dextro-rotation. This usage is no longer followed in carbohydrate chemistry. Fischer established the convention of calling ordinary glucose *d*-glucose and employed the prefix *d* in the configurational sense, to mean that a *d* compound is derivable from *d*-glucose, whereas an *l* compound is derivable from *l*-glucose. Hence, fructose was called *d*-fructose, although it exhibits a levorotation.

The Fischer system was modified by Rosanoff[10] to avoid certain ambiguities that had arisen. These ambiguities are illustrated by the series of transform-ations shown in the accompanying formulas. It is shown that either of the enantiomers of glucaric acid may be produced from ordinary glucose (D). Since the transformation of D-xylose (natural form) into a hexaric acid that is the mirror image of that obtained by the direct oxidation of the glucose was ob-served first, the natural xylose was originally called *l*-xylose by Fischer; had the conversion of glucose into xylose through D-glucuronic acid been observed first, the natural sugar would probably have been termed *d*-xylose.

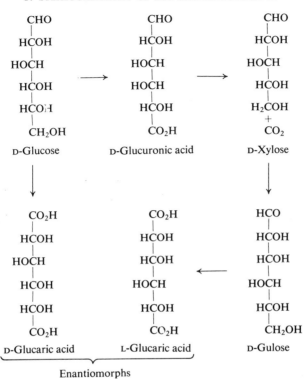

The system proposed by Rosanoff placed the use of the symbols *d* and *l* (later changed to D and L to avoid confusion) on a logical, genetic basis. His system is now universally accepted by carbohydrate chemists. It starts with the definition that the glyceraldehyde having formula **25** shall be called D-glyceraldehyde and that that having formula **26** shall be called L-glyceraldehyde.

CHO   CHO
| |
HCOH   HOCH
| |
CH₂OH   CH₂OH

D-Glyceraldehyde   L-Glyceraldehyde

25   **26**

According to Rosanoff, all the higher sugars that might conceivably be derived from D-glyceraldehyde by successive application of the cyanohydrin synthesis shall be called D sugars. Similarly, all those obtained in this way from L-glyceraldehyde shall be called L sugars.

* References start on p. 65.

$$
\begin{array}{ccccc}
 & & \text{CHO} & & \text{CHO} \\
 & & | & & | \\
 & & \text{HCOH} & & \text{HOCH} \\
\text{CHO} & & | & & | \\
| & \longrightarrow & \text{HCOH} & + & \text{HCOH} \\
\text{HCOH} & & | & & | \\
| & & \text{CH}_2\text{OH} & & \text{CH}_2\text{OH} \\
\text{CH}_2\text{OH} & & & & \\
\text{D-Glyceraldehyde} & & \text{D-Erythrose} & & \text{D-Threose}
\end{array}
$$

$$
\begin{array}{ccccc}
 & & \text{CHO} & & \text{CHO} \\
 & & | & & | \\
\text{CHO} & & \text{HOCH} & & \text{HCOH} \\
| & \longrightarrow & | & + & | \\
\text{HOCH} & & \text{HOCH} & & \text{HOCH} \\
| & & | & & | \\
\text{CH}_2\text{OH} & & \text{CH}_2\text{OH} & & \text{CH}_2\text{OH} \\
\text{L-Glyceraldehyde} & & \text{L-Erythrose} & & \text{L-Threose}
\end{array}
$$

As a new asymmetric center is produced on addition of an extra carbon atom (through the cyanohydrin synthesis), two epimers are produced from each glyceraldehyde. A continuation of this process with each of the four-carbon aldoses gives four D-aldopentoses and four L-aldopentoses; application of the cyanohydrin synthesis to the aldopentoses produces, in turn, eight D- and eight L-aldohexoses. Although this entire process has not yet been carried out experimentally, interconversions have been performed in sufficient number for the allocation of the configurations of all the possible aldoses through the aldohexose stage and for many of the higher aldoses.

In general, a compound may be defined as belonging to the D family when the asymmetric carbon atom most remote from the reference group (for example, aldehyde, keto, and carboxyl group) has the same configuration as that in D-glyceraldehyde; if this carbon atom has the same configuration as that in L-glyceraldehyde, the compound belongs to the L family. When the compound is written in a Fischer projection, with the reference group towards the top, the allocation to the D or L series is made on the basis of the con-figuration of the lowermost asymmetric carbon atom, usually the penultimate carbon atom. Compounds in the D series have the hydroxyl group lying on the right, and those in the L series, on the left. When two possible reference groups are present in the same molecule, the choice of the reference group is usually in the following order: CHO, CO$_2$H, CO (ketone); for example, in D-glucuronic acid, the reference group is the aldehyde group, not the carboxyl group.

This classification leads to assignment of two equally correct names for certain optically active, like-ended compounds. Thus, of those having six carbon atoms, D-glucitol may also be called L-gulitol. In this situation, the name now accepted is that which has alphabetic preference, and thus D-glucitol

The D-Family of Aldoses Having Three to Six Carbon Atoms

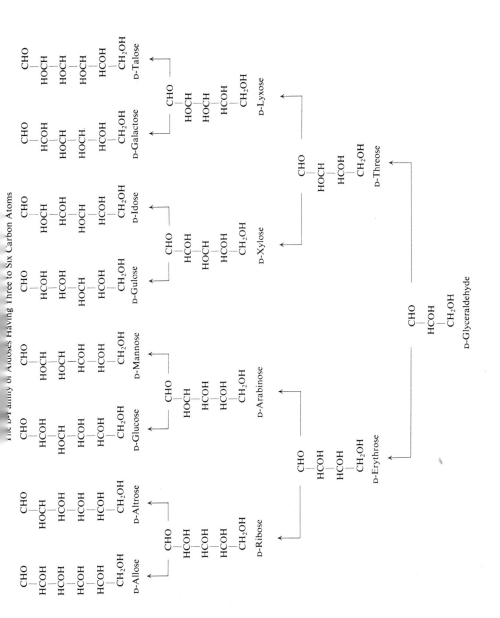

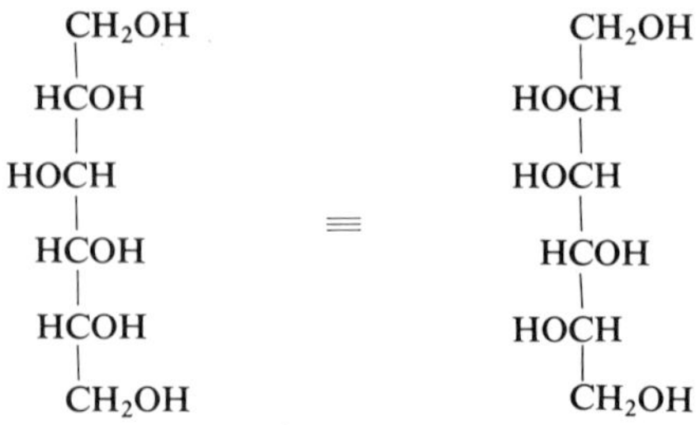

$$
\begin{array}{cc}
\text{CH}_2\text{OH} & \text{CH}_2\text{OH} \\
| & | \\
\text{HCOH} & \text{HOCH} \\
| & | \\
\text{HOCH} & \text{HOCH} \\
| \qquad \equiv & | \\
\text{HCOH} & \text{HCOH} \\
| & | \\
\text{HCOH} & \text{HOCH} \\
| & | \\
\text{CH}_2\text{OH} & \text{CH}_2\text{OH} \\
\text{D-Glucitol} & \text{L-Gulitol}
\end{array}
$$

(obsolete trivial name, sorbitol) is the name used. Likewise, D-altritol and (D)talitol are the same compound; the former is the accepted name.

The configurations of the family of D-aldoses having three to six carbon atoms are shown in the accompanying diagram. A genetic relationship to D-glyceraldehyde is shown. This relationship is chemically feasible by use of the cyanohydrin synthesis. Dextrorotatory D-glyceraldehyde was shown to be related to D-glucose.[11] In the diagram showing the formulas of the D-ketoses, the relationships cannot be provided by direct reactions, and the configurations were usually derived from those of the corresponding aldoses.

Because many optically active substances can be related to the tetraric acids, it is desirable to relate the configurations of the sugars to these acids. This correlation was first accomplished by Fischer,[12] but it will be illustrated by the conversions conducted by Hockett.[13]

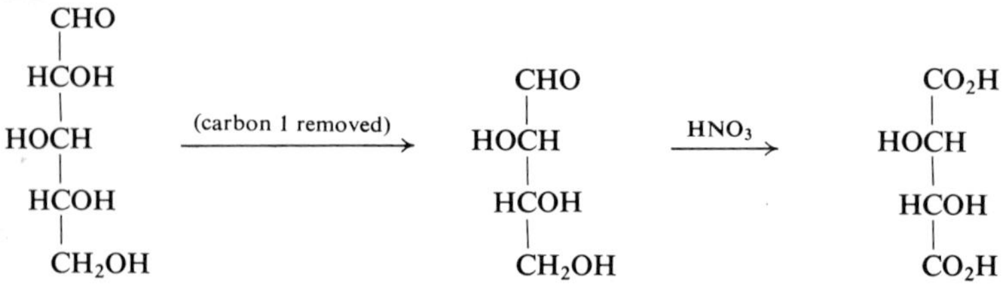

$$
\begin{array}{ccc}
\text{CHO} & & \\
| & & \\
\text{HCOH} & \text{CHO} & \text{CO}_2\text{H} \\
| & | & | \\
\text{HOCH} \xrightarrow{\text{(carbon 1 removed)}} & \text{HOCH} \xrightarrow{\text{HNO}_3} & \text{HOCH} \\
| & | & | \\
\text{HCOH} & \text{HCOH} & \text{HCOH} \\
| & | & | \\
\text{CH}_2\text{OH} & \text{CH}_2\text{OH} & \text{CO}_2\text{H} \\
\text{D-Xylose} & \text{D-Threose} & \text{D-Threaric acid} \\
& & [(-)\text{-tartaric acid}] \\
& & \text{(levorotatory)}
\end{array}
$$

The configuration of levorotatory tartaric acid is established by this process. In conformity with the Rosanoff system, it could be termed D-tartaric acid, but its original name was (*l-levo*)-tartaric acid, given because of its levorotation. The naturally occurring form is the dextrorotatory L-tartaric acid, earlier known as *d*-tartaric acid. This confusion is eliminated when the accepted nomenclature for these compounds as the D- and L-threaric acids (Vol. IB, Chapter 22) is used. The important acid known as sarcolactic acid, earlier called *d*-lactic acid from its dextrorotation, is L-lactic acid. The common *l*-malic acid should be termed 2-deoxy-D-*glycero*-tetraric acid.

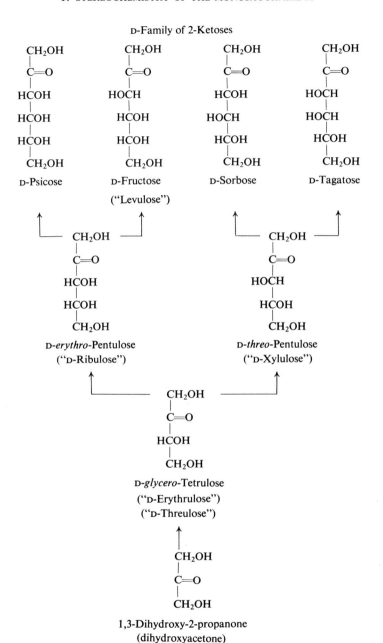

D-Family of 2-Ketoses

D-Psicose   D-Fructose ("Levulose")   D-Sorbose   D-Tagatose

D-*erythro*-Pentulose ("D-Ribulose")   D-*threo*-Pentulose ("D-Xylulose")

D-*glycero*-Tetrulose ("D-Erythrulose") ("D-Threulose")

1,3-Dihydroxy-2-propanone (dihydroxyacetone)

* *References start on p. 65.*

In the α-amino acid series, D-glyceraldehyde is also the configurational reference compound. The NH_2 group of a serine replaces the OH group at C-2 of a glyceric acid.[14] The projectional formulas of the amino acids are viewed with the carboxyl group at the top, and the assignment of configuration is made according to the position of the α-NH_2 group, D if it is to the right and

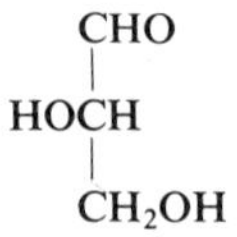

L-Glyceraldehyde

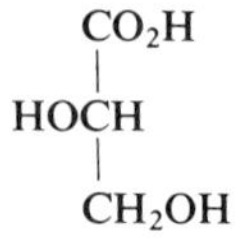

L-Glyceric acid

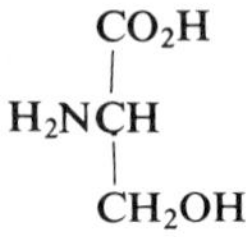

L-Serine

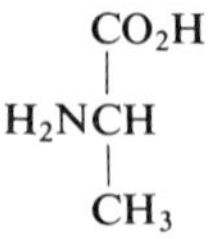

L-Alanine

L if it is to the left. All of the amino acids that occur in normal proteins are now allocated to the L series, but some of these compounds were earlier indicated as *d* isomers.

When a second asymmetric carbon atom is in the molecule, as in a threonine,

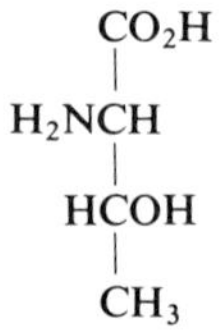

L_s-Threonine
D_g-Threonine

additional considerations are necessary. For example, D_g-threonine indicates that the compound has been named as a derivative of D-threose (in the manner of the sugars) from the relationship of D-glyceraldehyde to the configuration of the highest-numbered, asymmetric carbon atom. L_s-Threonine indicates the same compound, named (in the manner of the amino acids) from the relationship of the configuration of the α-carbon atom to a secondary standard, L-serine.

L-Glyceraldehyde was directly correlated by Wolfrom *et al.*,[15] with L-alanine, derived from 2-amino-2-deoxy-D-glucose. The configuration of the amino-bearing carbon atom in 2-amino-2-deoxy-D-glucose had already been established through syntheses involving known Walden inversions and by X-ray crystallography.

In view of the early confusion in use of the small *d* and *l* for both the optical rotations and the configurational relationships, the use of D and L is now wholly restricted to a configurational significance based on D-glyceraldehyde. Early *d*- and *l*- prefixes for the hydroxy acids and amino acids, particularly, must be translated into modern nomenclature only after careful consideration of the later conventions.[14]

The D,L system is a valuable "local" convention for configuration in the sugar series and in various related types of molecule. It is applicable whenever a backbone chain of asymmetric centers (not necessarily contiguous) can be recognized in the molecule, and the configurational prefixes *glycero, erythro, threo, arabino, lyxo, ribo, xylo, allo, altro, galacto, gluco, gulo, ido, manno,* and *talo* are used, together with D or L as appropriate, next to the root name of a molecule to specify the stereochemistry of one, two, three, or four asymmetric centers viewed according to the Fischer projection (see also p. 53). Combinations of prefixes may be used (see Vol. IIB, Chapter 46) when more than four asymmetric centers are involved. The system can be used when atoms other than oxygen are attached to the carbon atoms of the chain. The following examples illustrate the use of these configurational prefixes, where group A is

A	A	A	A
CH$_2$	HOCH	HCOH	HOCH
HOCH	HCSH	HCF	CH$_2$
CH$_2$	HCOH	HOCH	HCNH$_2$
HCOH	HCOH	HOCH	HCOH
B	B	B	B
D-*threo*	D-*altro*	L-*manno*	D-*arabino*

considered to be the principal function.† The system may also be applied with certain types of branched-chain sugar, by using the atom (at the branch point) of the higher atomic number as the atom considered in assigning the configurational symbol. Thus, an order of preference would be O before C, and C before H, as in the examples on page 30.

The D,L system breaks down when no principal chain can be recognized in the molecule, and it may then be necessary to use, at the individual chiral

---

† These prefixes are given in accordance with the British–American "Rules of Carbohydrate Nomenclature" (Vol. IIB, Chapter 46). The IUPAC–IUB "Tentative Rules for Carbohydrate Nomenclature"[1] allow the use of separate stereo designators when the sequence of asymmetric centers is interrupted by a non-asymmetric center. Thus the IUPAC–IUB rules would permit, as acceptable, longer alternatives for the first and fourth examples, the prefixes D-*glycero*-L-*glycero* (instead of D-*threo*) and D-*erythro*-L-*glycero* (instead of D-*arabino*). These more cumbersome forms are seldom encountered in the literature.

* *References start on p. 65.*

centers, the symbols (R) and (S), assigned according to the sequence rule of Cahn *et al.*[16] For example, (R) or (S) may be used if it is desired to specify the chirality at the acetalic carbon atom in the cyclic benzylidene or ethylidene acetals of sugars. The (R), (S) system was not devised to replace the D,L system,

<table>
<tr><td align="center">A
|
HOCH
|
CH$_3$—COH
|
HCOH
|
HCOH
|
B

D-*altro*</td><td align="center">A
|
HCOH
|
HC—CH$_3$
|
HOCH
|
HCOH
|
B

D-*gulo*</td><td align="center">A
|
HCOH
|
C$_2$H$_5$—CH
|
CH$_2$
|
HOCH
|
B

L-*arabino*</td><td align="center">A
|
CH$_3$—COH
|
CH$_2$
|
ClCH
|
HCOH
|
B

D-*xylo*</td></tr>
</table>

but to provide a rational system of stereochemical nomenclature for systems that cannot be named conveniently and unambiguously by the D,L system.

D. Conformational Representation of Acyclic Sugar Derivatives

By rotation about sigma bonds it is possible for the atoms in the molecule of a sugar (or some sugar derivatives) to adopt numerous, theoretically possible relative orientations in space. However, other considerations exclude many of them. These spatial orientations or shapes are termed conformations (in German, *Konstellation*). Various potential-energy functions govern torsion about sigma bonds, steric and electronic interactions of groups, bond-angle strain, and related factors, so that some conformations lie at potential-energy minima, whereas others are in high-energy states. For sugars and their derivatives in the liquid state, or in solution at room temperature, the energy barriers are usually small enough to allow rapid interconversion between conformational states (conformers), although, in the crystalline solid, lattice forces generally hold the molecule in one fixed conformation.

The Fischer projection, useful as it is for depicting configurational relationships in acyclic sugar chains, represents the molecule in a most improbable conformation having the largest groups eclipsed along each carbon–carbon bond. For depicting the molecule in its various conformational states it is useful to employ as a structural formula a formalized drawing that shows the relative dihedral angles of the substituents at each end of a carbon–carbon bond. Three types of representation are illustrated with reference to erythritol and D-threitol. The sawhorse and Newman (Böeseken) types of formula[17] clearly illustrate the dihedral angles in the molecules, although the Newman (Böeseken) formula is somewhat unsatisfactory for depicting eclipsed conformations; but the zigzag type of representation has the advantage of showing the relative orientations of groups along a chain of several atoms that lie approximately in a plane (for example, the plane of the paper). The sawhorse representation can be considered to be a view of the molecule at an angle of about 45° to the central carbon–carbon bond, whereas in the zigzag representation this bond is viewed at a right angle, in the plane of the paper. In the Newman type of formula the line of view is directly along the central bond.

Based on the premise that the most stable conformation of the four-carbon derivatives is the one having the largest groups antiparallel,[18] a vicinal diol group that is formally *cis* in the Fischer projection (as in the 2,3-diol group of erythritol) has, in fact, these hydroxyl groups antiparallel in the favored conformation. In D-threitol the 2,3-diol group, formally *trans* in the Fischer projection, actually has these groups in a gauche orientation (ideal dihedral angle of 60°) in the conformation in which C-1 and C-4 are antiparallel.

High-energy (eclipsed) conformations

Low-energy (staggered) conformations

Configurational and conformational representations of erythritol.

* *References start on p. 65.*

High-energy (eclipsed) conformations

Low-energy (staggered) conformations

Configurational and conformational representations of D-threitol.

Application of the concept of a favored antiparallel disposition of the largest pairs of groups along each carbon–carbon bond leads, with the longer-chain, acyclic sugar derivatives, to a conformation having an approximately planar backbone of carbon atoms arranged in a zigzag, with the substituent groups above and below this plane. Such a conformation is illustrated for galactitol.

Galactitol in a zigzag conformation having
the carbon atoms coplanar.

Conformations approximating the planar, zigzag form have been documented for acyclic sugar derivatives as being the energetically favored forms (in solution by n.m.r. spectroscopy,[19] and in the solid state by X-ray crystallography[20]). However, such destabilizing factors as interaction between 1,3-disposed oxygen atoms on the same side of the chain may cause the chain to adopt a conformation having one of the carbon atoms out of the plane; for acyclic sugar derivatives in solution, the term sickle conformation has been used,[19] as illustrated for xylitol.[20]

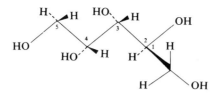

Xylitol in a "sickle" conformation derived from the planar, zigzag form by rotation about C-2–C-3 (to alleviate the 1,3-*syn*-interaction between the hydroxyl groups on C-2 and C-4).

VII. RING STRUCTURES OF THE SUGARS

A. NECESSITY FOR RING STRUCTURES

Soon after the formulation of glucose as a polyhydroxy aldehyde and of fructose as a polyhydroxy ketone, it became evident that the acyclic formulas would not account for all the reactions of these sugars. Thus, the sugars give a negative test with the Schiff reagent (fuchsin and sulfurous acid) under the usual conditions of test, although, under milder conditions, positive results are obtained.[21]

Also, the aldehyde and ketone structures do not account for the change in optical rotation that may be observed for the freshly prepared, aqueous solutions of many sugars. This phenomenon, now called mutarotation, was observed by Dubrunfaut in 1846 for solutions of D-glucose.

When the hydroxyl groups of D-glucose are esterified by treatment with acetic anhydride and a catalyst, two isomeric pentaacetates are formed. Similarly, isomeric methyl D-glucosides are formed by treatment of D-glucose with methanol and hydrogen chloride. The existence of two D-glucosides[22] of the same alcohol and of two pentaacetates[23] could not be predicted on the basis of the aldehyde formula, a conclusion stated by Fischer in the case of the methyl glucosides (which he discovered), and even earlier by Colley and Tollens.

The isolation of crystalline isomers of a single sugar provided additional evidence for the inadequacy of the aldehyde formulas. As early as 1856, two different crystalline kinds of lactose, now designated α- and β-lactose, were prepared by Erdmann;[24] he discovered their mutarotation to a common equilibrium rotation. Tanret[25] in 1895 reported the isolation of three forms of D-glucose, which he described as α-, β-, and γ-glucose having the following rotations:

α-Glucose		"β-Glucose"		"γ-Glucose"
+106°	⟶	+52.5°	⟵	+22.5°

* *References start on p. 65.*

When dissolved in water, the α-glucose mutarotated downward and the "γ-glucose" upward, to the same constant specific rotation of +52.5°. Tanret's "β-glucose" exhibited no mutarotation and was later considered to be a mixture of the two other forms in their equilibrium proportions. The name β-D-glucose is now given to the form that he termed "γ-glucose." The common form is the α-D-isomer.

Even before the various isomers of D-glucose and some of its derivatives had been isolated, the lack of some typical aldehyde reactions by D-glucose had been explained by Colley (1870) and by Tollens (1883) as arising from an alteration of the aldehyde group by the formation of an inner hemiacetal type of linkage. The formulas proposed by Colley and by Tollens are illustrated.

Colley formula	Tollens formula

The ring forms of the sugars are intramolecular hemiacetals. An aldehyde reacts with an alcohol with the formation of a hemiacetal and an acetal. For a

sugar, hemiacetal (ring) formation takes place by reaction of the aldehyde group with a hydroxyl group in the same molecule. Each of the ring formulas possible for D-glucose allows two isomers, which differ only in the configuration of the hemiacetal group, as C-1 is asymmetrically substituted in any ring form. Such isomers are distinguished by the symbols α and β—for example, α-glucose and β-glucose—and are termed *anomers*. The hemiacetal carbon atom is known as the *anomeric* or *reducing* carbon atom. The existence of isomeric D-glucoses, penta-O-acetyl-D-glucoses, and methyl D-glucosides became explicable when ring structures were assigned to the sugar and some of its derivatives.

B. PROOF OF RING STRUCTURE

Subsequent to the proposal of the ring structures for the sugars and their derivatives, acceptance by carbohydrate chemists[26] gradually ensued. However, it was not until the period 1920 to 1930 that conclusive proof could be offered for the position of the ring-forming hydroxyl group. Prior to this time, the rings were usually considered to be of the 1,4 type shown in the Tollens formula—that is, with ring formation between C-1 and O-4 of aldoses. This type of structure was based mainly on an analogy with the acid series, for which it was known that γ-hydroxy acids could be converted into inner esters (lactones) that have the 1,4 or γ structure.

Methods are now available for the unequivocal determination of the ring structures of the glycosides of the sugars. The glycosides may be prepared by condensing a sugar with an alcohol in the presence of a mineral acid. (For a detailed discussion of the preparation of glycosides and of the details of determination of their structures, see Chapter 9.) Originally, the structures of

these glycosides were demonstrated by oxidation of the glycosides to fragments, which were then identified. To prevent the oxidation from proceeding too far, the four unsubstituted hydroxyl groups were first etherified with methyl groups. (Details of this method are given in Chapter 9.) An easier and more direct method involves oxidation of the glycosides with periodic acid. As shown in the accompanying formula, this reagent cleaves the linkage between two adjacent, hydroxyl-bearing carbon atoms and leads to net oxidation of each CHOH group to a C=O group. A hydroxymethyl (CH_2OH) group yields formaldehyde; a secondary hydroxyl group (CHOH) gives rise to an aldehyde group or, if flanked by two secondary hydroxyl groups, to formic acid. The reaction is practically quantitative, and the consumption of periodate is a direct measure of the number of adjacent hydroxyl groups in a compound. (See Vol. IB, Chapter 25.) The structure is revealed by the nature of the oxidation products, together with the proportion of oxidant that is consumed.

* *References start on p. 65.*

$$
\begin{array}{c}
\overset{\frown}{\text{HCOMe}} \\
| \\
(\text{CHOH})_n \\
| \\
\text{HCO}\underline{} \\
| \\
\text{CH}_2\text{OH}
\end{array}
\quad \xrightarrow[(n > 1)]{(n-1)\ \text{HIO}_4} \quad
\begin{array}{c}
\overset{\frown}{\text{HCOMe}} \\
| \\
\text{CHO} \\
| \\
\text{CHO} \\
| \\
\text{HCO}\underline{} \\
| \\
\text{CH}_2\text{OH}
\end{array}
\quad
\begin{array}{l}
+\ \text{H}_2\text{O} + (n-1)\ \text{HIO}_3 \\
\quad + (n-2)\ \text{HCO}_2\text{H}
\end{array}
$$

The structures possible for the methyl α-D-glucosides are given in formulas **27** to **31**, in which the brackets indicate the adjacent hydroxyl groups. One mole

27	28	29	30	31	
3	2	2	2	3	Moles of oxidant
2	1	0	1	2	Moles of HCO_2H
1	1	1	0	0	Moles of HCHO

of ordinary methyl α-D-glucoside consumes two moles of periodic acid, and no formaldehyde is produced. Hence, the structure must be that shown in **30**, which has a 1,5 oxygen bridge.

In most instances, the evidence obtained confirmed the structures indicated by the earlier methylation–oxidation studies. The method of periodic acid oxidation is now widely used because of its simplicity. As a result of the application of the methylation–oxidation technique and of the periodic acid method, it was found that the ring most commonly present in the synthetic glycosides then known is of the six-membered type connecting carbon atoms 1 and 5. However, rings formed between C-1 and C-4 were found in some glycosides. Sugars and derivatives that have the 1,5 ring may be considered to be related to tetrahydropyran, and those with the 1,4-ring to be related to tetrahydrofuran. These relationships are shown in the accompanying formulas. The sugars having a tetrahydropyran ring are known as pyranoses, and the corresponding glycosides as pyranosides. Those having tetrahydrofuran rings are called furanoses and furanosides, respectively. The terms "pyranose" and

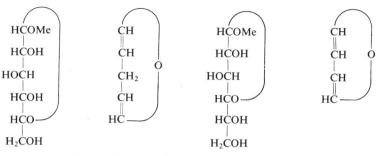

Methyl α-D-glucopyranoside   Pyran   Methyl α-D-glucofuranoside   Furan

"furanose" may be regarded as operators that specify the position of ring formation from the reducing (anomeric) atom to an oxygen atom elsewhere on the chain.

Absolute methods are now available for establishing the ring structures of glycosides, but the classical methods used for the sugars (before the advent of n.m.r. spectral and X-ray crystallographic methods) were indirect. For many glycosides, the rings are quite stable under alkaline and neutral conditions. However, with the free sugars, determination of ring structure is complicated by the ease with which ring changes may take place as soon as dissolution of the sugar occurs. The methods that are applicable to the determination of the ring forms of the sugars must be such that ring changes do not precede the subsequent reactions necessary. In the following methods, this proviso is assumed.

One method for locating the position of the ring in an unsubstituted aldose entails oxidation to the corresponding acid or lactone. As shown in the following formulas, the ring compound might be oxidized (dehydrogenated) by bromine to the corresponding lactones, whereas the free aldehyde forms would give the corresponding acids.

The oxidation is conducted in solution, and the nature of the oxidation products establishes the structure of the original sugar, unless ring shifts took place prior to the oxidation. By application of this method,[27] it was shown that the common form of D-glucose (the α anomer) gives D-glucono-1,5-lactone; β-D-glucose gave the same compound. Hence, both have pyranose (1,5) rings; otherwise the 1,4-lactone or the free acid would have been produced. The method has not been widely applied. A crystalline addition compound of D-mannose with calcium chloride yields D-mannono-1,4-lactone, and the sugar moiety appears to have the furanoid structure.[28]

By the bromine oxidation method, the structure of a sugar can be correlated with that of the corresponding lactone and acid. The proof requires that the structure of the lactone be known. In general, the method depends on a correlation of the properties of the lactone with those of the methylated

* References start on p. 65.

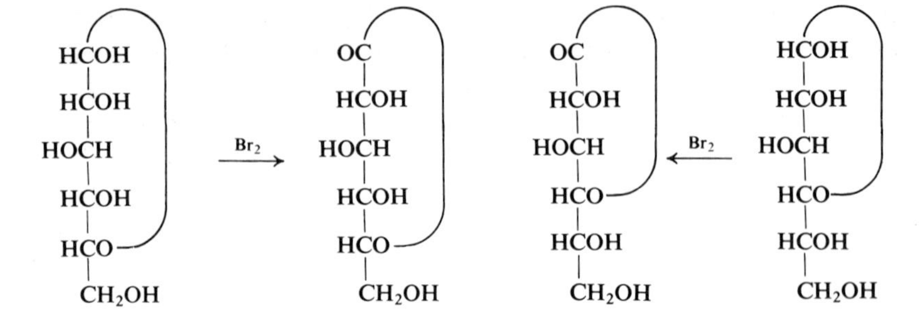

HCOH   OC   OC   HCOH
HCOH   HCOH   HCOH   HCOH
HOCH Br₂ HOCH   HOCH Br₂ HOCH
HCOH ——→ HCOH   HCO—— ←—— HCO——
HCO——   HCO——   HCOH   HCOH
CH₂OH   CH₂OH   CH₂OH   CH₂OH

α-D-Glucopyranose D-Glucono-   D-Glucono-   α-D-Glucofuranose
       1,5-lactone   1,4-lactone

OCH   CO₂H
HCOH   HCOH
HOCH   HOCH
HCOH   HCOH
HCOH   HCOH
CH₂OH   CH₂OH

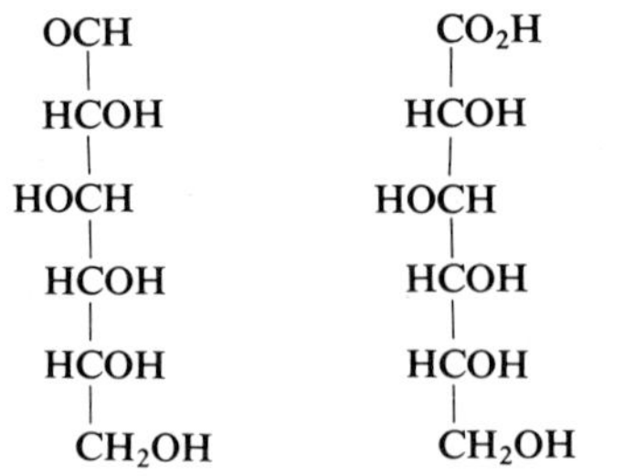

aldehydo-D-Glucose D-Gluconic acid

derivative obtained by methylation and oxidation of a glycoside of known structure.

Another method for establishing of the ring structure of D-glucose (and other sugars) involves correlation of the optical rotation of the sugar with those of the glycoside. This method, although not absolute, was developed and widely applied by C. S. Hudson, and has much value for this purpose. It is considered later (Chapter 4).

The D-glucopyranosides are hydrolyzed to D-glucose by certain enzymes (see Vol. IIA, Chapter 33). Identification of the anomeric form of the sugar that is released provides a method for the correlation of the glycoside with the crystalline forms of the sugar.[29] The product formed by enzymic hydrolysis of methyl α-D-glucopyranoside appears to be the α-D anomer; that from methyl β-D-glucopyranoside appears to be the β-D anomer. Hence, unless ring changes took place very rapidly, the α and β forms of D-glucopyranose would appear to have the same (pyranose) structures as the corresponding D-glucopyranosides.

The foregoing methods for the determination of the structures of the unsubstituted sugars are rather unsatisfactory as absolute methods, because of the possibility of ring shifts. However, the evidence available indicates that most of the crystalline sugars have pyranoid ring structures. In addition to the double compound of D-mannose with calcium chloride, which probably has the furanoid structure,[28] the monoketodisaccharide lactulose may exist as the furanoid modification when in the crystalline state.[30] Otherwise, crystalline, free furanoses are seldom encountered,[30a] although the furanoid ring is of frequent occurrence in compounds in which ring shifts are impossible (glycosides, disaccharides, etc.) or in compounds in which the hydroxyl group that could form the pyranose ring is blocked by substitution with a stable group.

The ring size of the free aldoses in solution can frequently be established by n.m.r. spectroscopy.[31] By use of solvents in which mutarotation is slow, it is possible to ascertain the tautomeric composition in the crystalline material, as well as in the equilibrated solution (see Chapter 4 and Vol. IB, Chapter 27, and ref. 31a, for more details).

C. CONFIGURATION AT THE ANOMERIC CARBON ATOM

For each of the ring modifications of a free sugar, two isomers (α and β isomers or anomers) can exist, because a new asymmetric center is created by ring closure at the reducing carbon atom.

As previously noted, the existence of such isomers was one of the most important reasons for the formulation of ring structures. The isomeric α- and β-D-glucopyranose have quite different solubilities, melting points, and optical rotations. The isomeric pentaacetates and methyl D-glucosides also exhibit differences in properties.

Historically, the terms α and β referred to the order of isolation of the individual anomers. Thus "α-glucose" was the common, first-isolated isomer. The terms now refer to absolute configurations and must be used with the D or L symbol (see p. 45). Early methods used for establishing the configuration at the anomeric carbon atoms of sugars and derivatives were optical rotatory relationships (see p. 45), changes in the conductivity of solutions of sugars and their boric esters,[32] and optical rotatory changes occurring during enzymic hydrolysis of glycosides of known configuration. The optical rotatory method is still the most convenient when suitable reference compounds are available.

The absolute method is crystallographic analysis by X-ray or neutron diffraction studies of the crystalline compounds.[33] (See Vol. IB, Chapter 27.) This method, although difficult and time-consuming, allows for the calculation of interatomic spacings and atomic positions. These data have, in turn, been correlated with n.m.r. data and theory, so that n.m.r. measurements may often be used for determining the absolute configuration (Vol. IB, Chapter 27). These measurements can be carried out for sugars in solution and provide a rapid and convenient method for establishing the configurations of many compounds. Infrared measurements (Vol. IB, Chapter 27) have had some use but have not generally been satisfactory for the purpose.

Periodic acid oxidation provides a means for correlating the configuration of the anomeric carbon atoms of the glycosides (see also Vol. IB, Chapter 25). As shown in the accompanying formulas, representative of the aldohexo-pyranosides, C-3 is removed in the process (as formic acid), and the asymmetry at C-2 and C-4 is removed. In the "dialdehyde" 33 only two asymmetric centers remain, and these are derived from C-1 and C-5 of the original D-glucopyranoside (32). Hence, all of the D-aldohexopyranosides should yield the same dialdehyde 33 as does the corresponding α- or β-D-glucopyranoside.

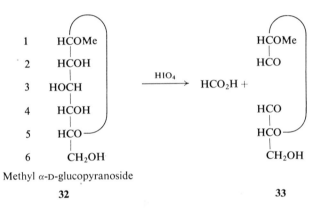

Methyl α-D-glucopyranoside

32   33

The configuration at C-1 of each of the glycosidic derivatives of the aldohexoses may be correlated with those of the D-glucopyranosides in this way.[34]

Evidence for the configuration at C-1 of some phenyl D-glucopyranosides has been obtained by the conversion of the β-glucopyranoside into a 1,6-anhydro derivative (see Chapter 13), and by the stability of the α-D anomer to strong alkali.[35]

D. DEPICTION OF THE RING STRUCTURES OF THE SUGARS

In the preceding discussion, the structure and configurations of the two isomeric D-glucopyranoses have been developed. The structure and configuration may be represented by the cyclic form of the Fischer formula, as in 34 for α-D-glucopyranose, and as in 35 for methyl α-D-glucopyranoside. However, the cyclic Fischer–Tollens projection formula has several shortcomings. Thus, the molecule is depicted as an extended chain of carbon atoms connected by an oxygen bridge between C-1 and C-5. Obviously, an extended linear chain is impossible, for carbon atoms 1 and 5 must be close enough for the existence of the oxygen bridge. Also, the configuration of C-5 as given by the cyclic Fischer formula is correct but does not give a true picture of the steric relationships between the terminal primary hydroxyl group and the hydroxyl groups attached to the ring carbon atoms. A formula of the type of **36**, derived by

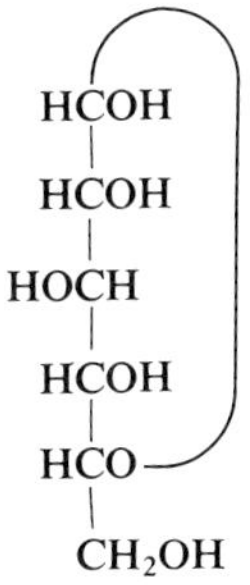

34 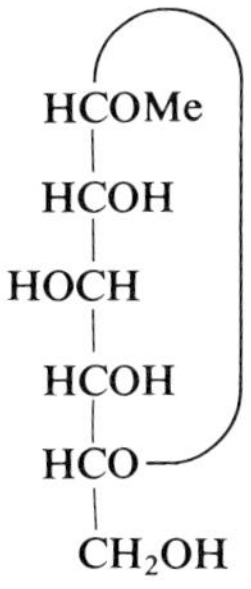   **35** 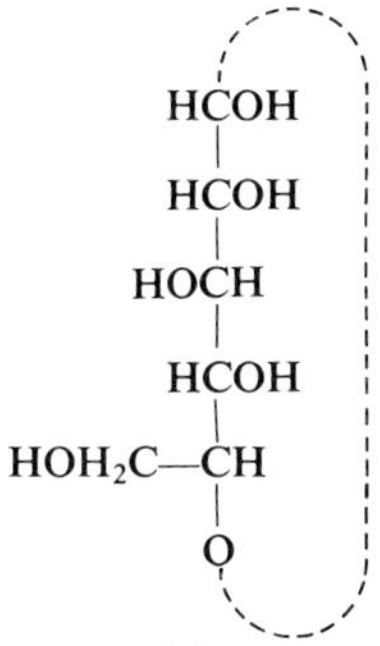   **36**

rotating C-5 clockwise by one-third of a revolution about C-4–C-5, gives a truer depiction of the spatial arrangement at C-5. In this hybrid depiction the hydroxymethyl group has a *trans* relationship to the hydroxyl groups on carbon atoms 1, 2, and 4.

To provide a better picture of the structure and configuration of the cyclic sugars, Böeseken[36] proposed a perspective representation that was later modified and popularized by Haworth.[37] Böeseken's original representation of α-D-glucofuranose and Haworth's depiction[37] of α-D-glucopyranose are shown below.

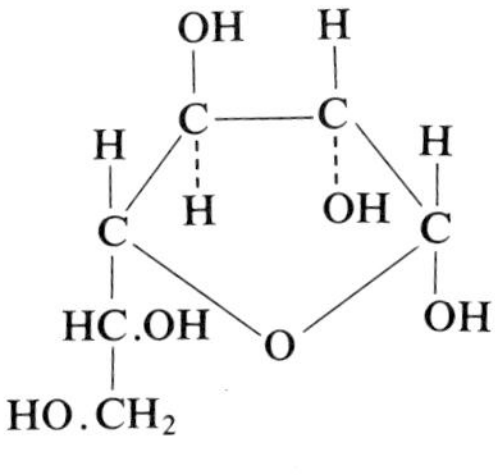

Böeseken formula (1913) 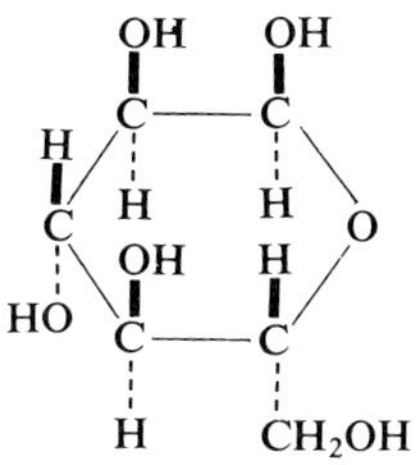   Haworth formula (1926)

* *References start on p. 65.*

These formulas are to be regarded as conventionalized perspective drawings of a three-dimensional model. In Haworth's representation, the fundamental pyranoid ring is considered, for the purpose of emphasizing configurational relationships, as a planar ring viewed from above at an angle, with the front edge tilted toward the observer. This orientation was emphasized by shading the bonds in the lower edge of the ring, and the representation was simplified by omitting the carbon atoms from the ring, as depicted below.

Development of the Haworth representation.

In essence, the Haworth convention sets up three ring planes having a common axis. One plane is that of the oxygen-containing heterocycle. Above that plane is a ring plane of the substituents above the heterocyclic ring, and below it is another plane of the substituents below the heterocyclic ring. (This representation bears an analogy to the common mode of depicting aromatic ring systems to show the plane of the ring atoms and the planes of the π-electron clouds above and below the ring.) Any reorientation of the Haworth representation must take into account the simultaneous reorientation of these three planes (see p. 55 for further discussion of this point).

Although the Haworth formula for a given pyranoid sugar can be correctly depicted in twelve different ways (ten for a furanoid sugar), it has become customary[38] to use only one of these orientations wherever possible, so that stereochemical differences between various structures are more quickly recognizable. The "standard" orientation has the anomeric (reducing) carbon atom on the extreme right of the formula, and the numbering of the ring proceeds in a clockwise direction.

Haworth representations of pyranoid and furanoid rings in the "standard" orientation. The arrow denotes the direction of advancing ring-numbering.

To relate the Haworth formula correctly to the Fischer representation, it is necessary first to orient the tetrahedra used to develop the Fischer formula so that all the ring atoms lie in the same vertical plane.[37] In such a modified

Fischer structure, all groups lying to the left of the ring plane are in the upper plane of substituents in the Haworth formula (in the standard orientation), and all groups on the right are in the lower plane of substituents in the Haworth formula. These transpositions may be illustrated with respect to α-D-gluco-pyranose and α-L-*xylo*-hexulose).

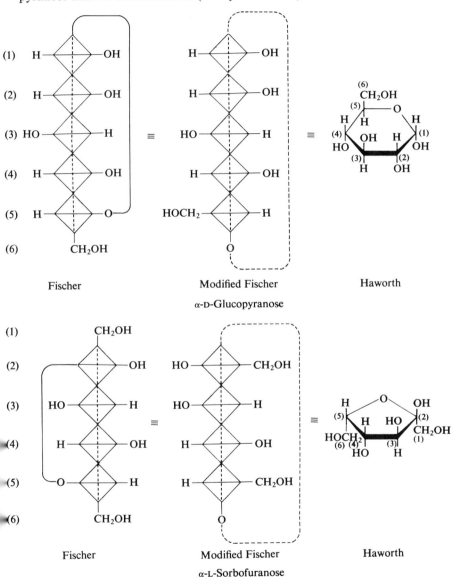

α-D-Glucopyranose

α-L-Sorbofuranose

References start on p. 65.

It may be noted that the same correlation applies for any of the other possible orientations of the Haworth ring that have clockwise numbering, but if the Haworth ring is "turned over," so that the ring numbering runs counter-clockwise, the groups on the left of the modified Fischer formula now appear below the plane of the counterclockwise-numbered Haworth ring, and those on the right appear above the plane of the ring. These relationships are readily recognized by use of simple cut-out models (see p. 58). The enantiomer of a sugar in a Haworth representation can be drawn by use of a hypothetical mirror plane placed in any desired orientation or, more simply, by reversing all pairs of substituents at each carbon atom of the ring. For object and mirror image to be in the standard orientation, the mirror plane must be horizontal. This relationship is shown for β-D-glucopyranose and its enantiomer, β-L-glucopyranose.

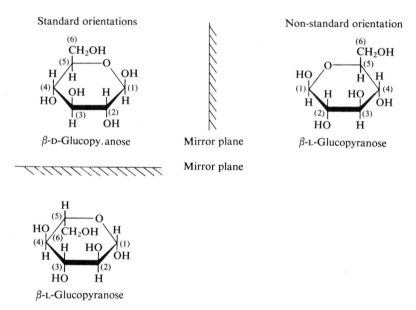

It is sometimes useful to depict the ring in orientations other than the standard one when bulky substituents exist in the molecule, or when linkages between two or more rings are to be represented, as in oligo- and poly-saccharides.

If there is a "tail" group attached to the ring and if it contains one or more asymmetric centers, as in the heptopyranoses and hexofuranoses and their corresponding glycosides, the normal Fischer convention can be used to represent the configuration at the exocyclic asymmetric centers. This practice, already illustrated in Böeseken's cyclic formula (p. 41), may be further

Fischer	Modified Fischer	Haworth

α-D-Glucopyranose

exemplified by the currently used representation[38] of α-D-glucofuranose. Note that the "tail" group is "up" in the normal Haworth formula when the ring of the standard Fischer formula is closed on the right; when the ring of the standard Fischer formula is closed on the left, the "tail" group is down in the normal Haworth formula.

E. NOMENCLATURE OF ANOMERS (α,β NOMENCLATURE)

The anomeric (α and β) isomers of the sugars are named by the system of Hudson.[39] Hudson[39] defined this system as follows: "In the D-series, the more dextrorotatory member of an α,β-pair of anomers is to be named α-D-, the other being β-D-. In the L-series, the more levorotatory member of such a pair is given the name α-L- and the other β-L-." A correct application of the system requires knowledge that the compounds being considered are truly anomeric— that is, that they differ only in the configuration at the hemiacetal carbon atom. If the compounds being considered mutarotate and are of the D series, the α anomer is usually the form that mutarotates to a rotational value less positive than the initial value. Particularly in the case of compounds that exhibit complex mutarotation, the mutarotational data must be interpreted with caution, for such data can be used for the naming of the sugars only when the mutarotation results from an α to β interconversion.

The configurational significance of the rules, as interpreted by Hudson,[40] is expressed in the skeleton stereostructures for the methyl α- and β-pyranosides as follows:

* *References start on p. 65.*

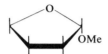

Methyl α-D-glycosides
and β-L-glycosides

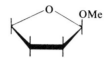

Methyl β-D-glycosides and
α-L-glycosides

The skeleton formulas for the anomeric furanose forms are:

Methyl α-D-glycosides
and β-L-glycosides

Methyl β-D-glycosides and
α-L-glycosides

It should be noted that α-D and β-L (also β-D and α-L) refer to the same absolute configuration of the anomeric carbon atom. Thus, β-L-arabinose and α-D-galactose have the same absolute configuration at carbon atom 1 and at the other asymmetric atoms and hence exhibit many similarities.

The anomeric designators α and β are now accepted as referring to absolute structural relationships; the designator α refers to that anomer which, in the Fischer projection, has the anomeric group and the group at the reference carbon atom in a formal *cis* relationship, and the designator β refers to that arrangement in which these two groups are *trans*-related.

α anomer   β anomer

Assignment of anomeric configurational symbols. The lower carbon atom shown is the reference atom (see Vol. IIB, Chapter 46); it need not necessarily be the position at which the ring is closed.

The Hudson rules of rotation, although built on a "principle" of optical superposition advanced by Van't Hoff that is definitely unsound, give configurational relationships that can be justified by more sophisticated interpretations based on conformational asymmetry. There is no doubt that the anomeric symbols assigned by Hudson's rules, for anomeric pairs of all the sugars and the overwhelming majority of their derivatives, correspond to the absolute relationships just stated, and these assignments are fully supported by X-ray crystallographic and n.m.r. spectral evidence in a number of examples.

However, there do exist[41] some anomeric pairs of sugar derivatives for which the α-D anomeric form (as defined by the *cis* relationship of the anomeric group and the group at the reference atom) has a lower specific rotation than the β-D anomeric form (as defined by the *trans* relationship of the anomeric group and the group at the reference atom), in violation of Hudson's rule. Because of the existence of such exceptions to Hudson's rules, the absolute configuration should be used as a basis for defining the anomeric symbolism, not optical rotatory data, despite the great utility of the latter in assigning anomeric configuration in many cases.

Anomeric forms of derivatives of acyclic sugars have been prepared. Thus, two 1-chloro-*aldehydo*-D-galactose hexaacetates (**37** and **38**) are known.[42]

<table>
<tr><td>H</td><td>H</td></tr>
<tr><td>|</td><td>|</td></tr>
<tr><td>ClCOAc</td><td>AcOCCl</td></tr>
<tr><td>|</td><td>|</td></tr>
<tr><td>HCOAc</td><td>HCOAc</td></tr>
<tr><td>|</td><td>|</td></tr>
<tr><td>AcOCH</td><td>AcOCH</td></tr>
<tr><td>|</td><td>|</td></tr>
<tr><td>AcOCH</td><td>AcOCH</td></tr>
<tr><td>|</td><td>|</td></tr>
<tr><td>HCOAc</td><td>HCOAc</td></tr>
<tr><td>|</td><td>|</td></tr>
<tr><td>CH_2OAc</td><td>CH_2OAc</td></tr>
<tr><td>**37**</td><td>**38**</td></tr>
</table>

These forms have been named phenomenonologically α and β according to whether they mutarotate downward (α) or upward (β) in acetyl chloride solution containing zinc chloride,[43] but no correlation of these symbols with the standard anomeric terminology was implied.

VIII. HOMOMORPHOUS SUGARS

A. HOMOMORPHOLOGY

Earlier in this Chapter, the genetic relationship of the various sugars to D- and L-glyceraldehyde was demonstrated. A relationship of considerably more importance for the correlation of the properties and reactions is based on the similarity of compounds that have the same configuration for each atom that is part of the pyranose ring. Since for the aldohexopyranoses the number of asymmetric carbon atoms is just sufficient to make each carbon atom in the pyranose ring asymmetric, the aldohexopyranoses may be considered to be the fundamental types for all sugars that can form pyranose rings. On the other hand, the fundamental types of the furanoses are the aldopentoses, which have just sufficient asymmetric carbon atoms to make each carbon atom in the furanose ring asymmetric. The aldopentoses and higher sugars can formally be obtained from the aldohexoses by substitution of the CH_2OH group of the aldohexoses by H or by $(CHOH)_n$–CH_2OH, respectively. The various types of aldohexose are illustrated in the accompanying formulas, which also show some of the members of each series. Although 32 aldohexo-pyranoses are theoretically possible, only the formulas for the eight D types are written, and the α,β configuration is not indicated. Because of the lack of asymmetry at C-5 of the aldopentoses, each of these sugars is related to a pair of aldohexoses.

Although each aldopentose is homomorphous to two aldohexoses that differ in the configuration at C-5, the aldopentose usually resembles one more than the other in its properties. This hexose is the one in which the 3-OH and the 5-CH_2OH groups are equatorial in the $C1$ (D) conformation (see p. 60). Thus, L-arabinopyranose is more homomorphous to D-galactopyranose, D-xylopyranose to D-glucopyranose, D-lyxopyranose to D-mannopyranose, and L-ribopyranose to D-talopyranose than to the other aldohexopyranose of each pair shown in the accompanying listings. The effect of the 3-OH group is attributed to a dipolar repulsive force between this OH and the ring-oxygen atom.[44]

As would be expected from the identity of the configuration on the pyranose or furanose rings, the members of each homomorphous series show marked chemical and physical similarities,[45] and it is often possible to predict the properties of unknown members from those of the fundamental type. As might be anticipated, the greatest differences are found between the aldopento-pyranoses and the corresponding aldohexopyranoses. It appears that enzymes that hydrolyze the hexoside members of each series also hydrolyze the glyco-sides of the other members of each series.[46] Thus, the enzyme α-D-mannosidase of almond emulsin hydrolyzes the α-D-lyxosides, as well as the α-D-mannosides.

Aldopyranose Types

D-Idose type

X = —H, L-xylose
X = —CH_2OH, D-idose
X = —CH_3, 6-deoxy-D-idose
X = —CHOH—CH_2OH (two aldoheptoses)

D-Gulose type

X = —H, L-lyxose
X = —CH_2OH, D-gulose
X = —CH_3, 6-deoxy-D-gulose
X = —CHOH—CH_2OH (two aldoheptoses)

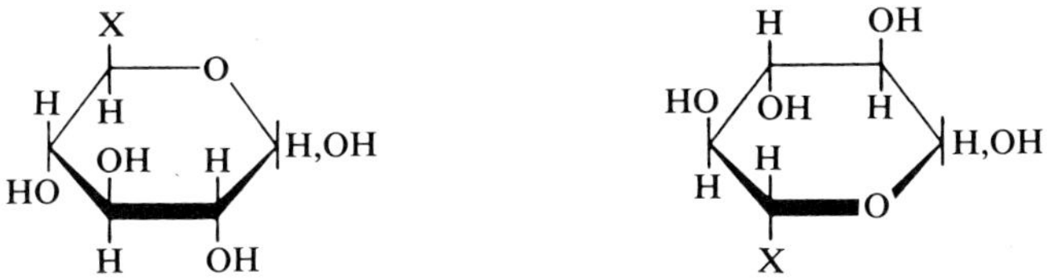

D-Glucose type

X = —H, D-xylose
X = —CH_2OH, D-glucose
X = —CH_3, D-quinovose
X = —CHOH—CH_2OH (two aldoheptoses)

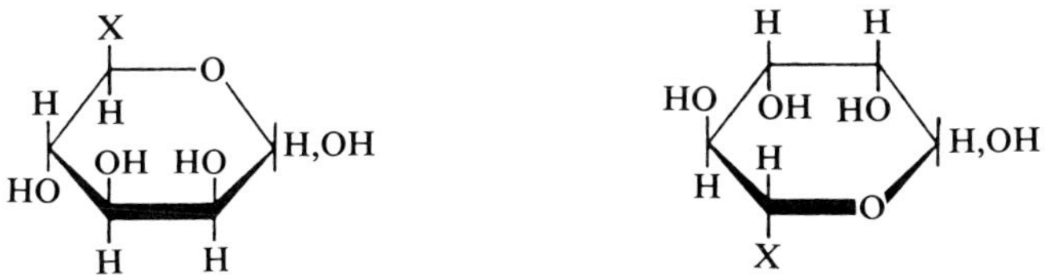

D-Mannose type

X = —H, D-lyxose
X = —CH_2OH, D-mannose
X = —CH_3, D-rhamnose
X = —CHOH—CH_2OH (two aldoheptoses)

* *References start on p.* 65.

Aldofuranose Types

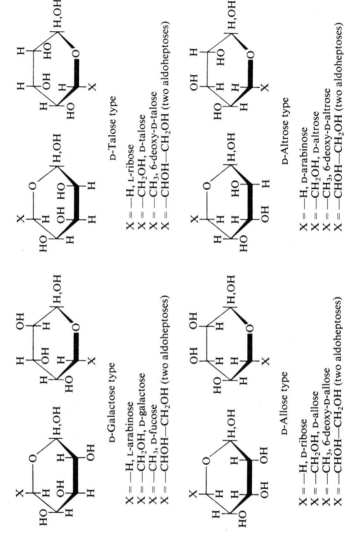

D-Galactose type

X = —H, L-arabinose
X = —CH₂OH, D-galactose
X = —CH₃, D-fucose
X = —CHOH—CH₂OH (two aldoheptoses)

D-Talose type

X = —H, L-ribose
X = —CH₂OH, D-talose
X = —CH₃, 6-deoxy-D-talose
X = —CHOH—CH₂OH (two aldoheptoses)

D-Allose type

X = —H, D-ribose
X = —CH₂OH, D-allose
X = —CH₃, 6-deoxy-D-allose
X = —CHOH—CH₂OH (two aldoheptoses)

D-Altrose type

X = —H, D-arabinose
X = —CH₂OH, D-altrose
X = —CH₃, 6-deoxy-D-altrose
X = —CHOH—CH₂OH (two aldoheptoses)

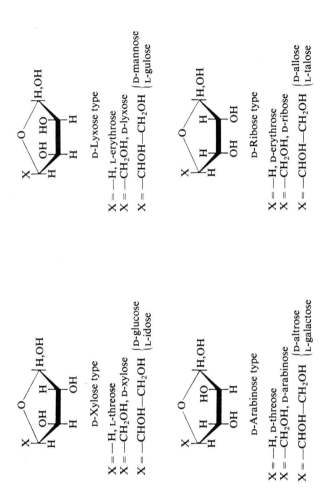

D-Lyxose type

X = —H, L-erythrose
X = —CH₂OH, D-lyxose
X = —CHOH—CH₂OH {D-mannose
                  {L-gulose

D-Ribose type

X = —H, D-erythrose
X = —CH₂OH, D-ribose
X = —CHOH—CH₂OH {D-allose
                  {L-talose

D-Xylose type

X = —H, L-threose
X = —CH₂OH, D-xylose
X = —CHOH—CH₂OH {D-glucose
                  {L-idose

D-Arabinose type

X = —H, D-threose
X = —CH₂OH, D-arabinose
X = —CHOH—CH₂OH {D-altrose
                  {L-galactose

B. Nomenclature for Higher Sugars and for Compounds Having Numerous Asymmetric Atoms in a Carbon Chain

In the development of the stereochemistry of the sugars, trivial names, as shown on p. 25, were assigned to sugars having two, three, and four CH(OH) groups. In earlier usage and recognized in the Carbohydrate Nomenclature Rules[1] (Vol. IIB, Chapter 46), these names of the C_3 to C_6 sugars have taken on a more fundamental aspect as the basis for indicating the configuration of consecutive CH(OH) groups in a chain of carbon atoms. These names and the configurations they describe are shown in Table II. The names are used as italicized prefixes before the chemical name, as in

$$\text{D-}\textit{gluco}\text{-pentahydroxypentyl, } HOCH_2\text{—}\overset{\displaystyle H}{\underset{\displaystyle OH}{C}}\text{—}\overset{\displaystyle H}{\underset{\displaystyle OH}{C}}\text{—}\overset{\displaystyle OH}{\underset{\displaystyle H}{C}}\text{—}\overset{\displaystyle H}{\underset{\displaystyle OH}{C}}\text{—}$$

If the sequence of asymmetric centers is broken by a nonasymmetric group, this group is passed over,* as in

$$\text{D-}\textit{ribo}\text{-1,3,4,5-tetrahydroxypentyl, } HOCH_2\text{—}\overset{\displaystyle H}{\underset{\displaystyle OH}{C}}\text{—}\overset{\displaystyle H}{\underset{\displaystyle OH}{C}}\text{—}CH_2\text{—}\overset{\displaystyle H}{\underset{\displaystyle OH}{C}}\text{—}$$

and

$$\text{D-}\textit{erythro}\text{-pentulose, } HOCH_2\text{—}CO\text{—}\overset{\displaystyle OH}{\underset{\displaystyle H}{C}}\text{—}\overset{\displaystyle OH}{\underset{\displaystyle H}{C}}\text{—}CH_2OH$$

The same prefixes are used when one or more CH(OH) groups are replaced by $CH(NH_2)$, CH(OMe), CH(Cl), CH(OAc), CMe(OH), and similar groups (see p. 29).

For a sequence of more than four asymmetric carbon atoms, two or more prefixes are used. The sequence of asymmetric centers is so divided that there is a four-carbon prefix for the carbon atoms closest to the principal function, and so that the other prefixes contain the maximal possible number of asymmetric centers. In the actual name, the order of citation of the prefixes starts with the grouping farthest removed from the principal function. The most common examples of compounds requiring this type of compound prefix are the sugars and alditols having seven or more carbon atoms, two examples of which are given on p. 54.

The nomenclature of the higher sugars and of the corresponding alditols and acids has undergone considerable modification. When synthesized from

* But see footnote, p. 29.

TABLE II

CONFIGURATIONAL PREFIXES

Carbons	Configuration and name[a]
1	H X—C—Y OH D-*glycero*

2	H H
	X—C—C—Y
	HO OH
	D-*erythro* H OH
	                X—C—C—Y
	                OH H
	                D-*threo*

3	H H OH
	X—C—C—C—Y
	OH OH H
	D-*arabino* H H H
	             X—C—C—C—Y
	             OH OH OH
	             D-*ribo* H OH H
	                       X—C—C—C—Y
	                       OH H OH
	                       D-*xylo* H OH OH
	                                X—C—C—C—Y
	                                OH H H
	                              D-*lyxo*

4	H H OH H
	X—C—C—C—C—Y
	OH OH H OH
	D-*gluco* H H OH OH
	            X—C—C—C—C—Y
	            OH OH H H
	            D-*manno* H OH OH H
	                       X—C—C—C—C—Y
	                       OH H H OH
	                       D-*galacto* H OH OH OH
	                                X—C—C—C—C—Y
	                                OH H H H
	                              D-*talo*
	
	H OH H OH
	X—C—C—C—C—Y
	OH H OH H
	D-*ido* H OH H H
	          X—C—C—C—C—Y
	          OH H OH OH
	          D-*gulo* H H H OH
	                    X—C—C—C—C—Y
	                    OH OH OH H
	                    D-*altro* H H H H
	                            X—C—C—C—C—Y
	                           OH OH OH OH
	                         D-*allo*

[a] The group Y is the main functional group such as CHO or CO_2H. Group Y is written at the top when the carbon chain is vertical. (X and Y cannot be hydrogen.)

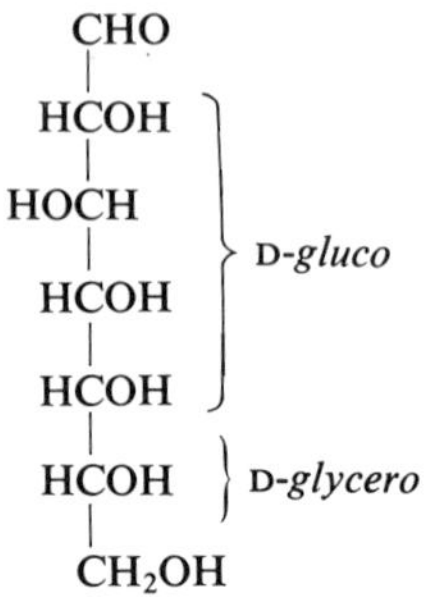

$$
\begin{array}{ll}
\text{CHO} & \\
\text{HCOH} & \\
\text{HOCH} & \left.\phantom{\begin{array}{c}a\\a\\a\end{array}}\right\} \text{D-}gluco \\
\text{HCOH} & \\
\text{HCOH} & \\
\text{HCOH} & \left.\right\} \text{D-}glycero \\
\text{CH}_2\text{OH} &
\end{array}
$$

D-*glycero*-D-*gluco*-Heptose

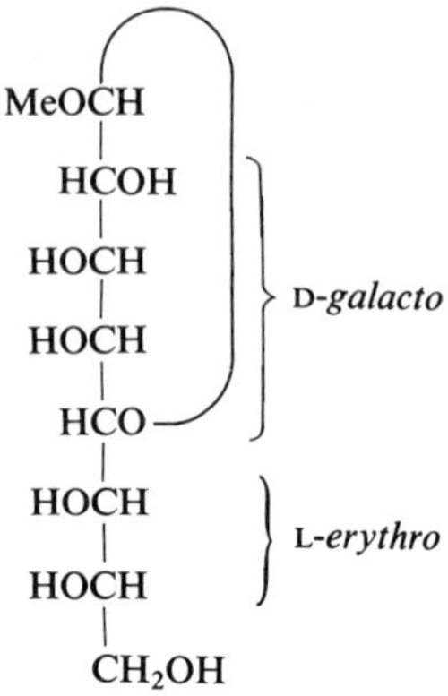

$$
\begin{array}{ll}
\text{MeOCH} & \\
\text{HCOH} & \\
\text{HOCH} & \\
\text{HOCH} & \left.\right\} \text{D-}galacto \\
\text{HCO} & \\
\text{HOCH} & \\
\text{HOCH} & \left.\right\} \text{L-}erythro \\
\text{CH}_2\text{OH} &
\end{array}
$$

Methyl L-*erythro*-β-D-*galacto*-
octopyranoside

the aldohexoses by the cyanohydrin synthesis of Kiliani, two aldoheptoses are derived from each aldohexose. Emil Fischer adopted the convention of naming the first isomer isolated the α-heptonic acid, and the second the β-heptonic acid. This process gave rise to such names as D-α-glucoheptonic and D-β-glucoheptonic acids for the acids derived from D-glucose. Isbell[47] later gave this usage a configurational significance, whereas Hudson[48] developed a system similar to that given above, except that overlapping prefixes were used.

For aldohexofuranoses and aldoheptopyranoses and higher sugars, the designation as a D or L sugar obviously does not determine the homomorphic ring series to which the sugar belongs. An example is given in the accompanying formulas for sialic acid (*N*-acetylneuraminic acid).

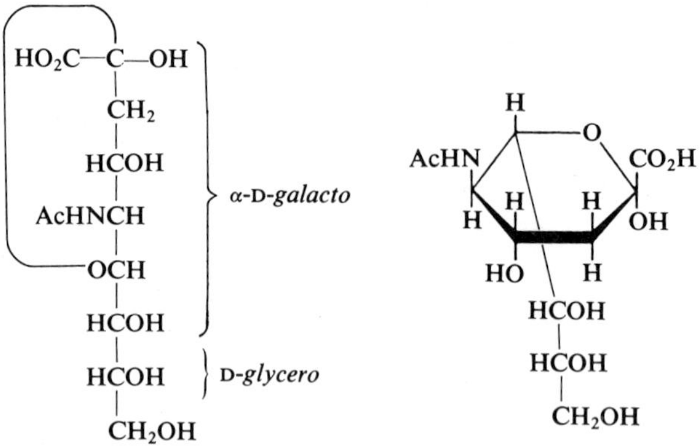

$$
\begin{array}{ll}
\text{HO}_2\text{C—C—OH} & \\
\text{CH}_2 & \\
\text{HCOH} & \\
\text{AcHNCH} & \left.\right\} \alpha\text{-D-}galacto \\
\text{OCH} & \\
\text{HCOH} & \\
\text{HCOH} & \left.\right\} \text{D-}glycero \\
\text{CH}_2\text{OH} &
\end{array}
$$

N-Acetylneuraminic acid (sialic acid)
5-Acetamido-3,5-dideoxy-D-*glycero*-α-D-*galacto*-nonulopyranosonic acid

This compound is named[1] α-D-*galacto*-, the D being based on C-7, as specified in the British–American Rules of Carbohydrate Nomenclature (see Vol. IIB, Chapter 46). Alternative names could conceivably be: 5-acetamido-3,5-dideoxy-α-D-*glycero*-D-*galacto*-nonulopyranosonic acid (by using C-8, the highest-numbered asymmetric center, as the reference atom), or 5-acetamido-3,5-dideoxy-D-*erythro*-β-L-*arabino*-nonulopyranosonic acid (by using C-6, the point of ring closure, as the reference atom). All three names are unambiguous. A minor difference between the British–American and the IUPAC–IUB Rules[1] is that the latter specify the highest-numbered asymmetric center as the reference atom regardless of chain length, so that the second name is favored by the IUPAC–IUB system; similarly the octoside depicted on p. 54 would be named methyl α-L-*erythro*-D-*galacto*-octopyranoside.

IX. CONFORMATIONAL REPRESENTATION OF CYCLIC SUGARS

The Haworth perspective formulas express more closely than the Fischer projection formulas the actual bond lengths in the cyclic sugars and their derivatives. The convention of establishing one plane for the atoms in the ring, and planes above and below the ring for the substituents, facilitates the recognition of configurational relationships. However, the Haworth formulas are not to be interpreted as indicating that the rings are planar in the actual molecule; rarely, if ever, do saturated ring systems having more than three atoms adopt a coplanar arrangement of the ring atoms. By rotation about sigma bonds it is possible for molecules of a cyclic sugar, having a given structure and configuration, to adopt an infinite number of relative orientations (conformations[37]) of the component atoms in three-dimensional space. The relative energies of the various possible conformations are determined by the net effect of various destabilizing factors, including bond-angle strain (Baeyer strain), bond-torsional strain (Pitzer strain), steric interactions (Van der Waals interactions), dipolar interactions, hydrogen-bonding effects, solvation effects, crystal-lattice forces, and the like[49] (for detailed considerations, see Chapter 5). The "favored conformation" is the conformation having the minimum net energy. For molecules within a crystal lattice, a single conformation is normally adopted, and its geometry can be determined quite exactly by crystallographic methods. For freely mobile molecules (as in solution or in the molten state), a conformational equilibrium is established in which the favored conformation is populated in equilibrium with other, higher-energy forms according to the classical thermodynamic distribution: $\Delta G° = -RT\ln K_{eq}$, where $\Delta G°$ is the difference in free energy between the favored conformation (A) and a higher-energy form (B) in equilibrium (equilibrium constant, $K_{eq} = [B]/[A]$) with it. The thermal energy in sugar molecules at room temperature is normally sufficient to establish a freely mobile system in conformational equilibrium[50] (see Chapter 5 for further details) if the molecule is capable of assuming more than one conformation.

Pyranoid sugars and their derivatives can be formulated in two energetically nonequivalent, chairlike conformations that are free from bond-angle strain. These conformations have potential-energy minima as compared with the various conformations theoretically possible. The idealized depictions given for the two possible chairlike shapes of β-D-glucopyranose show the axial and equatorial dispositions of the bonds to the various substituents. Interconversion

All-equatorial conformation   All-axial conformation

Chairlike conformational depiction of β-D-glucopyranose.

between the two chair conformers by "ring inversion" takes place readily at room temperature, so that isolation of the individual conformers is not possible. One of the two chairlike conformers is usually more stable than the other; for example, the all-equatorial form of β-D-glucopyranose is much more stable than the all-axial one. The conformational populations of pyranoid sugars and many of their derivatives can be measured by n.m.r. spectroscopy, and, at low temperatures, it is with certain derivatives possible to observe the spectra of the individual conformers.[50]

Conformational formulas of the type shown were also introduced for the sugars by Haworth;[37] they are idealized perspective drawings based on a model of cyclohexane. To avoid confusion they are simply termed "conformational formulas." They are readily related to the Haworth perspective formulas, and to the modified Fischer projection formulas used to develop the Haworth perspective formulas, by so orienting both the Haworth perspective and the conformational formulas that the ring numbering is clockwise (with the nearer edge of the ring shaded). In this orientation, all groups on the left of the modified Fischer formula are "up" in either the Haworth perspective or

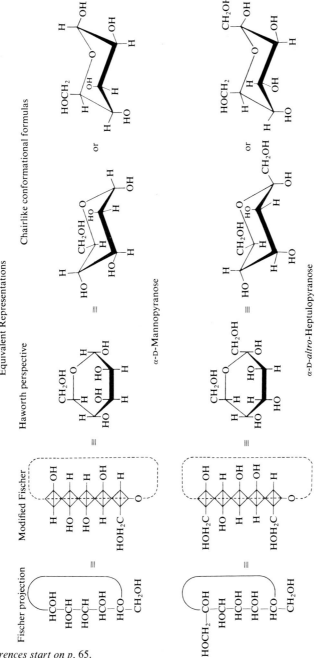

Equivalent Representations

Chairlike conformational formulas

Haworth perspective

Modified Fischer

Fischer projection

α-D-Mannopyranose

α-D-altro-Heptulopyranose

the conformational formulas, and groups that are on the right of the modified Fischer projection are "down" in both the Haworth perspective and the conformational formulas. Because only configurational relationships are correlated in this operation, the axial or equatorial dispositions in the conformational formulas are disregarded; the criterion in correlating the configurations is whether the groups are "up" or "down." These relationships are illustrated with reference to the Fischer projection, modified Fischer, Haworth perspective, and the two chairlike conformational depictions of α-D-mannopyranose (an aldohexopyranose) and β-D-*altro*-heptulopyranose (β-sedoheptulopyranose, a 2-ketoheptopyranose).

For a given anomer of an aldohexopyranose, it is possible to write twelve equally correct Haworth formulas. Six of them have clockwise numbering and are generated by rotating the three planes simultaneously through intervals of one-sixth of a revolution. The other six are generated in the same way, but by starting from the molecule drawn with counterclockwise numbering. There are also 372 additional variations, all of which represent other configurations of the parent structure. Because of the possibilities for confusion, it has, as already mentioned (see p. 42), become customary to adopt a standard ring orientation, in which the reducing carbon atom is on the extreme right and the ring is numbered clockwise. A simple paper model (Fig. 1) provides a convenient

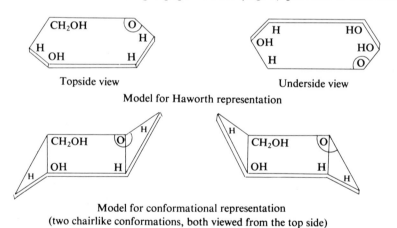

Topside view   Underside view
Model for Haworth representation

Model for conformational representation
(two chairlike conformations, both viewed from the top side)

FIG. 1. Cut-out paper or cardboard models for Haworth perspective and conformational structures. The example given is for α-D-glucopyranose.

device for easy recognition of the various correct orientations of the groups in a Haworth perspective formula. Models of this type, advocated by C. S. Hudson, consist of paper cut-outs of the ring system. After the ring-oxygen atom has been marked on one corner, on the top, and on the underside of the model, the ring is held in the conventional orientation and the various groups on the left of the modified Fischer projection are marked on the appropriate corners on the top of the model, and those groups on the right side of the modified Fischer projection are marked on the corresponding corners of the underside. The model can then be turned to any of the twelve possible orientations, and the Haworth structure having the correct orientations of all substituents can readily be drawn. The technique can be adapted for ring systems of any size, and it can also be extended to allow conformational depiction by appropriate folding and tilting of the corners of the model (Fig. 1), thus permitting drawing of configurationally correct depictions of the molecule in different conformations, in any desired orientation.

For depicting the relationships between dihedral angles of substituents along selected pairs of bonds on the pyranoid ring, the "Newman" type of formula[17] is useful, as illustrated for α-D-mannopyranose depicted in the favored chairlike form (A), and in the Newman depictions along C-1–C-2 and C-5–C-4 (B), along C-2–C-3 and O-5–C-5 (C), and along C-3–C-4 and C-1–O-5 (D).

Newman-type depictions of a chair form of α-D-mannopyranose.

There exists as yet no officially approved system of nomenclature for naming the conformers of sugars. A suitable structural formula can be used to express conformation without ambiguity, but it is frequently desirable to refer to a particular conformation by use of a suitable symbol as part of a name, in comparison in a table, or in a textual description. In practical terms, the major issue centers around the naming of the two chairlike forms of pyranoid sugars and their derivatives. A question of logic arises as to whether a symbol assigned should be independent of chirality (in the spirit of the α,β nomenclature for anomers), or whether it should refer to a particular ring shape and require the independent specification of chirality before the conformational symbol becomes meaningful. By the first system, a specific symbol is used to represent a given axial–equatorial disposition of substituent groups, and this applies to both enantiomers; the chirality must then be specified before a unique conformational structure for one enantiomer can be drawn. The second method allows the ring system to be drawn directly from the symbol, but leads to different symbols for mirror-image conformations (that is, enantiomorphs in chair conformations having the same axial–equatorial dispositions of substituents).

The system of conformational terminology that has thus far been utilized most widely was introduced by Reeves.[51] It is based on the second system just mentioned—that is, it assigns symbols for specified ring shapes, as shown below:

C1   *1C*

Reeves' symbols for pyranoid rings[51]

The symbols are defined as *C1* and *1C* for the ring shapes shown, with the understanding that the ring numbering runs clockwise (as the unsubstituted rings are superimposable). For these symbols to be meaningful, the chirality of the molecule must be specified, because, as shown on p. 61 for the two chairlike conformers of α-D-glucopyranose and α-L-glucopyranose, the symbol is changed when mirror-image conformations are named.

The conformational symbol by this system has no meaning unless the chirality of the molecules that it refers to is specified; it is correct to state "α-D-glucopyranose in the *1C* conformation," or "α-D-glucopyranose-*1C*," but if the chiral symbol is not given in the description of the compound, the appropriate chiral designation must be specified together with the conformational symbols. Thus, if it is desired to indicate a particular axial–equatorial relationship of the substituent groups in a sugar without specifying its chirality, two symbols must be used—for example, α-glucopyranose in the *C1* (D) or *1C* (L) conformation. This terminology, although explicit and unambiguous, is somewhat inconvenient to use in conjunction with such widely utilized physical tools as n.m.r. spectroscopy and X-ray crystallography, which specify information on the relative dispositions of the groups of the molecule, but seldom give chiral information directly.

Various alternatives to the Reeves system for the nomenclature of con-

α-D-Glucopyranose

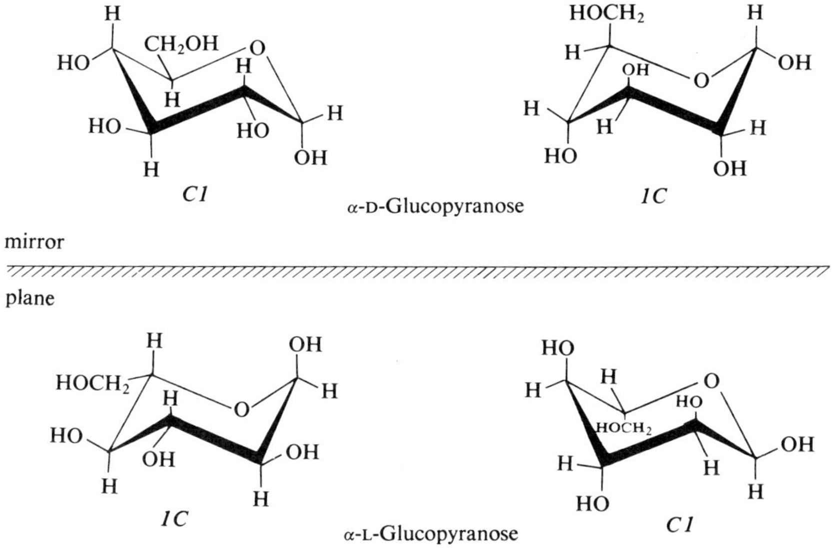

mirror

plane

α-L-Glucopyranose

Conformational designators by the Reeves system.

formers in pyranoid sugar rings have been proposed.[52-54] The system of Isbell and Tipson[52] embodies the principle of achiral, conformational symbols. To designate pyranoid chair conformations, the symbols CE and CA are used. For the purpose of assigning the conformational symbol, the particular anomeric configuration of the molecule under consideration is disregarded. That chair conformation which, in the α anomeric form, would have the anomeric substituent axial is assigned the symbol CA, and that chair conformation which, in its α anomeric form, would have the anomeric group equatorial is given the symbol CE. The use of these symbols is illustrated with reference to the conformational designation of both chairlike conformations of α-D-glucopyranose and α-L-glucopyranose. It is seen that mirror-image conformations receive the same symbol, thus following the system used for the anomeric designators α and β, which also remain the same, regardless of chirality.

Various arguments for and against both types of systems have been advanced. In support of the system of achiral conformational symbols, the obvious advantages of having a single symbol to describe a molecular shape independent of chirality is readily apparent; physical methods that give information concerning molecular geometry, but which do not differentiate between mirror images, are tools very frequently used with carbohydrate molecules, and when the Reeves terms are employed the symbolism required in such situations is clumsy. On the other hand, there is justification for the point of view that recognizes a particular molecular shape; such a system follows the general

* References start on p. 65.

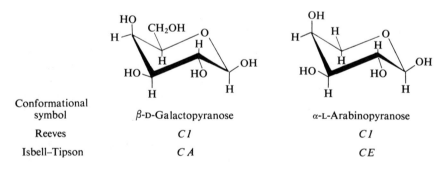

Conformational designators by the Isbell–Tipson system[52]

usage in the areas of steroid and terpene chemistry, and the Reeves symbols readily indicate the homomorphous relationship between, for example, β-D-galactopyranose and α-L-arabinopyranose when each of these is formulated in that chair conformation having the greater number of substituents equatorial. By the Reeves system, both of these molecules receive the same conformational symbol (*C1*), whereas by the system based on achiral conformational symbols, the symbols are different (*CA* and *CE*, respectively)

Conformational symbol	β-D-Galactopyranose	α-L-Arabinopyranose
Reeves	*C1*	*C1*
Isbell–Tipson	*CA*	*CE*

because the two compounds belong to different chiral series. The desirability of pointing out similarities between these molecules—as, for example, in recognition of a structure binding to an enzyme-acceptor site—provides justification for devising a system of conformational nomenclature in which these two ring systems receive the same conformational symbol.

The Isbell–Tipson system of conformational nomenclature has been much less widely used in the literature than the Reeves system, but there are numerous examples in which the Reeves system has been used in an ambiguous fashion because the chiral series was not specified. In the present book, the system adopted has been to employ the Reeves symbolism with the added chiral designators whenever the chirality is not immediately evident from the context; the four symbols used are $C1$ (D), $1C$ (D), $C1$ (L), and $1C$ (L).

In addition to the chair conformations, the pyranoid ring may also be formulated in a flexible cycle,[52] where it is also free from bond-angle strain. The cycle includes six boat forms and six skew forms; each boat form is a maximum-energy transition state between two skew forms. For unsubstituted sugars, nonbonded interactions render the skew, and more especially the boat, forms less stable than the favored chair form. The boat forms are insignificant contributors to the conformational population of free (unsubstituted) pyranoid sugars at room temperature.[49, 50] The energy required for conversion from the favored chair form into the flexible cycle is small compared with that required for most chemical reactions, and consequently the skew and boat forms may be significant intermediates in reactions of the sugars. Furthermore the skew or boat form may be the stable conformation for certain pyranoid sugar derivatives that are conformationally "locked" by bridging substituents.

Skew   Boat

Skew and boat ring shapes for a six-membered ring.

When four of the atoms of the pyranoid ring are constrained to coplanarity, as by a double bond or epoxide ring, two half-chair forms thereof are possible, one of which will usually be favored. When five of the six atoms are held coplanar, as by an unsaturated lactone ring, the single exoplanar atom leads to a shape that has been termed[52b, 55] a sofa conformation.

A generalized system of conformational nomenclature based on a proposal by Schwarz[54] gives a letter symbol (C, chair; S, skew; HC, half-chair, and so on); the symbols denote the (somewhat idealized) molecular shape by defining a plane that contains as many ring atoms as possible (four for a chair, boat, or skew form). The orientation of the remaining ring atoms as above or below this plane is identified by superscript or subscript numbers.

Half-chair conformations
(illustrated for D-glucal)

Sofa conformation
(for an unsaturated, pyranoid lactone)[56]

This type of system could be adapted, according to the way in which the terms are defined, to conform to either the achiral (Isbell–Tipson type) or chiral (Reeves type) concept of conformational symbol, and has been utilized[56a] in the latter manner. Furthermore, it can be used with furanoid and other ring systems, as well as pyranoid rings.

Furanoid rings are not planar because the relief of bond-angle (Baeyer) strain that exists in a planar ring would be outweighed by the excessive torsional (Pitzer) strain that a planar ring would generate between adjacent, eclipsed substituents. In furanoid sugar derivatives, the ring commonly adopts either an envelope conformation having four atoms (including the hetero atom) approximately coplanar, with the fifth atom out of the plane, or else a twist conformation wherein three adjacent atoms are coplanar and the other two (adjacent) lie above and below the plane of the other three. Such conformations for cyclopentane have been termed[57] C_s and C_2, respectively. For any furanoid ring there are, in principle, ten possible envelope forms and ten possible twist forms.[58] The energy barriers between these forms are low, and rapid interconversion undoubtedly takes place at room temperature by a "pseudorotation" type of process.

Envelope (V)   Twist (T)
(C_s type)   (C_2 type)
Conformations of furanoid rings.

REFERENCES

1. Rules of Carbohydrate Nomenclature, *J. Chem. Soc.*, 5307 (1962); *J. Org. Chem.* **28**, 281 (1963); see also IUPAC–IUB "Tentative Rules for the Nomenclature of Carbohydrates," *Biochemistry*, **10**, 3983 (1971); *Biochim. Biophys. Acta*, **244**, 223 (1971), and Vol. IIB, Chapter 46.

2. For more details of the history and earlier work, the reader is referred to the following references from which the present discussion was abstracted: E. O. von Lippmann, "Geschichte des Zuckers," 2nd ed., Berlin, 1929; "Beilstein's Handbuch der organischen Chemie," Vol. 31, Springer, Berlin, 1938; N. Deerr, "The History of Sugar," Chapman & Hall, London, 1949–1950.

3. For references see "Beilstein's Handbuch der organischen Chemie," Vol. 31, p. 83, Springer, Berlin, 1938.

4. For a detailed general treatment of stereochemistry, see E. L. Eliel, "Stereochemistry of Carbon Compounds," McGraw-Hill, New York, 1962.

5. E. Fischer, *Ber.*, **24**, 1836, 2683 (1891); see also, C. S. Hudson, *Advan. Carbohyd. Chem.*, **3**, 1 (1948).

6. See E. R. Alexander and A. G. Pinkus, *J. Amer. Chem. Soc.*, **71**, 1786 (1949); E. L. Eliel, *ibid.*, **71**, 3970 (1949); A. Streitwieser, *ibid.*, **75**, 5014 (1953).

7. A. G. Ogston, *Nature*, **162**, 963 (1948).

8. C. S. Hudson, *J. Chem. Educ.*, **18**, 353 (1941).

9. A. F. Peerdemann, A. J. van Bommel, and J. M. Bijvoet, *Nature*, **168**, 271 (1951); J. M. Bijvoet, *Endeavour*, **14**, 71 (1955); J. Trommel and J. M. Bijvoet, *Acta Crystallogr.*, **7**, 703 (1954).

10. M. A. Rosanoff, *J. Amer. Chem. Soc.*, **28**, 114 (1906); C. S. Hudson, *Advan. Carbohyd. Chem.*, **3**, 12 (1948).

11. A. Wohl and F. Momber, *Ber.*, **50**, 455 (1917).

12. E. Fischer, *Ber.*, **29**, 1377 (1896).

13. R. C. Hockett, *J. Amer. Chem. Soc.*, **57**, 2260 (1935).

14. Definitive Rules for the Nomenclature of Amino Acids and Related Substances, *J. Amer. Chem. Soc.*, **82**, 5575 (1960); *J. Org. Chem.*, **28**, 291 (1963).

15. M. L. Wolfrom, R. U. Lemieux, and S. M. Olin, *J. Amer. Chem. Soc.*, **71**, 2870 (1949).

16. R. S. Cahn, C. K. Ingold, and V. Prelog, *Experientia*, **12**, 81 (1956); *Angew. Chem. Intern. Ed. Engl.*, **5**, 385 (1966); R. S. Cahn, *J. Chem. Educ.*, **41**, 116 (1964).

17. M. S. Newman, *J. Chem. Educ.*, **32**, 344 (1955); similar formulas were used earlier by J. Böeseken and R. Cohen, *Rec. Trav. Chim.*, **47**, 839 (1928).

18. E. L. Eliel, N. L. Allinger, S. J. Angyal, and G. A. Morrison, "Conformational Analysis," Wiley (Interscience), New York, 1965, Chapter 1.

19. D. Horton and M. J. Miller, *J. Org. Chem.*, **30**, 2457 (1965); H. El Khadem, D. Horton, and T. F. Page, Jr., *ibid.*, **33**, 734 (1968); D. Horton and J. D. Wander, *Carbohyd. Res.*, **10**, 279 (1969); **15**, 271 (1970); H. El Khadem, D. Horton, and J. D. Wander, *J. Org. Chem.*, **37**, in press (1972).

20. G. A. Jeffrey and H. S. Kim, *Carbohyd. Res.*, **14**, 207 (1970); G. A. Jeffrey and E. J. Fasiska, *ibid.*, **21**, 187 (1972).

21. A. Villiers and M. Fayolle, *Bull. Soc. Chim. Fr.*, [3] **11**, 692 (1894); W. C. Tobie, *Ind. Eng. Chem.*, **14**, 405 (1942).

22. E. Fischer, *Ber.*, **28**, 1145 (1895).

23. E. Erwig and W. Koenigs, *Ber.*, **22**, 1464, 2207 (1889).

24. E. O. Erdmann, *Ber.*, **13**, 2180 (1880).

25. C. Tanret, *Compt. Rend.*, **120**, 1060 (1895).

26. See, for example, E. Fischer and K. Zach, *Ber.*, **45**, 456 (1912), footnote on p. 461.
27. H. S. Isbell and W. W. Pigman, *J. Res. Nat. Bur. Stand.*, **10**, 337 (1933); H. S. Isbell and C. S. Hudson, *ibid.*, **8**, 327 (1932); H. S. Isbell, *ibid.*, **8**, 615 (1932).
28. H. S. Isbell, *J. Amer. Chem. Soc.*, **55**, 2166 (1933).
29. E. F. Armstrong, *J. Chem. Soc.*, **83**, 1305 (1903).
30. H. S. Isbell and W. W. Pigman, *J. Res. Nat. Bur. Stand.*, **20**, 773 (1938).
30a. A furanose form has been established by X-ray crystallography for D-*altro*-3-heptulose (coriose): T. Okuda, K. Osaki, and T. Taga, *Chem. Commun.*, 851 (1969); see also p. 102.
31. S. J. Angyal and V. A. Pickles, *Carbohyd. Res.*, **4**, 269 (1967).
31a. W. Pigman and H. S. Isbell, *Advan. Carbohyd. Chem.*, **23**, 11 (1968); W. Pigman and H. S. Isbell, *Advan. Carbohyd. Chem. Biochem.*, **24**, 13 (1969).
32. For a summary, see J. Böeseken and H. Couvert, *Rec. Trav. Chim.*, **40**, 354 (1921); R. Verschuur, *ibid.*, **47**, 123, 423 (1928).
33. G. A. Jeffrey and R. D. Rosenstein, *Advan. Carbohyd. Chem.*, **19**, 7 (1964); G. Strahs, *ibid.*, **25**, 53 (1970).
34. E. L. Jackson and C. S. Hudson, *J. Amer. Chem. Soc.*, **59**, 994 (1937); M. Abdel-Akher, F. Smith, J. E. Cadotte, J. W. Van Cleve, R. Montgomery, and B. A. Lewis, *Nature*, **171**, 474 (1953).
35. E. M. Montgomery, N. K. Richtmyer, and C. S. Hudson, *J. Amer. Chem. Soc.*, **65**, 3 (1943).
36. J. Böeseken, *Ber.*, **46**, 2612 (1913); Böeseken did not make it clear that his formula had to be regarded as a view from below the plane of the ring, with the ring-oxygen atom remote from the observer.
37. H. D. K. Drew and W. N. Haworth, *J. Chem. Soc.*, 2303 (1926); compare, W. N. Haworth, "The Constitution of Sugars," Edward Arnold & Co., London, 1929.
38. C. G. Anderson, W. Charlton, and W. N. Haworth, *J. Chem. Soc.*, 1329 (1929).
39. C. S. Hudson, *J. Amer. Chem. Soc.*, **31**, 66 (1909); *Advan. Carbohyd. Chem.*, **3**, 15 (1948).
40. C. S. Hudson, *J. Amer. Chem. Soc.*, **60**, 1537 (1938). The Haworth-type perspective formulas given here have been turned-over by 180° in space, to be in conformity with the formulas as written in the present book. See previous discussion.
41. D. Horton, *J. Org. Chem.*, **29**, 1776 (1964); S. Guberman and D. Horton, *ibid.*, **32**, 294 (1967).
42. M. L. Wolfrom and R. L. Brown, *J. Amer. Chem. Soc.*, **63**, 1246 (1941).
43. For a further discussion of the naming of this type of compound, see M. L. Wolfrom, M. Konigsberg, and F. B. Moody, *J. Amer. Chem. Soc.*, **62**, 2343 (1940); R. J. Dimler and K. P. Link, *ibid.*, **62**, 1216 (1940).
44. H. S. Isbell, *Nat. Bur. Stand. Tech. Note* 405, p. 37 (Sept. 30, 1966).
45. R. M. Hann, A. T. Merrill, and C. S. Hudson, *J. Amer. Chem. Soc.*, **57**, 2100 (1935); R. M. Hann and C. S. Hudson, *ibid.*, **59**, 548 (1937); H. S. Isbell, *J. Res. Nat. Bur. Stand.*, **18**, 505 (1937); H. S. Isbell and W. W. Pigman, *ibid.*, **18**, 141 (1937). Many earlier workers had also noticed the resemblances in the structures for the members of the various series.
46. W. W. Pigman, *J. Res. Nat. Bur. Stand.*, **26**, 197 (1941).
47. H. S. Isbell, *J. Res. Nat. Bur. Stand.*, **18**, 529 (footnote) (1937).
48. C. S. Hudson, *J. Amer. Chem. Soc.*, **60**, 1537 (1938); *Advan. Carbohyd. Chem.*, **1**, 28 (1945).
49. See ref. 18, chapters 2, 4, and 6.
50. N. S. Bhacca and D. Horton, *J. Amer. Chem. Soc.*, **89**, 5993 (1967); P. L. Durette, D. Horton, and N. S. Bhacca, *Carbohyd. Res.*, **10**, 565 (1969); P. L. Durette and D. Horton, *Chem. Commun.*, 1608 (1971); *Advan. Carbohyd. Chem. Biochem.*, **26**, 49 (1971).

51. R. E. Reeves, *J. Amer. Chem. Soc.*, **71**, 215 (1949); *Advan. Carbohyd. Chem.*, **6**, 107 (1951).
52. H. S. Isbell and R. S. Tipson, (a) *Science*, **130**, 793 (1959); (b) *J. Res. Nat. Bur. Stand.*, **64A**, 171 (1960).
53. R. D. Guthrie, *Chem. Ind.* (London), 1593 (1958); D. F. Shaw, *Tetrahedron Lett.*, 1 (1965).
54. J. C. P. Schwarz, cited in L. Hough and A. C. Richardson, "Rodd's Chemistry of Carbon Compounds" (S. Coffey, Ed.), Elsevier, Amsterdam, Vol. I, part F, p. 90, 1967.
55. A. S. Smith, *Chem. Ind.* (London), 353 (1955); E. M. Philbin and T. S. Wheeler, *Proc. Chem. Soc.*, 167 (1958).
56. E. F. L. J. Anet, *Carbohyd. Res.*, **1**, 348 (1966).
56a. J. F. Stoddart, "Stereochemistry of Carbohydrates," Wiley—Interscience, New York, 1971.
57. J. E. Kilpatrick, K. S. Pitzer, and R. Spitzer, *J. Amer. Chem. Soc.*, **69**, 2483 (1947).
58. L. D. Hall, *Chem. Ind.* (London), 950 (1963).

2. OCCURRENCE, PROPERTIES, AND PREPARATION OF NATURALLY OCCURRING MONOSACCHARIDES (INCLUDING 6-DEOXY SUGARS)

ROBERT SCHAFFER

I. INTRODUCTION

The sugars that are found in naturally occurring materials are of particular importance because of interest in their biological function and in their present or potential industrial application. To the chemist, these sugars are of value in providing, along with the alditols and uronic acids, starting materials for the preparation both of the numerous monosaccharides known only by synthesis and of those natural monosaccharides which (because of inadequate sources or difficulties attending their isolation) are more readily obtained by transformation of other, abundant carbohydrates.

A. STRUCTURES OF THE NATURAL MONOSACCHARIDES

The chain length of the natural monosaccharides currently known ranges from 3 to 9 carbon atoms. Mainly, they are stereoisomeric members of the homologous series of sugars having ordinary structures; that is, they have a normal chain of carbon atoms, a carbonyl group at either C-1 or C-2, and one hydroxyl group on each remaining carbon atom. Structures in which a hydrogen

69

atom or an amino group replaces one hydroxyl group are termed deoxy sugars and amino sugars, respectively, and a number of such sugars occur. Further departures of natural monosaccharides from the ordinary structure are exemplified by a few sugars that have branched carbon chains or two carbonyl groups, and there is at least one natural sugar having its carbonyl group at C-3. Of the monosaccharides that occur in Nature, this chapter deals in detail only with aldoses and ketoses having a normal carbon chain, and with 6-deoxy-hexoses. Naturally occurring amino sugars are treated in Chapter 16; other deoxy sugars and the branched-chain sugars are discussed in Chapter 17. Enzymes, which catalyze reactions that provide some of the sugars found in Nature, are often capable of effecting similar reactions with substrates not normally available to these enzymes. However, if sugars result, they are not included in this chapter unless they have also been isolated from a natural product.

B. Monosaccharides in Nature

1. *Combined Monosaccharides*

The natural sugars commonly occur in glycosidic combinations (of two or more of the same or different sugars) that may extend in size to polymers—that is, as molecules ranging from oligosaccharides (disaccharides, trisaccharides, etc.) to polysaccharides, which may be composed of thousands of glycosidically linked sugar residues. (In some of these carbohydrates, uronic acid moieties are present as well; see Chapters 30, 39, 40, and 42.) Numerous substances occur in Nature in which saccharide combinations and sugars are linked glycosidically to a wide variety of nonsugar organic residues (see Chapter 32). Some sugars bear phosphate substituents; phosphorylated D-ribose and 2-deoxy-D-*erythro*-pentose ("2-deoxy-D-ribose") found in the polymeric nucleotides (nucleic acids) are notable in this respect. Sulfate esters of carbohydrates, and methyl ethers of many deoxy sugars, are some of the additional modifications of sugars found in glycosidic combination. Later chapters of this book are devoted to the occurrence, structure, and formation of many classes of complex substances in which natural sugars are found.

2. *Free Monosaccharides*

Prior to the advent of modern methods of separation and identification, the only sugars known to occur in botanic materials as free monosaccharides were two aldoses (D-glucose and L-arabinose) and four ketoses (D-fructose, L-sorbose, D-*manno*-heptulose, and sedoheptulose). Now, however, chromatographic techniques applied to extracts of plants have revealed traces of three free aldoses, already known to occur commonly in glycosidic combinations—namely, D-xylose, D-galactose, and D-mannose—and twelve other free sugars (hexoses to nonuloses), nine of which had not previously been recognized as naturally occurring. As regards the occurrence of free sugars in animals, D-glucose is a normal constituent of blood, lymph, and other body fluids. The blood of the newborn has been found to contain D-fructose; this ketose is also found in semen. Trace amounts of several C_5 and C_6 sugars in normal human urine have been detected by use of paper chromatography. Increased levels of two sugars, L-*threo*-pentulose (L-"xylulose") and D-glucose, are excreted in cases of essential pentosuria and diabetes, respectively.

II. NATURALLY OCCURRING MONOSACCHARIDES

A. PREPARATION AND ISOLATION

The preferred source materials for monosaccharide preparations, where they are available, are the polysaccharides built of repeated residues of a single sugar. A more complex natural product serves if the desired monosaccharide can be selectively liberated by hydrolysis or if other sugars present in the hydrolyzate can be effectively eliminated—for example, by fermentation; otherwise, the more complex source can provide at least a limited amount of a desired monosaccharide through use of some of the sensitive chromatographic separation methods that have been developed (see Chapter 28).

In the isolation of unsubstituted monosaccharides, care must be taken to prevent their being altered or destroyed—for example, as a result of excessive heating during concentration, or of keeping them in the presence of even a small proportion of alkali, which can effect profound changes in free sugars (see Chapter 4). Often, an effective procedure includes the preparation of a crystalline derivative from which purer sugar can be regenerated. Ultimately, however, isolation and purification require crystallization, preferably of the sugar itself. Most of the known sugars have been crystallized, and crystallizations of new preparations of them usually proceed without difficulty, although frequently this is facilitated by nucleating with a few crystals of the desired product. Impurities may impede crystallization; even small proportions of ionic contaminants can interfere, and ion-exchange treatments are commonly employed to remove their last traces.

B. PROPERTIES AND IDENTIFICATION

Sugars are often identified by comparison of their infrared or nuclear magnetic resonance spectra with those of known sugars, or identifications are based on likenesses in migration on chromatography in several solvent systems. Although these properties are highly significant for establishing identity with a known sugar, conclusive proof requires crystallization, derivatization, and

* *References start on p. 107.*

polarimetric measurements which should correspond with previous descriptions. Table I lists the natural sugars that are described further in this chapter and includes physical properties and a characteristic derivative for each sugar to the extent that these are known.

C. THE MONOSACCHARIDES

1. *Trioses*

D-Glyceraldehyde

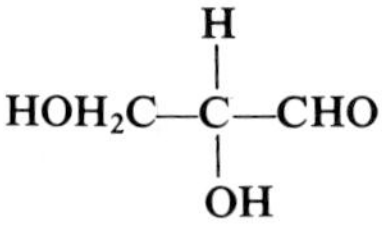

$$HOH_2C—\overset{\displaystyle H}{\underset{\displaystyle OH}{C}}—CHO$$

Synonym. D-Glycerose.

Occurrence. D-Glyceraldehyde occurs as the 3-phosphate in the intermediate stages of cellular metabolism and in related biological processes.

Preparation. D-Glyceraldehyde was first synthesized by Wohl and Momber,[1] who prepared it from resolved 3-amino-3-deoxy-DL-glyceraldehyde dimethyl acetal by deamination and hydrolysis. This preparation is of historical importance because their (*dextro*)-glyceraldehyde (D-glyceraldehyde) was converted into (*levo*)-tartaric acid (D-threaric acid), which provided a final experimental basis for the Rosanoff classification of sugar configurations. DL-Glyceraldehyde, known, in contrast to the optically active forms, as a crystalline material (dimer, m.p. 142°), has been resolved by condensation with (*dextro*)-α-(2-hydroxy-1-naphthyl)benzylamine.[2] Fischer and Baer introduced the glycol cleavage of 1,2:5,6-di-O-isopropylidene-D-mannitol, followed by acid hydrolysis,[3] to provide D-glyceraldehyde. However, the most convenient preparation is that of Perlin and Brice,[4] who oxidized D-fructose with a limited proportion of lead tetraacetate, and hydrolyzed the resulting derivative. The triose exhibits a "mutarotation," probably due to intermolecular hemiacetal bonding.[3] D-Glyceraldehyde 3-phosphate has been synthesized.[5]

1,3-Dihydroxy-2-propanone

$$HOH_2C—\overset{\displaystyle }{\underset{\displaystyle O}{\overset{\|}{C}}}—CH_2OH$$

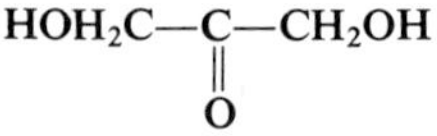

Synonyms. Oxantin, s-dihydroxyacetone.

Occurrence. As the 1-phosphate, 1,3-dihydroxy-2-propanone occurs in the intermediate stages of cellular metabolism and in the fermentation of hexoses.

Preparation. The oxidation of glycerol in the presence of *Acetobacter xylinum* gives the ketotriose, which can be separated as the insoluble sodium bisulfite addition compound, and then recovered on acidification. The dimer, m.p. 82°, is obtained on crystallization from ethanol,[6] but, on standing, higher-melting products are deposited from the mother liquor.[7] Distillation of the dimer gives the monomeric ketotriose,[8] m.p. 72°. The 1-phosphate has been synthesized.[9]

2. *Aldotetrose*

D-Erythrose

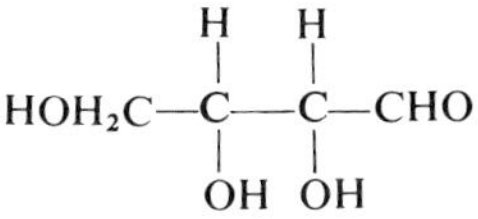

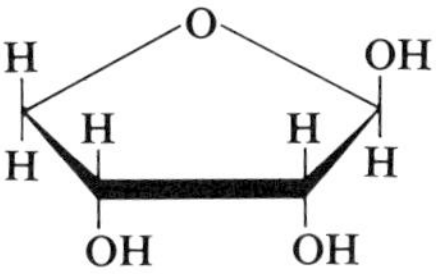

β-D-Erythrofuranose

Occurrence. D-Erythrose 4-phosphate is an intermediate in enzymic transformations of carbohydrates.

Preparation. D-Erythrose was first synthesized by the Wohl degradation;[10] subsequently, other methods usual for degrading an aldose (in this case, D-arabinose to D-erythrose) were employed. The Ruff procedure[11] had been recommended after an evaluation of the many routes to D-erythrose;[12] however, oxidative glycol-cleavage reactions provide better approaches to the tetrose. A high overall yield is obtained by oxidation of D-glucose with a limited proportion of lead tetraacetate.[13] D-Erythrose of the highest purity, as evidenced by maximal optical rotation,[14] results on hydrolysis of crystalline 2,4-*O*-ethylidene-D-erythrose. The tetrose is not fermentable by yeast. Synthesis of the 4-phosphate has been reported.[15]

3. *Aldopentoses*

D-Arabinose

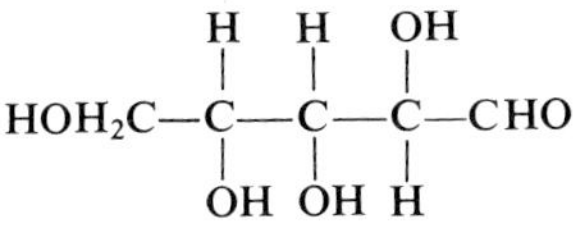

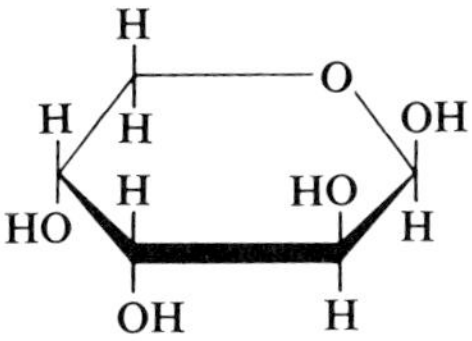

β-D-Arabinopyranose

TABLE I

PHYSICAL PROPERTIES AND DERIVATIVES OF SOME NATURAL MONOSACCHARIDES[1]

Sugar	Melting point[a] (degrees)	$[\alpha]_D^{20-25}$ (degrees, final in H_2O)	Characteristic derivative	Melting point (degrees)	$[\alpha]_D$ (degrees, final)
Aldotriose					
D-Glyceraldehyde[2]	Syrup	+8.7	Dimedon compound	196–8	+210 (EtOH)
Ketotriose					
1,3-Dihydroxy-2-propanone[3]	81 (dimer)	0	(p-Nitrophenyl)hydrazone[4]	160	0
Aldotetrose					
D-Erythrose[5]	Syrup	−41	(2,5-Dichlorophenyl)hydrazone[6]	110–2	−12.5 (MeOH)
Aldopentoses					
D-Arabinose[7,8]	159–60	−105	2-Benzyl-2-phenylhydrazone[9]	177–8	+14.4 (MeOH)
L-Arabinose[10,11]	160	+105	2,2-Diphenylhydrazone[12]	204–5	+14.9 (C5H5N)
L-Lyxose[11,13]	105	+13.8	(p-Bromophenyl)hydrazone[13]	157	
D-Ribose[14,15]	87	−23.7	(p-Bromophenyl)hydrazone[14]	172–3	−5.7 (EtOH)
D-Xylose[11,16]	145	+18.8	Di-O-benzylidene-, dimethyl acetal[17]	211–2	−7.0 (CHCl3)
Ketopentoses					
D-erythro-Pentulose[18]	Syrup	−16.3	(o-Nitrophenyl)hydrazone	168–9	−48 (MeOH)
D-threo-Pentulose[19]	Syrup	−33	(p-Bromophenyl)hydrazone[19]	128–9	−31 (C5H5N)
			2,3-O-Isopropylidene-[20]	70–1	+1.7 (Me2CO)
L-threo-Pentulose[21]	Syrup	+34.8	(p-Bromophenyl)hydrazone[21,22]	128	+31.5 (C5H5N)
Aldohexoses					
Allose[b]					
D-Galactose[11,23,24]	167 118–20 (monohydrate)	+80.2	2-Benzyl-2-phenylhydrazone[25]	157–8	−14.6 (C5H5N)

Sugar	[α]	mp (°C)	Derivative	mp (°C)	[α] (solvent)
L-Galactose[23,26]	-81	165	2-Methyl-2-phenylhydrazone[26]	189	-23 (MeOH)
D-Glucose[11,27]	+52.7	146	2-Benzyl-2-phenylhydrazone[25]	163-4	-48 (C₅H₅N)
		83-6 (monohydrate)	(p-Bromophenyl)hydrazone[25]	164-6	+19.8 (C₅H₅N)
D-Mannose[11,28]	+14.6	132	Phenylhydrazone[25,29]	199-200	+33.8 (C₅H₅N)
			Anhydro-, phenylhydrazone[30]	123-4	+12 (C₅H₅N)
D-Talose[31]	+21	120-1	2-Methyl-2-phenylhydrazone[31,32]	154	
Ketohexoses					
D-Fructose[33]	-92.4	102-4	(2,5-Dichlorophenyl)hydrazone[34]	154	+5.3 (C₅H₅N)
D-Psicose[35]	+3.1	Syrup	1,2:3,4-Di-O-isopropylidene-, furanose	57-8	-98 (Me₂CO)
L-Sorbose[36]	-43	159-61	(2,5-Dichlorophenyl)hydrazone[34]	117	-32.7 (C₅H₅N)
D-Tagatose[37,38]	-5	134-5	1,2:3,4-Di-O-isopropylidene-, pyranose[37]	65-6	+72 (H₂O)
			1,2,3,4,6-Pentaacetate[39]	132	+30.2 (CHCl₃)
6-Deoxyaldohexoses[c]					
6-Deoxy-D-allose[40]	+1.2	151-2	(p-Bromophenyl)hydrazone	145-6	-11.6 (C₅H₅N)
6-Deoxyaltrose[b]					
6-Deoxy-D-glucose[41]	+29.7	139-45	Phenylosazone	186-7	-95 (C₅H₅N, white light)
6-Deoxy-L-glucose[41a]			Phenylosazone	180-2	+44.8 (EtOH)
6-Deoxy-D-gulose[42]	-42.4	130-1	(p-Bromophenyl)hydrazone	133-5	
6-Deoxy-D-talose[43]	+20.6	129-31	Phenylosazone	176-8	
6-Deoxy-L-talose[44]	-20.9	116-8	(p-Bromophenyl)hydrazone	145-7	+3.8 (EtOH)
D-Fucose[45]	+76.3	140-45	2-Benzyl-2-phenylhydrazone[47]	178-9	-14.9 (MeOH)
L-Fucose[45,46]	-76	145	2,2-Diphenylhydrazone	198	-15.8 (C₅H₅N)
D-Rhamnose[48,49]	-8.2	90-1 (monohydrate)	(p-Bromophenyl)hydrazone[49]	167	
L-Rhamnose[11,50]	+8.2	93-4 (monohydrate)	Phenylhydrazone[51]	160	+27 (80% EtOH)

(continued)

TABLE I (*Continued*)

Sugar	Melting point[a] (degrees)	$[\alpha]_D^{20-25}$ (degrees, final in H_2O)	Characteristic derivative	Melting point (degrees)	$[\alpha]_D$ (degrees, final)
Aldoheptoses					
D-*glycero*-D-*galacto*-Heptose[52]	107 (monohydrate)	+64.7	Phenylosazone[53]	200	+27.2 (AcOH)
D-*glycero*-D-*gluco*-Heptose[54]	156–7	+46.2	α-Hexaacetate[55]	181–2	+107 (CHCl₃)
			β-Hexaacetate[54]	133–4	+19.6 (CHCl₃)
D-*glycero*-D-*manno*-Heptose[55]	Syrup		Hexaacetate	138–9	+66.5 (CHCl₃)
L-*glycero*-D-*manno*-Heptose[56,57]	Syrup		Phenylosazone[58]	202	−50.1 (C₅H₅N–EtOH)
7-Deoxy-D-*glycero*-D-*gluco*-heptose[58a]	180–90	+38	Phenylosazone	178	−75(C₅H₅N–EtOH)
Ketoheptoses					
D-*allo*-Heptulose[59]	130–2	+52.8	Phenylosazone	164–7	
D-*altro*-3-Heptulose[54]	~165	~+20			
D-*manno*-Heptulose[60]	152	+29.2	(*p*-Bromophenyl)hydrazone[60]	179	
			α-Hexaacetate[61]	110	+39 (CHCl₃)
Sedoheptulose[62,63]	Syrup	+8.2	Sedoheptulosan[62]	155	−146 (H₂O)
			Sedoheptulosan monohydrate[64]	101–2	−134 (H₂O)
			Sedoheptulosan tetrabenzoate[65]	166	−188 (CHCl₃)
D-*talo*-Heptulose[59]	135–7	+12.9	Phenylosazone	178	−75(C₅H₅N–EtOH)
7-Deoxy-D-*glycero*-D-*altro*-heptulose[58a]	160–70	+4			
Keto-octoses					
D-*glycero*-L-*galacto*-Octulose[66]	Syrup	−61	(2,5-Dichlorophenyl)hydrazone	178–80	
D-*glycero*-D-*manno*-Octulose[67]	Syrup	+20	(2,5-Dichlorophenyl)hydrazone	169–70	
			Phenylosazone	188–90	
Keto-nonoses					
D-*erythro*-L-*galacto*-Nonulose[68]	Syrup	−36.2	(2,5-Dichlorophenyl)hydrazone	247–9	
D-*erythro*-L-*gluco*-Nonulose[69,70]	Syrup	−47.2	(2,5-Dichlorophenyl)hydrazone[69]	248–50	

[a] The melting point is that of the usual crystalline modification.

[b] There is no evidence for assigning configuration to the natural sugar isolated. For the properties of synthetic D- and L-allose, see p. 84; for 6-deoxy-D- and L-altrose, see p. 95.

[c] For the naturally occurring methyl ethers, see Chapter 12.

1. For additional information and details, it is suggested that the following references in particular be consulted:

H. Vogel and A. Georg, "Tabellen der Zucker und ihrer Derivate," Springer Verlag, Berlin, 1931; H. Elsner, Ed., Tollens–Elsner, "Kurzes Handbuch der Kohlenhydrate," 4th ed., Barth, Leipzig, 1935 (Photo-lithoprint Reproduction, Edwards Brothers, Ann Arbor, Michigan, 1943); "Beilstein's Handbuch der organischen Chemie," Vol. 31. Springer Verlag, Berlin, 1938; F. J. Bates and Associates, *Nat. Bur. Stand. (U.S.) Circ.* **C440** (1942); M. L. Wolfrom, G. G. Maher, and R. G. Pagnucco in "Biology Data Book," P. L. Altman and D. S. Dittmer, Eds., Federation of American Societies for Experimental Biology, Washington, 1944, pp. 351–367; F. E. Hammerslag and B. Colesnick, *ibid.*, pp. 368–371; D. Horton in "Handbook of Biochemistry and Biophysics," H. C. Damm, Ed., Handbook House, Cleveland, Ohio, 1966, pp. 133–147; R. L. Whistler, M. L. Wolfrom, J. N. BeMiller, and F. Shafizadeh, Eds., "Methods in Carbohydrate Chemistry," Vol. 1. Academic Press, New York, 1962.

2. H. O. L. Fischer and E. Baer, *Helv. Chim. Acta*, **17**, 622 (1934); **19**, 519 (1936); *J. Amer. Chem. Soc.*, **61**, 761 (1939); A. S. Perlin and C. Brice, *Can. J. Chem.*, **34**, 85 (1956).

3. G. Bertrand, *Ann. Chim. Phys.*, **3** [8], 181 (1904).

4. H. O. L. Fischer and C. Taube, *Ber.*, **57**, 1502 (1924); H. P. den Otter, *Rec. Trav. Chim. Pays-Bas*, **56**, 474 (1937).

5. O. Ruff, *Ber.*, **32**, 3672 (1899); R. Schaffer, *J. Amer. Chem. Soc.*, **81**, 2838 (1959).

6. A. S. Perlin and C. Brice, *Can. J. Chem.*, **36**, 1216 (1955).

7. O. Ruff, *Ber.*, **32**, 550 (1899).

8. C. S. Hudson and E. Yanovsky, *J. Amer. Chem. Soc.*, **39**, 1013 (1917).

9. O. Ruff and G. Ollendorf, *Ber.*, **32**, 3234 (1899); E. Fischer, M. Bergmann, and H. Schotte, *ibid.*, **53**, 509 (1920).

10. C. Schiebler, *Ber.*, **1**, 108 (1868).

11. H. S. Isbell and W. W. Pigman, *J. Res. Nat. Bur. Stand.*, **18**, 141 (1937).

12. C. Neuberg, *Ber.*, **33**, 2253 (1900); A. Müther and B. Tollens, *ibid.*, **37**, 311 (1904).

13. W. Alberda van Ekenstein and J. J. Blanksma, *Chem. Weekbl.*, **11**, 189 (1914).

14. P. A. Levene and W. A. Jacobs, *Ber.*, **42**, 1198, 2102 (1909).

15. F. P. Phelps, H. S. Isbell, and W. W. Pigman, *J. Amer. Chem. Soc.*, **56**, 747 (1934).

16. H. I. Wheeler and B. Tollens, *Ann.*, **254**, 304 (1889).

17. L. J. Breddy and J. K. N. Jones, *J. Chem. Soc.*, 738 (1945); L. E. Wise and E. K. Ratliff, *Anal. Chem.*, **19**, 694 (1947).

18. C. Glatthaar and T. Reichstein, *Helv. Chim. Acta*, **18**, 80 (1935).

19. O. T. Schmidt and R. Treiber, *Ber.*, **66**, 1765 (1933).

20. P. A. Levene and R. S. Tipson, *J. Biol. Chem.*, **115**, 731 (1936).
21. P. A. Levene and F. B. LaForge, *J. Biol. Chem.*, **18**, 319 (1914); I. Greenwald, *ibid.*, **88**, 1 (1930); **89**, 501 (1930).
22. L. von Vargha, *Ber.*, **68**, 24 (1935).
23. E. Fischer and J. Hertz, *Ber.*, **25**, 1247 (1892).
24. H. Fudakowski, *Ber.*, **11**, 1069 (1878).
25. A. Hoffman, *Ann.*, **366**, 277 (1909).
26. C. Araki, *Nippon Kagaku Zasshi*, **59**, 424 (1938).
27. R. F. Jackson and C. G. Silsbee, *Sci. Papers Bur. Stand.* **17**, 715 (1922); O. Hesse, *Ann.*, **277**, 302 (1893); B. Tollens, *Ber.*, **17**, 2234 (1884).
28. W. Alberda van Ekenstein, *Rec. Trav. Chim. Pays-Bas*, **15**, 221 (1896).
29. C. L. Butler and L. H. Cretcher, *J. Amer. Chem. Soc.*, **53**, 4358 (1931).
30. M. L. Wolfrom and M. G. Blair, *J. Amer. Chem. Soc.*, **68**, 2110 (1946).
31. P. A. Levene and R. S. Tipson, *J. Biol. Chem.*, **93**, 631 (1931).
32. J. J. Blanksma and W. Alberda van Ekenstein, *Chem. Weekbl.*, **5**, 771 (1908).
33. C. S. Hudson and D. H. Brauns, *J. Amer. Chem. Soc.*, **38**, 1216 (1916); H. S. Isbell and W. W. Pigman, *J. Res. Nat. Bur. Stand.*, **20**, 773 (1938).
34. I. Mandl and C. Neuberg, *Arch. Biochem. Biophys.*, **35**, 326 (1952).
35. M. Steiger and T. Reichstein, *Helv. Chim. Acta*, **19**, 184 (1936).
36. H. H. Schlubach and J. Vorwerk, *Ber.*, **66**, 1251 (1933).
37. T. Reichstein and W. Bosshard, *Helv. Chim. Acta*, **17**, 753 (1934).
38. E. L. Totton and H. A. Lardy, *J. Amer. Chem. Soc.*, **71**, 3076 (1949).
39. Y. Khouvine and Y. Tomada, *Compt. Rend.*, **205**, 736 (1937); Y. Khouvine, G. Arragon, and Y. Tomada, *Bull. Soc. Chim. Fr.*, **6** [5], 354 (1939).
40. P. A. Levene and J. Compton, *J. Biol. Chem.*, **116**, 169 (1936).
41. E. Fischer and K. Zach, *Ber.*, **45**, 3761 (1912); K. Freudenberg and K. Raschig, *ibid.*, **62**, 373 (1929).
41a. I. F. Makarevich and D. G. Kolesnikov, *Khim. Prir. Soedin.*, **5**, 190 (1969).
42. P. A. Levene and J. Compton, *J. Biol. Chem.*, **111**, 335 (1935).
43. A. Markowitz, *J. Biol. Chem.*, **237**, 1767 (1962).
44. J. Schmutz, *Helv. Chim. Acta*, **31**, 1719 (1948).
45. E. Votoček, *Ber.*, **37**, 3859 (1904).
46. B. Tollens and F. Rorive, *Ber.*, **42**, 2009 (1909); J. Minsaas, *Rec. Trav. Chim. Pays-Bas*, **50**, 424 (1933).
47. A. Müther and B. Tollens, *Ber.*, **37**, 306 (1904).
48. W. T. Haskins, R. M. Hann, and C. S. Hudson, *J. Amer. Chem. Soc.*, **68**, 628 (1946).

49. E. Votoček and F. Valentin, *Compt. Rend.*, **188**, 62 (1926).
50. L. Behrend, *Ber.*, **11**, 1353 (1878).
51. E. Fischer and J. Tafel, *Ber.*, **20**, 2566 (1887); C. Tanret, *Bull. Soc. Chim. Fr.*, **27** [3], 395 (1902).
52. H. S. Isbell, *J. Res. Nat. Bur. Stand.*, **18**, 505 (1937); **20**, 97 (1938).
53. E. Fischer and F. Passmore, *Ber.*, **33**, 2226 (1890).
54. R. Begbie and N. K. Richtmyer, *Carbohyd. Res.*, **2**, 272 (1966); T. Okuda and K. Konishi, *Yakugaku Zasshi*, **88**, 1329 (1968).
55. D. A. Rosenfeld, N. K. Richtmyer, and C. S. Hudson, *J. Amer. Chem. Soc.*, **73**, 4907 (1951).
56. W. Weidel, *Hoppe-Seyler's Z. Physiol. Chem.*, **299**, 253 (1955); M. Teuber, R. D. Bevill, and M. J. Osborn, *Biochemistry*, **7**, 3303 (1968).
57. Crystalline, synthetic D-*glycero*-L-*manno*-heptose: see R. M. Hann, A. T. Merrill, and C. S. Hudson, *J. Amer. Chem. Soc.*, **56**, 1644 (1934); **57**, 2100 (1935).
58. From perseulose: R. M. Hann and C. S. Hudson, *J. Amer. Chem. Soc.*, **61**, 336 (1939).
58a. T. Ito, N. Ezaki, T. Tsuruoka, and T. Niida, *Carbohyd. Res.*, **17**, 375 (1971).
59. J. W. Pratt and N. K. Richtmyer, *J. Amer. Chem. Soc.*, **77**, 6326 (1955).
60. F. B. LaForge, *J. Biol. Chem.*, **28**, 511 (1917).
61. E. M. Montgomery and C. S. Hudson, *J. Amer. Chem. Soc.*, **61**, 1654 (1939).
62. F. B. LaForge and C. S. Hudson, *J. Biol. Chem.*, **30**, 61 (1917).
63. N. K. Richtmyer and J. W. Pratt, *J. Amer. Chem. Soc.*, **78**, 4717 (1956).
64. J. W. Pratt, N. K. Richtmyer, and C. S. Hudson, *J. Amer. Chem. Soc.*, **74**, 2200 (1952).
65. W. T. Haskins, R. M. Hann, and C. S. Hudson, *J. Amer. Chem. Soc.*, **74**, 2198 (1952).
66. H. H. Sephton and N. K. Richtmyer, *J. Org. Chem.*, **28**, 1691 (1963).
67. A. J. Charlson and N. K. Richtmyer, *J. Amer. Chem. Soc.*, **82**, 3428 (1960).
68. H. H. Sephton and N. K. Richtmyer, *Carbohyd. Res.*, **2**, 289 (1966).
69. H. H. Sephton and N. K. Richtmyer, *J. Org. Chem.*, **28**, 2388 (1963).
70. M. L. Wolfrom and H. B. Wood, Jr., *J. Amer. Chem. Soc.*, **77**, 3096 (1955).

Occurrence. This sugar is encountered infrequently. Cathartic-acting glycosides (aloins), such as barbaloin, homonataloin, isobarbaloin, and nataloin from plants of the genus *Aloe* (*A. barbadensis*), yield D-arabinose.[16] The sugar occurs in the furanose modification as a constituent of the polysaccharide fraction of tubercle bacilli.[17]

Preparation. D-Arabinose has the same configuration as the lower three asymmetric carbon atoms of D-glucose; therefore, all the methods for removing C-1 from D-glucose lead to D-arabinose or a derivative thereof. A convenient method is the oxidation of the readily obtainable calcium D-gluconate by hydrogen peroxide and ferric acetate.[18] In a novel procedure, *tert*-butyl tetra-*O*-acetyl-β-D-glucopyranoside gives D-arabinose quantitatively on treatment with dilute sodium methoxide.[19] D-Arabinose is not fermentable by yeast.

L-Arabinose

β-L-Arabinopyranose

Occurrence. The sugar occurs free in the heartwood of many coniferous trees. In a combined state, it is very widely distributed in plant products, being found in bacterial polysaccharides, gums, hemicelluloses, and pectic materials. Several glycosides yield the sugar on hydrolysis.

Preparation. Gum from mesquite (*Prosopis juliflora*), a plant common in the southwestern United States, and also cherry gum are good natural sources of L-arabinose. Mesquite gum consists of L-arabinose, D-galactose, and 4-*O*-methyl-D-glucuronic acid residues in combination, and cherry gum has, in addition, some D-xylose and D-mannose residues. The L-arabinose is present as readily hydrolyzed L-arabinofuranose residues, and, by controlled hydrolysis, most of the pentose is removed without hydrolyzing the other constituents to any great extent.[20] The L-arabinose is then partially purified by dialysis[21] or ion-exchange procedures,[22] and crystallized from ethyl alcohol. Australian black-wattle gum, peach gum, rye and wheat bran, and spent beet pulp have been utilized for the preparation of the pentose.

General Discussion. Although calcium chloride compounds of both the α- and β-L anomers have been crystallized,[23] only one crystalline anomer of the sugar itself is known, and this has usually been designated as the β-L anomer, following the nomenclature of Hudson. L-Arabinose is not fermentable by yeast.

L-Lyxose

$$HOH_2C \overset{\overset{\displaystyle OH}{|}}{\underset{\underset{\displaystyle H}{|}}{C}} \overset{\overset{\displaystyle H}{|}}{\underset{\underset{\displaystyle OH}{|}}{C}} \overset{\overset{\displaystyle H}{|}}{\underset{\underset{\displaystyle OH}{|}}{C}} CHO$$

α-L-Lyxopyranose

Occurrence. L-Lyxose is rare in Nature. It has been identified among the products of hydrolysis of the antibiotic curamycin.[24]

Preparation. The Ruff degradation of calcium L-galactonate, first employed by Alberda van Ekenstein and Blanksma for preparing L-lyxose, provides a useful route to the sugar.[25] Another route is by treatment of L-*galacto*-heptulose with periodic acid to cause selective C-2—C-3 cleavage.[25a]

D-Ribose

$$HOH_2C \overset{\overset{\displaystyle H}{|}}{\underset{\underset{\displaystyle OH}{|}}{C}} \overset{\overset{\displaystyle H}{|}}{\underset{\underset{\displaystyle OH}{|}}{C}} \overset{\overset{\displaystyle H}{|}}{\underset{\underset{\displaystyle OH}{|}}{C}} CHO$$

β-D-Ribopyranose

Occurrence. D-Ribose and 2-deoxy-D-*erythro*-pentose ("2-deoxy-D-ribose") are the carbohydrate constituents of nucleic acids, which are found in all plant and animal cells. D-Ribose is also a constituent of several coenzymes. In these natural products, the sugar occurs in the furanose modification. On hydrolysis of bacterial polysaccharides isolated from *Salmonella* and Arizona O groups, D-ribose was found as one of the resulting monosaccharides.[26]

Preparation. The best methods for laboratory preparation involve the stepwise hydrolysis of yeast nucleic acid.[27] The original procedure of Levene and Clark, which requires the action of ammonia at elevated temperatures and pressures, has been greatly improved by Phelps, who used magnesium oxide as the hydrolytic agent. The hydrolytic products, consisting of a mixture of nucleosides, are then further hydrolyzed by a mineral acid to produce D-ribose from the purine nucleosides.[28]

A similar method is based on the enzymic hydrolysis of yeast nucleic acid.[29] Emulsins prepared from alfalfa seeds, sweet almonds, and many sprouted seeds hydrolyze polynucleotides (nucleic acids) to the nucleosides. Guanosine (9-β-D-ribofuranosylguanine) is produced almost quantitatively, and adeno-

---

* *References start on p. 107.*

sine picrate (picrate of 9-β-D-ribofuranosyladenine) is obtained in high yield. As in the earlier methods, the purine nucleosides are then hydrolyzed by a mineral acid to give D-ribose. Crude D-ribose may be purified by preparing *N*-phenyl-D-ribosylamine, from which the free sugar is liberated on treatment with an aldehyde.[30]

D-Ribose may be synthesized from D-arabinose by alkaline isomerization, by the glycal synthesis, or through the pyridine-catalyzed epimerization of D-arabinonic acid followed by reduction. The sugar has also been prepared by the oxidative degradation of calcium D-altronate[31] and by the nitromethane synthesis from D-erythrose.[32] A practical synthesis has been described from 1,2:5,6-di-*O*-isopropylidene-D-glucose. The 3-methanesulfonate is degraded to 4-*O*-formyl-2-*O*-(methylsulfonyl)-D-arabinose, which gives D-ribose on treatment with alkali.[33]

Solutions of D-ribose in certain solvents contain considerable proportions of the furanose form (see Chapter 4). The mutarotation in water is complex and exhibits a minimum. D-Ribose is not fermentable by ordinary yeasts. The chemistry of D-ribose has been reviewed.[34]

D-Xylose

α-D-Xylopyranose

Synonyms. Wood sugar; in earlier literature, *l*-xylose.

Preparation. The sugar is prepared from corncobs (or many other woody materials) by boiling with acid, fermenting out the D-glucose with yeast, and crystallizing the D-xylose from the evaporated solution.[27, 35]

General Discussion. The presence of combined D-xylose in considerable quantities in many important agricultural wastes has stimulated interest in this sugar and its preparation. Corncobs, cottonseed hulls, pecan shells, and straw have been investigated as sources of the sugar, and several large-scale preparations[36, 37] have been performed. The sugar crystallizes fairly readily and could be made cheaply, but insufficient uses have been developed to make the manufacture of the sugar of commercial interest. Since it is not fermentable by ordinary yeasts or utilizable by many animals, the value of the sugar is considerably limited. Nevertheless, many bacteria and certain yeasts are able to ferment the sugar, with the formation of important compounds. Lactic acid and acetic acid in yields of 85 to 96% are formed[38] by the action of certain *Lactobacilli*.

Torula and *Monilia* yeasts grow well on hydrolyzed corncobs and straw, and the products are good cattle feeds.[37]

4. Ketopentoses

D-*erythro*-Pentulose

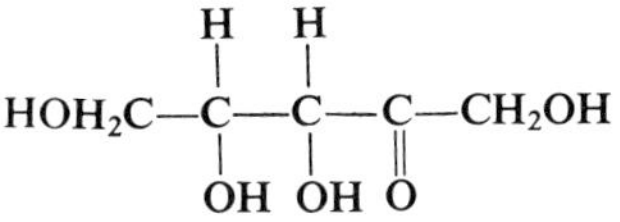

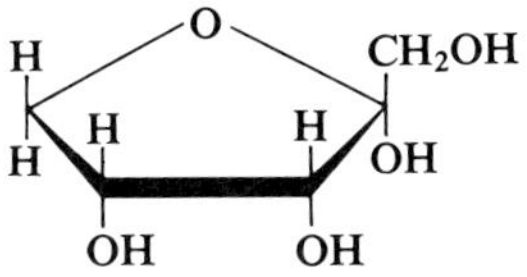

HOH$_2$C—C—C—C—CH$_2$OH

with H, H above the first two carbons and OH, OH, O below

α-D-*erythro*-Pentulofuranose

Synonyms. D-Arabinoketose, D-arabinulose, D-riboketose, D-ribulose.

Occurrence. Phosphorylated D-*erythro*-pentulose is an intermediate in the oxidative pathway of D-glucose metabolism by yeast or animal tissue, and is an early product of photosynthesis in plants.

Preparation. The ketopentose has been synthesized from D-arabinose by isomerization with pyridine followed by isolation as the crystalline (*o*-nitro-phenyl)hydrazone.[39] A useful photochemical route, involving extrusion of carbon monoxide from a derivative of D-fructose, has also been described.[39a] The free sugar has not been crystallized.

D-*threo*-Pentulose

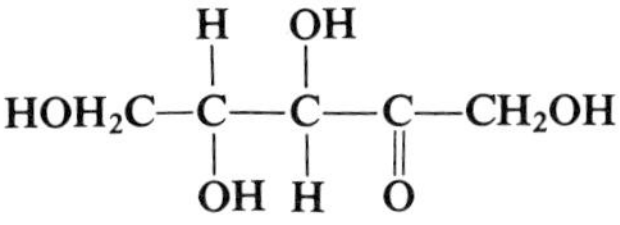

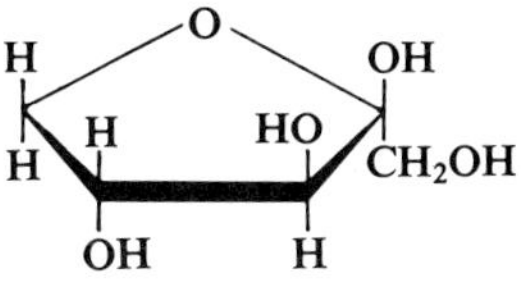

HOH$_2$C—C—C—C—CH$_2$OH

with H, OH above and OH, H, O below

β-D-*threo*-Pentulofuranose

Synonyms. D-Lyxoketose, D-lyxulose, D-xyloketose, D-xylulose.

Occurrence. D-*threo*-Pentulose 5-phosphate is an intermediate in the metabolism of D-xylose in bacterial extracts and is formed from D-*erythro*-pentulose 5-phosphate in the presence of epimerase. The 5-phosphate of D-*threo*-pentulose (but not of D-*erythro*-pentulose) is a substrate for transketolase.[40]

Preparation. D-*threo*-Pentulose is synthesized by isomerization of D-xylose in hot pyridine.[41] Better yields are obtained by the oxidation of D-arabinitol with *Acetobacter xylinum*.[42] The ketose is usually isolated as the distillable 2,3-*O*-isopropylidene-D-*threo*-pentulose,[43] which crystallizes readily.

* *References start on p.* 107.

L-*threo*-Pentulose

$$HOH_2C-\overset{\overset{\displaystyle OH}{|}}{\underset{\underset{\displaystyle H}{|}}{C}}-\overset{\overset{\displaystyle H}{|}}{\underset{\underset{\displaystyle OH}{|}}{C}}-\overset{\displaystyle}{\underset{\underset{\displaystyle O}{\|}}{C}}-CH_2OH$$

α-L-*threo*-Pentulofuranose

Synonyms. L-Arabinoketose, L-arabinulose, L-xyloketose, L-xylulose, urine pentose.

Occurrence. The ketopentose is found in the urine of many cases of pentosuria.

Preparation. The sugar has been synthesized by boiling L-xylose with pyridine, removing unchanged L-xylose by crystallization, and isolating the L-*threo*-pentulose as the (*p*-bromophenyl)hydrazone.[44]

General Discussion. The abnormal presence of pentoses in urine was known for a considerable time before the identification of the sugar as L-*threo*-pentulose by Levene and LaForge.[45] The precursor of the pentose is believed to be D-glucuronic acid, since administration of this substance induces the appearance of the ketopentose in the urine.[46]

5. *Aldohexoses*

Allose

$$HOH_2C-\overset{\overset{\displaystyle H}{|}}{\underset{\underset{\displaystyle OH}{|}}{C}}-\overset{\overset{\displaystyle H}{|}}{\underset{\underset{\displaystyle OH}{|}}{C}}-\overset{\overset{\displaystyle H}{|}}{\underset{\underset{\displaystyle OH}{|}}{C}}-\overset{\overset{\displaystyle H}{|}}{\underset{\underset{\displaystyle OH}{|}}{C}}-CHO$$

D-Allose   β-D-Allopyranose

Occurrence. An allose was obtained as a radioactive syrup by chromatography of extracts of *Ochromas malhamensis* that had been obtained after a prolonged photosynthesis by the organism in $^{14}CO_2$, but no evidence was obtained for assigning the sugar to the D or L series.[47]

Preparation. Both D-[48] and L-allose[49] have been prepared (from D- and L-ribose, respectively) by the Fischer–Kiliani cyanohydrin synthesis. The crystalline β-D- and β-L-allopyranoses are obtained (m.p. 129°, and a dimorph having m.p. 141°; at equilibrium, $[\alpha]_D^{20}$ +14° and −14°, respectively). A convenient, large-scale route to D-allose is from 1,2:5,6-di-*O*-isopropylidene-α-D-glucofuranose through configurational inversion at C-3 by various methods.[49a]

D-Galactose

$$HOH_2C-\overset{\overset{\displaystyle H}{|}}{C}-\overset{\overset{\displaystyle OH}{|}}{\underset{\underset{\displaystyle OH}{|}}{C}}-\overset{\overset{\displaystyle OH}{|}}{\underset{\underset{\displaystyle H}{|}}{C}}-\overset{\overset{\displaystyle H}{|}}{\underset{\underset{\displaystyle H}{|}}{C}}-CHO$$

α-D-Galactopyranose

Synonyms. Cerebrose, brain sugar.

Occurrence. The sugar is a constituent of several oligosaccharides, notably lactose, melibiose, and raffinose. Polysaccharides that yield D-galactose on hydrolysis include agar, gum arabic, mesquite gum, western larch gum, and many other plant gums and mucilages. The K or gel fraction of carrageenan has been shown to consist mainly of D-galactopyranose and 3,6-anhydro-D-galactopyranose residues.[50] A few glycosides have also been reported to yield D-galactose on hydrolysis (idaein, myrtillin, the cerebrosides). D-Galactose occurs glycosidically combined with *myo*-inositol[51] in sugar beets, and with glycerol[52] in certain algae. Crystalline D-galactose has been observed on ivy berries. Traces of the free sugar have been detected in fruits[53] and in the heartwood of *Chamaecyparis lawsonia*.[54]

D-Galactose is one of the few sugars, other than D-glucose and 2-deoxy-D-*erythro*-pentose, which is found distributed to any great extent in the animal kingdom. In combination with D-glucose, as the disaccharide lactose, it is an important constituent of the milk of mammals. D-Galactose polysaccharides from animal sources include the galactans of beef lung, frog spawn, and the albumin gland of the snail (*Helix pomatia*). Many of the cerebrosides and gangliosides, occurring in brain and nerve tissue, are D-galactosides (see Chapter 44). It is a common constituent of glycoproteins.

Preparation. The method most frequently used involves the hydrolysis of lactose by acids, and fractional crystallization of the D-galactose liberated.[55] A modification of the method entails removal of the D-glucose by fermentation with yeasts, and crystallization of the D-galactose remaining. Water-soluble gums extractable from the western or eastern larch also serve as sources of the sugar.[56]

General Discussion. The usual crystalline modification of the sugar is α-D-galactopyranose, although the β-D anomer is obtained by crystallization from cold, alcoholic solution. D-Galactose and D-glucose differ only in the configuration of C-4, and this difference is accompanied by a greater tendency for D-galactose to give furanose derivatives. As a result, the mutarotation of the D-galactose isomers does not follow the first-order equation. Considerable

* *References start on p. 107.*

proportions of furanose isomers are formed when the sugar is acetylated directly. The fermentation of D-galactose by D-galactose-adapted and lactose-fermenting yeasts has received much study.[57]

L-Galactose

$$HOH_2C-\overset{\overset{\displaystyle OH}{|}}{\underset{\underset{\displaystyle H}{|}}{C}}-\overset{\overset{\displaystyle H}{|}}{\underset{\underset{\displaystyle OH}{|}}{C}}-\overset{\overset{\displaystyle H}{|}}{\underset{\underset{\displaystyle OH}{|}}{C}}-\overset{\overset{\displaystyle OH}{|}}{\underset{\underset{\displaystyle H}{|}}{C}}-CHO$$

α-L-Galactopyranose

Occurrence. Several polysaccharides including agar-agar, chagual gum, and flaxseed mucilage afford L-galactose on hydrolysis, and, as D-galactose is usually present, some DL-galactose is isolated. Galactan from snails also gives D- and L-galactose on hydrolysis.[58]

Preparation. Synthetic methods are the most convenient, although preparation from agar and from flaxseed mucilage has been described.[59] The separation of L-galactose from natural or synthetic DL mixtures is accomplished by fermentation of the D-galactose by D-galactose-adapted yeasts, or by resolution of the hydrazones formed from 1-[(+)-2-pentyl]-1-phenylhydrazine.[60]

Reduction of the readily available D-galacturonic acid to L-galactonic acid and of this to L-galactose may be employed for the preparation of this sugar.[61]

D-Glucose

$$HOH_2C-\overset{\overset{\displaystyle H}{|}}{\underset{\underset{\displaystyle OH}{|}}{C}}-\overset{\overset{\displaystyle H}{|}}{\underset{\underset{\displaystyle OH}{|}}{C}}-\overset{\overset{\displaystyle OH}{|}}{\underset{\underset{\displaystyle H}{|}}{C}}-\overset{\overset{\displaystyle H}{|}}{\underset{\underset{\displaystyle OH}{|}}{C}}-CHO$$

α-D-Glucopyranose

Synonyms. Blood sugar, corn sugar, dextrose, grape sugar.

Occurrence. This sugar, in free or combined form, is not only the most common of the sugars but also is probably the most abundant organic compound. It occurs free in blood, cerebrospinal fluid, fruits, honey, lymph, plant juices, and urine, and is a major component of many oligosaccharides, polysaccharides (particularly cellulose, glycogen, and starch), and D-glucosides, including sucrose.

Preparation. D-Glucose (often called dextrose commercially, because of its dextrorotation) is manufactured on a large scale from starch,[62] usually potato starch in Europe and corn starch in the United States. An aqueous suspension of starch, containing 0.25 to 0.5 % of hydrochloric acid (by weight of starch), is put in a converter. Steam is passed into the converter, and a pressure of about 40 pounds per square inch is maintained until a 90 to 91 % conversion into D-glucose has been achieved. The acid solution is then run into tubs, and brought to pH 4.8 with sodium carbonate. Fatty materials from the starch are removed by centrifugation, and protein and insoluble carbohydrates are then filtered off. Alternatively, fats and proteins are removed from the initial hydrolyzate by coagulation with bentonite, and the clear solution is purified by ion exchange. A cation exchanger removes metal ions, and then an anion exchanger removes acids. The sugar solution from either process is decolorized and purified by passage through bone black (animal charcoal) and, after evaporation to approximately 30° Be. (about 55% by weight), is refiltered through bone black. The filtrate is then evaporated in a vacuum pan. Subsequent treatment depends on the product desired.

The last stage in the process is the most difficult to conduct on a large scale, because for economy an aqueous solution should be used for the crystallization, and the crystals should be homogeneous (at least three forms are possible) and easily centrifuged and washed. The conditions under which the pyranose forms of D-glucose and the hydrate are stable are illustrated in the phase diagram[63] of the system D-glucose–water (Fig. 1).

Below 50° (C), α-D-glucose·H_2O is the stable crystalline phase, but above 50°, the anhydrous form is obtained. At higher temperatures, β-D-glucose forms the solid phase. Although at any temperature it is usually possible to obtain any form by the addition of the proper seed crystals, this is usually not desirable, since the introduction of seed crystals of the more stable modification will result in a change to the latter if equilibrium conditions are attained. In the commercial process, these conditions are met for the hydrate by cooling the liquid at a concentration of about 40° Be. (about 77% by weight) to a temperature of about 50°, and after seeding heavily with the hydrate, allowing it to crystallize while the mass is stirred and slowly cooled. The crystals are then separated by centrifugation and passed through driers.

For preparation of the anhydrous material, the crystals are developed at higher temperatures in the vacuum pan while the evaporation is taking place. This is done by first evaporating about 15 to 20% of the total batch to a thick syrup (90% dry substance) and allowing crystals to form spontaneously. The remainder of the batch is then used to dilute the seed formed, and the evaporation is continued. When the crystals have developed to the desired point, the mass is passed into a centrifuge and the mother liquors are removed. During this final stage and during the washing, the proper temperatures are maintained to prevent spontaneous formation of the hydrate. The crystals are finally dried by filtered warm air.

* *References start on p. 107.*

β-D-Glucose has proved of some technical interest because of its initial solubility. It has been prepared[64] by dissolution in hot pyridine and crystallization at 0°. The accompanying molecule of pyridine is removed at 105°. The β-D anomer is also prepared[65] by crystallization from hot acetic acid and recrystallization from water and alcohol. At temperatures above about 115°, β-D-glucose is the stable form in contact with a saturated aqueous solution

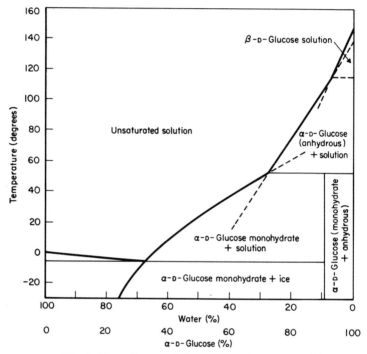

Fig. 1. Phase diagram of the system D-glucose–water.[63]

(see Fig. 1). Because of the high solubility at these temperatures, very concentrated solutions must be used. It is possible to work at somewhat lower temperatures (100°) if the seed of the α-D anomer is excluded. β-D-Glucose may be prepared by seeding a concentrated D-glucose solution at 100° with β-D-glucose and evaporating it at this temperature to a solid mass.[66] Spray-drying of a hot, concentrated solution of D-glucose produces a mixture of the α- and β-D anomers.[67]

In industrial circles, the term "glucose" is used to describe a partially hydrolyzed starch product that consists of dextrins, oligosaccharides, maltose, and D-glucose. The material is also designated as C.S.U. (corn syrup, unmixed).

The commercial material is made by autoclaving aqueous starch suspensions with acid. It usually has a reducing power in the range of 40 to 45 % of the same weight of D-glucose; the concentration of solid material lies in the range of 78 to 85 %.

"Hydrol" is the mother liquor remaining from the preparation of D-glucose, and it corresponds to the "molasses" of cane-sugar refining. The sugar content of a typical hydrol consists of about 65 % of D-glucose and 35 % of disaccharides and higher oligosaccharides. Disaccharides that have been identified[68] in hydrol include cellobiose, gentiobiose, maltose, 5-O-β-D-glucopyranosyl-D-glucose, 6-O-α-D-glucopyranosyl-D-glucose ("isomaltose," "brachiose"), and α,α-trehalose. Except for the 6-O-α-D-glucopyranosyl-D-glucose, which arises principally as a residuum of the original starch structure,[69] these disaccharides are formed mainly by acid reversion of the D-glucose.[70] Amylopectin, hydrolyzed under conditions leading to negligible reversion, yields 3-O-α-D-glucopyranosyl-D-glucose, thus indicating the presence of this linkage in starch.[71] Other oligosaccharides in hydrol are isomaltotriose, "isopanose," 3-O-maltosyl-D-glucose, panose, and a trisaccharide containing an α-D-(1 → 2) and an α-D-(1 → 6) linkage.[72] D-Glucose is fermented by yeast.

D-Mannose

α-D-Mannopyranose

Synonyms. Seminose, carubinose.

Occurrence. There are many polysaccharides that yield D-mannose on hydrolysis. For preparative purposes, the most important source is the seed of the tagua palm,[73] *Phytelephas macrocarpa,* also known as vegetable ivory. Salep mucilage from tubers of Orchidaceae, the seed of *Phoenix canariensis,* and white spruce hemicellulose are rich enough sources of D-mannose that they have been used for the preparation of this sugar. Konjac flour, which is commonly available in Japan from corms of *Amorphophallus konjac,* provides another useful source of the hexose.[74] Other mannans are proliferated by yeasts and by the red alga *Porphyra umbilicalis.*[75] D-Mannose has also been reported as a constituent of blood serum globulins, ovomucoid, and tubercle

References start on p. 107.

bacilli. Traces of the free sugar in apple and peach fruits have been revealed by chromatography.[53]

Preparation. Shavings obtained as by-products from the manufacture of buttons from the ivory nut (*Phytelephas macrocarpa*) are considered the best source. The shavings are hydrolyzed with mineral acid, and, by fractionation with alcohols, D-mannose is separated from other substances and crystallized directly from alcoholic solution or, alternatively, is converted into the readily crystallizable methyl α-D-mannopyranoside. Direct crystallization of D-mannose is a considerable improvement over the earlier methods, which separated the sugar as its phenylhydrazone.[76]

General Discussion. Both anomers of the pyranose sugar are known, and either may be obtained from aqueous solution by adding nucleating crystals of the desired form to a supersaturated solution. The importance of having seed crystals is well illustrated by this sugar. The single anomer known for many years was β-D-mannose, but in laboratories in which the α-D anomer had been obtained, it became very difficult to obtain the more soluble β-D form. β-D-Mannose can be obtained only by very careful exclusion of the seed of the α-D anomer.[76a]

D-Mannose forms with calcium chloride a readily crystallizable compound[77] of the formula $C_6H_{12}O_6 \cdot CaCl_2 \cdot 4H_2O$, which exhibits complex mutarotation having a maximum. The addition compound appears to contain the furanose modification of the sugar. D-Mannose is fermented by yeast.

D-Talose

α-D-Talopyranose

Occurrence. D-Talose is obtained on hydrolysis of the antibiotic hygromycin.[78] The natural occurrence of the hexose was first reported by Hesse, who isolated a hydrate (m.p. 89–90°, $[\alpha]_D^{15}$ +18.8°) as a hydrolysis product of cocacitrin.[79]

Preparation. D-Talose was first prepared by E. Fischer[80] by sodium-amalgam reduction of the D-talono-1,4-lactone obtained by epimerization of D-galactonic acid. The sugar is also obtained by oxidation of D-galactal,[81] or by configurational-inversion procedures from derivatives of D-mannose or D-galactose.[81a] D-Talose is fermented by yeast.

6. Ketohexoses

D-Fructose

$$HOH_2C-\overset{\overset{\displaystyle H}{|}}{\underset{\underset{\displaystyle OH}{|}}{C}}-\overset{\overset{\displaystyle H}{|}}{\underset{\underset{\displaystyle OH}{|}}{C}}-\overset{\overset{\displaystyle OH}{|}}{\underset{\underset{\displaystyle H}{|}}{C}}-\overset{\overset{\displaystyle}{\|}}{\underset{\underset{\displaystyle O}{}}{C}}-CH_2OH$$

β-D-Fructopyranose

Synonyms. D-*arabino*-Hexulose, levulose, fruit sugar.

Occurrence.[82] D-Fructose is found, usually accompanied by sucrose, in an uncombined form in fruit juices and honey. Apples and tomatoes are said to contain particularly large proportions of the sugar. Sucrose consists of D-fructose and D-glucose residues in glycosidic union. Plants of the family Compositae contain polysaccharides of D-fructose (inulins). It is of interest that common weeds, such as Jerusalem artichoke, burdock, dandelion, and goldenrod, as well as chicory and dahlias utilize inulins as reserve polysaccharides. The sugar is a frequent constituent of oligosaccharides, often combined with D-glucose as a sucrose residue, but it rarely occurs in glycosides other than oligosaccharides.

Preparation. The abundance and wide distribution of D-fructose in natural proc icts, its sweetness, and its poor crystallizing property have stimulated considerable experimental work on methods of preparation. Most methods for isolation of the sugar depend on the formation of a difficultly soluble calcium "levulate" or "fructosate" in which one molecule of the sugar is combined with one of lime. The compound is washed free from impurities, such as other sugars and inorganic salts, and decomposed to D-fructose and (insoluble) calcium carbonate by carbonation.[27]

The best source of D-fructose for large-scale purposes is probably the inversion of sucrose by acids or invertase. The separation of the ketose from the concomitant D-glucose may then be accomplished by direct crystallization, by removal of the D-glucose after oxidation with bromine to D-gluconic acid (the ketose is not affected), or by precipitation of the calcium "fructosate." Hydrolysis of the natural inulins mentioned above may also serve for the preparation of D-fructose, which is isolated from the hydrolyzate by precipitation of the lime complex. Conditions have been patented for preparing D-fructose by the action of alkali on D-glucose.[83] D-Fructose is fermented by yeast.

General Discussion. Only one crystalline isomer of the sugar is known, and this is probably the β-D-pyranose form. In solution, however, as indicated by evidence obtained from mutarotation studies, a considerable proportion of a

* References start on p. 107.

furanose modification (possibly also of β-D configuration) is present. When combined in natural products, the sugar is always found as a furanose.

Most tests have shown D-fructose to be the sweetest of the sugars, although the actual ratios between the various sugars depend to a considerable extent on the taster and on the methods and conditions adopted for the comparison. Compared to a sweetness value for sucrose of 100, that for D-fructose has been reported as varying from 103 to 173. In Table II are given the alleged relative sweetnesses of some sugars and other organic compounds.

D-Psicose

$$\begin{array}{c}\text{H} \quad \text{H} \quad \text{H}\\ | \quad | \quad |\\ \text{HOH}_2\text{C}-\text{C}-\text{C}-\text{C}-\text{C}-\text{CH}_2\text{OH}\\ | \quad | \quad | \quad ||\\ \text{OH} \quad \text{OH} \quad \text{OH} \quad \text{O}\end{array}$$

α-D-Psicopyranose

Synonyms. D-Allulose, D-altrulose, D-*ribo*-hexulose.

Occurrence. D-Psicose is produced in the hydrolysis of the antibiotic psicofuranine.[84] Although the sugar was isolated only as the phenylosazone, its identity was proved by synthesis of psicofuranine from tetra-O-acetyl-D-psicosyl chloride. The ketose also occurs in *Itea* plants, from which it is isolated as the diisopropylidene acetal.[85] Allitol accumulates during photosynthesis by the *Itea* plants: in the light, D-psicose is converted into allitol; in the dark, the reverse process is operative.

Preparation. Isomerization of D-allose in anhydrous pyridine gives syrupy D-psicose, which is isolated as the crystalline 1,2:3,4-diisopropylidene acetal.[86] Diazomethane plus D-ribonyl chloride tetraacetate gives a slightly more dextrorotatory preparation.[87] A promising, convenient synthesis is the isomerization of D-fructose by N,N'-dicyclohexylcarbodiimide in hot methanol to D-psicose, D-glucose, and D-mannose, followed by fermentation with yeast.[88] D-Psicose is not fermentable by yeast. A review of the chemistry of D-psicose has appeared.[89]

L-Sorbose

$$\begin{array}{c}\text{OH} \quad \text{H} \quad \text{OH}\\ | \quad | \quad |\\ \text{HOH}_2\text{C}-\text{C}-\text{C}-\text{C}-\text{C}-\text{CH}_2\text{OH}\\ | \quad | \quad | \quad ||\\ \text{H} \quad \text{OH} \quad \text{H} \quad \text{O}\end{array}$$

α-L-Sorbopyranose

TABLE II

RELATIVE SWEETNESS OF SOME ORGANIC COMPOUNDS[1]

Compound	Relative sweetness
Lactose	0.27
D-Glucitol	0.48
Glycerol	0.48
D-Glucose	0.5–0.6
Maltose	0.60
Invert sugar	0.8–0.9
Sucrose	1.0
D-Fructose	1.0–1.5
Sodium cyclamate (sodium cyclohexanesulfamate)[2]	30
Naringin dihydrochalcone[3]	70–230[a]
Dulcin [(p-ethoxyphenyl)urea]	70–350
Saccharin (1,2-benzisothiazolin-3-one 1,1-dioxide)	200–700
Stevioside[4]	300
Neohesperidin dihydrochalcone[3]	500–1600[a]
Perillaldehyde α-anti-oxime	2000
5-Nitro-2-propoxyaniline[5]	4100

[a] Sweetness had been measured relative to that of saccharin; the values listed are adjusted to that of sucrose.

1. C. F. Walton, Jr., "International Critical Tables," E. W. Washburn, Ed., McGraw-Hill, New York, Vol. I, 1926, p. 357; see also R. S. Shallenberger and T. E. Acree, in "Chemical Structure of Compounds and Their Sweet and Bitter Taste," Handbook of Sensory Physiology (L. M. Beidler, Ed.), Vol. 4, part 2, "Chemical Senses ,Taste," Springer-Verlag, Heidelberg, 1972.
2. L. F. Audrieth and M. Sveda, J. Org. Chem., 9, 89 (1944).
3. R. M. Horowitz, in "Biochemistry of Phenolic Compounds," J. B. Harborne, Ed., Academic Press, New York, 1964, p. 545.
4. E. Thomas, Bull. Assoc. Chim., 54, 844 (1937).
5. J. J. Blanksma and P. W. M. Van der Weyden, Rec. Trav. Chim., 59, 629 (1940); P. E. Verkade, C. P. Van Dijk, and W. Meerberg, ibid., 65, 346 (1946).

Synonyms. L-xylo-Hexulose, sorbinose; in earlier literature, d-sorbose.

Occurrence. L-Sorbose has been reported in the enzymic hydrolyzate of a pectin from the skin of the passion fruit (Passiflora edulis).[90] Although L-sorbose is found in the fermented juice of mountain-ash berries (Sorbus aucuparia L.), it has been shown to be a secondary product, formed by the oxidation of D-glucitol by such bacteria as Acetobacter xylinum.[91]

Preparation. The biochemical oxidation of D-glucitol is the most convenient route to this sugar, which, as an intermediate in the commercial synthesis of

* References start on p. 107.

L-ascorbic acid, is prepared in large quantities by this method. Early researches by Bertrand showed that D-glucitol may be oxidized by sorbose bacteria (*Acetobacter xylinum* Brown) to L-sorbose. Yields of 50 to 75% were reported. By performing the fermentation with *Acetobacter suboxydans* in rotating drums, instead of utilizing surface cultures of Bertrand's organism, yields of over 90% are obtained.[92] The chemistry of L-sorbose has been reviewed.[89]

D-Tagatose

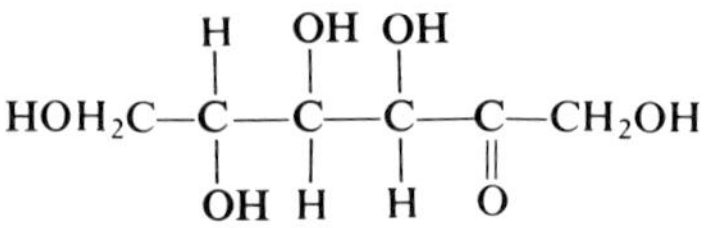

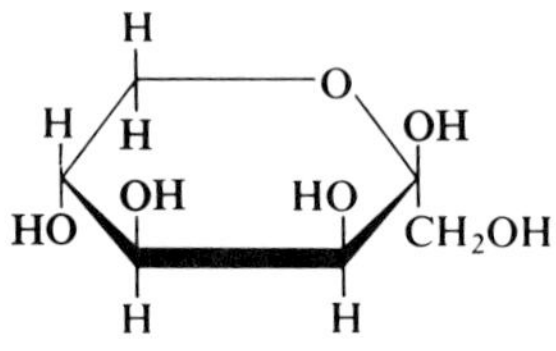

HOH$_2$C—C—C—C—C—CH$_2$OH

β-D-Tagatopyranose

Synonym. D-*lyxo*-Hexulose.

Occurrence. D-Tagatose has been obtained as a hydrolytic product from a gum exudate of the tropical tree *Sterculia setigera.*[93]

Preparation. D-Tagatose was originally synthesized by the alkaline isomerization of D-galactose.[94] It is more readily obtained from D-altritol (D-talitol) by oxidation with *Acetobacter suboxydans;*[95] however, this hexitol is not a readily available material. A newer route to D-tagatose is afforded by the isomerization of D-galactose with *N,N'*-dicyclohexylcarbodiimide.[88] The chemistry of D-tagatose has been reviewed.[89]

7. 6-Deoxyaldohexoses

6-Deoxy-D-allose

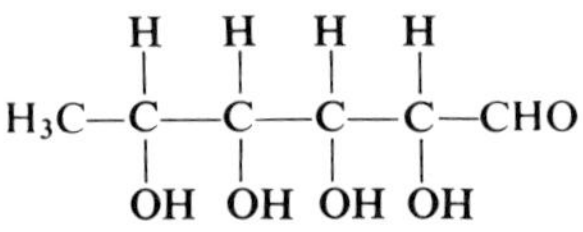

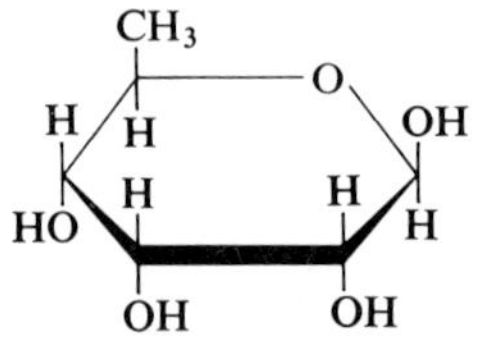

H$_3$C—C—C—C—C—CHO

6-Deoxy-β-D-allopyranose

Synonym. D-Allomethylose.

Occurrence.[96] The sugar is found in cardiac glycosides obtained, for example, from *Gomphocarpus fruticosus*[97] and *Digitalis canariensis* var. *isabelliana.*[98]

Preparation. 6-Deoxy-D-allose may be obtained through the series of configurational inversions that occur on treatment of methyl 2,3-*O*-isopropylidene-5-*O*-*p*-tolylsulfonyl-L-rhamnofuranoside with sodium methoxide.[99]

6-Deoxyaltrose

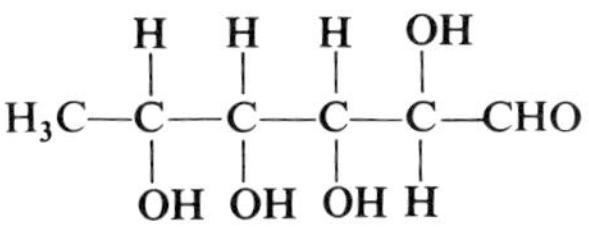

$$H_3C-\overset{\overset{\displaystyle H}{|}}{C}-\overset{\overset{\displaystyle H}{|}}{\underset{\underset{\displaystyle OH}{|}}{C}}-\overset{\overset{\displaystyle H}{|}}{\underset{\underset{\displaystyle OH}{|}}{C}}-\overset{\overset{\displaystyle OH}{|}}{\underset{\underset{\displaystyle OH}{|}}{C}}-\overset{}{\underset{\underset{\displaystyle H}{|}}{C}}-CHO$$

6-Deoxy-D-altrose 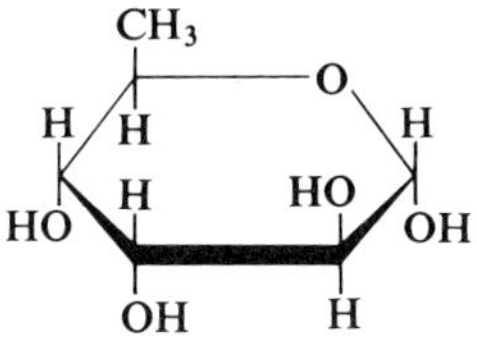   6-Deoxy-α-D-altropyranose

Synonym. Altromethylose.

Occurrence.[96] A cardiac glycoside from *Strophanthus gratus* is reported to contain the sugar, which was identified only by paper chromatography; hence, its D or L configuration is not known.

Preparation. Hygromycin A, which contains the 5,6-enol form of 6-deoxy-D-*arabino*-hexos-5-ulose,[100] gives 6-deoxy-D-altrose on reduction and hydrolysis. A related synthesis is the dehydrohalogenation of 6-deoxy-6-iodo-1,2:3,4-di-*O*-isopropylidene-D-galactose, followed by reduction and hydrolysis to give 6-deoxy-L-altrose plus 6-deoxy-D-galactose (D-fucose).[101] Other methods have been employed for preparing the deoxyaltrose.[102] 6-Deoxy-D- and -L-altrose have been obtained only as syrupy materials having $[\alpha]_D$ of $+16.2°$ and $-18°$, respectively.

6-Deoxy-D-glucose

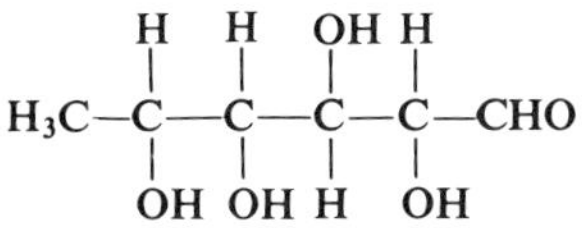

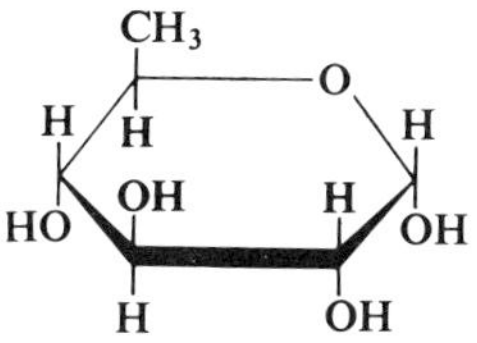

6-Deoxy-α-D-glucopyranose

Synonyms. Chinovose, D-epirhamnose, D-glucomethylose, D-isorhamnose, isorhodeose, D-quinovose.

Occurrence.[101, 103] The bark of many species of *Cinchona* contains a glycoside (quinovin or chinovin) which is extracted along with the quinine alkaloids. On treatment with ethanolic hydrogen chloride, an ethyl 6-deoxy-D-glucoside is obtained. Convolvulin is a mixture of glycosides, one of which yields 6-deoxy-D-glucose on hydrolysis[104] (see under D-Fucose). The sugar also occurs as a glycoside of digitoxigenin.[98] The Fischer–Zach preparation of 6-deoxy-D-glucose, by treating a 6-bromo-6-deoxy-D-glucoside with zinc and acetic acid, marks the earliest conversion of a sugar having an "ordinary" structure into a 6-deoxy sugar.[105]

* *References start on p. 107.*

The L-enantiomorph has been stated [105a] to occur as a constituent of alloside, a cardiac glycoside.

6-Deoxy-D-gulose

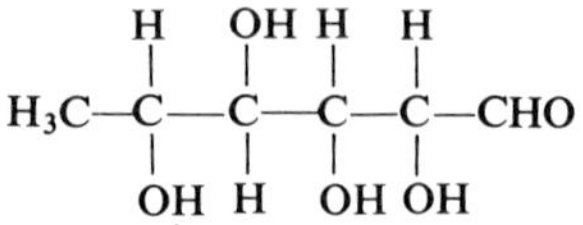

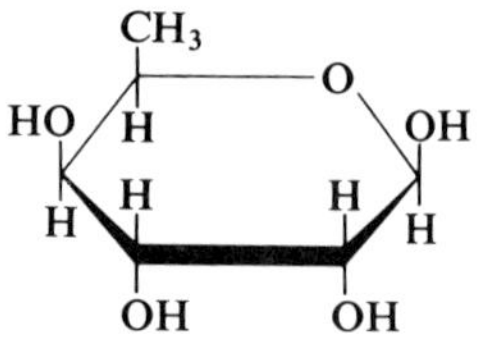

6-Deoxy-β-D-gulopyranose

Synonyms. D-Antiarose, D-gulomethylose.

Occurrence.[96] 6-Deoxy-D-gulose occurs in steroidal glycosides obtained, for example, from sap of the upas tree, *Antiaris toxicaria*,[106] from the wallflower, *Cheiranthus cheiri*,[107] or from *Digitalis canariensis* var. *isabelliana*.[98]

Preparation. Levene and Compton prepared the crystalline sugar by application of the cyanohydrin synthesis to 5-deoxy-D-xylose.[108]

6-Deoxy-D-talose

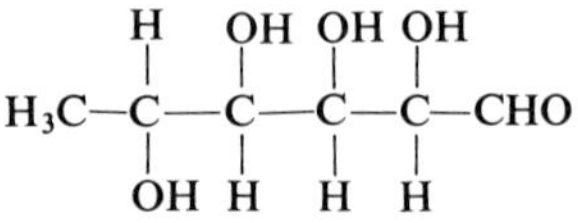

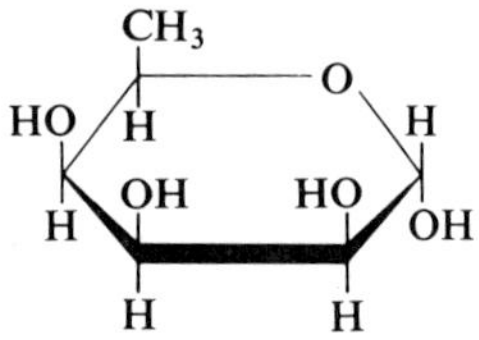

6-Deoxy-α-D-talopyranose

Synonym. D-Talomethylose.

Occurrence. This sugar is found, accompanied by D-rhamnose, among the products of hydrolysis of the capsular polysaccharide of gram-negative bacteria, strain GS.[109]

6-Deoxy-L-talose

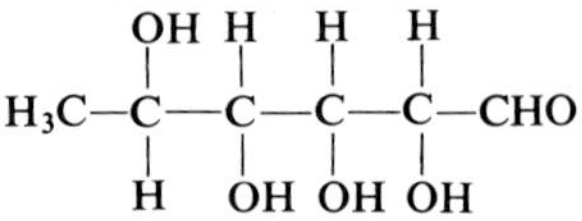

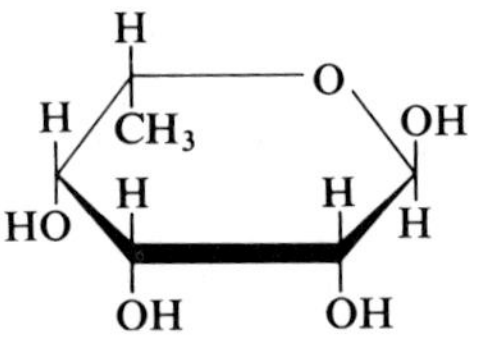

6-Deoxy-α-L-talopyranose

Synonym. L-Talomethylose.

Occurrence. Cardiac glycosides isolated from several plants contain 6-deoxy-L-talose as a component.[96] The sugar is also found in the cell wall carbohydrate of a strain of *Actinomyces bovis*,[110, 111] in the K antigen of *Pseudomonas pseudomallei*,[110] and in a group of specific glycolipids, from *Mycobacterium avium*, called mycoside C.[110, 112]

Preparation. 6-Deoxy-L-talose is synthesized by sodium-amalgam reduction of 6-deoxy-L-talonolactone, which is obtained by epimerization of L-fuconic acid.[113]

D-Fucose

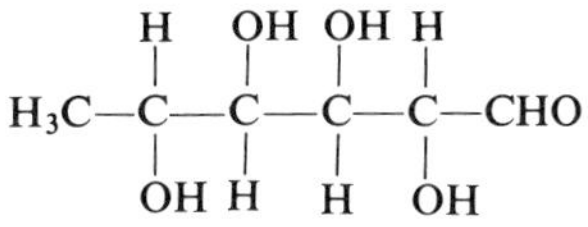

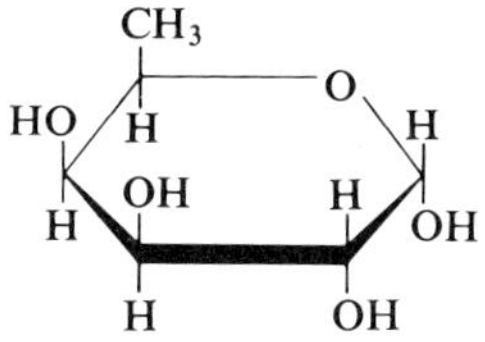

α-D-Fucopyranose

Synonyms. 6-Deoxy-D-galactose, D-galactomethylose, D-rhodeose.

Occurrence and Preparation. This sugar is found in the hydrolytic products of cardiac glycosides.[96] The roots of certain South and Central American plants (Convolvulaceae), used as purgatives, give resins of a glycosidic nature: Jalap resin (convolvulin) and Scammonium or Tampico jalap (jalapin) are obtained from *Tubera jalapae* and *Ipomoea orizabensis*, respectively. Jalapin and convolvulin give, among other products, L-rhamnose, 6-deoxy-D-glucose, and D-fucose.[114] (For synthetic D-fucose, see under 6-Deoxyaltrose.)

L-Fucose

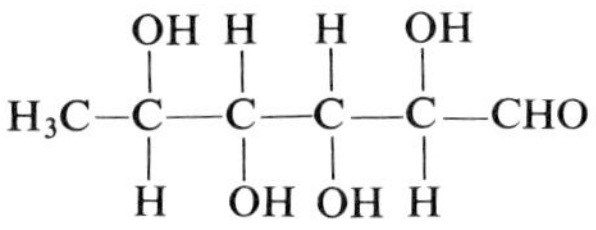

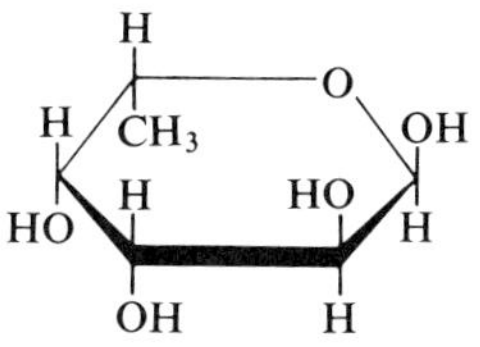

α-L-Fucopyranose

Synonyms. 6-Deoxy-L-galactose, L-galactomethylose, L-rhodeose.

Occurrence.[115] L-Fucose is a constituent of polysaccharides obtained from eggs of the sea urchin, frog spawn, gum tragacanth, and marine algae (seaweed). It is a common constituent of glycoproteins, including mucins and blood-group substances[116] and of milk oligosaccharides[117] (see Vol. IIA, Chap. 30, and Vol. IIB, Chap. 43).

Preparation.[115] Seaweed (*Fucus vesiculosus*) is hydrolyzed with acid. After deionization, and precipitation of impurities with methanol, the sugar is crystallized. To obtain the L-fucose remaining in the mother liquor, the phenyl-

---

* *References start on p. 107.*

hydrazone is prepared. Alternatively, the deionized hydrolyzate is fermented with D-galactose-adapted yeasts.[118]

D-Rhamnose

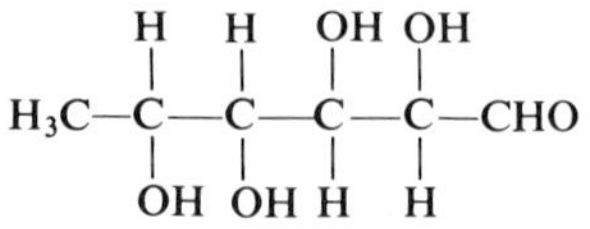

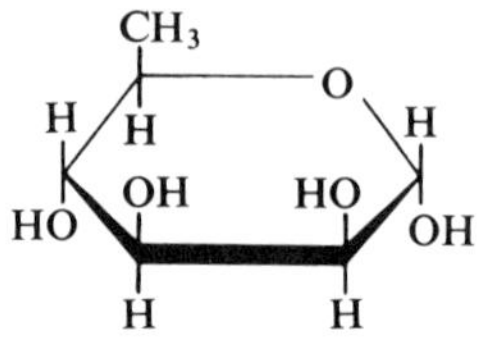

α-D-Rhamnopyranose

Synonyms. 6-Deoxy-D-mannose.

Occurrence. D-Rhamnose (accompanied by 6-deoxy-D-talose) is found among the hydrolysis products of the capsular polysaccharide of gram-negative bacteria strain GS.[109]

Preparation. D-Rhamnose is synthesized from 6-p-toluenesulfonic esters of D-mannose by displacement of the sulfonyloxy group.[119]

L-Rhamnose

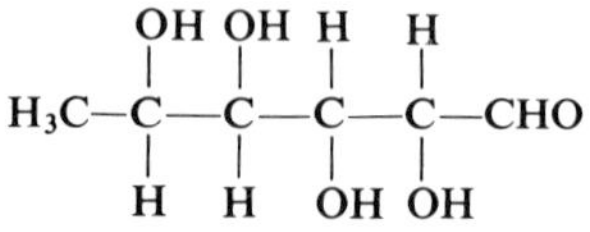

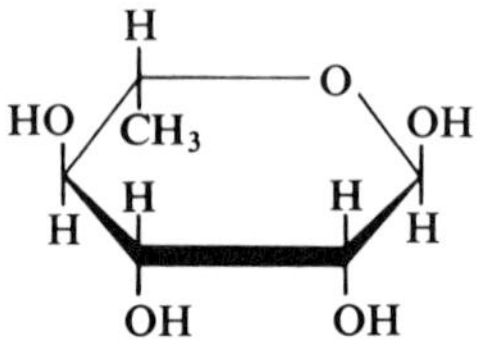

α-L-Rhamnopyranose

Synonyms. 6-Deoxy-L-mannose, "isodulcit," L-mannomethylose.

Occurrence. The sugar is a constituent of many glycosides, which provide its best source. It was isolated in the free state from extracts of the leaves and blossoms of poison ivy, *Rhus toxicodendron* L., where it is normally present as a glycoside; occurrence of the free sugar probably results from hydrolysis.[120] Some polysaccharides of gums and mucilages contain L-rhamnose. The sugar has also been detected in various immunological polysaccharides of bacterial origin[121] and in cardiac glycosides.[96]

Preparation. "Lemon flavin," a khaki dyestuff obtained from the bark of an oak species (*Quercus tinctoria* Mich.), provides an excellent source of the sugar. The main constituent of lemon flavin is the rhamnoside quercitrin, which, after hydrolysis, yields the aglycon (quercetin) and L-rhamnose. The lemon flavin is hydrolyzed by boiling acid, and, after neutralization of the solution and treatment with a considerable proportion of decolorizing carbon, the sugar crystallizes from the evaporated solution.[122]

The glycoside naringin, prepared readily from grapefruit-canning wastes, has also been suggested as a source of L-rhamnose.[123]

General Discussion. The sugar is known in both anomeric forms. Under ordinary conditions, α-L-rhamnose·H_2O crystallizes. Anhydrous acetone solutions, seeded with crystals of the β-L anomer, crystallize, giving β-L-rhamnose (anhydrous). Seed crystals of the β-L anomer are obtained by melting the rhamnose hydrate and allowing the melt to crystallize at high temperatures. A molecular compound, β-L-rhamnose·α-L-rhamnose, is reported,[124] but has been disproved.[125]

8. *Aldoheptoses*

D-*glycero*-D-*galacto*-Heptose

D-*glycero*-α-D-*galacto*-Heptopyranose

Synonyms. D-Manno-D-*gala*-heptose, *d*-α-mannoheptose.

Occurrence and Preparation. This heptose is a component of the specific polysaccharide of *Chromobacterium violaceum* (BN): the somatic antigen (extracted from dried cells) is hydrolyzed gently to separate the polysaccharide from lipoprotein, and the polysaccharide is then hydrolyzed; chromatography gives the sugar.[126] The heptose had previously been detected in three strains of *C. violaceum.*[127]

The heptose is found[128] among a number of sugars present in the ripe avocado (Californian Calavo, Hass variety): deionized, gum-freed extracts gave crystalline *myo*-inositol, perseitol (D-*glycero*-D-*galacto*-heptitol), D-*manno*-heptulose,† and D-*erythro*-D-*galacto*-octitol. Chromatography of the mother liquor on columns of cellulose (and hydrolysis of some of the fractions containing oligosaccharides, and rechromatography) gave arabinose, xylose, glucose, fructose, D-*glycero*-D-*galacto*-heptose, additional D-*manno*-heptulose, D-*glycero*-L-*galacto*-octulose,† and D-*glycero*-D-*manno*-octulose.† Further

---

† This sugar is discussed in a later section of this chapter.

* References start on p. 107.

fractionation gave D-*erythro*-L-*galacto*-nonulose* and D-*erythro*-L-*gluco*-nonulose.*

Synthetic D-*glycero*-D-*galacto*-heptose is obtained from D-mannose by the cyanohydrin[129] and nitromethane methods.[130]

D-*glycero*-D-*gluco*-Heptose

$$HOH_2C-\overset{\overset{\displaystyle H}{|}}{C}-\overset{\overset{\displaystyle H}{|}}{\underset{\underset{\displaystyle OH}{|}}{C}}-\overset{\overset{\displaystyle H}{|}}{\underset{\underset{\displaystyle OH}{|}}{C}}-\overset{\overset{\displaystyle OH}{|}}{\underset{\underset{\displaystyle H}{|}}{C}}-\overset{\overset{\displaystyle H}{|}}{\underset{\underset{\displaystyle OH}{|}}{C}}-CHO$$

D-*glycero*-α-D-*gluco*-Heptopyranose

Synonym. D-Altro-D-*gluco*-heptose.

Occurrence and Preparation. Begbie and Richtmyer[131] have shown that the dried root of the primrose (*Primula officinalis* Jacq.) contains the heptose among a number of sugars. Aqueous extracts of the root, after separation of volemitol (D-*glycero*-D-*manno*-heptitol), were treated with yeast. The non-fermented products were first fractionated by cellulose-column chromatography, and some fractions were rechromatographed on Dowex 50W-X8 (Ba^{2+}) ion-exchange resin. The following products were isolated and identified; glycerol, erythritol, xylitol, D-xylose, *myo*-inositol, D-*glycero*-D-*gluco*-heptitol (β-sedoheptitol), volemitol, D-*glycero*-D-*gluco*-heptose, D-*glycero*-D-*manno*-heptose,* D-*allo*-heptulose,* D-*altro*-3-heptulose,* D-*manno*-heptulose,* sedo-heptulose,* D-*glycero*-L-*galacto*-octulose,* D-*glycero*-L-*galacto*-octulose,* D-*glycero*-D-*manno*-octulose,* D-*erythro*-L-*galacto*-nonulose,* D-*erythro*-L-*gluco*-nonulose,* and α-primeverose (6-*O*-β-D-xylopyranosyl-α-D-glucopyran-ose). Regarding the discovery of D-*gluco*-heptose, D-*glycero*-D-*manno*-heptose, D-*allo*-heptulose, and D-*altro*-3-heptulose in the fractionation, Begbie and Richtmyer comment: "Although [these] compounds . . . have, at C-4 to C-7, the same configuration as sedoheptulose, and although it is conceivable that they were formed by isomerizations occurring at C-1 to C-3 during the course of their isolation, we do not believe them to be artifacts but, rather, that their formation was caused by the action of enzymes within the plant itself."

D-*glycero*-D-*gluco*-Heptose has been synthesized by application of the cyano-hydrin reaction to D-altrose.[132]

---

* This sugar is discussed in a later section of this chapter.

D-*glycero*-D-*manno*-Heptose

D-*glycero*-α-D-*manno*-Heptopyranose

Synonym. D-Altro-D-*manno*-heptose.

Occurrence and Preparation. Baker's yeast contains D-*glycero*-D-*manno*-heptose as the glycosyl component of a guanosine 5'-(glycosyl pyrophosphate).[133] D-*glycero*-D-*manno*-Heptose and L-*glycero*-D-*manno*-heptose were obtained on hydrolysis of cell wall lipopolysaccharides of enterobacteria.[134] The heptose is also present in the root of the primrose[123] (see under D-*glycero*-D-*gluco*-Heptose). The heptose is obtained from D-altrose by application of the cyanohydrin synthesis.[124]

L-*glycero*-D-*manno*-Heptose

L-*glycero*-β-D-*manno*-Heptopyranose

Occurrence and Preparation. On hydrolysis and chromatography, cell wall polysaccharide isolated from *Escherichia coli* yields the heptose,[135] which may also be present in polysaccharides from a strain of *Shigella somei*,[136] *S. flexneri*,[137] and *S. dysenteriae*.[138] Cell wall lipopolysaccharides of enterobacteria were found to contain D-*glycero*-D-*manno*-heptose and L-*glycero*-D-*manno*-heptose.[134] L-*glycero*-D-*manno*-Heptose has been synthesized by the nitromethane route.[138a] Some physical properties of the synthetic enantiomorph, known earlier as D-gala-L-*manno*-heptose or D-α-galaheptose,[139] are m.p. 77–78° and $[\alpha]_D^{20}$ −14.1°, at equilibrium.

* *References start on p.* 107.

9. Ketoheptoses

D-*allo*-Heptulose

α-D-*allo*-Heptulopyranose

Occurrence and Preparation. In its first known natural occurrence, Begbie and Richtmyer[131] isolated this heptulose from *Primula officinalis* Jacq. (see under D-*glycero*-D-*gluco*-Heptose). The ketose is obtained by alkaline isomerization of D-*glycero*-D-*allo*- and D-*glycero*-D-*altro*-heptose[140] and of D-*manno*-3-heptulose.[141] 1,3-Dihydroxy-2-propanone adds to D-erythrose in an alkaline medium to give D-*allo*-heptulose together with D-*altro*- and D-*gluco*-heptulose.[141] An *allo*-heptulose, probably of the D series, has been isolated from the avocado.[141a]

D-*altro*-3-Heptulose

β-D-*altro*-3-Heptulopyranose

Synonym. Coriose.

Occurrence and Preparation. This 3-heptulose was obtained from chromatographed extracts of the roots of *Primula officinalis* Jacq.[131] (see under D-*glycero*-D-*gluco*-Heptose), and it has also been isolated from the leaves and stem of *Coriana japonica* A. Gray.[142] D-*altro*-3-Heptulose is prepared by isomerization of sedoheptulose in pyridine,[131] or from 2,4-*O*-ethylidene-D-erythrose by an aldol reaction.[142]

General Discussion. D-*altro*-3-Heptulose is the first clearly characterized 3-heptulose found in a natural product. Several papers have appeared concerning an incompletely identified, syrupy substance said to be a 3-heptulose phosphate, which was isolated from a rat liver digest. Evidence was presented for its possessing the D-*arabino* configuration at the three adjacent asymmetric carbon atoms of the acyclic form,[143] but it is not D-*manno*-3-heptulose.

X-Ray crystallographic studies have shown that the sugar crystallizes as the α-furanose form.[143a]

D-*manno*-Heptulose

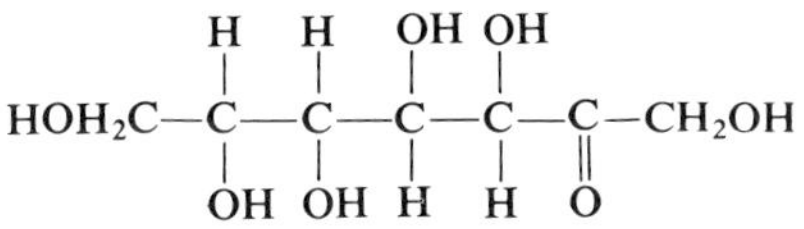

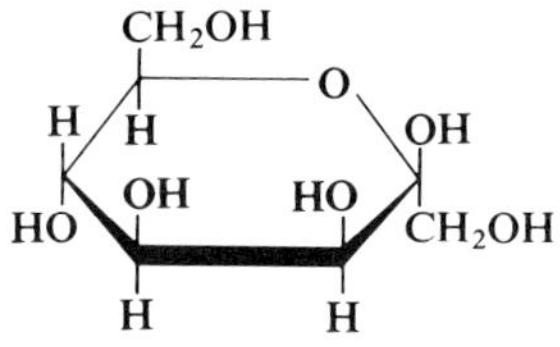

H   H   OH OH
| | | |
HOH₂C—C—C—C—C—C—CH₂OH
| | | | ‖
OH OH H   H   O

CH₂OH

β-D-*manno*-Heptulopyranose

Synonym. D-Mannoketoheptose.

Occurrence. LaForge reported the first isolation of a naturally occurring higher sugar, D-*manno*-heptulose, from the avocado (*Persea gratissima* Gaertn.).[144] Much earlier, the avocado had proved a source of perseitol[145] (D-*glycero*-D-*galacto*-heptitol).

Preparation. Avocado varieties differ greatly in their content of the heptulose; yields vary from 0.1 to 5%, based on the weight of wet pulp.[146] The purified extract of the avocado gives crystalline perseitol first. On nucleating, the mother liquor deposits the heptulose.[144, 146] (See also under D-*glycero*-D-*galacto*-Heptose for other sugars that have been isolated from the avocado.) Isomerization of D-*glycero*-D-*galacto*-heptose in alkali gives synthetic D-*manno*-heptulose plus D-*gluco*-heptulose; but, in pyridine, only D-*manno*-heptulose is allegedly obtained.[147] The nitroethanol[148] and aldol[149] syntheses may also be used.

Sedoheptulose

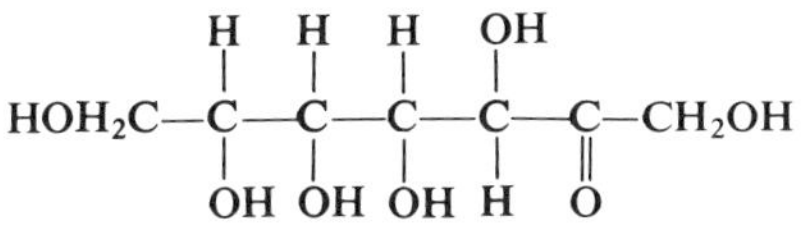

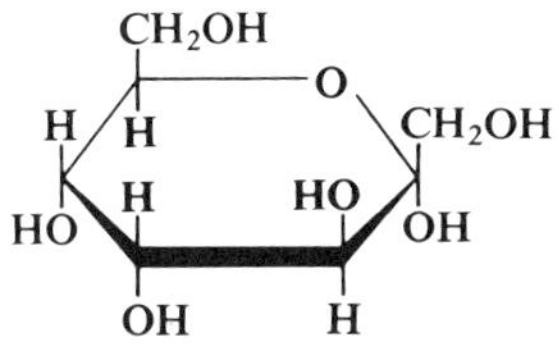

H   H   H   OH
| | | |
HOH₂C—C—C—C—C—C—CH₂OH
| | | | ‖
OH OH OH H   O

CH₂OH

α-D-Sedoheptulopyranose

Synonym. D-*altro*-Heptulose. "Volemulose,"[150] obtained by the oxidation of natural volemitol (D-*glycero*-D-*talo*-heptitol) with *Acetobacter xylinum*, has been identified as sedoheptulose,[151] possibly contaminated with D-*manno*-heptulose.[152] "Volemose," obtained by hypobromite oxidation of volemitol, is a complex mixture containing mainly the predictable aldoheptoses; the name should be abandoned.[153]

Occurrence. Sedoheptulose was found originally in *Sedum spectabile* Bor., a common herbaceous perennial plant used for decorative purposes.[154] It has since been detected in many of the succulent plants; indeed, as an intermediate of photosynthesis, it is probably present to some extent in all green plants.

* *References start on p. 107.*

Sedoheptulose plays an important role in the formation of hexoses from lower-carbon fragments, both in photosynthesis and in carbohydrate metabolism by animal tissues.

Preparation. The sugar is extracted by water from ground *Sedum* leaves and stems, and the extracts are evaporated to a thick syrup. The sedoheptulose is extracted by alcohol, which is then removed by evaporation. An aqueous solution of the syrup is purified with basic lead acetate. After removal of the excess of lead by precipitation with hydrogen sulfide, a crude solution of the sugar is obtained.[154, 155] Sedoheptulose has not yet been crystallized. It is usually isolated as the readily crystallized anhydride, sedoheptulosan monohydrate, whose structure is 2,7-anhydro-β-D-*altro*-heptulopyranose monohydrate.[156] Sedoheptulosan is formed by treating the sugar with acid.[154] The proportion of this anhydride varies from 84.5 to 91 %, according to the temperature employed to attain equilibrium[157] (see Chapter 13). The remainder consists mainly of unchanged sedoheptulose[157] and 2,7-anhydro-β-D-*altro*-heptulofuranose.[158] At 20° to 80°, minor proportions of 5-(D-*glycero*(?)-1,2-dihydroxyethyl)-2-furaldehyde are formed, but at 100° the proportion rapidly becomes considerable.[157]

General Discussion. The sugar has been suggested as a source for D-altrose and D-ribose, since it is readily oxidized by oxygen in alkaline solution to D-altronic acid; and calcium D-altronate is oxidized by hydrogen peroxide and ferric acetate to D-ribose.

D-*talo*-Heptulose

α-D-*talo*-Heptulopyranose

Occurrence and Preparation. The sugars present in the ripe avocado (Calavo, Fuerte variety) were first studied chromatographically by Charlson and Richtmyer, who obtained as one of the fractions a syrupy product having properties strongly suggestive of D-*talo*-heptulose.[159] The avocado extract (deionized, freed of gums, and fermented with yeast) gave perseitol, *myo*-inositol, and D-*erythro*-D-*galacto*-octitol (the first octitol discovered in Nature). Aldoses in the mother liquor were removed by bromine oxidation, and the remaining sugars were chromatographed on cellulose columns. In the fraction containing D-*talo*-heptulose, which could not be induced to crystallize, both D-fructose and D-*manno*-heptulose were detected.[141a] Another fraction gave D-*glycero*-D-*manno*-octulose.†

Oxidation of D-*glycero*-D-*altro*-heptitol by *Acetobacter suboxydans* gives D-*talo*-heptulose, which has been crystallized.[140]

10. Keto-octoses

D-*glycero*-L-*galacto*-Octulose

D-*glycero*-α-L-*galacto*-Octulopyranose

Occurrence and Preparation. In the first report of its occurrence,[128] D-*glycero*-L-*galacto*-octulose was obtained from the pulp of ripe avocados (Calavo, Hass variety) by extraction and chromatography (see under D-*glycero*-D-*galacto*-Heptose). The octulose was also isolated from the dried roots of *Primula officinalis* Jacq. (see under D-*glycero*-D-*gluco*-Heptose). Chromatography has shown the octulose to be present in several genera of Crassulaceae.[160] The synthetic octulose was prepared from D-gulose and 2-nitroethanol.[128]

D-*glycero*-D-*manno*-Octulose

D-*glycero*-α-D-*manno*-Octulopyranose

Occurrence and Preparation. D-*glycero*-D-*manno*-Octulose is the first naturally occurring octulose to have been isolated.[159] Charlson and Richtmyer obtained this ketose by chromatography of extracts of the ripe avocado (Calavo, Fuerte variety; see under D-*talo*-Heptulose), and also from extracts of *Sedum spectabile*. For the preparation of the octulose from *Sedum*, fermented extracts were heated in acid, and sedoheptulosan monohydrate crystallized. The mother liquor was freed from aldoses by oxidation with bromine, and the

---

† This sugar is discussed in a later section of this chapter.

* References start on p. 107.

unoxidized products were fractionated by cellulose-column chromatography. D-Mannitol, *myo*-inositol, additional sedoheptulosan, D-*glycero*-D-*gluco*-heptitol (β-sedoheptitol), and D-*glycero*-D-*manno*-octulose were isolated. The octulose is also present in *Primula officinalis* Jacq.[131] (see under D-*glycero*-D-*gluco*-Heptose).

Charlson and Richtmyer synthesized D-*glycero*-D-*manno*-octulose by isomerization of D-*erythro*-D-*galacto*-octose.[159]

11. *Keto-nonoses*

D-*erythro*-L-*galacto*-Nonulose

D-*erythro*-α-L-*galacto*-Nonulopyranose

Occurrence and Preparation. D-*erythro*-L-*galacto*-Nonulose is obtained as a hygroscopic syrup in very low yield by chromatography of extracts of the avocado[160] (*Persea gratissima* Gaertn., family Lauraceae; see under D-*glycero*-D-*galacto*-Heptose) and from dried roots of *Primula officinalis* Jacq.[131] (see under D-*glycero*-D-*gluco*-Heptose). It is also present in several genera of the Crassulaceae, including *Sedum*.[160] Sephton and Richtmyer prepared the nonulose by using Sowden's 2-nitroethanol procedure and by the diazomethane synthesis.

D-*erythro*-L-*gluco*-Nonulose

D-*erythro*-α-L-*gluco*-Monulopyranose

Occurrence and Preparation. This nonulose was obtained from the ripe avocado (Calavo, Hass variety) by extraction and cellulose chromatography[161] (see under D-*glycero*-D-*galacto*-Heptose). It is also present in *Sedum*.[160] Dried roots of *Primula officinalis* Jacq. provide another source of the nonulose[131] (see under D-*glycero*-D-*gluco*-Heptose). The synthetic nonulose was prepared by the diazomethane method.[161, 162]

REFERENCES

1. A. Wohl and F. Momber, *Ber.*, **50**, 455 (1917).
2. M. Betti and P. Pratesi, *Biochem. Z.*, **274**, 1 (1934).
3. H. O. L. Fischer and E. Baer, *Helv. Chim. Acta*, **17**, 622 (1934); **19**, 519 (1936); *J. Amer. Chem. Soc.*, **61**, 761 (1939).
4. A. S. Perlin and C. Brice, *Can. J. Chem.*, **34**, 85 (1956).
5. C. E. Ballou and H. O. L. Fischer, *J. Amer. Chem. Soc.*, **77**, 3329 (1955).
6. G. Bertrand, *Ann. Chim.*, **3** [8], 181 (1904).
7. H. H. Strain and W. H. Dore, *J. Amer. Chem. Soc.*, **56**, 2649 (1934).
8. H. O. L. Fischer and H. Milband, *Ber.*, **57**, 707 (1924).
9. C. E. Ballou and H. O. L. Fischer, *J. Amer. Chem. Soc.*, **78**, 1659 (1956); C. E. Ballou, *Biochem. Prep.*, **7**, 45 (1960).
10. A. Wohl, *Ber.*, **26**, 743 (1893).
11. O. Ruff, *Ber.*, **32**, 3672 (1899).
12. W. G. Overend, M. Stacey, and L. F. Wiggins, *J. Chem. Soc.*, 1358 (1949).
13. A. S. Perlin and C. Brice, *Can. J. Chem.*, **33**, 1216 (1955).
14. R. Schaffer, *J. Amer. Chem. Soc.*, **81**, 2838 (1959).
15. C. E. Ballou, H. O. L. Fischer, and D. L. MacDonald, *J. Amer. Chem. Soc.*, **77**, 5967 (1955).
16. M. E. Léger, *Ann. Chim.* (Paris), **8** [9], 265 (1917); C. S. Gibson and J. L. Simonsen, *J. Chem. Soc.*, 553 (1930).
17. M. Maxim, *Biochem. Z.*, **223**, 404 (1930); E. Chargaff and R. J. Anderson, *Hoppe-Seyler's Z. Physiol. Chem.*, **191**, 172 (1930); W. N. Haworth, P. W. Kent, and M. Stacey, *J. Chem. Soc.*, 1211, 1220 (1948).
18. R. C. Hockett and C. S. Hudson, *J. Amer. Chem. Soc.*, **56**, 1632 (1934); H. G. Fletcher, Jr., H. W. Diehl, and C. S. Hudson, *ibid.*, **72**, 4546 (1950); compare W. C. Griffiths, T. Galkowski, R. W. Kocan, and K. M. Reardon, *Carbohyd. Res.*, **13**, 177 (1970).
19. M. Schulz and H. Steinmaus, *Angew. Chem.*, **75**, 918 (1963).
20. T. S. Harding, *Sugar*, **24**, 656 (1922); E. Anderson and L. Sands, *J. Amer. Chem. Soc.*, **48**, 3172 (1926); *Org. Syn. Coll. Vol.*, **1**, 67 (1941).
21. E. V. White, *J. Amer. Chem. Soc.*, **69**, 715 (1947).
22. C. S. Hudson, *J. Amer. Chem. Soc.*, **73**, 4038 (1951); F. B. Cramer, *J. Franklin Inst.*, **256**, 93 (1953).
23. W. C. Austin and J. P. Walsh, *J. Amer. Chem. Soc.*, **56**, 934 (1934); J. K. Dale, *ibid.*, **56**, 932 (1934); H. S. Isbell and W. W. Pigman, *J. Res. Nat. Bur. Stand.*, **18**, 141 (1937).
24. O. L. Galmarini and V. Deulofeu, *Tetrahedron*, **15**, 76 (1961).
25. W. Alberda van Ekenstein and J. J. Blanksma, *Chem. Weekbl.*, **11**, 189 (1914).
25a. P. Fleury, J. E. Courtois, and D. Darzens-Souloumiac, *Ann. Pharm. Fr.*, **28**, 17 (1970).
26. F. Kauffmann, B. Jann, L. Kruger, O. Lüderitz, and O. Westphal, *Zentr. Bakteriol. Parasitenk. Abt. I*, **180**, 509 (1962).
27. F. J. Bates and Associates, *Nat. Bur. Stand. (U.S.) Circ.*, **C440** (1942).
28. P. A. Levene and E. P. Clark, *J. Biol. Chem.*, **46**, 19 (1921); F. P. Phelps, U.S. Patent 2,152,662 (1939); L. Laufer and J. Charney, U.S. Patent 2,379,913, 2,379,914 (1945).
29. H. Bredereck, M. Köthnig, and E. Berger, *Ber.*, **73**, 956 (1940).
30. J. Lee, U. V. Solmassen, and L. Berger, U.S. Patent 2,384,102, 2,384,103 (1945).
31. C. S. Hudson and N. K. Richtmyer, U.S. Patent 2,162,721 (1939).
32. J. C. Sowden, *J. Amer. Chem. Soc.*, **72**, 808 (1950).
33. D. C. C. Smith, *Chem. Ind.* (London), 92 (1955); H.-H. Stroh, D. Dargel, and R. Häussler, *J. Prakt. Chem.*, **23** [4], 309 (1964).

34. See R. W. Jeanloz and H. G. Fletcher, Jr., *Advan. Carbohyd. Chem.*, **6**, 135 (1951).
35. C. S. Hudson and T. S. Harding, *J. Amer. Chem. Soc.*, **40**, 1601 (1918).
36. W. T. Schreiber, N. V. Geib, B. Wingfield, and S. F. Acree, *Ind. Eng. Chem.*, **22**, 497 (1930).
37. N. A. Sytchev, *Compt. Rend. Acad. Sci. URSS*, **29**, 384 (1940).
38. M. Iwasaki, *Nippon Nogei Kagaku Kaishi*, **16**, 148 (1940).
39. C. Glatthaar and T. Reichstein, *Helv. Chim. Acta*, **18**, 80 (1935).
39a. P. M. Collins and P. Gupta, *Chem. Commun.*, 1288 (1969).
40. P. A. Srere, J. Cooper, V. Klybas, and E. Rocker, *Arch. Biochem. Biophys.*, **59**, 535 (1955); B. L. Horecker, J. Hurwitz, and P. Z. Smyrniotis, *J. Amer. Chem. Soc.*, **78**, 692 (1956).
41. O. Th. Schmidt and R. Treiber, *Ber.*, **66**, 1765 (1933).
42. R. Prince and T. Reichstein, *Helv. Chim. Acta*, **20**, 101 (1937).
43. P. A. Levene and R. S. Tipson, *J. Biol. Chem.*, **115**, 731 (1936); R. S. Tipson and R. F. Brady, Jr., *Carbohyd. Res.*, **10**, 549 (1969).
44. L. von Vargha, *Ber.*, **68**, 18 (1935).
45. P. A. Levene and F. B. LaForge, *J. Biol. Chem.*, **18**, 319 (1914); I. Greenwald, *ibid.*, **88**, 1 (1930).
46. M. Enklewitz and M. Lasker, *J. Biol. Chem.*, **110**, 443 (1935).
47. H. Kauss, *Z. Pflanzenphysiol.*, **53**, 58 (1965).
48. F. P. Phelps and F. J. Bates, *J. Amer. Chem. Soc.*, **56**, 1250 (1934); J. W. Pratt and N. K. Richtmyer, *ibid.*, **77**, 1906 (1955).
49. W. C. Austin and F. L. Humoller, *J. Amer. Chem. Soc.*, **56**, 1152 (1934).
49a. R. Ahluwahlia, S. J. Angyal, and M. H. Randall, *Carbohyd. Res.*, **4**, 478 (1967); D. Horton and C. G. Tindall, Jr., *ibid.*, **15**, 215 (1970).
50. A. N. O'Neill, *J. Amer. Chem. Soc.*, **77**, 2837 (1955).
51. R. J. Brown and R. F. Serro, *J. Amer. Chem. Soc.*, **75**, 1040 (1953); E. A. Kabat, D. L. MacDonald, C. E. Ballou, and H. O. L. Fischer, *ibid.*, **75**, 4507 (1953).
52. H. Colin and E. Guéguen, *Compt. Rend.*, **191**, 163 (1930); R. C. Bean, E. W. Putman, R. E. Trucco, and W. Z. Hassid, *J. Biol. Chem.*, **204**, 169 (1953).
53. L. Genevois, G. Vitte, and C. Guichard, *Compt. Rend.*, **240**, 1150 (1955).
54. G. Kritchevsky and A. B. Anderson, *J. Amer. Chem. Soc.*, **77**, 3391 (1955).
55. T. S. Harding, *Sugar*, **25**, 175 (1923); E. P. Clark, *Bur. Stand. Sci. Papers*, **17**, 227 (1921); G. Mougne, *Bull. Soc. Chim. Biol.*, **4**, 206 (1922); M. L. Wolfrom and A. Thompson, *Methods Carbohyd. Chem.*, **1**, 120 (1962).
56. A. W. Schorger and D. F. Smith, *Ind. Eng. Chem.*, **8**, 494 (1916); L. E. Wise, P. L. Hamer, and F. C. Peterson, *ibid.*, **25**, 184 (1933).
57. L. E. Wise, *Methods Carbohyd. Chem.*, **1**, 404 (1962).
58. D. J. Bell and E. Baldwin, *Nature*, **146**, 559 (1940).
59. E. Anderson, *J. Biol. Chem.*, **100**, 249 (1933); E. Anderson and H. J. Lowe, *ibid.*, **168**, 289 (1947); C. Araki, *Nippon Kagaku Zasshi*, **59**, 424 (1938); C. Araki and K. Arai, *Methods Carbohyd. Chem.*, **1**, 122 (1962).
60. C. Neuberg and M. Federer, *Ber.*, **38**, 872 (1905).
61. R. S. Tipson, *J. Biol. Chem.*, **125**, 341 (1938); H. S. Isbell, *J. Res. Nat. Bur. Stand.*, **33**, 45 (1944); H. L. Frush and H. S. Isbell, *Methods Carbohyd. Chem.*, **1**, 127 (1962).
62. W. B. Newkirk, *Ind. Eng. Chem.*, **16**, 1173 (1924); **28**, 760 (1936); **31**, 18 (1939); G. R. Dean and J. B. Gottfried, *Advan. Carbohyd. Chem.*, **5**, 127 (1950).
63. W. B. Newkirk, *Ind. Eng. Chem.*, **28**, 764 (1936).
64. R. Behrend, *Ann.*, **377**, 220 (1910); A. W. Mangam and S. F. Acree, *J. Amer. Chem. Soc.*, **39**, 965 (1917).

65. C. S. Hudson and J. K. Dale, *J. Amer. Chem. Soc.*, **39**, 323 (1917).
66. R. L. Whistler and B. F. Buchanan, *J. Biol. Chem.*, **125**, 557 (1938); C. Tanret, *Bull. Soc. Chim. Fr.*, **13** [3], 733 (1895).
67. A. T. Harding, U.S. Patent 2,369,231 (1945).
68. H. Berlin, *J. Amer. Chem. Soc.*, **48**, 1107, 2627 (1926); E. M. Montgomery and F. B. Weakley, *J. Assoc. Offic. Agr. Chem.*, **36**, 1096 (1953); J. C. Sowden and A. S. Spriggs, *J. Amer. Chem. Soc.*, **76**, 3539 (1954); **78**, 2503 (1956).
69. A. Thompson, M. L. Wolfrom, and E. J. Quinn, *J. Amer. Chem. Soc.*, **75**, 3003 (1953).
70. W. R. Fetzer, E. K. Crosby, C. E. Engel, and L. C. Kirst, *Ind. Eng. Chem.*, **45**, 1075 (1953); A. Thompson, K. Anno, M. L. Wolfrom, and M. Inatome, *ibid.*, **76**, 1309 (1954).
71. M. L. Wolfrom and A. Thompson, *J. Amer. Chem. Soc.*, **77**, 6403 (1955).
72. A. Sato and H. Ono, *Kogyo Gijutsuin, Hakko Kenkyusho Kenkyu Hokoku*, **22**, 147 (1962).
73. R. Reiss, *Ber.*, **22**, 609 (1889).
74. T. Fujita and T. Sato, *Bull. Chem. Soc. Jap.*, **33**, 353 (1960).
75. J. K. N. Jones, *J. Chem. Soc.*, 3292 (1950).
76. T. S. Harding, *Sugar*, **25**, 583 (1923); E. P. Clark, *J. Biol. Chem.*, **51**, 1 (1922); C. S. Hudson and E. L. Jackson, *J. Amer. Chem. Soc.*, **56**, 958 (1934); H. S. Isbell, *J. Res. Nat. Bur. Stand.* **26**, 47 (1941).
76a. S. Levine, R. G. Hansen, and H. M. Sell, *Carbohyd. Res.*, **6**, 382 (1968); **10**, 468 (1969).
77. J. K. Dale, *Bur. Stand. J. Res.*, **3**, 459 (1929); H. S. Isbell, *J. Amer. Chem. Soc.*, **55**, 2166 (1933); H. S. Isbell and W. W. Pigman, *J. Res. Nat. Bur. Stand.*, **18**, 141 (1937).
78. P. F. Wiley and M. V. Segal, *J. Amer. Chem. Soc.*, **80**, 1010 (1958).
79. O. Hesse, *J. Prakt. Chem.*, **66** [2], 407 (1902).
80. E. Fischer, *Ber.*, **24**, 3622 (1891).
81. P. A. Levene and R. S. Tipson, *J. Biol. Chem.*, **93**, 631 (1931); W. W. Pigman and H. S. Isbell, *J. Res. Nat. Bur. Stand.*, **19**, 189 (1937).
81a. D. Horton and J. S. Jewell, *Carbohyd. Res.*, **5**, 149 (1967); G. J. F. Chittenden, *ibid.*, **15**, 101 (1970).
82. See C. P. Barry and J. Honeyman, *Advan. Carbohyd. Chem.*, **7**, 53 (1952); L. M. J. Verstraeten, *ibid.*, **22**, 229 (1967); I. R. Siddiqui, *Advan. Carbohyd. Chem. Biochem.*, **25**, 285 (1970).
83. S. M. Cantor and K. C. Hobbs, U.S. Patent 2,354,664 (1944).
84. T. E. Eble, H. Hoeksema, G. A. Boyack, and G. M. Savage, *Antibiot. Chemotherapy*, **9**, 419 (1959); J. J. Vavra, A. Dietz, B. W. Churchill, P. Siminoff, and H. J. Koepsell, *ibid.*, **9**, 427 (1959); W. Schroeder and H. Hoeksema, *J. Amer. Chem. Soc.*, **81**, 1767 (1959); E. R. Garrett, *ibid.*, **82**, 827 (1960).
85. L. Hough and B. E. Stacey, *Phytochemistry*, **2**, 315 (1963); **5**, 171, 215 (1966).
86. M. Steiger and T. Reichstein, *Helv. Chim. Acta*, **19**, 184 (1936).
87. M. L. Wolfrom, A. Thompson, and E. F. Evans, *J. Amer. Chem. Soc.*, **67**, 1793 (1945).
88. S. Passeron and E. Recondo, *J. Chem. Soc.*, 813 (1965); cf. R. Grünnagel and H. J. Haas, *Ann.*, **721**, 234 (1969).
89. See J. V. Karabinos, *Advan. Carbohyd. Chem.*, **7**, 99 (1952).
90. C. M. Martin and F. H. Reuter, *Nature*, **164**, 407 (1949).
91. G. Bertrand, *Bull. Soc. Chim. Fr.*, **15** [3], 627 (1896).
92. P. A. Wells, J. J. Stubbs, L. B. Lockwood, and E. T. Roe, *Ind. Eng. Chem.*, **29**, 1385 (1937).
93. E. L. Hirst, L. Hough, and J. K. N. Jones, *Nature*, **163**, 177 (1949).

94. C. A. Lobry de Bruyn and W. Alberda van Ekenstein, *Rec. Trav. Chim.* **16**, 265 (1897); E. L. Totton and H. A. Lardy, *Methods Carbohyd. Chem.*, **1**, 155 (1962).

95. E. L. Totton and H. A. Lardy, *J. Amer. Chem. Soc.*, **71**, 3076 (1949).

96. See T. Reichstein and E. Weiss, *Advan. Carbohyd. Chem.*, **17**, 65 (1962).

97. M. Keller and T. Reichstein, *Helv. Chim. Acta*, **32**, 1607 (1949); A. Hunger and T. Reichstein, *ibid.*, **35**, 1073 (1952).

98. R. Rees, C. R. Gavilanes, W. Meier, A. Fürst, and K. Meyer, *Helv. Chim. Acta*, **44**, 1607 (1961).

99. P. A. Levene and J. Compton, *J. Biol. Chem.*, **116**, 169 (1936).

100. A. D. Elbein, H. Koffler, and H. R. Garner, *Biochim. Biophys. Acta*, **56**, 165 (1962).

101. K. Freudenberg and K. Raschig, *Ber.*, **62**, 373 (1929).

102. K. Iwadare, *Bull. Chem. Soc. Jap.*, **17**, 296 (1942); M. Gut and D. A. Prins, *Helv. Chim. Acta*, **29**, 1555 (1946); D. A. Rosenfeld, N. K. Richtmyer, and C. S. Hudson, *J. Amer. Chem. Soc.*, **70**, 2201 (1948).

103. C. Liebermann and F. Giesel, *Ber.*, **16**, 935 (1883); E. Fischer and C. Liebermann, *ibid.*, **26**, 2415 (1893).

104. E. Votoček, *Ber.*, **43**, 476 (1910).

105. E. Fischer and K. Zach, *Ber.*, **45**, 3761 (1912).

105a. I. F. Makarevich and D. G. Kolesnikov, *Khim. Prir. Soedin.*, **5**, 190 (1969).

106. H. Kiliani, *Arch. Pharm.* (Weinheim), **234**, 438 (1896); *Ber.*, **46**, 667 (1913).

107. J. A. Moore, C. Tamm, and T. Reichstein, *Helv. Chim. Acta*, **37**, 755 (1954).

108. P. A. Levene and J. Compton, *J. Biol. Chem.*, **111**, 335 (1935).

109. A. Markowitz, *J. Biol. Chem.*, **237**, 1769 (1962).

110. A. P. MacLennan and D. A. L. Davies, *Bull. Soc. Chim. Biol.*, **42**, 1373 (1960).

111. A. P. MacLennan, *Biochim. Biophys. Acta*, **48**, 600 (1961).

112. A. P. MacLennan, *Biochem. J.*, **82**, 394 (1962); P. Jollès, F. Bigler, T. Gendre, and E. Lederer, *Bull. Soc. Chim. Biol.*, **43**, 177 (1961).

113. J. Schmutz, *Helv. Chim. Acta*, **31**, 1719 (1948).

114. E. Votoček and F. Valentin, *Collect. Czech. Chem. Commun.*, **1**, 46, 606 (1929); F. B. Power and H. Rogerson, *J. Chem. Soc.*, **101**, 1 (1912); L. A. Davies and R. Adams, *J. Amer. Chem. Soc.*, **50**, 1749 (1928); C. Mannich and P. Schumann, *Arch. Pharm.* (Weinheim), **276**, 211 (1938).

115. E. E. Percival, *Methods Carbohyd. Chem.*, **1**, 195 (1962).

116. H. G. Bray, H. Henry, and M. Stacey, *Biochem. J.*, **40**, 124 (1946); E. A. Kabat, H. H. Baer, A. E. Bezer, and V. Knaub, *J. Exp. Med.*, **88**, 43 (1948).

117. See H. H. Baer, *Fortschr. Chem. Forsch.*, **3**, 822 (1958).

118. E. P. Clark, *J. Biol. Chem.*, **54**, 65 (1922); R. C. Hockett, F. P. Phelps, and C. S. Hudson, *J. Amer. Chem. Soc.*, **61**, 1658 (1939).

119. W. T. Haskins, R. M. Hann, and C. S. Hudson, *J. Amer. Chem. Soc.*, **68**, 628 (1946); W. W. Zorbach and C. O. Tio, *J. Org. Chem.*, **26**, 3543 (1961).

120. S. F. Acree and W. A. Syme, *Amer. Chem. J.*, **36**, 309 (1906).

121. S. M. Partridge, *Biochem. J.*, **42**, 251 (1948); G. Pon and A. M. Staub, *Bull. Soc. Chim. Biol.*, **34**, 1132 (1952); M. McCarty, *J. Exp. Med.*, **96**, 569 (1952).

122. C. F. Walton, Jr., *J. Amer. Chem. Soc.*, **43**, 127 (1921).

123. G. N. Pulley and H. W. von Loesecke, *J. Amer. Chem. Soc.*, **61**, 175 (1939).

124. E. L. Jackson and C. S. Hudson, *J. Amer. Chem. Soc.*, **59**, 1076 (1937); E. Fischer, *Ber.*, **28**, 1162 (1895); T. Purdie and C. R. Young, *J. Chem. Soc.*, **89**, 1194 (1906).

125. R. S. Tipson and H. S. Isbell, *J. Res. Nat. Bur. Stand.*, **66A**, 31 (1962).

126. A. P. MacLennan and D. A. L. Davies, *Biochem. J.*, **66**, 562 (1957).

127. D. A. L. Davies, *Biochem. J.*, **59**, 696 (1955).

128. H. H. Sephton and N. K. Richtmyer, *J. Org. Chem.*, **28**, 1691 (1963).

129. E. Fischer and F. Passmore, *Ber.*, **23**, 2226 (1890); H. S. Isbell, *J. Res. Nat. Bur. Stand.*, **18**, 505 (1937); **20**, 97 (1938); E. M. Montgomery and C. S. Hudson, *J. Amer. Chem. Soc.*, **64**, 247 (1942).

130. J. C. Sowden and R. Schaffer, *J. Amer. Chem. Soc.*, **73**, 4662 (1951).

131. R. Begbie and N. K. Richtmyer, *Carbohyd. Res.*, **2**, 272 (1966).

132. D. A. Rosenfeld, N. K. Richtmyer, and C. S. Hudson, *J. Amer. Chem. Soc.*, **73**, 4907 (1951).

133. V. Ginsberg, P. J. O'Brien, and C. W. Hall, *J. Biol. Chem.*, **237**, 497 (1962).

134. G. Bagdian, W. Dröge, K. Kotelko, O. Lüderitz, and O. Westphal, *Biochem. Z.*, **344**, 197 (1966).

135. W. Weidel, *Hoppe-Seyler's Z. Physiol. Chem.*, **299**, 253 (1955).

136. M. A. Jesaitis and W. F. Goebel, *J. Exp. Med.*, **96**, 409 (1952).

137. M. W. Slein and G. W. Schnell, *Proc. Soc. Exp. Biol. Med.*, **82**, 734 (1953).

138. D. A. L. Davies, W. T. J. Morgan, and B. R. Record, *Biochem. J.*, **60**, 290 (1955); D. A. L. Davies, *Biochim. Biophys. Acta*, **26**, 151 (1957).

138a. M. Teuber, R. D. Bevill, and M. J. Osborn, *Biochemistry*, **7**, 3303 (1968).

139. R. M. Hann, A. T. Merrill, and C. S. Hudson, *J. Amer. Chem. Soc.*, **56**, 1644 (1934); **57**, 2100 (1935).

140. J. W. Pratt and N. K. Richtmyer, *J. Amer. Chem. Soc.*, **77**, 6326 (1955).

141. R. Schaffer, *J. Org. Chem.*, **29**, 1471, 1473 (1964).

141a. I. Johansson and N. K. Richtmyer, *Carbohyd. Res.*, **13**, 461 (1970).

142. T. Okuda and K. Konishi, *Chem. Commun.*, 553 (1968); cf. *ibid.*, 1117 (1969); *Yakugaku Zasshi*, **88**, 1329 (1968); *Tetrahedron*, **24**, 6907 (1968).

143. V. N. Nigam, H.-G. Sie, and W. H. Fishman, *J. Amer. Chem. Soc.*, **82**, 1007 (1960); H.-G. Sie and G. P. Mathau, *Biochim. Biophys. Acta*, **64**, 497 (1962); H.-G. Sie, *Proc. Soc. Exp. Biol. Med.*, **113**, 733 (1963).

143a. T. Okuda, K. Osaki, and T. Taga, *Chem. Commun.*, 851 (1969).

144. F. B. LaForge, *J. Biol. Chem.*, **28**, 511 (1917).

145. J. B. Avequin, *J. Chim. Med. Pharm. Toxicol.*, **7** [1], 467 (1831).

146. N. K. Richtmyer, *Methods Carbohyd. Chem.*, **1**, 173 (1962).

147. E. M. Montgomery and C. S. Hudson, *J. Amer. Chem. Soc.*, **61**, 1654 (1939).

148. J. C. Sowden, *J. Amer. Chem, Soc.*, **72**, 3325 (1950); J. C. Sowden and D. R. Strobach, *ibid.*, **80**, 2532 (1958).

149. R. Schaffer and H. S. Isbell, *J. Org. Chem.*, **27**, 3268 (1962).

150. G. Bertrand, *Bull. Soc. Chim. Fr.*, **19** [3], 348 (1898).

151. L. C. Stewart, N. K. Richtmyer, and C. S. Hudson, *J. Amer. Chem. Soc.*, **71**, 3532 (1949).

152. V. Ettel, J. Liebster, M. Tadra, and M. Kulhánek, *Collect. Czech. Chem. Commun.*, **16**, 696 (1951).

153. W. T. Haskins and C. S. Hudson, *J. Amer. Chem. Soc.*, **69**, 1370 (1947); V. Ettel, J. Liebster, and M. Tadra, *Chem. Listy*, **46**, 445 (1952).

154. F. B. LaForge and C. S. Hudson, *J. Biol. Chem.*, **30**, 61 (1917).

155. N. K. Richtmyer, R. M. Hann, and C. S. Hudson, *J. Amer. Chem. Soc.*, **61**, 343 (1939).

156. J. W. Pratt, N. K. Richtmyer, and C. S. Hudson, *J. Amer. Chem. Soc.*, **73**, 1876 (1951); **74**, 2200 (1952).

157. N. K. Richtmyer and J. W. Pratt, *J. Amer. Chem. Soc.*, **78**, 4717 (1956).

158. L. P. Zill and N. E. Tolbert, *J. Amer. Chem. Soc.*, **76**, 2929 (1954).

159. A. J. Charlson and N. K. Richtmyer, *J. Amer. Chem. Soc.*, **82**, 3428 (1960).

160. H. H. Sephton and N. K. Richtmyer, *Carbohyd. Res.*, **2**, 289 (1966).

161. H. H. Sephton and N. K. Richtmyer, *J. Org. Chem.*, **28**, 2388 (1963).

162. M. L. Wolfrom and H. B. Wood, Jr., *J. Amer. Chem. Soc.*, **77**, 3096 (1955).

3. SYNTHESIS OF MONOSACCHARIDES

L. Hough and A. C. Richardson

I. INTRODUCTION

Efficient procedures are available for the synthesis of the less readily available monosaccharides by starting with readily available monosaccharides that occur naturally. Total synthesis of certain sugars has been achieved from simple noncarbohydrate substances by specific and nonspecific processes. Although generally of little preparative value, they have been of interest in biosynthetic studies. Chemical modifications of accessible carbohydrates are concerned in the main with processes for lengthening or shortening of the carbon chain, and for controlled modification of structure and configuration at individual carbon atoms. Experimental details of many of these procedures are given in "Methods in Carbohydrate Chemistry."[1]

II. TOTAL SYNTHESIS

A. The Formose Reaction

The base-catalyzed self-condensation of formaldehyde was first observed in 1861 by A. Butlerow,[2] who obtained a sugar-like syrup, now termed "formose," by the action of calcium hydroxide on "trioxymethylene." Similar products resulted when formaldehyde was condensed in other mildly alkaline media such as $CaCO_3$, $Mg(OH)_2$, or $Ba(OH)_2$. Monosaccharides were detected in formose by E. Fischer and J. Tafel,[3] who isolated two hexulose phenylosazones in approximately 13 % yield after treatment of the syrup with phenylhydrazine. They termed these α- and β-acrosazone, and they showed that α-acrosazone was DL-*arabino*-hexulose phenylosazone, a derivative that can be derived from DL-glucose, DL-mannose, or DL-fructose. By chemical and enzymic manipulations (see p. 116) they achieved total syntheses of D-glucose, D- and L-mannose, and D- and L-*arabino*-hexulose (D- and L-fructose).

β-Acrosazone was later identified[4, 5] as DL-*xylo*-hexulose phenylosazone, a derivative that can be derived from DL-idose, DL-gulose, or DL-*xylo*-hexulose (DL-sorbose). Fischer and Tafel[6] also obtained formose-like mixtures from 2,3-dibromopropionaldehyde (**1**) and from DL-glyceraldehyde (**2**) on treatment with mild alkali. The product was termed "acrose" and was thought to arise from the mixed aldol condensation between DL-glyceraldehyde (**2**) with 1,3-dihydroxy-2-propanone ("dihydroxyacetone," **3**), since the latter is formed from DL-glyceraldehyde by base-catalyzed isomerization. This reaction scheme is supported by the observations of H. O. L. Fischer and E. Baer,[7] who found that an equimolar mixture of D-glyceraldehyde (**4**) and 1,3-dihydroxy-2-propanone (**3**) in $0.01 M$ barium hydroxide gave equal amounts of D-*arabino*-

hexulose (D-fructose, **5**) and D-*xylo*-hexulose (D-sorbose, **6**), in high yield. The results suggested that the reaction was stereospecific, since the two newly formed asymmetric centers at C-3 and C-4 had the *threo* configuration.

Chromatographic studies of formose have shown that it contains a complex mixture of aldoses and ketoses, of which glucose, mannose, fructose, sorbose, the four aldopentoses, and the two pentuloses have all been detected.[8,9] The composition of "formose" varies with the time of reaction; shorter reaction times favor a preponderance of glycolaldehyde and the pentuloses, whereas longer times lead to isomerization of the ketoses and aldoses to give a more complex mixture. The primary products of the formose reaction are glycolaldehyde, glyceraldehyde, and 1,3-dihydroxy-2-propanone, which, by means of mixed aldol condensations together with epimerization and related rearrangements, give rise to all the products so far detected.[10,11] The formation of glycolaldehyde is difficult to explain, since this compound is unlikely to arise at an appreciable rate from the direct condensation of two units of formaldehyde.

* *References start on p. 158.*

Fischer and Tafel's Complete Synthesis of Hexose Sugars

"α-ACROSE" (from formaldehyde, glyceraldehyde, or acrolein dibromide)

$\downarrow$ PhNHNH$_2$

DL-*arabino*-Hexulose phenylosazone (DL-glucosazone, α-phenylacrosazone)

$\downarrow$ conc. HCl

DL-*arabino*-Hexosulose (DL-glucosone, α-acrosone)

$\downarrow$ Zn, HOAc

DL-Fructose $\xrightarrow{\text{fermentation with yeast}}$ L-FRUCTOSE

$\downarrow$ Na—Hg (reduction)

DL-Mannitol $\xleftarrow{\text{Na—Hg}}$ DL-Mannose

$\downarrow$ HNO$_3$ oxidation; purification through phenylhydrazone

DL-Mannose $\xrightarrow{\text{fermentation with yeast}}$ L-MANNOSE

$\downarrow$ Br$_2$ oxidation

DL-Mannonic acid

$\downarrow$ separated by fractional crystallization of strychnine salts

D-Mannonic acid $\xrightarrow{\text{Na—Hg}}$ D-MANNOSE ⎯ L-Mannonic acid

$\downarrow$ Heat with quinoline

D-Gluconic acid

$\downarrow$ Na—Hg (reduction)

D-GLUCOSE $\xrightarrow{\text{PhNHNH}_2}$ D-*arabino*-Hexulose phenylosazone (D-glucosazone)

$\downarrow$ conc. HCl

D-*arabino*-Hexosulose (D-glucosone)

$\downarrow$ Zn, HOAc

D-FRUCTOSE

Kinetic studies have revealed that a long induction period precedes an ensuing rapid reaction, suggestive of an autocatalytic process. The addition of small proportions of glycolaldehyde, glyceraldehyde, or 1,3-dihydroxy-2-propanone overcomes the induction period, and the reactions proceed at a rate equal to that of the rapid part of the normal formose reaction. R. Breslow[12] has rationalized this by suggesting that two processes are operative for the formation of glycolaldehyde in the formose reaction. Initially a very slow direct condensation of two molecules of formaldehyde occurs to give glycolaldehyde. There follows a fast overall reaction, which is initiated by condensation of formaldehyde with the small amount of glycolaldehyde present and is terminated by reverse aldol cleavage of the aldotetroses formed (Eq. 1).

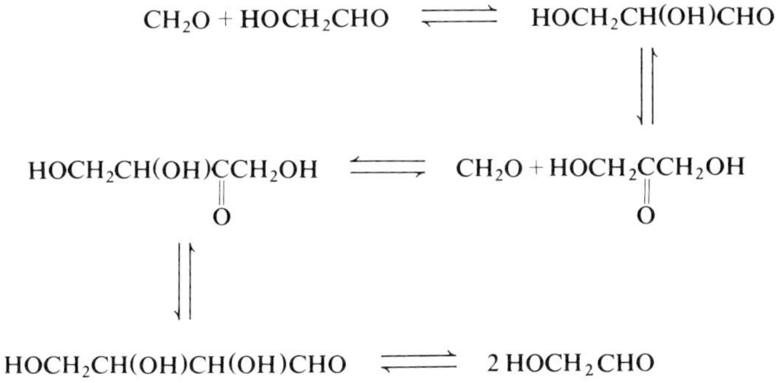

$$CH_2O + HOCH_2CHO \rightleftharpoons HOCH_2CH(OH)CHO$$

$$HOCH_2CH(OH)\underset{\underset{O}{\|}}{C}CH_2OH \rightleftharpoons CH_2O + HOCH_2\underset{\underset{O}{\|}}{C}CH_2OH$$

$$HOCH_2CH(OH)CH(OH)CHO \rightleftharpoons 2\ HOCH_2CHO \qquad (1)$$

The rate of the formose reaction is also dependent on the base cation used. The reaction is very rapid with thallium or lead hydroxides, relatively fast with calcium hydroxide, and of negligible rate with sodium or potassium hydroxide.[10] This effect is related to the well-known stabilizing effect of chelating ions on enediols, which are intermediates in the tautomerization and condensation steps of the above reaction sequence.[12]

The stereospecific aldol condensation between aldoses and 1,3-dihydroxy-2-propanone, to give products having the *threo* configuration at the two newly formed asymmetric centers, is used as a method of extending the chain of sugars (Section III, G) and synthesizing branched-chain sugars.[13]

B. SYNTHESIS FROM ALKENE AND ALKYNE DERIVATIVES

Various DL-sugars have been synthesized from simple precursors by methods involving alkyne intermediates and hydroxylation of double bonds. Iwai *et al.*[14] have described the total synthesis of all four DL-pentoses; thus the Grignard reagent derived from 3-(tetrahydropyran-2-yloxy)propyne (7) was condensed with 2,2-diethoxyacetaldehyde (8) to give 1,1-diethoxy-5-(tetrahydropyran-

* *References start on p. 158.*

2-yloxy)-3-pentyn-2-ol (**9**). Subsequent reduction with lithium aluminum hydride afforded the *trans*-alkene (**10**), whereas catalytic hydrogenation gave the *cis* alkene (**11**). Acetylation of the *trans* isomer followed by *cis* hydroxylation of the double bond gave, after acid hydrolysis of the acetal groups, a mixture of DL-lyxose and DL-xylose. By similar reactions the *cis* isomer gave a mixture of DL-arabinose and DL-ribose. A stereoselective, total synthesis has been used to obtain DL-glucose in 34% overall yield from an acrolein precursor.[14a]

III. ASCENT OF THE SERIES

A. THE FISCHER–KILIANI CYANOHYDRIN SYNTHESIS

The base- or acid-catalyzed reaction of hydrocyanic acid with aldehydes and ketones gives two isomeric 2-hydroxynitriles, hydrolysis of which affords 2-hydroxy acids. This sequence is used to lengthen the carbon chain (Eq. 2).

$$\tag{2}$$

Originally this reaction was applied by H. Kiliani to D-*arabino*-hexulose (D-fructose),[15] D-glucose,[16] L-arabinose,[17] and D-galactose.[18] The intermediary cyanohydrins are not usually isolated, but are hydrolyzed *in situ* to the corresponding aldonic acids. The formation of a new asymmetric center gives rise to two 2-epimers, usually in unequal amounts. The proportions vary according to the stereochemistry of the monosaccharide used and the reaction conditions. The two resultant acids can usually be separated by fractional crystallization of the derived metal salts, lactones, phenylhydrazides, or amides.

When aqueous hydrocyanic acid containing a little ammonia is used, some stereoselectivity may be observed in the reaction with aldoses, inasmuch as the 2,4-*threo* product preponderates over the 2,4-*erythro* isomer.[19] This selectivity is explicable on conformational grounds, since in the zigzag conformation the β-hydroxyl groups come into unfavorable, eclipsed proximity in the 2,4-*erythro* isomer (12) but not in the 2,4-*threo* isomer (13). However, alteration of

the reaction conditions can reverse the stereoselectivity; for example, L-arabinose yields mainly L-mannonic acid (2,4-*threo*) at pH <7, as would be predicted by Maltby's generalization,[19] whereas under alkaline conditions L-gluconic acid (2,4-*erythro*) preponderates.[20] Militzer[21] found that for all simple sugars the optimum reaction occurs at pH 9.0 to 9.1, even in the presence of a stoichiometric amount of cyanide.

The extension of this reaction to the ascent of the aldose series was made possible by Emil Fischer's observation[22] that aldonolactones, derived from the aldonic acids by heating, are reduced to aldoses by sodium amalgam under mildly acidic conditions. Careful control of the pH of the reaction mixture is essential; the optimum pH is 3.0 to 3.5. The reaction can be controlled either by the portionwise addition of mineral acid or, better, by the incorporation of

* *References start on p.* 158.

sodium hydrogen oxalate or benzoic acid into the reaction mixture. At other pH values, extensive over-reduction can give rise to substantial proportions of alditol impurities.

Reduction of aldonolactones to aldoses has also been achieved at pH 3 to 4 with aqueous sodium borohydride[23] at 0°, with sodium borohydride in ether–acetic acid,[24] and by catalytic hydrogenation.[25] Acylated aldonolactones are reduced readily to the acylated aldoses by bis(3-methyl-2-butyl)borane.[25a] R. Kuhn and his co-workers[26] have described a method by which cyanohydrins may be converted directly into the aldoses by hemihydrogenation over palladium on barium sulfate; the method has been found especially useful for synthesis of 2-amino-2-deoxyaldoses (see Chapter 16).

The cyanohydrin synthesis is widely used for the preparation of isotopically labeled sugars (Section VIII). Theoretically this synthesis permits the preparation of all the higher aldoses, but there are practical limitations and so far the method has not been extended beyond the decoses.[27] The sugars to which this reaction has been applied have been tabulated by J. Staněk et al.;[28] the subject has been reviewed by Hudson,[29] and practical details have been given by N. K. Richtmyer.[30]

B. THE NITROMETHANE SYNTHESIS[31]

The carbon chain of an aldose may be extended by a base-catalyzed aldol type of addition reaction with nitromethane. The electron-withdrawing nitro group of nitromethane resembles the carbonyl group in that it facilitates formation of a resonance-stabilized carbanion ($^-CH_2NO_2$). The latter attacks the electron-deficient carbon atom of the carbonyl group to give a 1-deoxy-1-nitroalditol. Subsequent decomposition of the *aci* form of the nitroalcohol in strongly acidic solution, known as the Nef reaction, gives the hydroxyaldehyde (Eq. 3).

$$(3)$$

The nitromethane reaction can be effected conveniently with unsubstituted aldoses. Since the reaction is reversible, a large excess of nitromethane is used with 1 to 2 molar equivalents of alkali.[31] Best yields are obtained by using a solvent in which the starting material is soluble but from which the resulting sodium salt of the *aci*-nitroalcohol precipitates. In most cases methanolic sodium methoxide is used, although the use of methyl sulfoxide as a solvent is reported to give improved yields.[32]

As in the cyanohydrin synthesis, a new asymmetric center is created at C-2, giving rise to two epimers, which generally can be separated by fractional crystallization of the 1-deoxy-1-nitroalditols. Usually some stereoselectivity is observed,[33] so that one epimer preponderates. Hough and Shute[33, 33a] have observed that the major product is that predicted by the Maltby rule (Section III,A)—that is, the isomer in which the 2,4-hydroxyl groups have a *threo* relationship. Thus D-galactose yields the D-*glycero*-L-*manno* epimer in higher yield than the D-*glycero*-L-*gluco* epimer.[33a]

1-Deoxy-1-nitroalditols are decomposed by acidification of their sodium salts with relatively concentrated sulfuric acid to give the corresponding aldoses. The resultant aldoses can usually be isolated in good yield after deionization of the reaction mixture.

The 1-deoxy-1-nitroalditols are used extensively as synthetic intermediates because their acetates readily lose the elements of acetic acid on treatment with sodium hydrogen carbonate in benzene to give 1-alkenes,[34] which can undergo a variety of addition reactions to the double bond (see Chapter 19) to give suitable precursors of deoxy sugars (Chapter 17), 2-O-methylaldoses (Chapter 12), and amino sugars (Chapter 16). The cyclization of the "dialdehydes" produced by the periodate oxidation of pyranosides and furanosides with nitromethane is an important method of synthesizing 3-amino-3-deoxy sugars (Chapter 16).

The method can be adapted for the preparation of ketoses, by treating an aldose with 2-nitroethanol, resulting in the addition of two carbon atoms to the chain (Eq. 4). The resulting 2-deoxy-2-nitroalditols undergo the Nef reaction to give 2-ketoses.[35, 36] Similarly, the reaction of nitroethane with aldoses affords 1-deoxyketoses.

$$
\begin{array}{ccc}
CH_2OH & CH_2OH & CH_2OH \\
| & | & | \\
H_2CNO_2 & CHNO_2 & C=O \\
\overset{+}{} & | & | \quad\quad (4)\\
H\diagdown_{C}\diagup O & CHOH & CHOH \\
| & | & | \\
R & R & R \\
\end{array}
$$

Practical details of the nitromethane condensation are given in "Methods in Carbohydrate Chemistry."[37]

C. THE DIAZOMETHANE SYNTHESIS

This method is used for the synthesis of 2-ketoses and their 1-deoxy derivatives and proceeds from the acetylated aldonic acid (14) to the acetylated aldonyl chloride (15) and then to the acetylated 1-deoxy-1-diazo-*keto* sugar (16), which is converted into either the ketose (17) or the deoxyketose (18). The diazoketones (16) are produced by treatment of poly-*O*-acetylaldonyl chlorides (15) with diazomethane.[38, 39] The conversion of the diazoketone (16) into the

ketose (17) can be accomplished in one step by direct hydrolysis with mineral acid, but it is usually accomplished by the action of acetic acid containing a catalytic amount of cupric acetate to give 20, with subsequent deacetylation. An acetylated diazoketone can react[40] with dry hydrogen chloride (or bromide) to give 1-deoxy-1-halogenoketones (19), which also arise as side products in the synthesis of the diazoketone (16) unless sufficient diazomethane is added to react with the hydrogen halide formed during the reaction.[41] Reduction of the diazoketones (16) with either hydriodic acid[40] or aluminum amalgam[42] affords 1-deoxyketoses. Treatment of diazoketones with cupric oxide causes elimination of nitrogen and head-to-head dimerization of the resultant α-oxo carbene to give an unsaturated dicarbonyl sugar.[42a]

The method may be adapted to the synthesis of 2-deoxy sugars by the Wolff rearrangement[43] of the diazoketone—that is, by treatment with silver oxide in water, followed by deacetylation and reduction of the resulting poly-*O*-acetyl-2-deoxyaldonolactone. The only recorded example[41] is the conversion of *keto*-1-deoxy-1-diazo-D-*gluco*-heptulose pentaacetate (21) into 2-deoxy-D-*gluco*-heptonic acid (22).

Poly-*O*-acetyl aldonic acids are obtained directly from cadmium aldonates by acetylation in the presence of hydrogen chloride,[44] from poly-*O*-acetyl aldonamides by deamination with nitrosyl chloride,[45] or by oxidation of poly-*O*-acetyl-*aldehydo*-aldoses derived from the corresponding dithioacetals

$$
\begin{array}{c}
\text{CHN}_2 \\
| \\
\text{C}\!=\!\text{O} \\
| \\
\text{HCOAc} \\
| \\
\text{AcOCH} \\
| \\
\text{HCOAc} \\
| \\
\text{HCOAc} \\
| \\
\text{CH}_2\text{OAc} \\
\mathbf{21}
\end{array}
\xrightarrow[\text{H}_2\text{O}]{\text{Ag}_2\text{O}}
\quad\quad
\xrightarrow{[\text{H}]}
\quad\quad
\begin{array}{c}
\text{CHO} \\
| \\
\text{CH}_2 \\
| \\
\text{HCOH} \\
| \\
\text{HOCH} \\
| \\
\text{HCOH} \\
| \\
\text{HCOH} \\
| \\
\text{CH}_2\text{OH} \\
\mathbf{22}
\end{array}
$$

(Chapter 10). Conversion into the aldonyl chloride is effected by conventional treatment of the acid with either thionyl chloride or phosphorus pentachloride. The diazomethane synthesis can also be used with protected uronic acids as precursors.[45a]

The relatively large number of steps involved in the diazomethane synthesis invariably results in a low overall yield.

Aldehydo sugars react with diazomethane to give 1-deoxy-*keto* derivatives initially. These can then react further to give 1,2-dideoxy-3-*keto* derivatives (Eq. 5). Thus *aldehydo*-D-glucose pentaacetate affords 1-deoxy-*keto*-D-*gluco*-heptulose pentaacetate.[46] On the other hand, reaction of *keto*-D-fructose

$$
\begin{array}{c}
\text{H}\diagdown \\
\quad\text{C}\!=\!\text{O} \\
| \\
\text{R}
\end{array}
\xrightarrow{\text{CH}_2\text{N}_2}
\begin{array}{c}
\text{CH}_3\diagdown \\
\quad\text{C}\!=\!\text{O} \\
| \\
\text{R}
\end{array}
\xrightarrow{\text{CH}_2\text{N}_2}
\begin{array}{c}
\text{CH}_3 \\
| \\
\text{CH}_2 \\
| \\
\text{C}\!=\!\text{O} \\
| \\
\text{R}
\end{array}
\qquad (5)
$$

pentaacetate (**23**) with diazomethane gives the epoxide (**24**) and hence provides a route to branched-chain sugars (Chapter 17).

$$
\begin{array}{c}
\text{AcOCH}_2 \\
| \\
\text{C}\!=\!\text{O} \\
| \\
\text{R} \\
\mathbf{23}
\end{array}
\xrightarrow{\text{CH}_2\text{N}_2}
\begin{array}{c}
\text{AcOCH}_2 \quad\text{O} \\
| \diagup\ \diagdown \\
\text{C}\!-\!\!-\!\!-\text{CH}_2 \\
| \\
\text{R} \\
\mathbf{24}
\end{array}
$$

D. Syntheses Based on Malonic Esters and Related Compounds

Aldehydo sugars undergo the Knoevenagel condensation with ethyl malonate, ethyl acetoacetate, ethyl cyanoacetate, and ethyl benzoylacetate to extend the carbon chain and give unsaturated derivatives (Eq. 6).[47] Kochetkov and

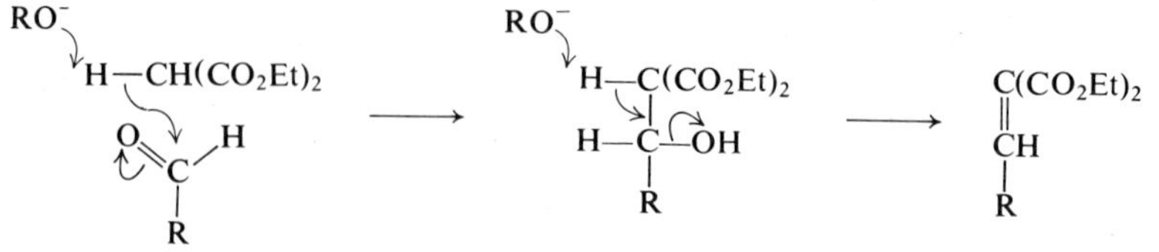

$$
\begin{array}{c}
RO^- \\
\searrow H-CH(CO_2Et)_2 \\
\underset{R}{O=C}-H
\end{array}
\longrightarrow
\begin{array}{c}
RO^- \\
\searrow H-C(CO_2Et)_2 \\
\underset{R}{H-C-OH}
\end{array}
\longrightarrow
\begin{array}{c}
C(CO_2Et)_2 \\
\| \\
\underset{R}{CH}
\end{array}
\qquad (6)
$$

Dmitriev[48] have extended this method by hydroxylation of the double bond, thereby ascending the series by two carbon atoms at a time. Thus the unsaturated derivative (26) obtained from the condensation of 2,3:4,5-di-*O*-isopropylidene-*aldehydo*-L-arabinose (25) with ethyl malonate was hydrolyzed

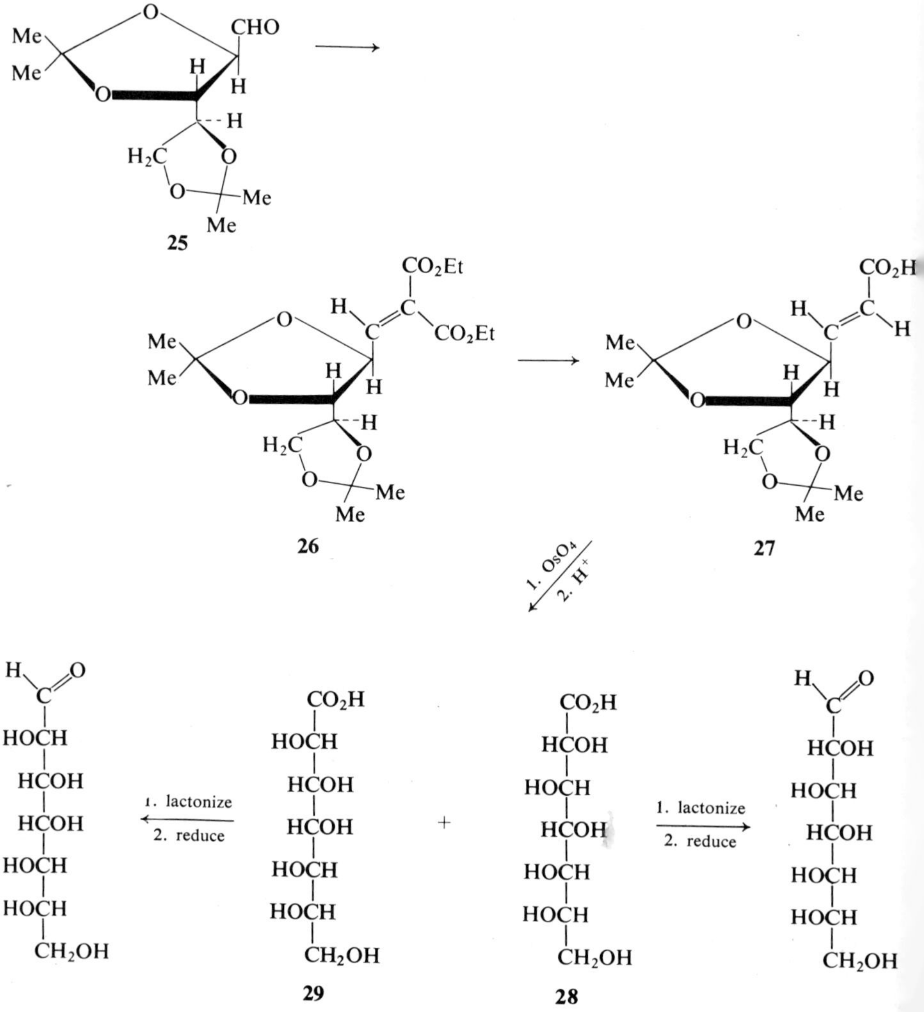

25

26

27

1. OsO₄
2. H⁺

H–CHO		CO₂H	CO₂H	H–CHO
HOCH		HOCH	HCOH	HCOH
HCOH		HCOH	HOCH	HOCH
HCOH	1. lactonize	HCOH	HCOH	HCOH
HOCH	2. reduce	HCOH	HCOH	HCOH
HOCH		HOCH	HOCH	HOCH
CH₂OH		HOCH	HOCH	HOCH
		CH₂OH	CH₂OH	CH₂OH

29　　+　　**28**

1. lactonize
2. reduce

and decarboxylated to give the *trans* form of the unsaturated aldonic acid (27). Subsequent *threo* hydroxylation of the double bond with osmium tetraoxide and acid hydrolysis of the acetal groups gave L-*glycero*-L-*galacto*- (29) and L-*glycero*-L-*ido*-heptonic acids (28), which yielded the corresponding aldoses on reduction of their lactones.

Suitable aldehydo sugars are not always available, and the formation of *cis* and *trans* forms of the unsaturated acid is always possible, so that four possible stereoisomeric aldonic acids could arise. Reduction of the olefinic double bond in 27 gives a precursor of a 2,3-dideoxyaldose.

A related procedure involving condensation of a poly-*O*-acetyl aldonyl chloride with benzyl malonate affords a *keto* derivative, which on catalytic reduction followed by heat-induced decarboxylation is converted into the 1-deoxyketose (Eq. 7).[49]

(7)

E. THE WITTIG REACTION

Reducing monosaccharides undergo the Wittig reaction with ethoxy-carbonylmethylenetriphenylphosphorane ($Ph_3P=CHCO_2Et$) and other phosphoranes to give the corresponding α,β-unsaturated derivatives[48, 50] in good yields (40 to 60%). Hydroxylation of the double bond, as in the case cited in Section III,D, affords a mixture of aldonic acid esters, which may be converted into aldoses by standard methods. For example, D-galactose reacts with ethoxycarbonylmethylenetriphenylphosphorane to give ethyl D-*galacto*-2,3-dideoxy-oct-2-enonate (**30**) in 43% yield. These unsaturated esters show some tendency to cyclize, in a manner analogous to that of the disulfones discussed in Section IV,D. Hence some 2,5-anhydro derivative (**31**) is formed together with the unsaturated ester.[48]

F. Ethynylation

Aldehydo derivatives of monosaccharides undergo addition of ethynyl-magnesium bromide (prepared from acetylene and ethylmagnesium bromide) to the carbonyl group to give a mixture of epimeric acetylenic alcohols (Eq. 8). Partial reduction of the triple bond of each of these derivatives followed by

$$
\begin{array}{c}
\text{H}\diagdown_{\text{C}}\diagup^{\text{O}} \\
\underset{\text{R}}{|}
\end{array}
+
\begin{array}{c}
\text{CMgBr} \\
\text{|||} \\
\text{CH}
\end{array}
\longrightarrow
\begin{array}{c}
\text{CH} \\
\text{|||} \\
\text{C} \\
\text{|} \\
\text{CHOH} \\
\text{|} \\
\text{R}
\end{array}
\longrightarrow
\begin{array}{c}
\text{CH}_2 \\
\text{||} \\
\text{CH} \\
\text{|} \\
\text{CHOH} \\
\text{|} \\
\text{R}
\end{array}
\xrightarrow{\text{O}_3}
\begin{array}{c}
\text{CHO} \\
\text{|} \\
\text{CHOH} \\
\text{|} \\
\text{R}
\end{array}
\quad (8)
$$

ozonolysis of the resulting alkenes gives the next higher aldehydo derivatives. The method is limited by the availability of suitable starting materials, but it has, for example, been applied to 2,3:4,5-di-*O*-isopropylidene-L-arabinose (**25**) for the synthesis of L-glucose and L-mannose.[51, 52]

G. Aldol Condensation Reaction

Aldol condensation reactions between aldoses and 1,3-dihydroxy-2-propa-none or its derivatives provide a method of ascending the series by three carbon atoms to give ketoses (Eq. 9).[8] The reaction may be catalyzed either by

$$
\text{HO}^-\diagdown
\begin{array}{c}
\text{CH}_2\text{OH} \\
\text{|} \\
\text{C}=\text{O} \\
\text{|} \\
\text{H}-\text{CHOH} \\
\text{|} \\
\text{H}\diagdown_{\text{C}}\diagup^{\text{O}} \\
\text{|} \\
\text{R}
\end{array}
\underset{(\longleftarrow)}{\longrightarrow}
\begin{array}{c}
\text{CH}_2\text{OH} \\
\text{|} \\
\text{C}=\text{O} \\
\text{|} \\
\text{HOCH} \\
\text{|} \\
\text{HCOH} \\
\text{|} \\
\text{R}
\end{array}
+
\begin{array}{c}
\text{CH}_2\text{OH} \\
\text{|} \\
\text{C}=\text{O} \\
\text{|} \\
\text{HCOH} \\
\text{|} \\
\text{HOCH} \\
\text{|} \\
\text{R}
\end{array}
\quad (9)
$$

alkali or, if 1,3-dihydroxy-2-propanone 1-phosphate is used, by enzymes. The *threo* configuration is usually favored for two new asymmetric centers that are created, but the extent of stereospecificity is dependent on the catalyst employed. With alkali the reaction is less stereospecific and is of limited utility. Only the lower aldoses, glyceraldehyde and erythrose, condense satisfactorily with 1,3-dihydroxy-2-propanone in the presence of alkali (under carefully controlled conditions) to give ketoses. D-Glyceraldehyde yields a mixture of D-*arabino*- and D-*xylo*-hexuloses in high yield (Section II,A), whereas D-erythrose gives a mixture of D-*gluco*-, D-*altro*, and D-*allo*-heptuloses in about 37% total yield (6.6:3.4:1, respectively), together with traces of D-*gluco*-L-*glycero*-3-octulose and "dendroketose".[53] The latter is a branched-chain sugar that arises from the combination of two moles of 1,3-dihydroxy-2-propanone.[54] The failure of higher aldoses to undergo reaction can be attributed to their existence in cyclic forms and to their isomerization and degradation by alkali.[55]

Enzyme-catalyzed reactions with pea or muscle aldolase and 1,3-dihydroxy-2-propanone 1-phosphate (**32**) are stereospecific, giving ketose 1-phosphates (**33**) having the D-*threo* configuration at the two newly created asymmetric centers. If the enzyme preparation also contains a phosphatase, dephosphorylated products are obtained. For example, incubation of a pea aldolase

$$
\begin{array}{ccc}
\text{CH}_2\text{OPO}_3\text{H}_2 & & \text{CH}_2\text{OPO}_3\text{H}_2 \\
| & & | \\
\text{C}\!\!=\!\!\text{O} & & \text{C}\!\!=\!\!\text{O} \\
| & \xrightarrow{\text{aldolase}} & | \\
\text{CH}_2\text{OH} & & \text{HOCH} \\
+ & & | \\
\text{CHO} & & \text{HCOH} \\
| & & | \\
\text{R} & & \text{R} \\
& & \mathbf{33}
\end{array}
$$

preparation with D-erythrose and a hexose 1,6-diphosphate (a source of 1,3-dihydroxypropanone 1-phosphate by reversal of the aldolase reaction) gives D-*altro*-heptulose.[56] Under similar conditions D- and L-threose yield D-*ido*- and L-*galacto*-heptulose, respectively,[57,58] The stereospecificity of the aldolase reaction is further emphasized by similar application to the four aldopentoses, all of which give octuloses having the 3,4-D-*threo* configuration.[59]

IV. DESCENT OF THE SERIES

A. THE RUFF DEGRADATION[60]

This reaction is one of the most widely used of the degradative methods. A soluble salt of an aldonic acid is made to undergo oxidative decarboxylation by the action of hydrogen peroxide in the presence of ferric ions, to give the next

* *References start on p. 158.*

lower aldose (Eq. 10). The reaction is generally assumed to proceed by way of a 2-keto acid, which undergoes decarboxylation.

$$
\begin{array}{c}
\overset{-O}{\underset{}{}}\overset{O}{\underset{}{}}
\end{array}
\xrightarrow{H_2O_2/Fe^{3+}}
\quad
\xrightarrow{-CO_2}
\quad
\tag{10}
$$

The reaction is fairly specific, since little oxidation occurs beyond the pentose stage in the degradation of hexoses. However, ferrous ions interfere by causing oxidation of the aldose to an aldosulose (see Chapter 23). The mechanism of the degradation is obscure but possibly involves the hydroperoxy radical (HO_2^-) (Eqs. 11 and 12) and not the hydroxyl radical, since the latter would probably attack each alcoholic hydroxyl group, and a more complex reaction would then be observed.

$$
HO—OH \;\; \rightleftharpoons \;\; H^+ + HO—O^- \tag{11}
$$

$$
HO—O^- + Fe^{3+} \;\; \rightleftharpoons \;\; Fe^{2+} + HO—O. \tag{12}
$$

An effective preparative method for 2-deoxy-D-*erythro*-pentose is by the Ruff degradation of calcium 3-deoxy-D-*arabino*-hexonate ("D-glucosaccharinic acid") (see Chapter 23).

In its original form, the Ruff degradation often gives low yields because crystallization of the aldoses is impeded by the gross amounts of inorganic material present. Sometimes the aldose crystallizes easily, as in the case of the preparation of D-arabinose (50% yield from calcium D-gluconate). Recent improvements by using ion-exchange resins for the removal of inorganic material have resulted in much better yields; the yield of D-lyxose from calcium D-galactonate has been raised[61] in this way from 17% to 41%. The conversion of D-galactaric acid 6-monoamide into D-lyxuronamide proceeds satisfactorily.[62] Unfortunately, the method is unsatisfactory for the preparation of tetroses from pentoses. Practical details of the Ruff degradation have been given by Fletcher,[63] by Whistler and BeMiller,[64] and by Richards.[65]

B. THE WOHL DEGRADATION

This method is essentially the reverse of the Fischer–Kiliani cyanohydrin synthesis (Section III, A). The aldose (34) is converted into its oxime (35), which, on treatment with acetic anhydride in the presence of sodium acetate, undergoes simultaneous acetylation and dehydration to form the poly-*O*-acetyl nitrile

(36). When the nitrile is treated with ammoniacal silver oxide, loss of hydrogen cyanide together with O-deacetylation occurs to give the next lower aldose. The latter reacts with ammonia as it is formed, and subsequent O → N acetyl migrations occur to give the corresponding 1,1-bis(acetamido) derivative (37), from which the aldose (38) may be obtained by hydrolysis with dilute acid.[66]

Various modifications have been described. The use of acetic anhydride in pyridine for acetylation and dehydration of the oxime is reported to give better results than acetic anhydride–sodium acetate, but the mildness of the former reagent favors acetylation of the oxime without dehydration. At low temperatures, oximes of only the aldopentoses, D-mannose, 3-deoxy-L-mannose, and 2-amino-2-deoxy-D-glucose have been shown to form nitriles. Higher temperatures, however, favor dehydration to the nitrile with all of the sugar oximes.[67] The deacetylation–degradation stage can be effected in aqueous ammonia alone.[68, 69] A marked improvement in this stage of the degradation may be achieved by use of sodium methoxide, which yields the lower aldose directly and obviates the need for separate hydrolysis of the intermediary 1,1-bis-(acetamido) derivative. Deulofeu[70] has shown that this modification is preferable for hexoses and higher monosaccharides, whereas the original procedure of Wohl is better for the lower sugars. The topic has been reviewed by Deulofeu.[71]

The fully acetylated aldehydo sugar is probably formed initially on treatment of the acetylated nitrile derivative with ammonia. The 1,1-bis(acetamido) derivative could conceivably arise by participation and migration of the acetyl groups of C-2 and C-3; this intramolecular reaction has been studied by using

* References start on p. 158.

[14]C-labeled O-benzoyl groups, but it was found that the largest migration occurred from C-3 and C-4, whereas that from C-2 was the smallest.[72] These observations have been rationalized on the basis of the stability of the cyclic intermediates formed from esters in their zigzag conformations.

A related procedure, in which an oxime of a higher aldose is degraded to the lower aldose in one step by the action of 2,4-dinitrofluorobenzene, has been developed by Weygand *et al.*[73] (Eq. 13).

$$\text{(13)}$$

C. The Weerman Degradation

This adaptation of the Hoffmann degradation of amides to primary amines has not found wide use in descending the series. The higher aldose is first converted into the aldonamide. Treatment with alkaline hypochlorite or hypobromite transforms the amide into the intermediary isocyanate, which disproportionates to give the lower aldose (Eq. 14).[74]

$$\text{(14)}$$

The method is of limited value because the aldonamides are rather inaccessible. Furthermore, the aldose that is formed undergoes oxidation by hypochlorite readily. However, the method has been used to demonstrate whether the 2-hydroxyl group of an aldose is free or substituted, since in the latter instance a cyclic carbamate results. In the case of 2,3,4,6-tetra-O-methyl-D-gluconamide (**39**) the reaction proceeds as far as the isocyanate (**40**), which then undergoes cyclization with the 5-hydroxyl group to give the cyclic carbamate (**41**). The degradation is completed[75] by alkaline hydrolysis of the carbamate to give 2,3,5-tri-O-methyl-D-arabinofuranose (**42**).

By this method D-lyxuronic acid (44) has been prepared[62] from the mono-
amide of D-galactaric acid 1-amide (43).

D. THE DISULFONE METHOD

Aldose diethyl dithioacetals (45) are oxidized by monoperoxyphthalic acid,[76]
peroxypropionic acid,[77] or hydrogen peroxide in the presence of molybdate
anions,[78] to the corresponding disulfones (46), which undergo base-catalyzed
cleavage of the bond between C-1 and C-2 with dilute aqueous ammonia at
room temperature giving bis(ethylsulfonyl)methane (47) and the lower aldose.
The good yields and high purity of the latter make this a valuable method for
descending the aldose series.

* References start on p. 158.

When applied to ketoses, this method effects degradation of the mono-saccharide by two carbon atoms, since the action of base on the derived disul-fone (**48**) causes simultaneous cleavage of the bonds between C-1 and C-2 and between C-2 and C-3, to give a lower aldose, formaldehyde, and 1,1,3,3-tetra-kis(ethylsulfonyl)propane (**49**). The latter arises from the action of alkali on a mixture of formaldehyde and bis(ethylsulfonyl)methane.[78] The disulfones

derived from the aldoses may have one of three possible structures (**51, 52,** or **53**), according to the nature of the starting material and the reaction conditions. Thus, oxidation of D-mannose diethyl dithioacetal (**50**) with peroxypropionic acid gives two disulfone derivatives, one being the saturated acyclic compound (**51**), which is readily isolated because it crystallizes from the reaction mixture. The other disulfone is derived from **51** by loss of the elements of water and cyclization to a pyranoid anhydride (**53**). On the other hand, oxidation of D-glucose diethyl dithioacetal afforded[79] only the cyclic disulfone (**53**). The cyclic compounds arise from the saturated sulfones (such as **51**) by dehydration to the 1-hexene derivative (for example, **52**), the double bond of which suffers attack by the 6-hydroxyl group to form the pyranoid ring. In the case of the pentose disulfones derived from aldopentoses, the 1-pentene structure was originally assigned,[80] but now they are known to have a cyclic furanoid struc-ture.[81] In solution they behave as an equilibrium mixture of unsaturated and furanoid forms. Oxidation of fully acetylated aldose diethyl dithioacetals yields unsaturated disulfones, since ring closure is prevented by the ester groups. The saturated acyclic derivatives are usually converted into the cyclic disulfones by heating in dilute acetic acid, although some reverse aldol-type reaction has been observed even under acidic conditions, to give bis(ethyl-sulfonyl)methane and the lower aldose, together with the cyclic disulfone.[82]

All three types of disulfone structure undergo base-catalyzed cleavage of the bond between C-1 and C-2, to give bis(ethylsulfonyl)methane and the lower aldose; the structure of the disulfone does not, therefore, affect the final product of the degradation sequence. However, the rates of cleavage of the three

types of disulfone are significantly different. In dilute aqueous ammonia at pH ~10 at room temperature the saturated acyclic disulfones are cleaved rapidly, whereas the cleavage of the pentofuranoid derivatives is comparatively slow, 5 to 30 hr usually being required for complete degradation.

The pyranoid disulfones are degraded even more slowly than the other two types, 5 to 7 days being necessary for complete reaction. An investigation of the mechanism of cleavage of bis(ethylsulfonyl)-(α-D-lyxopyranosyl)methane (54) to D-lyxose (56) has suggested that the reaction proceeds by a unimolecular mechanism[83] involving a cyclic oxonium ion (55), but an alternative or competitive mechanism by way of the acyclic unsaturated structure (57) cannot at this stage be ruled out.

* References start on p. 158.

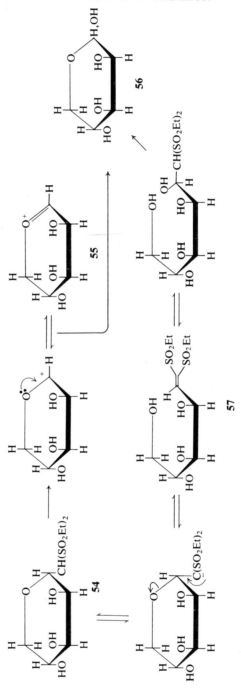

E. CONTROLLED OXIDATION WITH GLYCOL-CLEAVING REAGENTS

Oxidation cleavage of suitable vicinal diols with sodium metaperiodate, lead tetraacetate, and the like generates aldehyde groupings; hence oxidation of suitable protected monosaccharides can be used to descend the aldose series. As an example,[84] 2,4-O-benzylidene-D-glucitol (58) undergoes oxidation between C-5 and C-6 with lead tetraacetate to give 2,4-O-benzylidene-L-xylose (59), which on acid hydrolysis gives L-xylose in an overall yield of 80%.

Free monosaccharides are usually oxidized by these reagents in their pyranose or furanose ring forms; consequently, in the absence of a free 3-hydroxyl group, aldoses undergo selective cleavage of the bond between C-1 and C-2 to give 4-O-formyl derivatives (61), which at about pH 3.5 are fairly resistant to hydrolysis and thus to further oxidation. After removal of the inorganic material, hydrolysis of the formic ester by dilute base liberates the next lower

aldose. This method has been applied[85] to 3-O-benzyl-D-glucose (60), which gives 2-O-benzyl-D-arabinose (62), and 2-deoxy-D-erythro-pentose has been prepared from 3-deoxy-D-ribo-hexose.[86] Similarly, oxidation of 3-acetamido-

* References start on p. 158.

3-deoxy-D-mannose gives rise to 2-acetamido-2-deoxy-D-arabinose,[87] but the method fails for 3-acetamido-3-deoxy-hexoses having the *altro* and *gulo* configurations, presumably because they do not react predominantly in their pyranose modifications.[88, 89]

　　Exocyclic diols react more rapidly with periodate than *trans*-diols attached to five-membered furanoid rings. Consequently, methyl D-glucofuranosides (**63**) and methyl D-galactofuranosides (**64**) undergo preferential cleavage of the bond between C-5 and C-6 to give, after reduction with sodium borohydride and hydrolysis with acid, D-xylose (**65**) and L-arabinose (**66**), respectively.[90]

63　　　　　　　　　　　　　　　　　　　　　**65**

64　　　　　　　　　　　　　　　　　　　　　**66**

　　Oxidation of unprotected aldoses with lead tetraacetate has been exploited as a method for descending the series.[91] Reaction with one mole of lead tetraacetate occurs principally with cleavage of the bond between C-1 and C-2 to give the next lower aldose directly, after hydrolysis of the intermediary formic ester (**68**). However, the reaction is not highly specific, and impure products are usually obtained that require purification by chromatography or through a crystalline derivative. By using this method, D-galactose (**67**) affords mainly D-lyxose (**69**), together with small proportions of D-threose and starting material.[92]

67　　　　　　　　　　　　　　　　　　　　　**68**

69

F. The Hunsdiecker Reaction

Application of this reaction to silver salts of poly-*O*-acetyl-aldonic acids has been described. The silver salt (**70**) of penta-*O*-acetyl-D-gluconic acid on

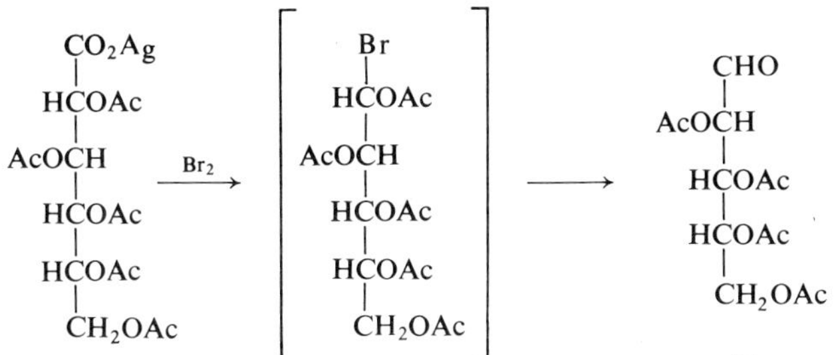

$$
\begin{array}{ccc}
\begin{array}{l}
\text{CO}_2\text{Ag} \\
\mid \\
\text{HCOAc} \\
\mid \\
\text{AcOCH} \\
\mid \\
\text{HCOAc} \\
\mid \\
\text{HCOAc} \\
\mid \\
\text{CH}_2\text{OAc} \\
\mathbf{70}
\end{array}
&
\xrightarrow{\text{Br}_2}
\left[
\begin{array}{l}
\text{Br} \\
\mid \\
\text{HCOAc} \\
\mid \\
\text{AcOCH} \\
\mid \\
\text{HCOAc} \\
\mid \\
\text{HCOAc} \\
\mid \\
\text{CH}_2\text{OAc}
\end{array}
\right]
\longrightarrow
&
\begin{array}{l}
\text{CHO} \\
\mid \\
\text{AcOCH} \\
\mid \\
\text{HCOAc} \\
\mid \\
\text{HCOAc} \\
\mid \\
\text{CH}_2\text{OAc} \\
\mathbf{71}
\end{array}
\end{array}
$$

treatment with bromine affords[93] tetra-*O*-acetyl-*aldehydo*-D-arabinose (**71**). Penta-*O*-acetyl-D-gluconyl chloride is similarly degraded by the combined action of silver oxide and bromine.[94]

G. Other Methods

Oxidative degradations of aldoses and ketoses either with oxygen in alkaline solution or with sodium hypochlorite have found preparative applications. In the presence of alkali, aldoses are oxidized by air or oxygen to the lower aldonic acids.[95, 96] Since ketoses yield similar products, it is generally assumed that the reaction proceeds by way of an enediol to a 2-hexulosonic acid, which then undergoes decarboxylation (Eq. 15), although other possibilities have been outlined.[97] Lactonization and reduction of the lactone then yields the lower aldose.

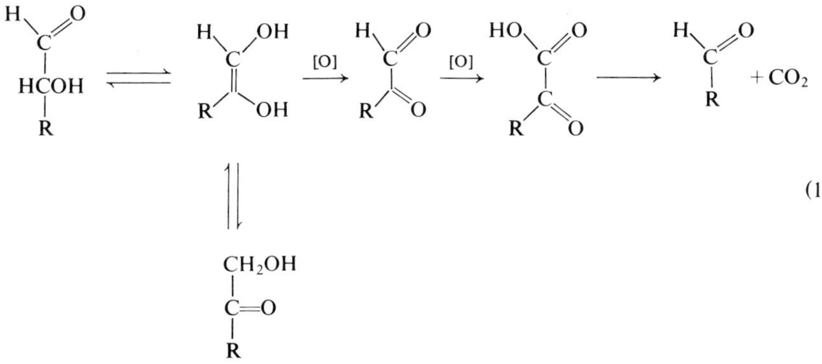

$$(15)$$

Aldonic acids are oxidized by two moles of hypochlorite at pH 4.5 to 5.0 to give the lower aldoses in 30 to 40% yield.[98] This reaction can be applied directly to the aldose itself by initial oxidation with hypochlorite at pH 11, followed by adjustment to pH 4.5 to 5.0 and the addition of further oxidant. In this way D-glucose is converted into D-arabinose by a two-stage single-batch process in about 35% yield.[99]

Electrolytic oxidation of aldoses and aldonic acids can also be used to shorten the chain.[100] Aldonic acids are reported to be converted into the lower aldoses by heating with mercuric oxide.[101, 102]

V. CHANGES IN CONFIGURATION AT ASYMMETRICALLY SUBSTITUTED CARBON ATOMS

A. BASE-CATALYZED EPIMERIZATIONS AND ISOMERIZATIONS

Aldonic acids undergo epimerization at C-2 in the presence of base. An aldose such as D-galactose (**72**) may thus be converted into D-talose (**76**) by a three-stage process, through oxidation to the aldonic acid (**73**), epimerization, isolation of the epimeric aldonic acid (**74**) as its lactone (**75**), and subsequent reduction of the latter.[103, 104] The epimerization is usually effected in pyridine,

although quinoline, inorganic bases such as calcium and barium hydroxides, or basic ion-exchange resins are used occasionally. Low overall yields are usually obtained, but in cases where an aldose is abundantly available and the epimer is not, this method can sometimes be advantageous. Epimerization of aldoses at C-2 is not usually feasible because they undergo the Lobry de Bruyn–

Alberda van Ekenstein transformation[105] and isomerize via enediolate ions to a mixture of ketoses and aldoses (Eq. 16). However, the method is frequently

$$(16)$$

applied to the preparation of ketoses, since the aldoses can be removed conveniently. Thus the four pentuloses can each be prepared by isomerization of the appropriate aldopentoses with hot, moist pyridine, followed by isolation of the ketoses through crystalline derivatives such as isopropylidene acetals.[106]

Substituted aldoses, in which the 2-hydroxyl group is not free, undergo epimerization only, since isomerization to a ketose is prevented. For example, 2,3,4,6-tetra-O-methyl-D-glucose can be readily isomerized to the D-mannose isomer.[107] Similarly 2-acetamido-2-deoxyaldoses equilibrate rapidly with their 2-epimers in basic solution, and the composition of the resulting mixture of two epimers is dependent on the relative thermodynamic stabilities of the epimers. The rate-determining step in the epimerization is the removal of the hydrogen substituent from C-2, and a strongly electron-withdrawing group at this carbon atom, such as acetamido, greatly facilitates the process.[108]

The influence of steric factors is illustrated by conversion of the conformationally unstable L-idose and 6-deoxy-L-idose (77) into L-sorbose and 6-deoxy-L-*xylo*-hexulose (78, 6-deoxy-L-sorbose), respectively, in the presence of acid,[109-111] or upon being kept for a long period of time. Similarly, it was shown[111a] that D-*threo*-L-*ido*-octose undergoes a slow, spontaneous conversion into D-*glycero*-L-*gluco*-octulose.

77   78

* References start on p. 158.

B. Replacement Reactions of Sulfonic Esters

Direct replacement of sulfonyloxy substituents at secondary, asymmetrically substituted carbon atoms of carbohydrates, by nucleophilic reagents such as sodium acetate in acetic anhydride[109] and sodium benzoate in N,N-dimethylformamide,[112] proceeds with inversion of configuration when the stereochemistry and electronic factors favor a bimolecular reaction.[112a] The nucleophilic displacement is usually facile when the sulfonic ester group is part of an acyclic molecule or is attached to an exocyclic side chain. Thus the 5,6-di-O-p-tolylsulfonyl derivative of 1,2-O-isopropylidene-α-D-glucofuranose (**79**) is readily converted[113] into L-idose derivatives (**80**). The primary p-tolylsulfonyloxy groups are replaced more rapidly than secondary p-toluenesulfonates, because of increased crowding in the transition state with the latter.

        79   **80**

There must be no hindrance to the incoming nucleophile, and therefore direct replacement of a sulfonyloxy group attached to a pyranoid or furanoid ring is critically dependent on steric factors. Even under favorable steric circumstances the use of N,N-dimethylformamide, or other high-boiling aprotic solvents of high dielectric constant such as hexamethylphosphoric triamide, appears to be essential for these displacement reactions to occur. Replacement reactions occur fairly readily at C-4 of both α- and β-D-gluco- and α-D-galacto-pyranosides, so that methyl 2,3,6-tri-O-benzoyl-α-D-galactopyranoside 4-methanesulfonate is readily converted into the 2,3,4,6-tetrabenzoate of methyl α-D-glucopyranoside by sodium benzoate in N,N-dimethylformamide.[114] Conversely, methyl 2,3-di-O-benzoyl-α-D-glucopyranoside 4,6-dimethanesulfonate (**81**) affords[115] the tetrabenzoate (**83**) of methyl α-D-galactopyranoside. These nucleophilic substitutions proceed by a bimolecular mechanism which involves a transition state (**82**) where the incoming nucleophile is relatively unhindered in the *gluco* and *galacto* series, but in the *manno* series the formation of the transition state is hindered by the axial 2-substituent. Thus the 4-methanesulfonate of methyl α-D-mannopyranoside 2,3,6-tribenzoate does not undergo direct substitution.[116] When direct substitution of an equatorial sulfonyloxy group is not possible because of the presence of a *β-trans*-axial group, as in the mannoside, then ring contraction may be observed. For example,[117, 118] methyl 6-deoxy-2,3-*O*-isopropylidene-α-D-(or L)-mannopyranoside 4-methanesulfonates (**84**) undergo reactions with various nucleophiles to give 5-substituted derivatives of methyl 6-deoxy-2,3-*O*-isopropylidene-α-D-talo- (or β-L-allo)furanoside (**85**).

Replacement reactions of 3-sulfonates of methyl D-glucopyranosides are similarly hindered in the α series by the *β-trans*-axial methoxyl group, whereas

---

* *References start on p. 158.*

the β-D-anomers (**86**) undergo replacement to give[119] allopyranoside derivatives (**87**). In the α-D series the replacement is very slow[119a] and there is a

possibility of ring contraction. For example, the 3-*p*-nitrobenzenesulfonates
of methyl α-D-glucopyranoside (**88**) and the corresponding α-D-manno-
pyranoside undergo solvolysis to give[120] ring-contracted, branched-chain
products (**89**).

Displacement reactions are also hindered by an adjacent electronegative
axial substituent[120a] due to a polar effect. The polar and steric factors which
control these reactions has been outlined by Richardson.[120b]

Despite the wide use made in cyclohexane chemistry of neighboring-group
participation in the displacement of sulfonyloxy substituents—as, for example,[121] by a *trans*-acyloxy group (**90**) through an intermediary acyloxonium

cation (91)—such participation has rarely been demonstrated in a practical way with pyranoside derivatives.[115, 122] However, solvolysis of methyl 2-O-benzoyl-β-L-arabinopyranoside 3-p-toluenesulfonate (92) and 3,4-p-toluenesulfonate (93) in moist N,N-dimethylformamide containing sodium fluoride gave, after hydrolysis of the ester groups, methyl β-L-lyxopyranoside (94) and methyl α-D-ribopyranoside (95), respectively.[123] These products must arise from the formation of intermediary benzoxonium ions and subsequent hydrolysis. In contrast, a *trans*-acylamido group is a good participating group and has been utilized extensively in synthesis of amino sugars (see Chapter 16).

Replacement of secondary sulfonyloxy groups on furanoid systems is also dependent on steric and electronic factors. Thus, neither anomer of 3-*O*-*p*-tolylsulfonyl-D-glucofuranose 1,2,5,6-tetrabenzoate reacts with sodium benzoate in *N*,*N*-dimethylformamide, whereas under similar conditions 1,2-*O*-isopropylidene-3,5-di-*O*-*p*-tolylsulfonyl-α-D-ribofuranose (**96**) is converted[124] into the D-xylofuranose derivative (**97**). In bicyclic systems such as **96**, replacement of the *endo* 3-sulfonate group is more favored than replacement of an *exo* substituent. Thus, 1,2:5,6-di-*O*-isopropylidene 3-*O*-*p*-tolylsulfonyl-α-D-glucofuranose (**98**) is resistant to sodium azide and sodium benzoate in *N*,*N*-dimethylformamide,[125, 125a] probably because of a field effect repelling the charged nucleophile, whereas "neutral" nucleophiles such as ammonia and hydrazine react[126, 127] relatively readily to give the allofuranose derivatives (**99**).

Sulfonyloxy substituents situated *trans* to a vicinal hydroxyl or acyloxy group undergo intramolecular base-induced displacements to give the corresponding epoxides with inversion of configuration. Ring opening of epoxides by aqueous alkali takes place with inversion of configuration. When the epoxide is asymmetrically substituted, the reaction can give a mixture of two *trans*-diols, and one of these often preponderates because of steric factors in the transition state. If the epoxide ring is fused to a pyranoid ring, the latter is constrained to a half-chair conformation and ring opening occurs stereospecifically to give the *trans*-diaxial diol as the major product, according to the Fürst–Plattner rule.[128] This stereospecificity is convincingly illustrated by the action of aqueous alkali on methyl 2,3-anhydro-4,6-*O*-benzylidene-α-D-allopyranoside (**100**) and -α-D-mannopyranoside (**101**). The former undergoes attack at the 2-position and the latter at the 3-position to give the same major product— namely, methyl 4,6-*O*-benzylidene-α-D-altropyranoside (**102**), in which the 2- and 3-hydroxyl groups are *trans*-diaxial.[129] In cases where the pyranoside ring is not held rigidly in one conformation, the epoxide may react in both half-chair conformations, giving rise to a mixture of products.[129]

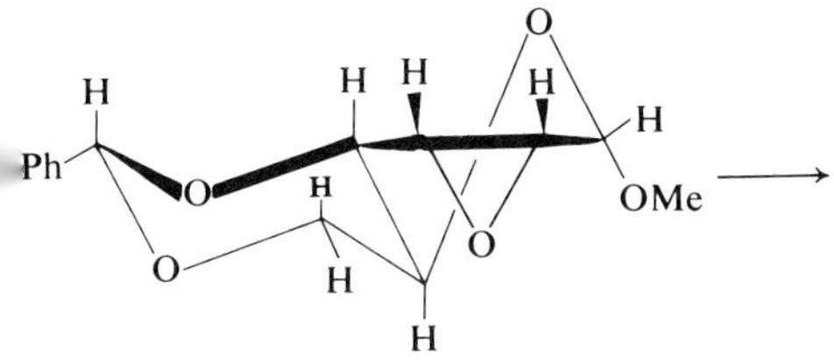

100

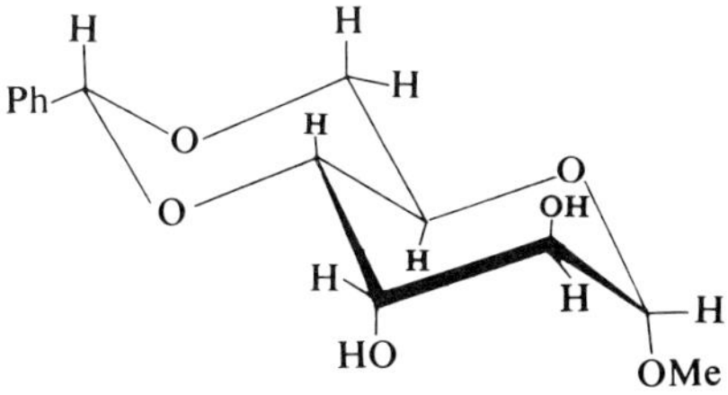

102

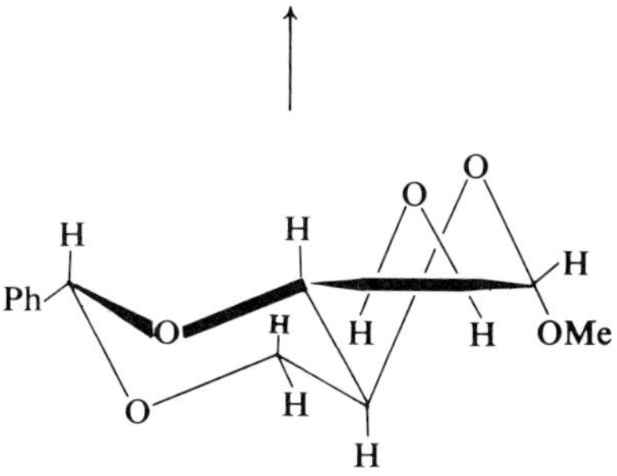

101

When the epoxide ring is situated at a terminal position, as in 5,6-anhydro-1,2-*O*-isopropylidene-β-L-idofuranose (**104**), stereospecific attack occurs at the primary carbon atom; consequently a 5-*O*-*p*-tolylsulfonyl-D-glucofuranose derivative (**103**) can be converted into the L-idofuranose derivative (**105**) by way of this epoxide.[130]

* *References start on p. 158.*

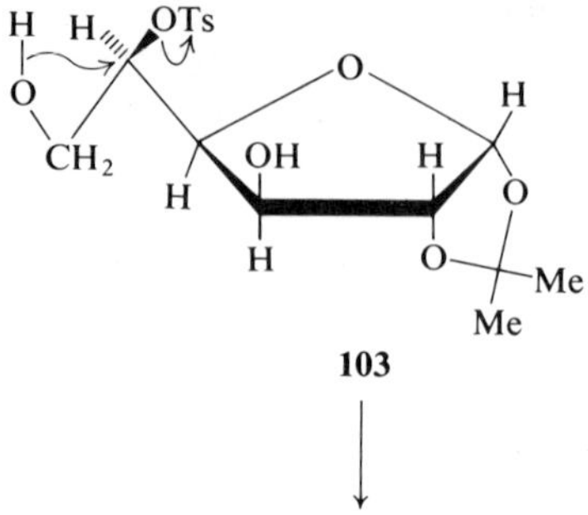

103

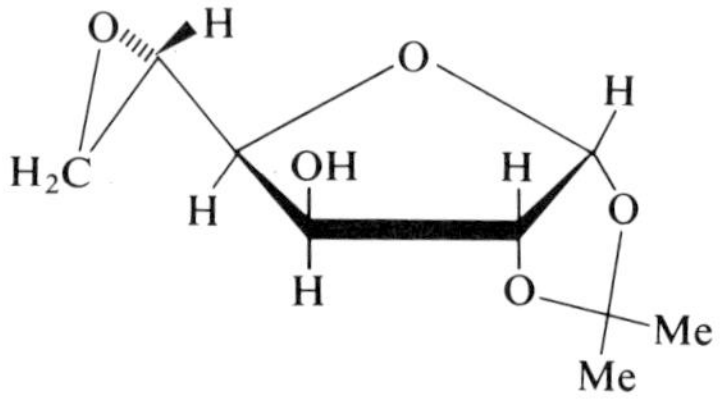

104 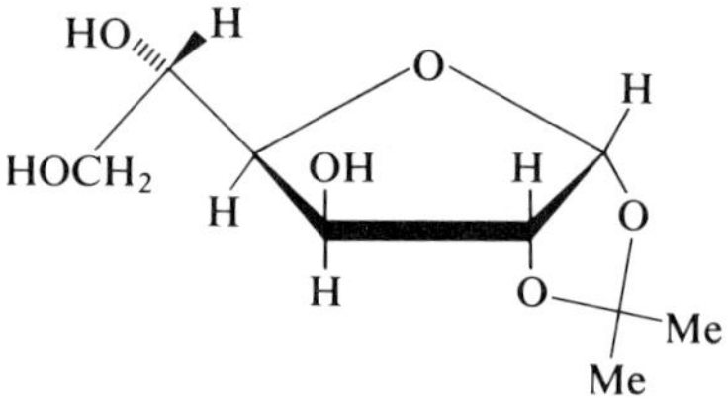　　　　　　　　　**105**

2-Sulfonic ester substituents in arabinose, xylose, and 6-deoxygalactose (fucose) are readily displaced by titration of the sugar derivatives with base, and the 2-epimers of the aldoses are formed in high yield[131]. The process appears to proceed by participation of the *trans*-anomeric hydroxyl group (106) to give, with inversion of configuration at C-2, the highly reactive 1,2-epoxide (107), which then undergoes stereospecific attack at C-1, thus affording the 2-epimer. By this method 2-*O*-methylsulfonyl-D-arabinose (106) can be converted[132] into

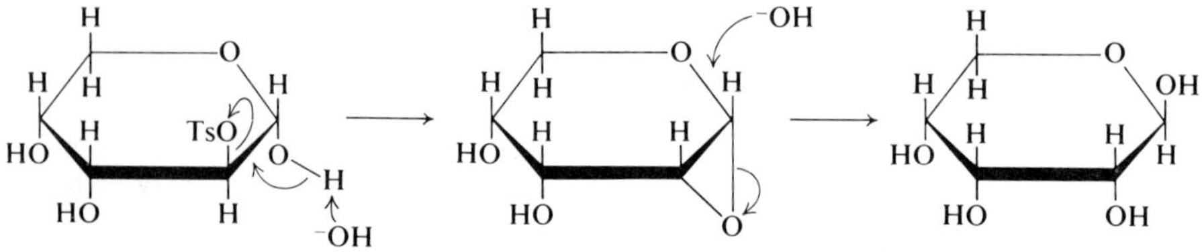

106　　　　　　　　　**107**　　　　　　　　　**108**

D-ribose (108), and 3-*O*-methylsulfonyl-D-fructose is conveniently converted into D-*ribo*-hexulose.

C. MOLECULAR INVERSION

Chemical manipulation of an aldose so that the two terminal groups of the aldose chain are interconverted may result in change of configuration, total inversion of configuration, or no net change, according to the starting configuration. Those hexoses that would be unaffected by such an exchange are the mannoses and the idoses. The interconversion is effected through the uronic acid; consequently the method is practical only if the uronic acid is readily available from the aldose. D-Galactose (**109**), for example, may be converted into D-galacturonic acid (**110**) by a several-step synthesis; reduction of the aldehydic function then affords L-galactonic acid (**111**), from which L-galactose

$$
\begin{array}{ccc}
\text{CHO} & \text{CHO} & \\
| & | & \\
\text{HCOH} & \text{HCOH} & \\
| & | & \\
\text{HOCH} & \text{HOCH} & \\
| \longrightarrow & | \longrightarrow & \\
\text{HOCH} & \text{HOCH} & \\
| & | & \\
\text{HCOH} & \text{HCOH} & \\
| & | & \\
\text{CH}_2\text{OH} & \text{CO}_2\text{H} & \\
\text{D-Galactose (109)} & \textbf{110} &
\end{array}
$$

$$
\begin{array}{ccc}
\text{CH}_2\text{OH} & \text{CO}_2\text{H} & \text{CHO} \\
| & | & | \\
\text{HCOH} & \text{HOCH} & \text{HOCH} \\
| & | & | \\
\text{HOCH} & \text{HCOH} & \text{HCOH} \\
| \quad \equiv & | \longrightarrow & | \\
\text{HOCH} & \text{HCOH} & \text{HCOH} \\
| & | & | \\
\text{HCOH} & \text{HOCH} & \text{HOCH} \\
| & | & | \\
\text{CO}_2\text{H} & \text{CH}_2\text{OH} & \text{CH}_2\text{OH} \\
\textbf{111} & & \text{L-Galactose (112)}
\end{array}
$$

(**112**) is obtained by reduction of the lactone.[133, 134] Application of this process to D-glucose gives L-gulose by way of D-glucuronic acid.[135]

D. SEQUENTIAL OXIDATION AND REDUCTION

Improved methods for the oxidation of suitably protected derivatives[135a] have made the *keto*-pyranosides and *keto*-furanosides readily available, and this has greatly increased the scope of specific reductions to effect inversion of configuration of secondary alcoholic groups. The oxidizing agents commonly used for the preparation of the *keto* intermediates are ruthenium tetraoxide[136] or methyl sulfoxide in the presence of either acetic anhydride,[137, 138] phosphorus pentaoxide,[139] or N,N'-dicyclohexylcarbodiimide–phosphoric acid (Pfitzner–Moffatt reagent),[140] and catalytic oxidation.[141] The use of dimethyl sulfoxide–sulfur trioxide–pyridine–triethylamine for oxidation of certain partially acylated derivatives results in oxidation followed by subsequent β-elimination of an acyloxy substituent.[141a] *keto*-Pyranosides undergo equa-

* *References start on p. 158.*

torial attack by borohydride in their favored conformation to give the product having the hydroxyl group axial,[141, 142] so that methyl α-D-glucopyranoside (113) can be converted into the allopyranoside (115) by oxidation to the 3-keto derivative (114) and subsequent reduction. Similarly, 1,2:5,6-di-O-isopropylidene-D-*ribo*-3-hexulofuranose (116) affords D-allose (117) on reduction

with lithium aluminum hydride or sodium borohydride followed by acid hydrolysis.[139] Furthermore, methyl 4,6-O-benzylidene-2-O-p-tolylsulfonyl-α-D-glucopyranoside is readily oxidized by the Pfitzner–Moffatt reagent to the

3-*keto* derivative, which on reduction affords the D-allopyranoside.[140] The sequence also provides convenient routes to D-*ribo*-hexulose (D-psicose), starting from the 1,2:4,5-diisopropylidene acetal of D-fructopyranose[142a] to D-*lyxo*-hexulose (D-tagatose),[142b] and D-gulose (starting from D-galactose derivatives).[142c]

E. INVERSION OF CONFIGURATION BY WAY OF CARBOXYLIC ESTER DERIVATIVES

1,2,3,4-Tetra-*O*-benzoyl-β-L-arabinopyranose and 2,3,4-tri-*O*-benzoyl-β-L-arabinosyl fluoride are readily converted into 3,4-di-*O*-benzoyl-β-L-ribosyl fluoride by the action of hydrogen fluoride.[143] Pederson[144] has shown that D-glucose pentaacetate gives rise to D-mannose and D-altrose derivatives when similarly treated. Similar rearrangements of fully acetylated monosaccharide derivatives that are catalyzed by aluminum chloride and phosphorus pentachloride resemble these transformations.[145] When 1,2,3,4-tetra-*O*-acetyl-6-*O*-*p*-tolylsulfonyl-hexopyranoses are treated with acetic anhydride–zinc chloride, they give mixtures containing all the possible diastereoisomeric 1,1,2,3,4,5,6-hepta-*O*-acetyl-*aldehydo*-hexoses.[146] Neighboring acetoxy-group participation to form cyclic carbonium ions is implicated in all these processes.[146a]

An attempt to prepare 3,4,6-tri-*O*-acetyl-α-D-glucopyranose 1,2-acetoxonium hexachloroantimonate (**118**) by the action of antimony pentachloride on β-D-glucopyranose pentaacetate, resulted in a series of epimerizations to the 2,3- (**119a**), the 3,4- (**119b**), and finally to the insoluble 4,6-acetoxonium (**119c**) ion. The action of sodium acetate on **119c** gave a mixture of α-D-ido-pyranose 1,2,3,6-tetraacetate (**120a**) and the 1,2,3,4-tetraacetate (**120b**) in fair yield. This reaction constitutes a convenient synthesis of the difficultly accessible sugar D-idose. Similarly D-talose was obtained from D-galactose derivatives.[147]

VI. SYNTHESIS FROM UNSATURATED DERIVATIVES

A. HYDROXYLATION

Glycals undergo hydroxylation of the double bond with peroxybenzoic acid to give the product having the 2,3-*cis* arrangement (for example, **121**), provided the 3-hydroxyl group is not substituted. However, when the 3-hydroxyl group is substituted, by an acetyl group for example, then the 2,3-*trans* product, such as **122**, is obtained.[148] *cis*-Hydroxylation of various 2,3-dideoxy-α-D-*erythro*-

hex-2-enopyranosides (for example, **123**, R and R′ = alkyl groups) with osmium tetraoxide has been studied with a view to developing a convenient synthesis of D-allose derivatives. However, hydroxylation occurs *trans* to the substituents at C-1 and C-4 to give the D-mannoside derivatives almost exclusively. Only in one case, when the anomeric configuration was β-D, did an allopyranoside result.[149, 150] *cis*-Hydroxylation of D-galactal with hydrogen peroxide in the presence of molybdenum trioxide gives the *cis*-2,3-substituted product almost exclusively, and thus provides a convenient route to D-talose.[150a]

B. HYDROBORATION

cis-Hydration of unsaturated carbohydrates of the enol-ether type can be achieved by hydroboration with diborane, followed by decomposition of the adduct. The hydration occurs in an anti-Markownikoff fashion, so that 3-deoxy-1,2:5,6-di-*O*-isopropylidene-α-D-*erythro*-hex-3-enofuranose (124) yields 1,2:5,6-di-*O*-isopropylidene-α-D-galactofuranose (125) by addition to

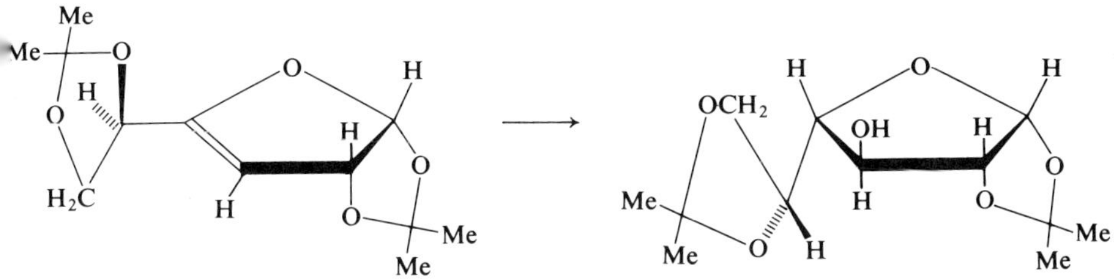

124   →   **125**

the least-hindered side of the molecule.[151,152] Similarly methyl 5,6-dideoxy-2,3-*O*-isopropylidene-β-D-*erythro*-hex-4-enofuranoside (126) gave[153] methyl 6-deoxy-2,3-*O*-isopropylidene-β-D-gulofuranoside (127). This reaction has

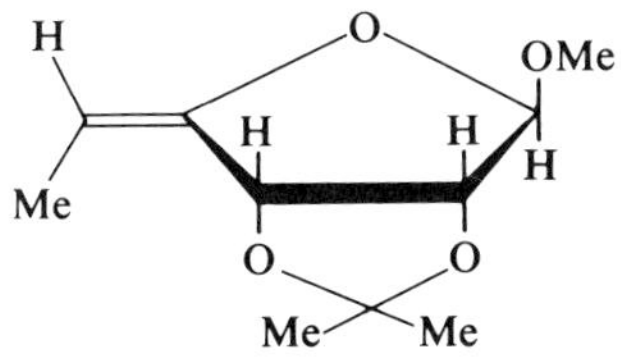

126 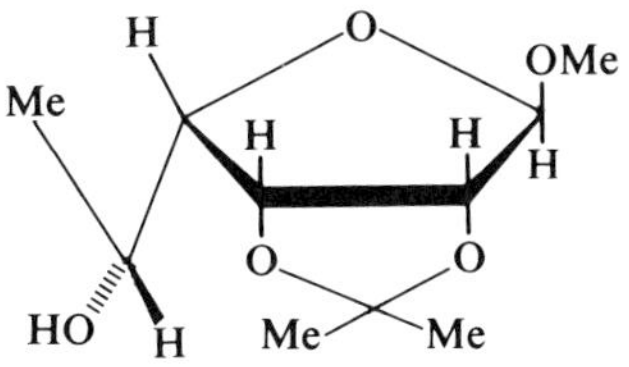   **127**

been used for the selective tritiation of monosaccharides.[151]

VII. OXIDATION OF ALDITOLS

A. CHEMICAL METHODS

Oxidation of a suitably protected alditol in which only one hydroxyl group remains free is of particular use in the synthesis of ketoses other than 2-ketoses. Thus a *meso-erythro*-3-pentulose derivative (129) is synthesized from 1,5-di-*O*-benzoyl-2,4-*O*-benzylidenexylitol (128) by oxidation with chromium trioxide

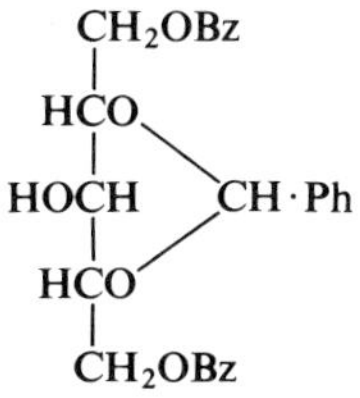

CH_2OBz
|
HCO
|
$HOCH$ $CH \cdot Ph$
|
HCO
|
CH_2OBz
128

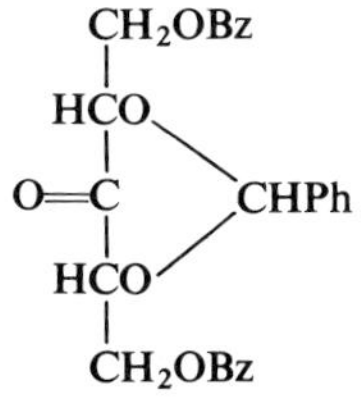

CH_2OBz
|
HCO
|
$O{=}C$ $CHPh$
|
HCO
|
CH_2OBz
129

followed by removal of substituents.[154] Oxidation of ribitol with mercuric acetate occurs mainly at the secondary hydroxyl groups, affording a mixture of meso-erythro-3-pentulose and DL-erythro-2-pentulose.[155]

Chromium trioxide is not always satisfactory in these oxidations, and several repeated treatments are usually required. Better results have been obtained with ruthenium tetraoxide,[136] which is generated[156] conveniently in situ by using a small amount of ruthenium dioxide and adding sodium metaperiodate solution dropwise at pH 6 to 7. Methyl sulfoxide containing an acid anhydride or acid chloride is normally an excellent reagent for these oxidations.[137-139] (see ref. 135a, p. 147, and Chap. 24). 1,2:5,6-Di-O-isopropylidene-3-O-(methylsulfonyl)-D-mannitol is readily oxidized to the D-arabino-3-hexulose derivative in 74% yield by the Pfitzner–Moffatt reagent.[140] The Pfitzner–Moffatt reagent is also effective for oxidizing "isolated" primary alcohol groups to aldehydes, as in the conversion of 1,2:3,4-di-O-isopropylidene-α-D-galactopyranose into the corresponding 6-aldehyde;[51, 156a] methyl sulfoxide–acetic anhydride, however, gives mainly the (methylthio)methyl ether of the parent alcohol.[156b]

Oxidation of alditols with hydrogen peroxide and a ferrous salt (Fenton's reagent)[157] gives a mixture of products containing aldoses and ketoses that arise by competitive reaction of the hydroxyl groups of the alditol with the hydroxyl radicals that are generated (Eq. 17).[158] D-Glucitol gives only a 2%

$$Fe^{2+} + H_2O_2 \longrightarrow Fe^{3+} + OH^- + OH\cdot \qquad (17)$$

yield of D-glucose,[159] whereas D-mannitol is reported to give a relatively high yield of D-mannose.[160] The similarity between the irradiation of aqueous solutions of alditols with ionizing radiation, the self-decomposition of ^{14}C-labeled D-glucoses, and the action of Fenton's reagent on the pure substances has been noted[161, 162] (see Chapter 26). Fenton's reagent reacts with 3,4-di-O-methyl-D-mannitol, yielding 3,4-di-O-methyl-D-mannose and, in addition, methylated derivatives of pentoses, uronic acids, D-mannose, and D-mannitol, including their monomethyl ethers.[163] The nonselective reactivity of Fenton's reagent and its ability to de-etherify complicates the use of this reagent.

B. Biological Methods

Various strains of Acetobacter oxidize alditols to give ketoses in high yield, providing a convenient method for the preparation of difficultly accessible ketoses. The organisms are highly specific and, in the case of Acetobacter xylinum, the penultimate hydroxyl group of an alditol is oxidized to a carbonyl group when there is present a vicinal, secondary hydroxyl group of the same configuration (the two hydroxyl groups being D- or L-erythro) together with a vicinal, primary hydroxyl group (as in 130 or 131, respectively).[164]

$$
\begin{array}{c}
\text{HCOH} \\
\text{HCOH} \\
\text{CH}_2\text{OH}
\end{array}
\qquad \longrightarrow \qquad
\begin{array}{c}
\text{HCOH} \\
\text{C=O} \\
\text{CH}_2\text{OH}
\end{array}
$$

130

$$
\begin{array}{c}
\text{HOCH} \\
\text{HOCH} \\
\text{CH}_2\text{OH}
\end{array}
\qquad \longrightarrow \qquad
\begin{array}{c}
\text{HOCH} \\
\text{C=O} \\
\text{CH}_2\text{OH}
\end{array}
$$

131

When allitol (**132**) is oxidized by this organism, L-*ribo*-hexulose (**133**) is obtained, and clearly the oxidation of the D-*erythro* system is favored over the L-*erythro* system when both arrangements are present in the same molecule.[165]

$$
\begin{array}{c}
\text{CH}_2\text{OH} \\
\text{HCOH} \\
\text{HCOH} \\
\text{HCOH} \\
\text{HCOH} \\
\text{CH}_2\text{OH}
\end{array}
\qquad \longrightarrow \qquad
\begin{array}{c}
\text{CH}_2\text{OH} \\
\text{HCOH} \\
\text{HCOH} \\
\text{HCOH} \\
\text{C=O} \\
\text{CH}_2\text{OH}
\end{array}
$$

132   **133**

Only D-*erythro* configurations (**130**) are oxidized by *Acetobacter suboxydans*, L-*glycero*-tetrulose being obtained from erythritol,[166] D-*threo*-pentulose from D-arabinitol,[166] L-*ribo*-hexulose (L-allulose, L-psicose) from allitol,[167] L-*xylo*-hexulose (L-sorbose) from D-glucitol (sorbitol),[168] L-*galacto*-heptulose (persulose) from L-*glycero*-D-*manno*-heptitol (perseitol),[169] and D-*glycero*-L-*gluco*-octulose from D-*threo*-L-*gulo*-octitol.[111a] The replacement of one of the terminal hydrogen atoms by a methyl group does not affect the oxidation pattern; thus 1-deoxy-D-galactitol (L-fucitol, **134**) behaves as if it were a derivative of D-arabinitol ($\equiv$ D-lyxitol) and gives[170] 1-deoxy-D-*xylo*-3-hexulose (**135**). Similarly 1-*S*-ethyl-1-thio-D-glucitol and -D-arabinitol, prepared by partial desulfurization of the corresponding dithioacetals, are oxidized to 6-*S*-ethyl-6-thio-L-*xylo*-hexulose and 5-*S*-ethyl-5-thio-D-*threo*-pentulose, respectively.[171] Also, 5-acetamido-5-deoxy-L-*xylo*-hexulose is obtained from 2-acetamido-2-deoxy-D-glucitol,[172] and derivatives of D-*threo*-5-pentulose, D-*xylo*-5-hexosulose, and D-*lyxo*-5-hexosulose can be prepared by oxidation of aldose dimethyl acetals and dithioacetals.[173]

* *References start on p. 158.*

$$
\begin{array}{ccc}
CH_2OH & & CH_2OH \\
| & & | \\
HOCH & & HOCH \\
| & & | \\
HCOH & \longrightarrow & C{=}O \\
| & & | \\
HCOH & & HCOH \\
| & & | \\
HOCH & & HOCH \\
| & & | \\
CH_3 & & CH_3 \\
134 & & 135
\end{array}
$$

VIII. ISOTOPICALLY LABELED CARBOHYDRATES

Investigations of biosynthetic pathways and reaction mechanisms in carbohydrate chemistry can be facilitated by the use of isotopically labeled precursors. For this purpose the radioactive isotopes carbon-14 and tritium are commonly used. (See Chapter 45 for details of the use of radioactive isotopes in analytical work with carbohydrates.) Derivatives labeled with deuterium (^{2}H), a nonradioactive isotope, are also used, particularly in proton magnetic resonance studies (see Chapter 27). Sugars specifically labeled with isotopes of hydrogen have been discussed in detail.[173a]

A. CARBON-14 LABELING

1. *Biological Methods*

All carbohydrates occurring in plants are derived from carbon dioxide by photosynthesis, and hence any plant, or an isolated leaf from that plant, which has been allowed to photosynthesize from $^{14}CO_2$ produces uniformly labeled carbohydrates. The common sugars D-glucose, D-fructose, and sucrose, normally found in the leaves of most plant species, are thus readily available, labeled uniformly with ^{14}C, by this method. Less common sugars such as D-galactose and D-*ribo*-hexulose (D-psicose, D-allulose) may be obtained by using red algae[174] (*Iridea laminaroides*) and leaves of *Itea* species,[175] respectively. The advantage of this method is that compounds having high specific activities can easily be obtained, and uniform distribution of the label is assured, a condition not easily fulfilled with other methods.

Carbohydrates labeled specifically at certain positions are more useful for most purposes, and most biological methods for their preparation are based on well-known enzymic reactions, performed with substrates that are already labeled specifically. An exception is provided by the observation that animals, such as rats, when fed with sodium hydrogen carbonate-^{14}C, succinic acid-^{14}C, or carboxyl-labeled acetate or pyruvate, synthesize glycogen (in the liver) containing mostly residues of D-glucose-*3,4*-^{14}C. This result is in agreement with established transformations in the glycolytic and citric acid cycles.[176, 177]

Biological preparations, such as enzymes, selected tissues, or even intact organisms, can be used to synthesize a specifically labeled compound as a metabolite from a specifically labeled substrate, provided the biosynthetic process is known and well understood. For example, sucrose phosphorylase from the bacterium *Pseudomonas saccharophila* synthesizes sucrose from α-D-gluco-pyranosyl phosphate and D-fructose, and so by the use of suitably labeled precursors, sucrose labeled in any desired position may be obtained.[178]

Acetobacter suboxydans has proved convenient for the preparation of labeled ketoses, the bacterium being specific for the oxidation of the D-*erythro* configuration of hydroxyl groups in alditols (see Section VII,B). The symmetry of the D-mannitol molecule allows oxidation by *A. suboxydans* to take place equally at C-2 and C-5 to give D-*arabino*-hexulose (D-fructose). Hence the use of D-mannitol-*1*-^{14}C gives a 1:1 mixture of D-fructose-*1*-^{14}C and D-fructose-*6*-^{14}C, which to all intents and purposes can be used as D-fructose-*1,6*-^{14}C. However, the distribution is unequal when D-mannitol-*2*-^{14}C is used because oxidation at C-2 is slower on account of an isotope effect, and a mixture of D-fructose-*2*-^{14}C (48.4%) and D-fructose-*5*-^{14}C (51.6%) is obtained.[179] Other hexitols that are oxidized by this organism lead to ketoses labeled in one position only.

2. Chemical Methods

Most of the chemical methods involve lengthening of the carbon chain, usually by the Fischer–Kiliani cyanohydrin synthesis (Section III,A) on a semimicro scale, although the Sowden–Fischer method (Section III,B) with nitromethane[180] is sometimes used.

1-^{14}C-Labeled aldoses are prepared[181] from the lower aldose and cyanide-^{14}C. In this way D-glucose-*1*-^{14}C and the D-mannose analog are conveniently prepared from D-arabinose. The radiochemical yield of D-glucose is usually about 50%. 2-Amino-2-deoxy-D-glucose-*1*-^{14}C and 2-amino-2-deoxy-D-galactose-*1*-^{14}C are prepared by addition of $^{14}CN^-$ to N-benzyl-D-arabino-pyranosylamine and N-benzyl-D-lyxopyranosylamine, respectively.[182]

2-^{14}C-Labeled aldoses are synthesized from aldoses-*1*-^{14}C and inactive cyanide. For example D-arabinose-*1*-^{14}C yields D-glucose-*2*-^{14}C and the D-mannose analog by this method.

3-^{14}C-Labeled aldoses have been prepared from aldoses-*1*-^{14}C by two successive cyanohydrin syntheses.[183]

ω-^{14}C-Labeled aldoses have been prepared by methods of chain ascent and descent. Pentoses-*5*-^{14}C can be prepared by degradation of the corresponding hexoses-6-^{14}C. The oxidation of D-fructose-*1,6*-^{14}C with molecular oxygen in alkaline solution leads to the formation of D-arabinose-*5*-^{14}C. Hexoses-6-^{14}C

* *References start on p. 158.*

can be prepared from 5-*aldehydo* derivatives of pentoses by application of the cyanohydrin synthesis. The 5-*aldehydo* derivatives are best obtained by periodate oxidation of suitable hexofuranoid compounds having an exocyclic 5,6-diol grouping. Thus, 1,2-*O*-isopropylidene-α-D-glucofuranose gives 1,2-*O*-isopropylidene-5-*aldehydo*-α-D-*xylo*-pentodialdo-1,4-furanose, which, by ascent of the series, yields[184] D-glucose-6-[14]*C* and L-idose-6-[14]*C*. Sodium D-glucuronate-6-[14]*C* and D-glucurono-3,6-lactone-6-[14]*C* can also be synthesized by this method.

[14]C-Labeled ketoses can be prepared from acetylated aldonyl chlorides by the diazomethane method (Section III,C). The use of diazomethane-[14]*C* would yield the corresponding ketose-1-[14]*C* whereas prior labeling of the aldonyl chloride would lead to a corresponding label in the final product. Both D- and L-ascorbic acid-1-[14]*C* have been synthesized from D- and L-*erythro*-pentosulose by treatment with [14]CN.[185, 186]

[14]C-Labeled carbohydrates undergo self-induced decomposition during storage (see Chapter 26). The decomposition is probably caused by hydroxyl free radicals formed by interaction of α-particles with nonbonded water. Thus, after 26 months of storage in the dark, D-glucose-[14]*C* had decomposed to the extent of 20%, with the formation of 36 new compounds. Decomposition is lessened if the product is stored at low temperatures and specific activities are kept low.[187]

B. TRITIATED CARBOHYDRATES

Tritium (^{3}H) is usually introduced into a molecule by reduction of a carbonyl group with NaBT$_4$, LiAlT$_4$, LiBT$_4$, or sodium amalgam in tritiated water. Aldoses-*1-t* are prepared[188] by reduction of the corresponding aldonolactones with tritiated reagents; for example, D-glucono-1,4-lactone affords α-D-glucose-*1-t*. An alternative procedure, used for the preparation of D-ribose-*1-t*, is by reduction of the fully acetylated aldonyl chloride with lithium tri-*tert*-butoxy-aluminohydride-*t*.[189] Complete reduction of D-glucose-*1-t* to the alditol gives D-glucitol-*1,1-t*, whereas alditols-*1-t* can be prepared by reduction of the corresponding unlabeled aldose.[188] Alditols-*2-t* are similarly prepared by reduction of ketoses, although a mixture of two alditols is normally obtained. Reduction of 1,2-*O*-isopropylidene-5-*aldehydo*-α-D-*xylo*-pentodialdo-1,4-furanose (136) with NaBT$_4$ followed by hydrolysis yields D-xylose-*5-t* (137). D-Galactose-*6-t* can be obtained by oxidation of D-galactose derivatives with D-galactose oxidase, and reduction of the resultant 6-aldehydes with NaBT$_4$.[190, 191] Oxidation of D-glucose with *Acetobacter suboxydans* to give D-*xylo*-5-hexulosonic acid, followed by reduction with borotritiide, provides a route to D-glucose-*5-t*.[192] D-Glucose-*6,6'-t* is prepared by reduction of D-glucurono-3,6-lactone with NaBT$_4$.

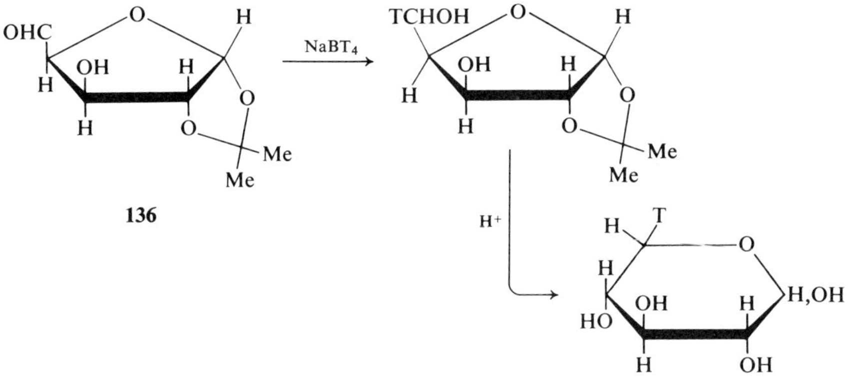

OHC

$NaBT_4$

136

TCHOH

H^+

137

The ready availability of *keto*-pyranosides and related compounds due to the development of methyl sulfoxide as an oxidizing agent increases the ease of synthesizing tritiated monosaccharides enormously (see Section V,D). By this route syntheses of D-ribose-3-*t*,[193] D-glucose-4-*t*, and D-galactose-4-*t*,[194] have been effected.

Application of hydroboration to unsaturated carbohydrates of the enol-ether type (such as **138**) by using B_2T_6 (Section VI,B) affords a method for selective tritiation at C-5 with pyranoid derivatives and at C-4 with furanoid derivatives.[151] For example, methyl 6-deoxy-α-D-*xylo*-hex-5-enopyranoside (**138**) is converted into a mixture of methyl α-D-glucopyranoside-5-*t* (**139**) and the corresponding β-L-idopyranoside (**140**) by this method. The triacetate of **138**

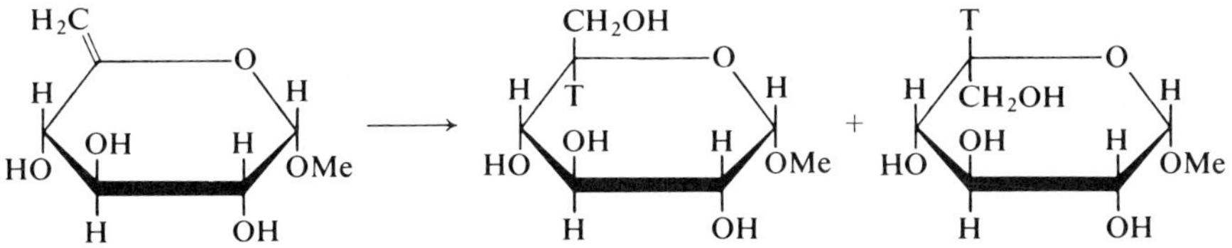

138   **139**   **140**

can be tritiated at C-6 by treatment with tritiated water in pyridine, in the presence of silver fluoride.[195]

By using the Wilzbach method of exposure to tritium gas, tritium labeling of D-glucose and D-ribose occurred to the extent of 90 to 95% at C-3, whereas tetra-*O*-acetyl-D-ribofuranose became labeled[196] at C-4 to the extent of 93%.

C. DEUTERIUM-LABELED CARBOHYDRATES

Deuterated compounds are prepared by methods used for tritiated derivatives (Section VIII,B), but with sodium borodeuteride or lithium aluminum

* *References start on p. 158.*

deuteride for reduction purposes. Introduction of deuterium at a position α to a carbonyl group has been achieved by base-catalyzed exchange in deuterium oxide. Thus D-arabinose-*2-d* is prepared from 2-*O*-benzyl-D-arabinose by treatment with sodium deuteroxide in deuterium oxide followed by removal of the benzyl group by hydrogenolysis.[197] Furthermore, 1,6-anhydro-2,3-*O*-iso-propylidene-β-D-*lyxo*-hexopyrano-4-ulose (141) undergoes exchange at C-3 only to give 142, because enolization at C-4 is not possible because of steric factors. Since reduction of the 4-keto derivative (141) with either sodium borohydride or sodium borodeuteride gives the *talo*-isomer, D-talose-*3-d* (143) and D-talose-*3,4-d* (144) are both readily obtained by this method.[198]

REFERENCES

1. R. L. Whistler and M. L. Wolfrom, Eds., "Methods in Carbohydrate Chemistry," Vol. 1, Academic Press, New York, 1962.
2. A. Butlerow, *Compt. Rend.*, **53**, 145 (1861).
3. E. Fischer and J. Tafel, *Ber.*, **23**, 2114 (1890).
4. W. Kuster and F. Schoder, *Hoppe-Seyler's Z. Physiol. Chem.*, **141**, 110 (1924).
5. E. Schmitz, *Ber.*, **46**, 2327 (1913).
6. E. Fischer and J. Tafel, *Ber.*, **20**, 1088, 2566, 3384 (1887).
7. H. O. L. Fischer and E. Baer, *Helv. Chim. Acta*, **19**, 519 (1936); **20**, 1213 (1937).
8. L. Hough and J. K. N. Jones, *J. Chem. Soc.*, 1122, 3191 (1951); *Advan. Carbohyd. Chem.* **11**, 185 (1956).
9. E. Mariani and G. Torraca, *Int. Sugar J.*, **55**, 309 (1953).
10. E. Pfeil and G. Schroth, *Ber.*, **85**, 293 (1952); E. Pfeil and H. Ruckert, *Ann.*, **641**, 121 (1961).
11. A. H. Weiss and J. Shapira, *Hydrocarbon Process.*, **49**, 119 (1970).

12. R. Breslow, *Tetrahedron Lett.*, **22** (1959).
13. R. Schaffer and H. S. Isbell, *Methods Carbohyd. Chem.*, **1**, 273 (1962).
14. I. Iwai and T. Iwashige, *Chem. Pharm. Bull.* (Tokyo), **9**, 316 (1961); I. Iwai and K. Tomita, *ibid.*, **10**, 976 (1962); **11**, 184 (1963); I. Iwai, T. Iwashige, M. Asai, K. Tomita, T. Hiroaka, and J. Ide, *ibid.*, **11**, 188 (1963).
14a. U. P. Singh and R. K. Brown, *Can. J. Chem.*, **48**, 1791 (1970).
15. H. Kiliani, *Ber.*, **18**, 3066 (1885); **19**, 221, 772 (1886).
16. H. Kiliani, *Ber.*, **19**, 767, 1128 (1886).
17. H. Kiliani, *Ber.*, **19**, 3029 (1886); **20**, 282, 339 (1887).
18. H. Kiliani, *Ber.*, **21**, 915 (1888); **22**, 521 (1889).
19. J. G. Maltby, *J. Chem. Soc.*, 2769 (1929).
20. H. S. Isbell, *J. Res. Nat. Bur. Stand.*, **48**, 163 (1962).
21. W. Militzer, *Arch. Biochem. Biophys.*, **21**, 143 (1949).
22. E. Fischer, *Ber.*, **22**, 2204 (1889).
23. M. L. Wolfrom and H. B. Wood, *J. Amer. Chem. Soc.*, **73**, 2933 (1951); M. L. Wolfrom and A. Thompson, *Methods Carbohyd. Chem.*, **2**, 65 (1963).
24. R. K. Hulyalker, *Can. J. Chem.*, **27**, 1594 (1966).
25. J. W. E. Glattfeld and E. M. Shaver, *J. Amer. Chem. Soc.*, **49**, 2305 (1927); **57**, 2204 (1935).
25a. P. Kohn, R. H. Samaritino, and L. M. Lerner, *J. Amer. Chem. Soc.*, **86**, 1457 (1964); **87**, 5475 (1965); P. Kohn, L. M. Lerner, A. Chan, Jr., S. D. Ginocchio, and A. Zitrin, *Carbohyd. Res.*, **7**, 21 (1968).
26. R. Kuhn and H. Grassner, *Ann.*, **612**, 55 (1958); R. Kuhn and P. Klesse, *Ber.*, **91**, 1989 (1958).
27. L. H. Philippe, *Ann. Chim. Phys.*, **26** [8], 393 (1912).
28. J. Staněk, M. Černý, J. Kocourek, and J. Pacák, "The Monosaccharides," Academic Press, New York, 1965, p. 144.
29. C. S. Hudson, *Advan. Carbohyd. Chem.*, **1**, 1 (1945).
30. N. K. Richtmyer, *Methods Carbohyd. Chem.*, **1**, 161 (1962).
31. J. C. Sowden, *Advan. Carbohyd. Chem.*, **6**, 291 (1951).
32. L. Hough and S. H. Shute, *J. Chem. Soc.*, 4633 (1962).
33. J. Kovař and H. H. Baer, *Can. J. Chem.*, **48**, 2377 (1970).
33a. L. Hough and S. H. Shute, unpublished results, 1962.
34. E. Schmidt and G. Rutz, *Ber.*, **61**, 2142 (1928).
35. J. C. Sowden and D. R. Strobach, *J. Amer. Chem. Soc.*, **80**, 2532 (1958).
36. J. K. N. Jones, *J. Chem. Soc.*, 3643 (1954); cf. B. A. McFadden, L. L. Barden, N. W. Rokke, M. Uyeda, and T. J. Siek, *Carbohyd. Res.*, **4**, 254 (1967).
37. J. C. Sowden, *Methods Carbohyd. Chem.* **1**, 132 (1962); R. L. Whistler and J. N. BeMiller, *ibid.*, p. 137; J. C. Sowden and M. L. Oftedahl, *ibid.*, p. 235.
38. M. L. Wolfrom and A. Thompson, *Methods Carbohyd. Chem.*, **1**, 118 (1962).
39. J. M. Webber, *Advan. Carbohyd. Chem.*, **17**, 20 (1962).
40. M. L. Wolfrom and R. L. Brown, *J. Amer. Chem. Soc.*, **65**, 1516 (1943).
41. M. L. Wolfrom, S. W. Waisbrot, and R. L. Brown, *J. Amer. Chem. Soc.*, **64**, 1701 (1942).
42. M. L. Wolfrom and J. B. Miller, *J. Amer. Chem. Soc.*, **80**, 1678 (1958).
42a. Y. A. Zhdanov, V. I. Kornilov, and G. V. Bogdanova, *Carbohyd. Res.*, **3**, 139 (1966).
43. L. Wolff, *Ann.*, **394**, 23 (1912).
44. M. L. Wolfrom and P. W. Cooper, *J. Amer. Chem. Soc.*, **71**, 2668 (1949); **72**, 1345 (1950).
45. M. L. Wolfrom and H. B. Wood, Jr., *J. Amer. Chem. Soc.*, **73**, 730 (1951).
45a. S. David and M.-O. Popot, *Carbohyd. Res.*, **5**, 234 (1967).

46. M. L. Wolfrom, J. D. Crum, J. B. Miller, and D. I. Weisblat, *J. Amer. Chem. Soc.*, **81**, 243 (1959).
47. H. Zinner, E. Wittenburg, and G. Rembarz, *Ber.*, **92**, 1614 (1959).
48. N. K. Kochetkov and B. A. Dmitriev, *Tetrahedron*, **21**, 803 (1965).
49. E. J. Reist, P. A. Hart, B. R. Baker, and L. Goodman, *J. Org. Chem.*, **27**, 1722 (1962).
50. Y. A. Zhdanov, G. N. Dorofeenko and L. A. Uzlova, *Zh. Obshch. Khim.*, **25**, 181 (1965); compare N. K. Kochetkov, B. A. Dmitriev, and L. V. Backinowsky, *Carbohyd. Res.*, **11**, 193 (1969); Y. A. Zhdanov, Y. E. Alexeev, and V. G. Alexeeva, *Advan. Carbohyd. Chem. Biochem.*, **27**, in press (1972).
51. D. Horton, J. B. Hughes and J. M. J. Tronchet, *Chem. Commun.*, 481 (1965); D. Horton, J. B. Hughes, and J. K. Thomson, *J. Org. Chem.*, **33**, 728 (1968).
52. D. Horton and J. M. J. Tronchet, *Carbohyd. Res.*, **2**, 315 (1966); J. L. Godman, D. Horton, and J. M. J. Tronchet, *ibid.*, **4**, 392 (1967).
53. R. Schaffer, *J. Org. Chem.*, **29**, 1471 (1964).
54. L. M. Utkin, *Dokl. Akad. Nauk. SSSR*, **67**, 301 (1949); *Chem. Abstr.* **44**, 3910 (1950).
55. R. Schaffer and H. S. Isbell, *J. Amer. Chem. Soc.*, **81**, 2178 (1959).
56. L. Hough and J. K. N. Jones, *J. Chem. Soc.*, 342 (1953).
57. P. A. J. Gorin and J. K. N. Jones, *J. Chem. Soc.*, 1537 (1953).
58. J. K. N. Jones and N. K. Matheson, *Can. J. Chem.*, **37**, 1784 (1959).
59. J. K. N. Jones and H. H. Sephton, *Can. J. Chem.* **38**, 753 (1960).
60. O. Ruff, *Ber.*, **31**, 1573 (1898); O. Ruff and G. Ollendorf, *ibid.*, **33**, 1798 (1900); O. Ruff, *ibid.*, **34**, 1362 (1901).
61. H. G. Fletcher, Jr., H. W. Diehl, and C. S. Hudson, *J. Amer. Chem. Soc.*, **72**, 4546 (1950).
62. M. Bergmann, *Ber.*, **54**, 1362 (1921).
63. H. G. Fletcher, Jr., *Methods Carbohyd. Chem.*, **1**, 77 (1962).
64. R. L. Whistler and J. N. BeMiller, *Methods Carbohyd. Chem.*, **1**, 79 (1962).
65. G. N. Richards, *Methods Carbohyd. Chem.*, **1**, 180 (1962).
66. A. Wohl, *Ber.*, **26**, 730 (1893).
67. E. Restelli de Labriola and V. Deulofeu, *J. Amer. Chem. Soc.*, **62**, 1611 (1940).
68. L. Maquenne, *Compt. Rend.*, **130**, 1402 (1900).
69. R. C. Hockett, *J. Amer. Chem. Soc.*, **57**, 2265 (1935).
70. V. Deulofeu, *J. Chem. Soc.*, 2602 (1930).
71. V. Deulofeu, *Advan. Carbohyd. Chem.*, **4**, 119 (1949).
72. E. G. Gros and V. Deulofeu, *J. Org. Chem.*, **29**, 3647 (1964).
73. F. Weygand and R. Lowenfeld, *Ber.*, **83**, 559, 563 (1950).
74. R. A. Weerman, *Rec. Trav. Chim.*, **37**, 16 (1917).
75. W. N. Haworth, S. Peat, and J. Whetstone, *J. Chem. Soc.*, 1975 (1938).
76. D. L. MacDonald and H. O. L. Fischer, *J. Amer. Chem. Soc.*, **74**, 2087 (1952).
77. D. L. MacDonald and H. O. L. Fischer, *Biochim. Biophys. Acta*, **12**, 203 (1953).
78. E. J. Bourne and R. W. Stephens, *J. Chem. Soc.*, 4009 (1954).
79. L. Hough and T. J. Taylor, *J. Chem. Soc.*, 970 (1956).
80. L. Hough and T. J. Taylor, *J. Chem. Soc.*, 1212 (1955).
81. A. Farrington and L. Hough, *Carbohyd. Res.*, **16**, 59 (1971).
82. L. D. Hall, L. Hough, S. H. Shute, and T. J. Taylor, *J. Chem. Soc.*, 1154 (1965).
83. L. Hough and A. C. Richardson, *J. Chem. Soc.*, 1019, 1024 (1962).
84. E. Dimant and M. Banay, *J. Org. Chem.*, **25**, 475 (1960).
85. J. C. P. Schwarz and M. McDougall, *J. Chem. Soc.*, 3065 (1956).
86. G. Rembarz, *Ber.*, **95**, 1565 (1962).
87. H. H. Baer and H. O. L. Fischer, *J. Amer. Chem. Soc.*, **82**, 3709 (1960).
88. B. Coxon and L. Hough, *J. Chem. Soc.*, 1463 (1961).
89. A. C. Richardson and H. O. L. Fischer, *J. Amer. Chem. Soc.*, **83**, 1132 (1961).

90. O. Kjølberg, *Acta Chem. Scand.*, **14**, 1118 (1960).
91. A. S. Perlin and C. Brice, *Can. J. Chem.*, **34**, 552 (1956).
92. A. S. Perlin, *Advan. Carbohydr. Chem.*, **14**, 36 (1959).
93. F. A. H. Rice and A. R. Johnson, *J. Amer. Chem. Soc.*, **78**, 428 (1956).
94. F. A. H. Rice, *J. Amer. Chem. Soc.*, **78**, 3173 (1956).
95. J. U. Nef, *Ann.*, **403**, 204 (1914).
96. O. Spengler and A. Pfannenstiel, *Z. Wirtschafsgruppe Zuckerind.* **85**, 547 (1935); *Chem. Abstr.*, **30**, 4470 (1936).
97. J. Dubourg and P. Naffa, *Bull. Soc. Chim. Fr.*, 1353 (1959).
98. R. L. Whistler and K. Yagi, *J. Org. Chem.*, **26**, 1050 (1961).
99. R. L. Whistler and R. Schweiger, *J. Amer. Chem. Soc.*, **81**, 5190 (1959).
100. C. Neuberg, *Biochem. Z.*, 527 (1907).
101. M. Guerbet, *Bull. Soc. Chim. Fr.*, **3** [4], 427 (1908).
102. K. H. Boddener and B. Tollens, *Ber.*, **43**, 1645 (1910).
103. E. Fisher, *Ber.*, **24**, 3622 (1891).
104. C. Glatthaar and T. Reichstein, *Helv. Chim. Acta*, **21**, 3 (1938).
105. C. A. Lobry de Bruyn and W. Alberda van Ekenstein, *Rec. Trav. Chim. Pays-Bas*, **14**, 156, 203 (1895); **15**, 92 (1896); **16**, 241, 257, 267, 274, 282 (1897); **18**, 147 (1899); **19**, 1 (1900).
106. L. Hough and R. S. Theobald, *Methods Carbohyd. Chem.*, **1**, 94 (1962).
107. N. Prentice, L. S. Cuendet, and F. Smith, *J. Amer. Chem. Soc.*, **78**, 4439 (1956).
108. B. Coxon and L. Hough, *J. Chem. Soc.*, 1577 (1961).
109. L. Vargha, *Ber.*, **87**, 1351 (1954).
110. M. L. Wolfrom and S. Hanessian, *J. Org. Chem.*, **27**, 1800 (1962).
111. M. Müller and T. Reichstein, *Helv. Chim. Acta*, **21**, 263 (1938).
111a. N. K. Richtmyer, *Carbohyd. Res.*, **17**, 401 (1971).
112. B. R. Baker and A. H. Haines, *J. Org. Chem.*, **28**, 438 (1963).
112a. D. H. Ball and F. W. Parrish, *Advan. Carbohyd. Chem., Biochem.*, **24**, 139 (1969).
113. D. H. Buss, L. D. Hall, and L. Hough, *J. Chem. Soc.*, 1616 (1965).
114. E. J. Reist, R. R. Spencer, and B. R. Baker, *J. Org. Chem.*, **24**, 1618 (1959).
115. J. Hill, L. Hough, and A. C. Richardson, *Carbohyd. Res.*, **8**, 7 (1968).
116. J. M. Williams and A. C. Richardson, *Tetrahedron*, **23**, 1369 (1967).
117. S. Hanessian, *Chem. Commun.*, 796 (1966).
118. C. L. Stevens, R. P. Glinski, K. G. Taylor, P. Blumbergs and F. Sirokman, *J. Amer. Chem. Soc.*, **88**, 2073 (1966).
119. N. A. Hughes and P. R. H. Speakman, *J. Chem. Soc.*, 2236 (1965).
119a. R. Ahluwahlia, S. J. Angyal, and M. H. Randall, *Carbohyd. Res.*, **4**, 478 (1967).
120. F. W. Austin, J. G. Buchanan, and R. M. Saunders, *Chem. Commun.*, 146 (1965).
120a. Y. Ali and A. C. Richardson, *J. Chem. Soc.* [C], 320 (1969).
120b. A. C. Richardson, *Carbohyd. Res.*, **10**, 395 (1969).
121. S. Winstein, H. V. Hess, and R. A. Buckles, *J. Amer. Chem. Soc.*, **64**, 2796 (1942); S. Winstein, C. Hanson, and E. Grunwald, *ibid.*, **70**, 812 (1948); S. Winstein and R. Heck, *ibid.*, **74**, 5585 (1952).
122. R. W. Jeanloz and D. A. Jeanloz, *J. Amer. Chem. Soc.*, **80**, 5692 (1958); compare L. Goodman, *Advan. Carbohyd. Chem.*, **22**, 109 (1967).
123. E. J. Reist, L. V. Fischer, and D. E. Gueffroy, *J. Org. Chem.*, **31**, 226 (1966).
124. N. A. Hughes and P. R. H. Speakman, *Carbohyd. Res.*, **1**, 341 (1966).
125. M. L. Wolfrom, J. Bernsmann, and D. Horton, *J. Org. Chem.*, **27**, 4505 (1962).
125a. R. L. Whistler and L. W. Doner, *J. Org. Chem.*, **35**, 3562 (1970).
126. R. U. Lemieux and P. Chu, *J. Amer. Chem. Soc.*, **80**, 4745 (1958).
127. B. Coxon and L. Hough, *J. Chem. Soc.*, 1643 (1961).

128. J. A. Mills, *Advan. Carbohyd. Chem.*, **10**, 1 (1955); N. R. Williams, *Advan. Carbohyd. Chem. Biochem.*, **25**, 109 (1970).
129. F. H. Newth, *Quart. Rev.*, **13**, 30 (1959).
130. H. Ohle and K. Tessmar, *Ber.*, **71**, 1843 (1938).
131. J. K. N. Jones and W. H. Nicholson, *J. Chem. Soc.*, 3050 (1965).
132. D. C. C. Smith, *Chem. Ind.* (London), 92 (1953).
133. C. Glatthaar and T. Reichstein, *Helv. Chim. Acta*, **20**, 1537 (1937).
134. R. A. Pizzarello and W. Freudenberg, *J. Amer. Chem. Soc.*, **61**, 611 (1939).
135. E. Fischer and O. Piloty, *Ber.*, **24**, 521 (1891).
135a. R. F. Butterworth and S. Hanessian, *Synthesis*, 70 (1971).
136. P. J. Beynon, P. M. Collins, P. T. Doganges, and W. G. Overend, *J. Chem. Soc.* (*C*), 1131 (1966); *Carbohyd. Res.*, **6**, 431 (1968).
137. W. Sowa and G. H. S. Thomas, *Can. J. Chem.*, **44**, 836 (1966).
138. D. Horton and J. S. Jewell, *Carbohyd. Res.*, **2**, 251 (1966); **5**, 149 (1967).
139. K. Onodera, S. Hirano, and N. Kashimura, *J. Amer. Chem. Soc.*, **87**, 4651 (1965); *Carbohyd. Res.*, **6**, 276 (1968).
140. B. R. Baker and D. H. Buss, *J. Org. Chem.*, **30**, 2304 (1965); J. S. Brimacombe and A. Husain, *Carbohyd. Res.*, **6**, 491 (1968).
141. O. Theander, *Advan. Carbohyd. Chem.*, **17**, 223 (1962).
141a. G. M. Cree, D. W. Mackie, and A. S. Perlin, *Can. J. Chem.*, **47**, 511 (1969).
142. E. E. Grebner, R. Durbin, and D. S. Feingold, *Nature*, **201**, 419 (1964).
142a. E. J. McDonald, *Carbohyd. Res.*, **5**, 106 (1967); K. James, A. R. Tatchell, and P. K. Ray, *J. Chem. Soc.* [C], 2681 (1967).
142b. A. A. H. Al-Jabore, R. D. Guthrie, and R. D. Wells, *Carbohyd. Res.*, **16**, 474 (1971).
142c. G. J. F. Chittenden, *Carbohyd. Res.*, **15**, 101 (1970).
143. C. Pedersen and H. G. Fletcher, Jr., *J. Amer. Chem. Soc.*, **82**, 945 (1960).
144. C. Pedersen, *Acta Chem. Scand.*, **16**, 1831 (1962); **17**, 673 (1963).
145. N. K. Richtmyer, *Advan. Carbohyd. Chem.*, **1**, 37 (1945).
146. F. Micheel and R. Bohm, *Tetrahedron Lett.*, 107 (1962).
146a. H. Paulsen, *Advan. Carbohyd. Chem. Biochem.*, **26**, 127 (1971).
147. H. Paulsen, W.-P. Trautwein, F. Garrido Espinosa, and K. Heyns, *Ber.*, **100**, 2822 (1967); **101**, 179, 186, 191 (1968); H. Paulsen and C. P. Herold, *ibid.*, **103**, 2450 (1970); H. Paulsen, C. P. Herold, and F. Garrido Espinosa, *ibid.*, **103**, 2463 (1970).
148. P. A. Levene and R. S. Tipson, *J. Biol. Chem.*, **93**, 631 (1931).
149. S. McNally and W. G. Overend, *J. Chem. Soc.* (*C*), 1978 (19
150. C. L. Stevens, J. B. Filippi, and K. G. Taylor, *J. Org. Chem.*, **31**, 1292 (1966).
150a. V. Bilík and Š. Kučár, *Carbohyd. Res.*, **13**, 311 (1970).
151. J. Lehmann, *Carbohyd. Res.*, **2**, 1 (1966).
152. H. Paulsen and H. Behre, *Carbohyd. Res.*, **2**, 80 (1966).
153. H. Arzoumanian, E. M. Acton, and L. Goodman, *J. Amer. Chem. Soc.*, **86**, 74 (1964).
154. A. Sera, *Bull. Chem. Soc. Jap.*, **35**, 2033 (1962).
155. R. J. Stoodley, *Can. J. Chem.*, **39**, 2593 (1961).
156. V. M. Parikh and J. K. N. Jones, *Can. J. Chem.*, **43**, 3452 (1965); B. T. Lawton, W. A. Szarek, and J. K. N. Jones, *Carbohyd. Res.*, **10**, 456 (1969).
156a. D. Horton, M. Nakadate, and J. M. J. Tronchet, *Carbohyd. Res.*, **7**, 56 (1968); cf. A. Kampf, A. Felsenstein, and E. Dimant, *ibid.*, **6**, 220 (1968).
156b. J. L. Godman and D. Horton, *Carbohyd. Res.*, **6**, 229 (1968).
157. H. J. H. Fenton and H. Jackson, *J. Chem. Soc.*, **75**, 1 (1899).
158. G. J. Moody, *Advan. Carbohyd. Chem.*, **19**, 156 (1964).
159. J. D. Anderson, P. Andrews, and L. Hough, *Biochem. J.*, **84**, 140 (1962).
160. F. Haber and W. Weiss, *Proc. Roy. Soc. Ser.*, **A147**, 332 (1934).

161. M. L. Wolfrom, *Radiation Res.*, **10**, 37 (1959).
162. H. Weigel, D. H. Hutson, and E. J. Bourne, *J. Chem. Soc.*, 5155 (1960).
163. J. K. N. Jones, B. Fraser-Reid, and M. B. Perry, *Can. J. Chem.*, **39**, 555 (1961).
164. G. Bertrand, *Ann. Chim.* (France), **3** [vii], 181 (1904).
165. M. Steiger and T. Reichstein, *Helv. Chim. Acta*, **18**, 790 (1935).
166. R. M. Hann, E. B. Tilden, and C. S. Hudson, *J. Amer. Chem. Soc.*, **60**, 1201 (1938).
167. L. Hough and B. E. Stacey, unpublished results, 1965.
168. P. A. Wells, L. B. Lockwood, J. J. Stubbs, and E. T. Roe, *Ind. Eng. Chem.*, **31**, 1518 (1939); U.S. Patent 2,121,533 (1938).
169. R. M. Hann and C. S. Hudson, *J. Amer. Chem. Soc.*, **61**, 336 (1939).
170. N. K. Richtmyer, L. C. Stewart, and C. S. Hudson, *J. Amer. Chem. Soc.*, **72**, 725 (1950).
171. L. Hough, J. K. N. Jones, and D. L. Mitchell, *Can. J. Chem.*, **37**, 725, 1561 (1959).
172. J. K. N. Jones, M. B. Perry, and J. C. Turner, *Can. J. Chem.*, **39**, 965 (1961).
173. J. K. N. Jones and D. T. Williams, *Can. J. Chem.*, **43**, 955 (1965).
173a. J. E. G. Barnett and D. L. Corina, *Advan. Carbohyd. Chem. Biochem.*, **27**, in press (1972).
174. R. C. Bean, E. W. Putman, R. E. Trucco, and W. Z. Hassid, *J. Biol. Chem.*, **204**, 169 (1953).
175. L. Hough and B. Stacey, *Phytochemistry*, **5**, 215 (1966).
176. H. G. Wood, N. Lifson, and V. Lorber, *J. Biol. Chem.*, **159**, 475 (1945).
177. S. Abraham and I. L. Charkoff, *Arch. Biochem. Biophys.*, **41**, 143 (1952).
178. H. Wolochow, E. W. Putman, M. Doudoroff, W. Z. Hassid, and H. A. Barker, *J. Biol. Chem.*, **180**, 1237 (1949).
179. H. L. Frush and L. J. Tregoning, *Science*, **128**, 59 (1958).
180. J. C. Sowden, *J. Biol. Chem.*, **180**, 55 (1949).
181. H. S. Isbell, N. B. Holt, and H. L. Frush, *Methods Carbohyd. Chem.*, **1**, 276 (1962).
182. R. Kuhn and W. Kirschenlohr, *Ann.*, **600**, 115, 126 (1956).
183. H. L. Frush, L. T. Sniegoski, N. B. Holt, and H. S. Isbell, *J. Res. Nat. Bur. Stand.*, **A69**, 535 (1965).
184. R. Schaffer and H. S. Isbell, *Methods Carbohyd. Chem.*, **1**, 281 (1962).
185. A. C. Neish and A. C. Blackwood, *Can. J. Biochem. Physiol.*, **33**, 323 (1955).
186. J. K. Hamilton and F. Smith, *J. Amer. Chem. Soc.*, **74**, 5162 (1952).
187. E. J. Bourne, D. H. Hutson, and H. Weigel, *J. Chem. Soc.*, 5153 (1960).
188. H. S. Isbell, H. L. Frush, B. N. Holt, and J. D. Moyer, *J. Res. Nat. Bur. Stand.*, **A64**, 177, 359 (1960).
189. R. J. Suhadolnik, T. Uematsu, and R. M. Ramer, *Carbohyd. Res.*, **5**, 479 (1967).
190. G. Avigad, *Carbohyd. Res.*, **3**, 430 (1967).
191. J. E. G. Barnett, *Carbohyd. Res.*, **4**, 267 (1967).
192. J. E. G. Barnett and D. L. Corina, *Carbohyd. Res.*, **3**, 134 (1966).
193. H. P. C. Hogenkamp, *Carbohyd. Res.*, **3**, 239 (1966).
194. O. Gabriel, *Carbohyd. Res.*, **6**, 319 (1968).
195. J. Lehmann, *Carbohyd. Res.*, **4**, 196 (1967).
196. H. Simon, *Z. Naturforsch.*, **18b**, 360 (1963).
197. R. U. Lemieux and J. D. Stevens, *Can. J. Chem.*, **44**, 539 (1966).
198. D. Horton and J. S. Jewell, *Carbohyd. Res.*, **3**, 255 (1966); cf. D. Horton and E. K. Just, *ibid.*, **9**, 129 (1969); D. Horton, J. S. Jewell, E. K. Just, and J. D. Wander, *ibid.*, **18**, 49 (1971).

4. MUTAROTATIONS AND ACTIONS OF ACIDS AND BASES

Ward Pigman and E. F. L. J. Anet

I. MUTAROTATIONS AND CHANGES OCCURRING IN THE ABSENCE OF STRONG ACIDS AND BASES[*][1]

A. Introduction

Although fairly stable when in the crystalline condition, the sugars undergo many transformations when dissolved in water, particularly in the presence of acids or alkalies. Initially, these changes usually involve the carbon atom carrying the aldehyde or ketone groups. Hence, when these groups are blocked, as in the nonreducing compound sugars (for example, sucrose or glycosides), the compounds are more stable and do not undergo isomerizations until the blocking groups are removed.

In solution, the polar groups of sugars are highly solvated. With water as the solvent, the hydrogen atom of each hydroxyl group is rapidly exchanged with hydrogen atoms of the solvent. Within a few seconds at room temperature, these hydrogen atoms are exchanged completely with the deuterium atoms of heavy water.[2] The carbon-bound hydrogen and oxygen atoms

---

* This section was prepared by Ward Pigman.

(except for the "carbonyl" oxygen atom) are bound much more firmly and require bases, acids, or heat to effect their removal. One oxygen atom, presumably that of the carbonyl or hemiacetal hydroxyl group is much more active than the others. Thus, when glucose is kept in ^{18}O-labeled water, one oxygen atom is exchanged after 100 hr at 55°.[3]

Interconversions between α and β anomers, and between ring tautomers, take place under the mildest possible condition of acidity and temperature. Such changes are manifested by the change of optical rotation with time which may be observed for freshly prepared sugar solutions. This change of rotation is known as mutarotation. Mutarotations may arise from changes other than interconversions between α and β anomers and between ring tautomers, but for neutral or slightly acid or slightly alkaline solutions of the sugars they arise most often from such changes. The phenomenon was observed first by Dubrunfaut (1846), who noted that the optical rotation of freshly dissolved D-glucose changes with time and that after a number of hours the rotation becomes constant. The ordinary form of glucose (α-D-glucopyranose) mutarotates downward, and the β-D anomer mutarotates upward; in both cases the same equilibrium value is reached.

$$\alpha\text{-D-Glucopyranose} \; \rightleftharpoons \; \text{equilibrium} \; \rightleftharpoons \; \beta\text{-D-Glucopyranose}$$
$$+112° \; \rightarrow \; +52.7° \; \leftarrow \; +18.7°$$

As mentioned in Chapter 1, the mutarotation of glucose and other sugars showed that the original aldehyde structure for glucose was not adequate for explaining the properties of the sugar. The separation of isomers (anomers) of lactose (Erdmann, 1880), and later of D-glucose (Tanret, 1896), which mutarotated to the same equilibrium value, provided good evidence that the observed mutarotations result from an interconversion of the various modifications.

B. KINETICS

The mutarotation of α-D-glucose may be represented by the equation for a first-order reversible reaction.

$$\alpha \underset{k_2}{\overset{k_1}{\rightleftharpoons}} \beta \tag{1}$$

$$-\frac{d\alpha}{dt} = k_1[\alpha] - k_2[\beta] \tag{2}$$

Equation (2) gives the rate of change of the α into the β form at the time t. The reaction constant for $\alpha \rightarrow \beta$ is k_1, and for $\beta \rightarrow \alpha$ is k_2. The concentrations of the α and β form at the time t are represented by $[\alpha]$ and $[\beta]$.

Equation (2) may be integrated and expressed in terms of the optical rotations in the form of Eq. (3).[4]

$$k_1 + k_2 = \frac{1}{t} \log \frac{r_0 - r_\infty}{r_t - r_\infty} \qquad (3)$$

In Eq. (3), r_0 = the rotation at $t = 0$; r_∞ = the final equilibrium rotation; and r_t = the rotation at the time t. The rotations may be expressed as observed or as specific rotations. The specific rotations are calculated from the observed rotations by the relation

$$\text{Specific rotation} = [\alpha] = \frac{\alpha \times 100}{l \times c} \qquad (4)$$

where α = observed rotation (in degrees).

l = length of column of solution (in decimeters).

c = concentration of active substance (grams per 100 ml of solution).

If the rotations are read on a saccharimeter, the values observed (^0S) are multiplied by the factor 0.3462 to give α.

The rotation varies with the wavelength of the light source, and usually the sodium D line is employed. Most rotations are measured at 20°. The solvents most commonly employed are water and chloroform.

The mutarotation coefficient, $k_1 + k_2$, should be the same for the α and β anomers of each sugar. Hudson[5] demonstrated that the α and β anomers of lactose and of some other sugars give identical values for $k_1 + k_2$ and that the mutarotations follow the first-order equation. Table I lists the mutarotation coefficients for several sugars.[6]

The mutarotations of the sugars listed in Table I and those for many other sugars follow the first-order equation. The activation energy averages about 17,000 cal mole^{-1}; this value corresponds to an increase in rate of 2.5 times for a 10° rise in temperature. The conformity of the mutarotation data to the first-order equation makes it probable that the main constituents of the equilibrium solution are the α- and β-pyranose modifications. The actual composition may be calculated from the optical rotations of the equilibrium solution when the rotations of the pure α and β anomers are known. Data of this type are included in Table I. Independent confirmation of the composition of the equilibrium solutions is provided by studies of the rates of bromine oxidation of the sugars, the results of which are also found in Table I, and also by n.m.r. spectroscopy (see Chapter 5).

A number of important sugars exhibit mutarotations that do not follow the first-order equation (see Fig. 1). A striking case[6] is presented by the

* *References start on p. 191.*

TABLE I

TABLE I

Mutarotation Coefficients and Activation Energies for Some Sugars

Sugar	$k_1 + k_2$ (20°)	Q (cal)	Composition of equilibrium solution (%)	
			From rotations	From oxidation studies
α-D-Glucose	0.00632	17,200	{α—36.2	37.4
β-D-Glucose	0.00625	17,200	{β—63.8	62.6
α-D-Mannose	0.0173	16,700	{α—68.8	68.9
β-D-Mannose	0.0178	17,100	{β—31.2	31.1
α-D-Xylose	0.0203	16,800		
α-D-Lyxose	0.0568	15,300	{α—76.0	79.7
β-D-Lyxose	0.0591	15,700	{β—24.0	20.3
α-Lactose·H₂O	0.00471	17,300	{α—36.8	37.5
β-Lactose	0.00466	17,600	{β—63.2	62.5
β-Maltose·H₂O	0.00527	17,500		

pentose ribose; the specific rotation of freshly dissolved L-ribose decreases from an initial value of +23.4° to a minimum of +18.2° and then rises to a constant value of +23.2°. Some other sugars, such as α- and β-galactose, α- and β-talose, and α- and β-arabinose, exhibit similar but less-striking deviations

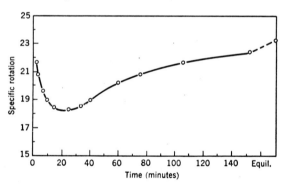

Fig. 1. Mutarotation of L-ribose in water at 0°.

from the first-order equation. In Fig. 2, $\log(r_t - r_\infty)$ versus time is plotted for α-D-glucose and α-D-talose. Although the curve for α-D-glucose is linear and follows the first-order equation, that for α-D-talose deviates greatly from a straight line during the initial period. This deviation indicates the failure of the mutarotation of talose to conform with the first-order equation.

In general, those mutarotations that cannot be expressed by the first-order equation conform to equations derived by assuming that three components are present in the equilibrium mixture. The equilibrium involved may be formulated:

$$\alpha \rightleftharpoons \mu \rightleftharpoons \beta \tag{5}$$

Equations fitting this condition were derived by Riiber and Minsaas[7] and by Smith and Lowry.[8] The Smith and Lowry type of equation is represented by Eq. (6).

$$[\alpha] = A \times 10^{-m_1 t} + B \times 10^{-m_2 t} + C \tag{6}$$

In this equation, C is the equilibrium rotation, A is the total change in optical rotation due to the slowly mutarotating component, and B is $(r_0 - r_\infty) - A$. Methods for applying these equations are described elsewhere.[6] The constants m_1 and m_2 are functions of the velocity constants for the various reactions represented in Eq. (5).

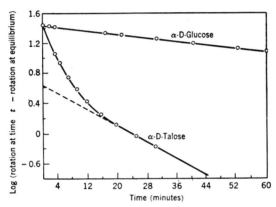

FIG. 2. "Simple" and "complex" mutarotations.

Changes in other properties such as the solution volume, the refractive index, and the heat content have been shown by Riiber and his associates[7] to parallel the changes in rotations.

Mutarotations which cannot be expressed by the first-order equation but which are expressed by Eq. (6) must represent the establishment of equilibria in which three or more components are present in appreciable proportions. Hence, equilibrated solutions of sugars such as galactose, arabinose, talose, and, particularly, ribose must have appreciable quantities of isomers other than the pyranose modifications. The ease of conversion of galactopyranose into furanose and free-aldehyde (or aldehydrol) modifications is shown by the formation of appreciable quantities of such isomers in the products of the acetylation (see Chapt. 6, p. 221; also Chapt. 10).

* *References start on p. 191.*

C. Composition of Solutions

Measurements involving nuclear magnetic resonance (n.m.r.) have considerably clarified the problem of the composition of sugar solutions.[9] Solutions in water, deuterium oxide, and methyl sulfoxide have been used. Data provided by Angyal[9e] are given in Table II for those aldohexoses and aldopentoses that have appreciable amounts of furanose forms at equilibrium.

TABLE II

Equilibrium Composition[9e] of Sugars by N.M.R. in Deuterium Oxide at 31°

Sugar	α-Pyranose (%)	β-Pyranose (%)	α-Furanose (%)	β-Furanose (%)
Allose	14	77.5	3.5	5
Altrose	30	41	18	11
Galactose	~29	64	~3	~4
Glucose	38	62	—	—
Gulose (at 44°)	16	78	6	
Idose	38.5	36	11.5	14
Mannose (at 44°)	65.5	34.5	—	—
Talose (at 44°)	~37	32	~17	14
Arabinose	60	35.5	2.5	2
Lyxose	70	28	1.5	0.5
Ribose	21.5	58.5	6.5	13.5
Xylose	36.5	63	<1	

Gas–liquid chromatographic analysis of mixtures of sugars, after their conversion into the corresponding trimethylsilyl ethers, provides another powerful tool for investigating the mutarotation reaction;[9g] it can be applied, especially in combination with mass spectrometry, for the analysis of sub-milligram quantities of mixtures.[9h]

Mackie and Perlin[9b] found that methylation of D-arabinose, D-galactose, and D-altrose at C-2 and C-3 increased the tendency for formation of furanose forms. An additional enhancement occurred when methyl sulfoxide was the solvent.

In agreement with optical rotatory data, the principal anomer in water was shown[9c, 9d] by n.m.r. spectroscopy to be the β-pyranose for D-galactose and D-glucose and the α-pyranose for D-mannose. Horton and associates[9c] showed that an amino group at C-2 of these sugars caused an increase in the proportion of the other pyranose anomer. The aqueous solutions of these sugars had little or no detectable furanose forms. For the corresponding 2-acetamido-2-deoxy sugars, the α-pyranose was the major isomer; no detectable furanose forms could be found in their aqueous solutions.

The mutarotation reactions that follow Eq. (6) may be considered to involve two simultaneous or consecutive reactions, one of which is slow and the other of which is rapid. The values of m_1 (which represent the reaction constant for the slowest reaction) are about the same as those for $k_1 + k_2$ for glucose, and the activation energies also have almost the same value as for glucose.[6] It is probable then that the slower reactions are $\alpha \rightleftharpoons \beta$-conversions between pyranose anomers. The reactions represented by m_2 are 5 to 10 times as rapid, and the activation energy is much smaller (about 13,200 cal mole^{-1} as compared with 17,000 for the normal mutarotations). For the rapid mutarotation reactions of galactose, talose, and ribose, the magnitude of the reaction constant, the small activation energy, and the influence of pH on the rate of mutarotation are similar to those for the mutarotation of the furanose modification of fructose. Since the mutarotation of fructose probably represents mainly a pyranose–furanose change,[10] the fast mutarotations of the other sugars also may represent pyranose–furanose interconversions.

It is usually considered that the interconversion of the α and β anomers and of pyranose and furanose forms takes place through the intermediate formation of the *aldehydo* or *keto* forms of the sugars (see Chapters 1 and 10).

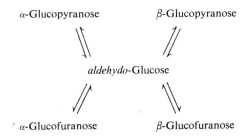

There is no direct proof for the existence of the open-chain forms. However, small quantities of acetylated open-chain forms are obtained along with the ring forms when some sugars are acetylated (see Chapter 6). Sugar solutions contain tautomers that are reducible at the dropping-mercury electrode of the polarograph.[11] The proportions of the reducible form present in 0.25 M solutions of several aldoses at pH 7.0 and 25° are shown in Table III. As may be seen from the table, the amount of the reducible form in solutions of D-glucose is very small (0.024 mole %). Solutions of other sugars, particularly ribose, contain fairly large proportions. The proportion of reducible material increases rapidly as the pH becomes greater. These data, obtained by Cantor and Peniston, are said to agree with those reported by Lippich[12] for the amount of material in solution which reacts "instantaneously" with hydrocyanic acid.

* *References start on p. 191.*

It has been suggested by Isbell and co-workers[12a] that, in the mutarotation reaction, the sugar ring may be opened momentarily, forming a pseudo-acyclic intermediate having a conformation similar to that of the parent sugar. Such an intermediate would then pass through characteristic transition states to pyranoses, furanoses, and other products.

TABLE III

QUANTITY OF REDUCIBLE FORM PRESENT
IN SOLUTIONS OF SEVERAL SUGARS

Sugar	Reducible forms (mole % of total sugar)
Glucose	0.024
Mannose	0.064
Galactose	0.082
Allose	(1.38)
Xylose	0.17
Arabinose	0.28
Lyxose	0.40
Ribose	8.5 (0.1M)

The mutarotation of 2-deoxy-D-erythro-pentose has been studied in detail.[126b]

D. MECHANISMS

The mutarotation reactions are catalyzed by both hydrogen and hydroxyl ions. The rate of mutarotation of glucose and galactose is at a minimum between the pH limits 3.0 to 7.0. At pH values greater than 7.0 and less than 3.0, the velocity increases rapidly. The curve for mutarotation velocity versus pH is represented by a catenary. The influence of hydrogen and hydroxyl ions on the rate was found by Hudson to be expressible by equations of the type

$$k_1 + k_2 = A + B[H^+] + C[OH^-] \tag{7}$$

where A, B, and C are constants. For glucose at 20° the equation is[10]

$$k_1 + k_2 = 0.0060 + 0.18[H^+] + 16,000[OH^-] \tag{8}$$

According to Eq. (8), glucose mutarotates most slowly at pH 4.61. Acids and alkalies influence the mutarotation of fructose and some other sugars much more markedly than glucose, although the minimum for fructose occurs near that for glucose. As may be shown from Eqs. (7) and (8), at pH 4.6 the portion of the catalysis that is due to the water (term A) is much greater than that caused by the hydrogen and hydroxyl ions. In turn, the hydroxyl ions are much more effective catalysts than the hydrogen ions (compare values for *B* and *C*).

Equations (7) and (8) are special cases for aqueous solutions of the equation for generalized acid–base catalysis. As shown by Lowry,[13] the mutarotations of sugars are reactions involving simultaneous catalysis by both acids and bases, in the generalized concept of acids and bases proposed by Lowry and by Brönsted. Water functions as a complete catalyst because of its amphoteric dissociation into ions: $H_2O \rightleftharpoons H^+ + OH^-$. Acids or bases alone are not effective catalysts but in mixture are complete catalysts.

Thus in a mixture of pyridine and cresol, tetra-*O*-methyl-D-glucose was found to mutarotate, whereas in either pyridine or cresol mutarotation was insignificant.[14] Lowry therefore proposed that the mutarotation of sugars is basically a ternary reaction involving simultaneous acid–base catalysis. The ternary reaction involves the simultaneous transfer of a proton from the acid catalyst to the sugar in the same step that a proton is transferred from the sugar to the base catalyst, yielding the *aldehydo*-sugar (or aldehydrol) directly.[15, 16]

Strong evidence exists for the intermediary role of the free aldehyde in the mutarotation reaction in the observation that no carbon-bound hydrogen atoms and no oxygen atoms exchange with water during the mutarotation reaction. Since one oxygen atom per mole is exchanged only slowly, the possible hydrated aldehyde is formed too slowly to be the intermediate. The dependence of mutarotations on acid and base catalysis accounts for the observation that the rate of mutarotation decreases in aqueous methanol and ethanol solutions as the alcohol concentration is increased,[17] since the alcohols are less amphoteric than water. In heavy water, the mutarotation of glucose proceeds more slowly than in ordinary water.[2]

The findings of Swain and Brown[16] support the ternary mechanism proposed by Lowry for the mutarotation of tetra-*O*-methyl-D-glucose. It was found

* *References start on p.* 191.

that the mutarotation of tetra-O-methyl-D-glucose in benzene in the presence of both an acid (phenol) and a base (pyridine) followed third-order kinetics but was first-order with respect to each component: tetra-O-methyl-D-glucose, pyridine, and phenol.[16] 2-Hydroxypyridine was found to be a very effective bifunctional catalyst, and since both acid and base functions were in the same molecule, the mutarotation in benzene followed second-order kinetics. Its catalytic action was essentially independent of the other acid and base species present. Although 2-hydroxypyridine is a much weaker acid or base than either phenol or pyridine, its catalysis of the mutarotation of tetra-O-methylglucose in benzene was much greater than that of either pyridine or phenol, or a mixture of both.[16]

Although basically a ternary reaction, mutarotation may appear to follow first- or second-order kinetics. As indicated earlier (p. 166), under the usual experimental conditions, which involve water as the solvent in large excess and a fixed hydrogen-ion concentration, (pseudo) first-order kinetics are followed. Actually, the reaction usually appears to be second-order in an aqueous system when the concentration of the catalyst is taken into consideration. This special situation has led to the interpretation that two consecutive bimolecular reactions are involved; in one a proton is added, and in the other a proton is removed in a separate step.[18, 19]

Isbell and Wade[20] studied the mutarotations of D-glucose and D-fructose in water and deuterium oxide over a wide pH range. Although that for glucose is an anomeric interconversion of pyranose isomers and that for fructose a furanose–pyranose interchange, the isotope effects were similar for both types of reaction. The k_H/k_D ratios were always greater than unity and were highest (4) in the pH region of minimal rate and lowest (1.3) in highly acid solutions.

A ratio greater than unity discounts a mechanism involving a cyclic carbonium ion and supports an acyclic intermediate. However, aldehydes are known to react with water and alcohols rapidly with formation of the aldehydrol or acetal. This reaction occurs rapidly for both aliphatic aldehydes[7] and aldehydo-sugar acetates, as shown by Wolfrom and associates, and would lead to exchange of the C-1 oxygen with the solvent. However, such an exchange does not occur during mutarotations. Probably stable acyclic ions are the actual intermediates, with the aldehydo-sugar being formed more slowly.[1] The small proportion of this aldehydo-sugar in solution is also an indication of its thermodynamic instability and of the rapidity of ring-closure reactions.[12a]

E. MUTAROTASE[1]

Enzymes exist which catalyze the mutarotations of some sugars and are called mutarotases.[21] A mutarotase found originally[22] in preparations of

D-glucose oxidase from *Penicillium notatum* has been purified extensively and studied.[23] A similar enzyme has been found distributed widely throu animal tissues.[24]

The enzyme acts on sugars having D-glucose and D-galactose moieties, including cellobiose, maltose, and lactose.[23–25] Mutarotations taking place in the presence of mutarotase follow Michaelis–Menton kinetics,[26] and the mechanism appears to be the same as for nonenzyme-catalyzed reactions.[23] Keston[24] has provided evidence that the enzyme may be involved in the active transport of sugars through biological membranes.

II. ACTION OF ACIDS AND ALKALIES†

Under the mildest of conditions, acids and alkalies induce the sugars to mutarotate, the only carbon atom involved in this reaction being the anomeric carbon atom; with increasingly vigorous conditions, the α-carbon atom is attacked (enolization) and then the β-carbon atom (β elimination). The products of further degradation then depend on whether the solution is acid or alkaline. Other reactions, such as fragmentation of the carbon chain, may occur also.

A. ENOLIZATION

Enolization is a general reaction, acid–base catalyzed, of carbonyl compounds having an α-hydrogen atom (1). Since the acyclic carbonyl form of sugars is generally assumed to exist in solution in equilibrium with the cyclic forms, enolization of sugars should take place under suitable conditions. However, the enolic forms (2) of the sugars have not been isolated, and have probably not been detected in solution, although this has been claimed.[27] The main evidence for enolization is indirect, and comes from the Lobry de Bruyn–Alberda van Ekenstein transformation (see also the related Amadori and Heyns rearrangements, Chapter 20).[28, 29]

The Lobry de Bruyn–Alberda van Ekenstein transformation includes the epimerization of both aldoses and ketoses as well as aldose–ketose isomerization. These reactions are subject to acid–base catalysis; they proceed readily in alkaline solution and still occur under acid or neutral conditions.[29] Estimates of comparative rates of enolization of a number of sugars have been obtained by Isbell and co-workers from the uptake of tritium by sugars from alkaline water-*t* solution.[12a]

---

† This section was prepared by E. F. L. J. Anet.

$$\text{Ketose (4)} \quad \begin{array}{c} CH_2OH \\ | \\ C \quad O \\ | \\ R \end{array} \quad \rightleftharpoons \quad \begin{array}{l} \text{Epimerization at C-3,} \\ \text{formation of 3-ketoses, etc.} \end{array}$$

$$\begin{array}{ccc}
\begin{array}{c} HC \quad O \\ | \\ HOCH \\ | \\ R \end{array} & \rightleftharpoons &
\begin{array}{c} HCOH \\ \| \\ COH \\ | \\ R \end{array} & \rightleftharpoons &
\begin{array}{c} HC=O \\ | \\ HCOH \\ | \\ R \end{array} \\
\text{Aldose (1)} & & \begin{array}{c}\text{Enediol (2)} \\ \text{(or ionic forms)}\end{array} & & \text{Aldose (3)}
\end{array}$$

$$(cis \; + \; trans)$$

Ketose (4) formation is precluded when an aldose (1, 3) contains a stable 2-substituent (or an alkali-stable group at C-2, as in the case of 2-deoxy-2-acetamidohexoses).[30, 31] Lewis and co-workers[32, 33] studied the reactions of 2,3,4,6-tetra-O-methyl-D-glucose (5) and 2,3,4,6-tetra-O-methyl-D-mannose (6) with saturated lime-water and found that in both instances the product consisted essentially of a mixture of these two epimeric aldoses in the same proportion.

$$\begin{array}{ccc}
\begin{array}{c} HC=O \\ | \\ HCOMe \\ | \\ MeOCH \\ | \\ HCOMe \\ | \\ HCOH \\ | \\ CH_2OMe \end{array} & \rightleftharpoons &
\begin{array}{c} HCO^\ominus \\ \| \\ COMe \\ | \\ MeOCH \\ | \\ HCOMe \\ | \\ HCOH \\ | \\ CH_2OMe \end{array} & \rightleftharpoons &
\begin{array}{c} HC=O \\ | \\ MeOCH \\ | \\ MeOCH \\ | \\ HCOMe \\ | \\ HCOH \\ | \\ CH_2OMe \end{array} \\
\mathbf{5} & & & & \mathbf{6}
\end{array}$$

In the absence of protecting groups, the Lobry de Bruyn–Alberda van Ekenstein transformation gives mixtures of aldoses and ketoses. Wolfrom and Lewis[32] showed that when D-glucose was treated with saturated lime-water at 35° for 10 days the composition of the mixture was glucose 63.5%, fructose 31%, and mannose 2.5%. By means of an isotope-dilution method, Sowden and Schaffer[34] showed that these three sugars constituted only about 80% of the final mixture, and that the remainder was unfermentable.

The main preparative use of the Lobry de Bruyn–Alberda van Ekenstein transformation is for the synthesis of ketoses, because, although the yields are frequently low, the aldose may be readily available and can easily be removed from the mixture of products, usually by oxidation with hypoiodite, as in the preparation of D-*gluco*-heptulose[35] or of lactulose (4-*O*-α-D-galactopyranosyl-D-fructose).[36] In the preparation of D-tagatose from D-galactose, the aldose can be removed by fermentation.[37] To prevent side reactions, the hydroxide ion concentration must be kept low; pyridine or calcium hydroxide is usually employed. The calcium ion appears to have an extra catalytic effect on the reaction.[34] A large increase in the proportion of ketose in the equilibrium mixture is found when sodium or potassium aluminate is used as the base; thus starting from D-glucose the proportion[37a] of D-fructose reaches 70%.

By further keto–enediol tautomerizations, the carbonyl group of the sugar can eventually "move" right down the carbon chain.[38–40] This mechanism explains the formation of D-psicose from D-glucose,[41] the interconversion of D-sorbose and D-galactose,[39] and the occurrence of allitol[42] among the products of the electrolytic reduction of D-glucose in mildly alkaline solution. From the treatment of a solution of D-glucose with a strongly basic resin, Blair and Sowden[43] isolated DL- and D-sorbose. It is evident that, under suitable conditions, any hexose may be converted into any other hexose,[39] although other mechanisms, such as dealdolization and recombination, may also be involved to some extent.

The mechanism of the Lobry de Bruyn–Alberda van Ekenstein transformation requires enolization.[29] In an alkaline medium, enolization of **7**, **8**, and **10** to the hybrid anions occurs after an attack by the Brönsted base (A⊖) at the α-hydrogen atom; the actual formation of the enediol (**9**) is not necessary.[29] In acid solution, the carbonyl group is first protonated (11) and the α-proton is abstracted to yield the enediol, thus causing loss of asymmetry at the α-carbon atom; on de-enolization, the carbonyl group can appear in the original position or at the α-carbon atom.[44]

As acids are far less effective catalysts than alkalies for enolization, sugars show their maximum stability in acid conditions. Thus the optimal pH for stability of D-glucose[45] and of D-fructose[46] lies between pH 3 and 4.

The enolization of a ketose can occur by the removal of a proton from one of the two α-carbon atoms. In acid conditions the 2-ketoses lose a proton more easily from C-1.[47] In alkaline solution the same position seems to hold, a proton being removed more readily from C-1 than from C-3, but the proportion of isomeric sugars resulting from repeated enolization and deenolization depends on the extent of the reaction. In the early stages of the reaction, the composition depends on the rate of formation of the various sugars; in later stages, the proportion of the sugars present in the mixture depends on their relative thermodynamic stability.[48]

B. Beta Elimination

β Elimination involves the elimination of a hydroxyl or alkoxyl group in a β position relative to a carbonyl group. The dehydration of aldol to croton-aldehyde is a typical example. Since prior or simultaneous enolization is required, the overall reaction is acid–base catalyzed. With sugars, the initial products are usually degraded rapidly by acid or alkali, exceptions being compounds such as 2-O-methylaldoses in alkaline solution, where further reaction is largely prevented.

In lime-water, 2,3,4,6-tetra-O-methyl-D-glucose (12) enolizes rapidly to a mixture of the epimeric aldoses.[32] The enolate ion (13) present in the equilibrium mixture, then slowly eliminates a methoxide ion to form the unsaturated sugar (14), which exists in cyclic forms (3-deoxy-2,4,6-tri-O-methyl-α,β-D-erythro-hex-2-enopyranose), the α-D anomer being obtained crystalline in good yield.[49, 50]

In the absence of the blocking 2-*O*-methyl group, the unsaturated sugar tautomerizes to a 3-deoxyaldosulose, which is readily degraded. Nevertheless, 3-deoxy-D-*erythro*-hexosulose (16) was isolated, albeit in low yield, by treating D-fructose (15) with acid[51] and 3-*O*-benzyl-D-glucose (17) with alkali.[52] The elimination of the hydroxyl group at C-3 from D-fructose is

CH$_2$OH	HC=O	HC=O
C=O	C=O	HCOH
HOCH	CH$_2$	PhCH$_2$OCH
HCOH	HCOH	HCOH
HCOH	HCOH	HCOH
CH$_2$OH	CH$_2$OH	CH$_2$OH
D-Fructose (15)	16	3-*O*-Benzyl-D-glucose (17)

regarded as a β elimination because the reaction involves C-1 and C-3 of 19, the enolic form common to D-fructose, D-glucose, and D-mannose. 3-Deoxy-aldosuloses can be conveniently isolated from reaction mixtures as their sparingly soluble bis(benzoylhydrazones).[52a]

Detailed mechanisms involving β elimination were proposed by Isbell[44] for many degradations of aldoses long before deoxydicarbonyl sugars were characterized. An extension of Isbell's mechanism to ketoses leads to the reaction sequence 18 → 21 for the formation of compound 16 from D-fructose.[47] In acid, the β elimination (19 → 20) is an allylic rearrangement and takes place readily, the rate-limiting step being the enolization (18 to 19).[47]

CH$_2$OH	HC—O—H	HC=O	HC=O
C=O	COH	COH	C=O
HOCH	H$_2$O—CH	CH	CH$_2$
R	R	R	R
18	19	20	21

In alkaline solution (22–25) the position is reversed; enolization is rapid, but β elimination is slow and takes place from the enediol (23) ionized at C-1.[44,47] The rate of the elimination reaction depends on the substituent group R'; it is high with a benzyl group, moderate with an alkyl or glycosyl group, and slow[53] with a hydroxyl group at C-3. It has been suggested[53a] that these reactions involve a hydride shift, but this has been refuted.[53b]

In most instances of β eliminations, several products are formed, because by the process of enolization the carbonyl group can first migrate down the carbon chain, and also because elimination can be from either of two positions if the enediol is not terminal. The 2,3-enediol (26) can be derived by a single enolization from a 2- or 3-hexulose.[39, 40] If elimination then takes place at C-1, the intermediate 27 is formed, and this tautomerizes to the dicarbonyl sugar (28), which may be degraded further, depending on the conditions.

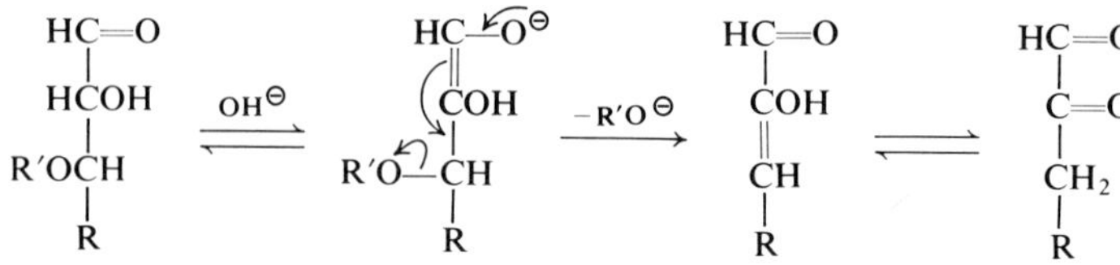

| | 22 | 23 | 24 | 25 |

Under alkaline conditions, saccharinic (2-C-methylaldonic) acids are formed.[53c] Maltol, isomaltol, and Hodge's reductones may also be produced from 28, following a β elimination of the amine from N-substituted 1-amino-1-deoxy derivatives of the enediol (26) (see Chapter 20). If elimination occurs at C-4 the product (29) tautomerizes to compound 30, which is converted by acid into 2-(2-hydroxyacetyl)furan (see p. 184),[47] and by alkali into iso-saccharinic acids. The dicarbonyl derivative 28 has been synthesized.[47a]

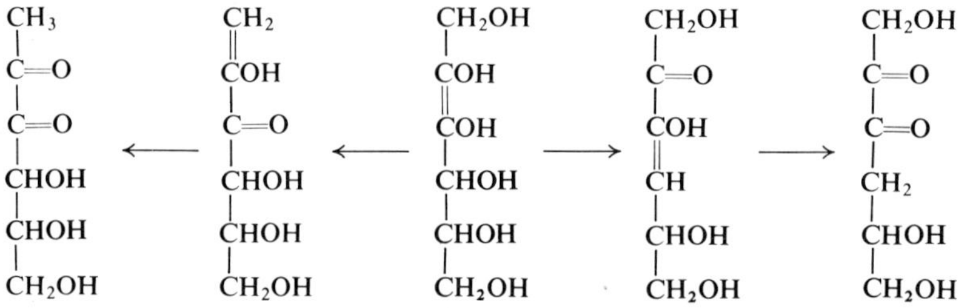

| | 28 | 27 | 26 | 29 | 30 |

Whether the group at C-1 or that at C-4 is eliminated from the enediol (26) depends on their relative ease of elimination.[47] This is illustrated by the following reactions in alkaline solution. Elimination from C-1 is favored (as indicated by the isolation of saccharinic acid [54]) from 1-O-methyl-D-fructose, whereas C-4 elimination occurs, with the formation of isosaccharinic acid, from 4-O-methyl- or 4-O-glycosylhexoses.[55] Further support for this mechanism was obtained by Machell and Richards by their isolation of 4-deoxy-D-glycero-hexo-2,3-diulose (30) from the treatment of maltose with sodium hydroxide.[56]

"Activated" glycosides can be cleaved by alkali by a β-elimination reaction. This is especially important for determining the linkage between oligosaccharides and the protein core of glycoproteins (see Chapters 9 and 43).

C. FURTHER ACTION OF ACIDS

Sugars exhibit moderate stability to acids, but under vigorous conditions the enolizations that occur lead to β elimination and the formation of furan compounds (with hexoses the degradation proceeds further to yield levulinic acid and formic acid). Other possible reactions are fragmentation and glycosidation. Products of fragmentation, such as methylglyoxal[57] and lactic acid,[58] are isolated in trace amounts only and are believed to be formed by a dealdolization mechanism (compare fragmentation under alkaline conditions).

1. Glycosidation (Including Reversion and Formation of Anhydro Sugars)

Aldoses and ketoses react with alcohols under acid catalysis to form glycosides (mixed acetals from an internal hydroxyl group and the alcohol; see Chapter 9). If the alcohol is another sugar molecule, a disaccharide and eventually polysaccharides are formed, the process being called reversion; whereas if the alcohol is a part of the sugar molecule itself, an internal bicyclic acetal arises (a type of anhydro sugar; see Chapter 13).

2. Formation of Furan Compounds

a. 2-Furaldehydes.—When treated with acid, most carbohydrates and their derivatives are degraded to 2-furaldehydes. Good yields of 2-furaldehyde (**32**, R = H) are obtained by the distillation of solutions of pentoses (**31**, R = H) in hydrochloric acid and of pentosans, the method being used for their determination.[59, 60] Under similar conditions 6-deoxyhexoses (**31**, R = Me) give 5-methyl-2-furaldehyde (**32**, R = Me).[60]

The preparation of 5-(hydroxymethyl)-2-furaldehyde (**34**) from hexoses is more difficult because the product cannot be removed from the reaction mixture by distillation as it is formed, and also because it is less stable than 2-furaldehyde.[61, 62] It may be obtained by treating sucrose or hexuloses with dilute oxalic acid, the fructose portion only of sucrose being converted.[63]

Aldohexoses are stable under the above conditions,[61] and under more forcing conditions the 5-(hydroxymethyl)-2-furaldehyde is converted into levulinic acid.[64] However, good yields of 5-(hydroxymethyl)-2-furaldehyde can be obtained by raising the pH of the reaction mixture, the yield being

46% when D-glucose is heated at 200° to 238° for 20 minutes in a solution of aqueous pyridine–phosphoric acid.[65]

The hydroxyl group of 5-(hydroxymethyl)-2-furaldehyde is very reactive, being easily removed (as in levulinic acid formation) or replaced by either another alkoxyl group to yield the condensation product (35) or a mixed ether,[66] or by halogen (in concentrated hydrochloric or hydrobromic acid) to give a 5-(halomethyl)-2-furaldehyde.[62]

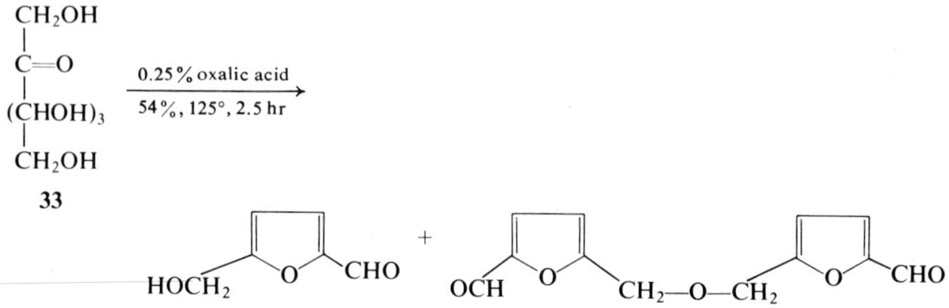

$$CH_2OH$$
$$|$$
$$C=O$$
$$|$$ $\xrightarrow[54\%,\,125°,\,2.5\ \text{hr}]{0.25\%\ \text{oxalic acid}}$
$$(CHOH)_3$$
$$|$$
$$CH_2OH$$

33

HOCH$_2$ O   CHO   +   OCH   O   CH$_2$—O—CH$_2$ O   CHO

      34   **35**

The first mechanism for the formation of 2-furaldehydes, proposed by Nef[67] and later by Haworth and Jones,[63] was based on the ease of conversion of 2,5-anhydroaldoses,[61] into 2-furaldehydes. Although the yield from 2,5-anhydro-D-mannose (chitose) is greater than that from aldoses, it is lower than that from ketoses.[61] These anhydro sugars are more reactive than aldoses, probably because their carbonyl groups are not inactivated by being in cyclic hemiacetal forms.[47]

Other mechanisms proposed by Hurd and Isenhour,[68] by Isbell,[44] and by Wolfrom et al.[69] all applied to aldoses only and involved the well-established β-elimination reaction. Since β elimination requires a preliminary enolization to the enediol, the same mechanisms can apply to ketoses.[47] Evidence for the

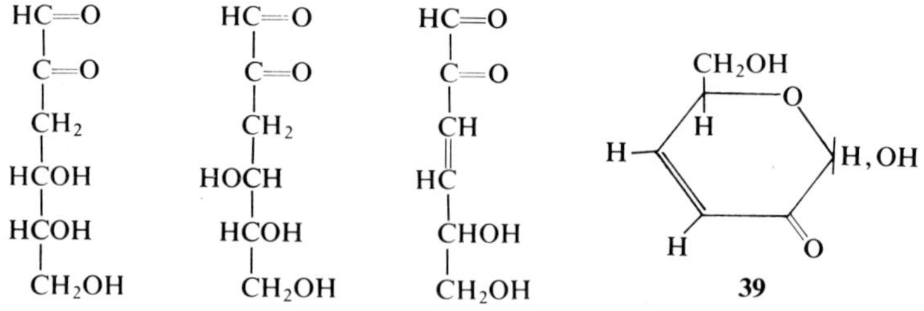

      36   **37**   **38**   **39**

β-elimination mechanism was obtained by Anet who, from the action of oxalic or acetic acids on D-fructose,[47, 51] isolated (as osazones) the deoxy-hexosuloses (**36–39**). L-Sorbose behaved similarly.[51] The mechanism of

2-furaldehyde formation is therefore complex, and the following scheme has been suggested.[47, 51]

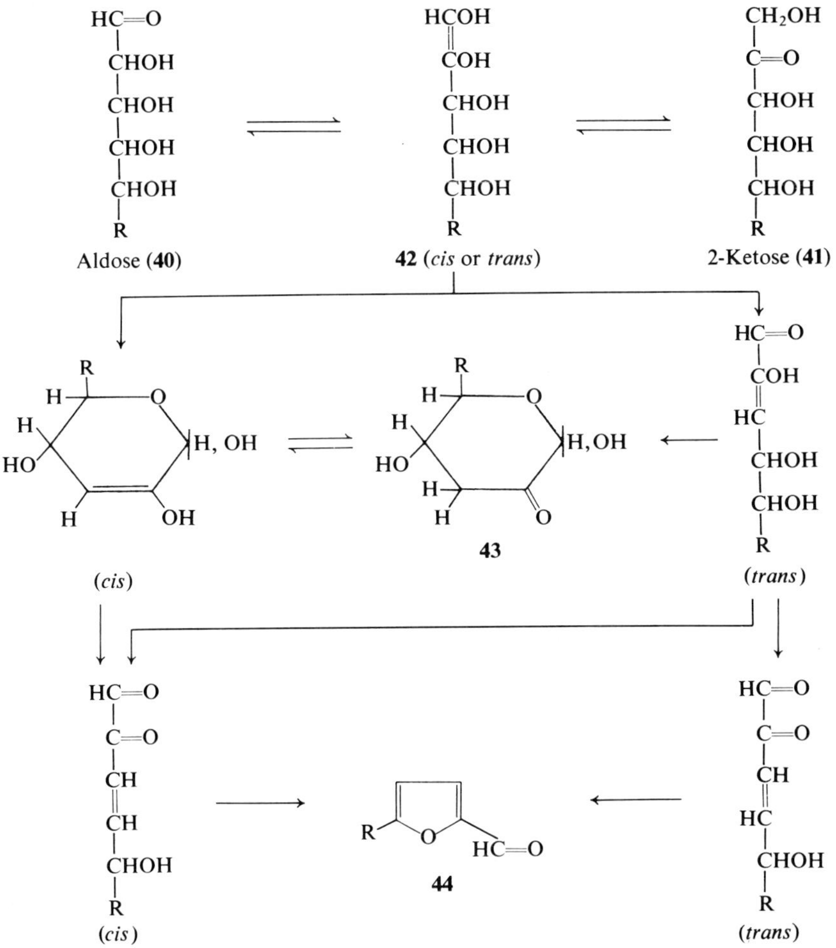

```
   HC=O              HCOH              CH₂OH
   |                 ‖                 |
   CHOH              COH               C=O
   |                 |                 |
   CHOH              CHOH              CHOH
   |                 |                 |
   CHOH     ⇌        CHOH     ⇌        CHOH
   |                 |                 |
   CHOH              CHOH              CHOH
   |                 |                 |
   R                 R                 R
 Aldose (40)    42 (cis or trans)   2-Ketose (41)
```

```
                                          HC=O
                                          |
                                          COH
                                          ‖
                                          HC
                                          |
                                          CHOH
                                          |
                                          CHOH
                                          |
                                          R
   (cis)              43                (trans)
```

```
   HC=O                              HC=O
   |                                 |
   C=O                               C=O
   |                                 |
   CH                                CH
   ‖          →        44      ←     ‖
   CH                                HC
   |                                 |
   CHOH                              CHOH
   |                                 |
   R                                 R
   (cis)                            (trans)
```

In this scheme, since the rate-limiting step is the formation of the enediol (42), the rate of formation of the 2-furaldehyde (44) depends on the ease of enolization of the sugar.[47] The saturated 3-deoxyhexosuloses (43) are not necessary intermediates; it appears that with D-fructose over half of the material transformed bypasses these 3-deoxyhexosuloses.[51] In deuterium oxide solution less than 20 % of the reaction proceeds via the 3-deoxyhexosuloses,[71a] but this finding cannot be extended to aqueous solution because of probable isotope effects.

* References start on p. 191.

b. 2-(2-Hydroxyacetyl)furan.—When D-fructose is treated with acid there is obtained, besides the large amount of 5-(hydroxymethyl)-2-furaldehyde, a by-product—namely, 2-(2-hydroxyacetyl)furan.[70, 71] This furan (**47**) is probably derived from the dicarbonyl intermediates (**45** and **46**), the mechanism of formation[47] being similar to that for 5-(hydroxymethyl)-2-furaldehyde (**40–44**), except that 2,3-enolization takes place instead of 1,2-enolization, the hydroxyl group at C-4 of the 2,3-enediol then undergoing a β elimination to yield[47] **45**. Since 2-(2-hydroxyacetyl)furan is a minor product, 2,3-enolization of D-fructose is not as important as 1,2-enolization.[47]

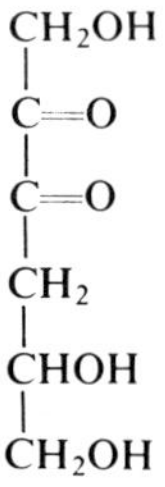

45 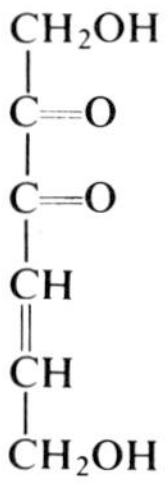   46 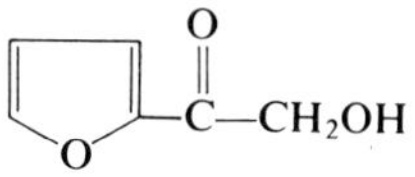   2-(2-Hydroxyacetyl)furan (47)

c. 3(2H)-Furanones and 3-hydroxyfurans.—The 3(2H)-furanones (**48**) and 3-hydroxyfurans (**49**) are keto–enol tautomers and normally exist in the furanone form unless the furan ring is stabilized by conjugation and chelation of the hydroxyl groups, as in isomaltol (**49**, R' = H, R = Ac) (see Chapter 20). Sugars having no potential hydroxyl group at C-5 do not yield 2-furaldehydes on acid treatment, 3(2H)-furanones or 3-hydroxyfurans being formed

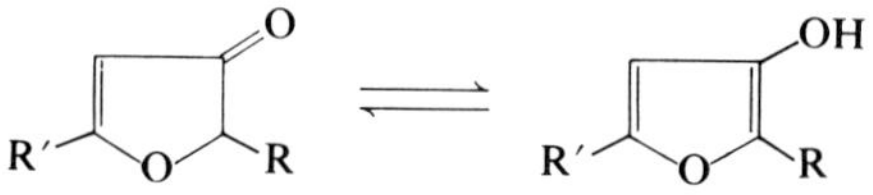

48   49

instead.[72] In methanolic hydrogen chloride, 6-deoxy-L-*lyxo*-hex-5-ulosonic acid is converted into the methyl ether of **49** (R = Me, R' = CO$_2$Me),[73] whereas in aqueous acid **50** is degraded[72, 74] to the furanone **51**.

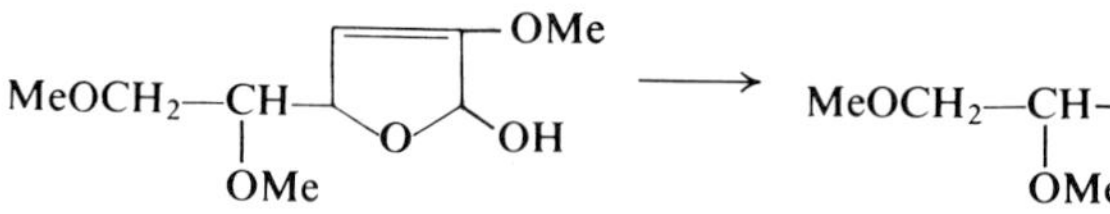

50   51 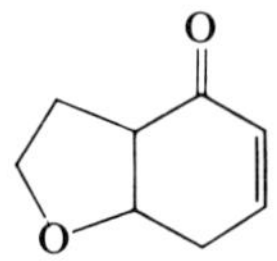   52

From D-erythrose a small yield of a dimerized product, "erythropyrone" (**52**), has been isolated.[75]

3. Formation of Levulinic Acid

Levulinic acid (**54**) is a product of the acid degradation of 5-(hydroxy-methyl)-2-furaldehyde, arabinal, furfuryl alcohol, and 2-deoxypentoses.[76, 77] Therefore hexoses, which are degraded initially to 5-(hydroxymethyl)-2-furaldehyde, also yield levulinic acid. Sowden[78] has shown that the levulinic acid formed from D-glucose-1-^{14}C is devoid of activity, the radioactivity being all present in the other product of the reaction, formic acid. Several mechanisms[44, 79-81] have been proposed for this reaction, but the following (**53–54**) by Isbell[44] seems the most probable.

Arabinal, furfuryl alcohol, and the 2-deoxypentoses react with methanolic hydrogen chloride to yield the dimethyl acetal of 5-methoxylevulinic alde-hyde.[82] Isbell[44] has proposed a mechanism for the conversion of 2-deoxy-pentoses into levulinic acid, and his scheme includes the enolic form of 5-hydroxylevulinic aldehyde as an intermediate.

Methanolic hydrogen chloride reacts with D-glucal (**54a**) or with methyl 4,6-*O*-benzylidene-2,3-dideoxy-α-D-*erythro*-hex-2-enopyranoside (**54c**) to cause a stepwise series of transformations,[82a] with an initial fast reaction leading to 2-(D-*glycero*-1,2-dihydroxyethyl)furan (**54b**). A slower step involves formation of the racemic 1′-methyl ether (**54e**), and the 1′,2′-dimethyl ether (**54f**) is produced at a still lower rate. The latter product undergoes reversible conversion into 2,6-dimethoxy-4-oxohexanal dimethyl acetal (**54g**) to give an equilibrium mixture of **54g** and **54f**. A side product in the reaction from D-glucal is the 2-deoxyhexose (**54d**).

* *References start on p. 191.*

D. FURTHER ACTION OF ALKALIES

Although reducing sugars are very labile in alkaline solution, they show a surprising stability at 5° in 0.04M sodium hydroxide. Isbell et al.[48] found that, under these conditions, D-glucose, D-mannose, and D-fructose are largely unchanged after nine months, 85 to 90 % of the starting sugar remaining intact. At higher temperatures sugars are rapidly isomerized by enolization (Lobry de Bruyn–Alberda van Ekenstein transformation), and β elimination eventually leads to saccharinic acids. However, large proportions of fragmentation products such as lactic acid and some polymeric material may also be formed.[38] The proportion of the various products in these degradations depends on many factors, such as the sugar and its substituent, temperature, duration of the reaction, nature of the alkali, and its concentration.[83, 84]

All of the reactions discussed in this section are those taking place in the absence of oxygen. In the presence of oxygen the formation of oxidation products further complicates the reaction.[38, 53c]

1. Formation of Saccharinic Acids

Alkalies were shown by Peligot[85] to degrade D-glucose to acids, which were later found by Kiliani and Nef to be a complex mixture consisting largely of saccharinic acids (see Chapter 22). The accepted mechanism for the formation of saccharinic acids is that of Nef–Isbell:[44] enolization of the sugar, β elimination to give a deoxydicarbonyl intermediate, which undergoes a benzilic acid rearrangement to give saccharinic acids. The mechanisms for enolization and β elimination have been discussed earlier in this chapter.

$$
\begin{array}{ccc}
\text{R—C=O} \\
| \\
\text{R'—C=O} \\
\textbf{55}
\end{array}
\xrightarrow{\ominus \text{OH}}
\begin{array}{c}
\text{:R—:COH} \\
| \\
\text{R'—C=O} \\
\textbf{56}
\end{array}
\longrightarrow
\begin{array}{c}
\text{CO}_2\text{H} \\
| \\
\text{CO} \\
\text{R} \quad \text{R'} \\
\textbf{57}
\end{array}
\longrightarrow
\begin{array}{c}
\text{CO}_2\text{H} \\
| \\
\text{COH} \\
\text{R} \quad \text{R'} \\
\textbf{58}
\end{array}
$$

From the mechanism **55–58** for the benzilic acid rearrangement it can be seen that the same acid would be obtained if the other carbonyl group were attacked and R' migrated, unless one of the carbonyl carbon atoms was labeled. In the rearrangement of D-glyceraldehyde-3-^{14}C to lactic acid, the migrating group can be either the hydrogen atom at C-1 or the methyl group of the intermediate deoxydicarbonyl compound, pyruvaldehyde. The group that migrates depends on the alkali; in sodium hydroxide it is the hydrogen atom, but in lime-water migration of the methyl group also occurs[86] to the extent of 26%.

$$
\begin{array}{l}
\text{HC=O} \\
| \\
\text{C=O} \\
| \\
\text{CH}_2 \\
| \\
\text{CHOH} \\
| \\
\text{CHOH} \\
| \\
\text{CH}_2\text{OH} \\
\textbf{59}
\end{array}
\longrightarrow
\begin{array}{l}
\text{CO}_2\text{H} \\
| \\
\text{CHOH} \\
| \\
\text{CH}_2 \\
| \\
\text{CHOH} \\
| \\
\text{CHOH} \\
| \\
\text{CH}_2\text{OH} \\
\text{Metasaccharinic} \\
\text{acid (60)}
\end{array}
\qquad
\begin{array}{l}
\text{CH}_3 \\
| \\
\text{C=O} \\
| \\
\text{C=O} \\
| \\
\text{CHOH} \\
| \\
\text{CHOH} \\
| \\
\text{CH}_2\text{OH} \\
\textbf{61}
\end{array}
\longrightarrow
\begin{array}{l}
\text{CO}_2\text{H} \\
| \\
\text{C(CH}_3)\text{OH} \\
| \\
\text{CHOH} \\
| \\
\text{CHOH} \\
| \\
\text{CH}_2\text{OH} \\
\text{Saccharinic} \\
\text{acid (62)}
\end{array}
$$

$$
\begin{array}{l}
\text{CH}_2\text{OH} \\
| \\
\text{C=O} \\
| \\
\text{C=O} \\
| \\
\text{CH}_2 \\
| \\
\text{CHOH} \\
| \\
\text{CH}_2\text{OH} \\
\textbf{63}
\end{array}
\longrightarrow
\begin{array}{l}
\text{CO}_2\text{H} \\
| \\
\text{C(OH)CH}_2\text{OH} \\
| \\
\text{CH}_2 \\
| \\
\text{CHOH} \\
| \\
\text{CH}_2\text{OH} \\
\text{Isosaccharinic} \\
\text{acid (64)}
\end{array}
\qquad
\begin{array}{l}
\text{CH}_2\text{OH} \\
| \\
\text{CHOH} \\
| \\
\text{C=O} \\
| \\
\text{C=O} \\
| \\
\text{CH}_2 \\
| \\
\text{CH}_2\text{OH} \\
\textbf{65}
\end{array}
\longrightarrow
\begin{array}{l}
\text{CO}_2\text{H} \\
| \\
\text{C—OH} \\
\text{CH}_2 \quad \text{CHOH} \\
| \qquad | \\
\text{CH}_2\text{OH} \quad \text{CH}_2\text{OH} \\
\text{"Parasaccharinic acid" (66)}
\end{array}
$$

* *References start on p.* 191.

In the hexose series, four types of saccharinic acid are possible, since four types of deoxydicarbonyl sugar could be formed by enolization, de-enolization, and β elimination (59–66).[44] Parasaccharinic acids (66) have not yet been definitely isolated.

The formation of saccharinic acid (62) is not straightforward, and apparently it occurs by several paths. Sowden and co-workers[87] found that the "α"-D-glucosaccharinic acid formed from D-mannose-1-^{14}C was 39%, labeled at the methyl carbon atom, and 57% at the tertiary carbon atom. Several mechanisms have been proposed, but they do not fully account for the experimental results. According to the latest mechanism, saccharinic acid (62) may also be formed from hexoses by way of the 3-deoxydicarbonyl sugar (59).[48] However, this intermediate (59) reacts readily[47, 52, 53c] with alkali to yield metasaccharinic acids, but not the saccharinic acids (62).

2. Fragmentation

The fragmentation of sugar molecules to substances having fewer carbon atoms is a reaction especially important in alkaline solution. The most important fragmentation product, DL-lactic acid, has been obtained[88] in yields of 60%, and the following acids in some instances in 10 to 30% yield: formic acid,[89] glycolic acid,[55] DL-2,4-dihydroxybutyric acid,[38] and D-3,4-dihydroxy-butyric acid.[55] A 7% yield of acetic acid has been reported,[90] but the other fragmentation products—acetol,[91] methylglyoxal,[91, 92] pyruvic acid,[93, 94] reductone,[95, 96] butanedione,[97] and methylcyclopentenolone[98]—have been detected in smaller proportions only. Nef[38] proposed the splitting of a 3,4-enediol of a hexose to account for three-carbon fragments, but this is most unlikely. The mechanism operating in most instances of fragmentation involves either dealdolization of sugars or the splitting of dicarbonyl degradation products. The dicarbonyl compound 61 is stable only in acid solution; under non-acid conditions it is rapidly split between C-2 and C-3, D-erythronic acid having been isolated as the main product.[47a]

a. Dealdolization.—The aldol condensation of D-glyceraldehyde (68) with 1,3-dihydroxy-2-propanone (69) proceeds rapidly in weakly alkaline solution with the formation, in 90 to 95% yield, of D-fructose and D-sorbose.[99] Further examples of this reaction are given in Chapter 2. Aldolization is a reversible reaction, and although it may be catalyzed by acids or alkalies, alkalies are far more efficient catalysts. The splitting of a hexulose (67) to the two trioses (68 and 69) is a dealdolization. In alkaline conditions the trioses are quickly converted into DL-lactic acid, which can be considered as the metasaccharinic acid derived from 1,3-dihydroxypropanone or glyceraldehyde by a β elimination to methylglyoxal and then a benzilic acid rearrangement. DL-Lactic acid is the major product from the action of hot N alkali on D-glucose.[88] Nef[38] showed that besides lactic acid and six-carbon saccharinic acids, DL-2,4-

dihydroxybutyric acid is obtained in about 15% yield from the treatment of D-glucose with $8M$ sodium hydroxide. This butyric acid is the metasaccharinic acid derived from tetroses and probably arises by dealdolization of the hexose to glycolaldehyde and a tetrose. A different dihydroxybutyric acid (72) has also been reported (see the next section, Splitting of Dicarbonyl Intermediates).

$$
\begin{array}{ccc}
\text{CH}_2\text{OH} & \left[\text{CH}_2\text{OH}\right. & \text{CH}_2\text{OH} \\
| & | & | \\
\text{C}=\text{O} & \text{C}=\text{O} \longleftrightarrow & \text{C}-\text{O}^{\ominus} \\
| & | & \| \\
\text{CHOH} & {}^{\ominus}\text{CHOH} & \text{CHOH} \\
| & & \\
\text{CH}-\text{O}-\text{H} & \text{HC}=\text{O} & \text{CH}_2\text{OH} \\
| & | & | \\
\text{CHOH} & \text{CHOH} + & \text{C}=\text{O} \\
| & | & | \\
\text{CH}_2\text{OH} & \text{CH}_2\text{OH} \left.\right] & \text{CH}_2\text{OH} \\
\mathbf{67} & \mathbf{68} & \mathbf{69}
\end{array}
$$

The formation of D-sorbose and L-sugars by the treatment of D-glucose by alkali was originally thought to involve dealdolization and recombination of the three-carbon fragments.[43, 100] It was later shown that dealdolization could account for only a very small proportion of the isomerization.[39]

The accepted mechanism[84] for aldolization and dealdolization in alkaline solution is shown in formulas 67–69. In an earlier mechanism, suggested for the formation of trioses from hexoses, the 1,2-enediol was proposed as an intermediate rather than the hexulose,[101] but an enediol cannot lead to the dealdolization of a hexose to a tetrose and glycolaldehyde. That the enediol mechanism is not important is further suggested by the fact that lactic acid is produced from 3-O-methyl-D-fructose by treatment with alkali, but not from 3-O-methyl-D-glucose.[102]

Alkaline solutions of sugars develop reducing properties typical of reductones. Earlier workers believed that sugar enediols were responsible, but it now appears[95, 96] that these reducing properties are due to reductone (70).

$$
\text{Hexose} \longrightarrow \left[\begin{array}{cc}
\text{HC}=\text{O} & \text{CH}_2\text{OH} \\
| & | \\
\text{CHOH} + & \text{C}=\text{O} \\
| & | \\
\text{CH}_2\text{OH} & \text{CH}_2\text{OH}
\end{array}\right] \longrightarrow
\begin{array}{c}
\text{HC}=\text{O} \\
| \\
\text{COH} \\
\| \\
\text{HCOH}
\end{array}
$$

Reductone (70)

Reductone is prepared, in about 1% yield, by treating hot solutions of D-glucose with sodium hydroxide in the presence of lead acetate and traces of potassium cyanide and copper acetate. Euler[95] proposed that fragmentation of one molecule of D-glucose to one molecule of reductone and one molecule

* *References start on p. 191.*

of glycerol occurred. However, no glycerol is produced in the reaction, and, furthermore, the reductone arises almost equally from the two halves of the D-glucose molecule.[96, 103] The reductone is probably formed by oxidation of the triose sugars formed by dealdolization of the hexose.[96]

b. *Splitting of dicarbonyl intermediates.*—Several fragmentation products of sugars are believed to arise by the splitting of dicarbonyl compounds. Glycolic acid and D-3,4-dihydroxybutyric acid are the major products of the degradation of maltose and cellulose by cold, dilute sodium hydroxide.[55] These acids are also formed from the deoxydicarbonyl sugar (71),[56] which is derived by β elimination of the glycosyloxy group from the terminal reducing sugar of maltose or from (1 → 4)-linked polysaccharides.

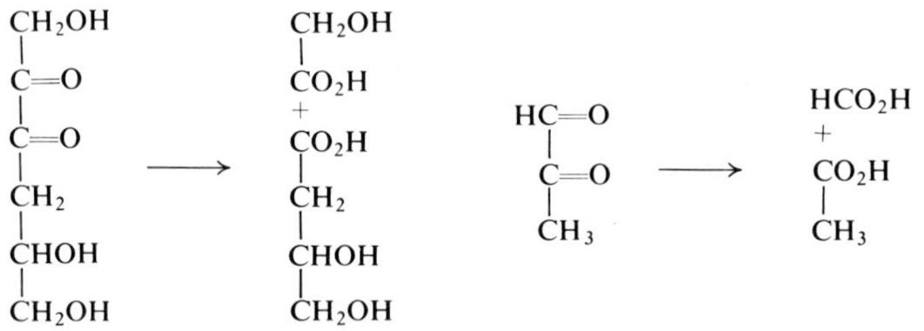

$$
\begin{array}{ccccc}
\text{CH}_2\text{OH} & \text{CH}_2\text{OH} & & & \\
| & | & & & \\
\text{C}=\text{O} & \text{CO}_2\text{H} & & & \text{HCO}_2\text{H} \\
| & + & \text{HC}=\text{O} & & + \\
\text{C}=\text{O} & \text{CO}_2\text{H} & | & & \text{CO}_2\text{H} \\
| & | & \text{C}=\text{O} & \longrightarrow & | \\
\text{CH}_2 & \text{CH}_2 & | & & \text{CH}_3 \\
| & | & \text{CH}_3 & & \\
\text{CHOH} & \text{CHOH} & & & \\
| & | & & & \\
\text{CH}_2\text{OH} & \text{CH}_2\text{OH} & & & \\
\mathbf{71} & \mathbf{72} & \mathbf{73} & &
\end{array}
$$

In a similar way, formic acid and acetic acid could arise from the splitting of pyruvaldehyde (73) by alkali.[90] Formic acid is formed[52] also from 3-deoxy-D-*erythro*-hexosulose (16). These reactions explain the formation of acids, but, since the splitting is assumed to be hydrolytic, equal amounts of the corresponding aldehydes should be formed and be further degraded to acids and alcohols. However, only acids have been detected.[47a, 56]

The splitting of β-dicarbonyl sugars has been proposed by Hayami[104] to explain the production of acetol (77) from pentoses and hexoses. This reaction occurs in the pH range 3 to 11, especially in the presence of phosphate buffers. By using ^{14}C-labeled sugars, the methyl group of the acetol was shown to be derived equally from the two terminal carbon atoms of the sugar.[105] The following scheme was suggested[105] for the formation of acetol from pentoses:

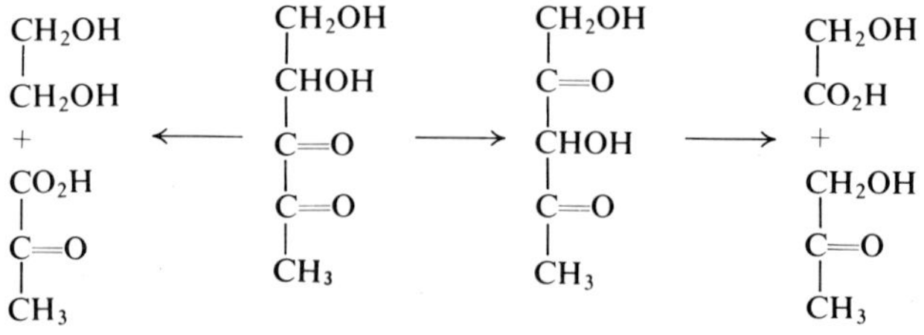

$$
\begin{array}{cccc}
\text{CH}_2\text{OH} & \text{CH}_2\text{OH} & \text{CH}_2\text{OH} & \text{CH}_2\text{OH} \\
| & | & | & | \\
\text{CH}_2\text{OH} & \text{CHOH} & \text{C}=\text{O} & \text{CO}_2\text{H} \\
+ & | & | & + \\
\text{CO}_2\text{H} & \text{C}=\text{O} & \text{CHOH} & \text{CH}_2\text{OH} \\
| & | & | & | \\
\text{C}=\text{O} & \text{C}=\text{O} & \text{C}=\text{O} & \text{C}=\text{O} \\
| & | & | & | \\
\text{CH}_3 & \text{CH}_3 & \text{CH}_3 & \text{CH}_3 \\
\mathbf{75} & \mathbf{74} & \mathbf{76} & \mathbf{77}
\end{array}
$$

isomerization to a 3- or 4-ketose, β elimination of the C-1 or C-5 hydroxyl group to give **74**, isomerization of **74** to **76**, and hydrolytic cleavage to acetol (**77**) and glycolic acid. Hayami[104] also explained the concurrent formation of pyruvic acid [94] (**75**) as being due to the cleavage of **74**.

REFERENCES

1. For a more extensive discussion and list of references, see: W. Pigman and H. S. Isbell, *Advan. Carbohyd. Chem.*, **23**, 11 (1968); **24**, 14 (1969); T. M. Lowry and E. F. Smith, "Rapports sur les hydrates de carbone," 10th Conf. Int. Union Chem., Liège, 1930.
2. H. Fredenhagen and K. F. Bonhoeffer, *Z. Phys. Chem.*, **A181**, 392 (1938).
3. K. Goto and T. Titani, *Bull. Chem. Soc. Jap.*, **16**, 172, 403 (1941).
4. T. M. Lowry, *J. Chem. Soc.*, **75**, 211 (1899); H. Trey, *Z. Phys. Chem.*, **18**, 198 (1895).
5. C. S. Hudson, *Z. Phys. Chem.*, **44**, 487 (1903).
6. H. S. Isbell and W. W. Pigman, *J. Res. Nat. Bur. Stand.*, **18**, 141 (1937).
7. C. N. Riiber and J. Minsaas, *Ber.*, **59**, 2266 (1926); N. A. Sørensen, *Kgl. Norske Videnskab Selskab Skrifter*, No. 2 (1937).
8. G. F. Smith and T. M. Lowry, *J. Chem. Soc.*, 666 (1928).
9. (a) M. Rudrum and D. F. Shaw, *J. Chem. Soc.*, 52 (1965); (b) W. Mackie and A. S. Perlin, *Can. J. Chem.*, **44**, 2039 (1966); (c) D. Horton, J. S. Jewell, and K. D. Philips, *J. Org. Chem.*, **31**, 4022 (1966); (d) R. U. Lemieux and J. D. Stevens, *Can. J. Chem.*, **44**, 249 (1966); (e) S. J. Angyal and V. A. Pickles, *Aust. J. Chem.*, **25**, in press (1972); (f) T. E. Acree, R. S. Shallenberger, and L. R. Mattick, *Carbohyd. Res.*, **6**, 498 (1968); compare (g) T. E. Acree, R. S. Shallenberger, C. Y. Lee, and J. W. Einset, *ibid.*, **10**, 355 (1969); A. S. Hill and R. S. Shallenberger, *ibid.*, **11**, 541 (1969); G. G. S. Dutton, *Advan. Carbohyd. Chem. Biochem.*, **27**, in press (1972); (h) I. M. Campbell and R. Bentley, *Abstr. Papers Amer. Chem. Soc. Meeting*, **162**, CARB (1971).
10. H. S. Isbell and W. Pigman, *J. Res. Nat. Bur. Stand.*, **20**, 773 (1938).
11. S. M. Cantor and Q. P. Peniston, *J. Amer. Chem. Soc.*, **62**, 2113 (1940); J. M. Los and K. Wiesner, *ibid.*, **75**, 6346 (1953); J. M. Los, L. B. Simpson, and K. Wiesner, *ibid.*, **78**, 1564, (1956).
12. F. Lippich, *Biochem. Z.*, **248**, 280 (1932).
12a. H. S. Isbell, H. L. Frush, C. W. R. Wade, and C. E. Hunter, *Carbohyd. Res.*, **9**, 163 (1969); H. S. Isbell, K. Linek, and K. E. Hepner, *ibid.*, **19**, 319 (1971).
12b. R. U. Lemieux, L. Anderson, and A. H. Conner, *Carbohyd. Res.*, **20**, 59 (1971).
13. T. M. Lowry and E. M. Richards, *J. Chem. Soc.*, **127**, 1385 (1925).
14. T. M. Lowry and I. J. Faulkner, *J. Chem. Soc.*, **127**, 2883 (1925).
15. T. M. Lowry, *J. Chem. Soc.*, 2554 (1927).
16. C. G. Swain and J. F. Brown, Jr., *J. Amer. Chem. Soc.*, **74**, 2534, 2538 (1952).
17. H. H. Rowley and W. N. Hubbard, *J. Amer. Chem. Soc.*, **64**, 1010 (1942).
18. K. J. Pedersen, *J. Phys. Chem.*, **38**, 581 (1934).
19. C. G. Swain, *J. Amer. Chem. Soc.*, **72**, 4578 (1950).
20. H. S. Isbell and C. W. R. Wade, *J. Res. Nat. Bur. Stand.*, **71A**, 137 (1967).
21. D. Keilin and E. F. Hartree, *Biochem. J.*, **50**, 341 (1952).
22. R. Bentley and A. Neuberger, *Biochem. J.*, **45**, 584 (1949).
23. R. Bentley and D. S. Bhate, *J. Biol. Chem.*, **235**, 1225 (1960).
24. A. S. Keston, *Science*, **120**, 356 (1954); **143**, 698 (1964); *J. Biol. Chem.*, **239**, 3241 (1964).
25. G. B. Levy and E. S. Cook, *Biochem. J.*, **57**, 50 (1954).
26. L. Lin, A. M. Chase, and S. L. Lapedes, *J. Cell Comp. Physiol.*, **64**, 283 (1964).
27. A. Kusin, *Ber.*, **69**, 1041 (1936).

28. C. A. Lobry de Bruyn and W. Alberda van Ekenstein, *Rec. Trav. Chim.*, **14**, 156, 203 (1895); **15**, 92 (1896); **16**, 241, 257, 262, 274, 282 (1897); **18**, 147 (1899); **19**, 1 (1900); W. Alberda van Ekenstein and J. J. Blanksma, *ibid.*, **27**, 1 (1908).

29. For a review see J. C. Speck, Jr., *Advan. Carbohyd. Chem.*, **13**, 63 (1958).

30. C. T. Spivak and S. Roseman, *J. Amer. Chem. Soc.*, **81**, 2403 (1959).

31. B. Coxon and L. Hough, *J. Chem. Soc.*, 2403 (1959).

32. M. L. Wolfrom and W. L. Lewis, *J. Amer. Chem. Soc.*, **50**, 837 (1928).

33. R. D. Green and W. L. Lewis, *J. Amer. Chem. Soc.*, **50**, 2813 (1928).

34. J. C. Sowden and R. Schaffer, *J. Amer. Chem. Soc.*, **74**, 499 (1952).

35. J. W. Pratt, N. K. Richtmyer, and C. S. Hudson, *J. Amer. Chem. Soc.*, **74**, 2210 (1952).

36. E. M. Montgomery and C. S. Hudson, *J. Amer. Chem. Soc.*, **52**, 2101 (1930).

37. T. Reichstein and W. Bosshard, *Helv. Chim. Acta*, **17**, 753 (1934); E. L. Totton and H. A. Lardy, *Methods Carbohyd. Chem.*, **1**, 155 (1962).

37a. E. Haack, F. Braun, and K. Kohler, Ger. Pat. 1,163,307, *Chem. Abstr.*, **60**, 14598a (1964).

38. J. U. Nef, *Ann. Chem.*, **376**, 1 (1910).

39. J. C. Sowden and R. T. Thompson, *J. Amer. Chem. Soc.*, **80**, 1435 (1958).

40. R. Schaffer, *J. Org. Chem.*, **29**, 1473 (1964).

41. L. Hough, J. K. N. Jones, and E. L. Richards, *J. Chem. Soc.*, 3854 (1952).

42. M. L. Wolfrom, B. W. Lew, and R. M. Goepp, Jr., *J. Amer. Chem. Soc.*, **68**, 1443 (1946).

43. M. G. Blair and J. C. Sowden, *J. Amer. Chem. Soc.*, **77**, 3323 (1955).

44. H. S. Isbell, *J. Res. Nat. Bur. Stand.*, **32**, 45 (1944).

45. W. Kröner and H. Kothe, *Ind. Eng. Chem.*, **31**, 248 (1939); E. J. McDonald, *J. Res. Nat. Bur. Stand.*, **45**, 200 (1950).

46. J. A. Mathews and R. F. Jackson, *Bur. Stand. J. Res.*, **11**, 619 (1933).

47. E. F. L. J. Anet, *Advan. Carbohyd. Chem.*, **19**, 181 (1964).

47a. A. Ishizu, B. Lindberg, and O. Theander, *Carbohyd. Res.*, **5**, 329 (1967).

48. H. S. Isbell, H. L. Frush, R. Schaffer, C. W. R. Wade, and R. A. Peterson, *Nat. Bur. Stand. (U.S.) Tech. Note*, **274**, 36 (1965).

49. E. F. L. J. Anet, *Chem. Ind.* (London), 1035 (1963); *Aust. J. Chem.*, **18**, 837 (1965).

50. A. Klemer, H. Lukowski, and F. Zerhusen, *Ber.*, **96**, 1515 (1963).

51. E. F. L. J. Anet, *Chem. Ind.* (London), 262 (1962); *Aust. J. Chem.*, **18**, 240 (1965).

52. G. Machell and G. N. Richards, *J. Chem. Soc.*, 1938 (1960).

52a. H. El Khadem, D. Horton, M. H. Meshreki, and M. A. Nashed, *Carbohyd. Res.*, **17**, 183 (1971); in press (1972).

53. J. Kenner and G. N. Richards, *J. Chem. Soc.*, 3019 (1957).

53a. C. Fodor and J. Sachetto, *Tetrahedron Lett.*, 401 (1968).

53b. E. F. L. J. Anet, *Tetrahedron Lett.*, 3525 (1968).

53c. R. M. Rowell and J. W. Green, *Carbohyd. Res.*, **15**, 197 (1970).

54. J. Kenner and G. N. Richards, *J. Chem. Soc.*, 1784 (1954).

55. G. Machell and G. N. Richards, *J. Chem. Soc.*, 1924 (1960); compare R. M. Rowell, P. J. Somers, S. A. Barker, and M. Stacey, *Carbohyd. Res.*, **11**, 17 (1969).

56. G. Machell and G. N. Richards, *J. Chem. Soc.*, 1932 (1960).

57. H. G. Sento, J. C. Underwood, and C. O. Willits, *Food Res.*, **25**, 750 (1960).

58. R. Weidenhagen, *Zucker*, **12**, 244 (1959).

59. W. E. Stone and B. Tollens, *Ann. Chem.*, **249**, 227 (1888).

60. F. J. Bates and Associates, "Polarimetry, Saccharimetry and the Sugars," National Bureau of Standards, Circular C440, U.S. Government Printing Office, Washington, D.C., p. 241.

61. W. Alberda van Ekenstein and J. J. Blanksma, *Ber.*, **43**, 2355 (1910).
62. F. H. Newth, *Advan. Carbohyd. Chem.*, **6**, 83 (1951).
63. W. N. Haworth and W. G. M. Jones, *J. Chem. Soc.*, 667 (1944).
64. W. W. Moyer, U.S. Patent 2,2270,328 (1942).
65. M. L. Mednick, *J. Org. Chem.*, **27**, 398 (1962).
66. C. J. Moye and Z. S. Krzeminski, *Aust. J. Chem.*, **16**, 258 (1963).
67. Ref. 38, p. 117.
68. C. D. Hurd and L. L. Isenhour, *J. Amer. Chem. Soc.*, **54**, 317 (1932).
69. M. L. Wolfrom, R. D. Schuetz, and L. F. Cavalieri, *J. Amer. Chem. Soc.*, **71**, 3518 (1949).
70. R. E. Miller and S. M. Cantor, *J. Amer. Chem. Soc.*, **74**, 5236 (1952).
71. E. F. L. J. Anet, *J. Chromatogr.*, **9**, 291 (1962).
71a. M. S. Feather and J. F. Harris, *Carbohyd. Res.*, **15**, 304 (1970).
72. E. F. L. J. Anet, *Tetrahedron Lett.*, 1649 (1966).
73. E. Votoček and S. Malachta, *Collection Czech. Chem. Commun.*, **4**, 87 (1932).
74. E. F. L. J. Anet, *Carbohyd. Res.*, **2**, 448 (1966).
75. F. Catala, J. Defaye, P. László and E. Lederer, *Bull. Soc. Chim. Fr.*, 3182 (1964).
76. P. A. Levene and T. Mori, *J. Biol. Chem.*, **83**, 803 (1929).
77. W. Alberda van Ekenstein and J. J. Blanksma, *Chem. Weekblad*, **6**, 717 (1909); **7**, 387 (1910); *Ber.*, **43**, 2355 (1910).
78. J. C. Sowden, *J. Amer. Chem. Soc.*, **71**, 3568 (1949).
79. R. Pummerer and W. Gump, *Ber.*, **56**, 999 (1923); R. Pummerer, O. Guyot, and L. Birkofer, *ibid.*, **68**, 480 (1935).
80. H. P. Teunissen, *Rec. Trav. Chim.*, **49**, 784 (1930); **50**, 1 (1930).
81. F. Leger and H. Hibbert, *Can. J. Res.*, **16B**, 68 (1938).
82. L. Birkofer and R. Dutz, *Ann.*, **608**, 7 (1957); K. G. Lewis, *J. Chem. Soc.*, 531 (1957); R. E. Deriaz, M. Stacey, E. G. Teece, and L. F. Wiggins, *ibid.*, 1222 (1949).
82a. E. Albano, D. Horton, and T. Tsuchiya, *Carbohyd. Res.*, **2**, 349 (1966); D. Horton and T. Tsuchiya, *ibid.*, **3**, 257 (1967); D. Horton and T. Tsuchiya, *Chem. Ind.* (London), 2011 (1966).
83. J. C. Sowden, *Advan. Carbohyd. Chem.*, **12**, 35 (1957).
84. R. L. Whistler and J. N. BeMiller, *Advan. Carbohyd. Chem.*, **13**, 289 (1958).
85. E. Peligot, *Compt. Rend.*, **7**, 106 (1838).
86. J. C. Sowden and E. K. Pohlen, *J. Amer. Chem. Soc.*, **80**, 242 (1958).
87. J. C. Sowden, M. G. Blair, and D. J. Kuenne, *J. Amer. Chem. Soc.*, **79**, 6450 (1957).
88. P. A. Shaffer and T. E. Friedeman, *J. Biol. Chem.*, **86**, 345 (1930).
89. G. N. Richards and H. H. Sephton, *J. Chem. Soc.*, 4492 (1957).
90. W. L. Evans, R. H. Edgar, and G. P. Hoff, *J. Amer. Chem. Soc.*, **48**, 2665 (1926).
91. A. Emmerling and G. Loges, *Ber.*, **16**, 837 (1883).
92. J. Groot, *Biochem. Z.*, **146**, 72 (1924).
93. R. Nodzu, K. Matsui, R. Goto, and S. Kunichika, *Mem. Coll. Sci. Kyoto Imp. Univ.* **20**, 197 (1937); *Chem. Abstr.*, **32**, 8365 (1938).
94. S. Otani, *Bull. Chem. Soc. Jap.*, **38**, 1873 (1965).
95. H. von Euler and C. Martius, *Svensk Kem. Tidskr.*, **45**, 73 (1933).
96. H. F. Bauer and C. Teed, *Can. J. Chem.*, **33**, 1824 (1955).
97. R. Nodzu and R. Goto, *Bull. Chem. Soc. Jap.*, **11**, 381 (1936).
98. T. Enkvist, *Acta Chem. Scand.*, **8**, 51 (1954).
99. H. O. L. Fischer and E. Baer, *Helv. Chim. Acta*, **19**, 519 (1936).
100. M. L. Wolfrom and J. N. Schumacher, *J. Amer. Chem. Soc.*, **77**, 3318 (1955).
101. O. Schmidt, *Chem. Rev.*, **17**, 137 (1935).

102. J. Kenner and G. N. Richards, *J. Chem. Soc.*, 278 (1954).
103. F. Weygand and G. Billek, *Z. Naturforsch.*, **12b**, 601 (1957).
104. J. Hayami, *Bull. Chem. Soc. Jap.*, **34**, 927 (1961).
105. J. Hayami, *Bull. Chem. Soc. Jap.*, **34**, 924 (1961).

5. CONFORMATIONS OF SUGARS

S. J. ANGYAL

I. INTRODUCTION

The term "conformation" was introduced into organic chemistry by Haworth[1] in his book "The Constitution of Sugars." He considered the various possible conformations of sugar molecules and predicted "that these considerations open up a large field of inquiry into the conformation of groups as distinct from structure or configuration." In 1929, the time was not yet ripe: there was insufficient theoretical knowledge, and the necessary data were not available. In more recent years, however, the study of the conformations of sugars in solution has acquired major importance.

All of the monosaccharides belong to a few families of diastereoisomers; within these families, they differ only in their steric arrangement, and these differences can therefore be accounted for by conformational factors. If these factors could be exactly evaluated, th edifferent chemical and physical proper-

195

ties of the various sugars could be predicted. Even without exact evaluation, semiquantitative application of conformational analysis can explain facts which, for many years, have been presented without any explanation: for example, why the aqueous equilibrium solution of D-lyxose consists mostly of the α-pyranose, but that of D-xylose of the β-pyranose form; why the aqueous solution of D-altrose contains substantial proportions of furanose forms, whereas in the solution of D-glucose they are not detectable by our present methods; why D-idose is extensively converted in acid solution into a 1,6-anhydride, whereas D-galactose is not. These facts are now readily explained by a study of the conformations involved.

II. DETERMINATION OF CONFORMATIONS

Six-membered ring systems (pyranoses) will be considered first, because most free sugars exist and react mainly in this form, and because the conformations of this ring system are clearly defined and readily recognized. The conformations of furanoses are discussed in Section V.

Conformational analysis of the pyranoid sugars is based on the assumption that the geometry of the pyranoid ring is the same as that of cyclohexane:[2] the substitution of two carbon–oxygen bonds for the slightly longer carbon–carbon bonds causes only a slight distortion of the regular cyclohexane structure, but the flexibility of the ring system is somewhat increased. Like substituted cyclohexane, the substituted pyranoid ring may be in one of two chair forms or in the less-stable, flexible forms (boat and skew forms). In solution, all of these forms coexist in equilibrium, but usually one or two conformers preponderate; the others may not be detectable by present techniques, but they may be the intermediates through which some reactions proceed.

The conformations of sugars have been determined by several methods. The pioneering work of Reeves[3] was carried out by studying the formation of complexes from sugars in cuprammonia solution. X-Ray crystallography, polarimetry, optical rotatory dispersion, and infrared spectroscopy have been used, but the best source of information is nuclear magnetic resonance (n.m.r.) spectroscopy. The conformations of the sugars have been reviewed several times.[4-7b]

A. Complexing with Cuprammonia

Reeves[3] determined the prevalent conformations of many methyl glycosides in aqueous solution by studying the complexes formed with cuprammonia reagent. This method, like several others that have been used for the same purpose, determines the relative positions of hydroxyl groups. The exact structure of the complexes is not known, but it is certain that two hydroxyl groups, mostly on adjacent carbon atoms, are involved. By studying the formation of complexes from compounds having a rigid conformation, Reeves found that a small dihedral angle between two participating hydroxyl groups (as in *cis*-tetrahydrofuran-2,3-diol) results in very strong complexing; a compound having a dihedral angle of 60° (cellulose) is also reactive, but with angles of 120° (2,5-anhydro-D-glucitol) and 180° (methyl 4,6-*O*-benzylidene-α-D-altropyranoside), no complexing occurs. Occasionally, complex formation was observed between two *syn*-axial hydroxyl groups (as in 1,6-anhydro-3-*O*-methyl-β-D-glucopyranose). Reeves concluded that, for complexing to occur, the distance between two participating oxygen atoms must be less than 3.45 Å.

Complex formation is observed as a decrease in the electrical conductivity of the cuprammonia solution; it also causes a change in the optical rotation, which yields further information on the relative positions of the oxygen atoms involved. If the hydroxyl groups are in a true *cis* relationship, or in a *syn*-axial arrangement, the rotational change is small, because the conformation of the molecule is not substantially altered on complex formation. However, if the dihedral angle is about 60°, formation of a copper-containing ring requires deformation of the molecule and causes a large rotational change, the sign of which is determined by the sign of the dihedral angle. Thus, in suitably protected derivatives of methyl β-D-glucopyranoside (**1**), complexing at the C-2, and C-3 hydroxyl groups causes a levorotatory shift, but at the C-3 and C-4 hydroxyl groups a dextrorotatory shift. When formation of both complexes is possible in the same molecule, the change in rotation may be small, but the decrease in conductivity indicates the extent of complex formation. Little or no complex formation occurs elsewhere in the methyl β-D-glucopyranoside molecule, as is shown by the nonreactivity of the 3-methyl ether. The behavior of these compounds clearly indicates that the β-D-glucopyranosides are in the *C1* (D) conformation.

Careful study of the D-glucopyranosides, D-galactopyranosides, and 1,6-anhydro-D-hexopyranosides showed that the boat form plays no significant role in the equilibrium of sugar conformations. On the assumption that only chair forms need be considered, the prevalent conformations of many glycopyranosides were determined, and most of them were found in the *C1* (D) or in the enantiomorphic *1C* (L) conformation. Only methyl α-D-idopyranoside and methyl 4,6-*O*-benzylidene-α-D-idopyranoside appeared to exist in the *1C* (D) conformation. The D-lyxo- and D-altro-pyranosides showed a behavior intermediate between those expected for the two chair forms, and it was concluded that both chair conformations contribute significantly to the equilibrium.

A glance at the two chair forms of methyl β-D-glucopyranoside (**1**) clearly shows why the *C1* conformation is favored; in this form, all the bonds to the

* *References start on p. 213.*

substituent groups are equatorial. The other extreme is shown in the two chair forms of methyl α-D-idopyranoside (**2**); here, the distribution of bonds to the substituent groups appears to be more favorable in the *1C* form. A more detailed consideration of this aspect is given in Section III,A.

1, *C1* (D)   **1**, *1C* (D)

2, *C1* (D)   **2**, *1C* (D)

In principle, against Reeves' method[3] the objection can be raised that formation of a complex, like other chemical reactions applied to similar systems, alters the equilibrium between the conformations. If the sugar can form a complex in one conformation only, the presence of that conformer will be indicated by the method, even though it is not normally the preponderant one. Thus Reeves' claim that the 3-methyl ether of methyl β-D-idopyranoside is in the *C1* form but its 4,6-benzylidene acetal exists in the *1C* form is not justified; the former can form a complex in the *C1* form only, and the latter in the *1C* form only.

In practice, however, the method works well. This is partly due to the fact that nonformation of a complex is important in Reeves' derivation of conformations, and the above criticism does not here apply. Most of Reeves' results have been confirmed by other methods.

Although the cuprammonia method was applied to methyl glycosides, the results are applicable to the free sugars in aqueous solution, because replacement of the hydroxyl group by the methoxyl group has only a small effect on the conformation.

B. X-RAY CRYSTAL STRUCTURE ANALYSIS

The most definite and most accurate method for determining the conformation of a sugar is by complete X-ray crystal structure analysis.[8] It is not necessarily true, however, that the conformation that occurs in a crystal is the one that is favored in solution; and although the method yields valuable information on bond lengths, angles, and interatomic distances, this information may not necessarily be valid for the molecules in solution.

So far, all of the pyranoid monosaccharides that have been studied by X-ray analysis, either as free sugars or as derivatives (and which are listed in Table I), have been found in the conformations postulated by Reeves. Further details of X-ray crystal structure analysis of carbohydrates and their derivatives are given in Vol. IB, Chapter 27.

C. INFRARED SPECTROSCOPY

Infrared spectroscopy can often yield valuable information on conformations;[9] for example, axial and equatorial hydrogen atoms at the anomeric position can be distinguished.[10] The main disadvantage of the method is that it is not readily applicable to aqueous solutions. The conformations of sugar derivatives that are soluble in carbon tetrachloride can, however, be studied by Kuhn's method,[11] which makes use of the strengths of the intramolecular hydrogen bonds involving hydroxyl groups in each molecule, to provide information about the distance between the hydroxyl group and the nearest oxygen atom. This method has been used for determining the conformations of several methyl 4,6-O-benzylidene-D-aldohexosides[12] in carbon tetrachloride solution. The same conformation would, however, not necessarily be favored in aqueous solution, since the hydrogen bond itself may alter the conformation (Section IV,B).

D. OPTICAL ROTATION

Whiffen[13] described a method for calculating the molecular rotations of cyclic compounds, especially carbohydrates, by assigning rotational increments to conformational units in the structures. This approach has been extended by Brewster[14] to apply to a wide range of acyclic and cyclic compounds. The values obtained are not always in close agreement with the experimental data, the deviation being as much as 40° in some cases, and it is not always possible to decide from the rotation which conformation is present.[15] However, when the calculated values for two alternative conformations differ considerably from each other, and one of them agrees well with the experimentally determined molecular rotation, it is justified to regard this as evidence for that conformation. For example, the calculated molecular rotation of methyl β-D-arabinopyranoside is −75° for the *C1*, and −400° for the *1C* conformation;

* *References start on p. 213.*

the observed value is −403°, indicating that the latter is the favored conformation. This method is applicable to aqueous solutions of the sugars, but, unfortunately, the rotation of the *IC* (D) conformation can at present be calculated only for the pentoses, as there are no authentic models from which to derive the numerical constants needed for the *IC* (D) or *CI* (L) forms of hexopyranoses.

Optical rotatory dispersion has recently shown promise as a method for determining the conformations of sugars.[16]

E. NUCLEAR MAGNETIC RESONANCE SPECTROSCOPY

In recent years, the most important information on the conformations of sugars has been obtained by nuclear magnetic resonance (n.m.r.) spectroscopy[17] (see Chapter 27). The value of this method for the determination of conformations was first shown for the acetates of inositols[18] and sugars,[19] but it is also applicable to free sugars.[20-22] The technique is very valuable not only because it can be applied to aqueous solutions, but also because it often permits the determination, in equilibrium solutions, of the conformation of anomers that cannot be isolated.

Two kinds of data given by n.m.r. spectroscopy yield information on conformations: the coupling constants and the chemical shifts. Of these two, the former are the more valuable. The coupling constants for hydrogen atoms on neighboring carbon atoms vary with the dihedral angle according to the Karplus equation:[23] $J = J_0 \cos^2 \phi + K$, where J is the coupling constant between two hydrogen atoms attached to adjacent carbon atoms at a dihedral angle of ϕ, and J_0 and K are constants. Although it is now realized that this equation is only approximate and that the coupling constants also depend on the nature, and even on the configuration, of substituents, the coupling constants can often distinguish clearly between two possible conformations—for example, between one having two axial and one having two equatorial hydrogen atoms in a six-membered ring.

The anomeric proton of aldoses, being on a carbon atom attached to two oxygen atoms, always resonates at lower field than the other ring protons, and its signal is clearly visible in the spectrum. The $J_{1,2}$ coupling constant of this one proton alone will distinguish between the two possible chair forms of those sugars in which H-1 and H-2 are *trans* to each other; if both protons are axial, the coupling is 7 to 8 Hz, if they are both equatorial, the coupling is 1 to 2 Hz. An intermediate value, as in α-D-altrose and α-D-lyxose, is an indication that both conformers are present in substantial proportions in equilibrium (see p. 201). The splitting of the signal for the anomeric proton therefore defines the conformations of β-D-allose, α-D-altrose, β-D-galactose, β-D-glucose, β-D-gulose, α-D-idose, α-D-mannose, α-D-talose, α-D-arabinose, α-D-lyxose, β-D-ribose, and β-D-xylose, and their enantiomorphs.

When H-1 and H-2 are *cis*, the coupling constant does *not* offer a clear distinction between conformations, although the coupling for *eq.*H-1,*ax.*H-2 is usually 2.5 to 3.5 Hz, and for *ax.*H-1,*eq.*H-2 1.0 to 1.5 Hz. If signals of other ring protons can be distinguished in the spectrum, or if their coupling constants can be determined by analysis of the spectrum, the conformation may be defined. The aldo-hexoses and -pentoses whose conformations have been ascertained by n.m.r. spectroscopy are listed in Table I; the conformations of numerous fully and partially substituted sugars and deoxy sugars have also been determined by this method.[7a, 17]

The assumption that coupling constants intermediate between those expected from each of the chair forms signifies a conformational mixture has been shown valid, for instance, the 220-MHz n.m.r. spectrum of tetra-*O*-acetyl-β-D-ribopyranose at −84° shows signals of both chair forms, whereas at room temperature only signals having time-averaged coupling constants are observed.[23a] The use of low-temperature n.m.r. spectroscopy has permitted determination of the conformational populations of several complete series of acylated aldopentopyranose derivatives.[7a, 23a]

The chemical shift of a proton is also characteristic of its environment and may yield information that defines the conformation. The early rule[18] that an equatorial proton will produce a signal to lower field than a chemically similar but axial proton appears to be generally valid, provided that the compounds compared are epimeric and both in the same conformation.[20] For D-pyranoses in the *C1* conformation, therefore, the anomeric proton of the α-D anomer resonates at lower field than that of the β-D-anomer. In the cases of D-ribose, D-altrose, and D-idose, where the opposite is true, the α-D forms are equilibrium mixtures of the *C1* and *1C* conformations.[22]

The chemical shifts of ring protons are strongly dependent on configurational changes at positions other than the neighboring ones. Axial protons, for example, are deshielded not only by neighboring but also by *syn*-axial hydroxyl groups. Lemieux and Stevens established simple empirical rules for predicting the chemical shifts of the ring protons in free[20] and acetylated[19] sugars; comparison of the predicted values for each chair conformation with the chemical shift found experimentally can sometimes distinguish between the conformations.

In the preceding discussions, it has been assumed that only the two chair forms of the pyranoses need to be considered. In no case has a monocyclic pyranose been shown to exist in other than a chair form. Nevertheless, the possibility cannot be excluded that some of the pyranose sugars exist predominantly in a skew form; α-D or L-altropyranose and β-D or L-idopyranose would have the most favorable skew forms. Evidence at present available does not definitely exclude this possibility, although penta-*O*-acetyl-α-D-altro-

* *References start on p. 213.*

pyranose has been shown to exist, in several solvents, in the *Cl* form;[24] the same is true for penta-*O*-acetyl-α-D-idopyranose,[24a] but the β-D anomer shows couplings intermediate between those anticipated for the two chair conformations.[24b] The n.m.r. spectrum provides definite proof of a chair form only if several coupling constants can be determined; this is the case with β-D-xylopyranose, α-D-arabinopyranose, β-D-ribopyranose, and α- and β-D-allopyranose.

III. CALCULATION OF CONFORMATIONAL FREE ENERGIES

The preponderance of one or other of the chair forms, as shown in Table I, can be explained by conformational analysis. Reeves[3] introduced arbitrarily chosen "instability factors," but better results are obtained by calculating the conformational free energies of the molecules[25] or the atomic overlap of non-bonded atoms.[26] Although these calculations, owing to the approximations involved, do not give accurate results, they can serve as a general guide to the behavior of pyranoid sugars, and they allow an approximate prediction of the predominant conformation, of the α:β ratio at equilibrium, etc. The calculation of conformational free energies will, therefore, be reviewed, and the results tabulated.

A. NONBONDED INTERACTION ENERGIES

All of the monosaccharides belong to a few families of diastereoisomers. The differences in their free energies are caused mainly by different interactions between nonbonded atoms; in a comparatively rigid system, such as the six-membered ring of pyranoses, the interaction energies can be evaluated and totaled.[25]

The calculations of nonbonded interactions are based on two assumptions: (1) that the pyranoid ring has the same geometry as that of cyclohexane, and (2) that the free energies of conformational isomers are additive functions of energy terms associated with the presence of nonbonded interactions—that is, that the occurrence of one interaction does not affect the magnitude of another one. Both of these assumptions are only approximations, but the results obtained seem to indicate that they introduce only minor errors.

The values of interaction energies used in these calculations have been obtained, in aqueous solution, from the equilibria of cyclitols with their borate complexes[27] and from equilibria of sugars.[28] The following values were used: interaction between an axial hydrogen atom and an axial oxygen atom, 0.45; between an axial hydrogen atom and an axial methyl (or hydroxymethyl) group, 0.9; between two axial oxygen atoms, 1.5; between an axial oxygen

TABLE I

PREPONDERANT CONFORMATIONS OF D-ALDOPYRANOSES IN AQUEOUS SOLUTION[a]

| Aldose | Conformation found by | | | | Calculated interaction energies, kcal mole^{-1} | |
	Cuprammonia complexing[b]	N.m.r.[c]	Other methods	Calculation	C1	1C
α-D-Allose		C1		C1	3.9	5.35
β-D-Allose		C1		C1	2.95	6.05
α-D-Altrose	C1, 1C	C1, 1C		C1, 1C	3.65	3.85
β-D-Altrose	C1	C1		C1	3.35	5.35
α-D-Galactose	C1	C1		C1	2.85	6.3
β-D-Galactose	C1	C1		C1	2.5	7.75
α-D-Glucose	C1	C1	C1[d]	C1	2.4	6.65
β-D-Glucose	C1	C1	C1[e]	C1	2.05	8.0
α-D-Gulose	C1			C1	4.0	4.75
β-D-Gulose		C1		C1	3.05	5.45
α-D-Idose	1C	C1, 1C		C1, 1C	4.35	3.85
β-D-Idose				C1	4.05	5.35
α-D-Mannose	C1	C1		C1	2.5	5.55
β-D-Mannose	C1	C1		C1	2.95	7.65
α-D-Talose		C1		C1	3.55	5.9
β-D-Talose				C1	4.0	8.0
α-D-Arabinose	1C	1C	1C[f]	1C	3.2	2.05
β-D-Arabinose	1C		1C[f,g]	C1, 1C	2.9	2.4
α-D-Lyxose	C1, 1C	C1, 1C		C1, 1C	2.05	2.6
β-D-Lyxose	C1, 1C	C1	C1[h]	C1	2.5	3.55
α-D-Ribose	C1	C1, 1C	C1, 1C[f]	C1, 1C	3.45	3.55
β-D-Ribose	C1	C1, 1C		C1, 1C	2.5	3.1
α-D-Xylose	C1	C1	C1[f]	C1	1.95	3.6
β-D-Xylose	C1	C1		C1	1.6	3.9

[a] The conformational symbols C1 and 1C refer to the D sugars. The given symbols are reversed (1C and C1) when L sugars are considered (see Chapter I).

[b] For the methyl glycosides, see R. E. Reeves, J. Amer. Chem. Soc., 71, 215 (1949); 72, 1499 (1950).

[c] R. U. Lemieux and J. D. Stevens, Can. J. Chem., 44, 249 (1966); S. J. Angyal and V. A. Pickles, Aust. J. Chem., 25, in press (1972).

[d] X-ray crystallography, T. R. R. McDonald and C. A. Beevers, Acta Crystallogr., 5, 654 (1952).

[e] X-ray crystallography of cellobiose, R. A. Jacobson, J. A. Wunderlich, and W. N. Lipscomb, Acta Crystallogr., 14, 598 (1961).

[f] Optical rotation of the methyl glycoside, D. H. Whiffen, Chem. Ind. (London), 964 (1956).

[g] X-ray crystallography, A. Hordvik, Acta Chem. Scand., 15, 16 (1961); S. H. Kim and G. A. Jeffrey, Acta Crystallogr., 22, 537 (1967).

[h] X-ray crystallography, A. Hordvik, Acta Chem. Scand., 15, 1781 (1961).

* References start on p. 213.

atom and an axial methyl group, 2.5; between two *gauche* oxygen atoms, 0.35; and between an oxygen atom and a methyl (or hydroxymethyl) group when they are *gauche* to each other, 0.45 kcal mole^{-1}. All other nonbonded interactions are believed to be negligible. These values were obtained in aqueous solutions at room temperature and are valid only under these conditions.

B. THE ANOMERIC EFFECT

Besides the interaction energies, another factor has to be taken into account in calculating the free energies of pyranoses. Most 1-substituted derivatives of α-D-glucose (and many other sugars) are more stable than the corresponding β-D anomers, although the former have the bond to the substituent in the axial position, and the latter, in the equatorial one. There appears to be an effect, first discussed by Edward[29] and named "anomeric effect" by Lemieux,[5, 30] which makes an equatorially attached group on the anomeric carbon atom less stable than it would be at other positions on the ring. An explanation of this effect was given[29] in terms of the unfavorable interaction between the equatorial carbon–oxygen dipole and the dipole formed by the resultant of the unshared electron pairs on the ring-oxygen atom. These dipoles form a small angle when the substituent is equatorial, and a large one when it is axial (Fig. 1). In accordance with this explanation, the anomeric effect varies inversely with the dielectric constant of the solvent[30a] and is greatest when the effective charge

FIG. 1. Anomeric effect: direction of dipole moments.

density on the substituent atom directly attached to C-1 is high.[31] Thus, the more stable anomer of the acetylated aldopyranosyl halides is the one in which the bond to the halogen atom adopts the axial orientation in the favored chair conformation.[32, 33] When a positively charged atom is attached to the anomeric carbon atom, as in the pyridinium glycosides, the direction of the dipole moment on the anomeric carbon atom will be opposite to the usual one. The resulting "reverse anomeric effect" will favor an equatorial, rather than an axial, orientation of the positively charged group.[34]

The anomeric effect of a hydroxyl group in aqueous solution is comparatively small; it can be readily evaluated by considering the differences between the free energies of anomers.[25] The free energy of α-D-glucopyranose should be 0.9 kcal mole^{-1} higher than that of the β-D anomer, owing to the presence of an axial anomeric hydroxyl group in the former. In fact, the equilibrium solution contains 36% of α-D-glucose and 64% of β-D-glucose, corresponding to a free-energy difference of only 0.35 kcal mole^{-1}. The difference between the two values, 0.55 kcal mole^{-1}, is due to the anomeric effect. A similar calculation for the D-mannopyranoses gives a difference of 1.0 kcal mole^{-1}, for the 2-deoxy-*arabino*-hexoses, 0.85 kcal mole^{-1} (Fig. 2). The magnitude of the anomeric

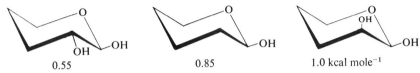

0.55   0.85   1.0 kcal mole^{-1}

FIG. 2. Values of anomeric effect in aqueous solution.

effect, therefore, depends on the presence and on the configuration of the hydroxyl group on C-2. The presence or absence of a hydroxyl group on C-6 (and probably in other positions) also modifies the anomeric effect,[30] but in aqueous solution this modification is small enough to be neglected. In these calculations, the anomeric effect will be taken as 0.55 kcal mole^{-1} when the oxygen atom on C-2 is attached equatorially, and as 1.0 kcal mole^{-1} when it is attached axially. These values are valid only for aqueous solutions: in other solvents, the anomeric effect is greater, but its exact value is not yet known.

Reeves[3] introduced the term "Δ2" to describe the configuration in which the hydroxyl group on C-2 is axial and the one on C-1 is equatorial (see the last formula in Fig. 2). He regarded the "Δ2 condition" as the most unfavorable interaction in pyranoses. The term is still often used, but it can be seen from the values in Fig. 2 that the "Δ2 condition" represents only a minor variation in the value of the anomeric effect. It would, therefore, be better to discontinue the use of this term.

C. CONFORMATIONAL FREE ENERGIES

To calculate the free energy of a sugar in a particular chair conformation, all of the interaction energies and the anomeric effect (if present) are totaled. The values listed in Table I for all of the aldohexoses and aldopentoses have been obtained in this way. The conformation that has the lower free energy is, of course, the more stable one. When the free energies of the two chair forms differ by 0.6 kcal mole^{-1} or less, both conformations are listed, since they would both be present in substantial amounts. The table shows good agreement between the calculated and the experimental data.

* *References start on p.* 213.

Besides indicating which conformation is preponderant, the free-energy values in the table also show the relative stability of the α-D and β-D anomers; the α:β ratio in equilibrium can be predicted approximately.[25] The relative stabilities of the different sugars are also indicated.

A different method for assessing the relative stabilities of the pyranose chair forms has been advanced by Barker and Shaw.[26] They calculated the total amount of atomic overlap of nonbonded atoms, taking into account the exact shape of the pyranose ring, but not the anomeric effect. These calculations lead, in most cases, to the same conclusions as those of Reeves[3] and of Angyal,[25] but the overlaps (expressed in Ångstrom units) are not readily translated into free-energy values and therefore cannot be used to predict the composition of equilibrium mixtures.

Conformational energies in good agreement with the values estimated in Table I have been determined by potential-energy calculations that take into account non-bonded and electrostatic interactions in the aldohexopyranoses and aldopentopyranoses.[34a]

IV. THE CONFORMATIONS OF PYRANOSES

A. IN AQUEOUS SOLUTION

As the data in Table I show, the conformations of the aldohexopyranoses in aqueous solution are governed mainly by the disposition of the hydroxymethyl group. This, being the bulkiest of the substituents, tends to assume the equatorial position. All of the β-D anomers (except, perhaps, β-D-idose) are predominantly in the $C1$ (D) form, because the $1C$ form involves a large interaction between the axial hydroxymethyl group and the axial anomeric hydroxyl group. This interaction is absent from the α-D anomers, but most of these also favor the $C1$ form; α-D-idose, however, exists mainly in the $1C$ (D) conformation, and aqueous solutions of α-D-altrose and, probably, α-D-gulose have substantial proportions of both chair forms in equilibrium.

As the hydroxymethyl group at C-5 is absent, the conformations of the aldopentopyranoses are governed by the dispositions of the hydroxyl groups, the anomeric effect favoring the conformation in which the anomeric hydroxyl group is axially attached. Thus the D-arabinoses favor the $1C$ (D) form, α-D-lyxose and α-D-ribose are each mixtures, and the other D-aldopentoses are mostly in the $C1$ (D) form.

Not much is known about the conformations of the ketohexopyranoses; n.m.r. spectroscopy offers no easy answers, because these compounds have no anomeric hydrogen atoms. Calculation of the conformational free energies indicates that here, again, the disposition of the hydroxymethyl group controls the conformation adopted.[4] It would then be inferred that only β-D-sorbose

(3) would be found mainly in that conformation [$C1$ (D)] in which the hydroxymethyl group is attached axially (all the hydroxyl groups then being attached equatorially); β-D-tagatose would be a mixture of the two chair forms, and the other ketohexoses would favor that conformer having the hydroxymethyl group equatorial—that is, β-D-fructose (4) and β-D-psicose would be in the $1C$ (D) and all the α-D anomers in the $C1$ (D) chair form.

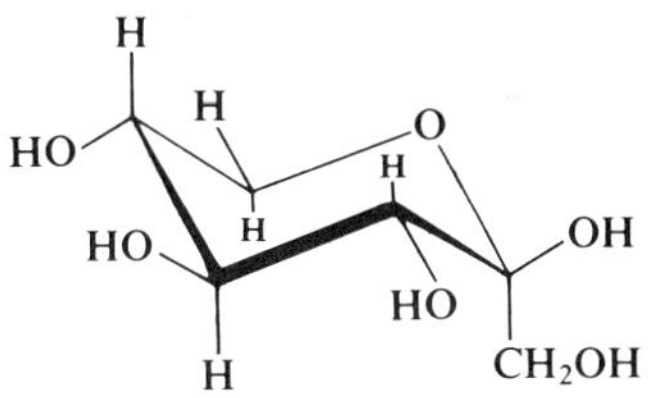

3

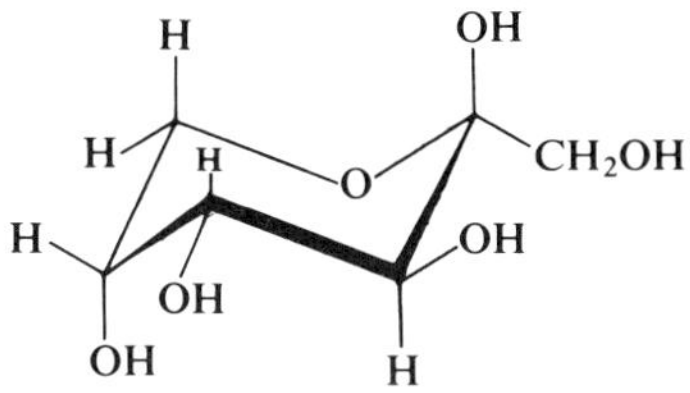

4

B. The Effect of Solvents and Substituents

A substituent, even a bulky one, separated by at least one atom from the six-membered ring has little effect on its conformation.[35] Hence, O-substitution in pyranoses should not markedly alter the conformational equilibrium. Substitution on the anomeric hydroxyl group affects the conformation, because it changes the magnitude of the anomeric effect. For example, in aqueous solution, the anomeric effect of a methoxyl or acetoxyl group is greater than that of a hydroxyl group, but quantitative data are not yet available. Replacement of the anomeric hydroxyl group by a halogen atom greatly increases the anomeric effect, whereas substitution by an amino group decreases it.[36] When a quaternary ammonium group is attached to the anomeric carbon atom, the "reverse anomeric effect" might be greater than the steric interactions. Thus, in aqueous solution, N-(tetra-O-acetyl-α-D-glucopyranosyl)-4-methylpyridinium bromide is found in the $1C$ form (5), with all of the acetoxy groups attached axially.[34]

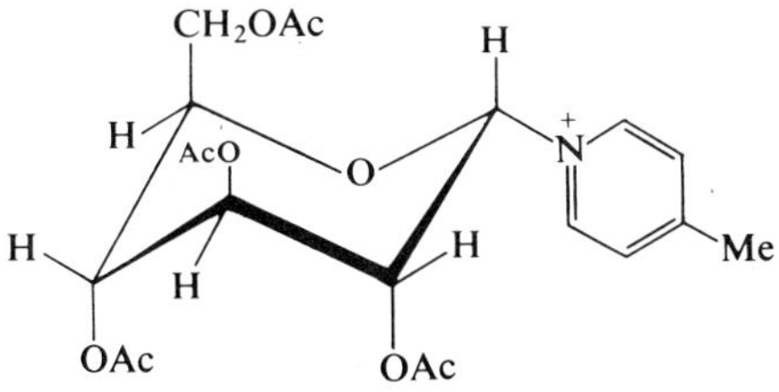

5

The nature of the solvent also has an effect on the conformation. The interaction energies (except those of hydroxyl and of amino groups) seem to vary

* *References start on p. 213.*

little from one solvent to another, but the anomeric effect increases as the dielectric constant of the solvent decreases. Thus, Lemieux and Chü[5, 30] estimated the following values for acetylated sugars in 1:1 acetic acid–acetic anhydride: interaction between an axial acetoxy group and an axial hydrogen atom, 0.18; between two *syn*-axial acetoxy groups, 2.08; between acetoxy groups on adjacent carbon atoms, 0.55; and the anomeric effect of an acetoxy group, 1.3 kcal mole^{-1} for pentopyranoses and 1.5 kcal mole^{-1} for hexopyranoses.

An increase in the anomeric effect shifts the equilibrium between the chair conformations toward the *C1* form in α-D anomers and toward the *1C* form in β-D anomers. Thus, α-D-lyxose consists of a higher proportion of the *C1* (D) form in methyl sulfoxide than in water.[37] The acetates of α-D-altrose, α-D-lyxose, and α-D-ribose, dissolved in chloroform, are predominantly in the *C1* form[19] whereas the free sugars in aqueous solution appear as mixtures of the two conformations. On the other hand, tetra-*O*-acetyl-β-D-ribopyranose in acetone is a mixture of chairs in which the *1C* form preponderates,[23a] whereas β-D-ribopyranose in aqueous solution is mainly in the *C1* form. The anomeric effect of a bromine atom is greater (probably over 2 kcal mole^{-1}) than that of an acetoxy group; n.m.r. spectroscopy shows that, in chloroform solution, the tri-*O*-acetyl-β-D-pentopyranosyl halides are predominantly in the *1C* conformation.[33] Even the tri-*O*-acetyl-β-D-xylopyranosyl halides are[7a, 23a, 38] in the *1C* form, with all substituents oriented axially (6).

X – Cl or I⁻

6

There is another way in which the solvent may affect the conformation. Intramolecular hydrogen bonding, as already stated, is negligible in hydrogen-bonding solvents; however, when a partially substituted sugar is dissolved in a nonbonding solvent, intramolecular hydrogen bonds may form, and their energy may stabilize a conformation which would otherwise be less favored. An example is methyl 4,6-*O*-benzylidene-α-D-idopyranoside; in aqueous solution it complexes[3] with cuprammonia in the *1C* conformation, but in dilute carbon tetrachloride solution the infrared spectrum shows hydrogen bonding between axial oxygen atoms,[12] which is possible only in the *C1* or a non-chair conformation. The n.m.r. spectra confirm this change in conformation.[39] Another example is methyl 2-deoxy-α-D-*erythro*-pentopyranoside, shown by its n.m.r. spectrum to be predominantly in the *C1* form (7) in chloroform, but mainly in the *1C* form (8) in water; presumably, hydrogen bonding from O-3 to O-1 occurs in the nonhydroxylic solvent.[40]

V. THE CONFORMATIONS OF FURANOSES

The conformations of furanoses are not as easily established as those of pyranoses. The furanoid ring is not planar, and either one or two of its atoms are out of the plane containing the others;[5] in order to avoid extensive eclipsing, one of the atoms is out of the plane formed by the other four (envelope form), or two adjacent atoms are on opposite sides of the plane formed by the other three (twist form). For example, the β-D-fructofuranose moiety of sucrose in the crystals of sucrose sodium bromide dihydrate[41] is in an envelope conformation (9), whereas the preponderant conformation of β-D-lyxofuranose[42] is a twist form (10). The different envelope and twist conformations are of similar free-energy content, and the barrier to their interconversion is much smaller than that between the two chair forms of pyranoses. Hence, it would be expected that, in solution, furanoses might be mixtures of these conformations.

The shape of the furanoid ring has been determined by X-ray crystallography in many derivatives of D-ribofuranose.[43] One definite conformation is found, of course, for each compound, but there is considerable variation in the shape

* *References start on p. 213.*

of the furanoid ring from compound to compound. Nuclear magnetic resonance spectroscopy has also been applied to the determination of furanoid conformations,[42,44-46] and sharp spectra, suggestive of a single conformation, have been obtained; it is possible, however, that in some cases these represent averages of several conformations.

Because eclipsing of a carbon atom and an oxygen atom requires less energy than that of two carbon atoms, the oxygen atom in the furanoid ring tends to occupy the least puckered part of the ring; hence, usually either C-2 or C-3, or both, will be out of the plane. In most cases, the conformation is one intermediate between the ideal envelope and twist forms, and is governed by the disposition of the substituents. Analysis[43] of the crystallographically determined conformations of β-D-ribofuranose derivatives shows that the average dihedral angles are: O:C-1, 8°; C-1:C-2, 29°; C-2:C-3, 37°; C-3:C-4, 32°; and C-4:O, 14°. Other furanoses have different dihedral angles.[42] The section of the ring containing C-2 and C-3 resembles the half-chair form of pyranoses, and the substituents occupy *quasi*-axial and *quasi*-equatorial positions. The preponderant conformation appears to be the one in which bulky substituents are staggered as much as possible, and assume *quasi*-equatorial positions on the most staggered carbon atoms. The substituents therefore tend to move into the plane of the furanose ring. On the other hand, owing to the operation of the anomeric effect,[42] the oxygen atom on C-1 tends to take up a *quasi*-axial position—that is, it tends to move out of the plane of the ring. Hence, *trans*-substituents at C-3 and C-4 usually form[42] a dihedral angle of less than 120°, those at C-1 and C-2, more than 120°. On the basis of these tendencies it is often possible to predict the favored conformation of a furanose, but a quantitative treatment, similar to that of the pyranoses, would probably be difficult.

The main interactions in furanoses are those between atoms or groups on adjacent carbon atoms, the strongest one being between the hydroxymethyl group and a neighboring *cis* hydroxyl group, on C-4 and C-3; hence, sugars of the *xylo* and *lyxo* configuration, which contain this interaction, do not readily form furanoses.[4] Two hydroxyl groups on adjacent carbon atoms are more stable when *trans* rather than *cis*; in equilibrium, therefore, the anomer having O-1 and O-2 *trans* predominates (α for the lyxoses and the arabinoses, β for the xyloses and the riboses). The free-energy difference between the anomers is 0.3 to 0.7 kcal mole^{-1}, depending on the configuration of the other substituents.[47] However, *cis*-1,3 interactions also play some destabilizing role; the two anomers of glucofuranose and of idofuranose are of approximately equal stability.[22] Accumulation of *cis* substituents probably leads to high free energies: the β-D or L-lyxofuranose configuration should be particularly unfavorable.[28] The conformational behavior of furanoses has been discussed in considerable detail by Lemieux, based on the simplifying assumption that they all assume an envelope conformation.[5]

VI. SUGARS HAVING FUSED RINGS AND UNSATURATED SUGARS

Fusion of another ring to a pyranose or furanose ring often restricts the number of possible conformations. For example, the 4,6-O-benzylidene derivatives of D-aldopyranoses in which O-4 and C-6 are *trans* (glucose, mannose, allose, and altrose) cannot exist in the $1C$ (D) form. In contrast to α-D-altrose, methyl 4,6-O-benzylidene-α-D-altropyranoside (**11**) exists wholly in the $C1$ (D) form;[48] the n.m.r. spectrum clearly excludes skew forms. The 1,6-anhydro-D-aldohexopyranoses cannot be in the $C1$ (D) form: the n.m.r. spectra are consistent with $1C$ (D) forms (**12**) but not with skew forms.[49]

11   **12**

In the above cases, there is no substantial deformation of the regular chair form. However, when a five-membered ring is fused to a pyranose ring, as in vicinal cyclic acetals, the ring cannot retain its undistorted chair form. At one time, it was assumed that the 1,3-dioxolane ring of the acetals was planar and that the four centers adjacent to the bridgehead of the dioxolane ring *cis*-fused to a pyranose ring would have to be coplanar; this would force six-membered sugar rings into a so-called half-chair or a boat conformation. Nuclear magnetic resonance spectroscopy, however, proved that the 1,3-dioxolane ring is *not* planar, the angle between the bridgehead hydrogen atoms[46,50-52] being about 40°. The pyranose ring, in most cases, remains in a chair form which is only slightly distorted.[53] Only in the case of 1,2-O-alkylidene acetals of α-D-gluco-pyranose was it claimed that the pyranoid ring is in a skew conformation;[52] in the crystalline state, however, one of these acetals was found to be in the $C1$ (D) form.[54] When there are two acetal rings *cis*-fused to a six-membered ring, however, the most stable form is a skew; this conformation was first established for some di-O-isopropylidene derivatives of inositols[55] but was later found[56] to be also predominant in derivatives of 1,2:3,4-di-O-isopropyli-dene-β-L-arabinopyranose (**13**).

The presence of a double bond in a six-membered ring is incompatible with a chair form, since it requires the coplanarity of four atoms. The so-called half-

* *References start on p. 213.*

chair form is the most stable conformation of cyclohexene, and the same form is found in unsaturated sugars. High-resolution n.m.r. techniques involving double-resonance experiments have shown that tri-*O*-acetyl-D-glucal adopts a conformation which approximates to a so-called half-chair conformation derived from the *Cl* (D) form.[57] The same half-chair form has been found to be the conformation of four hex-2-enopyranoses.[58] The anomeric effect is operative in these cases, too, and, in addition, the so-called "allylic effect" which causes a substituent in the allylic positions to favor (by 0.8 to 1.3 kcal mole^{-1}) the *quasi*-axial to the *quasi*-equatorial orientation.[59] As a result of the operation of the anomeric and the allylic effects, tetra-*O*-acetyl-3-deoxy-β-D-*erythro*-hex-2-enopyranose is, in chloroform solution, in the conformation (**14**) in which all substituent groups are axial or *quasi*-axial.[60]

Epoxides (and also episulfides and ethylenimines) usually have the same geometry as the corresponding unsaturated compounds. Several epoxides of pyranoses have been studied by n.m.r. spectroscopy and were found to adopt half-chair forms.[60a]

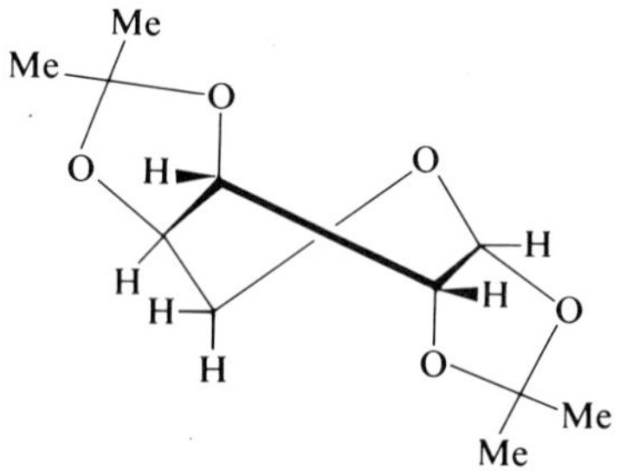

13 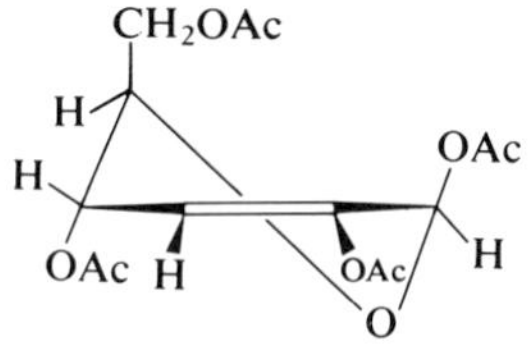　　　　　　　　　　　　　　　　　　14

VII. ACYCLIC SUGAR DERIVATIVES

The favored conformation of acyclic sugar derivatives, like that of other molecules having unbranched chains, has been assumed to be a planar zigzag arrangement of the carbon atoms, in which the largest groups along each carbon–carbon bond are in antiparallel disposition. X-ray crystallographic analysis has shown[61] that in the crystal the D-gluconate ion (**15**) is present as a nearly planar zigzag carbon chain with the seven oxygen atoms above and below

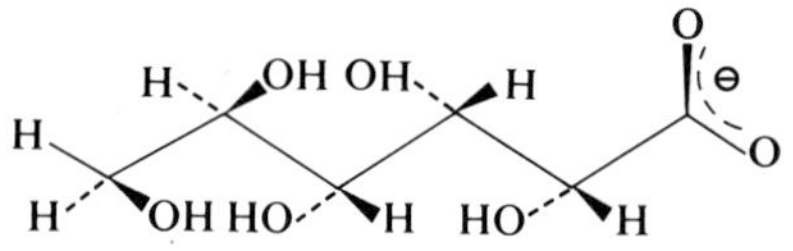

15

the plane formed by the carbon atoms. The L-arabinonate ion has a similar conformation.[62]

Data presented as evidence for the predominance of the zigzag conformation in solution have been provided by chemical studies. Formation of cyclic acetals gives mainly derivatives of the zigzag form,[63] although the outcome of this reversible reaction is governed by the stability of the products rather than by the conformation of the starting materials.[64] Similarly, periodate splits polyols preferentially between vicinal *threo*-hydroxyl groups (which appear as formally *trans* in the Fischer projection), as would be expected if the polyol reacted in the zigzag conformation;[65] but here, again, the relative stability of the intermediate cyclic periodate ester, and not the initial conformation, will determine the rate of the reaction.

Detailed, comparative investigations by n.m.r. spectroscopy and by X-ray crystallography have permitted considerable advances in our understanding of the conformations of acyclic sugar derivatives, both substituted and unsubstituted.[7a]

The n.m.r. spectra of several series of acyclic sugar derivatives have shown[66] that, in a five-carbon sequence, only the *arabino* and *lyxo* isomers are preponderantly in the planar zigzag form. In those cases (*xylo*, *ribo*) in which there are *erythro* substituents on two carbon atoms once removed from each other, parallel 1,3-interactions between these substituents decrease the stability of the zigzag form. In these cases rotation around one of the carbon–carbon bonds to a *gauche* arrangement results in the preponderance of a non-planar chain. The position of the bond around which rotation to the *gauche* conformation occurs depends on the relative size of the end-groups. In the crystalline state it has been shown[67] that the alditols all adopt the planar zigzag form unless a parallel, 1,3-interaction of oxygen groups would result. In such a case a more favorable conformation, formed by rotation about one carbon–carbon bond, is observed.

REFERENCES

1. W. N. Haworth, "The Constitution of Sugars," Arnold, London, 1929, p. 90.
2. O. Hassel and B. Ottar, *Acta Chem. Scand.*, **1**, 929 (1947).
3. R. E. Reeves, *J. Amer. Chem. Soc.*, **71**, 215 (1949); **72**, 1499 (1950); *Advan. Carbohyd. Chem.*, **6**, 107 (1951).
4. E. L. Eliel, N. L. Allinger, S. J. Angyal, and G. A. Morrison, "Conformational Analysis," Wiley (Interscience), New York, 1965, pp. 351–432.
5. R. U. Lemieux, in "Molecular Rearrangements," P. de Mayo, Ed., Wiley (Interscience), New York, 1963, pp. 713–769.
6. R. J. Ferrier and W. G. Overend, *Quart. Rev.* (London), **13**, 265 (1959); B. Capon and W. G. Overend, *Advan. Carbohyd. Chem.*, **15**, 11 (1960).
7. S. J. Angyal, *Angew. Chem.*, **81**, 172 (1969); *Angew. Chem. Intern. Ed.*, **8**, 157 (1969); H. S. Isbell and W. Pigman, *Advan. Carbohyd. Chem.*, **24**, 14 (1969).

7a. P. L. Durette and D. Horton, *Advan. Carbohyd. Chem. Biochem.*, **26**, 49 (1971).

7b. J. F. Stoddart, "Stereochemistry of Carbohydrates," Wiley—Interscience, New York, 1971.

8. For reviews see G. A. Jeffrey and R. D. Rosenstein, *Advan. Carbohyd. Chem.*, **19**, 7 (1964); G. Strahs, *Advan. Carbohyd. Chem. Biochem.*, **25**, 53 (1970).

9. For reviews, see W. B. Neely, *Advan. Carbohyd. Chem.*, **12**, 13 (1957); H. Spedding, *ibid.*, **19**, 23 (1964); see also Vol. IB, Chapter 27.

10. S. A. Barker, E. J. Bourne, M. Stacey, and D. H. Whiffen, *J. Chem. Soc.*, 171 (1954); S. A. Barker, E. J. Bourne, R. Stephens, and D. H. Whiffen, *ibid.*, 3468, 4211 (1954).

11. L. P. Kuhn, *J. Amer. Chem. Soc.*, **74**, 2492 (1952); **76**, 4323 (1954).

12. H. Spedding, *J. Chem. Soc.*, 3617 (1961).

13. D. H. Whiffen, *Chem. Ind.* (London), 964 (1956).

14. J. H. Brewster, *J. Amer. Chem. Soc.*, **81**, 5475, 5483 (1959).

15. For a general discussion, see ref. 4, pp. 384–390; see also R. U. Lemieux and J. C. Martin, *Carbohyd. Res.*, **13**, 139 (1970).

16. I. Listowsky, G. Avigad, and S. England, *J. Amer. Chem. Soc.*, **87**, 1765 (1965); N. Harada, H. Sato, and K. Nakanishi, *Chem. Commun.*, 1691 (1970).

17. For a review, see L. D. Hall, *Advan. Carbohyd. Chem.*, **19**, 51 (1964).

18. R. U. Lemieux, R. K. Kullnig, H. J. Bernstein, and W. G. Schneider, *J. Amer. Chem. Soc.*, **79**, 1005 (1957).

19. R. U. Lemieux and J. D. Stevens, *Can. J. Chem.*, **43**, 2059 (1965).

20. R. U. Lemieux and J. D. Stevens, *Can. J. Chem.*, **44**, 249 (1966).

21. M. Rudrum and D. F. Shaw, *J. Chem. Soc.*, 52 (1965).

22. S. J. Angyal and V. A. Pickles, *Aust. J. Chem.*, **25**, in press (1972).

23. M. Karplus, *J. Chem. Phys.*, **30**, 11 (1959); *J. Amer. Chem. Soc.*, **85**, 2870 (1963).

23a. P. L. Durette, D. Horton, and N. S. Bhacca, *Carbohyd. Res.*, **10**, 565 (1969); P. L. Durette and D. Horton, *ibid.*, **18**, 57, 289, 389, 403, 419 (1971); *J. Org. Chem.*, **36**, 2658 (1971).

24. B. Coxon, *Carbohyd. Res.*, **1**, 357 (1966).

24a. N. S. Bhacca, D. Horton, and H. Paulsen, *J. Org. Chem.*, **33**, 2484 (1968).

24b. N. S. Bhacca, D. Horton, and H. Paulsen, to be published.

25. S. J. Angyal, *Aust. J. Chem.*, **21**, 2737 (1968); S. J. Angyal and K. Dawes, *ibid.*, **21**, 2747 (1968).

26. G. R. Barker and D. F. Shaw, *J. Chem. Soc.*, 584 (1959).

27. S. J. Angyal and D. J. McHugh, *Chem. Ind.* (London), 1147 (1956).

28. S. J. Angyal, V. A. Pickles, and R. Ahluwahlia, *Carbohyd. Res.*, **1**, 365 (1966).

29. J. T. Edward, *Chem. Ind.* (London), 1102 (1955); see also S. Wolfe, A. Rauk, L. M. Tel, and I. G. Csizmadia, *J. Chem. Soc. B*, 136 (1971); R. U. Lemieux, *Pure Appl. Chem.*, **25**, 527 (1971).

30. R. U. Lemieux and N. J. Chü, *Abstr. Papers Amer. Chem. Soc. Meeting*, **133**, 31N (1958).

30a. See, however, R. U. Lemieux, A. A. Pavia, J. C. Martin, and K. A. Watanabe, *Can. J. Chem.*, **47**, 4427 (1969).

31. B. Coxon, *Tetrahedron*, **22**, 2281 (1966).

32. L. J. Haynes and F. H. Newth, *Advan. Carbohyd. Chem.*, **10**, 207 (1955).

33. D. Horton and W. N. Turner, *J. Org. Chem.*, **30**, 3387 (1965).

34. R. U. Lemieux and A. R. Morgan, *Can. J. Chem.*, **43**, 2205 (1965).

34a. V. S. R. Rao, K. S. Vijayalapshami, and P. R. Sundarajan, *Carbohyd. Res.*, **17**, 341 (1971).

35. See Ref. 4, p. 458.

36. B. Capon and B. E. Connett, *J. Chem. Soc.*, 4497 (1965).

37. W. Mackie and A. S. Perlin, *Can. J. Chem.*, **44**, 2039 (1966).
38. L. D. Hall and J. F. Manville, *Carbohyd. Res.*, **4**, 512 (1967); C. V. Holland, D. Horton, and J. S. Jewell, *J. Org. Chem.*, **32**, 1818 (1967).
39. S. J. Angyal, unpublished data (1968).
40. R. U. Lemieux and S. Levine, *Can. J. Chem.*, **42**, 1473 (1964).
41. C. A. Beevers and W. Cochran, *Proc. Roy. Soc.* (London), *Ser. A.*, **190**, 257 (1947).
42. J. D. Stevens and H. G. Fletcher, Jr., *J. Org. Chem.*, **33**, 1799 (1968).
43. M. Sundaralingam, *J. Amer. Chem. Soc.*, **87**, 599 (1965).
44. R. U. Lemieux, *Can. J. Chem.*, **39**, 116 (1961).
45. C. D. Jardetzky, *J. Amer. Chem. Soc.*, **82**, 229 (1960); **83**, 2919 (1961); **84**, 62 (1962).
46. R. J. Abraham, L. D. Hall, L. Hough, and K. A. McLaughlan, *J. Chem. Soc.*, 3699 (1962).
47. C. T. Bishop and F. P. Cooper, *Can. J. Chem.*, **41**, 2743 (1963).
48. B. Coxon, *Tetrahedron*, **21**, 3481 (1965).
49. L. D. Hall and L. Hough, *Proc. Chem. Soc.*, 382 (1962); K. Heyns and J. Weyer, *Ann.*, **718**, 224 (1968).
50. R. U. Lemieux, J. D. Stevens, and R. R. Frazer, *Can. J. Chem.*, **40**, 1955 (1962).
51. A. S. Perlin, *Can. J. Chem.*, **41**, 399 (1963).
52. B. Coxon and L. D. Hall, *Tetrahedron*, **20**, 1685 (1964).
53. L. D. Hall, L. Hough, K. A. McLauchlan, and K. G. R. Pachler, *Chem. Ind.* (London), 1465 (1962).
54. J. Trotter and J. K. Fawcett, *Acta Crystallogr.*, **21**, 366 (1966); R. G. Rees, A. R. Tatchell, and R. D. Wells, *J. Chem. Soc.* (*C*), 1768 (1967).
55. S. J. Angyal and R. M. Hoskinson, *J. Chem. Soc.*, 2991 (1962).
56. C. Cone and L. Hough, *Carbohyd. Res.*, **1**, 1 (1965).
57. L. D. Hall and L. F. Johnson, *Tetrahedron*, **20**, 883 (1964).
58. E. F. L. J. Anet, *Carbohyd. Res.*, **1**, 348 (1966).
59. R. J. Ferrier and G. H. Sankey, *J. Chem. Soc.* (*C*), 2345 (1966).
60. R. U. Lemieux and R. J. Bose, *Can. J. Chem.*, **44**, 1855 (1966); R. J. Ferrier, W. G. Overend, and G. H. Sankey, *J. Chem. Soc.*, 2830 (1965).
60a. J. G. Buchanan, R. Fletcher, K. Parry, and W. A. Thomas, *J. Chem. Soc.* (*B*), 377 (1969).
61. C. D. Littleton, *Acta Crystallogr.*, **6**, 775 (1953); G. A. Jeffrey and E. J. Fasiska, *Carbohyd. Res.*, **21**, 187 (1972).
62. S. Furberg and S. Helland, *Acta Chem. Scand.*, **16**, 2373 (1962).
63. S. A. Barker, E. J. Bourne, and D. H. Whiffen, *J. Chem. Soc.*, 3865 (1952).
64. J. A. Mills, *Advan. Carbohyd. Chem.*, **10**, 1 (1955).
65. J. C. P. Schwarz, *J. Chem. Soc.*, 276 (1957); D. H. Hutson and H. Weigel, *ibid.*, 1546 (1961).
66. H. S. El Khadem, D. Horton, and T. F. Page, Jr., *J. Org. Chem.*, **33**, 734 (1968); D. Horton and J. D. Wander, *Carbohyd. Res.*, **10**, 279 (1969); **15**, 271 (1971); J. B. Lee and B. F. Scanlon, *Tetrahedron*, **25**, 3413 (1969); S. J. Angyal and K. James, *Aust. J. Chem.*, **23**, 1223 (1970); P. L. Durette, D. Horton, and J. D. Wander, *Advan. Chem. Ser.*, in press (1972).
67. G. A. Jeffrey and H. S. Kim, *Carbohyd. Res.*, **14**, 207 (1970).

6. ESTERS

MELVILLE L. WOLFROM AND WALTER A. SZAREK

I. INTRODUCTION

The esters of the acyclic carbohydrate structures, such as the alditols and the aldehydo and keto forms of the reducing sugars, present no unusual facets. These derivatives involve primary and secondary alcohol groups that are unusual to the organic chemist only in their presenting an array of adjacent functions. The esters of the cyclic structures, which do offer special problems, will be emphasized. These esters again involve primary and secondary alcohol groups, but in addition they constitute substituents in heterocyclic rings. Tertiary alcohol groups are found in some of the branched-chain sugars. These groups predictably resist esterification, a finding encountered in the streptose moiety of streptomycin.[1] The unusual property of the sugar ring esters is the presence therein of a mixed function on the reducing carbon atom (C-1 of an aldose or C-2 of a ketose) consisting of an ester and an acetal (internally cyclized), for which the name esteral can be proposed. This function produces asymmetry and leads to anomerism. The chemical properties of the esteral ester

217

are different from those of the esters of the secondary and primary alcohol groups. The sulfonic esters of alcohols have some similarities to the carboxylic esters and have become very useful to the carbohydrate chemist.

Esterification is accomplished by treatment of the carbohydrate with an acid halide or an acid anhydride and a catalyst. The catalyst may be: (1) an acid, such as perchloric acid, sulfuric acid, trifluoroacetic anhydride, or zinc chloride; or (2) a base, such as pyridine or sodium acetate. Transesterification is occasionally employed. Esters of inorganic acids are described in Chapters 7 and 8.

II. REACTIVITY OF SACCHARIDE HYDROXYL GROUPS TOWARD ESTERIFICATION

A knowledge of the relative reactivities of hydroxyl groups in carbohydrates is of considerable value in synthesis.[2] In general, primary hydroxyl groups are esterified more readily than are secondary hydroxyl groups. The selective sulfonylation of primary hydroxyl groups is established. Thus, 6-p-toluenesulfonates of methyl α- and β-D-glucopyranosides,[3a] the 6,6'-di-p-toluenesulfonates of methyl β-cellobioside,[3b] and 1,6-di-O-p-tolylsulfonyl-D-fructose[4] have been prepared. Selective esterification at O-6 of methyl α-D-glucopyranoside can be effected by base-catalyzed transesterification from the methyl esters of long-chain fatty acids.[5] Secondary hydroxyl groups may also show some selectivity toward esterification. Methyl α-D-glucopyranoside has been converted into the 2,6-di-O-benzoyl,[6] 2,6-di-O-palmitoyl,[7] 2,6-di-O-acryloyl,[8] and 2,6-di-O-methylsulfonyl[9] derivatives. Methyl α-D-galactopyranoside shows a like reactivity toward p-toluenesulfonylation and methanesulfonylation. The enhanced reactivity of the C-2 hydroxyl group in α-D-glycosides toward sulfonylation has not been observed for the methyl β-D-glycosides of glucose and galactose,[10] which have equatorial, rather than axial, methoxyl groups at C-1, in the favored (C1) conformations. Thus, more than one factor controls the relative reactivities of the ring hydroxyl groups.

Differences in the conformational orientation (axial or equatorial) of certain hydroxyl groups may allow selective esterification. Axial substituents are sterically more crowded than equatorial substituents, and this steric factor influences reactivity. With simple alkyl-substituted cyclohexanols the equatorial alcohols are acetylated more readily than the axial alcohols.[11] Examples of such a difference toward esterifying agents have been furnished in the carbohydrate field. In compounds of fixed conformation, as 1,6-anhydro-2-O-benzoyl-β-D-altropyranose[12] (1) and methyl 4,6-O-ethylidene-α-D-mannopyranoside[13] (2), preferential sulfonylation occurred at the equatorial (C-3) positions.

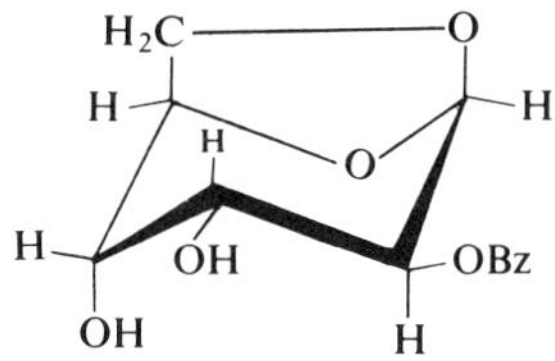

1

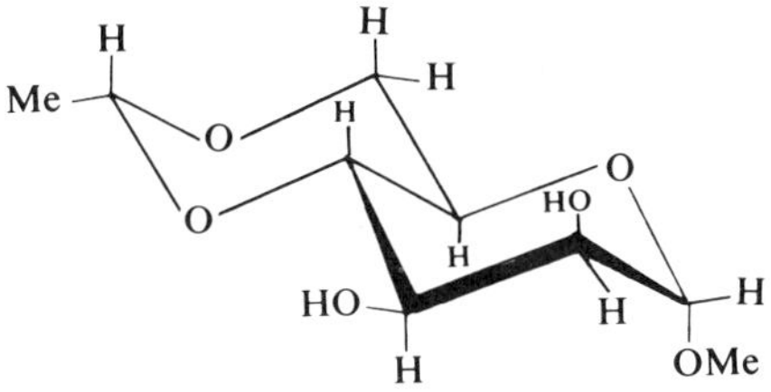

2

These observations indicate that steric factors can dominate the electronic effect of the anomeric center. The preferential benzoylation of equatorial, secondary hydroxyl groups in the presence of an axial secondary hydroxyl group has been described[14, 15] in the preparation of methyl 2,3,6-tri-*O*-benzoyl-α-D-galactopyranoside. In the favored *C1* conformation of methyl α-D-galactopyranoside (**3**), the C-4 hydroxyl group is axial, whereas the remaining secondary hydroxyl groups at C-2 and C-3 are equatorial. The preferential

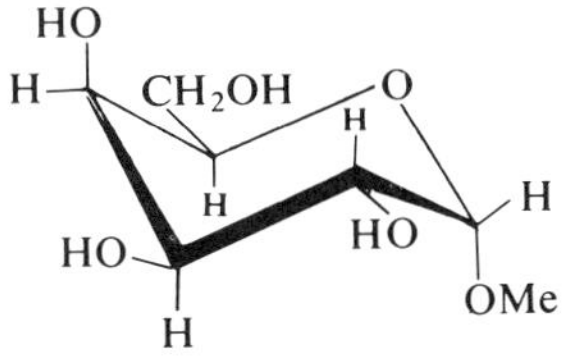

3

sulfonylation of an equatorial over an axial hydroxyl group was demonstrated by the selective sulfonylation of the 3-hydroxyl group of methyl 2-*O*-benzoyl-β-L-arabinopyranoside (**4**) to give the 3-*p*-toluenesulfonate[16] (**5**).

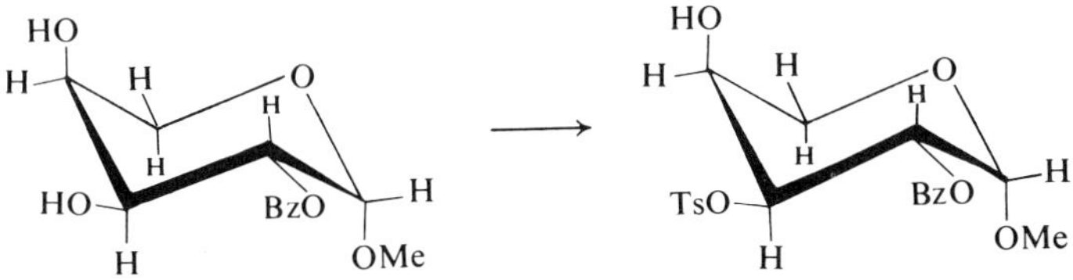

4 5

Methyl α-D-mannopyranoside in its favored conformation has the 2-hydroxyl group in an axial orientation. Partial benzoylation with 3.1 moles of benzoyl chloride gave methyl 2,3,6-tri-*O*-benzoyl-α-D-mannopyranoside and not the expected 3,4,6-isomer.[15] Clearly, the order of reactivity of the secondary hydroxyl groups of pyranosides is not dependent solely on whether a particular hydroxyl group is axial or equatorial, and inductive effects are certainly also a

* *References start on p.* 235.

factor. More examples must be investigated before complete generalizations are possible.

III. ACETATES

The acetyl derivatives of the sugars have been employed extensively as intermediates in sugar syntheses and for the isolation and identification of the sugars. Their value for these purposes arises from their ease of preparation and crystallization and from the fact that the acetyl groups are easily removed. In general, the methods evolved for the acetates may be adapted for other carboxylic esters. Thiolacetates (R–SAc) are known but are best considered as derivatives of thio sugars (Chapter 18).

A. FORMATION

1. *By Direct Esterification*

The hydroxyl groups of a carbohydrate react readily with acetic anhydride in the presence of an acidic catalyst or of a basic catalyst such as sodium acetate or pyridine. The acidic catalysts may be zinc chloride, hydrogen chloride, sulfuric acid, trifluoroacetic anhydride, perchloric acid, or an acidic ion-exchange resin.[17] Assuming that no tertiary hydroxyl groups are present, the complete acetylation of nonreducing sugars and other derivatives that consist of a single modification can be effected by any method that does not affect glycosidic linkages. Acetylation of the reducing sugars is complicated, however, by the fact that they exist in solution as equilibrium mixtures of tautomers. The isomer obtained depends on the catalyst used in the acetylation and on the temperature. Acetylation of α-D- or β-D-glucopyranose with acetic anhydride and pyridine[18] at 0° occurs without appreciable anomerization. The production of the acetylated β-D anomer is favored at higher temperatures with sodium acetate[19] as catalyst, an acetylating system first established with phenols.[20] Under these conditions β-D-glucopyranose pentaacetate is not isomerized to the more stable α-D anomer. The formation of β-D-glucopyranose pentaacetate is explained by the fact that the rate of acetylation is much lower than the rate of mutarotation of D-glucose, and also that the equatorial 1-hydroxyl group of β-D-glucopyranose undergoes acetylation considerably more rapidly than does the axial 1-hydroxyl group of α-D-glucopyranose. For acetylation of ketoses, low temperatures and acidic catalysts, especially zinc chloride, are favored; acyclic keto acetates are frequently obtained. Complete acetylation of D-fructose is best effected in two steps.[21] Acetylation with hot sodium acetate is not applicable to the sensitive ketoses. Acetylation of tertiary hydroxyl groups in branched-chain sugars requires forcing conditions.[21a]

The ring size of the cyclic acetates formed by the common acetylation procedures is normally pyranoid. In some cases, as with D-galactose,[22, 23] furanoid structures may be produced simultaneously. The publication of Hudson[22] is noteworthy in that it provided definitive evidence, on a crystalline basis, that a sugar could possess more than one ring form. Both anomeric forms of penta-O-acetyl-D-galactoseptanose are known also.[24] These seven-membered ring forms are prepared indirectly, however.

2. By Acetate Anomerization

This reaction was established as a preparative method by Erwig and Koenigs[25] and was correctly interpreted by Fischer.[26] It has been applied to the acetylated D-galactofuranoses.[27] Acetylated aldoses readily undergo anomeric interconversion,[28, 29] usually under acidic conditions, although acetylated aldoses have been rearranged with solid sodium hydroxide in an anhydrous, inert solvent.[30] The major product of the equilibrium mixtures is therefore the thermodynamically more stable anomer. The more stable anomeric form of the common aldohexopyranose peracetates is α (D or L) in each case. Thus, Jungius[31] reported that interconversion of the D-glucopyranose pentaacetates in acetic anhydride with zinc chloride as catalyst yielded an equilibrium mixture consisting of 90% of the α anomer and 10% of the β anomer. According to Painter,[32] the composition at equilibrium was approximately 87% of the α form and 13% of the β form, with either sulfuric or perchloric acid as catalyst, in nearly all mixtures of acetic acid and acetic anhydride.

The acetoxy group (or other polar group) at the anomeric center of an aldopyranose derivative tends to adopt the axial orientation (anomeric effect, see Chapters 5 and 7). In the aldohexopyranose peracetates, the acetoxy group in the α form has indeed the axial orientation. This effect has been studied in detail with the eight D-aldopentopyranose tetraacetates.[33] The effect on free hydroxyl groups in aqueous solution is small because of the high dielectric constant of water,[33a] and probably because of preferential solvation of equatorial hydroxyl groups.

It has long been known that sugars having 2,3-*trans*- and 2,3-*cis*-hydroxyl groups differ markedly in reactivity. The rates of exchange of 1,2-trans sugar acetates with ^{14}C-labeled stannic trichloride acetate in chloroform have been measured by Lemieux and Brice.[34] The sugars having *cis*-2,3-acetoxy groups (β-D-ribopyranose, α-D-lyxopyranose, β-D-allopyranose, and α-D-mannopyranose) reacted markedly more slowly than those in which these two groups are trans (β-D-xylopyranose, α-L-arabinopyranose, β-D-glucopyranose, and α-D-altropyranose). In the former group the formation of a cyclic acetoxonium ion (6) at C-1 and C-2 is hindered by the adjacent *cis*-acetoxy group (x) at C-3.

---

* *References start on p. 235.*

Cyclic acetoxonium ions of the type **6**, as generated by the action of antimony pentafluoride on sugar derivatives having a 1,2-*trans* diacetate group, have been studied extensively by Paulsen and coworkers.[34a] When an adjacent acetoxy group is trans to the acetoxonium ion (at *a* in **6**) it can attack with inversion at C-2 to give a new, cyclic, acetoxonium ion. Since this process can be repeated as long as another *trans*-acetoxy group is present, the reaction provides the opportunity for preparation of acetates differing in stereochemistry from the starting material. By this route α-D-idopyranose pentaacetate is readily prepared from β-D-glucopyranose pentaacetate.

6

3. From Glycosyl Halides

Mercuric acetate in acetic acid is an excellent reagent for replacing halogen in acetylated glycopyranosyl halides[35] (Chapter 7). The reaction has been utilized for the formation of β-D-acetates from the acetylated α-D-glycosyl halides. Wolfrom and Groebke[36] have reported, however, that treatment of tetra-*O*-acetyl-β-D-galactopyranosyl chloride and tetra-*O*-acetyl-β-D-glucopyranosyl chloride with the reagent yielded the β-D-glycopyranose pentaacetate. It appears then that the 1,2-*trans*-acetates are formed regardless of the anomeric nature of the glycosyl halide (see Chapter 7). Tetra-*O*-acetyl-β-D-galactofuranosyl chloride reacts similarly to form β-D-galactofuranose pentaacetate.[36]

B. DEACETYLATION

Acetates are hydrolyzed by both acids and bases but much more readily by bases. *O*-Deacetylation is generally effected by bases. Since the unsubstituted reducing sugars are sensitive to bases, special methods are required to obtain them from their acetates in pure form. The ketoses are especially base-sensitive, and the only suitable preparative method for them is saponification in the cold with aqueous barium hydroxide.[37] Protective complexes with barium hydroxide are probably formed with the hydroxyl groups. Acetylated aldoses may be saponified by various methods. The most useful of these are methanolic ammonolysis in the cold and transesterification with methanol containing catalytic amounts of sodium or barium methoxide. Acetylated amino sugars may be *N*-deacetylated with dilute hydrochloric acid (see Chapter 16).

C. ANALYSIS

It is of interest that Emil Fischer had no confidence in any group analysis for acetyl. The same distrust can still be shown toward most of the current microanalytical methods for the acetyl group. The macro or semimicro Kunz[38] method is a satisfactory alkaline saponification procedure; it consists in the dropwise addition of dilute, aqueous sodium hydroxide to a cold acetone solution of the acetate, maintenance in the cold for a short time, and back-titration with acid. This is suitable for O-acetyl but not for N-acetyl. Total O- and N-acetyl are evaluated by an acidic method, and thus N-acetyl may be determined by difference.[39] A recommended acidic method, suitable for micro techniques, is that of Chaney and Wolfrom.[40] Total acetyl can also be determined by n.m.r. spectroscopy (see Chapter 27).

IV. ORTHO ESTERS[41]

While not restricted to the acetates, these structures were first encountered in the reactions of acetylated glycosyl halides of the trans type[42] (see Chapter 7), and most of the known examples are 1,2-(orthoacetates). This structure was suggested by Freudenberg,[43] and supporting spectral evidence was provided by Braun.[43, 44] At about the same time, Haworth and co-workers[45] offered methylation evidence for such a structure. Like acetals, the ortho esters are resistant to alkali and are very acid-sensitive. Helferich and co-workers[46, 47] showed that a 1,2-cis glycosyl halide, tetra-O-acetyl-α-D-glucopyranosyl bromide (7), underwent reaction with an alcohol and sym-collidine to yield a 1,2-(orthoacetate) (9). Lemieux and Morgan[48] improved the yields in this reaction by the addition of tetraethylammonium bromide. They considered that the trans-β-D-halide (8) was an intermediate. By using a (methoxycarbonyl)oxy group on C-2 they obtained, with methanol, an interesting 1,2-(dimethyl orthocarbonate). The 1,2-(orthoacetate) structure is readily recognized by n.m.r. spectroscopy.[48, 48a]

An unusual method for obtaining ortho esters from 1,2-cis glycosyl halides has been reported.[49] Here the acetylated glycosyl halide was refluxed in ethanol and ethyl acetate with lead carbonate and calcium sulfate, followed by treatment at room temperature with silver carbonate and acetone.

Hydrogenolysis of ortho esters with lithium aluminum hydride–aluminum trichloride leads to replacement of the alkoxy group by hydrogen. This reaction, which presumably passes by way of an intermediate dioxolanium cation, provides a route to convert an ortho ester into an alkylidene (or arylidene) acetal.[49a]

* *References start on p. 235.*

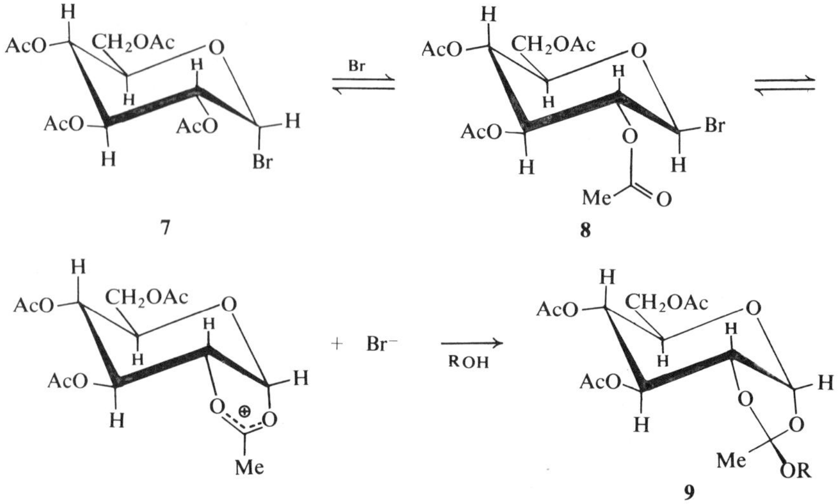

7

8

9

The ortho esters are useful starting compounds in the synthesis of glycosides (see Chapter 9), and this route of glycosylation has been successfully applied in the synthesis of oligo- and poly-saccharides.[49, 49b]

V. BENZOATES

Sugar benzoates have been widely used in a large variety of synthetic processes, since these esters are as easily obtained in crystalline form as are the acetates, and the benzoates are probably more stable than the acetates. Since benzoic anhydride is unreactive, fully benzoylated sugars are usually prepared by the use of benzoyl chloride in pyridine[50] (quinoline was first employed[51]) according to the general procedure of Behrend and Roth[18] for the acetylation of sugars. With a limited amount of reagent, primary hydroxyl groups are preferentially benzoylated.[5] Benzoylation of some sugars in hot pyridine may lead to the isolation of glycofuranose benzoates.[52] Crystalline benzoates of D-fructose have been prepared by Brigl and Schinle.[53, 54] These workers obtained 1,3,4,5-tetra-O-benzoyl-D-fructopyranose, 1,3,4,6-tetra-O-benzoyl-D-fructofuranose, and penta-O-benzoyl-keto-D-fructose, illustrative of the highly tautomeric nature of this ketose. These results would furthermore seem to indicate that the acyclic form is a substantial component of the complex equilibria present.

Prolonged treatment of benzoic esters in liquid hydrogen fluoride may produce partial debenzoylation, accompanied, in some instances, by Walden inversion or a change in ring size.[34a, 55] Acyloxonium-ion rearrangements of benzoates of sugars have been studied in detail.[34a]

The presence of the benzene ring in benzoates allows a number of aryl substituent variations to be prepared. Among these are the *p*-nitrobenzoates. Others are known. Phenylazobenzoates, *p*-PhN=NC$_6$H$_4$(C=O)OR, were the first derivatives to be used in the column chromatography of sugars.[56] The (*p*-benzamido)benzoyl[57] and (*p*-nitrophenylazo)benzoyl[58] esters have been recommended for the characterization of sugars.

A thionobenzoate, Ph(C=S)OR, substituted on the 2-hydroxyl group of a partially benzoylated anhydroalditol was prepared, as yellow crystals, from the acid chloride and pyridine.[59]

Several partially benzoylated sugars have been found in Nature. A mono-*O*-benzoyl-D-glucose (vaccinin) has been isolated from the juice of blueberries (*Vaccinium vitisidaea* L.). It was shown by Ohle[60] to be 6-*O*-benzoyl-D-glucose. Populin, which is found in the bark of a species of poplar (*Populus pyramidalis*), was demonstrated by Richtmyer and Yeakel[61] to be 6-*O*-benzoylsalicin (**10**). 1-*O*-Benzoyl-β-D-glucopyranose (periplanetin) has been found in the cockroaches *Periplanata americana* L. and *Blatta orientalis* L.[62] From *Daviesia latifolia* there was isolated a disaccharide dibenzoate.[63]

Populin (**10**)

VI. ESTERS OF OTHER CARBOXYLIC ACIDS

The sugars and their derivatives have been esterified with many other organic acids. Of considerable interest are the esters of trifluoroacetic acid.[64] Trifluoroacetylation may be effected[65] by warming the hydroxy compound with trifluoroacetic anhydride in the presence of dry sodium trifluoroacetate under anhydrous conditions. The trifluoroacetates are stable when pure and dry, but are hydrolyzed readily in the presence of water or anhydrous methanol.

Galloyl esters of D-glucose are related to the gallotannins, a group of hydrolyzable vegetable tannins. The gallotannins, including Turkish (*Quercus infectoria*, galls) and Chinese (*Rhus semialata*, galls) gallotannins and sumach tannins (*Rhus coriaria* and *typhina*), yield D-glucose and phenolic acids, predominately gallic, on hydrolysis with acid, alkali, or enzymes. By using hydrolytic, synthetic, and methylation procedures, it was concluded that

* *References start on p. 235.*

Chinese gallotannin was a mixture of isomers and closely related compounds whose average composition corresponded to that of a penta-O-(m-digalloyl)-β-D-glucose[66, 67] (11). Subsequent work[68] has shown that Chinese gallotannin

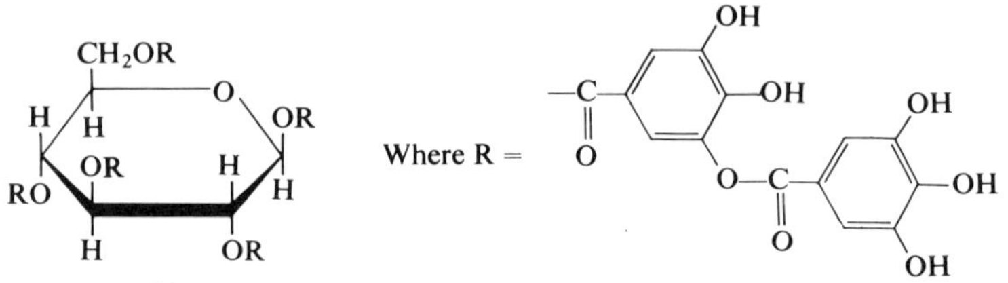

11

contains a penta-O-galloyl-β-D-glucose core to which at least three additional galloyl groups are attached by depside (R in 11) linkages. Evidence suggests[69] that the gallotannin can be formulated as a 1,3,4,6-tetra-O-galloyl-β-D-glucose with a m-trigalloyl chain (on an average) at the 2-position.

A significant finding was that pendunculagin, a tannin belonging to the ellagen group, contained D-glucose esterified at C-2,3 and C-4,6, respectively, with the dibasic (−)-hexahydroxydiphenic acid[70](12).

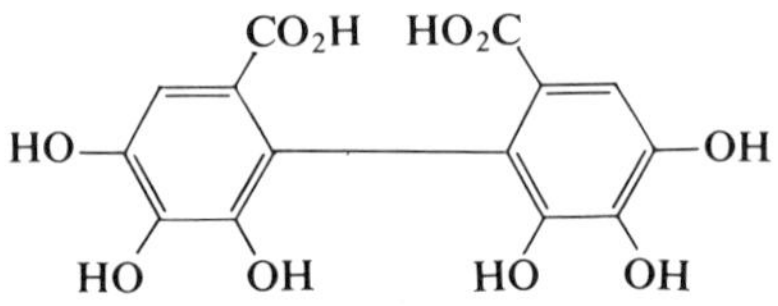

12

The preparation of fatty acid esters of carbohydrates has been reviewed.[71] Transesterification has been useful for this purpose.[5]

Polymerizable esters of sugars can be made by use of methylacrylic anhydride and pyridine.[72]

Carbanilates are excellent derivatives of carbohydrates, since they are, in general, easily formed and readily crystallized. Phenyl isocyanate in pyridine solution is commonly used for their preparation.[73, 74] The carbanilate group is

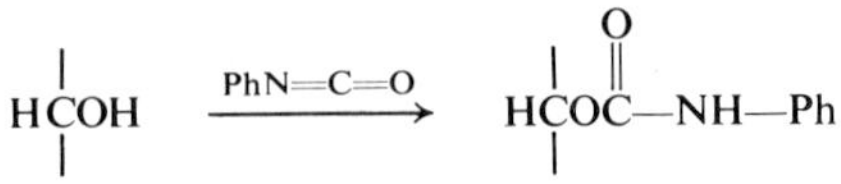

resistant to acid hydrolysis but is slowly and incompletely removed by alkali.[75] The group can be removed by sodium methoxide in methanol[76] or reductively by means of lithium aluminum hydride.[77]

Numerous fully and partially esterified derivatives of sucrose have been prepared and evaluated for possible commercial applications.[78]

VII. PARTIALLY ESTERIFIED SUGAR STRUCTURES

A number of examples of partially esterified sugars have been mentioned. The usual approach for their synthesis involves the use of removable protecting groups. Acid-sensitive protecting groups such as cyclic acetals (Chapter 11) and the 6-trityl ether (Chapter 12) have been widely used for this purpose. Hydrolysis of the acylated glycosyl halides leads to the acylated sugars having the anomeric hydroxyl group unsubstituted.

The (benzylthio)carbonyl group, $PhCH_2-S-CO-$, is a useful protecting group for the synthesis of partial esters.[79] The ester is stable in concentrated hydrochloric acid[80] and can be removed by oxidative cleavage with hydrogen peroxide in glacial acetic acid without cleavage of other ester groupings.[79] Treatment of 3-O-[(benzylthio)carbonyl]-1,2:5,6-di-O-isopropylidene-α-D-glucofuranose (13) with concentrated hydrochloric acid at room temperature gave crystalline 3-O-[(benzylthio)carbonyl]-β-D-glucopyranose[80] (14).

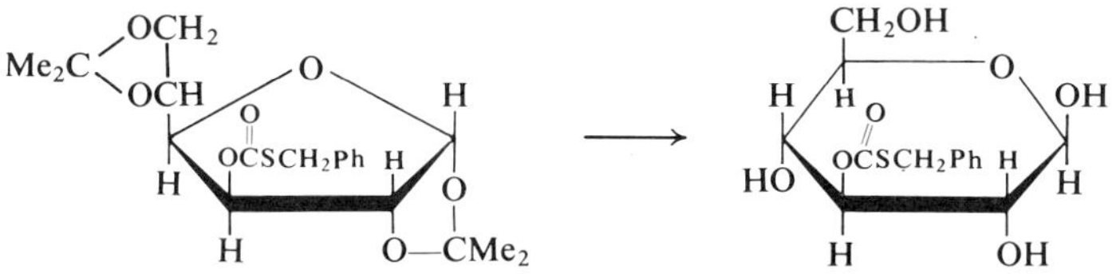

13   14

1-O-Acylaldoses have been synthesized by various methods. Zervas[81] obtained a sodium salt of 4,6-O-benzylidene-α-D-glucopyranose which, on treatment with benzoyl chloride and subsequent removal of the benzylidene group (by hydrogenolysis), yielded 1-O-benzoyl-β-D-glucopyranose.

Hydrolysis with methanolic ammonia of 2,3,4,6-tetra-O-acetyl-1-O-mesitoyl-β-D-glucose (15) gives 1-O-mesitoyl-β-D-glucopyranose[82] (16). The sterically hindered mesitoyl group is sufficiently stable to survive the deacetylation. The corresponding α-D anomer was prepared[83] by the action of silver mesitoate on ethyl 1-thio-β-D-glucopyranoside.

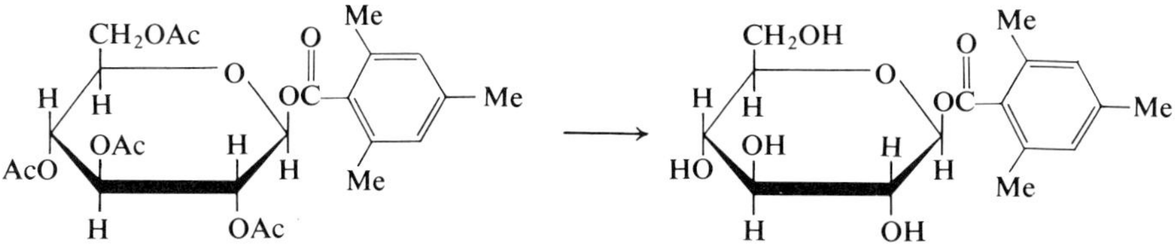

15   16

* *References start on p. 235.*

1-O-Benzoyl-α-D-talopyranose has been obtained by treatment of D-galactal with peroxybenzoic acid.[84]

Treatment[85] of methyl 2,6-dideoxy-α-D-*lyxo*-hexopyranoside with trimethyl orthoacetate in the presence of an acid catalyst, followed by hydrolysis of the resultant 3,4-O-methoxyethylidene derivative, gave a mixture of the 3- and 4-acetates in the approximate ratio 1:2.

VIII. MIGRATION OF ACYL GROUPS

Partially substituted esters are subject to intramolecular transesterification. Fischer[86] first observed an acyl group migration and postulated the mechanism currently held.[87] The migrations are considered to proceed through an intermediate cyclic orthoacid rather than by an actual hydrolysis and recombination of the wandering group. The reaction is acid–base catalyzed.

If C-1 is substituted, the migrating acetyl group tends to move down the chain. Thus, the methylation of 1,2,3,6-tetra-O-acetyl-β-D-glucopyranose with methyl iodide and silver oxide gave methyl tetra-O-acetyl-β-D-glucopyranoside.[88] If C-1 is not substituted, migration from C-2 to C-1 may occur (see Chapter 7). The reverse has also been noted.[89] It is generally considered that benzoyl groups do not migrate as readily as acetyl groups. The subject of acyl migration has been reviewed.[2]

The migration of acetyl groups can be monitored by n.m.r. spectroscopy if compounds specifically deuterated in individual acetyl groups are employed, because the signals for acetate groups at various positions can be differentiated and identified.[89a]

IX. MIXED ESTERS

Some examples of esters derived from more than one carboxylic acid have been described above. Partial esters may be further esterified with another acid. The acylated glycosyl halides may be replaced on C-1 with a different acyl group. Methyl α-D-glucopyranoside tetrakis(penta-O-acetyl-D-gluconate), $C_7H_{10}O_6[C_6H_6O_6(Ac)_5]_4$, has been described.[90] A 1,6′ ester condensation-polymer of low molecular weight has been prepared from 2,4:3,5-di-O-methylene-D-gluconic acid.[91]

X. CARBONATES AND THIOCARBONATES

The carbohydrate esters of carbonic and thiocarbonic acids have been applied widely in synthetic carbohydrate chemistry.[92] The principal methods of preparation of carbonates involve condensation of the carbohydrate with phosgene or with a chloroformic ester. The type of derivative obtained depends on the stereochemistry of the carbohydrate and the reaction conditions. When phosgene is used with a pyridine solution of carbohydrates, adjacent *cis*-hydroxyl groups normally react to give a cyclic carbonate. In some cases the

$$
\begin{array}{c}
| \\
HCO \\
| \\
HCO \\
|
\end{array}
\!\!\! >\!\! C\!\!=\!\!O
$$

carbohydrate changes from its normal pyranoid ring into the furanoid form to give dicarbonates. Thus, D-mannose yields D-mannofuranose 2,3:5,6-dicarbonate.[93] If it is sterically unfavorable for cyclic carbonates to be formed, as with *trans*-glycols fixed in a ring, and if the phosgene concentration is relatively low, intermolecular carbonates are generally formed.

Cyclic carbonates from *trans*-glycols fixed in a ring can be produced under forcing conditions; thus treatment of methyl 4,6-O-benzylidene-α-D-glucopyranoside with a large excess of ethyl chloroformate in the presence of triethylamine gives the 2,3-cyclic carbonate in good yield.[93a]

The direct synthesis of intermolecular carbonates from carbohydrate derivatives by the action of phosgene in pyridine has been reported. Thus, bis-(1,2,3,4-tetra-O-acetyl-β-D-glucopyranose) 6,6'-carbonate (18) was prepared by treatment of 1,2,3,4-tetra-O-acetyl-β-D-glucopyranose (17) with phosgene in pyridine.[94]

17   18

If phosgene is condensed with a carbohydrate in the presence of acetone, both a carbonate ester and an O-isopropylidene group may be introduced in a single step. Under these conditions D-xylose[95] and D-glucose[96] react in their

* References start on p. 235.

furanose forms to give 1,2-O-isopropylidene-α-D-xylofuranose 3,5-carbonate
and 1,2-O-isopropylidene-α-D-glucofuranose 5,6-carbonate, respectively,
whereas D-galactose[95] reacts in the pyranose form to give 6-O-(chloroformyl)-
1,2:3,4-di-O-isopropylidene-α-D-galactose. In all these cases the 1,2-O-iso-
propylidene acetal is formed in preference to the 1,2-carbonate.

The sugars react with chloroformic esters in the presence of pyridine to yield
alkoxycarbonyl derivatives; attack at other hydroxyl groups by further molecules of the alkyl chloroformate is favored over cyclization. Thus, treatment of
β-D-glucose in pyridine at 0° with methyl (or ethyl) chloroformate yields the
penta-O-(alkoxycarbonyl) derivative[97] (19).

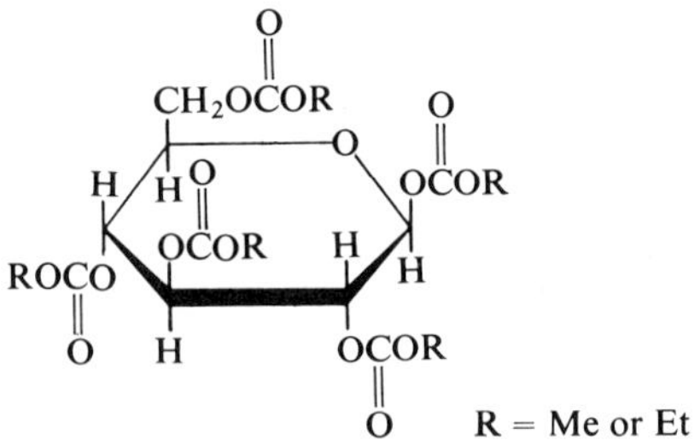

19

In the presence of aqueous alkali and chloroformic esters, mixed cyclic
carbonates and alkoxycarbonyl derivatives may be obtained.[98] Carbonates are
hydrolyzed easily by alkali but are relatively stable to acids. It is thus possible to
remove the acetone residue selectively from 1,2-O-isopropylidene-α-D-gluco-
furanose 5,6-carbonate with dilute acid.[96]

Aldgarose (**20**), a component of the macrolide antibiotic aldgamycin E, is a
cyclic carbonate of natural origin. The skeletal structure of this branched
3^2,4,6-trideoxyoctopyranose ester was deduced from nuclear magnetic resonance and mass spectra.[99] (See Chapter 31 for additional details.)

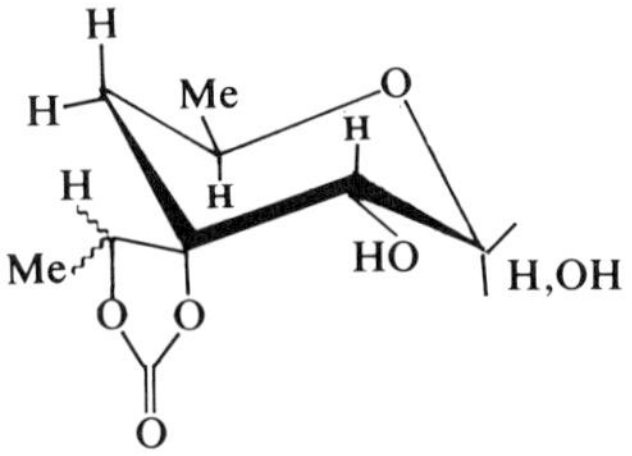

Aldgarose (**20**)

Of the types of carbohydrate thiocarbonates reported, the "xanthates" or
O-(metal thiolthiocarbonyl) derivatives have received the most attention.
These dithiocarbonates are prepared by use of carbon disulfide and alkali.

$$ROH + CS_2 + NaOH \longrightarrow RO\overset{\overset{\displaystyle S}{\|}}{C}S^-Na^+ + H_2O$$

The products are usually monosubstituted, although, by using high concentrations of alkali, or a tetraalkylammonium hydroxide,[100] more fully substituted derivatives have been obtained. Another method involves the action of carbon disulfide on sodium derivatives of carbohydrate alkoxides.[101] No reaction occurs in the total absence of water. The "xanthate" group migrates readily, especially from secondary to primary positions.[101a]

Treatment of the O-(metal thiolthiocarbonyl) compounds with methyl iodide yields O-[(methylthio)thiocarbonyl] derivatives (colorless).

$$ROH + NaH \xrightarrow{-H_2} RONa \underset{\longleftarrow}{\overset{CS_2}{\rightleftharpoons}}$$

$$RO\overset{\overset{\displaystyle S}{\|}}{C}S^-Na^+ + MeI \longrightarrow RO\overset{\overset{\displaystyle S}{\|}}{C}SMe + NaI$$

Repetition of the above reaction with alkyl halides higher than methyl led to the formation of the bis(thionocarbonate), presumably by reaction of some RONa still present because of the slower reaction of the higher halides with the O-(S-alkyl dithiocarbonate). The structure of the colorless product was estab-

$$RO\overset{\overset{\displaystyle S}{\|}}{C}SR' + RONa \longrightarrow RO\overset{\overset{\displaystyle S}{\|}}{C}OR + R'SNa$$

lished by its synthesis from thiophosgene.[102] O-[(Alkylthio)thiocarbonyl] esters of carbohydrates may rearrange, on pyrolysis at high temperature, to S-[(alkylthio)carbonyl] derivatives.[103] Thus, pyrolysis of 1,2:5,6-di-O-isopropylidene-3-O-[(methylthio)thiocarbonyl]-α-D-glucofuranose (21) gave rise to 1,2:5,6-di-O-isopropylidene-3-S-[(methylthio)carbonyl]-3-thio-α-D-glucofuranose (22).

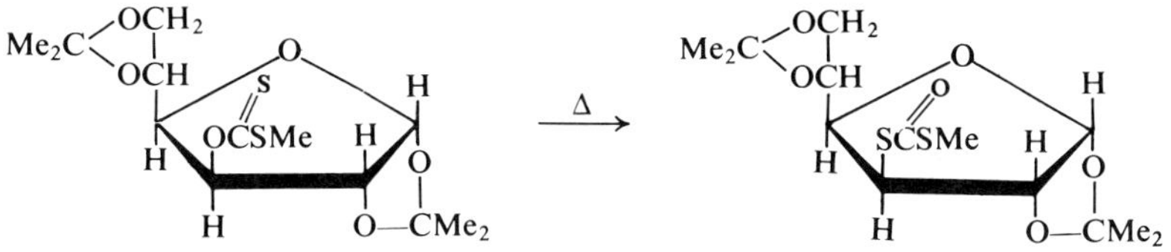

The thermal rearrangement presumably occurs by an SNi mechanism involving a four-membered cyclic intermediate (see Chapter 18), and it proceeds with retention of stereochemistry.[103a]

* *References start on p. 235.*

The N,N-dimethylthiocarbamates [RO(CS)NMe$_2$] of sugars are readily prepared by use of dimethylthiocarbamoyl chloride [Me$_2$N(CS)Cl] and pyridine, and resemble the O-[(methylthio)thiocarbonyl] derivatives in many respects.[103a]

The xanthates can be converted into their parent hydroxyl compounds by acidic or basic hydrolysis, by Raney nickel, or by oxidative fission.[87]

XI. SULFONATES

The sulfonic esters of the sugars are used extensively in synthetic work. The subject has been reviewed in detail.[104] A wide range of sulfonates, particularly aryl, is available to the chemist, but most of the work has been done with p-toluenesulfonic acid (a byproduct of saccharin manufacture), introduced by Freudenberg and Ivers.[105] The alkylsulfonic acids were for some time not readily available and had to be prepared for use. They are now commercially obtainable, and the simplest one, methanesulfonic acid,[106] has been widely utilized. The alkylsulfonic esters are perhaps more reactive than the aryl, particularly on secondary hydroxyl groups. In some respects the sulfonic acids resemble the carboxylic acids in reactivity, but the sulfonyloxy group more nearly resembles the halide, and this property is what makes these esters especially useful.

Esterification is customarily effected by the sulfonyl chloride in pyridine solution. As noted elsewhere (Chapter 7), prolonged action of these reagents may introduce chlorine at C-1 and C-6.

Removal of an O-arylsulfonyl group, without inversion, to regenerate the original sugar or sugar derivative can be effected in several ways. All these procedures are reductive in nature. Sodium amalgam and ethanol[107] cleaves an "isolated" (all other hydroxyl groups protected) sulfonic ester. Since this reaction involves alkali, anhydride formation will be a competing reaction if an unsubstituted hydroxyl group is sterically available. Raney nickel may replace[108] sodium amalgam with the same general results except that anhydro-ring formation, if possible, appears to be enhanced. Lithium aluminum hydride, in suitable organic solvents, is a valuable reagent for desulfonylation.[109] With primary sulfonates, reduction mainly to the hydrocarbon stage occurs. Secondary sulfonates are removed less readily, without inversion or C—O cleavage. This reagent thus operates, on secondary hydroxyl groups, principally by oxygen–sulfur fission and formation of a sulfinic acid, or, on primary hydroxyl groups, with carbon–oxygen fission and formation of a sulfonic acid.

The sulfonic esters are removable, without inversion, by bases. This is effected slowly with the O-arylsulfonyl group but is the preferred method for hydrolysis of a methanesulfonate.[110] A particularly effective, mild procedure for removal of O-p-tolylsulfonyl groups is by photolysis of the ester in methanolic solution, in the presence of an equimolar proportion of sodium methoxide.[110a]

$$\text{HCOSO}_2\text{R} \xrightarrow{\text{[H]}} \text{HCOH} + \text{RSO}_2\text{H}$$

$$\text{CH}_2\text{OSO}_2\text{R} \xrightarrow{\text{[H]}} \text{CH}_3 + \text{RSO}_2\text{OH}$$

A procedure useful in general organic chemistry is the nucleophilic SN2 replacement by acetate of a sulfonic ester by potassium acetate in boiling acetic anhydride. This reaction occurs with inversion of configuration. It is not always readily applicable with the secondary hydroxyl groups of the sugar heterocyclic ring. It was found, however, that sodium benzoate in N,N-dimethylformamide, on heating for 24 hours at 140°, would effect such a displacement.[111] This is a significant method for sugar interconversion, since the benzoate formed is readily convertible into the hydroxyl group.

Methyl 2,3-di-O-benzoyl-4,6-di-O-
p-tolylsulfonyl-α-D-galactoside

Methyl tetra-O-benzoyl-α-D-
glucopyranoside

A neighboring *trans*-benzoate is considered to accelerate the reaction by participation.[112]

Other SN2 nucleophilic displacements of isolated, secondary sulfonyloxy groups are known. These include displacements with hydrazine, ammonia, and azide. Displacement, with iodide, of secondary methylsulfonyloxy groups can be accomplished in some cases.[113]

Displacement of a primary sulfonyloxy group occurs more readily. As noted elsewhere (Chapter 7), a primary p-tolylsulfonyloxy group is replaced selectively by iodide. Alkylthio and thiocyanate groups can also be used in these primary-group displacement reactions, as well as the groups noted above that can replace secondary sulfonyloxy groups. Various nitrogen nucleophiles have been utilized as well as the acetylthio and benzylthio groups. The benzylthio group is especially noteworthy, as its reduction provides a synthesis for deoxy sugars.

As mentioned previously, the sulfonyloxy group undergoes attack under alkaline conditions by an unsubstituted hydroxyl group to give an anhydride. If the hydroxyl group is adjacent and configurationally trans, an epoxide ring is formed.[113a] If the epoxide ring is adjacent to another hydroxyl group it may be opened by the alkaline agent and re-formed with the new hydroxyl group, thus causing an essential migration of the epoxide ring. If a sterically available hydroxyl group is present in a nonadjacent position, a larger anhydro ring may be formed. If an adjacent hydroxyl group is configurationally cis, the sulfonic ester is saponified slowly without inversion. Reaction of the alkoxide in methyl sulfoxide can lead to the formation of an ether with retention of configuration.[114]

Many reactions of the sulfonyloxy group are accompanied by eliminations, which decrease yields. Contiguous primary and secondary sulfonyloxy groups in acyclic compounds are eliminated on heating with sodium iodide in a ketonic solvent.[115] Butadiene is formed from D-erythritol tetrakis(p-toluenesulfonate),[115] and iodine is liberated. The elimination probably involves nucleophilic

$$
\begin{array}{c}
\text{CH}_2\text{OTs} \\
| \\
\text{HCOTs} \\
| \\
\text{HCOTs} \\
| \\
\text{CH}_2\text{OTs}
\end{array}
\quad + \quad 4\text{NaI} \quad \longrightarrow \quad
\begin{array}{c}
\text{CH}_2 \\
\| \\
\text{CH} \\
| \\
\text{HC} \\
\| \\
\text{CH}_2
\end{array}
\quad + \quad 2\text{I}_2 \quad + \quad 4\text{NaOTs}
$$

displacement of the primary sulfonyloxy group, followed by a second attack by iodide ion on the iodine and elimination of the secondary sulfonyloxy group.[116] This reaction is also applicable to acyclic side chains as in the

$$
\begin{array}{c}
\text{I}^- \,\, \text{CH}_2 \text{—OTs} \\
| \\
\text{TsO—CH} \\
|
\end{array}
\quad \longrightarrow \quad
\begin{array}{c}
\text{CH}_2\text{—I} \,\, \text{I}^- \\
| \\
\text{TsO—CH} \\
|
\end{array}
\quad \longrightarrow \quad
\begin{array}{c}
\text{CH}_2 \\
\| \\
\text{CH} \\
|
\end{array}
\quad + \quad \text{I}_2
$$

formation of a 5,6-olefinic linkage **(24)** from 1,2-*O*-isopropylidene-5,6-di-*O*-
p-tolylsulfonyl-α-D-glucofuranose[110] **(23)**. A modification suitable for the

elimination of contiguous secondary sulfonyloxy groups in acyclic and cyclic
structures has been described.[117]

R = Ms or Ts

REFERENCES

1. R. U. Lemieux and M. L. Wolfrom, *Advan. Carbohyd. Chem.*, **3**, 361 (1948).
2. J. M. Sugihara, *Advan. Carbohyd. Chem.*, **8**, 1 (1953).
3a. J. Compton, *J. Amer. Chem. Soc.*, **60**, 395 (1938).
3b. J. Compton, *J. Amer. Chem. Soc.*, **60**, 1203 (1938).
4. W. T. J. Morgan and T. Reichstein, *Helv. Chim. Acta*, **21**, 1023 (1938).
5. G. N. Bollenback and F. W. Parrish, *Carbohyd. Res.*, **17**, 431 (1971).
6. T. Lieser and R. Schweizer, *Ann.*, **519**, 271 (1935).
7. J. Asselineau, *Bull. Soc. Chim. Fr.*, 937 (1955).
8. Z. Jedliński, *Bull. Acad. Polon. Sci., Ser. Sci. Chim.*, **9**, 103 (1961).
9. A. K. Mitra, D. H. Ball, and L. Long, Jr., *J. Org. Chem.*, **27**, 160 (1962).
10. R. C. Chalk, D. H. Ball, and L. Long, Jr., *J. Org. Chem.*, **31**, 1509 (1966).
11. E. L. Eliel and C. A. Lukach, *J. Amer. Chem. Soc.*, **79**, 5986 (1957).
12. F. H. Newth, *J. Chem. Soc.*, 441 (1956).
13. G. O. Aspinall and G. Zweifel, *J. Chem. Soc.*, 2271 (1957).
14. E. J. Reist, R. R. Spencer, D. F. Calkins, B. R. Baker, and L. Goodman, *J. Org. Chem.*, **30**, 2312 (1965).
15. A. C. Richardson and J. M. Williams, *Tetrahedron*, **23**, 1641 (1967).
16. E. J. Reist, L. V. Fisher, and D. E. Gueffroy, *J. Org. Chem.*, **31**, 226 (1966).
17. G. M. Christensen, *J. Org. Chem.*, **27**, 1442 (1962).
18. R. Behrend and P. Roth, *Ann.*, **331**, 359 (1904).
19. A. P. N. Franchimont, *Ber.*, **12**, 1940 (1879).
20. C. Liebermann and O. Hörmann, *Ber.*, **11**, 1618 (1878).
21. C. S. Hudson and D. H. Brauns, *J. Amer. Chem. Soc.*, **37**, 1283, 2736 (1915).
21a. G. B. Howarth, W. A. Szarek, and J. K. N. Jones, *Carbohyd. Res.*, **7**, 284 (1968); *Can. J. Chem.*, **46**, 3375 (1968).

22. C. S. Hudson, *J. Amer. Chem. Soc.*, **37**, 1591 (1915).
23. H. H. Schlubach and V. Prochownick, *Ber.*, **63**, 2298 (1930).
24. F. Micheel and F. Suckfüll, *Ann.*, **502**, 85 (1933).
25. E. Erwig and W. Koenigs, *Ber.*, **22**, 2207 (1889).
26. E. Fischer, *Ber.*, **26**, 2400 (1893).
27. C. S. Hudson and J. M. Johnson, *J. Amer. Chem. Soc.*, **38**, 1223 (1916).
28. R. U. Lemieux, *Advan. Carbohyd. Chem.*, **9**, 1 (1954).
29. B. Capon and W. G. Overend, *Advan. Carbohyd. Chem.*, **15**, 11 (1960).
30. M. L. Wolfrom and D. R. Husted, *J. Amer. Chem. Soc.*, **59**, 364 (1937).
31. C. L. Jungius, *Z. Physik. Chem.*, **52**, 101 (1905).
32. E. P. Painter, *J. Amer. Chem. Soc.*, **75**, 1137 (1953).
33. P. L. Durette and D. Horton, *J. Org. Chem.*, **36**, 2658 (1971).
33a. E. L. Eliel, N. L. Allinger, S. J. Angyal, and G. A. Morrison, "Conformational Analysis," Wiley (Interscience), New York, 1965, p. 376.
34. R. U. Lemieux and C. Brice, *Can. J. Chem.*, **34**, 1006 (1956).
34a. H. Paulsen, *Advan. Carbohyd. Chem. Biochem.*, **26**, 127 (1971).
35. B. Lindberg, *Acta Chem. Scand.*, **3**, 1355 (1949).
36. M. L. Wolfrom and W. Groebke, *J. Org. Chem.*, **28**, 2986 (1963).
37. C. S. Hudson and D. H. Brauns, *J. Amer. Chem. Soc.*, **38**, 1216 (1916).
38. A. Kunz and C. S. Hudson, *J. Amer. Chem. Soc.*, **48**, 1982 (1926).
39. M. L. Wolfrom, M. Konigsberg, and S. Soltzberg, *J. Amer. Chem. Soc.*, **58**, 490 (1936).
40. A. Chaney and M. L. Wolfrom, *Anal. Chem.*, **28**, 1614 (1956).
41. E. Pacsu, *Advan. Carbohyd. Chem.*, **1**, 77 (1945).
42. E. Fischer, M. Bergmann, and A. Rabe, *Ber.*, **53**, 2362 (1920).
43. K. Freudenberg and E. Braun, *Naturwissenschaften*, **18**, 393 (1930).
44. E. Braun, *Ber.*, **63**, 1972 (1930).
45. H. G. Bott and W. N. Haworth, *J. Chem. Soc.*, 1395 (1930).
46. B. Helferich, A. Doppstadt, and A. Gottschlich, *Naturwissenschaften*, **40**, 441 (1953).
47. B. Helferich and K. Weis, *Ber.*, **89**, 314 (1956).
48. R. U. Lemieux and A. R. Morgan, *Can. J. Chem.*, **43**, 2199 (1965).
48a. N. E. Franks and R. Montgomery, *Carbohyd. Res.*, **6**, 286 (1968).
49. A. Y. Khorlin, A. F. Bochkov, and N. K. Kochetkov, *Izv. Akad. Nauk SSSR, Ser. Khim.*, 2214 (1964); *Chem. Abstr.*, **52**, 9215 (1965); N. K. Kochetkov, A. Y. Khorlin, and A. F. Bochkov, *Tetrahedron*, **23**, 693 (1967).
49a. S. S. Bhattacharjee and A. S. Perlin, *Carbohyd. Res.*, **12**, 57 (1970).
49b. N. K. Kochetkov and A. F. Bochkov, *Carbohyd. Res.* **9**, 61 (1969).
50. P. A. Levene and G. M. Meyer, *J. Biol. Chem.*, **76**, 513 (1928).
51. E. Fischer and K. Freudenberg, *Ber.*, **45**, 2724 (1912).
52. H. G. Fletcher, Jr., *J. Amer. Chem. Soc.*, **75**, 2624 (1953).
53. P. Brigl and R. Schinle, *Ber.*, **66**, 325 (1933).
54. P. Brigl and R. Schinle, *Ber.*, **67**, 754 (1934).
55. C. Pedersen and H. G. Fletcher, Jr., *J. Amer. Chem. Soc.*, **82**, 941, 945 (1960).
56. W. S. Reich, *Compt. Rend.* **208**, 589, 748 (1939).
57. J. Kiss, *Chem. Ind.* (London), 32 (1964).
58. El S. Amin, *J. Chem. Soc.*, 5544 (1961).
59. E. J. Hedgley and H. G. Fletcher, Jr., *J. Org. Chem.*, **30**, 1282 (1965).
60. H. Ohle, *Biochem. Z.*, **131**, 611 (1922).
61. N. K. Richtmyer and E. H. Yeakel, *J. Amer. Chem. Soc.*, **56**, 2495 (1934).
62. A. Quilico, F. Piozzi, M. Pavan, and E. Mantica, *Tetrahedron*, **5**, 10 (1959).

63. B. Hansson, I. Johansson, and B. Lindberg, *Acta Chem. Scand.*, **20**, 2358 (1966).
64. T. G. Bonner, *Advan. Carbohyd. Chem.*, **16**, 59 (1961).
65. E. J. Bourne, C. E. M. Tatlow, and J. C. Tatlow, *J. Chem. Soc.*, 1367 (1950).
66. E. Fischer, *Ber.*, **52**, 809 (1919).
67. K. Freudenberg, "Tannin, Cellulose and Lignin," Springer, Berlin, 1933, p. 38.
68. R. Armitage, G. S. Bayliss, J. W. Gramshaw, E. Haslam, R. D. Haworth, K. Jones, H. J. Rogers, and T. Searle, *J. Chem. Soc.*, 1842 (1961).
69. G. Britton, P. W. Crabtree, E. Haslam, and J. E. Stangroom, *J. Chem. Soc. (C)*, 783 (1966).
70. O. Th. Schmidt, L. Würtele, and A. Harreus, *Ann.*, **690**, 150 (1965).
71. G. R. Ames, *Chem. Rev.*, **60**, 541 (1960).
72. R. H. Treadway and E. Yanovsky, *J. Amer. Chem. Soc.*, **67**, 1038 (1945).
73. L. Maquenne and W. Goodwin, *Bull. Soc. Chim. Fr.*, **31** [3], 430 (1904).
74. M. L. Wolfrom and D. E. Pletcher, *J. Amer. Chem. Soc.*, **62**, 1151 (1940).
75. W. M. Hearon, G. D. Hiatt, and C. R. Fordyce, *J. Amer. Chem. Soc.*, **65**, 833 (1943).
76. M. R. Salmon and G. Powell, *J. Amer. Chem. Soc.*, **61**, 3507 (1939).
77. R. L. Whistler and D. G. Medcalf, *Arch. Biochem. Biophys.*, **104**, 150 (1964).
78. V. Kollonitsch, "Sucrose Chemicals," International Sugar Research Foundation, Inc., 1970.
79. J. J. Willard and E. Pacsu, *J. Amer. Chem. Soc.*, **82**, 4347 (1960); J. J. Willard, *Can. J. Chem.*, **40**, 2035 (1962).
80. J. J. Willard, J. S. Brimacombe, and R. P. Brueton, *Can. J. Chem.*, **42**, 2560 (1964).
81. L. Zervas, *Ber.*, **64**, 2289 (1931).
82. H. B. Wood, Jr. and H. G. Fletcher, Jr., *J. Amer. Chem. Soc.*, **78**, 207 (1956).
83. C. Pedersen and H. G. Fletcher, Jr., *J. Amer. Chem. Soc.*, **82**, 3215 (1960).
84. H. B. Wood, Jr. and H. G. Fletcher, Jr., *J. Amer. Chem. Soc.*, **79**, 3234 (1957).
85. J. S. Brimacombe and D. Portsmouth, *Carbohyd. Res.*, **1**, 128 (1965).
86. E. Fischer, *Ber.*, **53**, 1621 (1920).
87. A. P. Doerschuk, *J. Amer. Chem. Soc.*, **74**, 4202 (1952).
88. B. Helferich and W. Klein, *Ann.*, **455**, 173 (1927).
89. H. B. Wood, Jr., and H. G. Fletcher, Jr., *J. Amer. Chem. Soc.*, **78**, 2849 (1956).
89a. D. Horton and J. H. Lauterbach, *J. Org. Chem.*, **34**, 86 (1969).
90. M. L. Wolfrom and P. W. Morgan, *J. Amer. Chem. Soc.*, **64**, 2026 (1942).
91. C. L. Mehltretter and R. L. Mellies, *J. Amer. Chem. Soc.*, **77**, 427 (1955).
92. L. Hough, J. E. Priddle, and R. S. Theobald, *Advan. Carbohyd. Chem.*, **15**, 91 (1960).
93. W. N. Haworth and C. R. Porter, *J. Chem. Soc.*, 649 (1930).
93a. W. M. Doane, B. S. Shasha, E. I. Stout, C. R. Russell, and C. E. Rist, *Carbohyd. Res.*, **4**, 445 (1967); **11**, 321 (1969).
94. D. D. Reynolds and W. O. Kenyon, *J. Amer. Chem. Soc.*, **64**, 1110 (1942).
95. W. N. Haworth, C. R. Porter, and A. C. Waine, *Rec. Trav. Chim.*, **57**, 541 (1938).
96. W. N. Haworth and C. R. Porter, *J. Chem. Soc.*, 2796 (1929).
97. G. Zemplén and E. D. László, *Ber.*, **48**, 915 (1915).
98. C. F. Allpress and W. N. Haworth, *J. Chem. Soc.*, **125**, 1223 (1924).
99. G. A. Ellestad, M. P. Kunstmann, J. E. Lancaster, L. A. Mitscher, and G. Morton, *Tetrahedron*, **23**, 3893 (1967).
100. T. Lieser and R. Thiel, *Ann.*, **522**, 48 (1936).
101. M. L. Wolfrom and M. A. El-Taraboulsi, *J. Amer. Chem. Soc.*, **75**, 5350 (1953).
101a. D. Trimnell, W. M. Doane, C. R. Russell, and C. E. Rist, *Carbohyd. Res.*, **5**, 166 (1967); **11**, 497 (1969).
102. A. B. Foster and M. L. Wolfrom, *J. Amer. Chem. Soc.*, **78**, 2493 (1956).

103. K. Freudenberg and A. Wolf, *Ber.*, **60**, 232 (1927).
103a. D. Horton and H. S. Prihar, *Carbohyd. Res.*, **4**, 115 (1967).
104. R. S. Tipson, *Advan. Carbohyd. Chem.*, **8**, 107 (1953); D. H. Ball and F. W. Parrish, *ibid.*, **23**, 233 (1968); *Advan. Carbohyd. Chem., Biochem.*, **24**, 139 (1969).
105. K. Freudenberg and O. Ivers, *Ber.*, **55**, 929 (1922).
106. B. Helferich and R. Hiltmann, *Ann.*, **531**, 160 (1937).
107. K. Freudenberg and F. Brauns, *Ber.*, **55**, 3233 (1922).
108. G. W. Kenner and M. A. Murray, *J. Chem. Soc.*, S178 (1949).
109. H. Schmid and P. Karrer, *Helv. Chim. Acta*, **32**, 1371 (1949).
110. J. K. N. Jones and J. L. Thompson, *Can. J. Chem.*, **35**, 955 (1957).
110a. S. Zen, S. Tashima, and S. Koto, *Bull. Chem. Soc.* (Japan), **41**, 3025 (1968); A. D. Barford, A. B. Foster, and J. H. Westwood, *Carbohyd. Res.*, **13**, 189 (1970).
111. E. J. Reist, R. R. Spencer, and B. R. Baker, *J. Org. Chem.*, **24**, 1618 (1959).
112. K. J. Ryan, H. Arzoumanian, E. M. Acton, and L. Goodman, *J. Amer. Chem. Soc.*, **86**, 2497 (1964); compare L. Goodman, *Advan. Carbohyd. Chem.*, **22**, 109 (1967).
113. B. Helferich and A. Gnüchtel, *Ber.*, **71**, 712 (1938).
113a. N. R. Williams, *Advan. Carbohyd. Chem. Biochem.*, **25**, 109 (1970).
114. C. D. M. Eades, D. H. Ball, and L. Long, Jr., *J. Org. Chem.*, **31**, 1159 (1966).
115. R. S. Tipson and L. H. Cretcher, *J. Org. Chem.*, **8**, 95 (1943).
116. A. B. Foster and W. G. Overend, *J. Chem. Soc.*, 3452 (1951).
117. R. S. Tipson and A. Cohen, *Carbohyd. Res.*, **1**, 338 (1965); E. L. Albano, D. Horton, and T. Tsuchiya, *ibid.*, **2**, 349 (1966); D. M. Clode, D. Horton, M. H. Meshreki, and H. Shoji, *Chem. Commun.*, 694 (1969).

7. HALOGEN DERIVATIVES

MELVILLE L. WOLFROM AND WALTER A. SZAREK

I. INTRODUCTION

This Chapter will treat briefly the halogen derivatives of the cyclic carbohydrate structures; those of the acyclic structures are discussed in Chapter 10. Synthesis, rather than reactivity, will be emphasized.

II. GLYCOSYL HALIDES

A. GENERAL CONSIDERATIONS

Glycosyl halides are saccharide derivatives in which the hydroxyl group of the anomeric center of an aldose or ketose is replaced by a halogen atom. They accordingly possess a mixed acetal–halide function wherein the acetal is cyclic and in which the halogen can be very reactive. The stem substituents are

generally esters, among which the acetate has been the most widely exploited; ether substituents, especially methyl or benzyl, have also been utilized. The latter group does not participate in replacement reactions at the anomeric center and is removed readily by hydrogenolysis. The benzoylated glycosyl halides are perhaps more stable than the acetylated forms and are frequently preferred for synthesis in the pentose series.

The stabilities of the acetylated glycosyl halides are in the normal order $F > Cl > Br > I$. The iodides decompose rapidly, whereas the fluorides are stable and may often be deacetylated without loss or isomerization of fluorine.[1] The acetylated glycosyl halides in the pyranose forms of the common sugars, including those of D-fructose, have been prepared by Brauns.[2] The bromides have been the most widely utilized halides in the pyranose series. The halides of the furanoses are less accessible synthetically and are more reactive and unstable. They have been employed especially in nucleoside synthesis wherein the acetylated glycofuranosyl chloride has sufficient reactivity for utilization. The chloride is also the favored form for the acetylated 2-amino-2-deoxyglycosyl halides. A ketopentofuranosyl fluoride structure is present in the antibiotic nucleocidin (see Vol. IIB, Chapter 31).

The glycosyl halides can predictably exist in anomeric forms, both of which have been prepared in many cases. The anomeric forms of the acetylated aldo-pyranosyl bromides or chlorides are known to differ sufficiently in stability that, under suitable conditions, rearrangement from the less-stable form to the other may occur. It is well established that the anomer having the halogen atom in the axial orientation in the "normal" conformation is the most stable. Thus, the preparation of a glycosyl halide under conditions of thermodynamic control provides that anomer which has the halogen atom in the axial orientation. For hexopyranose structures, this orientation normally corresponds to the α (D or L) configuration. The acetylated aldopyranosyl halides having the halogen in equatorial orientation are usually prepared, therefore, under kinetic control of the reaction, and, unless the conditions are very mild, the initially formed β (D or L) anomer rapidly isomerizes to the α (D or L) form. The anomerization of tetra-O-acetyl-β-D-glucopyranosyl chloride (**1**) by chloride ion has been depicted by Lemieux and Hayami[3] as a direct displacement through the transition state **2** to give the stable anomer **3**.

An investigation of the conformations in solution of a number of the acetylated glycosyl halides, by nuclear magnetic resonance spectroscopy, has established that the thermodynamically more stable anomer in every case had the halogen atom axial. The *C1* (D) conformation shown above (**3**) was found for the corresponding bromide (tetra-*O*-acetyl-α-D-glucopyranosyl bromide). The more stable form of the tri-*O*-acetyl-D-ribopyranosyl bromide is the β-D anomer, which assumes the *1C* (D) conformation[4] (**4**). The favored conformation in solution of tri-*O*-acetyl-β-D-xylopyranosyl chloride, the thermodynamically less-stable anomer, was found to be the all-axial *1C* (D) conformation[5] (**5**).

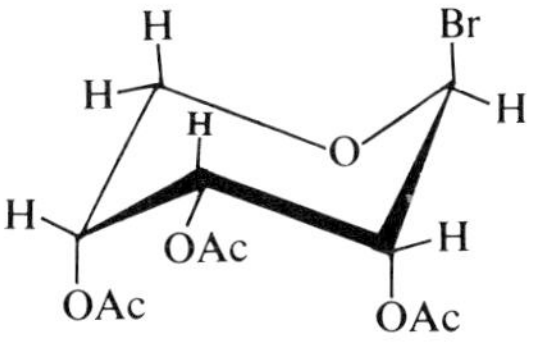

4 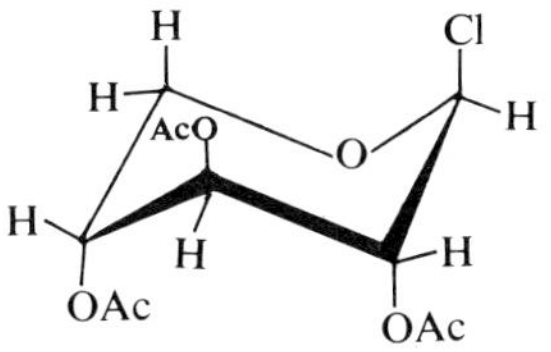   5

The preference for a halogen atom at C-1 to adopt the axial orientation, as the thermodynamically stable arrangement, has been termed the anomeric effect (see Chapter 5). It has been explained[6] in terms of an unfavorable parallel interaction between an equatorial C-1–halogen dipole and the atomic dipole formed by the resultant of the unshared *p* electrons on the ring oxygen atom.

B. SYNTHESIS

Most reactions for obtaining the acetylated glycosyl halides are acidic in nature and, accordingly, any of the unstable isomer formed will be isomerized to the stable variety.

The first synthesis of an acetylated glycosyl halide was effected by Colley,[7] who treated D-glucose with acetyl chloride and obtained the substance we now know to be tetra-*O*-acetyl-α-D-glucopyranosyl chloride. The reaction is difficult to control but is a recommended procedure for preparing 2-acetamido-3,4,6-tri-*O*-acetyl-2-deoxy-D-glucopyranosyl chloride.[8] The publication of Colley is notable for containing the first postulation of a ring form (ethylene oxide) in a sugar.

Koenigs and Knorr[9] first prepared tetra-*O*-acetyl-α-D-glucopyranosyl bromide by the action of acetyl bromide on D-glucose. Fischer and Armstrong[10] prepared the chloride and bromide, in nearly quantitative yield, by the action of the liquid hydrogen halide upon β-D-glucopyranose pentaacetate. A more convenient method is that utilizing an acetic acid solution of the hydrogen halide with either anomer of the acetylated sugar.[11] The reaction is effected at low temperatures, can be very rapid,[12] and consists in the replacement, by halogen, of the hemiacetal ester on the anomeric carbon atom. Several steps may be combined, and a number of modifications starting from the unsubstituted sugar have been reported. One of the best of these involves the following combination:[13]

Sugar + Ac_2O + $HClO_4$ + P + Br_2 + H_2O ⟶ acetylated glycopyranosyl bromide

The procedure employing liquid hydrogen halide still constitutes the preferred method for the synthesis of the fluorides;[14] alternatively the acetylated glycosyl fluorides may be obtained from the corresponding chlorides by the action of silver tetrafluoroborate.[14a]

The acylated glycosyl halides may be prepared from the corresponding methyl glycosides by methoxyl replacement with an acetic acid solution of hydrogen halide. This may be of special use for the synthesis of the acylated glycofuranosyl halides.[15] Preparation of the glycosyl halide of tetra-O-methyl-D-glucopyranose involves hydroxyl group replacement.

A mild and useful method for the preparation of acetylated glycosyl chlorides consists in replacement of the 1-acetoxy group, in the fully acetylated sugar, with titanium tetrachloride.[16] This procedure was utilized in the preparation of tetra-O-acetyl-D-galactoseptanosyl chloride.[17] Lemieux and Brice[18] have shown that, at 40°, penta-O-acetyl-α-D-glucopyranose is highly stable toward titanium tetrachloride, whereas the β-D anomer reacts extremely rapidly to give initially tetra-O-acetyl-β-D-glucopyranosyl chloride, which then anomerizes slowly to the α-D chloride.

An early synthesis of acetylated glycopyranosyl chlorides involved the action of a mixture of phosphorus pentachloride and aluminum chloride on the fully acetylated sugars.[19, 20] Some sugars (cellobiose and lactose) undergo inversion at both C-2 and C-3 of the reducing moiety during such a reaction. Completely acetylated mono- and disaccharides, of the β-D series, give with dichloro-(dibromo)methyl methyl ether, in the presence of catalytic amounts of zinc chloride, the corresponding acetylated α-D-glycosyl halides.[21] Zinc chloride–thionyl chloride is also a convenient reagent for preparing acetylated glycosyl chlorides from aldose peracetates.[22]

Glycosyl halides can be formed by addition of hydrogen halide to anhydro rings involving C-1 of an aldose or to 1,2-unsaturated sugars. With hydrogen bromide and acetic anhydride (acetylating conditions), 2,3,4-tri-O-acetyl-1,6-anhydro-β-D-glucopyranose forms tetra-O-acetyl-α-D-glucopyranosyl bromide.[23] Addition of hydrogen bromide or hydrogen chloride, in benzene, to the acetylated glycals furnishes the acetylated 2-deoxyglycosyl bromides or chlorides.[24]

A useful synthesis of glycosyl halides was provided by Bonner,[25] who converted phenyl 2,3,4,6-tetra-O-acetyl-1-thio-β-D-glucopyranoside into tetra-O-acetyl-α-D-glucopyranosyl bromide by treatment with bromine in an inert solvent. The reaction is general for 1-thioglucosides (6) with alkyl or aryl aglycons, and a mechanism was proposed[25] that involves S-bromination by positive bromine to give a bromonium ion (7) followed by heterolysis of the C-1–S bond to yield a glycosyl carbonium ion, which is attacked by bromide ion to give the halide (8). Chlorine may replace bromine,[26] and this modification

6

7   8

is especially useful for the preparation of glycofuranosyl chlorides. Since the reaction is essentially nonacidic, the thermodynamically unstable halides may arise. The anomeric form obtained is dependent on neighboring-group effects, conformational factors, and steric hindrance when bulky groups are present. A glycosyl chloride having an unusual nonacylated structure, 3-amino-3-deoxy-D-mannopyranosyl chloride hydrochloride (10), has been prepared[27] by the reaction of ethyl 3-amino-3-deoxy-1-thio-α(and β)-D-mannopyranoside (9) with chlorine.

9   10

There is a considerable variation in the thermodynamic stability of acylated glycosyl halides, and their stability would appear to depend on configurational, inductive, and neighboring-group effects. Thus, anomeric pairs of crystalline

* References start on p. 250.

benzoylated D-ribopyranosyl chlorides[28] and D-arabinofuranosyl bromides[29] are known with each anomer having considerable stability. The situation is quite otherwise, however, in the acetylated D-glucopyranosyl halides, where the β-D form is highly unstable and anomerizes readily to the α-D isomer. The first preparation of an unstable tetra-O-acetyl-β-D-aldopyranosyl chloride was effected by Schlubach[30] through the action of "active" silver chloride on tetra-O-acetyl-α-D-glucopyranosyl bromide. Lemieux and Hayami[3] replaced the silver chloride by tetraethylammonium chloride in acetonitrile and obtained the β-D chloride in high yield. Korytnyk and Mills[31] have shown that completely acetylated aldopyranoses of 1,2-*trans* configuration react with aluminum chloride in chloroform to give the unstable 1,2-*trans*-per-O-acetylglycosyl chlorides (12). Thus, the acetylated form of α-D-mannopyranose and the β-D forms of acetylated glucopyranose (11), galactopyranose, and galactofuranose all reacted readily and gave the O-acetylglycosyl chloride of the same configuration as the acetate used, whereas the anomeric acetate was recovered unchanged in every case. These results can be accommodated by the following mechanism, suggested by these authors, which involves displacement of the acetoxy group at C-1 by aluminum chloride with participation by the neighboring acetoxy group at C-2 (11):

11   12

Another method to increase the stability of the β-D-glucopyranosyl chloride structure is to substitute an electronegative (frequently nonparticipating) group on C-2. The nitrate ester is such a group. Brigl[32] treated β-D-glucopyranose pentaacetate (13) with phosphorus pentachloride and obtained the 2-O-trichloroacetyl chloride (14), from which the trichloroacetyl group could be removed with dilute ammonium hydroxide. Nitration[33] of the crystalline chloride (15) with acetyl nitrate yielded the crystalline 3,4,6-tri-O-acetyl-2-O-nitro-β-D-glucosyl chloride (16).

Another procedure for obtaining an unsubstituted C-2 position in some sugar structures is that found by Helferich and Zirner.[34] These investigators prepared tetra-O-acetyl-α-D-glucopyranosyl bromide directly from D-glucose according to the aforementioned method of Bárczai-Martos and Kőrösy,[13] but instead of isolating the bromide they added an aqueous solution of sodium acetate to the reaction mixture and so obtained 1,3,4,6-tetra-O-acetyl-α-D-glucose in 35 to 40% yield. This product probably resulted from hydrolysis of the halide followed by acetyl migration from C-2 to C-1. A similar acyl migration had been established by Ness and Fletcher[35] in the acidic or neutral hydrolysis

13   14

15   16

of tri-*O*-benzoyl-D-ribofuranosyl bromide. Helferich and Zirner[34] substituted an *O*-methylsulfonyl group on C-2 and converted the mixed ester into the bromide. In a related manner, Glaudemans and Fletcher[15] prepared 1,3,5-tri-*O*-(*p*-nitrobenzoyl)-β-D-arabinofuranose by hydrolysis of the corresponding bromide. The triester was converted into its 2-nitrate and this in turn into 2-*O*-nitro-3,5-di-*O*-(*p*-nitrobenzoyl)-α-D-arabinofuranosyl bromide. The latter, on treatment with silver chloride according to Schlubach,[30] gave an anomeric mixture of crystalline 2-*O*-nitro-3,5-di-*O*-(*p*-nitrobenzoyl)-D-arabinofuranosyl chlorides.

A method for obtaining a nonparticipating group at C-2 without selective substitution is by benzylation. The benzyl ethers are, however, generally noncrystalline. The benzyl group is easily removed by hydrogenolysis.

C. REACTIVITY

The glycosyl halide derivatives constitute reactive intermediates for a variety of syntheses in the carbohydrate series. Specific applications in synthesis are detailed in other chapters of this treatise. The following is an outline of the types of reactivity shown by this very reactive carbonyl-derived function, which may be designated as a halogenal or mixed halide–acetal function.

It has long been known that the glycosyl halides fall into two types in respect to their reactivity. The halides are classified as *cis* or *trans*, a designation refer-

* *References start on p. 250.*

ring to the configurational relationship of the halogen on C-1 to the hydroxyl-substituting group on C-2. The reactions of the halides are profoundly affected by the participating or nonparticipating nature of the group on C-2; 1,2-*trans*

cis   trans

products normally result when a participating C-2 substituent is present.

The glycosyl halides undergo replacement reactions at C-1 with hydroxyl compounds. Reactions with alcohols or phenols lead to glycosides. If the alcohol is that of another saccharide, the product is an oligosaccharide. If the reaction is effected intramolecularly, an anhydro sugar will result. These reactions are effected heterogeneously with silver salts or homogeneously with mercuric cyanide as catalysts and acid acceptors. The alkali salts of phenols are utilizable.[36] These conditions are adaptable to the formation of 1-thio- or 1-selenoglycosides. The glycosyl halides will react rapidly with the trityl ethers of alcohols in the presence of silver perchlorate.

$$RX + TrOCH_2R' \longrightarrow ROCH_2R' + TrX$$

The acylated *trans* halides (**17**) are susceptible to formation, in part, of ortho esters (**18**) in their reactions with alcohols. Ortho esters can be obtained in good

17   18

yield from acylated *cis* halides by treatment with the appropriate alcohol, in boiling ethyl acetate, in the presence of lead carbonate and anhydrous calcium sulfate.[36a]

The halides undergo hydrolysis. They undergo displacement reactions by anions with metal salts such as silver or mercuric acetate; they form Grignard reagents. They can be brought into reaction with the trimethylsilyl, silver, or chloromercury derivatives of heterocyclic bases to form glycosylamines, to which class belong the nucleosides. They can form simple glycosylamines with primary or secondary amines, although dehydrohalogenation (hydroxyglycal formation) may be a predominant competing reaction with the latter (see Chapter 19). The halides may be quaternized with tertiary bases, such as pyridine or nicotine.

Conditions are known for effecting reduction of the acylated glycosyl halides to the corresponding anhydroalditols or, with reductive dehydrohalogenation, to form glycals.

It is readily apparent that the (stem hydroxyl)-substituted glycosyl halides are of real value in carbohydrate synthetic procedures and in the study of the varied reactivity of carbohydrate structures.

III. NONANOMERIC HALIDES

A. GENERAL CONSIDERATIONS

This section deals with the formation and reactivity of carbohydrate halides where a hydroxyl group has been replaced, in the cyclic structures, by halogen at positions other than at C-1 of an aldose or C-2 of a ketose. These positions are sometimes designated stem positions.

Formation, through diazomethyl ketone reactivity, of halides at C-1 of an acyclic ketose derivative is discussed in Chapter 10 (p. 372). It may be noted that among these derivatives is an example of a 1,1-dihalide (diiodo). This, together with a 1,1-dichloride formed from an acyclic aldose structure (Chapter 10, p. 364) constitutes the first known examples in the carbohydrate series of a halide containing more than one halogen atom attached to the same carbon atom.

In general, the cyclic stem halides have not been as useful for synthetic purposes as have been the alkyl halides of simpler aliphatic structures. Their formation and general properties have been reviewed in detail.[37] The stem halides have been largely replaced, for synthetic purposes, by the more controllably introduced sulfonyloxy group, which is also an excellent leaving group. The halides are still useful in forming deoxy sugars by halide reduction.

Contrary to the reactivity of the simple aliphatic alcohols, the primary hydroxyl group in an aldose is more readily replaced by halogen than are the interior, secondary hydroxyl groups. This is probably because of the steric availability of the primary alcohol group as a side chain on the ring. Very little is known at present of halides formed from tertiary alcohol groups, such as occur in the rare branched-chain sugars. Knowledge of the ketose halides is also meager, no doubt because of the sensitivity of these substances toward acids; only their terminal (C-1 or C-6) primary halides are known.

B. SYNTHESIS

Synthetic methods fall into three main classes: (1) displacement reactions dependent on nucleophilic S_N1 or S_N2 attack of the halide ion; (2) the addition of halogen or hydrogen halide to a double bond; and (3) epoxide ring-opening

* References start on p. 250.

with hydrogen halides. As is well established, the nucleophilicities fall in the order $I^- > Br^- > Cl^- > F^-$. As is usual, special methods are required for the introduction of fluorine.[37a]

1. Group Replacement on a Substituted Terminal Primary Hydroxyl Group

Primary sulfonyloxy or nitrate[38] groups are selectively replaced by iodide with sodium iodide in acetone or 2,5-hexanedione.[39] Halogen exchange may be effected by this procedure. If a derivative is available where all hydroxyl groups, except the terminal one, are blocked with cyclic acetal or glycosidic groups, fluorine may be so introduced, but here hydroxylic solvents replace the ketonic.[40,41] Replacement of sulfonyloxy groups may be a side reaction on prolonged sulfonylation in pyridine, where the replacement may occur on the terminal position by reaction with the pyridinium chloride present.[41a]

Acetate replacement on C-6 was early encountered[42] in the formation of tri-O-acetyl-6-bromo-6-deoxy-α-D-glucopyranosyl bromide by the prolonged action of liquid hydrogen bromide on D-glucose pentaacetate. A convenient route to 6-bromo-6-deoxy derivatives of aldohexoses is by the action of N-bromosuccinimide on 4,6-O-benzylidene derivatives.[42a]

Primary trityloxy groups are displaced by phosphorus pentachloride.[43] When such groups are present in an acyclic derivative, displacement with hydrogen bromide occurs with remarkable ease.[44] When N,N-diethyl-P,P-dipropylphosphinous amide (Pr_2PNEt_2) is subjected to alcoholysis by a C-6 hydroxyl group in an otherwise protected (by cyclic acetals) hexose, there is obtained a 6-(dipropyl phosphinite), $ROPPr_2$. This phosphinite group can be displaced by halogens (Cl, Br, I) on reaction with alkyl halides. Fluorides may be obtained by treating the phosphinite with perfluoropropylene.[45]

2. Direct Replacement of Hydroxyl Groups

Probably the earliest reaction of this type was the formation, in an alditol, of crystalline 1,6-dibromo(and dichloro)-1,6-dideoxy-D-mannitol by heating D-mannitol with the halogen acid.[46]

Certain hydroxyl groups in otherwise protected (as by isopropylidene cyclic acetals) structures can be replaced by chlorine with triphenylphosphine in refluxing carbon tetrachloride.[47] Methyl 6-chloro-6-deoxy-α-D-glucopyrano-side has been obtained directly from the unsubstituted glucoside with sulfur monochloride.[48] Tetra-O-acetyl-6-chloro-6-deoxy-β-D-mannopyranose is formed from the corresponding tetraacetate on reaction with phosphoryl oxychloride.[49]

Helferich first used sulfuryl chloride to introduce chlorine into glycosides.[50] Jones and co-workers have studied this reaction.[51] Depending on the conditions there may be obtained mono-, di-, and trichloro substitution products, cyclic sulfates, and chlorosulfates. Again, the terminal, primary hydroxyl group is always replaced by chlorine.

Partially methylated or sulfonated glycosides may have their unsubstituted secondary hydroxyl groups replaced, with inversion, by iodine with triphenyl phosphite methiodide, $(PhO)_3PMeI$. Replacement of secondary hydroxyl groups, with inversion, was observed in some otherwise protected acyclic compounds.[52] Secondary sulfonyloxy groups in protected sugars can be displaced, with inversion, by fluorine with tetrabutylammonium fluoride in acetonitrile.[52a]

3. From Anhydro Sugars

Extensive literature exists on the opening of anhydro rings with hydrogen halides (see Chapter 13). The most effective and general route to the formation of secondary halides is by cleavage of epoxides with hydrogen or magnesium halides. Trans groupings are formed.[53] Epoxide migration may occur with consequent formation of several products. Direction of cleavage may be difficult to predict and is governed by a variety of factors. Epoxide ring opening by hydrogen chloride in a branched-chain alditol structure is mentioned in Chapter 10 (p. 379). Secondary fluorides may be obtained[54] by the opening of epoxy rings by hydrogen fluoride in p-dioxane at 120°, with potassium fluoride in acetamide at 200°, or by hydrogen tetrafluoridoborate in hydrogen fluoride[55] at −70°; related procedures have been explored in detail.[37a, 56]

4. By Addition to Olefinic Linkages

The addition of halogen to unsaturated sugar derivatives has been reviewed.[57, 58] This method was the first to yield secondary halides and is especially useful in obtaining halides on C-2 because of the availability of the acetylated derivatives unsaturated at the 1,2-position. Addition of halogen furnishes a glycosyl halide substituted at C-2, from which the halogen on C-1 is readily removed leaving the 2-halide. Here the attack, in the addition, is considered to be by a positive halogen. A mixed halide was obtained by the combined action of N-bromosuccinimide and hydrogen fluoride. The product is a 2-bromo-2-deoxyglycosyl fluoride from which the fluoride is removable to form the 2-bromide[59] This mixed halogen addition to a double bond was first established in the steroid field with N-bromoacetamide and hydrogen fluoride.[60, 61] The stereochemistry of thèse additions has been studied in detail.[62]

The addition of acetyl hypobromite to 2,3-unsaturated sugars yields trans-acetylated bromohydrins.[63] Addition of trifluoromethyl hypofluorite to glycals is a useful route to 2-deoxy-2-fluoro sugars.[64]

C. REACTIVITY

Nucleophilic attack on a saccharide stem halide can lead either to displacement, elimination, or anhydro ring formation. Most of the displacement reactions have been effected with bromo or iodo derivatives. Anhydro ring formation takes precedence over displacement or elimination, provided an unsubstituted hydroxyl group is available. An epoxide ring may migrate to give a final product at a position other than that originally involved. Other than epoxy, 3,6-anhydro ring formation is especially favored.

The halides are reducible to the hydrocarbon stage by a variety of reducing agents to form the deoxy sugars. Since the terminal primary and C-2 (aldose) halides are the most readily available, their reductions have furnished routes for the synthesis of the 6-deoxyhexoses and 2-deoxyaldoses. The iodide and bromide have been especially utilized for such purposes.

REFERENCES

1. B. Helferich, K. Bäuerlein, and F. Wiegand, *Ann.*, **447**, 27 (1926).
2. D. H. Brauns, *Bur. Stand. J. Res.*, **7**, 573 (1931); summarizing paper.
3. R. U. Lemieux and J. Hayami, *Can. J. Chem.*, **43**, 2162 (1965).
4. D. Horton and W. N. Turner, *J. Org. Chem.*, **30**, 3387 (1965); P. L. Durette and D. Horton, *Carbohyd. Res.*, **18**, 57 (1971).
5. C. V. Holland, D. Horton, and J. S. Jewell, *J. Org. Chem.*, **32**, 1818 (1967).
6. J. T. Edward, *Chem. Ind.* (London), 1102 (1955); S. Wolfe, A. Rauk, L. M. Tel, and I. G. Csizmadia, *J. Chem. Soc.*, (*B*), 136 (1971).
7. A. Colley, *Ann. Chim. Phys.* **21** [4], 363 (1870).
8. D. Horton and M. L. Wolfrom, *J. Org. Chem.*, **27**, 1794 (1962).
9. W. Koenigs and E. Knorr, *Ber.*, **34**, 957 (1901).
10. E. Fischer and E. F. Armstrong, *Ber.*, **34**, 2885 (1901).
11. E. Fischer, *Ber.*, **44**, 1898 (1911).
12. M. L. Wolfrom and D. L. Fields, *Tappi*, **40**, 335 (1957).
13. M. Bárczai-Martos and F. Kőrösy, *Nature*, **165**, 369 (1950).
14. D. H. Brauns, *J. Amer. Chem. Soc.*, **45**, 833 (1923).
15. C. P. J. Glaudemans and H. G. Fletcher, Jr., *J. Org. Chem.*, **29**, 3286 (1964).
16. E. Pacsu, *Ber.*, **61**, 1508 (1928).
17. F. Micheel and F. Suckfüll, *Ann.*, **507**, 138 (1933).
18. R. U. Lemieux and C. Brice, *Can. J. Chem.*, **30**, 295 (1952).
19. F. v. Arlt, *Monatsh. Chem.*, **22**, 144 (1901).
20. Z. H. Skraup and R. Kremann, *Monatsh. Chem.*, **22**, 375 (1901).
21. H. Gross and I. Farkas, *Ber.*, **93**, 95 (1960); I. Farkas, M. Menyhart, R. Bognar, and H. Gross, *Magy. Kém. Foly.*, **71**, 377 (1965); *Chem. Abstr.*, **65**, 7252 (1966).
22. L. P. Egan, T. G. Squires, and J. R. Vercellotti, *Carbohyd. Res.*, **14**, 263 (1970).
23. K. Freudenberg and K. Soff, *Ber.*, **69**, 1252 (1936).
24. J. Davoll and B. Lythgoe, *J. Chem. Soc.*, 2626 (1949); compare I. Lundt and C. Pedersen, *Acta Chem. Scand.*, **24**, 240 (1970).
25. W. A. Bonner, *J. Amer. Chem. Soc.*, **70**, 3491 (1948).
26. M. L. Wolfrom and W. Groebke, *J. Org. Chem.*, **28**, 2986 (1963).
27. M. L. Wolfrom, H. G. Garg, and D. Horton, *J. Org. Chem.*, **28**, 2989 (1963).
28. R. K. Ness, H. G. Fletcher, Jr., and C. S. Hudson, *J. Amer. Chem. Soc.*, **73**, 959 (1951).
29. R. K. Ness and H. G. Fletcher, Jr., *J. Amer. Chem. Soc.*, **80**, 2007 (1958).
30. H. H. Schlubach, *Ber.*, **59**, 840 (1926).
31. W. Korytnyk and J. A. Mills, *J. Chem. Soc.*, 636 (1959).
32. P. Brigl, *Hoppe Seyler's Z. Physiol. Chem.*, **116**, 1 (1921).
33. M. L. Wolfrom, A. O. Pittet, and I. C. Gillam, *Proc. Nat. Acad. Sci. U.S.*, **47**, 700 (1961).
34. B. Helferich and J. Zirner, *Ber.*, **95**, 2604 (1962).

35. R. K. Ness and H. G. Fletcher, Jr., *J. Amer. Chem. Soc.*, **78**, 4710 (1956).
36. A. Michael, *Amer. Chem. J.*, **1**, 305 (1879).
36a. N. K. Kochetkov, A. J. Khorlin, and A. F. Bochkov, *Tetrahedron*, **23**, 693 (1967).
37. J. E. G. Barnett, *Advan. Carbohyd. Chem.*, **22**, 177 (1967); S. Hanessian, *Advan. Chem. Ser.*, **74**, 159 (1968).
37a. A. B. Foster and J. H. Westwood, *Angew. Chem. Intern. Ed. Engl.*, **11**, in press (1972).
38. J. W. H. Oldham and J. K. Rutherford, *J. Amer. Chem. Soc.*, **54**, 366 (1932).
39. G. E. Murray and C. B. Purves, *J. Amer. Chem, Soc.*, **62**, 3194 (1940).
40. B. Helferich and A. Gnüchtel, *Ber.*, **74**, 1035 (1941).
41. N. F. Taylor and P. W. Kent, *J. Chem. Soc.*, 872 (1958); E. M. Bessell, A. B. Foster, J. H. Westwood, L. D. Hall, and R. N. Johnson, *Carbohyd. Res.*, **19**, 39 (1971).
41a. F. W. Parrish, F. H. Bissett, M. E. Evans, M. L. Bazinet, W. Yeomans, and L. Long, Jr., *Carbohyd. Res.*, **6**, 503 (1968); M. E. Evans, L. Long, Jr., and F. W. Parrish, *J. Org. Chem.*, **33**, 1074 (1968).
42. E. Fischer and E. F. Armstrong, *Ber.*, **35**, 833 (1902).
42a. S. Hanessian, *Carbohyd. Res.*, **2**, 86 (1966); S. Hanessian and N. R. Plessas, *J. Org. Chem.*, **34**, 1035 (1969); T. L. Hullar and S. B. Siskin, *J. Org. Chem.*, **35**, 228 (1970).
43. B. Helferich and H. Bredereck, *Ber.*, **60**, 1995 (1927).
44. M. L. Wolfrom, J. L. Quinn, and C. C. Christman, *J. Amer. Chem. Soc.*, **57**, 713 (1935).
45. K. A. Petrov, É. E. Nifant'ev, A. A. Shchegolev, and V. G. Terekhov, *Zh. Obshch. Khim.*, **34**, 1459 (1964); *J. Gen. Chem. USSR*, **34**, 1463 (1964) (Engl. transl.).
46. G. Bouchardat, *Ann. Chim. Phys.*, **6** [5], 100 (1875).
47. J. B. Lee and T. J. Nolan, *Can. J. Chem.*, **44**, 1331 (1966); C. R. Haylock, L. D. Melton, K. N. Slessor, and A. S. Tracey, *Carbohyd. Res.*, **16**, 375 (1971).
48. H. B. Sinclair, *J. Org. Chem.*, **30**, 1283 (1965).
49. B. Helferich and J. F. Leete, *Ber.*, **62**, 1549 (1929).
50. B. Helferich, *Ber.*, **54**, 1082 (1921).
51. P. D. Bragg, J. K. N. Jones, and J. C. Turner, *Can. J. Chem.*, **37**, 1412 (1959); and subsequent publications; compare B. T. Lawton, W. A. Szarek, and J. K. N. Jones, *Carbohyd. Res.*, **15**, 397 (1970).
52. N. K. Kochetkov and A. I. Usov, *Tetrahedron*, **19**, 973 (1963).
52a. A. B. Foster, R. Hems, and J. M. Webber, *Carbohyd. Res.*, **5**, 292 (1967); D. M. Marcus and J. H. Westwood, *ibid.*, **17**, 269 (1971).
53. S. Peat, *Advan. Carbohyd. Chem.*, **2**, 37 (1946).
54. N. F. Taylor, R. F. Childs, and R. V. Brunt, *Chem. Ind.* (London), 928 (1964); S. Cohen, D. Levy, and E. D. Bergmann, *ibid.*, 1802 (1964).
55. I. Johansson and B. Lindberg, *Carbohyd. Res.*, **1**, 467 (1966).
56. A. D. Barford, A. B. Foster, J. H. Westwood, L. D. Hall, and R. N. Johnson, *Carbohyd. Res.*, **19**, 49 (1971).
57. B. Helferich, *Advan. Carbohyd. Chem.*, **7**, 209 (1952).
58. R. J. Ferrier, *Advan. Carbohyd. Chem.*, **20**, 67 (1965); **24**, 199 (1969).
59. P. W. Kent, F. O. Robson, and V. A. Welch, *J. Chem. Soc.*, 3273 (1963); J. C. Campbell, R. A. Dwek, P. W. Kent, and C. K. Prout, *Carbohyd. Res.*, **10**, 71 (1969).
60. C. H. Robinson, L. Finckenour, E. P. Oliveto, and D. Gould, *J. Amer. Chem. Soc.*, **81**, 2191 (1959).
61. A. Bowers, *J. Amer. Chem. Soc.*, **81**, 4107 (1959).
62. L. D. Hall and J. F. Manville, *Chem. Commun.*, 35 (1968); L. D. Hall, R. N. Johnson, J. Adamson, and A. B. Foster, *Can. J. Chem.*, **49**, 118 (1971).
63. E. L. Albano, D. Horton, and J. H. Lauterbach, *Carbohyd. Res.*, **9**, 149 (1969).
64. J. Adamson, A. B. Foster, L. D. Hall, and R. H. Hesse, *Chem. Commun.*, 309 (1969); E. L. Albano, R. L. Tolman, and R. K. Robins, *Carbohyd. Res.*, **19**, 63 (1971).

8. PHOSPHATES AND OTHER INORGANIC ESTERS

Donald L. MacDonald

I. PHOSPHATE ESTERS

Of all the inorganic esters of carbohydrates, the phosphate esters are of the greatest importance because of their intimate involvement in the enzymic breakdown and synthesis of carbohydrates. Furthermore, their role as components of nucleic acids and a number of coenzymes has prompted the investigation of the properties of a variety of phosphorylated sugars, and elegant methods for the synthesis of a number of these compounds have been devised.[1]

The mono- and diphosphates of sugars are strong acids, usually stronger than orthophosphoric acid.[2] A representative comparison (Table I) is taken from the excellent review of Leloir.[3] The phosphates are usually isolated as their barium, potassium, sodium, or alkaloid salts. The cyclohexylammonium salts usually are crystalline, and nonreducing phosphates are often isolated in this form.

A. Synthesis of Phosphate Esters

Phosphate esters of carbohydrates may be obtained by isolation from natural sources or by enzymic or chemical synthesis. Except in a few instances, such as the preparation of D-fructose 1,6-diphosphate from yeast,[12] isolation from natural sources is not the method of choice. In some cases, such as in the preparation of certain glycosyl phosphates, enzymic synthesis with the appropriate phosphorylase provides the simplest preparation. In this fashion, α-D-glucopyranosyl phosphate is readily prepared by the phosphorolysis of starch or glycogen,[7, 13] and α-D-ribofuranosyl phosphate[14] and 2-deoxy-α-D-*erythro*-pentofuranosyl phosphate[15] can be prepared from the appropriate nucleosides.

In many instances chemical synthesis is to be preferred over other preparative methods. Synthesis may involve two distinctly different approaches. In the one case, the unprotected sugar is phosphorylated with a reagent such as polyphosphoric acid, and the product, which usually consists of a complex

TABLE I

DISSOCIATION CONSTANTS OF SOME CARBOHYDRATE PHOSPHATES[3]

Compound	pK_1'	pK_2'	Reference
Orthophosphoric acid	1.95–2.00	6.83–6.93	4–7
α-D-Xylopyranosyl phosphate	1.25	6.15	8
α-D-Glucopyranosyl phosphate	1.10	6.13	7
α-D-Galactopyranosyl phosphate	1.00	6.17	9, 10
α-Maltosyl phosphate	1.52	5.89	8
D-Glucose 3-phosphate	0.84	5.67	11
D-Glucose 6-phosphate	0.94	6.11	11
D-Fructose 6-phosphate	0.97	6.11	11
D-Fructose 1,6-diphosphate[a]	1.48	6.29	6

[a] The pK values are average values of the constants of the two phosphate groups.

mixture of polyphosphates, is then hydrolyzed to give the desired sugar phosphate. This method is usually successful only when the final product is a sugar which bears a phosphate group on its primary hydroxyl group; D-glucose 6-phosphate has been prepared in this manner,[16, 17] as have several ^{32}P-labeled sugar phosphates.[17a]

The other procedure involves monophosphorylation of a suitably protected sugar. In these instances, a wide variety of phosphorylating agents is available.[18] These include pyrophosphates such as tetra-*p*-nitrophenyl pyrophosphate,[19, 20] which is a useful reagent because the *p*-nitrophenyl groups on the resulting phosphate can be removed by alkaline hydrolysis. In a like manner, the use of cyanoethyl phosphate and *N,N'*-dicyclohexylcarbodiimide[21] gives an intermediate having an exceedingly labile protecting group; very mild alkaline treatment eliminates the cyanoethyl group as acrylonitrile. Phosphorylated derivatives of amylose have been prepared in this way,[22] and the reagent has been used for the preparation of the 6-phosphate of o-nitrophenyl β-D-galactopyranoside.[22a]

Phosphorus oxychloride, a reagent early used in phosphorylations,[23] has been utilized in the phosphorylation of benzyl 4,6-O-benzylidene-2-(benzyloxycarbonyl)amino-2-deoxy-α-D-glucopyranoside to give, after removal of protecting groups, 2-amino-2-deoxy-D-glucose 3-phosphate.[24] This synthesis is typical of many in that it involves the preparation of a suitably protected carbohydrate having free only that hydroxyl group which is to be phosphorylated; after phosphorylation the protecting groups are removed under conditions which do not affect the position of the phosphate group.

The most frequently used phosphorylating agents are dibenzylphosphorochloridate[25] and diphenylphosphorochloridate.[26] These reagents are normally used in pyridine solution, and the resulting nonpolar phosphate triesters can be isolated readily. The protecting groups on the phosphate are usually removed in the case of the dibenzyl esters with palladium and hydrogen or by anionic attack, and in the case of the diphenyl esters with platinum and hydrogen. When the sugar to be phosphorylated has two or more free hydroxyl groups, preferential phosphorylation of the primary hydroxyl group can often be accomplished. In this manner 2-amino-2-deoxy-D-glucose 6-phosphate[27] and D-glucose 6-phosphate[28] have been obtained. When two hydroxyl groups are available for phosphorylation, the initially formed phosphate triester may be induced to undergo attack by a suitably located hydroxyl group. A phenoxide ion is lost and a cyclic phosphate triester is formed. Further alkaline action leads to loss of the second phenoxide ion and then ring opening of the cyclic phosphate to produce a mixture of two phosphate monoesters.[19,29] Cyclic phosphates may also be prepared by phosphorylation with phenyl phosphorodichloridate, a reagent used extensively by Baer and co-workers for the synthesis of phospholipids.[30] In this way, for example, methyl 2-deoxy-β-D-$arabino$-hexopyranoside 4,6-(monophenyl phosphate) has been prepared by phosphorylation of methyl 2-deoxy-β-D-$arabino$-hexopyranoside.[31]

Other phosphorylating agents that have been employed for the preparation of carbohydrate phosphates include catechol cyclic phosphate[32] and hydrobenzoin cyclic phosphate.[33] The opening of epoxide rings by means of inorganic orthophosphoric acid is a useful method, when the appropriate epoxides are available, for the preparation of ^{32}P-labeled phosphates;[34] cyanoethyl phosphate has also been used for this purpose.[21] A phosphate group can be introduced at the primary hydroxyl group, by displacement of a suitable sulfonyloxy group by anions of phosphate diesters.[34a] Similar displacement reactions can be used to prepare analogs having a C–P bond.[34b]

Aldosyl phosphates can be prepared by the condensation of a per-O-acetyl-

* References start on p. 271.

glycosyl halide with trisilver phosphate, followed by controlled hydrolysis of the product with acid.[7, 13] Usually, but not always, the anomer formed has the phosphate group *cis* to the hydroxyl group on C-2 of the sugar. The use of "monosilver phosphate"[35, 36] (the silver salt present is actually disilver phosphate[37]) leads normally to the *trans* anomer. The reaction of silver dibenzyl phosphate with a per-*O*-acetylglycosyl halide normally gives, after hydrogenolysis and saponification, the *trans* anomer.[35, 36, 38] The isolation of a different anomer may be accomplished, however, by varying the reaction conditions.[39, 40] The use of silver diphenyl phosphate[39–43] may result in either the *cis* or the *trans* product, depending on the sugar employed. In certain instances, tertiary amine salts rather than silver salts of orthophosphoric acid[44] and of phosphate diesters[45, 46] have been used.

Glycosyl phosphates have also been prepared by fusion at moderate temperatures of orthophosphoric acid with fully acetylated aldoses and ketoses. By using β-D-glucopyranose pentaacetate, either α-D-glucopyranosyl phosphate[47] or β-D-glucopyranosyl phosphate[47a] can be obtained in good yield, and ^{32}P-labeled material can be prepared.[47b] Both anomers of 2-acetamido-2-deoxy-D-glucopyranosyl phosphate are formed from 2-acetamido-1,3,4,6-tetra-*O*-acetyl-2-deoxy-α-D-glucopyranose.[48] β-D-Fructopyranosyl phosphate has also been prepared in this manner.[49]

Examples of the preparation of a number of carbohydrate phosphates by isolation from natural sources as well as by enzymic and chemical synthesis are to be found in certain monographs.[50, 51]

B. Phosphate Migration

In acid solution, phosphate monoesters may undergo an intramolecular phosphate migration.[52] This phenomenon, which probably proceeds through an intermediate cyclic phosphate, necessitates the cautious use of hot acid during any isolation procedure. For example, the mixture obtained on treatment of glyceric acid phosphates with hot acid consists of four parts of the 3-ester to one part of the 2-ester.[53]

Phosphate migration may also be accomplished readily when a phosphate diester or phosphate triester having a sterically favorable hydroxyl group is hydrolyzed by alkali. In this manner the ribonucleic acids are hydrolyzed with alkali to a mixture of 2'- and 3'-ribonucleotides. In a like manner, 1,2-*O*-isopropylidene-D-xylofuranose 5-(diphenyl phosphate) is converted into a mixture of D-xylose 3-phosphate and D-xylose 5-phosphate by the series of reactions shown, through an intermediate 3,5-cyclic phosphate.[19]

Cyclic phosphates are most conveniently prepared by treatment of phosphate monoesters with trifluoroacetic anhydride[54] or with a carbodiimide.[55] The latter type of reagent, which may be used in aqueous pyridine, has been used to

prepare cyclic phosphates involving five-, six-, and seven-membered rings.[55,56] Treatment of certain aldosyl phosphates with N,N'-dicyclohexylcarbodiimide results in the almost quantitative formation of the 1,2-cyclic phosphates. Acid hydrolysis results in ring opening, and the 2-phosphate can be isolated in good yield.[57] In a like manner, D-fructose 1-phosphate can be converted into a mixture of 1,2-cyclic phosphates, alkaline hydrolysis of which gives D-fructofuranosyl phosphate and D-fructopyranosyl phosphate.[58]

C. HYDROLYSIS OF PHOSPHATES

With few exceptions, carbohydrate phosphate monoesters can be hydrolyzed by acid or by alkali, but the ease of hydrolysis varies considerably with the structure.[59]

In general the phosphate esters, particularly the glycosyl phosphates and other nonreducing phosphates, are reasonably stable to alkali. However, compounds such as glyceraldehyde 3-phosphate and D-glucose 3-phosphate readily undergo β-elimination. The decomposition of the latter compound is indicated[60] on p. 258 (see Chapter 4). The decomposition of D-glucose 6-phosphate in strong alkali has also been studied.[60a]

Hydrolysis of carbohydrate phosphates is usually accomplished with acid. The structure of the compound markedly influences the rate of hydrolysis, compounds such as D-glucose 6-phosphate[60b] or glycerol phosphate requiring

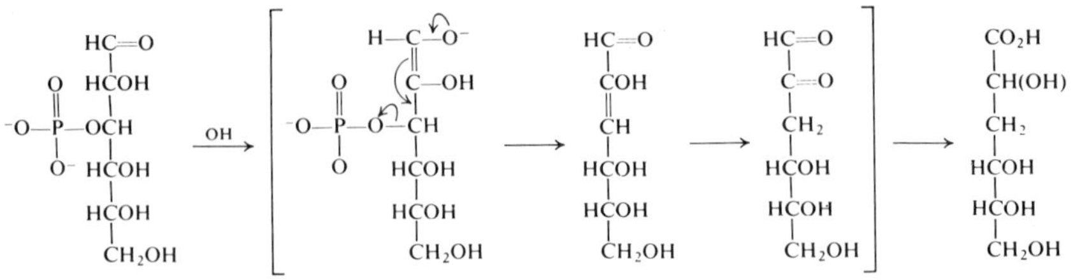

$$
\begin{array}{ccccc}
\begin{array}{l}\text{HC=O}\\ \text{HCOH}\\ \text{-O-P-OCH}\\ \text{HCOH}\\ \text{HCOH}\\ \text{CH}_2\text{OH}\end{array}
&\xrightarrow{\text{OH}}
&\left[\begin{array}{cccc}
\begin{array}{l}\text{H-C-O}^-\\ \text{C-OH}\\ \text{-O-P-O-CH}\\ \text{HCOH}\\ \text{HCOH}\\ \text{CH}_2\text{OH}\end{array}
&\begin{array}{l}\text{HC=O}\\ \text{COH}\\ \text{CH}\\ \text{HCOH}\\ \text{HCOH}\\ \text{CH}_2\text{OH}\end{array}
&\xrightarrow{}
\begin{array}{l}\text{HC=O}\\ \text{C=O}\\ \text{CH}_2\\ \text{HCOH}\\ \text{HCOH}\\ \text{CH}_2\text{OH}\end{array}
\end{array}\right]
&\xrightarrow{}
&\begin{array}{l}\text{CO}_2\text{H}\\ \text{CH(OH)}\\ \text{CH}_2\\ \text{HCOH}\\ \text{HCOH}\\ \text{CH}_2\text{OH}\end{array}
\end{array}
$$

much more strenuous conditions than those compounds where the phosphate group is α or β to the carbonyl function. Studies on D-glucose 2-phosphate have shown that the rate of phosphate migration in acid is greater than the rate of hydrolysis of the phosphate group,[61] and this situation probably obtains with other phosphates. Glycosyl phosphates are quite labile in acid, particularly the furanose forms. The hydrolysis of α-D-glucopyranosyl phosphate has been the subject of considerable study.[62, 63] Over the pH range 1 to 8, the hydrolysis was interpreted in terms of two reactions, one involving the monoanion and proceeding with phosphorus–oxygen bond fission, and the other involving the neutral species and proceeding with carbon–oxygen bond fission. In strongly acid solution the hydrolysis involves carbon–oxygen bond fission, but the two possibilities shown below cannot be kinetically distinguished, although

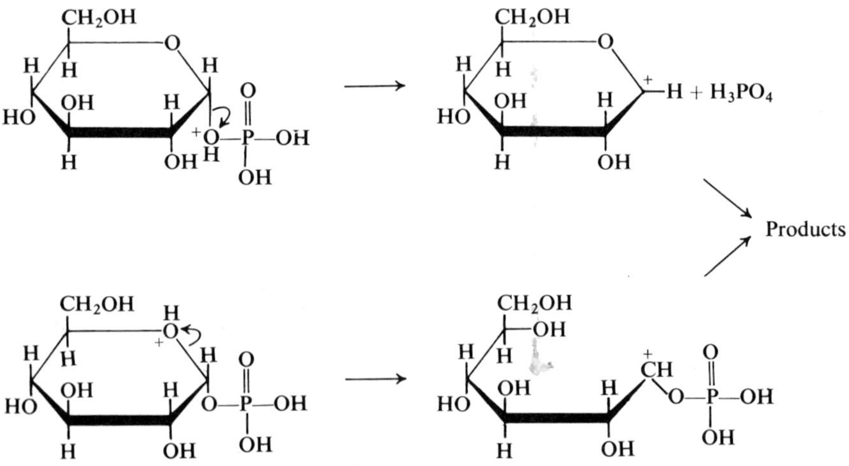

the reaction not involving ring opening is favored. The hydrolysis constants of a number of carbohydrate phosphates are found in Table II.

The alkaline hydrolysis of phosphates occurs with phosphorus–oxygen bond cleavage,[90] as does enzymic hydrolysis with prostatic acid phosphatase.[91]

TABLE II

OBSERVED ACID HYDROLYSIS CONSTANTS OF SOME CARBOHYDRATE PHOSPHATES[a]

Compound	Normality of acid	$T, °C$	$t_{1/2}, min$	$k \times 10^3$	Reference
Glycosyl phosphates					
2-Deoxy-α-D-*erythro*-pentosyl phosphate (pH 4)		25	12	57	64
α-D-Ribofuranosyl phosphate	0.5	25	2.5	275	14
β-D-Ribofuranosyl phosphate	0.1	26	240	2.9	45
α-D-Xylopyranosyl phosphate	0.38	36	110	6.21	8
α-D-Glucopyranosyl phosphate	0.01	61	310	2.2	65
	0.1	36	160	4.4	8
	0.25	37	230	3.0	7
	1	33	60	11.5	38
	1	37	97	7.1	66
	0.95	30	260	2.65	43
β-D-Glucopyranosyl phosphate	1	33	20	35	38
2-Amino-2-deoxy-α-D-glucopyranosyl phosphate	1	100	3	230	46
	1	100	ca. 4	ca. 170	67
2-Acetamido-2-deoxy-α-D-glucopyranosyl phosphate	1.33	26	570	1.2	48
	1	37	170	4.1	66
	1	37	190	3.7	46
2-Acetamido-2-deoxy-β-D-glucopyranosyl phosphate	1.33	26	8	86	48
α-D-Glucopyranosyluronic acid phosphate	0.01	61	3000	0.23	65
α-D-Galactopyranosyl phosphate	0.25	25	330	2.1	68
	0.25	37	50	14	68
	0.1	100	ca. 2	ca. 320	69
β-D-Galactopyranosyl phosphate	0.25	37	53	12	35
2-Amino-2-deoxy-α-D-galactopyranosyl phosphate	1	100	4.1	170	69
2-Acetamido-2-deoxy-α-D-galactopyranosyl phosphate	1	37	31	22	70
α-D-Mannopyranosyl phosphate	0.95	30	360	1.9	43
β-D-Fructopyranosyl phosphate (pH 4)		37	28	25	58
		37	20	34	49
β-D-Fructofuranosyl phosphate (pH 4)		37	8	86	58
α-Maltosyl phosphate	0.38	36	215	3.21	8
α-Lactosyl phosphate	1	37	250	2.7	71
β-Lactosyl phosphate	1	37	53	13	71

* *References start on p. 271.* (*continued*)

TABLE II (*Continued*)

OBSERVED ACID HYDROLYSIS CONSTANTS OF SOME CARBOHYDRATE PHOSPHATES[a]

Compound	Normality of acid	T, °C	$t_{1/2}$, min	$k \times 10^3$	Reference
α-D-Glucopyranose 1,6-diphosphate	0.25	37	970	0.71[b]	72
	1	30	880	0.78[b]	73
β-D-Glucopyranose 1,6-diphosphate	1	30	220	3.15[b]	73
α-D-Mannopyranose 1,6-diphosphate	0.95	30	1450	0.48[b]	43
Other Carbohydrate Phosphates					
Triose phosphates	1	100	8.1	85	74
D-*glycero*-Tetrulose 1-phosphate	1	100	30	22	75
	1	100	7	100	76
D-*glycero*-Tetrulose 4-phosphate	1	100	15	46	76
2-Deoxy-D-*erythro*-pentose 5-phosphate	1	100	ca. 6	ca. 115	77
D-Ribose 3-phosphate	0.01	100	ca. 180	ca. 3.9	78
	0.25	100	ca. 69	ca. 10	79
D-Ribose 5-phosphate	0.01	100	ca. 1000	ca. 0.7	78
	0.25	100	ca. 600	ca. 1.1	79
D-Xylose 5-phosphate	1	100	90	7.6	80
D-*erythro*-Pentulose 5-phosphate	1	100	ca. 60	ca. 11.5	81
D-*erythro*-Pentulose 1,5-diphosphate	0.1	100	12	57[b]	82
D-Glucose 2-phosphate	0.1	100	140	5.0	61
D-Glucose 6-phosphate	0.1	100	2300	0.3	83
	1	100	1400	0.5	84
D-Gluconic acid 6-phosphate	1	100		0.35–0.6	85
D-Fructose 1-phosphate	1	100	2.8	160	86
	0.1	100	33	21	86
D-Fructose 6-phosphate	1	100	70	10	84
D-Mannose 6-phosphate	0.1	100	2300	0.3	84
	1	100	1000	0.67	84
D-Mannonic acid 6-phosphate	1	100		0.25–0.46	85
D-Mannonolactone 6-phosphate	1	100	2500	0.28	85
L-Sorbose 6-phosphate	1	100	63	11	87
D-Fructose 1,6-diphosphate	1	100	6	115[b]	88
			75	9.2[c]	
Sedoheptulose 7-phosphate	1	100	75	9	82, 89

[a] Compiled by L. F. Leloir and C. E. Cardini, *Methods Enzymol.*, **3**, 840 (1957), with additions by the author. The k values are in min^{-1}, natural logarithms; $t_{1/2} = 0.69/k$.

[b] Constant for the phosphate on C-1.

[c] Constant for the phosphate on C-6.

D. Separation and Detection of Phosphate Esters

Carbohydrate phosphates are frequently fractionated as their barium salts. The more difficult separations are carried out by using paper chromatography and ion-exchange chromatography.[92] Thin-layer chromatography also has been applied.[93] The phosphate esters, after conversion into their dimethyl esters followed by trimethylsilylation, can be separated by gas–liquid chromatography.[94]

The quantitative determination of phosphate esters usually involves the formation of phosphomolybdate and its reduction to molybdenum blue.[95,96] Their detection on paper chromatograms is accomplished in the same manner,[97] the reduction being carried out photochemically with ultraviolet irradiation.[98] Irradiation has been reported to bring about dephosphorylation of carbohydrate phosphates at neutral pH.[99]

E. Specific Rotations of Natural and Synthetic Carbohydrate Phosphates

Table III lists the specific rotations of a number of carbohydrate phosphates that have been obtained by isolation from natural sources, or by enzymic or chemical synthesis.

TABLE III

Optical Rotation of Some Carbohydrate Phosphates[a]

Compound	Salt[b]	Solvent[c]	λ^d	$[\alpha]$, degrees	Reference
Trioses and Derivatives					
D-Glyceraldehyde 3-phosphate	F.A			+14	100
1,3-Dihydroxy-2-propanone phosphate					74, 101
D-Glyceric acid 3-phosphate		N HCl		−14.5	53, 102
		Molybdate		−745	53, 102
D-Glyceric acid 2-phosphate		N HCl		+13	53
		Molybdate		+5	53
D-Glyceric acid 1,3-diphosphate	F.A			−2.3	103
D-Glyceric acid 2,3-diphosphate	F.A			−2.3	104
L-Glycerol 3-phosphate		$2N$ HCl		−1.45	105
Tetroses and Derivatives					
D-Erythrose 4-phosphate	F.A			0	106
D-*glycero*-Tetrulose 1,4-diphosphate	F.A		400	−1.4 −11.7	107 107

* References start on p. 271.

(continued)

TABLE III (*Continued*)

OPTICAL ROTATION OF SOME CARBOHYDRATE PHOSPHATES[a]

Compound	Salt[b]	Solvent[c]	λ^d	$[\alpha]$, degrees	Reference
D-Erythritol 4-phosphate	CHA			−2.3	108
	F.A			+2.6	108
L-Erythritol 4-phosphate	CHA			+2.3	108
D-Erythronolactone 2-phosphate	CHA	N HCl		−55.0	109
D-Erythronic acid 4-phosphate	CHA			−20.0	109
4-Deoxy-D-erythronic acid 2-phosphate	F.A	N HCl		+15	110
4-Deoxy-D-erythronic acid 3-phosphate	F.A	N HCl		−14.5	110
	F.A	Molybdate		−737	110
Pentoses and Derivatives					
α-L-Arabinofuranosyl phosphate	Ba			+16.9	111
α-L-Arabinopyranosyl phosphate	CHA			+40.4	111
	CHA			+30.8	36
β-L-Arabinopyranosyl phosphate	CHA			+91	36
α-D-Arabinofuranosyl phosphate	Ba			+6.4	111
α-D-Arabinopyranosyl phosphate	CHA			−39.1	111
D-Arabinose 5-phosphate	Ba			−18.8	112
	Brucine	50% Pyridine		−48.6	112
α-D-Ribofuranose 1,5-diphosphate	CHA			+20.8	113
α-D-Ribofuranosyl phosphate	CHA			+40.3	114
	CHA			+53	115
β-D-Ribofuranosyl phosphate	CHA	Ethanol		−13.6	114
β-D-Ribopyranosyl phosphate	Ba	5% Acetic acid		−47.1	45
D-Ribose 3-phosphate	Na			−9.7	116
	Na	Half-saturated boric acid		+38	116
D-Ribose 5-phosphate	F.A			+16.5	78
	Ba			+5	117
2-Deoxy-α-D-*erythro*-pentosyl phosphate	CHA			+38.8	118
2-Deoxy-β-D-*erythro*-pentosyl phosphate	CHA			−15.8	44
2-Deoxy-D-*erythro*-pentose 5-phosphate	F.A			+19	119
	Ba			+16.5	120
	Ba			+10.8	121

TABLE III (*Continued*)

OPTICAL ROTATION OF SOME CARBOHYDRATE PHOSPHATES[a]

Compound	Salt[b]	Solvent[c]	λ[d]	[α], degrees	Reference
2-Deoxy-D-*threo*-pentose 5-phosphate	Ba			−35	122
3-Deoxy-D-*erythro*-pentose 5-phosphate	Ba			−10.6	123
D-*erythro*-Pentulose 5-phosphate	Ba	0.02N HCl		−40	124
2-Deoxy-D-*erythro*-pentitol 5-phosphate	Ba			−16.8	121
	CHA			−10.0	120
2-Deoxy-D-*erythro*-pentonic acid 5-phosphate	Ba			+1.95	121
α-D-Xylopyranosyl phosphate	Ba			+65	8
	Ba	5% Acetic acid		+70.9	39
	K			+76	8
	K			+75.5	39
	CHA			+58	36
β-D-Xylopyranosyl phosphate	Ba	5% Acetic acid		−13.3	39
	CHA			+0.8	36
D-Xylose 3-phosphate	Ba			+1.27	19
D-Xylose 5-phosphate	Na	50% Pyridine		+3.2	80
	Na	Half-saturated borax		+4	80
	Ba			+8	125
D-*threo*-Pentulose 5-phosphate	Brucine			−37.8	19, 126
D-Xylitol 5-phosphate	Ba			+1.27	19
Hexoses and Derivatives					
α-L-Fucopyranosyl phosphate	CHA			−77.8	126a
α-D-Galactopyranosyl phosphate	Ba		546	+113	68, 127
	Ba			+92	68, 127
	K			+98	68, 127
	CHA			+78	36
β-D-Galactopyranosyl phosphate	Ba			+31.2	35
	CHA			+21	36
D-Galactose 3-phosphate	K			+25.2	128
D-Galactose 6-phosphate				+36.5	128
α-D-Glucopyranosyl phosphate	F.A			+120	7
	Ba			+75	7
	K		546	+90	129
	Brucine			+0.5	38
	CHA			+64	36
	K			+78.5	129

* *References start on p. 271.*

(*continued*)

TABLE III (*Continued*)

OPTICAL ROTATION OF SOME CARBOHYDRATE PHOSPHATES[a]

Compound	Salt[b]	Solvent[c]	λ^d	$[\alpha]$, degrees	Reference
β-D-Glucopyranosyl phosphate	Brucine			−20	38
	Ba			+20.6	35
	CHA			+7.3	36
α-L-Glucopyranosyl phosphate	Ba			−73.2	130
	K			−78.2	130
α-D-Glucopyranose 1,6-diphosphate	F.A			+83	73
β-D-Glucopyranose 1,6-diphosphate	F.A			−19	73
D-Glucose 2-phosphate	K			+15	61
	K	0.1N H_2SO_4		+35	61
D-Glucose 3-phosphate	F.A		546	+39	131
	F.A			+39.5	132
	Brucine	50% Pyridine		−14.5	133
	Ba			+26.5	132
	Ba		546	+27	131
D-Glucose 4-phosphate	Brucine	Pyridine		−45.3	134
	Brucine	20% Ethanol		−9.8	134
D-Glucose 5-phosphate	Ba			+15	135
D-Glucose 6-phosphate	F.A			+35.1	83
	F.A		546	+41.4	83
	Ba			+18	83
	Ba		546	+21	83
	K			+21.2	136
3-O-Methyl-D-glucose 6-phosphate	CHA			+22	137
α-L-Idopyranosyl phosphate	CHA			−27.5	137a
5-Thio-α-D-glucopyranosyl phosphate	CHA			+136.6	137b
α-D-Mannopyranosyl phosphate	F.A			+58	10
	Ba			+36	10
	CHA			+28.7	138
α-D-Mannopyranose 1,6-diphosphate	K			+29.9	43
D-Mannose 6-phosphate	F.A		546	+15.1	84
	Ba		546	+3.6	84
α-L-Rhamnopyranosyl phosphate	CHA			−21.5	138a
D-Fructose 1-phosphate	F.A		546	−64.2	86
	Ba		546	−39	86
	Brucine		546	−52.1	86
β-D-Fructopyranosyl phosphate	Ba			−83.3	58
	CHA			−77.9	49

TABLE III (*Continued*)

OPTICAL ROTATION OF SOME CARBOHYDRATE PHOSPHATES[a]

Compound	Salt[b]	Solvent[c]	λ[d]	[α], degrees	Reference
β-D-Fructofuranosyl phosphate	Na			−53.6	58
D-Fructose 6-phosphate	Ba			+3.6	139
D-Fructose 1,6-diphosphate	F.A			+4.1	139
L-Sorbose 1-phosphate	K			−16.5	87
L-Sorbose 6-phosphate	Ba			−12.0	87
D-Tagatose 6-phosphate	Ba			+5.6	140
β-D-Galactopyranosyluronic acid phosphate	Benzyl-amine			−14	141
α-D-Glucopyranosyluronic acid phosphate	K			+51	142
	K			+53.6	65
D-Gluconic acid 6-phosphate	Ba		546	−1.5	83
	F.A		546	+0.2	83
D-Gluconolactone 6-phosphate			546	+21	83
D-Mannono-1,4-lactone 6-phosphate			546	+54.1	85
D-Mannono-1,5-lactone 6-phosphate			546	+60.6	85
2-Acetamido-2-deoxy-α- D-galactopyranosyl phosphate	Li		578	+189	143
	Li		578	+197	144
	K			+112.4	145
	F.A			+178	146
2-Acetamido-2-deoxy-α-D-galactopyranosyl phosphate 6-sulfate	Ba			+71.5	147
2-Amino-2-deoxy-α-D-galactopyranosyl phosphate	F.A			+142.6	145
2-Acetamido-2-deoxy-α-D glucopyranosyl phosphate	K			+79	46
	K			+76.1	48
	Ca			+107	66
	Li		578	+144	144
2-Acetamido-2-deoxy-β-D-glucopyranosyl phosphate	Na			−1.7	48, 148
2-Amino-3-O-(D-1-carboxy-ethyl)-2-deoxy-D-glucose 6-phosphate (muramic acid 6-phosphate)	F.A			+79	148a
2-Amino-2-deoxy-α-D-glucopyranosyl phosphate	K			+100	46
2-Amino-2-deoxy-D-glucose 3-phosphate	F.A			+70	149
	F.A			+79	24
2-Amino-2-deoxy-D-glucitol 3-phosphate	F.A			−20.5	24

* *References start on p. 271.*   (*continued*)

TABLE III (*Continued*)

OPTICAL ROTATION OF SOME CARBOHYDRATE PHOSPHATES[a]

Compound	Salt[b]	Solvent[c]	λ^d	[α], degrees	Reference
2-Deoxy-D-*arabino*-hexonic acid 6-phosphate	CHA	Ethanol		+6	31
2-Deoxy-D-*lyxo*-hexose 3-phosphate	F.A			+25	128
2-Deoxy-D-*lyxo*-hexose 6-phosphate	F.A			+41	128
3-Deoxy-D-*ribo*-hexose 6-phosphate	Ba			+6.6	121
	Ba			+3.8	150
	Brucine			−24.2	150
3-Deoxy-D-*ribo*-hexonic acid 6-phosphate	Brucine			−20.2	150
3-Deoxy-α-D-*xylo*-hexo-pyranosyl phosphate	Ba			+4	40
3-Deoxy-β-D-*xylo*-hexo-pyranosyl phosphate	Ba			−96	40
3,6-Dideoxy-α-D-*xylo*-hexo-pyranosyl phosphate	Ba			+1.5	41
3,6-Dideoxy-β-D-*xylo*-hexo-pyranosyl phosphate	Ba			−3.8	41
D-Glucosaccharinic acid 6-phosphate				+62	151
α,β-D-Glucometasaccharinic acid 6-phosphate	Ba			−5.5	137
α,β-D-Glucometasaccharinic acid 5-phosphate	Ba			−6.7	137
α-D-*xylo*-Hexopyranos-4-ulosyluronic acid phosphate				+30	152
Heptoses and Derivatives					
D-*glycero*-D-*galacto*-Heptose 6-phosphate	Ba			+26.8	153
L-*glycero*-α-D-*manno*-Hepto-pyranosyl phosphate	CHA			+32	153a
Sedoheptulose 7-phosphate	Ba		546	+8	82, 154
3-Deoxy-D-*gluco*-heptonic acid 7-phosphate	CHA			+9.2	155
3-Deoxy-D-*arabino*-heptulosonic acid 7-phosphate	K			+15.7	155
	F.A			+42	155
Disaccharides					
α-Lactosyl phosphate	Ba			+73.3	71
	F.A			+99.5	156

TABLE III (*Continued*)

OPTICAL ROTATION OF SOME CARBOHYDRATE PHOSPHATES[a]

Compound	Salt[b]	Solvent[c]	λ^d	$[\alpha]$, degrees	Reference
β-Lactosyl phosphate	Ba			+24.8	71
	F.A			+31.5	156
α-Maltosyl phosphate	Ba			+107	8
N-Acetyl-α-chondrosinyl phosphate	Tri-*n*-butyl amine			+66.4	157
Sucrose 6′-phosphate	Ba			+35.4	157a
Trehalose 6-phosphate	F.A		546	+185	158
	Ba		546	+132	158
	Brucine		546	+31	158
	Ba			+99	159
Trehalose 6,6′-diphosphate	CHA			+62	159

[a] Compiled by L. F. Leloir and C. E. Cardini, *Methods Enzymol.*, **3**, 840 (1957), with additions by W. Z. Hassid and C. E. Ballou, in "The Carbohydrates," 2nd ed. (W. Pigman, ed.), Academic Press, New York, 1957, p. 172, and by the author,

[b] Abbreviations used are F.A = free acid, CHA = cyclohexylamine. A blank indicates that the salt used was not specified by the author.

[c] The solvent is water unless otherwise indicated.

[d] The D-line of sodium, unless otherwise indicated in nm.

II. BORATE ESTERS

The enhanced optical rotation of sugars and alditols in boric acid solution provided early evidence for borate complexes of sugars.[160] The formation of such complexes has been of particular value in chromatographic separations (see Vol. IB, Chapter 28). Sugars[161] and their phosphorylated derivatives[162] can be separated on ion-exchange resins in the presence of borate ions, and by using paper chromatography in the presence of borate ions,[163] a number of separations have been achieved. Ionophoresis in the presence of borate[164] or sulfonated benzeneboronate[165] buffers has also led to some insight regarding the nature of the complex and the stereochemical factors governing its formation.

Although much has been written on complex formation, there are relatively few examples of crystalline esters of boric acid with sugars. Orthoboric acid reacts with 1,2-*O*-isopropylidene-α-D-glucofuranose in the presence of concentrated sulfuric acid to give a crystalline 3,5-orthoborate[166] (m.p. 90–110°), and during the reduction of *scyllo*-inosose with sodium borohydride, a crystalline scyllitol diborate monohydrate crystallizes from solution.[167] Crystalline derivatives of boric acid with sugars have also been prepared by heating trialkyl borates with substituted sugars. In this way a crystalline 1,2:5,6-di-O-iso-propylidene-α-D-glucofuranose 3-(dimethyl borate) has been obtained, and certain other O-isopropylidene and O-benzylidene derivatives give crystalline esters.[168]

Considerable interest in the benzeneboronate esters of carbohydrates arises from the fact that crystalline esters are readily formed under anhydrous conditions and that these, in turn, are readily hydrolyzed by reaction with water,[169] or by transesterification reactions involving propane-1,3-diol.[170] The cyclic derivatives formed, in which the benzeneboronate is usually a part of a five- or six-membered ring, are stable to the conditions of glycoside synthesis[171] and esterification[172] but may not be stable during methylation.[172] It has been shown that oxidation of methyl α-D-xylopyranoside 2,4-benzeneboronate with methyl sulfoxide–acetic anhydride[173] provides a convenient route to methyl α-D-erythro-pentopyranosid-3-ulose.[174] Butaneboronates of carbohydrates are useful derivatives for gas–liquid chromatography.[174a]

III. NITRATE ESTERS

Nitrate esters of carbohydrates, in particular of polysaccharides, have long been of considerable importance as explosives.[175] The preparation of such esters has been of interest, therefore, from a commercial as well as other points of view, and their formation as well as their decomposition has been well investigated. The controlled thermal decomposition of compounds such as cellulose nitrate has been investigated in a series of papers by Wolfrom et al.[176] In recent years, there has been much interest in the preparation and properties of nitrate esters of the simpler sugars, and these derivatives have been found to be useful intermediates in synthetic carbohydrate chemistry.[177]

A variety of conditions is available for the preparation of sugar nitrates, including absolute nitric acid, nitric acid with acetic anhydride and acetic acid, nitric acid with sulfuric acid, nitric acid with chloroform, and dinitrogen pentaoxide in chloroform. The nitrate group can also be introduced by displacement of a halogen atom by the action of silver nitrate; both secondary as well as primary halides may react. 1,4:3,6-Dianhydro-2,5-dideoxy-2,5-diiodo-L-iditol reacts with silver nitrate in dry acetonitrile to give dinitrate esters of the 1,4:3,6-dianhydrides of D-mannitol, D-glucitol, and L-iditol; a cyclic carbonium-ion intermediate is proposed, with the preferential formation of an exo-O-nitro group.[178] The halogen atom of acetylated glycopyranosyl halides can be replaced by using silver nitrate, giving rise to acetylated glycopyranosyl nitrates.[179]

The removal of nitrate groups is most frequently accomplished by reduction. Polynitrates are readily reduced, xylitol pentanitrate giving an almost quantative yield of xylitol on hydrogenolysis using palladium-on-carbon catalyst.[180] Hydrazine in boiling ethanol has also been used for smooth denitration of glycosides.[181] Lithium aluminum hydride is not an efficient reducing agent for sugar nitrates, but when reduction does occur, the parent alcohol is formed.[182] Sodium borohydride converts 2,3,4,6-tetra-O-acetyl-α-D-glucopyranosyl nitrate into 1,5-anhydro-D-glucitol in low yield, but is ineffective in reducing nitro groups on primary and secondary hydroxyl groups.[183] Denitration is sometimes selective, for treatment of methyl 4,6-O-benzylidene-α-D-altropyranoside 2,3-dinitrate with sodium nitrite in aqueous alcohol brings about removal of the 2-O-nitro group.[184]

Sugar nitrate groups are relatively stable to acid hydrolysis, but alkaline hydrolysis is complex and can lead to the formation of carbonyl compounds, ethers, and anhydro compounds. Thus tetra-O-acetyl-α-D-glucopyranosyl nitrate gives methyl β-D-glucopyranoside with sodium methoxide in methanol, whereas sodium hydroxide in aqueous dioxane converts it into 1,6-anhydro-β-D-glucopyranose.[185] However, 1,2:5,6-di-O-cyclohexylidene-D-glucofuranose 3-nitrate is smoothly denitrated by boiling sodium methoxide in methanol.[186] O'Meara and Shepherd[187] have studied the hydrolysis of mono- and oligonitrates of methyl β-D-glucopyranoside and interpreted the results in terms of substitution and elimination reactions.

One of the interesting properties of the nitrate group involves its use as a nonparticipating blocking group on C-2 during glycoside syntheses. This has led to successful syntheses of isomaltose[188] and panose[189] by using 3,4,6-tri-O-acetyl-2-O-nitro-β-D-glucopyranosyl chloride. A crystalline pair of anomers, having nonparticipating 2-O-nitro groups, is available in the α and β anomers of 2-O-nitro-3,5-di-O-p-nitrobenzoyl-D-arabinofuranosyl chloride.[190]

IV. SULFATE ESTERS

Sulfate esters occur naturally in a variety of polysaccharides of plant and animal origin, and considerable investigation of monosaccharide and oligosaccharide sulfates has been performed in order to determine their properties and to relate these findings to the biological significance of the sulfate group. A comprehensive review of this work has been published.[191]

Sulfating agents generally used are chlorosulfonic acid in pyridine, or a sulfur trioxide–pyridine complex. Reagents used less frequently include concentrated aqueous sodium bisulfite, which has been reported to react with D-glucose to give some D-glucose 6-sulfate.[192] In other cases sulfur is introduced as sulfite, as in the case of 1,2:5,6-di-O-isopropylidene-α-D-glucofuranose

3-(methyl sulfite), which can be oxidized to the 3-(methyl sulfate) with permanganate and hydrolyzed to the 3-sulfate.[193]

Starting materials for the preparation of sugar sulfates can be the free sugars, although this may lead to mixtures from which individual sulfates may be isolated only with difficulty. For example, sulfation of D-glucose or D-galactose with excess of pyridine–sulfur trioxide gives a mixture in which individual hexose molecules may contain as many as four sulfate groups.[194] However, in a number of cases the preferentially formed 6-sulfate may be isolated, as for instance D-glucose 6-sulfate and 2-acetamido-2-deoxy-D-glucose 6-sulfate.[195] A number of polysaccharide sulfates have been prepared by appropriate methods of sulfation. For example, by using chlorosulfonic acid in pyridine, synthetic polyglucose sulfates having different degrees of sulfation have been prepared.[196] Treatment of D-glucose, D-galactose, and D-mannose with concentrated sulfuric acid gives products considered to be the 1,3,6-trisulfates; the 1,2,4-trisulfate was obtained from D-fructose.[196a]

Definitive syntheses of sugar sulfates and of sulfates of glycosides are achieved from starting materials that are fully protected except on the position to be sulfated. In this manner D-galactose 4-sulfate has been prepared from benzyl 2,3-di-O-benzyl-6-O-trityl-β-D-galactopyranoside by sulfation followed by removal of protecting groups,[197] and 2-deoxy-2-sulfoamino-D-glucose results from sulfation of 1,3,4,6-tetra-O-acetyl-2-amino-2-deoxy-α-D-glucopyranose followed by deacetylation with sodium methoxide.[198] Protected starting materials such as 1,2:3,4-di-O-isopropylidene-α-D-galactopyranose and 1,2:5,6-di-O-isopropylidene-α-D-glucopyranose in N,N-dimethylformamide can be converted into the respective monosulfates by direct coupling with sulfuric acid through the action of N,N'-dicyclohexylcarbodiimide.[198a]

The reaction of sulfuryl chloride with sugars, glycosides, and substituted sugars[199] leads to complex mixtures which include sulfates, cyclic sulfates, and chlorodeoxy derivatives, as well as a variety of chlorosulfates. Observations in this field are to be found in a series of papers by J. K. N. Jones and co-workers.[200]

The removal of sulfate groups can be accomplished in a variety of ways. Acetolysis in absolute sulfuric acid[201] and reduction with lithium aluminum hydride[202] proceed without inversion on the carbon atom bearing the O-sulfate group. Acid hydrolysis of sulfate esters proceeds with retention of configuration,[203] whereas treatment with alkali results mainly in inversion with the formation of anhydro sugar derivatives.[204] Cyclic sulfate esters are cleaved by mild alkali to give the sulfate salts with retention of configuration, while acid hydrolysis can lead to the production of anhydro sugar derivatives (see Chapter 13) accompanied by inversion; 1,2:5,6-di-O-isopropylidene-D-mannitol 3,4-(cyclic sulfate) on treatment with hot dilute acid gives 1,4-anhydro-D-talitol in 88% yield.[205]

V. GLYCOSYL AZIDES

Glycosyl azides,[206] in which the glycosidic hydroxyl of a sugar has been replaced by an azide group, are useful intermediates in the preparation of glycosylamines (see Chapter 20). Starting materials for their preparation are the acylated glycosyl halides; by the action of sodium azide on 2,3,4,6-tetra-O-acetyl-α-D-glucopyranosyl bromide, the corresponding β-azide is readily formed.[207] The α-anomer can be obtained in good yield by using active silver azide and 3,4,6-tri-O-acetyl-2-O-trichloroacetyl-β-D-glucopyranosyl chloride in ether.[208] Azides of amino sugars are similarly prepared, silver azide with 3,4,6-tri-O-acetyl-2-amino-2-deoxy-N-p-tolylsulfonyl-β-D-glucopyranosyl fluoride giving the crystalline β-azide in excellent yield.[209] Furanosyl azides, such as the syrupy tri-O-benzoyl-α-D-lyxofuranosyl azide[210] and the crystalline 2,3,5-tri-O-benzoyl-β-D-ribofuranosyl azide,[211] are also known.

These azides can be deacylated using sodium methoxide in methanol without loss of the azide group, and β-D-ribofuranosyl azide[211] can be converted into the 2,3-O-isopropylidene derivative, which in turn can be phosphorylated and then reduced to give D-ribofuranosylamine 5-phosphate, a useful intermediate in nucleotide syntheses.[212]

Glycosyl azides react readily with acetylenes to give N-glycosyltriazoles.[213]

REFERENCES

1. For a pertinent summary of this work see H. G. Khorana, "Some Recent Developments in the Chemistry of Phosphate Esters of Biological Interest," Wiley, New York, 1961; A. B. Foster and W. G. Overend, *Quart. Rev.* (London), **11**, 61 (1957); L. Szabó, *Advan. Chem. Ser.*, **74**, 70 (1968).
2. W. D. Kumler and J. J. Eiler, *J. Amer. Chem. Soc.*, **65**, 2355 (1943).
3. L. F. Leloir, *Fortschr. Chem. Org. Naturstoffe*, **8**, 47 (1951).
4. D. D. Van Slyke, *J. Biol. Chem.*, **52**, 525 (1922).
5. H. T. S. Britton and R. A. Robinson, *Trans. Faraday Soc.*, **28**, 531 (1932).
6. O. Meyerhof and J. Suranyi, *Biochem. Z.*, **178**, 427 (1926).
7. C. F. Cori, S. P. Colowick, and G. T. Cori, *J. Biol. Chem.*, **121**, 465 (1937).
8. W. R. Meagher and W. Z. Hassid, *J. Amer. Chem. Soc.*, **68**, 2135 (1946).
9. H. W. Kosterlitz, *Biochem. J.*, **37**, 321 (1943).
10. S. P. Colowick, *J. Biol. Chem.*, **124**, 557 (1938).
11. O. Meyerhof and K. Lohmann, *Biochem. Z.*, **185**, 113 (1927).
12. C. Neuberg and H. Lustig, *J. Amer. Chem. Soc.*, **64**, 2772 (1942); I. Mandl and C. Neuberg, *Methods Enzymol.*, **4**, 162 (1957).
13. E. W. Putman, *Methods Carbohyd. Chem.*, **2**, 267 (1963).
14. H. M. Kalckar, *J. Biol. Chem.*, **167**, 477 (1947); R. E. Plesner and H. Klenow, *Methods Enzymol.*, **3**, 181 (1957).
15. M. Friedkin and H. M. Kalckar, *Methods Enzymol.*, **3**, 183 (1957).
16. J. E. Seegmiller and B. L. Horecker, *J. Biol. Chem.*, **192**, 175 (1951).
17. M. Viscontini and C. Olivier, *Helv. Chim. Acta*, **36**, 466 (1953).
17a. F. R. Zuleski and E. T. McGuinness, *J. Label. Compounds*, **5**, 371 (1970).

18. Reviews on phosphorylating agents include F. Cramer, *Angew Chem.*, **72**, 236 (1960); V. M. Clark, D. W. Hutchinson, A. J. Kirby, and S. G. Warren, *Angew. Chem. Int. Ed. Engl.*, **3**, 678 (1964); D. M. Brown, *Advan. Org. Chem.*, **3**, 75 (1963).

19. J. G. Moffatt and H. G. Khorana, *J. Amer. Chem. Soc.*, **79**, 1194 (1957).

20. T. Hashizume, K. Fujimoto, H. Unuma, K. Takinami, and K. Morimoto, *Bull. Inst. Chem. Res., Kyoto Univ.*, **38**, 70 (1960); *Chem. Abstr.*, **56**, 1516 (1962).

21. G. M. Tener, *J. Amer. Chem. Soc.*, **83**, 159 (1961).

22. J. L. Sannella and R. L. Whistler, *Arch. Biochem. Biophys.*, **102**, 226 (1963).

22a. W. Hengstenberg and M. L. Morse, *Carbohyd. Res.*, **7**, 180 (1968).

23. C. Neuberg and H. Pollak, *Biochem. Z.*, **23**, 515 (1910); **26**, 514 (1910); *Ber.*, **43**, 2060 (1910); E. Fischer, *Ber.*, **47**, 3193 (1914); E. Baer, *Biochem. Prep.*, **2**, 25, 31 (1952).

24. R. Lambert and F. Zilliken, *Ber.*, **96**, 2350 (1963).

25. F. R. Atherton, *Biochem. Prep.*, **5**, 1 (1957).

26. E. Baer, *Biochem. Prep.*, **1**, 51 (1949); **2**, 97 (1952).

27. F. Maley and H. A. Lardy, *J. Amer. Chem. Soc.*, **78**, 1393 (1956).

28. F. R. Atherton, H. T. Howard, and A. R. Todd, *J. Chem. Soc.*, 1106 (1948).

29. See for example, P. Rivaille and L. Szabó, *Bull. Soc. Chim. Fr.*, 712, 716 (1963).

30. E. Baer, *J. Amer. Oil. Chem. Soc.*, **42**, 257 (1965).

31. M. L. Wolfrom and N. E. Franks, *J. Org. Chem.*, **29**, 3645 (1964).

32. T. Ukita and K. Nagasawa, *Chem. Pharm. Bull.* (Tokyo), **9**, 544 (1961).

33. T. Ukita, N. Imura, K. Nagasawa, and N. Aimi, *Chem. Pharm. Bull.* (Tokyo), **10**, 1113 (1962).

34. G. P. Lampson and H. A. Lardy, *J. Biol. Chem.*, **181**, 693 (1949).

34a. A. K. Chatterjee and D. L. MacDonald, *J. Org. Chem.*, **33**, 1584 (1968).

34b. L. D. Hall and P. R. Steiner, *Chem. Commun.*, 84 (1971); compare K. Kumamoto, H. Yoshida, T. Ogata, and S. Inokawa, *Bull. Chem. Soc. Japan*, **42**, 3245 (1969); G. H. Jones, H. P. Albrecht, N. P. Damodaran, and J. G. Moffatt, *J. Amer. Chem. Soc.*, **92**, 5510 (1970).

35. F. J. Reithel, *J. Amer. Chem. Soc.*, **67**, 1056 (1945).

36. E. W. Putman and W. Z. Hassid, *J. Amer. Chem. Soc.*, **79**, 5057 (1957).

37. D. L. MacDonald and H. G. Fletcher, Jr., *J. Amer. Chem. Soc.*, **82**, 1832 (1960).

38. M. L. Wolfrom, C. S. Smith, D. E. Pletcher, and A. E. Brown, *J. Amer. Chem. Soc.*, **64**, 23 (1942); compare C. L. Stevens and R. E. Harmon, *Carbohyd. Res.*, **11**, 93, 99 (1969).

39. N. J. Antia and R. W. Watson, *J. Amer. Chem. Soc.*, **80**, 6134 (1958).

40. K. Antonakis, *Compt. Rend.*, **258**, 3511 (1964).

41. K. Antonakis, *Bull. Soc. Chim. Fr.*, 2112 (1965).

42. T. Posternak, *J. Amer. Chem. Soc.*, **72**, 4824 (1950).

43. T. Posternak and J. P. Rosselet, *Helv. Chim. Acta*, **36**, 1614 (1953).

44. D. L. MacDonald and H. G. Fletcher, Jr., *J. Amer. Chem. Soc.*, **84**, 1262 (1962).

45. R. S. Wright and H. G. Khorana, *J. Amer. Chem. Soc.*, **78**, 811 (1956).

46. F. Maley, G. F. Maley, and H. A. Lardy, *J. Amer. Chem. Soc.*, **78**, 5303 (1956).

47. D. L. MacDonald, *J. Org. Chem.*, **27**, 1107 (1962).

47a. D. L. MacDonald, *Carbohyd. Res.*, **3**, 117 (1966).

47b. W. A. Khan and K. E. Ebner, *Anal. Biochem.*, **22**, 338 (1968).

48. P. J. O'Brien, *Biochim. Biophys. Acta*, **86**, 628 (1964).

49. D. L. MacDonald, *J. Org. Chem.*, **31**, 513 (1966).

50. *Methods Enzymol.*, **3**, 129–223 (1957); **8**, 121–131 (1966).

51. E. W. Putman, *Methods Carbohyd. Chem.* **2**, 261–297 (1963).

52. M. C. Bailly, *Compt. Rend.*, **206**, 1902 (1938); **208**, 443 (1939).

53. C. E. Ballou and H. O. L. Fischer, *J. Amer. Chem. Soc.*, **76**, 3188 (1954).
54. D. M. Brown, D. I. Magrath, and A. R. Tood, *J. Chem. Soc.*, 2708 (1952).
55. H. G. Khorana, G. M. Tener, R. S. Wright, and J. G. Moffatt, *J. Amer. Chem. Soc.*, **79**, 430 (1957).
56. See, for instance, B. Zmudzka and D. Shugar, *Acta Biochim. Polon. (Engl. transl.)*, **11**, 509 (1964).
57. R. Piras, *Arch. Biochem. Biophys.*, **103**, 291 (1963).
58. H. G. Pontis and C. L. Fischer, *Biochem. J.*, **89**, 452 (1963).
59. See "Phosphoric Esters and Related Compounds," *Chem. Soc. Spec. Publ., No.* **8** (1957); and in J. R. Cox, Jr., and O. B. Ramsay, *Chem. Rev.*, **64**, 317 (1964); L. F. Leloir and C. E. Cardini, in "Comprehensive Biochemistry," M. Florkin and E. H. Stotz, Eds., Vol. V, Elsevier, Amsterdam, 1963, p. 113.
60. D. M. Brown, F. Hayes, and A. R. Todd, *Ber.*, **90**, 936 (1957).
60a. Ch. Degani and M. Halmann, *J. Amer. Chem. Soc.*, **90**, 1313 (1968).
60b. Ch. Degani and M. Halmann, *J. Amer. Chem. Soc.*, **88**, 4075 (1966); C. A. Bunton and H. Chaimovitch, *ibid.*, **88**, 4082 (1966).
61. K. R. Farrar, *J. Chem. Soc.*, 3131 (1949).
62. C. A. Bunton, D. R. Llewellyn, K. G. Oldham, and C. A. Vernon, *J. Chem. Soc.*, 3588 (1958); C. A. Bunton and E. Humeres, *J. Org. Chem.*, **34**, 572 (1969).
63. A. R. Osborn and E. Whalley, *Can. J. Chem.*, **39**, 597 (1961).
64. M. Friedkin, H. M. Kalckar and E. Hoff-Jørgensen, *J. Biol. Chem.* **178** 527 (1949); M. Friedkin, *ibid.*, **184**, 449 (1950).
65. S. A. Barker, E. J. Bourne, J. G. Fleetwood, and M. Stacey, *J. Chem. Soc.*, 4128 (1958).
66. L. F. Leloir and C. E. Cardini, *Biochim. Biophys. Acta*, **20**, 33 (1956).
67. D. H. Brown, *J. Biol. Chem.*, **204**, 877 (1953).
68. H. W. Kosterlitz, *Biochem. J.*, **33**, 1087 (1939).
69. C. E. Cardini and L. F. Leloir, *Arch. Biochem. Biophys.*, **45**, 55 (1953).
70. L. F. Leloir, C. E. Cardini, and J. M. Olavarria, *Arch. Biochem. Biophys.*, **74**, 84 (1958).
71. R. Sasaki and K. Taniguchi, *Nippon Nôgei Kagaku Kaishi*, **33**, 183 (1959); *Chem. Abstr.*, **54**, 308 (1960).
72. C. E. Cardini, A. C. Paladini, R. Caputto, L. F. Leloir, and R. E. Trucco, *Arch. Biochem. Biophys.*, **22**, 87 (1949).
73. T. Posternak, *J. Biol. Chem.*, **180**, 1269 (1949).
74. O. Meyerhof and K. Lohmann, *Biochem, Z.*, **271**, 89 (1934).
75. F. C. Charalampous and G. C. Mueller, *J. Biol. Chem.*, **201**, 161 (1953).
76. N. J. Chu and C. E. Ballou, *J. Amer. Chem. Soc.*, **83**, 1711 (1961).
77. E. Racker, *J. Biol. Chem.*, **196**, 347 (1952).
78. P. A. Levene and E. T. Stiller, *J. Biol. Chem.*, **104**, 299 (1934).
79. H. G. Albaum and W. W. Umbreit, *J. Biol. Chem.*, **167**, 369 (1947).
80. P. A. Levene and A. L. Raymond, *J. Biol. Chem.*, **102**, 347 (1933).
81. B. L. Horecker, P. Z. Smyrniotis, and J. E. Seegmiller, *J. Biol. Chem.*, **193**, 383 (1951).
82. A. A. Benson, in "Modern Methods of Plant Analysis," K. Paech and M. V. Tracey, Eds., Vol. II, Springer–Verlag, Berlin, 1955, p. 113.
83. R. Robison and E. J. King, *Biochem. J.*, **25**, 323 (1931).
84. R. Robison, *Biochem. J.*, **26**, 2191 (1932).
85. V. R. Patwardhan, *Biochem, J.*, **28**, 1854 (1934).
86. B. Tanko and R. Robison, *Biochem. J.*, **29**, 961 (1935).
87. K. M. Mann and H. A. Lardy, *J. Biol. Chem.*, **187**, 339 (1950).
88. M. Macleod and R. Robison, *Biochem. J.*, **27**, 286 (1933).
89. R. Robison, M. G. Macfarlane, and A. Tazelaar, *Nature*, **142**, 114 (1938).

90. E. Blumenthal and J. B. M. Herbert, *Trans. Faraday Soc.*, **41**, 611 (1945).
91. M. Cohn, *J. Biol. Chem.*, **180**, 771 (1949).
92. A. A. Benson, *Methods Enzymol.*, **3**, 110 (1957); D. L. MacDonald, *Carbohyd. Res.*, **6**, 376 (1968).
93. See, for example, C. P. Dietrich, S. M. C. Dietrich, and H. G. Pontis, *J. Chromatogr.*, **15**, 277 (1964).
94. W. W. Wells, T. Katagi, R. Bentley, and C. C. Sweeley, *Biochim. Biophys. Acta*, **82**, 408 (1964).
95. L. F. Leloir and C. E. Cardini, *Methods Enzymol.*, **3**, 840 (1957).
96. G. R. Bartlett, *J. Biol. Chem.*, **234**, 466 (1959).
97. C. S. Hanes and F. A. Isherwood, *Nature*, **164**, 1107 (1949).
98. R. S. Bandurski and B. Axelrod, *J. Biol. Chem.*, **193**, 405 (1951).
99. H. Trapmann and M. Devani, *Naturwissenschaften*, **52**, 208 (1965).
100. O. Meyerhof and R. Junowicz-Kocholaty, *J. Biol. Chem.*, **149**, 71 (1943); C. E. Ballou and H. O. L. Fischer, *J. Amer. Chem. Soc.*, **77**, 3329 (1955); see G. R. Gray and R. Barker, *Carbohyd. Res.*, **20**, 31 (1971) for the 3,3'-dideuterated analog.
101. C. E. Ballou and H. O. L. Fischer, *J. Amer. Chem. Soc.*, **78**, 1659 (1956).
102. O. Meyerhof and W. Schulz, *Biochem. Z.*, **297**, 60 (1938).
103. E. Negelein and H. Brömel, *Biochem. Z.*, **303**, 132 (1939).
104. E. Baer, *J. Biol. Chem.*, **185**, 763 (1950).
105. E. Baer, and H. O. L. Fischer, *J. Biol. Chem.*, **128**, 491 (1939).
106. C. E. Ballou, H. O. L. Fischer, and D. L. MacDonald, *J. Amer. Chem. Soc.*, **77**, 5967 (1955).
107. G. A. Taylor and C. E. Ballou, *Biochemistry*, **2**, 553 (1963).
108. D. L. MacDonald, H. O. L. Fischer, and C. E. Ballou, *J. Amer. Chem. Soc.*, **78**, 3720 (1956).
109. R. Barker and F. Wold, *J. Org. Chem.*, **28**, 1847 (1963).
110. C. E. Ballou, *J. Amer. Chem. Soc.*, **79**, 984 (1957).
111. R. S. Wright and H. G. Khorana, *J. Amer. Chem. Soc.*, **80**, 1994 (1958).
112. P. A. Levene and C. C. Christman, *J. Biol. Chem.*, **123**, 607 (1938).
113. G. M. Tener and H. G. Khorana, *J. Amer. Chem. Soc.*, **80**, 1999 (1958).
114. G. M. Tener, R. S. Wright, and H. G. Khorana, *J. Amer. Chem. Soc.*, **79**, 441 (1957).
115. D. H. Hayes and H. M. Kalckar, quoted in ref. 45.
116. P. A. Levene and S. A. Harris, *J. Biol. Chem.*, **101**, 419 (1933).
117. A. M. Michelson and A. R. Todd, *J. Chem. Soc.*, 2476 (1949).
118. H. L. A. Tarr, *Can. J. Biochem. Physiol.*, **36**, 517 (1958).
119. D. L. MacDonald and H. G. Fletcher, Jr., *J. Amer. Chem. Soc.*, **81**, 3719 (1959).
120. T. Ukita and K. Nagasawa, *Chem. Pharm. Bull.* (Tokyo), **7**, 655 (1959).
121. P. Szabó and L. Szabó, *J. Chem. Soc.*, 5139 (1964).
122. K. Antonakis, A. Dowgiallo, and L. Szabó, *Bull. Soc. Chim. Fr.*, 1355 (1962).
123. P. Szabó and L. Szabó, *J. Chem. Soc.*, 2944 (1965).
124. A. Kornberg, quoted in B. L. Horecker, *Methods. Enzymol.*, **3**, 190 (1957).
125. P. A. J. Gorin, L. Hough, and J. K. N. Jones, *J. Chem. Soc.*, 582 (1955).
126. J. L. Barnwell, W. A. Saunders, and R. W. Watson, *Can. J. Chem.*, **33**, 711 (1955).
126a. F. Schanbacher and D. R. Wilken, *Biochim. Biophys. Acta*, **141**, 646 (1967); D. H. Leaback, E. C. Heath, and S. Roseman, *Biochemistry*, **8**, 1351 (1969).
127. H. W. Kosterlitz, *Biochem. J.*, **37**, 318 (1943).
128. A. B. Foster, W. G. Overend, and M. Stacey, *J. Chem. Soc.*, 980 (1951).
129. M. L. Wolfrom and D. E. Pletcher, *J. Amer. Chem. Soc.*, **63**, 1050 (1941).
130. A. L. Potter, J. C. Sowden, W. Z. Hassid, and M. Doudoroff, *J. Amer. Chem. Soc.*, **70**, 1751 (1948).

131. K. Josephson and S. Proffe, *Ann.*, **481**, 91 (1930).
132. P. A. Levene and A. L. Raymond, *J. Biol. Chem.*, **89**, 479 (1930).
133. P. A. Levene and A. L. Raymond, *J. Biol. Chem.*, **91**, 751 (1931).
134. A. L. Raymond, *J. Biol. Chem.*, **113**, 375 (1936).
135. K. Josephson and S. Proffe, *Biochem. Z.*, **258**, 147 (1933).
136. H. A. Lardy and H. O. L. Fischer, *J. Biol. Chem.*, **164**, 513 (1946).
137. S. Lewak and L. Szabó, *J. Chem. Soc.*, 3975 (1963).
137a. P. Perchemlides, T. Osawa, E. A. Davidson, and R. W. Jeanloz, *Carbohyd. Res.*, **3**, 463 (1967).
137b. R. L. Whistler and J. H. Stark, *Carbohyd. Res.*, **13**, 15 (1970).
138. D. L. Hill and C. E. Ballou, *J. Biol. Chem.*, **241**, 895 (1966).
138a. G. R. Barber, *Biochim. Biophys. Acta*, **141**, 174 (1967); A. K. Chatterjee and D. L. MacDonald, *Carbohyd. Res.*, **6**, 253 (1968).
139. C. Neuberg, H. Lustig, and M. A. Rothenberg, *Arch. Biochem. Biophys.*, **3**, 33 (1944).
140. E. L. Totton and H. A. Lardy, *J. Biol. Chem.*, **181**, 701 (1949).
141. O. Touster and V. H. Reynolds, *J. Biol. Chem.*, **197**, 863 (1952).
142. C. A. Marsh, *J. Chem. Soc.*, 1578 (1952).
143. E. A. Davidson and R. W. Wheat, *Biochim. Biophys. Acta*, **72**, 112 (1963).
144. T. Y. Kim and E. A. Davidson, *J. Org. Chem.*, **28**, 2475 (1963).
145. D. M. Carlson, A. L. Swanson, and S. Roseman, *Biochemistry*, **3**, 402 (1964).
146. C. E. Cardini and L. F. Leloir, *J. Biol. Chem.*, **225**, 317 (1958).
147. A. H. Olavesen and E. A. Davidson, *Biochim. Biophys. Acta*, **101**, 245 (1965).
148. G. Baluja, B. H. Chase, G. W. Kenner, and A. R. Todd, *J. Chem. Soc.*, 4678 (1960).
148a. R. W. Jeanloz, Y. Konami, and T. Osawa, *Biochemistry*, **10**, 192 (1971).
149. O. Westphal and R. Stadler, *Angew. Chem.*, **75**, 452 (1963).
150. M. Dahlgard and E. Kaufmann, *J. Org. Chem.*, **25**, 781 (1960).
151. J. B. Lee, *J. Org. Chem.*, **28**, 2473 (1963).
152. D. B. E. Stroud and W. Z. Hassid, *Biochem. Biophys. Res. Commun.*, **15**, 65 (1964).
153. D. R. Strobach and L. Szabó, *J. Chem. Soc.*, 3970 (1963).
153a. M. Teuber, R. D. Bevill, and M. J. Osborn, *Biochemistry*, **7**, 3303 (1968).
154. R. Robison, M. G. Macfarlane and A. Tazelaar, *Nature*, **142**, 114 (1938).
155. D. B. Sprinson, J. Rothschild and M. Sprecher, *J. Biol. Chem.*, **238**, 3170 (1963).
156. F. J. Reithel and R. C. Young, *J. Amer. Chem. Soc.*, **74**, 4210 (1952).
157. A. H. Olavesen and E. A. Davidson, *J. Biol. Chem.*, **240**, 992 (1965).
157a. J. G. Buchanan, D. A. Cummerson, and D. M. Turner, *Carbohyd. Res.*, **21**, 283 (1972).
158. R. Robison and W. T. J. Morgan, *Biochem. J.*, **22**, 1277 (1928).
159. D. L. MacDonald and R. Y. K. Wong, *Biochim. Biophys. Acta*, **86**, 390 (1964).
160. J. Böeseken, *Advan. Carbohyd. Chem.*, **4**, 189 (1949).
161. J. X. Khym and L. P. Zill, *J. Amer. Chem. Soc.*, **74**, 2090 (1952).
162. J. X. Khym and W. E. Cohn, *J. Amer. Chem. Soc.*, **75**, 1153 (1953).
163. S. S. Cohen and D. B. M. Scott, *Science*, **111**, 543 (1950).
164. A. B. Foster, *J. Chem. Soc.*, 1395 (1957).
165. P. J. Garegg and B. Lindberg, *Acta Chem. Scand.*, **15**, 1913 (1961).
166. L. Vargha, *Ber.*, **66**, 704 (1933).
167. A. Weissbach, *J. Org. Chem.*, **23**, 329 (1958).
168. Y. Y. Makarov-Zemlyanskii and V. V. Gertsev, *Zh. Obshch. Khim.*, **35**, 272 (1965); *Chem. Abstr.*, **62**, 13215 (1965).
169. H. G. Kuivila, A. H. Keough, and E. J. Soboczenski, *J. Org. Chem.*, **19**, 780 (1954); M. L. Wolfrom and J. Solms, *ibid.*, **21**, 815 (1956); F. Shafizadeh, G. D. McGinnis, and P. S. Chin, *Carbohyd. Res.*, **18**, 357 (1971).
170. R. J. Ferrier, W. Prasad, A. Rudowski, and I. Sangster, *J. Chem. Soc.*, 3330 (1964).

171. R. J. Ferrier and D. Prasad, *J. Chem. Soc.*, 7429 (1965).
172. R. J. Ferrier, *J. Chem. Soc.*, 2325 (1961); R. J. Ferrier and D. Prasad, *ibid.*, 7425 (1965).
173. J. D. Albright and L. Goldman, *J. Amer. Chem. Soc.*, **87**, 4214 (1965).
174. B. Lindberg and K. N. Slessor, *Acta Chem. Scand.*, **21**, 910 (1967).
174a. F. Eisenberg, Jr., *Carbohyd. Res.*, **19**, 135 (1971); P. J. Wood and I. R. Siddiqui, *ibid.*, **19**, 283 (1971).
175. A general discussion is found in T. L. Davis, "Chemistry of Powder and Explosives", Vol. II, Wiley, New York, 1943, p. 191.
176. M. L. Wolfrom and G. P. Arsenault, *J. Amer. Chem. Soc.*, **82**, 2819 (1960) and earlier papers in this series.
177. J. Honeyman and J. W. W. Morgan, *Advan. Carbohyd. Chem.*, **12**, 117 (1957).
178. L. D. Hayward, M. Jackson, and I. G. Csizmadia, *Can. J. Chem.*, **43**, 1656 (1965).
179. H. H. Schlubach, P. Stadler, and I. Wolf, *Ber.*, **61**, 287 (1928).
180. L. P. Kuhn, *J. Amer. Chem. Soc.*, **68**, 1761 (1946); cf. I. G. Wright and L. D. Hayward, *Can. J. Chem.*, **38**, 316 (1960).
181. K. S. Ennor and J. Honeyman, *J. Chem. Soc.*, 2586 (1958).
182. E. G. Ansell and J. Honeyman, *J. Chem. Soc.*, 2778 (1952).
183. F. A. H. Rice and M. Inatome, *J. Amer. Chem. Soc.*, **80**, 4709 (1958).
184. K. S. Ennor, J. Honeyman, C. J. G. Shaw, and T. C. Stening, *J. Chem. Soc.*, 2921 (1958).
185. E. K. Gladding and C. B. Purves, *J. Amer. Chem. Soc.*, **66**, 76 (1944).
186. J. Honeyman and T. C. Stening, *J. Chem. Soc.*, 537 (1958).
187. D. O'Meara and D. M. Shepherd, *J. Chem. Soc.*, 3377 (1957).
188. M. L. Wolfrom, A. O. Pittet, and I. C. Gillam, *Proc. Nat. Acad. Sci. U.S.*, **47**, 700 (1961).
189. M. L. Wolfrom and K. Koizumi, *J. Org. Chem.*, **32**, 656 (1967).
190. C. P. J. Glaudemans and H. G. Fletcher, Jr., *J. Org. Chem.*, **29**, 3286 (1964); *J. Amer. Chem. Soc.*, **87**, 2456 (1965).
191. J. R. Turvey, *Advan. Carbohyd. Chem.*, **20**, 183 (1965).
192. D. L. Ingles, *Chem. Ind.* (London), 1159 (1960).
193. A. B. Foster and E. B. Hancock, *J. Chem. Soc.*, 968 (1957).
194. J. R. Turvey and T. P. Williams, *J. Chem. Soc.*, 2242 (1963).
195. A. G. Lloyd, *Nature*, **183**, 109 (1959); *Biochem. J.*, **75**, 478 (1960); *ibid.*, **83**, 455 (1962).
196. J. W. Wood and P. T. Mora, *J. Amer. Chem. Soc.*, **80**, 3700 (1958).
196a. K. Takiura, H. Yuki, S. Honda, Y. Kojima, and L.-Y. Chen, *Chem. Pharm. Bull.*, **18**, 429 (1970).
197. J. R. Turvey and T. P. Williams, *J. Chem. Soc.*, 2119 (1962); M. J. Harris and J. R. Turvey, *Carbohyd. Res.*, **9**, 397 (1969).
198. M. L. Wolfrom, R. A. Gibbons, and A. J. Huggard, *J. Amer. Chem. Soc.*, **79**, 5043 (1957); A. B. Foster, E. F. Martlew, M. Stacey, P. J. M. Taylor, and J. M. Webber, *J. Chem. Soc.*, 1204 (1961).
198a. R. O. Mumma, C. P. Hoiberg, and R. Simpson, *Carbohyd. Res.*, **14**, 119 (1970).
199. B. Helferich, *Ber.*, **54**, 1082 (1921).
200. S. S. Ali, T. J. Mepham, I. M. E. Thiel, E. Buncel, and J. K. N. Jones, *Carbohyd. Res.*, **5**, 118 (1967) and earlier papers in this series; compare B. T. Lawton, W. A. Szarek, and J. K. N. Jones, *Carbohyd. Res.*, **15**, 397 (1970).
201. M. L. Wolfrom and R. Montgomery, *J. Amer. Chem. Soc.*, **72**, 2859 (1950).
202. D. Grant and A. Holt, *Chem. Ind.* (London), 1492 (1959); G. Coleman, M. Higgs, A. Holt, and M. Mulvin, *ibid.*, 376 (1963).
203. M. J. Clancy and J. R. Turvey, *J. Chem. Soc.*, 2935 (1961).
204. E. G. V. Percival, *Quart. Rev.* (London), **3**, 369 (1949).
205. J. S. Brimacombe, M. E. Evans, A. B. Foster, and J. M. Webber, *J. Chem. Soc.*, 2735 (1964).

206. F. Micheel and A. Klemer, *Advan. Carbohyd. Chem.* **16**, 95 (1961).

207. A. Bertho, *Ber.*, **63**, 836 (1930).

208. A. Bertho and D. Aures, *Ann.*, **592**, 54 (1955).

209. F. Micheel and H. Wulff, *Ber.*, **89**, 1521 (1956).

210. M. Nys and J. P. Verheijden, *Bull. Soc. Chim. Belges*, **69**, 57 (1960).

211. J. Baddiley, J. G. Buchanan, R. Hodges, and J. F. Prescott, *J. Chem. Soc.*, 4769 (1957).

212. R. Carrington, G. Shaw, and D. V. Wilson, *J. Chem. Soc.*, 6864 (1965).

213. F. Micheel and G. Baum, *Ber.*, **90**, 1595 (1957); H. El Khadem, D. Horton, and M. H. Meshreki, *Carbohyd. Res.*, **16**, 409 (1971).

9. GLYCOSIDES

W. G. Overend

I. INTRODUCTION

The designation *glycoside* is used for the acetal derivatives of the cyclic forms of sugars in which the hydrogen atom of the hemiacetal hydroxyl group has been replaced by an alkyl, aralkyl, or aryl group. In this restricted sense, glycosides are mixed, monocyclic acetals. On complete hydrolysis they afford a mono- or polyhydric alcohol or phenol, and one or more monosaccharides. Glycosides derived from aldoses are referred to as aldosides, and those from ketoses are ketosides. In aldosides and ketosides the ring-oxygen atom is connected to C-1 and C-2, respectively, of the sugar.

Thioaldosides may be regarded as derivatives of 1-thio sugars, but frequently they are defined as 1-thioglycosides to differentiate them from glyco-

279

sides of a thio sugar in which the thiol group is located in the parent sugar at a site other than C-1.

For convenience, the alkyl, aralkyl, or aryl group is referred to as the "aglycon group," and the corresponding alcohol (or thiol) or phenol (or thiophenol) is called the "aglycon." The sugar residue is the "glycosyl" group ("glycofuranosyl" or "glycopyranosyl" for five- and six-membered rings, respectively).

The term glycoside is used in the generic sense, and specific glycosides are named by replacing the ending "ose" of the parent sugar by "oside" and by adding the name of the alkyl or other radical and the symbol α or β to designate the configuration of the glycosidic (anomeric) carbon—for example, methyl α-D-xylofuranoside or phenyl β-D-glucopyranoside.

For complex groups, it is sometimes more convenient to use the name of the alcohol or phenol rather than the radical, as in hydroquinone α-D-galactopyranoside. p-Hydroxyphenyl β-D-glucopyranoside may also be named catechol β-D-glucopyranoside. When several hydroxyl groups in polyhydric alcohols or phenols are linked glycosidically, the nomenclature is not uniform. For natural glycosides phytochemical names are used frequently, although chemical names are preferable because they indicate structure and facilitate classification. The trivial names have the advantage of brevity and indicate the source of the glycoside, as, for example, salicin (o-hydroxymethylphenyl β-D-glucopyranoside) from the bark of willow (*Salix helix*).

Di-, oligo-, and polysaccharides have glycosidic linkages, the aglycon group being a sugar residue. Many of these higher saccharides have trivial names.

The nomenclature of 1-thioglycosides is analogous to that given for glycosides in general.

This Chapter is limited to an account of aspects of the chemistry of simple glycosides of the type described, and, although such simple glycosides undergo numerous reactions at sites in the molecule other than the anomeric center, only reactions involving the glycosidic linkage will be described.

Mention should be made of usage of the term "glycoside" in a wider context nowadays than is covered by the subject matter of this Chapter. Glycosans, in which acetalation has taken place within an aldose molecule to produce an internal bicyclic acetal, are regarded as inner glycosides. Ketohexoses form bimolecular dianhydrides containing a central p-dioxane ring, in which the glycosidic center of each sugar residue is linked to an oxygen atom of the central ring.

Glycosides are widely distributed in Nature, particularly in plants. Because the chemistry of the natural glycosides resides to a considerable degree in the aglycon residue, and in biochemical relationships, they are described separately in Vol. IIA, Chapters 32 and 33.

II. METHODS FOR SYNTHESIS

Although many methods for the synthesis of glycosides have been studied, relatively few have found wide or general application. Most of the individual procedures are restricted to a certain aglycon type and often depend also on the nature of the glycose undergoing glycosidation. General preparative methods for methyl and phenyl glycosides have been reviewed.[1]

A. FISCHER METHOD

Aldehydes or ketones react in anhydrous alcoholic solutions of hydrogen chloride to form acetals. The simplest members of the sugar series, glycolaldehyde and glyceraldehyde, react similarly. In attempting to synthesize acetals of higher sugars by treating them with methanol and hydrogen chloride, Fischer[2] found that only one methyl group was introduced per mole of sugar and that a methyl glycoside was formed. The sugars in their cyclic forms (hemiacetals) establish an equilibrium in the reaction medium, in which

anomeric glycopyranosides and glycofuranosides preponderate. The formation and hydrolysis of glycosides is a reversible reaction, but, as carried out in practice, the reaction is forced in one direction as far as possible, by use of a large excess of alcohol or of water.

The Fischer synthesis, which is applicable with alcohols but not with phenols, is particularly suited to the preparation of glycosides with lower aliphatic alcohols. Disaccharides frequently undergo alcoholysis of the linkage between the constituent residues, and O-acetyl groups of acetylated sugars are hydrolyzed. The customary procedure is to heat a solution or suspension of the monosaccharide in the alcohol in the presence of a few percent of hydrogen chloride as catalyst. The furanoid forms of the sugars react most readily, but pyranosides are generally the principal constituents under equilibrium conditions, and so by appropriate selection of reaction conditions preponderant formation of either pyranosides or furanosides can be achieved (see p. 282).

* References start on p. 346.

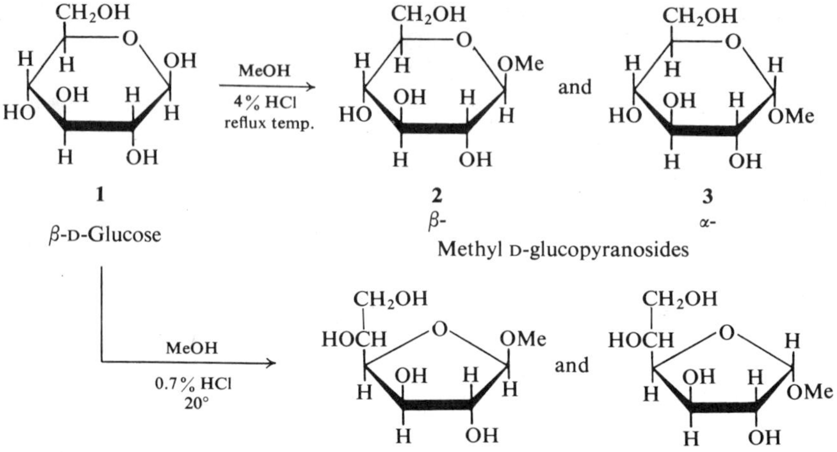

1
β-D-Glucose

2
β-
Methyl D-glucopyranosides

3
α-

4
β-
Methyl D-glucofuranosides

5
α-

This method is particularly useful for the preparation of α-D-hexopyrano-sides, which otherwise are less easily obtained. They can be separated from the reaction mixture as a result of their ready crystallization. As expected, reaction conditions may need to be modified for particular classes of sugars. For example, formation of the methyl glycosides of 2-deoxy sugars proceeds easily at room temperature.[3]

In addition to its value for the synthesis of alkyl glycopyranosides, the method has preparative value for alkyl glycofuranosides[4-8] (for example, 4 and 5 are methyl D-glucofuranosides), particularly when coupled with the method described by Baddiley and co-workers[9] for the separation of a furano-side mixture, usually syrupy, on a strongly basic anion-exchange resin or by a somewhat more tedious and less convenient separation on a column of powdered cellulose.[10] The furanosides have been the subject of a review by Green.[8]

Frequently, furanosides are prepared from sugar derivatives protected in the furanoid form. For example, D-glucose can be converted into an alkyl D-glucofuranoside via 1,2-O-isopropylidene-α-D-glucofuranose 5,6-carbonate. The latter, on treatment with an alcohol and hydrogen chloride, undergoes cleavage of the O-isopropylidene group and glycosidation to give the glyco-furanoside 5,6-monocarbonate. Subsequent treatment with alkali affords the D-glucofuranoside by hydrolysis of the 5,6-carbonate protecting group.[11, 12] Another route to D-glucofuranosides is based on the facile glycosidation of D-glucurono-3,6-lactone by alcohols under the influence of acid catalysts. The glycosides derived from the lactone are known to be furanoid[13] and are readily reduced by sodium borohydride[14] to afford anomeric forms of alkyl D-gluco-furanosides.

Various modifications of the general method have been reported. A cation-exchange resin in the acid form can be used as the acid catalyst, and isolation of the product is facilitated.[15, 16] The use of ion-exchange resins makes possible a commercial continuous process for the preparation of methyl α-D-gluco-pyranoside.

In recent years there have been attempts to obtain information about the mechanistic and conformational factors that influence the formation of glycosides in the Fischer reaction. Kinetic investigations have been hampered by lack of suitable methods to analyze the complex, multicomponent equilibrium mixture that is formed. In a paper published well ahead of its time in 1932, Levene and co-workers[17] described the results of an investigation of the formation of methyl glycofuranosides and -pyranosides from nine sugars. They examined changes in composition with time of solutions of D-glucose, D-galactose, D-mannose, L-rhamnose, D-fructose, L-arabinose, D-lyxose, D-xylose, and D-ribose in methanol containing 0.5% of hydrogen chloride. As the authors emphasized in the paper, the results are only approximate because the analytical methods then available did not distinguish clearly between furanosides and pyranosides. Nevertheless it was obvious that furanosides are formed in the first stage of all the reactions, but they decrease in quantity in the later stages and the proportion of pyranosides increases progressively with time. The amount of furanoside varies greatly with the nature of the sugar and seems to be particularly high for D-ribose. This faster formation of furano-sides is in accord with the generally faster closure of five-membered rings by comparison with those that are six-membered.[18, 19] Few obvious conformational correlations can be drawn from these workers' results, except that it can be noted that among the hexoses formation of furanoside was slowest in the case of D-mannose. For the pentoses, D-lyxose was the slowest to yield the furanoside. These are the sugars that have the least favorable *cis-cis* arrangement of substituents at C-2, C-3, and C-4 in the furanoid ring.

The first complete analyses of equilibrium mixtures of glycosides were reported by Bishop and Cooper,[20, 21] who used gas–liquid chromatography for the separations. Rates of methanolyses with D-xylose, D-arabinose, D-lyxose, and D-ribose were determined, and it was concluded that methanolysis of a pentose proceeds to equilibrium through four distinguishable, competing reactions: (1) pentose → furanosides; (2) anomerization of furanosides; (3) furanosides → pyranosides; (4) anomerization of pyranosides. Glycoside compositions at equilibrium were interpreted in terms of stabilities of each of the four glycosides from each sugar as influenced by steric and ionic

---

* *References start on p. 346.*

effects. The values shown in Table I refer to glycoside compositions at equilibrium for the pentoses and some of their *O*-methyl derivatives.

TABLE I[a]

GLYCOSIDE COMPOSITIONS[b] AT EQUILIBRIUM[21]

Sugar	α-Furano-side	α + β	β-Furano-side	α-Pyrano-side	α + β	β-Pyrano-side
D-Xylose	1.9		3.2	65.1		29.8
3-*O*-Methyl-D-xylose		9.0			91.0	
2-*O*-Methyl-D-xylose		12.8			87.2	
2,3-Di-*O*-methyl-D-xylose		16.4			83.6	
D-Arabinose	21.5		6.8	24.5		47.2
3-*O*-Methyl-D-arabinose		50.7			49.3	
2-*O*-Methyl-D-arabinose		66.7			33.3	
2,3-Di-*O*-methyl-D-arabinose		75.4			24.6	
D-Lyxose	1.4		c	88.3		10.3
D-Ribose	5.2		17.4	11.6		65.8

[a] Reproduced by permission of the National Research Council of Canada from *Can. J. Chem.*, **41**, 2743 (1963).   [b] Sugar (2%) in 1% methanolic hydrogen chloride at 35°.   [c] Not detected.

Several conclusions can be drawn from these results:

1. The proportion of furanosides, though small, is by no means negligible and is quite substantial in the case of the α-D-arabinofuranoside, which has the most favorable all-*trans* configuration. Methylation of D-xylose and D-arabinose increases the proportion of furanosides. The methylated D-arabinosides represent the unusual case in which the furanosides are more stable than the pyranosides. For these sugars the *trans* substituents at C-2–C-3 are closer to each other in the six-membered ring than is the case for the five-membered ring, and substitution at these positions should decrease the stability of the pyranosides more than the furanosides.

2. Furanosides in which the groups at C-1 and C-2 are *trans*-located (β forms for derivatives of D-xylose and D-ribose, α forms for derivatives of D-arabinose and D-lyxose) are considerably more stable than their anomers and are formed in greater proportion.

3. The anomer having an axial methoxyl group preponderates, which indicates that the anomeric effect on a methoxyl group in methanol is higher (on the average by 0.8 kcal mole^{-1}) than that on a hydroxyl group in water.

Column chromatography has been used also for the analysis of glycoside mixtures and provides more-definitive evidence than the techniques used by

Levene and co-workers,[17] but the method is more tedious and less convenient than the gas-chromatographic procedure.[20,21] By this method, values obtained for the methyl L-arabinosides agree well with those of Bishop and Cooper. For the methyl D-mannosides the results are 89% α-pyranoside and 7% β-pyranoside and about 2% of each furanoside.[22] An earlier investigation[23] of the glycosidation of D-galactose with methanol and hydrogen chloride, employing column chromatography for the separations, indicated that in the first stage of the reaction a mixture of β-D-furanosides and β-D-pyranosides is formed, which then shifts to a mixture of the α-D anomers. Similar results were obtained when a cation-exchange resin, instead of hydrogen chloride, was used as catalyst in the methanolysis.[24] However, because of the methods used for some of the estimations, caution must be exercised in evaluating both these results and those of Levene and co-workers.[17]

Although acyclic acetals are feasible intermediates in the glycosidation of sugars, Bishop and Cooper[20,21] obtained no evidence for the existence of such compounds. However, under the conditions they used for the chromatographic separations, it is possible that these acetals may have been masked by cyclic products, particularly as others have since reported the detection by radiochemical techniques of acyclic acetals in glycosidation reactions.[25,26] Heard and Barker[25] noted the presence of the dimethyl acetal in the products of the methanolysis of D-arabinose, and Ferrier and Hatton[26] reported the corresponding derivatives in the glycosidation of D-xylose and D-glucose. Acyclic hemiacetals have not been detected yet but cannot be discounted as possible reaction intermediates.

Ferrier and Hatton[26] found that the glycosidations of D-xylose and D-glucose follow a similar course, and in neither case does the acetal concentration exceed 2.5% at any time. In the methanolysis of D-xylose they found that the two furanosides are formed first, that the β form predominates over the α form, and that latterly ring expansion occurs to give pyranosides, which constitute 95% of the reaction products at equilibrium. These are all results that agree qualitatively with the findings of Bishop and Cooper, but there were quantitative differences between the results obtained by the two methods (g.l.c. and radiochemical methods).

The concentration of the dimethyl acetal builds up to 2.5% at the time when the furanosides are present in maximum amount, and with them it subsides as the pyranosides are formed. At equilibrium only 1% remains. Evidence suggests that the acetal is not the primary product from which the furanosides are derived.

Examination of the early stages of the methanolysis of D-glucose revealed that, as with D-xylose, the furanosides are formed first, with the α form preponderating, and that the acetal is formed in minor proportion and even in the early stages of reaction it is not a major component.

The extrapolated initial proportions (68:32) of α-:β-D-xylofuranoside (which compare well with values obtained by g.l.c. analysis), together with similar observations in the D-glucose series, point to the conclusion that the furanosyl carbonium ions (6, R = H or CH_2OH) are not involved in furano-side formation, since the hydroxyl group at C-2 would be expected to shield the "α side" of C-1 and cause preferential solvolytic attack to give β-glycosides.

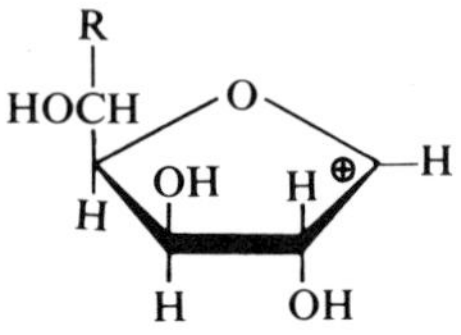

6

Several alternative mechanisms for formation of furanosides from furanoses can be visualized involving either an acyclic ion (7) as intermediate, or syn-chronous processes (10 → 8 → 9, 11 → 9). From the available evidence it is

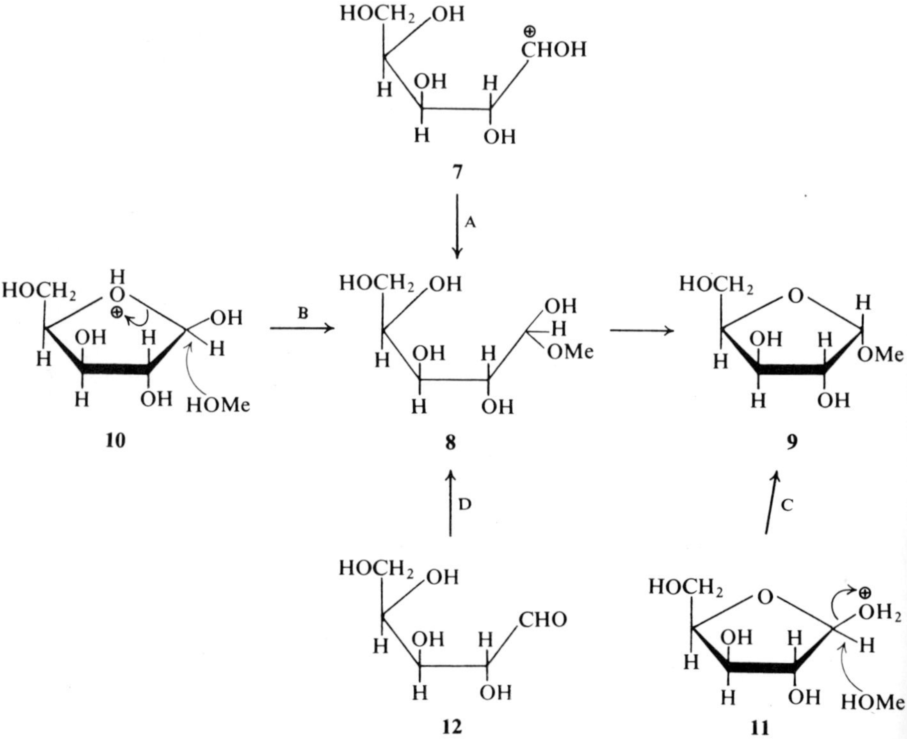

7

10   8   9

12   11

not possible to decide between them, although there is some support for the route **11** → **9**. It would satisfy the finding of initial preponderant amounts of the α-furanoside, since the β modification of the free furanose would be expected to preponderate in solution. In keeping with this possible pathway, it has been shown that the anomerization of methyl β-D-glucofuranoside in ^{14}C-methanol leads to labeled furanosides.[27] Pathway **12** → **8** → **9** cannot be eliminated and would result first in formation of the hemiacetal (**8**), which features, therefore, in three of the four possible routes to furanosides. At no stage in the work of Ferrier and Hatton was it detected, although it would have been separable from the other reaction intermediates and would have been detectable had its concentration in the final products reached ~0.3%. Its high reactivity in the methanolysis medium or during subsequent operations could account for its apparent absence.

On finding that the methyl α- and β-D-xylopyranosides were produced in the ratio of 1:1.7, which was the equilibrium ratio found for the furanosides, Bishop and Cooper[20] concluded that ring expansion occurs with retention of configuration at the anomeric center. This deduction has been criticized[27] on the grounds that anomerization of furanosides is rapid in comparison with the rate of the ring expansion, but it would be valid in the event of the rates of ring expansion of the two furanosides being identical. Because the methyl D-xylopyranosides were indistinguishable chromatographically by the conditions used by Ferrier and Hatton[26] evidence on this point was sought from an examination of the ethanolysis of radioactive D-xylose.

The overall glycosidation was closely similar to the methanolysis. The α- and β-furanosides were formed initially in the ratio 2.6:1; they anomerized to give an "equilibrium ratio" of 1:1.3; they then underwent ring expansion to α- and β-pyranosides, formed in the ratio of 1:1.9. Finally the pyranosides anomerized to give an equilibrium mixture. In this reaction, therefore, the α,β ratio changes appreciably during ring expansion, but no information on the stereochemistry of the process can be deduced from the results.

In spite of advances in recent years, further work is needed before this glycosidation reaction can be described in full detail.

B. FROM ACETALS AND THIOACETALS

By analogy with simple acetals, the hydrolysis of an aldose dimethyl acetal might be expected to yield the aldose and methanol, but in fact the reaction is more complicated. From a detailed investigation of the hydrolyses of the acyclic dimethyl acetals of D-glucose and D-galactose in dilute aqueous hydrochloric acid, Capon and Thacker[28] have shown that, in addition to the aldoses, the methyl furanosides are also produced in quantity by a concurrent ring closure; for example:

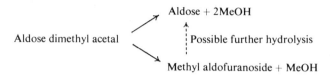

Pyranosides were formed in only very small proportion (see Table II).

The hydrolysis of these acetals was investigated first by Wolfrom and Wais-brot,[29] who concluded that their results were best interpreted on the basis of a

TABLE II

KINETICALLY CONTROLLED PRODUCTS AND RATE CONSTANTS[28] FOR THE REACTIONS OF SOME ALDOSE ACETALS IN 0.05M HYDROCHLORIC ACID AT 35°[a]

	Products (%)			
Dimethyl acetal of:	Aldose	Furanosides	Pyranosides	$10^4\,k_{total}\ (sec^{-1})$
D-Glucose	<2	>98	<0.5	17
D-Galactose	29	71	<0.5	1.58

SECOND-ORDER RATE CONSTANTS FOR THE ACID-CATALYZED HYDROLYSIS AND RING CLOSURE OF SOME DIMETHYL ACETALS AT 25°

Reaction	$10^4\,k\ l.\ mole^{-1}$	$10^4\,k_{calc}\ (sec^{-1})^b$	k/k_{calc}
Ring closure of D-glucose acetal	110	0.324	340
Ring closure of D-galactose acetal	9.3	0.324	29
Hydrolysis of D-glyceraldehyde acetal	1.77	2.48	0.71

[a] Reprinted from *J. Amer. Chem. Soc.*, **87**, 4199 (1965). Copyright (1965) by the American Chemical Society. Reprinted by permission of the copyright owner.   [b] Rate of ionization calculated by using Taft's $\rho^*\sigma^*$ equation based on a value for the second-order rate constant for the hydrolysis of acetaldehyde dimethyl acetal of 2.74×10^{-1} l. mole^{-1} sec^{-1}. The value of ρ^* was assumed to be -3.65, and σ^* for the first hydroxymethyl group to be 0.555.[32] The σ^* values for the other hydroxyl groups were calculated from this by assuming an alternation factor of 0.5 for each additional carbon atom between the hydroxyl group and the reaction center.[33]

very rapid initial reaction due to hydrolysis of the acetal followed by the formation of unstable, nonpyranoid glycosides which in turn are slowly converted into the stable pyranosides. Capon and Thacker,[28] however, were unable to detect any pyranosides by paper chromatography. From D-glucose dimethyl acetal the carbohydrate products are overwhelmingly furanosides, whereas the acetal of D-galactose (and of L-arabinose) yields significant amounts (about 30%) of the aldose. It was found[30] that the major furanoside produced is the thermodynamically less stable *cis*-1,2 isomer, but the proportion of *cis*:*trans*-1,2-furanosides was approximately equal from each of the acetals. Two mechanisms (I and II) are possible for the ring closure to the furanosides.

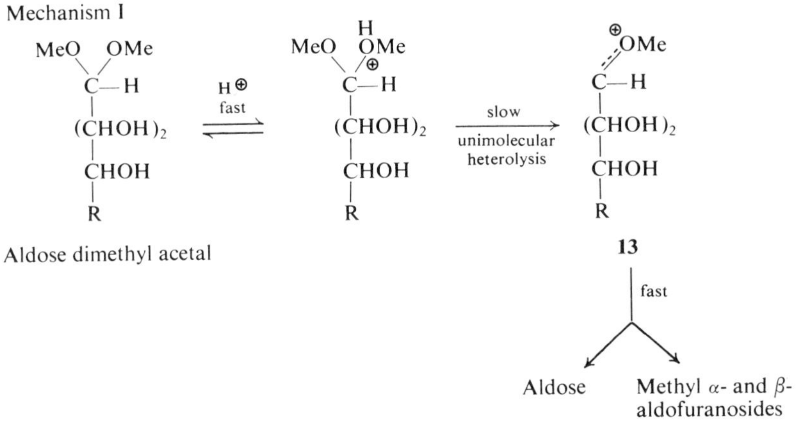

Mechanism I

$$
\begin{array}{ccccc}
\underset{|}{\overset{\text{MeO}\quad\text{OMe}}{\overset{\diagdown\diagup}{\text{C—H}}}} & \underset{\text{fast}}{\overset{\text{H}\oplus}{\rightleftharpoons}} & \underset{|}{\overset{\text{MeO}\quad\overset{\text{H}}{\overset{\oplus}{\text{OMe}}}}{\overset{\diagdown\diagup}{\text{C—H}}}} & \underset{\substack{\text{unimolecular}\\\text{heterolysis}}}{\overset{\text{slow}}{\longrightarrow}} & \underset{|}{\overset{\overset{\oplus}{\text{OMe}}}{\overset{}{\text{C—H}}}} \\
(\text{CHOH})_2 & & (\text{CHOH})_2 & & (\text{CHOH})_2 \\
\underset{|}{\text{CHOH}} & & \underset{|}{\text{CHOH}} & & \underset{|}{\text{CHOH}} \\
\text{R} & & \text{R} & & \text{R}
\end{array}
$$

Aldose dimethyl acetal — **13**

13 | fast

Aldose   Methyl α- and β-aldofuranosides

In mechanism I, the carbonium ion (**13**), of the type normally postulated to be involved in the hydrolysis of acetals,[31] is competed for by water and by the internal hydroxyl group. In mechanism II, ring closure is synchronous

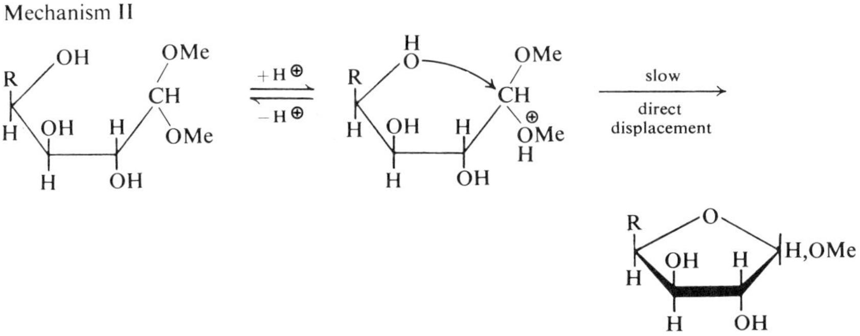

Mechanism II

with rupture of the acetal bond. The available evidence favors this mechanism.

In mechanism I the total rate of reaction is controlled by the heterolysis to give the ion (**13**), and this should be independent of the configuration at C-4.

On the other hand, for mechanism II the rate of ring closure in the synchronous process (and hence the total rate) could well depend on the configuration at C-4. Supporting mechanism II is the observation that k_{total} for the D-glucose acetal is about ten times as great as the value for the D-galactose acetal, and that the product from the D-glucose acetal contains a much higher proportion of furanosides. The anchimeric assistance associated with the ring closures is indicated by the results given in Table II.

The rate of ring closure of the D-glucose dimethyl acetal is 340 times, and the D-galactose acetal is 29 times, as great as the estimated unassisted rate of ionization of a pentahydroxyhexanal acetal, based on the $\rho^*\sigma^*$ relationship.

That D-glyceraldehyde dimethyl acetal is hydrolyzed at about the rate expected from the $\rho^*\sigma^*$ correlation[32] implies that it behaves as a simple acetal. Since hydroxyl substituents cause a decrease in the rates of hydrolysis of acetals through their inductive effect, the rate constant of this reaction sets an upper limit on the unassisted rate of ionization for the D-glucose and D-galactose acetals. It is concluded, therefore, that the ring closure of the dimethyl acetal of D-glucose and D-galactose involves a nucleophilic attack that is synchronous with rupture of the acetal bond. It was considered[30] unlikely that the observed increases in rate could arise from either solvation or steric effects, because differences in solvation of the transition states for the acetals of both hexoses would be at a distance from the reaction site, and, for the rate increases observed, appreciable relief of steric strain would be necessary. Any steric acceleration of the hydrolysis must be due to nonbonded interactions of the hydroxyl groups remote from the reaction site. Since the molecule is free to take up a "low-energy" staggered conformation, the steric acceleration from this source would be small.

The relative rates of reaction of the dimethyl acetals of D-glucose and D-galactose may derive from the unfavorable interaction between the cis-hydroxyl groups on C-2 and C-4 in the D-glucose acetal (15), which is absent in the D-galactose isomer[30] (14). (In solution these molecules might not take up the conformations shown, but similar interactions will be operative.) The difference in ratio is ascribed, in this way, to steric acceleration of the rate-controlling decomposition of the conjugate of the acetal.

Formation of furanosides in preference to pyranosides is explicable on the basis of the much faster ring closure to form five-membered rings in comparison to closure to form six-membered cyclic structures, as observed for other hydroxyl group-assisted hydrolyses.[34]

Thacker[30] has offered an explanation for the preferential formation of the less-stable 1,2-cis furanoside by considering somewhat idealized representations (16 and 17) of the two possible transition states for the cyclization. The diagrams represent the molecule as seen along the C-1–C-2 bond, and the formation of a hydrogen bond (or at least a partial bond) between the hydroxyl

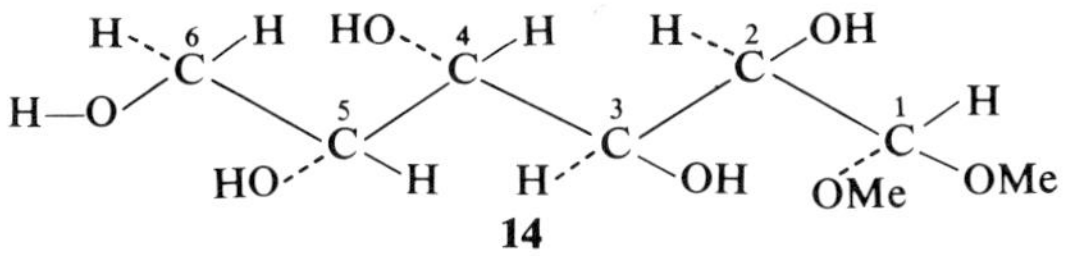

14

D-Galactose dimethyl acetal

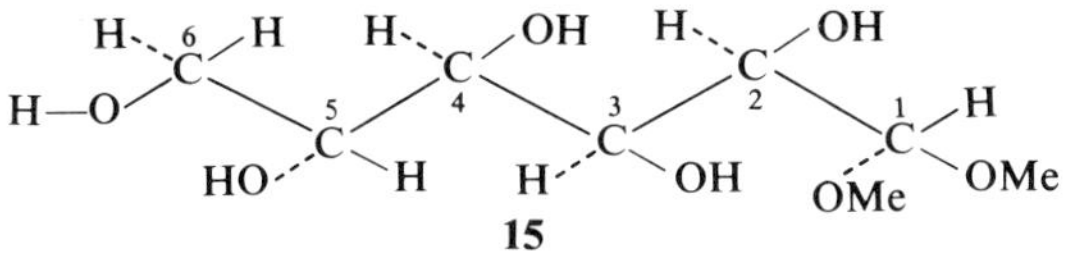

15

D-Glucose dimethyl acetal

group at C-2 and the nonprotonated methoxyl group at C-1 lowers the energy of **16** relative to **17** and leads to the observed products by determining the pathway of the reaction.

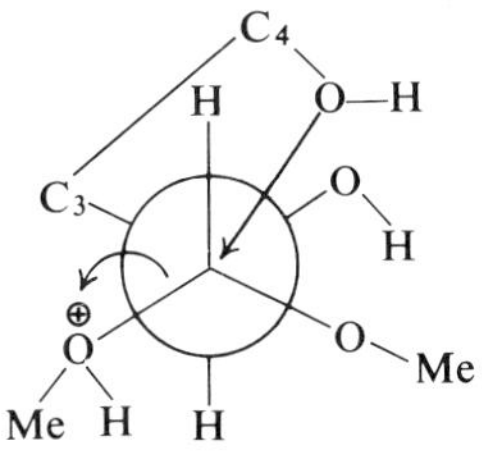

16 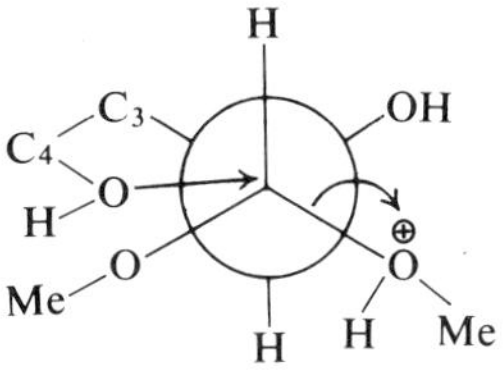   **17**

The reaction of an aldose dimethyl acetal in methanolic acid is less complex than in aqueous acid, since the only products are the methyl aldofuranosides. Of the two possible mechanistic pathways for the reaction, a synchronous bimolecular mechanism is favored.[30] When the reaction with D-galactose dimethyl acetal was effected in deuterated methanol, there was no incorporation to any marked extent of solvent in the products, as would be expected if the reaction were to involve a unimolecular ionization followed by capture of the carbonium ion by the hydroxyl group at C-4.

Aliphatic glycosides, especially furanosides, may be obtained from dithio-acetals of aldoses or ketoses.[35-43] By variation of the conditions for the hydrolysis of the alkylthio group, pyranosides, furanosides, 1-thioglycosides, or acetals may be obtained. The products also vary with the nature of the sugar structure. For formation of a glycoside, the reaction is conducted in a medium of the respective alcohol and in the presence of mercuric chloride, with or without mercuric oxide. Figure 1 illustrates the products that can be obtained from D-galactose diethyl dithioacetal under various conditions.

* References start on p. 346.

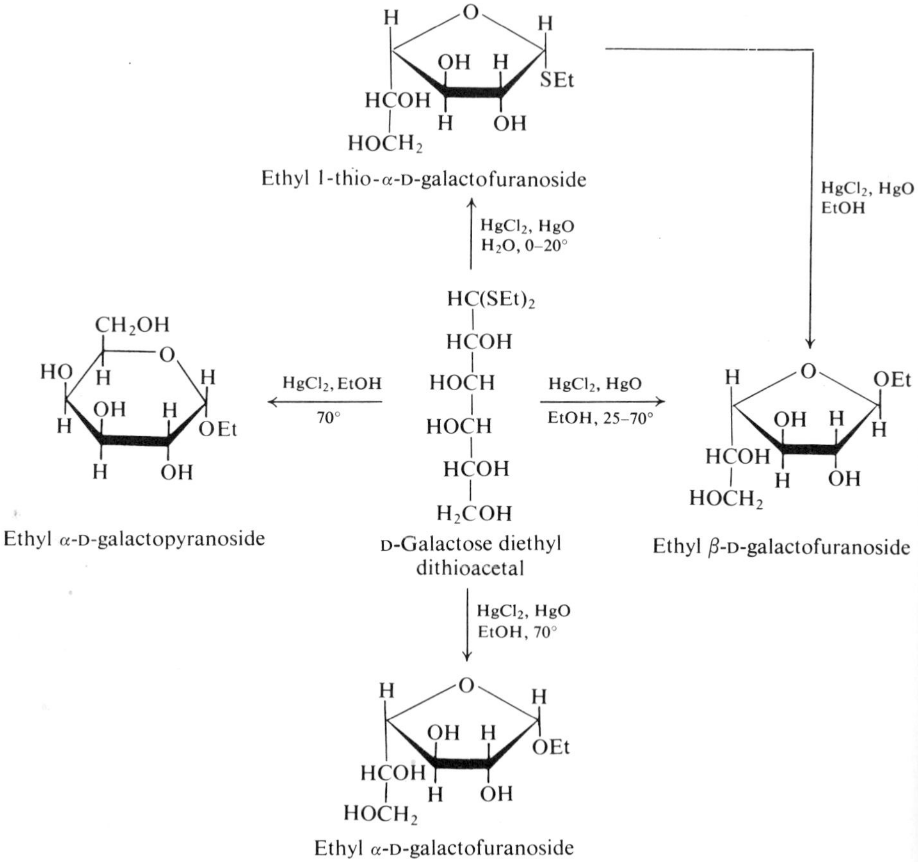

FIG. 1. Formation of glycosides and 1-thioglycosides from D-galactose diethyl dithioacetal.

C. ANOMERIC REPLACEMENTS

1. Helferich Method[44]

The acetoxy group at C-1 of acetylated aldopyranoses is more labile than the other acetoxy groups and is replaceable by a phenoxy group, when the acetylated sugar is heated with a phenol in the presence of an acid catalyst (usually zinc chloride or p-toluenesulfonic acid). This reaction is the basis of an important method for the synthesis of aromatic glycosides that was developed by Helferich and Schmitz-Hillebrecht.[45] Its advantage as a preparative procedure[44] lies in the use of the readily available, fully acetylated aldose. The weak p-toluenesulfonic acid catalyst and a short time of heating favor replacement with retention of configuration, but unless optimal conditions are found, considerable anomerization occurs and both isomers are produced, which is a shortcoming of the method. The stronger, aprotic zinc chloride and longer time of heating favor anomerization and production of the stable isomer.

Improved yields result from removal, under diminished pressure, of the acetic acid produced in the reaction, as well as any that may be added.[46,47] With some phenols, reaction proceeds better if xylene is present; in other cases benzene is recommended.[48]

Other catalysts have been employed successfully—for example, phosphorus oxychloride,[49] sulfuric acid,[50,51] and anhydrous stannic chloride,[52] all of which yield the β anomer.† Ion-exchangers have also been used.[53] Very good yields of aryl tetra-O-acetyl-β-D-hexosides were obtained with anhydrous aluminum chloride in the melt.[54] Another modification of the method consists in the application of aluminum phenolates.[55] When boron trifluoride was used in traces as catalyst for the reaction of penta-O-acetyl-β-D-glucose with phenol (4 moles) in benzene solution during 5 days, a 70% yield of the β-D-glucoside tetraacetate was obtained.[56] After numerous trials, a satisfactory, simple, and reproducible procedure was developed by Trevelyan[57] for the preparation of phenyl α-D-glucopyranoside and the corresponding p-nitrophenyl derivative. The phenyl glycoside is readily prepared, but the preparation of the p-nitrophenyl glycoside requires more care.

† R. U. Lemieux and W. P. Shyluk,[52] concluded that formation of β-glycosides was favored by the prominent role played by the carbonium ion generated by elimination of the acetoxy group at C-1 and stabilized by the anchimeric assistance of the acetoxy group at C-2. This view has been challenged by Bose and Ingle [J. L. Bose and T. R. Ingle, *Chem. Ind.* (*London*), 1451 (1967)] who found by thin-layer chromatographic examination of the products from several reactions that considerable amounts of the α-anomers are also produced. With the conditions employed by Lemieux and Shyluk for the preparation of phenyl β-D-glucopyranoside tetraacetate it was found by Bose and Ingle that in fact the α- and β-anomers were produced in the proportion respectively of approximately 3:7. The mixture obtained after the first fraction of β-form had crystallized from ethanol, could be separated conveniently by column chromatography or preparative-layer chromatography on silica gel. Other examples were cited and it is claimed that the reaction between acetylated sugar and a phenol in the presence of anhydrous stannic chloride constitutes a method for the synthesis of aryl α-glycosides. Clearly, in this case the anchimeric assistance from the acetoxy group at C-2 cannot be considered a significant factor in the reaction and it is more likely that the stabilized, open ion A is involved.

A

The method has been used also to prepare phenyl glycofuranosides by condensation of glycofuranose acetates with phenol in the presence of p-toluenesulfonic acid,[58, 59] although attempts to prepare o- and p-nitrophenyl arabinofuranoside in this way were unsuccessful. However, the method used by Feier and Westphal[60] to synthesize p-nitrophenyl α-L-arabinopyranoside from the pyranosyl acetate with mercuric cyanide as catalyst afforded, after deacetylation, p-nitrophenyl α-L-arabinofuranoside when the corresponding furanosyl acetate was used.[61]

Although the Helferich reaction is usually considered to relate to the synthesis of aryl glycosides, similar displacements have been used to prepare alkyl glycosides. In fact, in the first reference to the direct use of an acetylated sugar in glycoside synthesis, the aglycon was an alcohol.[63] Ethyl hepta-O-acetyl-α-cellobioside was prepared from α-cellobiose octaacetate.[62] Methyl tetra-O-acetyl-β-D-glucopyranoside has been obtained in fair yield by condensing equimolar amounts of penta-O-acetyl-β-D-glucose and methanol at 40° in benzene or chloroform solution with stannic chloride as catalyst;[52] the O-acetylglycosyl chloride was isolated as a by-product.

Other esters have also been displaced. For example, the ester group in a number of aldosyl (2,4,6-trimethylbenzoates) has been displaced by treatment with methanol containing methanesulfonic acid.[63–66] In a typical example, β-D-glucopyranosyl (2,4,6-trimethylbenzoate) is converted into methyl α-D-glucopyranoside.

Tetra-O-acetyl-α-D-glucopyranosyl nitrate will react with methanol under the influence of a variety of bases to give methyl tetra-O-acetyl-β-D-glucopyranoside.[67] Unlike tetra-O-acetyl-β-D-glucopyranosyl nitrate, β-D-glucopyranosyl pentanitrate[68] is stable in polar solvents and does not undergo facile anomerization. The pentanitrate therefore, might be considered a more suitable intermediate in glycoside synthesis, particularly because the nitrate ester group apparently does not participate in a displacement reaction at a neighboring carbon atom.[69, 70] Consequently, the pentanitrate might be expected to react with hydroxylic compounds with Walden inversion at C-1 to give rise to α-D-glucopyranosides. In fact, only methyl α-D-glucopyranoside was isolated when a methanolic solution of β-D-glucopyranose pentanitrate was boiled in the presence of silver carbonate and the product subsequently reductively denitrated.[71] The rate of this solvolytic reaction is low, and attempts to bring β-D-glucopyranose pentanitrate into reaction with hydroxylic compounds in inert solvents showed that the material is not sufficiently reactive to produce any detectable proportion of glycoside in reasonable periods of time. At elevated temperatures the rate of decomposition of the nitrate exceeds the rate of glycoside formation.

Acetylated glycosyl perchlorates react with alcohols to give glycosides.[72]

2. *Michael and Koenigs–Knorr Methods*

The first successful synthesis of a glycoside was accomplished by Michael[73] by the interaction of tetra-*O*-acetyl-α-D-glucopyranosyl chloride (**18**) and the potassium salt (**19**) of a phenol. Under the conditions of the reaction, the acetyl groups were also removed and the product was the β-D-glycoside (**20**). Considerably later, Koenigs and Knorr[74] introduced the use of the corresponding bromide. The utility of the method was further increased by performing the reaction in an alkaline aqueous–acetone solution of the phenol.[75–77]

The method is a convenient one for the preparation of aryl glycosides and takes place with inversion at the anomeric carbon atom. Yields are not always satisfactory, but good results are obtained with appropriately substituted phenols.[78] In the modified procedure the acetyl groups are not saponified and acetylated glycosides are obtained.

Although the Michael procedure can be applied only to prepare aryl glycosides, Koenigs and Knorr[74] devised a method, suitable for the synthesis of both alkyl and aryl glycosides, which has become an important one. It is used frequently for the synthesis of glycosides having complex aglycon groups. In general, the procedure involves treatment of an *O*-acetylglycosyl halide with the alcohol or phenol, in certain inert solvents when necessary, and in the presence of an excess of a heavy metal salt or organic base as an acid acceptor. The acceptor speeds up the reaction and prevents deacetylation of the product. Usually silver carbonate or silver oxide is used, as for example in the preparation of methyl tetra-*O*-acetyl-β-D-glucopyranoside (**23**) by treatment of tetra-*O*-acetyl-α-D-glucopyranosyl bromide (**21**) with methanol and silver carbonate. The reaction is considered to proceed via the ion (**22**) (see later). An interesting example is the condensation of hydroxymethylferrocene with the bromide (**21**) in the presence of silver oxide and calcium sulfate to give mainly ferrocenylmethyl 2,3,4,6-tetra-O-acetyl-β-D-glucopyranoside with some bis-(ferrocenylmethyl) ether.[78a]

The O-acetylglycosyl bromides react at a lower temperature than the corresponding chlorides and are to be preferred for most reactions. The longer-chain aliphatic alcohols do not react with the chlorides under the usual conditions. However, in some cases the chlorides are valuable, and α-D-glucosides have been obtained by a simple general method from 2,3,4,6-tetra-O-benzyl-D-glucosyl chloride.[79] Treatment of the chloride with an alcohol in the presence of silver carbonate and silver perchlorate gives a mixture of the corresponding α- and β-D-glucosides containing a high proportion of the α anomer; the mixture is readily resolved by chromatography on a resin column.[9] Likewise, 2,3,4,6-tetra-O-benzyl-D-galactosyl chloride (obtained by treatment of 2,3,4,6-tetra-O-benzyl-D-galactose with thionyl chloride under carefully controlled conditions) has been used to synthesize α- and β-D-galactopyranosides.[80] Since other methods for synthesizing α-D-galactosides either can be applied only in certain cases, or give products in very low yield, the reaction of 2,3,4,6-tetra-O-benzyl-D-galactosyl chloride with alcohols under specified conditions provides a method generally applicable for obtaining α-D-galactosides in good yield.

The O-benzoylglycosyl bromides often may be used advantageously in place of the acetyl analogs.[81, 82]

Numerous variations and improvements in the original method have been reported, and certain of these are particularly valuable.[83-85] The use of Drierite is often beneficial, and the presence of iodine may improve the yield.[83, 86] Zemplén and Csürös[87] were able to prepare some α-D-glycosides by replacing the silver carbonate with mercuric acetate or ferric chloride. More recently, mercuric salts have been used frequently in place of or in combination with the silver salt, and although there seems to be a general tendency for a reaction so conducted to proceed with Walden inversion, either anomer may be produced, depending on the experimental conditions.

It is reported[88] that use of dry quinoline at 100° as sole acid acceptor (instead of silver carbonate) is valuable in the preparation of phenyl α-D-glycosides. Although frequently used with silver compounds, pyridine was not recommended as sole acid acceptor;[89, 90] however, pyridine in ether was employed successfully in the synthesis of methyl tetra-O-acetyl-α-D-fructopyranoside.[91]

The method can be used to obtain glycosides other than pyranosides. For example, tetra-O-acetyl-D-galactofuranosyl chloride and tetra-O-acetyl-α-D-galactoseptanosyl chloride have been converted into ethyl β-D-galactofuranoside[92] and methyl β-D-galactoseptanoside,[93] respectively. Glaudemans and Fletcher[94] concluded, from a study of the rates of methanolysis of three 2-O-nitro-3,5-di-O-p-nitrobenzoyl-D-arabinofuranosyl halides and analyses by gas–liquid chromatography of the amounts of products, that an S$_N$1 mechanism operates in the methanolyses and that the stability of the intermediate ion pair has a strong influence on the configuration of the products. From the examples studied it seems that the solvolysis of pentofuranosyl halides lacking a participating group at C-2 favors the 1,2-*cis* product, and this may be true generally, irrespective of the configuration of the halide.

This investigation by Glaudemans and Fletcher[94] is but one example of a number of reports in recent years concerning the mechanism of the solvolysis of fully acylated glycosyl halides,[95–102] and it appears established that an SN1 type of reaction is normally involved but that with certain nucleophiles and with solvents of low polarity some SN2 character is manifest. Schroeter and co-workers[102] have concluded that primary alcoholyses of 2,3,4,6-tetra-*O*-acetyl-α-D-glucopyranosyl bromide proceed by an SN1 mechanism, but for secondary alcoholyses an SN2 mechanism is proposed.

Solvolyses of acylated glycosyl halides may be divided into two classes: (1) those that involve an "open-ion" intermediate (**24**) with no participation by a neighboring group, and (2) those having a "closed-ion" intermediate (**25**) resulting from participation by the neighboring acyloxy group.

24   **25**

Halides having an acyloxy group at C-2 *cis* to the halogen at C-1 normally react with inversion, whereas the corresponding *trans* halides react with predominant retention of configuration. Hence, from a practical point of view, aldose derivatives having a substituent at C-1 *trans* to that at C-2 are readily available no matter whether the parent halide is *cis* or *trans*.[102a]

The intermediate ion (**24**) presumably has a half-chair conformation in which C-2, C-1, C-5, and the oxygen atom of the ring will lie in one plane. In going from the chair conformation of the halide to the half-chair conformation of the ion (**24**), large equatorial substituents in the starting molecule will hinder the conversion whereas large axial ones will assist the change (for a discussion of this point see p. 326). An example of this is the much greater rate of methanolysis of 2,3,4-tri-*O*-acetyl-α-D-xylosyl bromide compared with 2,3,4,6-tetra-*O*-acetyl-α-D-glucosyl bromide. The latter has an equatorial CH_2OAc group at C-5 (see Table III).

When the acyloxy group at C-2 is *trans* to the halide it can participate in the displacement by attack on the back-face of C-1 as the halogen departs, thereby

TABLE III

Unimolecular Methanolysis of Acetylglycopyranosyl Halides[95,105]

Glycopyranosyl halide	$10^5 k_1$ (sec^{-1})	Temperature (degrees)
Tetra-O-acetyl-α-D-glucosyl bromide	2.80	21.2
Tetra-O-acetyl-α-D-galactosyl bromide	12.40	21.2
Tetra-O-acetyl-α-D-mannosyl bromide	30.00	21.2
Tri-O-acetyl-α-D-xylosyl bromide	139.00	21.2
Tetra-O-acetyl-α-D-glucosyl chloride	0.42	35
Tetra-O-acetyl-β-D-glucosyl chloride	265	23.7
Tetra-O-acetyl-α-D-mannosyl chloride	0.69	23.5

forming an orthoester carbonium ion.† With such anchimeric assistance possible, the orientation of the acyloxy group at C-3 is also significant. If it is *cis* to that at C-2, some hindrance to the formation of a five-membered ring intermediate may result, and the greater reactivity of 2,3,4,6-tetra-O-acetyl-β-D-glucosyl chloride compared with the α-D-mannosyl analog has been ascribed to this cause[105] (see Table III). It is doubtful whether the entire difference in reactivity can be attributed to this hindrance, and at least part of the difference may be due to a higher free energy of the initial state of the β-D-glucosyl halide. From the foregoing account it would seem possible to prepare some of the more difficultly accessible α-D-glycosides by starting with O-acetyl-β-D-glycopyranosyl halides rather than the common α-D anomers. The notable instances in which this procedure has succeeded was accomplished with 3,4,6-tri-O-acetyl-2-O-trichloroacetyl-β-D-glucopyranosyl chloride[106] and with 3,4,6-tri-O-acetyl-β-D-glucopyranosyl chloride.[107,108] Neither the O-trichloroacetyl group nor the hydroxyl group at C-2 participates in the substitution reaction, which allows simple replacement to occur without complications arising from formation of orthoesters.[109] The instability of ordinary O-acetyl-β-D-glycopyranosyl halides limits this extension of the method, and when the α-D-glycopyranoside is not obtained in this procedure it is probable that interconversion of the isomeric halides takes place faster than the replacement reaction. Interestingly, treatment of 2,3,5-tri-O-benzyl-D-ribosyl bromide (having a nonparticipating group at C-2) with methanol and silver carbonate gives mainly the methyl α-D-ribofuranoside derivative.[110]

† For a full discussion of this effect see ref. 103. It is noteworthy that formation of cyclic acetoxonium ions as a consequence of displacement of groups by neighboring acetoxyl groups was postulated in carbohydrate chemistry by Isbell[103] before Winstein's now classical work on neighboring group effects appeared [S. Winstein and R. E. Buckles, *J. Amer. Chem. Soc.*, **64**, 2780, 2787 (1942)]. See also ref. 104.

The methods described normally lead to the production of acylated glycosides, usually the acetates. Deacylation can be achieved, for example, by treatment with alkali or sodium methoxide, potassium methoxide or barium methoxide, or methanolic ammonia or dimethylamine.

3. *From Glycosyl Fluorides*

Glycosyl fluorides do not react under the conditions of the Koenigs–Knorr procedure but are easily converted into glycosides by the action of the corresponding alkoxides in boiling alcohol. When the fluorine atom and the hydroxyl group at C-2 are *trans* disposed, the glycoside has the same configuration as the fluoride and is produced via the 1,2-anhydride. When they are *cis* related or when *trans* but with the hydroxyl group at C-2 blocked, the glycoside has an opposite configuration to that of the fluoride (see Fig. 2). This reaction may be used to prepare some of the more difficultly accessible alkyl α-D-glycosides.

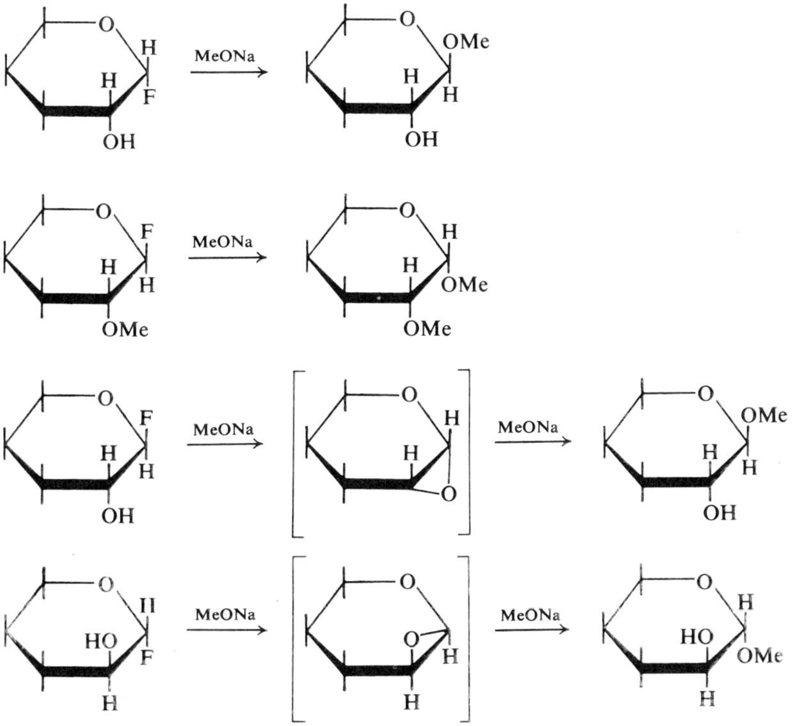

FIG. 2. Formation of glycosides from glycosyl fluorides.

* References start on p. 346.

4. *From Ortho Esters*

Under the slightly basic conditions of the Koenigs–Knorr reaction an alkoxide anion can attack the "closed-ion" (**26**) that results from participation of an acetoxy group at C-2 in replacement reactions with acetylglycosyl halides. This results in formation of a stable ortho ester (**27**); for example:

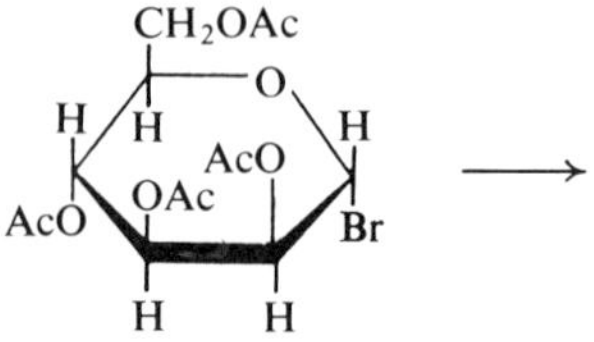

CH$_2$OAc

Tetra-*O*-acetyl-α-D-
mannopyranosyl bromide

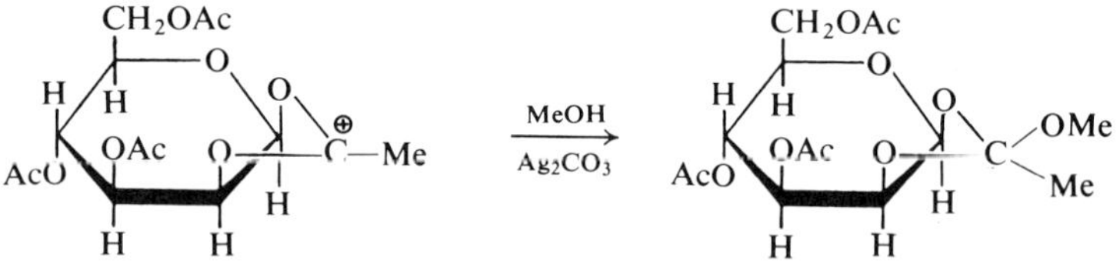

CH$_2$OAc

26

MeOH
Ag$_2$CO$_3$

CH$_2$OAc

OMe

27
Tri-*O*-acetyl-D-mannopyranose-
1,2-(methyl orthoacetate)

This "side-reaction" affects adversely the yield of desired product, particularly when formation of the glycosidic linkage involves an unreactive hydroxylic compound. Whereas, for example, tetra-*O*-acetyl-α-D-glucopyranosyl bromide (a *cis* halide) yields 90 to 95% of methyl tetra-*O*-acetyl-β-D-glucopyranoside when condensed with methanol in the presence of silver carbonate, the products from the corresponding α-D-mannopyranosyl bromide (a *trans* halide) vary according to the exact conditions of reaction.[111] In pure methanol at 20° and below, the product is 78% of 3,4,6-tri-*O*-acetyl-β-D-mannopyranose 1,2-(methyl orthoacetate) and 15% of methyl tetra-*O*-acetyl-β-D-manno-pyranoside, but no methyl tetra-*O*-acetyl-α-D-mannoside is formed, a result that indicates that conversion of the ortho ester ionic intermediate into the ortho ester is the favored course when methanol alone is present. In the presence of ether and small proportions of methanol, tetra-*O*-acetyl-α-D-mannosyl bromide yielded 7.5% of the orthoester, 23% of the acetylated β-D-glycoside, and 34% of the acetylated α-D-glycoside. The effect of benzene on the proportion of the products was similar to that of ether, but less pronounced.

When the temperature was raised to 50°, the proportion of the ortho ester dropped to 53%, and 8% of the acetylated α-D-mannoside was formed, together with 25% of the acetylated β-D-mannoside.

Several products have been isolated[112] from the reaction mixture resulting from treatment of tetra-*O*-acetyl-α-D-mannopyranosyl bromide with silver oxide in benzene. These include 2,3,4,6-tetra-*O*-acetyl-D-mannose, 3,4,6-tri-*O*-acetyl-β-D-mannopyranose 1,2-(2,3,4,6-tetra-*O*-acetyl-D-mannopyranosyl orthoacetate), and 10% of a "trisaccharide" that contains three residues of D-mannose linked together through the functional groups of a molecule of acetoacetic acid and considered[113] to have structure **28**.

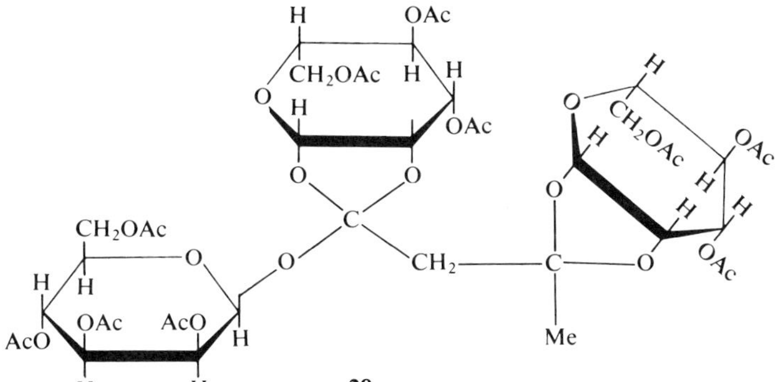

28

Although these side reactions complicate formation of glycosides by the Koenigs–Knorr reaction, the production of ortho esters is not detrimental to the synthesis of glycosides because ways have been devised for their conversion into glycosides. It has been demonstrated by Kochetkov *et al.*[114] that acetylated monosaccharide 1,2-(alkyl orthoacetates) will react with alcohols in the presence of catalytic amounts of mercuric bromide and *p*-toluenesulfonic acid (or sometimes in its absence) to give good yields of the 1,2-*trans* glycosides or of other ortho esters, depending on the reaction conditions—for example, **29** → **30** or **31**.

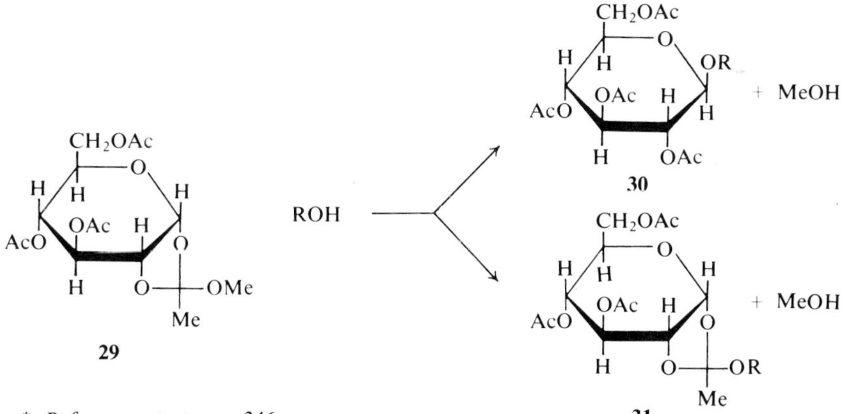

29

30

31

For example, 3,4,6-tri-O-acetyl-1,2-O-(methyl orthoacetyl)-α-D-gluco-pyranose and cholesterol, when boiled in nitromethane in the presence of mercuric bromide and p-toluenesulfonic acid, afford 3,4,6-tri-O-acetyl-α-D-glucopyranose 1,2-(cholesteryl orthoacetate) (26%) and cholesteryl 2,3,4,6-tetra-O-acetyl-β-D-glucopyranoside (15%). It was shown that the direction of the reaction is determined completely by the amount of mercuric bromide added to the reaction mixture, and by the nature of the solvent. By the use of dichloromethane and small proportions (0.001 mole) of mercuric bromide, selective formation of the ortho ester is achieved. Larger proportions (0.008 to 0.1 mole) of mercuric bromide and use of nitromethane as solvent lead to the production of acetylated 1,2-*trans*-glycopyranosides. Since Kochetkov *et al.* not only determined the conditions necessary for selective reaction but also devised a new route to ortho esters,† this method is a convenient one for the preparation of 1,2-*trans*-glycosides. The method has been extended to the synthesis of polysaccharides having predictable types of glycosidic linkage,[115] and variants of the procedure are useful for synthesis of oligosaccharides.[115a] Stereoselective ring opening of β-D-mannopyranose 1,2-(alkyl orthoacetate) has been examined also by Franks and Montgomery.[116] 3,4,6-Tri-O-benzyl-β-D-mannose 1,2-(methyl orthoacetate) in dichloromethane with p-toluene-sulfonic acid underwent complete rearrangement with almost exclusive forma-tion of methyl α-D-glycoside derivatives; methyl 2-O-acetyl-3,4,6-tri-O-benzyl-α-D-mannoside (82%) and methyl 3,4,6-tri-O-benzyl-α-D-mannoside (7%) were isolated and a small proportion of the β-D-anomer was identified by thin-layer chromatography. A similar reaction in nitromethane with mer-curic bromide as catalyst was slower, but otherwise gave the same results. The presence of methanol in the reaction mixture appeared to have little or no effect on the direction of ring opening, but there was more extensive loss of the acetyl group at C-2.

Orthoacetates in cold dilute methanolic hydrogen chloride undergo a rapid change in optical rotation. Dale,[117] who first reported this property, obtained a small yield of methyl 2,3,4,6-tetra-O-acetyl-β-D-mannoside as the only identifiable product formed by methanolysis of 3,4,6-tri-O-acetyl-β-D-man-nose 1,2-(methyl orthoacetate). Repetition of this experiment by Perlin[118] indicated that the main product was 3,4,6-tri-O-acetyl-D-mannose (80%) and small amounts in about equal proportion of the anomeric methyl 3,4,6-tri-O-acetyl-D-mannopyranosides. The methanolysis conditions are critical, and prolonged treatment yields products that correspond to di- and monoacetates as well as free D-mannose. Several concurrent reactions appear to take place, initiated by protonation at different positions on the orthoester ring.

---

† This route involves heating the readily available 1,2-*cis*-acetylglycosyl halides in boiling ethyl acetate with lead carbonate–calcium sulfate mixture as the acceptor of hydrogen halide.

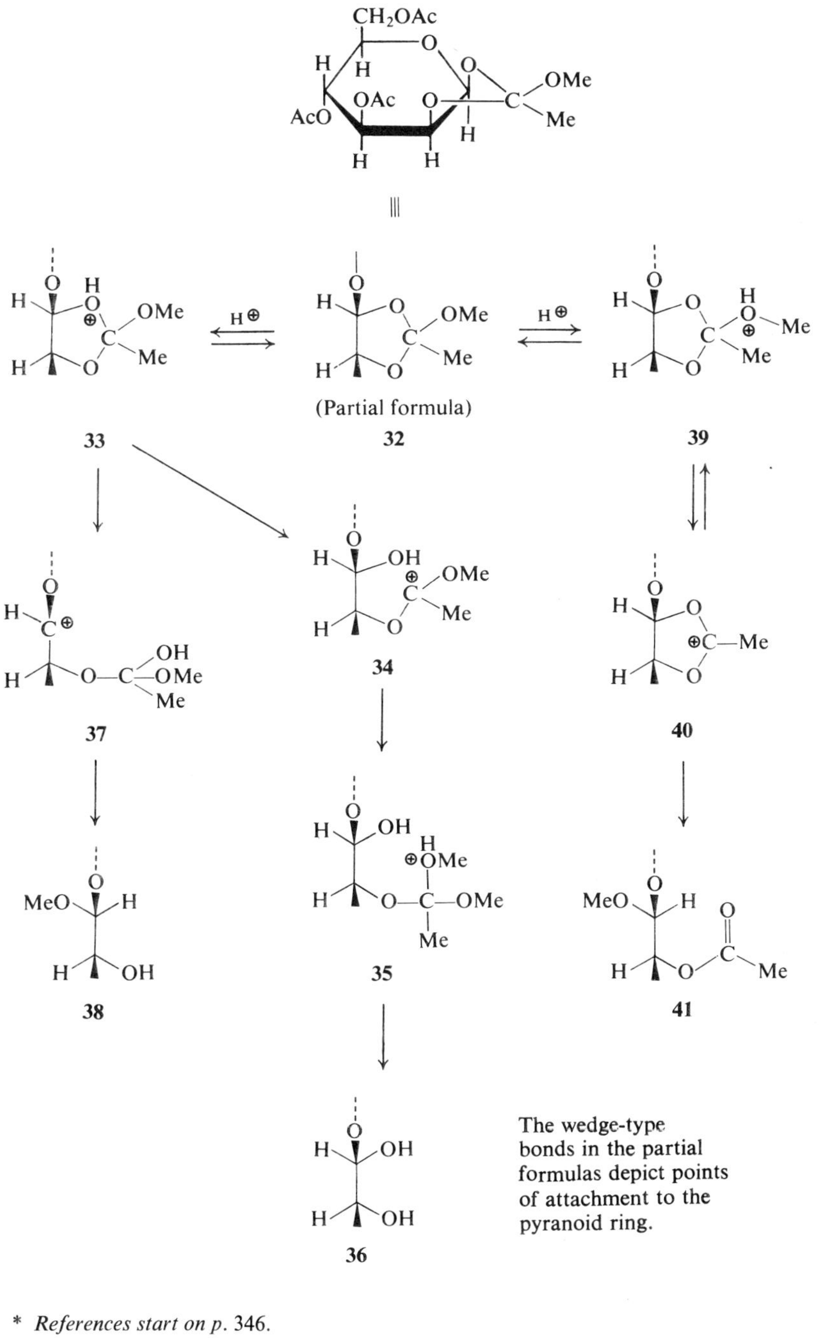

‖

33 **32** (Partial formula) **39**

37 **34** **40**

38 **35** **41**

36

The wedge-type bonds in the partial formulas depict points of attachment to the pyranoid ring.

Protonation of the orthoacetate (**32**) could lead successively to **33** and the rearranged ion (**34**), on which solvent attack can give the protonated entity (**35**). By a similar sequence involving protonation of the oxygen atom attached to C-2 in **35**, the acyclic ortho ester group can be removed as methyl orthoacetate, thereby affording the triacetate (**36**). As suggested by Pacsu,[119] a transesterification occurs between the ortho ester and the hydroxylic solvent. The overall reaction should be the same if the reaction were initiated by protonation of the oxygen at C-2 of **32** except that, in the intermediates corresponding to **34** and **35**, C-1 rather than C-2 would be substituted. If rearrangement of **33** were to occur with opening of the C-1–O bond, the orthoacid ion (**37**) would be formed, which on solvent attack and transesterification of the orthoacid group could yield the 3,4,6-triacetate of the methyl glycoside (**38**). Because of the 1,2-*cis* orientation of **33** and the 1,2-*trans* arrangement of the derived glycoside, ion **37** may be regarded as analogous to the ionic intermediate associated with nucleophilic substitution of 1,2-*cis* acyl halides.[104] Protonation of the *O*-methyl group as in **39** can lead to the cyclic ion (**40**), which by rearward solvent attack yields the 1,2-*trans* acyl glycoside (**41**) (that is, the α-mannoside derivative).

5. *Transglycosylation*

The aglycon group of a glycoside may be exchanged for another by what is obviously an intermolecular reaction. For example, ethyl α-D-glucopyranoside in methanol containing hydrogen chloride is transformed into methyl α-D-glucopyranoside. The methyl and benzyl α-D-fructofuranosides yield benzyl β-D-fructopyranoside when dissolved in benzyl alcohol under similar conditions.[120] Phenyl D-glucopyranosides are obtained by the action of the phenol on methyl α-D-glucopyranoside–boron trichloride.[121] Acetylated methyl aldopyranosides have been converted into the analogous acetylated phenyl glycosides by heating with a phenol and zinc chloride,[47] or with moist phosphorus oxychloride.[49] It is noteworthy that methyl α-D-glucofuranoside, on treatment with either *p*-nitro- or *p*-chloro-aniline in anhydrous methanol containing 0.01 to 0.1 mole/mole of hydrogen chloride, afforded, via an acyclic intermediate that could be isolated, the *N*-*p*-nitro- or *N*-*p*-chlorophenyl-D-glucopyranosylamine, whereas methyl α-D-glucopyranoside was recovered unchanged.[122] Similar results were obtained with the ethyl and propyl D-glucosides and with the corresponding D-galactosides.

The most economical method of preparing certain methyl glycosides is by methanolysis of polysaccharides—for example, the preparation of methyl α-D-mannopyranoside from D-mannan.[123]

D. OTHER METHODS

1. *Direct Alkylation*

Glycosides can be obtained by direct alkylation of a sugar, or certain derivatives thereof, with one equivalent of alkyl sulfate and alkali.[124] The method is restricted to the lowest aliphatic derivatives but is particularly useful for preparing methyl glycosides of the mannose type for which the Koenigs–Knorr reaction is complicated by formation of ortho esters; such selective methylation of D-mannose is a recognized way of making methyl β-D-mannopyranoside.[125] The mixture of anomers can be fractionated after acetylation, and the deacetylated β anomer is crystallized with one molecule of isopropyl alcohol of crystallization.

Diazomethane has been employed[126] for selective methyl glycosidation of a sugar. Alkylation of tetra-O-acetyl-β-D-fructopyranose with silver oxide and methyl iodide leads to methyl β-D-fructopyranoside tetraacetate.[127]

2. *From Anhydrides*

Brigl's anhydride (tri-O-acetyl-1,2-anhydro-α-D-glucopyranose) can be employed for forming α- or β-D-glucopyranosides, depending on the conditions used.[98, 104] Brigl[128] prepared methyl β-D-glucopyranoside 3,4,6-triacetate by evaporating to dryness a solution of the anhydride in methanol. Lemieux and Huber[129] have extended the method to the synthesis of sucrose (see Chapter 30).

Analysis of the fragmentation products of a polymer made by heating 1,6-anhydro-β-D-glucopyranose reveals that when heated the anhydride is capable of polymerizing to produce the usual glycosidic linkage found in a pyrodextrin.[130]

3. *From Glycals*

Enol ethers, in general, give acetals on treatment with alcohols in the presence of an acid[131] and by the same means a number of alkyl glycosides of 2-deoxypentoses and -hexoses have been prepared.[132] As in the hydration of glycals, competing reactions interfere.

Phenyl 3,4,6-tri-O-acetyl-2-deoxy-α-D-*lyxo*-hexopyranoside has been prepared in good yield by the addition of phenol to the double bond of tri-O-acetyl-D-galactal under the catalytic influence of p-toluenesulfonic acid,[133] but when a mixture of p-nitrophenol with tri-O-acetyl-D-glucal was treated under the same conditions extensive decomposition took place and the glycoside could be isolated only in low yield.[134] In the absence of the acid catalyst

---

* *References start on p. 346.*

these latter reagents gave unsaturated products.[135] When the unfractionated dibromides obtained by addition of bromine to tri-O-acetyl-D-glucal were treated with methanol in the presence of silver carbonate, high yields of the methyl 3,4,6-tri-O-acetyl-2-bromo-2-deoxy-β-D-glycosides were produced.[136] These were separated, and characterized configurationally by nuclear magnetic resonance spectroscopy.[137] Reduction of the mixture of glycosides, in the presence of palladium on charcoal, gave, after deacetylation, methyl 2-deoxy-β-D-*arabino*-hexopyranoside.

Halogenomethoxylation[137, 138] and methoxymercuration[139, 140] of acetylated glycals have also been used to provide intermediates that are convertible into methyl 2-deoxyglycosides.

III. DETERMINATION OF THE STRUCTURE OF GLYCOSIDES

Several methods have been developed to determine the size of the ring and the configuration at the anomeric center in a glycoside. The classical method of investigation of the ring size of glycosides involves successive methylation and oxidation procedures. Figure 3 illustrates this approach as applied to the methyl D-glucosides. Methylation of either anomer of methyl D-glucopyranoside and subsequent hydrolytic cleavage of the glycosidic methoxyl group led to the same crystalline tetra-O-methyl-D-glucose (**42**). This substance on oxidation with nitric acid yielded tri-O-methylxylaric acid (**43**), which can be characterized as its diamide or bis(N-alkylamide). This type of proof of ring structure was first established with D-xylose by Hirst and Purves[141] and later was extended to D-glucose.[142] Similar treatment of methyl α-D-glucofuranoside led to the identification of di-O-methyl-L-threaric acid (**44**) as the final oxidation product.[143] From the nature of the final substituted dibasic acid it was possible to deduce the size of the ring in the initial glycoside. The intermediate, methylated lactones (**45** and **46**) exhibited the hydrolytic behavior characteristic of 1,5- and 1,4-lactones, respectively. A similar oxidative degradation was employed by Micheel and Suckfüll[144] in establishing the 1,6 or septanoid ring in methyl tetra-O-methyl-β-D-galactoseptanoside, which afforded tetra-O-methylgalactaric acid.

Oxidative degradation of methyl ethers has been utilized in establishing the ring structures of glycosides derived from ketoses.[145]

Another method used extensively for determination of both ring size and anomeric configuration is based on the oxidative cleavage of glycol systems with either sodium periodate or lead tetraacetate, or, much less frequently, with sodium bismuthate (see Vol. IB, Chap. 25). The extent of substitution in a glycoside of known structure can also be estimated.

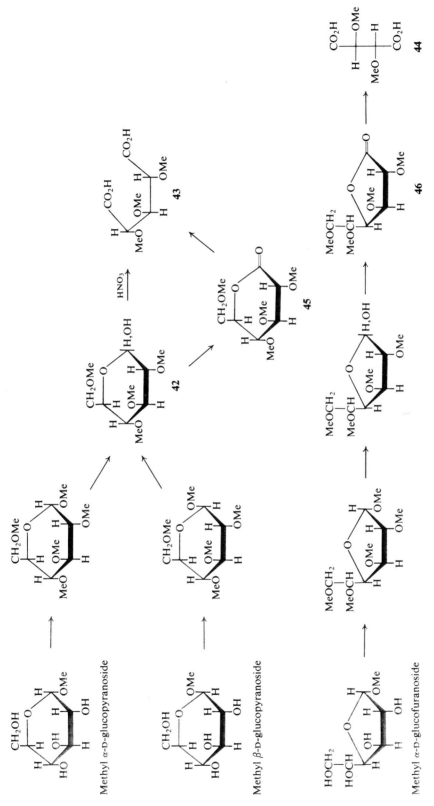

Fig. 3. Determination of structures of methyl D-glucosides.

Physical methods are used widely in structural and conformational assignments. Infrared and particularly n.m.r. spectral measurements are especially useful.

The infrared spectroscopy of carbohydrates has been reviewed[146, 147] and is discussed in detail in Vol. IB, Chap. 27. Tipson and Isbell and co-workers[148] have examined aldopyranosides, their acetates, and fully acetylated pyranoses, over a wide spectral range (usually 250 to 5000 cm^{-1}). They have demonstrated exceptions to the criteria given by earlier workers for characterizing the disposition at the anomeric carbon atom.

Since the anomeric proton of cyclic aldose derivatives usually resonates in the n.m.r. spectrum at lower field than the other ring protons, the chemical shift and splitting of the H-1 signal can often provide evidence for the configuration at this center (see Chapter 27). In particular, in cases where H-1 and H-2 of pyranoid compounds both have axial orientations, configurations can be readily assigned, since only in this condition are large $J_{1, 2}$ values observed. Fortunately stereochemical factors frequently favor structures that possess this feature, since the bulky C-1 and C-2 substituents are then equatorial. Many cases of anomeric configurational assignments have been reported,[149] including glycopyranosidic compounds.[150] The conformational equilibria in solution of the acylated methyl aldopentopyranosides have been studied in detail, and the conformational influence of the steric bulk of the aglycon has also been examined.[151]

To test the range of usefulness of n.m.r. spectral measurements for assigning the configurations to the anomeric hydrogen atoms of furanoid derivatives, Capon and Thacker[151a] measured the spectra of eight methyl aldofuranosides and the corresponding pyranosides for comparison (Table IV). For the furanosides measured, the distinction between anomers is clear, since the signals from the anomeric protons of the 1,2-*trans* compounds all have splittings of 2 Hz or less. There is little difference in the chemical shifts between anomers, and the differences in chemical shifts for the methoxyl groups of anomeric furanosides are also very small, thus making a clear distinction between anomers impossible on this basis.

Although the flexibility of the furanoside ring makes it possible that there would be no difference in the splitting of the anomeric proton signal for *cis* and *trans* 1,2-compounds, probably it is significant that this was not observed for the four anomeric pairs of furanosides studied. Thus, for these compounds the anomer having the smaller splitting of the anomeric signal is the *trans* 1,2-compound, and this might be expected to hold for other pairs of *simple* furanosides.

For the pyranosides the situation is different, since the α and β anomers may be distinguished on the basis of the splitting of the H-1 signal and also by the difference in chemical shift of the signal. There is a larger difference in the

TABLE IV[a]

THE NUCLEAR MAGNETIC RESONANCE SPECTRA OF SOME GLYCOSIDES[151a]

Glycoside	H-1		MeO	H-1		MeO
	τ	$J_{1,2}$ (Hz)	τ	τ	$J_{1,2}$ (Hz)	τ
	D-Glucose			D-Galactose		
Methyl α-furanoside	4.77	4.2	6.57	4.88	4.5	6.51
Methyl β-furanoside	4.80	0	6.63	4.61	2.0	6.52
Methyl α-pyranoside	4.87	3.5	6.57	4.79	3.5	6.51
Methyl β-pyranoside	5.27	7.0	6.41	5.32	8.0	6.37
	D-Xylose			L-Arabinose		
Methyl α-furanoside	4.87	4.0	6.57	4.72	1.0	6.60
Methyl β-furanoside	4.69	1.0	6.54	4.88	4.0	6.53
Methyl α-pyranoside	4.94	3.5	6.60	5.48	8.0	6.43
Methyl β-pyranoside	5.52	7.0	6.48	4.92	2.5	6.61

[a] Adapted by permission of the Editor of the Chemical Society and the senior author of the paper. All spectra were measured with pyridine solutions at 60 MHz.

chemical shifts of the methyl protons of the pyranosides than for the furanosides.

Crystal-structure analysis has been applied to the determination of carbohydrate structures (see Chapter 27) including glycosides[152]—for example, methyl β-D-xylopyranoside,[153] methyl α-D-lyxofuranoside,[154] and methyl 6-bromo-6-deoxy-α-D-galactopyranoside.[155] As so much of carbohydrate chemistry involves reactions in solution, it must be emphasized that the degree to which this type of structural information is relevant to the dynamic stereo relationships existing between sugar and solvent molecules in solutions of sugars depends on inferences that must be critically examined by other experimental methods.[150]

IV. PROPERTIES OF GLYCOSIDES

The glycosides are water-soluble substances except when the hydrocarbon portion of the aglycon becomes large enough to dominate the physical behavior of the compound. In the series of n-alkyl β-D-glucopyranosides, the D-glucosides become quite difficultly soluble in water when the aglycon has

* References start on p. 346.

more than nine carbon atoms. The higher members of the *n*-alkyl series of
β-D-glucosides are surface-active and form liquid crystals at the melting
point.[156, 157] The solubility of the surface-active materials in water is improved
by treating the glycosides with alkylene oxides (as ethylene oxides) in the presence of catalysts such as sodium hydroxide or an amine.[158] The conversion of
water-insoluble substances into water-soluble materials by glycosidic union
with a sugar is important for the detoxification of many phenolic compounds in
the animal organism by excretion in the urine as D-glucosiduronic acids.[159]
An analogous role—that is, to increase solubility and decrease toxicity—is
played by glycosidic derivatives of some drugs.

In contrast to the free sugars, glycosides (apart from a few exceptions like
methyl α-D-glucopyranoside) are not sweet, and for the greater part they taste
more or less bitter.

The optical rotations of *o*-nitrophenyl α- and β-D-glucopyranoside are un-
usually temperature-dependent,[160] probably because of electronic interaction
between the nitro group and the sugar. Optical rotatory dispersion studies of
some common glycosides have been reported.[161]

V. REACTIONS INVOLVING THE GLYCOSIDIC LINKAGE

A. ANOMERIZATION

It is well known[104] that, except for the hydroxyl group, polar aglycons tend
to assume an axial rather than an equatorial orientation.[162, 163] For example,
the equilibration of methyl 2,3,4,6-tetra-*O*-methyl-α- and -β-D-glucopyrano-
side in 5% methanolic hydrogen chloride yields about a 3:1 ratio of α and β
forms, showing that the axial orientation is favored for the aglycon. Judging
from the results obtained by Lemieux and Shyluk[164] in their study of the rearrangement of methyl β-D-glucopyranoside tetraacetate into the α anomer,
when the conversion is catalyzed either by boron trifluoride or titanium tetrachloride, the α form having the methoxyl group at C-1 axial is produced in over
90% yield. This phenomenon has been termed the "anomeric effect" and is
considered to arise from the more favorable dipole–dipole interactions that
exist with a polar, axial substituent at C-1. The conclusion that the favored
anomer had the larger substituent at C-1 axial required the assumption that
the sugar derivatives have the chair conformation with the largest number of
bulky groups equatorially disposed. That, in fact, many of these sugar derivatives did have the assumed conformations was demonstrated first by Lemieux
and co-workers.[165]

The "anomeric effect" is important in understanding the anomerization
(rearrangement of α and β forms) of glycosides. In 1928 Pacsu showed that

acetylated alkyl β-D-glycosides, in solution in absolute chloroform, are transformed by titanium tetrachloride[166] or stannic chloride[167] into the α-D anomers in high yield. Titanium tetrachloride was the more effective catalyst, and methods employing it have been used to prepare glycosides that are anomeric with those obtained from 1,2-cis-O-acetylglycosyl halides by the Koenigs–Knorr reaction. For the most part, analogous procedures with aromatic glycosides are unsatisfactory,[47,168] and anomerization is best achieved under the conditions of the Helferich reaction. According to some investigators,[169] anomerization proceeds better with benzoyl derivatives of β-glycosides. Lindberg[170] demonstrated that the O-acetylated ethyl β-glycosides of D-glucose and cellobiose are rearranged to their respective α anomers by heating a solution of each in benzene with hydrogen bromide and mercuric bromide. Also, he examined the effect of various acids and of boron trifluoride on acetylated alkyl glycosides in acetic acid or acetic anhydride and found that the rate of anomerization depends on the nature of the aglycon;[171] the rate followed the order tert-butyl > isopropyl > ethyl > methyl > allyl > benzyl glycoside.

Although a number of attempts have been made to elucidate the mechanism of anomerizations, further work is needed to resolve some of the differences reported in the literature, and the following account merely records the results announced from various laboratories.

In Fig. 4, three reasonable mechanisms are illustrated for the anomerization of methyl β-D-glucopyranoside. These involve as intermediate a cyclic ion (47), an acyclic ion (48), and an acetal (49), which undergoes ring closure stereospecifically with inversion. For this glycoside Capon[172] has produced evidence that mechanism 1 is the most likely. He distinguished between the three possibilities by following by n.m.r. spectroscopy the anomerization of methyl β-D-glucopyranoside in methanol-d_4 containing methanesulfonic acid at $70°$. The extent of anomerization was determined by the areas under the signals of the anomeric protons at τ 5.30 (α) and τ 5.79 (β). Since the signals of the methoxyl groups of the two D-glucosides at τ 6.59 (α) and τ 6.47 (β) are well resolved from each other and from that of CH_3OH (τ 6.64), it is easy to see whether α-D-glucoside formed on anomerization of β-D-glucoside contains a protonated methoxyl group. The results showed, after $45 \pm 5\%$ conversion into the α anomer, that less than 2% of the methyl α-D-glucoside formed had retained the methoxyl group of the original β-D-glucoside. Similar results were obtained when methyl α-D-glucoside was used at the outset, although the data obtained were less accurate because signals of certain ring protons overlapped the signal of the methoxyl group of methyl β-D-glucopyranoside.

These results exclude mechanism 2 and also any mechanism involving the acetal as an intermediate in which it is formed by nonstereospecific processes,

* References start on p. 346.

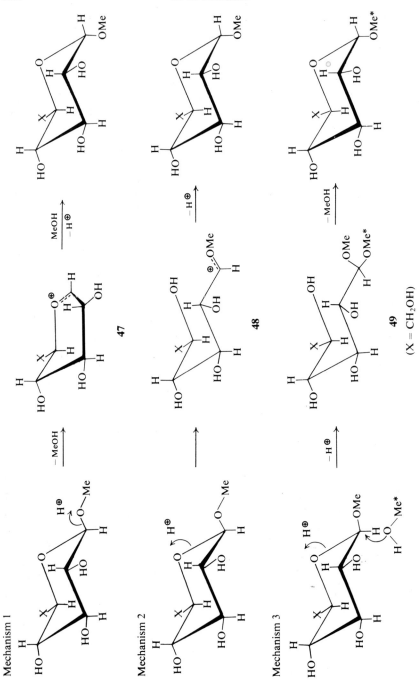

FIG. 4. Possible mechanisms for anomerization.

but they are consistent with mechanisms 1 and 3. Mechanism 3 could be excluded, since it was found that when the acetal (49) was subjected to the reaction conditions, the n.m.r. spectrum initially showed signals at τ 5.07 and 5.22 corresponding to the anomeric protons of methyl α- and β-D-glucofuranoside, and that these disappeared only after several hours, with concurrent appearance of the signals of the anomeric protons of the pyranosides. If the acetal (49) were an intermediate, on ring closure it would yield first furanosides, which are not converted rapidly into pyranosides under the reaction conditions. Since the former are not observed, the acetal (49) cannot be an intermediate.

From a study of the methanolysis of a wider range of glycosides, by paper chromatography and isotope exchange, evidence was obtained for both cyclic and acyclic ionic intermediates.[173] According to chromatographic analysis and the extent of isotope exchange (in ^{14}C-labeled methanol containing 1 % of hydrogen chloride), the glycosides examined were found to be divisible into two groups. The first group included methyl α-D-glucopyranoside and methyl α-D-mannopyranoside, and, for these, exchange was considered to proceed most probably by way of a cyclic carbonium ion. The cyclic-ion mechanism was also considered to be the principal pathway for the acid-catalyzed anomerization of ethyl D-xylopyranosides.[174] The other group, for which the exchange mechanism probably involves an acyclic ion, included methyl β-D-glucopyranoside, methyl α- and β-D-galactopyranoside, and methyl β-D-mannopyranoside.

Previous to the investigation of anomerization by n.m.r. and isotope-exchange methods, attempts had been made to explain the reaction in terms of both cyclic and acyclic ionic intermediates. Since the heterogeneous mixtures obtained from anomerizations of O-acetylated alkyl glycosides induced by the preferred catalysts (TiCl$_4$ or BF$_3$) were not amenable to kinetic measurement, Lindberg studied the kinetics of the reactions occurring in 10:3 mixtures of acetic anhydride and acetic acid with sulfuric acid as catalyst. For a series of β-glycosides he found that the rate of anomerization increased as the availability of electrons from the aglycon increased, and it was concluded[175] that the results can be explained in terms of a reaction in which there is coordination of the acid catalyst with the ring-oxygen atom of the 1,2-*trans* glycoside to give an open-chain intermediate (like 48). That the reaction was intramolecular was based on the fact that isopropyl tetra-O-acetyl-α-D-glucopyranoside and ethyl hepta-O-acetyl-α-D-cellobioside were isolated, in 66 % and 75 % yield, respectively, when a mixture of isopropyl tetra-O-acetyl-β-D-glucopyranoside and ethyl hepta-O-acetyl-β-D-cellobioside was anomerized with titanium tetrachloride in chloroform.

Lemieux[104] forwarded an alternative mechanism, which is analogous to that proposed to account for the anomerization of β-D-glycosyl chlorides in methanol. The first stage (see Fig. 5) is coordination of the catalyst (A) with the oxygen atom of the aglycon group (OR), which results in a stretching of the C-1–OR bond. In the case of O-acetylated glycosides, it was assumed that the free energy of activation for this stage is lowered by participation of the substituent at C-2 (for example, 53 and 54). Janson and Lindberg[176] have shown, however, that absence of an acetyl group at C-2 does not retard the anomerization reaction. In fact, replacement of the 2-O-acetyl group of isopropyl tetra-O-acetyl-β-D-glucopyranoside by a methyl group results in a higher rate of anomerization in an acetic anhydride–acetic acid–sulfuric acid mixture.

The evidence for the conclusion that the bond between C-1 and the ring-oxygen atom may cleave in preference to that joining C-1 to the alkoxyl group in glycopyranosides rests on the isolation of both anomeric forms of the O-acetylated methyl hemiacetal (50) in good and approximately equal yields from the product resulting from treatment of methyl tri-O-acetyl-β-D-arabino-pyranoside with either 8% of zinc chloride or 0.16% of sulfuric acid in a 7:3 acetic anhydride–acetic acid mixture.[177]

$$HC(OMe)OAc$$
$$|$$
$$AcOCH$$
$$|$$
$$HCOAc$$
$$|$$
$$HCOAc$$
$$|$$
$$CH_2OAc$$
$$\mathbf{50}$$

It seems reasonable to assume, if the group AOR is strongly nucleophilic (as would be expected when R is an alkyl group), that as the group moves outward the positive charge on the carbonium ion will shift to the neighboring ring-oxygen atom to afford the transition state (51). Winstein and co-workers[178] have presented evidence for the existence of ion-pair intermediates like 51 in a variety of rearrangements. If this ion pair should now collapse, the lactol carbon can assume the α configuration (52).

An interesting problem is presented by methyl 3,6-anhydro-2,4-di-O-methyl-α-D-glucoside, which anomerizes very rapidly in the presence of acids to afford a mixture in which the β-D anomer preponderates.[179] (If the hydroxyl group at C-4 is not protected by methylation, the change is from pyranoside into furanoside.) The corresponding α-D-galactoside undergoes the same change,[179] but the position of the equilibrium is not known. Under the conditions of the experiment the anomerization is nearly complete because the

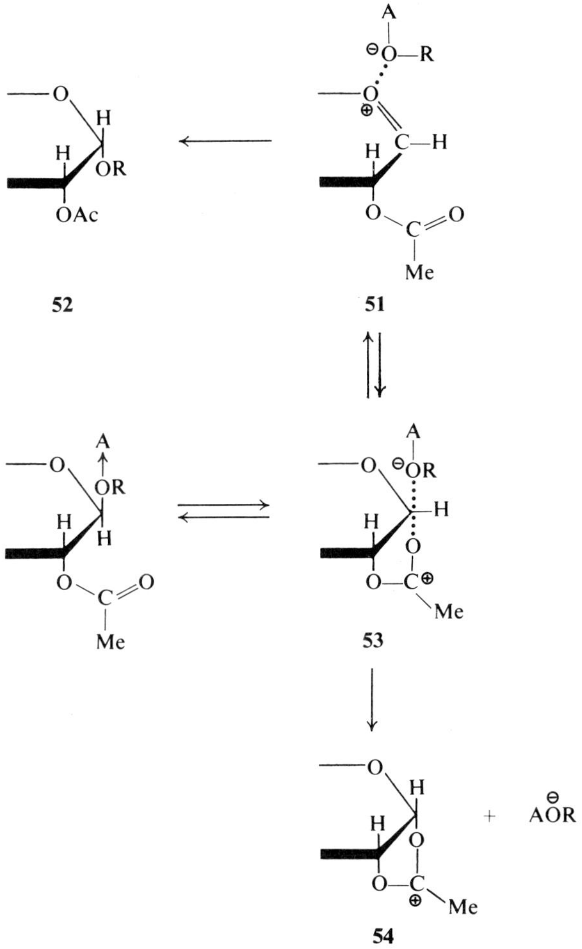

FIG. 5. Mechanism of anomerization proposed by Lemieux.[104]

syrupy α-D anomer changes into the crystalline β-D derivative. At first sight these results appear surprising,[180] since the change apparently leads to an increase in the number of axial substituents present (see **55** → **56** for conversion of the D-glucoside having the pyranoid ring in a chair conformation), but the electronic destabilization of an equatorial C-1 substituent, especially with an axial group at C-2 (see Chap. 5 for a discussion of the anomeric effect) undoubtedly outweighs any steric contribution and accounts for the greater stability of **56** (having the C-1 substituent axial).

* *References start on p. 346.*

α

55

β

56

B. Acidic Hydrolysis

Rates of hydrolysis of many glycosides have been measured with the aim of providing information about the reaction mechanism, and about the effect on the rate of steric and electronic changes in both the sugar and aglycon. The subject has been reviewed.[181] In 1894 Alberda van Ekenstein[182] prepared methyl β-D-glucopyranoside, and to distinguish it from the α anomer (prepared a year earlier by Fischer)[2] he compared their rates of acidic hydrolysis. Subsequently, quantitative information gradually accumulated. In 1908, Armstrong and Glover[183] pointed out that various glycosides are hydrolyzed by acids at very different rates, and twenty years later Moelwyn-Hughes[184] reported a more-extensive investigation of the acidic hydrolysis of certain D-glucosides. He noted that the rates of hydrolysis were more dependent on hydrogen-ion activity than on hydrogen-ion concentration, at least when such concentrations were 20% or higher. From rate measurements at two different temperatures, energies of activation were calculated. Another ten years elapsed before the rates of hydrolysis of glycosides derived from sugars other than D-glucose were studied.[185,186] Isbell and Frush[186] noted that, whereas the methyl β-D-glycopyranosides having the *gluco, manno, galacto,* and *xylo* configuration are hydrolyzed more rapidly than the corresponding α anomers, methyl α-D-gulopyranoside is hydrolyzed more rapidly than the β-D form. The main conclusions emerging from an examination of the hydrolysis of methyl or phenyl or benzyl α- and β-D-glucopyranoside in 0.1 M hydrochloric acid, and of methyl α- and β-D-fructopyranoside, benzyl β-D-fructopyranoside, and methyl α-D-fructofuranoside in 0.01 M hydrochloric acid, were that D-glucosides are hydrolyzed more slowly than D-fructosides and that pyranosides are hydrolyzed more slowly than furanosides.[187] Previously, Haworth[188] had concluded that aldofuranosides are hydrolyzed from 50 to 200 times as rapidly as the corresponding aldopyranosides, and Micheel and Suckfüll[93] found that the furanosides are hydrolyzed only slightly faster than the aldoseptanosides. A different behavior is exhibited by methyl α-D-fructofuranoside, which is hydrolyzed only about three times as fast as the corresponding pyranoside.[189]

1. Mechanism

Whatever mechanism is operative in the acid-catalyzed hydrolysis of an acetal, at some stage a carbon–oxygen bond must be cleaved. There are two ways in which this can occur:

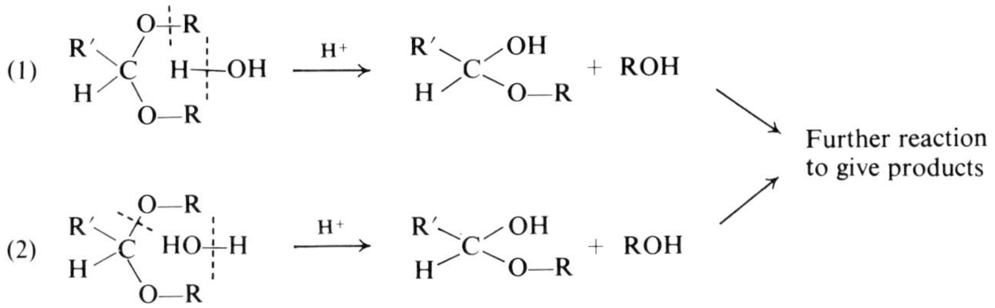

To distinguish between these possibilities for hydrolysis of glycosides, Bunton and co-workers[190] hydrolyzed methyl and phenyl α- and β-D-glucopyranoside in water enriched in ^{18}O and measured the excess abundance of the isotope in the methanol or phenol produced. The D-glucose formed in each reaction was isolated as the phenylosazone, and this also was analyzed for the isotope. The results showed that hydrolysis of the D-glucopyranoside involved fission of

$$(C_6H_{11}O_5){+}O{-}R + H_2^{18}O \longrightarrow (C_6H_{11}O_5){-}^{18}OH + HOR$$

the hexosyl–oxygen bond, as shown. A subsequent report[191] from the same laboratory indicated that the same pattern of bond cleavage occurred in the hydrolysis of o-(hydroxymethyl)phenyl β-D-glucopyranoside, methyl 2-deoxy-α- and β-D-arabino-hexopyranoside, lactose, and maltose. However, those glycosides that can lead to an alkyl carbonium ion that is much more stable than the glycosyl carbonium ion will undergo hydrolysis by cleavage of the bond between the alkyl group and the glycosidic oxygen atom. For example, tert-butyl β-D-glucopyranoside (57) is hydrolyzed in this way to produce D-glucose (58) and labeled tert-butyl alcohol via the tert-butyl cation. The hydrolysis of triethylmethyl β-D-glucopyranoside proceeds by similar bond

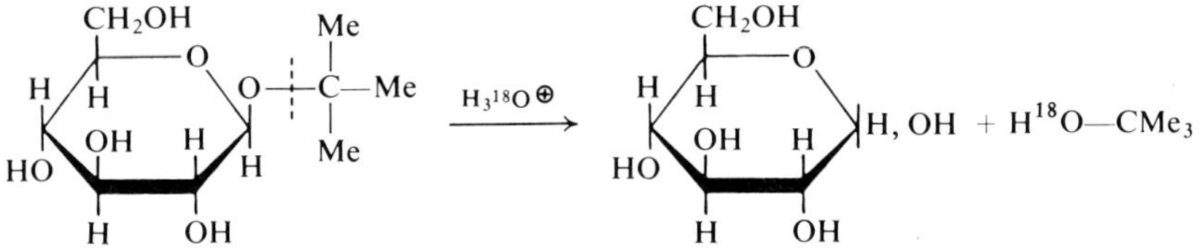

* References start on p. 346.

cleavage. The somewhat higher rate of hydrolysis of this compound relative to the *tert*-butyl analog probably is explicable on the basis of steric compression in the transition state.

The hydrolysis of aldofuranosides has been studied in less detail than that of the aldopyranosides, but similarly it has been shown that the ethanol produced in the hydrolysis of ethyl β-D-galactofuranoside (**59**) contains no ^{18}O-isotope,[192] and for this furanoside, at least, glycosyl–oxygen bond cleavage occurs during acidic hydrolysis, as shown.

In an early study of the hydrolysis of acetals, Brönsted and Wynne-Jones[193] demonstrated that in water the reaction exhibits specific hydronium-ion cata-lysis, and later studies have supported this conclusion.[194] The existence of specific hydronium-ion catalysis for a reaction implies that a pre-equilibrium of the substrate occurs, followed by decomposition of the conjugate acid to the products. In D_2O the concentration of the conjugate acid will be greater than in H_2O, and the rate of reaction should be higher in D_2O than in H_2O. Measurements of k_{D_2O}/k_{H_2O} (values in parentheses) for the hydrolysis of methyl 2-deoxy-α-D-*arabino*-hexopyranoside (2.5),[191] methyl α-D-glucopyran-oside (~2),[195] and methyl α-D-xylofuranoside (2.5)[30] indicated a deuterium isotope-effect (as had been found for simple acetals[196]). The evidence indicates that the hydrolysis of glycosides, like simple acetals, exhibits specific hydro-nium-ion catalysis† and proceeds by means of a rapid pre-equilibrium protonation to give the conjugate acid.

Before an overall mechanism can be postulated, a detailed knowledge is required of the breakdown of the conjugate acid to yield the hydrolysis pro-ducts, and therefore the molecularity of the rate-determining step has been examined. The consensus of the results is that the hydrolysis of glycopyrano-sides proceeds by an A-1 mechanism (Ingold terminology), and that of glyco-furanosides by an A-2 mechanism. Because hydrolysis by both A-1 or A-2 mechanisms will follow a first-order rate law (since the solvent is in such large

† Intramolecular catalysis, probably general acid catalysis, occurs in the hydrolysis of 2-carboxyphenyl β-D-glucopyranoside [B. Capon, *Tetrahedron Lett.*, **911** (1963)]. At pH 3.5 the rate of hydrolysis of 2-carboxyphenyl β-D-glucopyranoside is about 10^4 times as great as that estimated for the 4-carboxyphenyl isomer by extrapolation from results at higher acidities.

excess over the reactants), criteria other than the form of the rate law for the reaction have had to be used to deduce molecularity.

Bunton and co-workers[190] followed polarimetrically the rate of hydrolysis of the D-glucopyranosides mentioned previously, in perchloric acid solutions ranging from $0.465M$ to $3.72M$ at $72.9°$, and the first-order velocity coefficients (k_1) were found to be constant throughout each reaction.

Plots of $\log k_1$ against the Hammett acidity function, H_0, were nearly linear, having slopes $= 1$, whereas a plot of $\log k_1$ against pH was not linear, thereby indicating a unimolecular mechanism for these reactions;[197] this is a result analogous to that obtained for the acid-catalyzed hydrolysis of acetals.[198]

From a consideration of the steric requirements of a bimolecular reaction it seems likely that the more flexible five-membered furanoid ring will present less steric hindrance to such a reaction than will a six-membered pyranoid ring, and this is what is found. Hammett–Zucker plots ($\log k_1 \ v - H_0$) for the hydrolysis of furanosides yield straight lines of non-unit slope, which complicate interpretation of the data. However the w values (a parameter, the magnitude and sign of which yields information about the rate-determining step) obtained from plots of $\log_{10} k + H_0$ against $\log a_{H_2O}$ ($a_{H_2O} =$ activity of water)[199] are in the range required for a bimolecular reaction.

The first suggestion that hydrolysis of furanosides proceeds by an A-2 mechanism was made by Overend, and co-workers,[195] who found a value of -5.3 cal mole^{-1} for the entropy of activation (ΔS^+)† for the hydrolysis of ethyl β-D-galactofuranoside. This was confirmed by a subsequent study of more furanosides, for which negative values for ΔS^+ were reported. The fact that ΔS^+ for hydrolysis of furanosides is of the order of 20 cal mole^{-1} less than that for hydrolysis of the corresponding pyranosides is good evidence for a change of mechanism. In two extensive studies[195, 200] of the hydrolysis of glycopyranosides, positive values for ΔS^+ were found.

Neither of these criteria of molecularity is entirely satisfactory. Strong criticism has been made[31, 201] of the Bunnett relationship. Furthermore, in determinations of the entropy of activation (ΔS^+) the standard entropy change for the pre-equilibrium protonation of the substrate is frequently large and positive (and not zero, as has been assumed), and so a positive value for ΔS^+ must be treated with caution. Since conclusions regarding the molecularity for

---

† It is considered [F. A. Long et al., J. Amer. Chem. Soc., **79**, 2362 (1957)] that the activated complex for a reaction proceeding by an A-1 mechanism would have a less-ordered structure than the ground state of the reactant, with a consequent increase in entropy. For a bimolecular A-2 mechanism the reactants will be more ordered as the transition state is approached, causing a decrease of entropy. Hence ΔS^+, defined as the standard entropy of the transition state less the standard entropies of the reactants at the temperature of the reaction, should provide evidence for an A-2 mechanism if it has a negative value and for an A-1 mechanism if the value is positive.

* References start on p. 346.

hydrolysis of furanosides are based only on studies of the dependence of rate on acidity and on measurements of entropy of activation, they must be accepted with reserve until better evidence becomes available. If the evidence is accepted, the reaction can be considered to proceed as shown in Fig. 6.

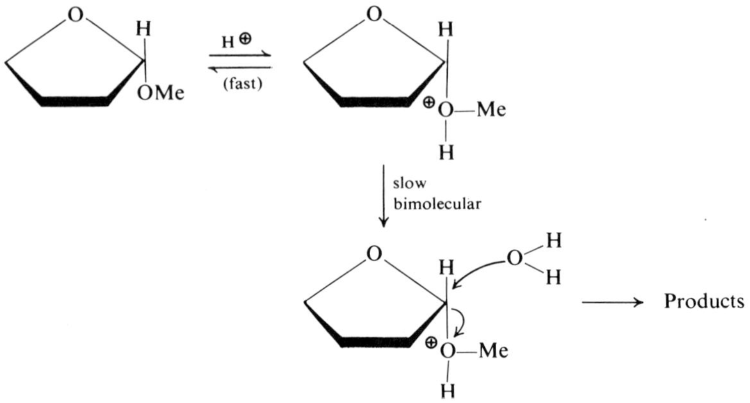

FIG. 6. Probable mechanism for hydrolysis of methyl aldofuranosides.[195]

For hydrolysis of methyl α-D-glucopyranoside, rather better evidence for the A-1 mechanism is available from a measurement of the volume of activation ($+5$ cm^3 mole^{-1}).[201, †]

From a study of the products of the hydrolysis of methyl α- or β-D-glucopyranoside, Overend and co-workers[195] found that, at the concentration of glycoside used, 5 to 7% of a disaccharide was formed (by a reversion[201a] process). If D-glucose were added before hydrolysis, the proportion of disaccharide formed was increased, but the overall rate of reaction remained unchanged. Thus the product-forming step is not the rate-controlling step, indicating an A-1 mechanism.

For glycopyranosides, details of the rate-determining unimolecular heterolysis are complicated by two possible reaction pathways, resulting according to whether protonation occurs at the ring-oxygen atom (pathway 1) or on the exocyclic oxygen atom of the aglycon (pathway 2) (see Fig. 7). It has proved to be difficult to devise experiments to distinguish unequivocally between these two possibilities, but fairly conclusive support for pathway 2 for the hydrolysis of methyl α-D-glucopyranoside has been provided by Banks and co-workers.[202]

† Measurements of the volume of activation for a reaction have been used by Whalley [E. Whalley, *Trans. Faraday Soc.*, **55**, 798 (1959); see also J. Koskikallio and E. Whalley, *ibid.*, **55**, 809 (1959)] for determination of the molecularity of a reaction. For a simple, unimolecular decomposition the volume of activation (ΔV^*) is positive, whereas for an A-2 solvolysis the value is negative. It is argued that these measurements provide a more reliable criterion of mechanism than do those for ΔS^*.

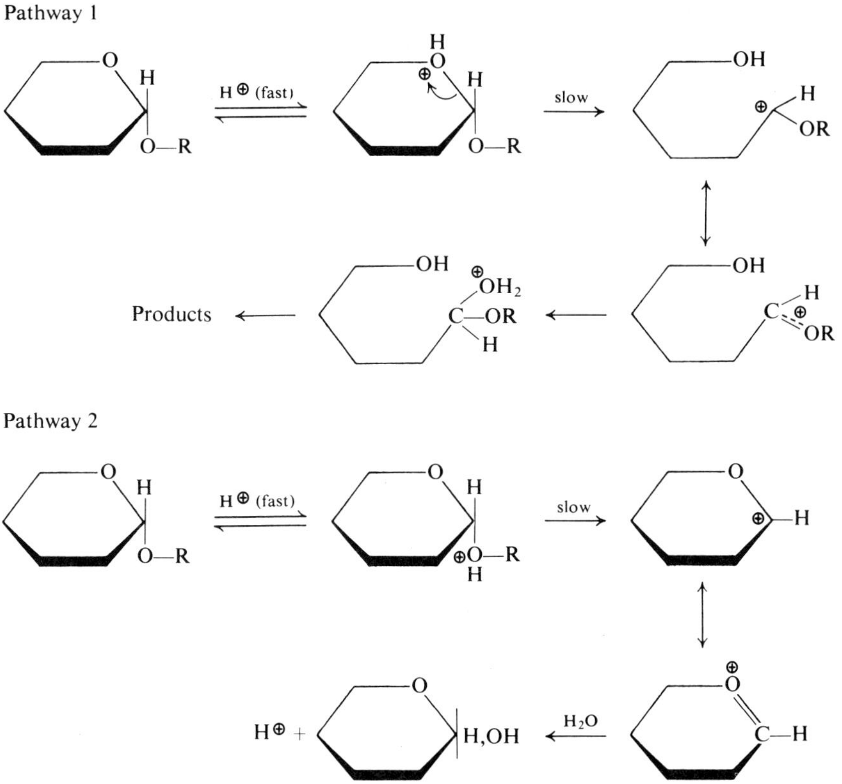

Pathway 1

Pathway 2

FIG. 7. Possible mechanisms for acid-catalyzed hydrolysis of alkyl aldopyranosides.

They showed that the hydrolysis of this glycoside was associated with an oxygen isotope-effect and that methanol near the beginning of the hydrolysis had a significantly lower content of ^{18}O than that isolated after complete hydrolysis. Reaction by pathway 1 should give no observable isotope effect. Shafizadeh[203] has outlined circumstantial evidence in favor of pathway 1 by analogy with a number of reactions in which open-chain derivatives have been isolated. Others[163,204,205] have discussed the relative rates of hydrolysis of glycopyranosides in terms of pathway 2. However, as pointed out by Capon and Overend,[206] the arguments forwarded by Shafizadeh are based on a mis-application of the principle of microscopic reversibility and on analogies of doubtful validity between glycoside formation and hydrolysis. Capon[172] has drawn attention to the analogy between the hydrolysis and methanolysis of glycopyranosides and has indicated that his evidence (see p. 311) for a cyclic

* References start on p. 346.

ionic intermediate in methanolysis makes it likely that the cyclic ion (that is, pathway 2) will be an intermediate in hydrolysis also.

2. Structural Effects

Structural influences may be divided into electronic and steric contributions.

a. Glycopyranosides.—As glycopyranosides are unsymmetrical cyclic acetals, predictions about their relative reactivities are somewhat complicated, particularly since conformational effects would be expected to be more significant than for simple acetals. Moreover, the effect of a substituent on reactivity will depend not only on its character but also on whether it occurs in the pyranoid ring or in the aglycon, and also on its proximity to the bond being cleaved in the rate-determining step.

The effect of polar substituents in the aglycon has been studied extensively[195,200,207-209] and, with a few exceptions, follows the general pattern found for acyclic acetals. For example, Nath and Rydon[207] found for a series of substituted phenyl β-D-glucopyranosides that the effect of substitution was as predicted: hydrolysis was facilitated by electron-releasing substituents in the phenyl group, but in many cases the increase in rate was rather small (see Table V). The Hammett reaction constant[210] is -0.66, and the first-order velocity constant for the fifteen β-D-glucosides studied varied between 5.5×10^{-6} sec^{-1} and 1.49×10^{-5} sec^{-1}.

TABLE V

HYDROLYSIS OF SUBSTITUTED PHENYL β-D-GLUCOPYRANOSIDES[207] IN $0.1M$ HCl AT 60°

Substituent:	None	o-Me	m-Me	p-Me	o-OMe	m-OMe	p-OMe	o-Cl	m-Cl	p-Cl
$k \times 10^6$ (sec^{-1}):	1.92	3.02	2.35	3.87	14.9	1.28	4.02	2.9	1.26	1.38

If the conclusion that the glycopyranosyl carbonium ion is an intermediate is accepted as a basis for discussion, it will be noted that substituents in the aryl group will affect the observed rate of reaction by their influence on both the formation of the conjugate acid and its subsequent "rate-determining" decomposition to the "onium" ion, and will affect these two processes in opposing ways. Electron-releasing substituents will facilitate the former and hinder the latter process, and such mutual, partial cancellation of opposing effects will lead to a low value for the Hammett reaction constant. It is reasonable to suppose, as did Bunton and co-workers,[190] that the low value of the constant found for the acid hydrolysis of aryl β-D-glucosides is, indeed, due to such partial cancellation.

The introduction of negative substituents into the aromatic ring of phenyl β-D-xylosides was found by Hibbert and associates[211] to have little effect on the rate of acid hydrolysis; from a study of the hydrolysis of some alkyl β-D-glucopyranosides in which the aglycon contained various electron-withdrawing polar groups, Timell[212] concluded that the rate increased only slightly as the electronegativity was increased.

Separation of an electron-attracting or -releasing group from the glycosyl oxygen atom by two or more carbon atoms results in very little effect on the rate of hydrolysis of the glycoside (see Table VI); only in the case of the carboxymethyl aglycon is there a slight increase in the rate.

TABLE VI

RATE CONSTANTS AND KINETIC PARAMETERS FOR THE HYDROLYSIS OF VARIOUS β-D-GLUCO-PYRANOSIDES IN 0.5M SULFURIC ACID[212] AT 60°

Aglycon:	Methyl	Carboxymethyl	2-Carboxyethyl	2-Hydroxyethyl
$k \times 10^6$ (sec^{-1}):	1.38	4.11	1.58	1.73

Aglycon:	2-Methoxyethyl	2-Chloroethyl	3-Hydroxypropyl
$k \times 10^6$ (sec^{-1}):	1.83	2.21	1.24

When Rydon and co-workers[208] extended their work to include an examination of the hydrolysis rates of a series of substituted phenyl α-D-glucosides, they found that the rates were unaffected by substitution in the phenyl group (see Table VII); the Hammett reaction constant is vanishingly small (-0.006),

TABLE VII[a]

HYDROLYSIS OF SUBSTITUTED PHENYL α-D-GLUCOSIDES IN ACID SOLUTION[208]
(0.001M IN 0.1M HCl at 70°)

Substituent:	None	o-Me	m-Me	p-Me	m-OMe
$k \times 10^6$ (sec^{-1}):	19.5	12.8	19.2	15.3	20.3
σ^b:	0.000	-0.054	-0.069	-0.170	$+0.115$

Substituent:	p-OMe	m-Cl	p-Cl	m-NO₂	p-NO₂
$k \times 10^6$ (sec^{-1}):	17.8	20	16.2	15.2	16.8
σ^b:	-0.268	$+0.373$	$+0.227$	$+0.710$	$+1.270$

[a] Reproduced by permission of the Editor of the Chemical Society and the senior author of the paper. [b] Values given by Hammett[210] for *meta* and *para* substituents and by Mamalis and Rydon[213] for *ortho* substituents.

* *References start on p. 346.*

and the most divergent first-order velocity constant (that for the *o*-Me compound) differs from the mean value (17.3×10^{-6} sec^{-1}) for the ten D-glucosides studied by only 26%.

This difference from the β series was ascribed to a difference in mechanism, and it was suggested[208] that the one operating in the α series involved a stereochemically controlled, diaxial elimination (Fig. 8). There is an apparent

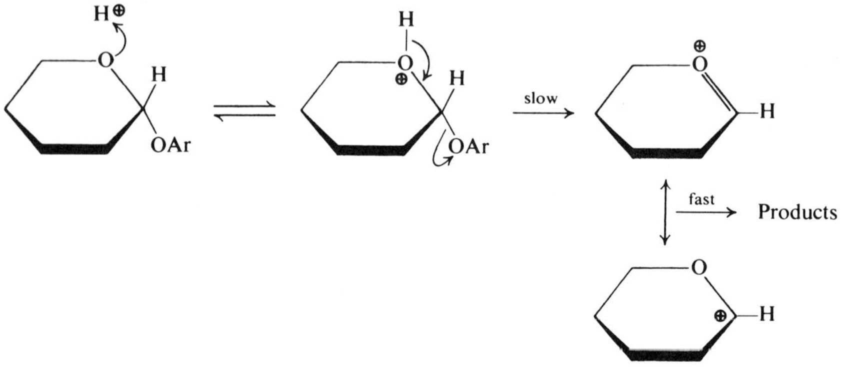

Fig. 8. Suggested mechanism for hydrolysis of aryl α-aldopyranosides.[208]

anomaly in that, whereas the aryl β-D-glucosides are assumed to be protonated on the glucosidic oxygen atom, the α anomers are assumed to be protonated on the ring-oxygen atom. Of these two atoms, the ring oxygen would be expected to be the more readily protonated, owing to the base-weakening effect of the aryl group on the glucosidic oxygen atom. It is claimed[208] that models indicate that, in the β-glucosides, the two oxygen atoms are so close together as to make it likely that the two conjugate acids (**61** and **62**) are but two contributory forms of a single mesomeric cation. It is difficult to reconcile the positive catalysis by acid of the hydrolysis of aryl α-D-glucopyranosides with the mechanism proposed. Prior protonation of the ring-oxygen atom would be expected to hinder and not facilitate the departure of the aryloxy group.

Variation in size of the aglycon, in general, leads only to a small steric effect, as expected for a unimolecular reaction (see Table VIII).

TABLE VIII

HYDROLYSIS OF SOME β-D-GLUCOPYRANOSIDES[200] IN 0.5N H$_2$SO$_4$ AT 60°[a]

Aglycon:	Ethyl	Propyl	Isopropyl	n-Butyl	Isobutyl
k × 10⁶ (sec⁻¹):	1.54	1.82	2.65	1.62	1.90

Aglycon:	neo-Pentyl	Cyclohexyl	Benzyl	Allyl	
k × 10⁶ (sec⁻¹):	2.41	3.25	1.56	1.83	

[a] Reproduced by permission of the National Research Council of Canada from *Can. J. Chem.*, **42**, 1456 (1964).

With some aglycons, facile hydrolysis is encountered. It is well known that the aryl glycosides hydrolyze more rapidly than the alkyl glycosides (see Table IX), supporting the hypothesis that an increase in the electronegativity of the aglycon enhances the rate by weakening the glycosyl–oxygen bond.

Much more pronounced is the facile hydrolysis encountered with *o*-carboxyphenyl β-D-glucopyranoside and *tert*-butyl β-D-glucopyranoside. Several investigators[207,211,214] have observed that *ortho*-substituted phenyl β-D-glucosides exhibit relatively high rates of hydrolysis compared with those containing *meta*- and *para*-substituted aglycons. The nature of this phenomenon is not, at present, clear. It has been suggested[207] that intramolecular association may be partly responsible. In the case of *o*-carboxyphenyl β-D-glucopyranoside, the rate of hydrolysis is 10⁴ times as great as that of the *para* isomer, and it is thought that a different mechanism from the usual A-1 type is operative.[215] In the case of the *tert*-butyl D-glucopyranoside, bond fission is of the alkyl–oxygen type leading to an alkyl carbonium ion rather than a D-glucopyranosyl cation[191] (see page 317).

The effects of substitution in the pyranoid ring and of variation of sugar configuration on the rates of hydrolysis of glycosides have been studied by several groups of workers. Representative results of two reports are shown in Tables IX and X. The results have been discussed mainly in terms of steric influences that operate in the conversion of a chair form (**63**) of the glycoside into a half-chair conformation (**64**) for the glycosyl ion.

63   64

Edward[163] forwarded the proposal that the free-energy difference between the ground state and the transition state arises from the relative ease with which the molecule can change from the chair conformation to the carbonium ion stabilized by conjugation with the ring-oxygen atom, for which C-2, C-1, the ring-oxygen atom, and C-5 must lie in one plane with the ion in the half-chair conformation. The conversion of a chair into a half-chair conformation will be hindered by the increased opposition of substituents on C-2 relative to C-3, and on C-5 relative to C-4. The larger the substituent, the greater will be the hindrance to the formation of the half-chair ion. This leads to the following predictions about the order of stability of glycosides:

1. The order of stability should be heptopyranosides > hexopyranosides > 6-deoxyhexopyranosides > pentopyranosides, when the nature and configuration of substituents elsewhere in the pyranoid ring are unaltered.

2. The conversion of the chair into the half-chair conformation will be aided by the recession of the axial substituents at C-2 and C-5 away from the C-4 and C-3 axial substituents, respectively. This effect will be more powerful as the size of these axial substituents increases. Consequently, for methyl D-glycopyranosides (all in the *C1* conformation), which differ only at C-2, C-3, and C-4, the order of reactivity will be D-*ido* (three axial substituents) > D-*altro*-, D-*gulo* (two axial substituents) > D-*allo*-, D-*manno*-, D-*galacto*- (one axial substituent) > D-*gluco*- (no axial substituent). Similarly, in the pentopyranoside series the order of reactivity will be D-*lyxo*- > D-*arabino*- > D-*ribo*- > D-*xylo*- for glycosides in the *C1* (D) conformation.

Reference to Tables IX and XI shows that, in general, these predictions are substantiated.

The results listed in Table X indicate that all of the monomethyl ethers of methyl β-D-glucopyranoside are hydrolyzed somewhat more slowly than the unsubstituted compound, the rate decreasing in the order 3-methyl > 2-methyl > 4-methyl > 6-methyl. That the 4-methyl ether is hydrolyzed slightly more slowly than the 2- and 3-methyl ether is to be expected, since a substituent at C-4 is in opposition not to a hydroxyl group but to the (larger) hydroxymethyl group at C-5. If conformational factors only were involved, a methoxyl group at either C-2 or C-3 would be expected to have the same influence on the rate. The fact that the 2-methyl ether is hydrolyzed slightly more slowly than the 3-methyl ether could be attributed possibly to steric interference of the 2-*O*-methyl group in the protonation of the exocyclic oxygen atom.

The above explanations are not entirely satisfactory, and anomalous examples may be cited. A slightly different interpretation, which appears to give a more satisfactory correlation of the rates of hydrolysis of various pyranosides, has been described by Feather and Harris.[217] The ease of conversion into the half-chair form depends not only on the ease of rotation about the C-2–C-3 and C-5–C-4 bonds but also on interactions of the 1,3-type, and it

TABLE IX

RATE COEFFICIENTS AND KINETIC PARAMETERS FOR THE HYDROLYSIS OF GLYCOSIDES IN 2.0*M* HYDROCHLORIC ACID[195]

Glycoside	10^5k, *at 60° (sec*$^{-1}$*)*	E (*kcal mole*$^{-1}$)	$\Delta S^{\ddagger}$ *at 60°* (*cal deg*$^{-1}$ *mole*$^{-1}$)
Methyl pyranoside			
α-D-*gluco-*	0.708	34.1 ± 1.0	+14.8
α-D-*galacto-*	3.55	34.0 ± 0.3	+17.7
6-Deoxy-α-D-*galacto-*	20.0	33.9 ± 0.6	+20.8
α-D-*manno-*	2.09	31.9 ± 0.4	+10.4
α-D-*altro-*	12.6	31.7 ± 1.0	+13.5
α-D-*xylo-*	2.69	33.5 ± 0.9	+15.7
α-L-*arabino-*	14.1	30.6 ± 0.4	+10.2
α-D-*lyxo-*	13.5	31.2 ± 1.0	+12.1
β-D-*gluco-*	1.26	34.3 ± 0.4	+16.5
β-D-*galacto-*	5.13	32.3 ± 0.6	+13.3
β-D-*xylo-*	5.89	33.6 ± 0.9	+17.5
β-D-*ribo-*	8.71	31.4 ± 0.9	+11.8
β-L-*arabino-*	9.55	32.5 ± 0.6	+15.2
Phenyl pyranoside			
α-D-*gluco-*	38.0	31.1 ± 2.0	+13.3
α-D-*galacto-*	128	30.2 ± 0.4	+13.5
β-D-*gluco-*	9.33	31.0 ± 1.2	+10.8
β-D-*galacto-*	24.5	28.1 ± 0.1	+ 4.1

TABLE X

RATE COEFFICIENTS AND KINETIC PARAMETERS FOR THE HYDROLYSIS OF METHYL MONO-*O*-METHYL-β-D-GLUCOPYRANOSIDES[216]

Position of substituent	k × 10^6 *at 60° (sec*$^{-1}$*)*	E (*kcal mole*$^{-1}$)	$\Delta S^{\ddagger}$ *at 60°* (*cal deg*$^{-1}$ *mole*$^{-1}$)
2-	1.19	33.4	+12.9
3-	1.27	34.0	+41.9
4-	1.15	33.8	+41.1
6-	0.84	34.9	+16.8
Unsubstituted	1.38		

* *References start on p.* 346.

has not been possible to find data suitable to evaluate the effect of the latter. It has been noted by Feather and Harris that the rate is decreased by increased opposition of substituents on C-2 with respect to those on C-3 when rotation is in the direction that tends to eclipse the equatorial groups. The rate is also decreased by increased opposition of substituents on C-5 relative to those on C-4, for the same reasons.

It should be noted that in the examples cited in Tables IX and X the fairly small differences in reactivity often result from changes in the entropy of activation. The idea of relief of repulsive strain energy in the transition state must, therefore, be an oversimplification. This becomes particularly obvious when the hydrolysis of deoxyglycosides is considered. The high rates of hydrolysis characteristic of the methyl 2- and 3-deoxyaldopyranosides[195] and the still greater lability of the 2,3-dideoxyaldopyranosides have been attributed to a decrease of the steric interactions that influence conversion of a chair conformation for a glycoside into the half-chair conformation of the glycosyl ion. It is more probable, however, that the lability of the glycosidic linkage in

TABLE XI

CORRELATION OF RELATIVE RATES OF ACID HYDROLYSIS OF METHYL D-GLYCOPYRANOSIDES IN THEIR MOST STABLE CONFORMATIONS WITH THE NUMBER OF AXIAL HYDROXYL GROUPS IN THE MOLECULE[195]

Configuration of glycoside	Position of axial hydroxyl groups	Relative rate
α-D-*gluco*-	None	1.0
β-D-*gluco*-	None	1.8
α-D-*manno*-	C-2	2.9
α-D-*galacto*-	C-4	5.0
β-D-*galacto*-	C-4	7.2
α-D-*altro*-	C-2, C-3	18.0
α-D-*ido*-	C-2, C-3, C-4	28–37
α-D-*xylo*-	None	3.8
β-D-*xylo*-	None	8.3
β-D-*ribo*-	C-3	12.3
α-D-*lyxo*-	C-2	19.0

these compounds has its origin only partly in steric influences, and electronic factors also operate. The extent of each type of influence will depend on the position of the deoxy group in the glycoside.

Overend and co-workers[195] have explained the reactivity of 2-deoxyaldopyranosides on the basis of removal of the large inductive effect of the hydroxyl group at C-2, together with a small steric effect. Marshall[218] has correlated the available data on the effect of substitution at C-2 of the sugar ring on the rate of hydrolysis of glycosides and concluded that, for a given aglycon, the largest differences in rate are due to the inductive effect at C-2. Richards[219] found that methyl 3-deoxy-α-D-*ribo*- and -*arabino*-hexopyranosides undergo hydrolysis, respectively, 7.4 and 11.5 times as fast as methyl α-D-glucopyranoside. These rates, although faster than that for the analogous hexoside, are considerably slower than those found for the 2-deoxy-D-*arabino*-hexosides, but are in the expected direction and of the right magnitude for rate changes due to inductive effects. On the other hand, it has been found[195] that the rate of hydrolysis of methyl 4-deoxy-α-D-*xylo*-hexoside (methyl "4-deoxy-α-D-glucoside") is faster than that of the 3-deoxy analog (methyl 3-deoxy-α-D-*ribo*-hexoside), and this is an increase not attributable to inductive effects. In this case it is quite reasonable to assume that steric effects are preponderant and that replacement of the hydroxyl group at C-4 by a hydrogen atom will result in removal of considerable steric strain because the hydroxyl group would interact with both the hydroxyl group at C-3 and the hydroxymethyl group at C-5. These interactions would be diminished in the 4-deoxyglycoside.

Likewise, it is difficult to attribute to an inductive effect the rate changes arising from the presence of a 6-deoxy group, because this group is too far removed from the active site and the rate changes are relatively large. Unfortunately, in this case explanations in terms of steric interactions are not wholly satisfactory either.

It is well known that replacement of the hydroxymethyl group in a hexoside can have a significant effect on the rate of hydrolysis of the glycoside. Extensive investigation of the kinetics of hydrolysis of glucosiduronic acids has been made.[212,220-222] Timell[220,223] has concluded that the effect is complex and that polar effects are of minor importance in the acidic hydrolysis of D-glyco-pyranosides substituted at C-5, and that conformational factors are more important. Apparently there is little or no correlation between the polarity of substituents at C-5 and the rate of hydrolysis of the glycosides. The carboxy-methyl and chloromethyl groups, for example, have the same polar substituent-constants, yet the 6-chloro-6-deoxy compound undergoes hydrolysis more slowly than the glucosiduronic acid by a factor of 6 or more. It is also note-worthy that the methyl 6-O-methyl-D-glucopyranosides are both hydrolyzed at a lower rate than the methyl glucosiduronic acids.

Table XII provides a comparison between the rate of hydrolysis of anomers and the orientation of the methoxyl group. The results demonstrate that the β-D-glycosides in which the methoxyl group is equatorial [*C1* (D) conformation] are hydrolyzed at a faster rate than the α anomers in which this group is axial. The ratio of the rates for anomers is approximately twofold, which is small compared with the large differences in rates between some pairs.

* *References start on p. 346.*

TABLE XII

RATES OF HYDROLYSIS OF SOME ANOMERIC GLYCOPYRANOSIDES[a]

| Pyranoside | Relative rate[b] | Conditions of comparison | |
		Acid	Temperature (°C)
Methyl α-D-glucoside	1.0	0.5M HCl	75
Methyl β-D-glucoside	1.9	0.5M HCl	75
Methyl α-D-mannoside	2.4	0.5M HCl	75
Methyl β-D-mannoside	5.7	0.5M HCl	75
Methyl α-D-galactoside	5.2	0.5M HCl	75
Methyl β-D-galactoside	9.2	0.5M HCl	75
Methyl α-D-xyloside	4.5	0.5M HCl	75
Methyl β-D-xyloside	9.1	0.5M HCl	75
Methyl α-D-glucosiduronic acid	0.47	0.47M H$_2$SO$_4$	75
Methyl β-D-glucosiduronic acid	0.62	0.47M H$_2$SO$_4$	75
Methyl 2-deoxy-α-D-*arabino*-hexoside	2090	0.01M HCl	58
Methyl 2-deoxy-β-D-*arabino*-hexoside	5125	0.01M HCl	58
Methyl 2,3,4,6-tetra-O-methyl-α-D-glucoside	0.16	0.01M HCl	95–100
Methyl 2,3,4,6-tetra-O-methyl-β-D-glucoside	0.40	0.01M HCl	95–100
Phenyl α-D-glucoside	61.5	0.05M HCl	60
Phenyl β-D-glucoside	18.0	0.05M HCl	60
p-Nitrophenyl α-D-glucoside	35.8	2.0M HCl	60
p-Nitrophenyl β-D-glucoside	4.1	2.0M HCl	60
Benzyl α-D-glucoside	2.0	0.05M HCl	60
Benzyl β-D-glucoside	2.97	0.05M HCl	60

[a] Taken from figures quoted by Feather and Harris.[217]   [b] Ratio of the rate constant for the hydrolysis of the methyl aldopyranoside to that of methyl α-D-glucopyranoside at the conditions specified (extrapolated when necessary).

The greater reactivity of glycosides containing the aglycon oriented equatorially is contrary to the usual order of conformational stability. Several explanations have been offered. It has been contended[203,224] that neighboring groups introduce hindrance to different extents to the α and β positions. Edward[163] has ascribed the greater reactivity of the β anomers to their higher free energy in the ground state, caused by polar interaction between the equatorial methoxyl group and the lone pair of electrons of the ring-oxygen atom. This repulsive interaction will be destroyed on protonation, and so from this viewpoint the concentration of conjugate acid would be greater for the β anomer, with the rate of its decomposition but little affected: indeed k_β may well be less than k_α, as would be predicted in the absence of this special interaction. There are exceptions to the above generalization; for example, methyl α-D-gulopyranoside is hydrolyzed faster than the β anomer, probably because of a much higher free energy in the ground state arising from interaction (when in the *C1* (D) conformation) between the axial methoxyl group and the axial hydroxyl group at C-3. In the *1C* (D) conformation the α-D anomer has O-1 equatorial and O-2 axial, a factor that would also introduce strain in the molecule (see Chapter 5).

Edward[163] recognized that, with a bulky aglycon, steric factors would again preponderate and the axial α anomer would be hydrolyzed more rapidly because of its higher free energy in the ground state. This is borne out by the relative rates reported[195] for phenyl α- and β-D-glucopyranoside and -galactopyranoside, *p*-nitrophenyl α- and β-D-glucopyranoside, and ethyl α- and β-D-galactopyranoside. The α anomers are all more reactive than the β forms. However, inspection of the results in Table XIII reveals the interesting

TABLE XIII

RATE COEFFICIENTS AND KINETIC PARAMETERS FOR THE HYDROLYSIS OF SOME ARYL GLYCOSIDES IN 2.0*M* HYDROCHLORIC ACID

Aldopyranoside	10^5k at 60° (sec^{-1})	E (kcal mole^{-1})	ΔS^+ at 60° (cal deg^{-1} mole^{-1})
Phenyl α-D-glucoside	38.0	31.1	13.3
Phenyl β-D-glucoside	9.33	31.0	10.8
p-Nitrophenyl α-D-glucoside	25.1	30.3	10.5
p-Nitrophenyl β-D-glucoside	2.88	30.3	6.4
Phenyl α-D-galactoside	128	30.2	13.5
Phenyl β-D-galactoside	24.5	28.1	4.1

feature that the differences in rate between the anomeric pairs result from differences in the entropy rather than the energy of activation. This suggests that, with other than the smallest aglycons, the α anomers are more highly oriented than the β anomers in the ground state and their increased reactivity is associated with greater loss of molecular order on passing to the transition state. That is, the important factor is not the increase in potential energy of the molecule owing to repulsive interaction between the aglycon and axial 3- and 5-substituents, but a decrease in entropy caused by restriction imposed upon rotation of the aglycon by these substituents.

Feather and Harris[217] have drawn attention to the fact that the aglycon may determine the conformation of the glycon in a reaction. In a phenyl α-D-glycoside in the *C1* (D) conformation, having an axial phenoxyl group, there are considerable axial–axial interactions, but these are largely removed if the (D) glycon assumes the *1C, B1, 2B,* or *3B* conformation with the aglycon in an equatorial position. The α-D anomer would then be expected to hydrolyze more rapidly than the β form, as the resistance to rotation about the C-2–C-3 and C-5–C-4 bonds of any of these conformers is less than that of the *C1* (D) conformer. In the case of the benzyl D-glucosides, the interaction between the glycon and aglycon portions of the molecule is much less than that for the phenyl D-glucosides because of the intervening methylene group. This is reflected in the smaller ratio of the hydrolysis rates of the α and β anomers (see Table XII).

b. *Glycofuranosides.*[8]—Relative rates of hydrolysis of some methyl aldo-furanosides are listed in Table XIV. These results indicate that, in an anomeric

TABLE XIV

RATES OF HYDROLYSIS OF METHYL ALDOFURANOSIDES[30] RELATIVE TO METHYL β-D-GALACTOFURANOSIDE $=1$

Aldofuranoside:	α-D-*xylo-*	β-D-*xylo-*	α-D-*gluco-*	β-D-*gluco-*
Relative rate:	98	65	147	52
Aldofuranoside:	β-L-*arabino-*	α-L-*arabino-*	α-D-*galacto-*	
Relative rate:	11	1.7	8.3	

pair of furanosides, the anomer having the *cis*-1,2 configuration for the hydroxyl group on C-2 relative to the methoxyl group on C-1 (in the Haworth formula) is less stable toward hydrolysis than the *trans*-1,2 anomer. Augestad and Berner[225] measured the rates of hydrolysis of a number of methyl aldofuranosides and found in all cases that the isomer having the substituents *cis*-located at C-1 and C-2—that is, the less stable anomer—is hydrolyzed fastest. For a series of alkyl aldofuranosides Capon and Thacker[225a] observed negative entropies of activation for acid-catalyzed hydrolysis, suggesting that the reaction involved either bimolecular displacement of the alcohol molecule from C-1 by water, or a ring-opening step.

C. ALKALINE HYDROLYSIS

The alkaline hydrolysis of glycosides has not been investigated kinetically as extensively as acid hydrolysis, and the reaction is less well understood. In dilute alkali some glycosides are stable at normal temperatures. It seems that alkali-sensitivity of the glycosidic linkage is mainly a function of the aglycon. Alkali-labile glycosides have been divided into three types:[226] glycosides of (1) phenols, (2) enols conjugated with a carbonyl group, and (3) alcohols substituted in the β position by a negative group. In fact the first two groups could be put under the third, which is the significant structural detail. An aglycon that imparts alkali-sensitivity acts by withdrawing electrons from the glycosidic linkage. With the notable exception of the phenyl glycosides, in which steric factors are important, the lability of the glycosidic linkage to alkali is but little affected by the sugar residue.

Fairly drastic conditions are needed to cleave aliphatic glycosides, although the glycofuranosides are considerably more labile than the pyranosides.[227] The mechanism of hydrolysis of such glycosides by 10% sodium hydroxide at 170° has been investigated by Lindberg and his associates.[228, 229] Their results show that a number of different mechanisms of comparable importance may be operative, making it difficult to distinguish between the polar and steric influences of the alkyl groups.

The alkyl glycosides are attacked by alkali only at temperatures higher than those required for phenyl glycosides.[230] Of the various glycopyranosides and -furanosides examined, the 1,2-*trans*-glycosides were more reactive than the 1,2-*cis* anomers (see Table XV), but the differences were not large. The re-

TABLE XV

HYDROLYSIS OF SOME METHYL GLYCOSIDES[227] IN 10% SODIUM HYDROXIDE AT 170°[a]

		10^3k (*in Briggs log, hour*)	
Configuration of glycoside	*Relationship between OMe at C-1 and OH at C-2*	*Pyranoside*	*Furanoside*
α-D-*gluco*-	cis	1.0	—
β-D-*gluco*-	trans	2.5	>100
α-D-*galacto*-	cis	1.0	7.8
β-D-*galacto*-	trans	5.7	28
α-D-*manno*-	trans	2.8	30
β-D-*manno*-	cis	1.1	—
α-D-*xylo*-	cis	1.2	8.1
β-D-*xylo*-	trans	5.8	>100
α-L-*arabino*-	trans	10.0	32
β-L-*arabino*-	cis	1.0	—

[a] Reproduced from *Acta Chem. Scand.*, by permission of the Editor-in-Chief.

activity of the glycopyranosides was found to increase as their conformational instability increased.

Methylation of the hydroxyl group at C-2 in methyl β-D-glucopyranoside diminishes the rate by more than half, whereas in the α series a methoxyl group

* *References start on p. 346.*

at C-2 produces a smaller effect. In a series of β-D-glucopyranosides the re-action rates increase in the order methyl < ethyl < isopropyl < *tert*-butyl (see Table XVI).

TABLE XVI

HYDROLYSIS OF SOME D-GLUCOPYRANOSIDES[229] IN 10% SODIUM HYDROXIDE AT 170°[a]

β Series

Aglycon:	Me	Me 2-*O*-Me-	Et	*iso*-Pr	*tert*-Bu	MeO(CH$_2$)$_2$-	HO(CH$_2$)$_2$-
10^3k^b:	2.5	1.2	2.9	3.9	12.0	1.7	4.5

α Series

Aglycon:	Me	Me 2-*O*-Me-	MeO(CH$_2$)$_2$-	HO(CH$_2$)$_2$-
10^3k^b:	1.0	0.8	1.0	1.8

[a] Reproduced from *Acta Chem. Scand.*, by permission of the Editor-in-Chief.   [b] In Briggs logs and hours.

tert-Butyl β-D-glucopyranoside and sodium methoxide in methanol at 170° gave only chromatographic traces of methyl β-D-glucopyranoside and 1,6-anhydro-β-D-glucopyranose. In similar experiments with ethyl or *tert*-butyl β-D-xyloside, only traces of methyl β-D-xyloside were detected; these results should be compared with the almost exclusive production of methyl β-D-xylopyranoside when 2,4,6-tribromophenyl β-D-xylopyranoside was treated with sodium methoxide in methanol.[231]

Apparently, the first example of the alkaline degradation of an enol glyco-side was theobromine β-D-glucopyranoside tetraacetate which, on attempted deacetylation with barium hydroxide solution, gave theobromine and D-glucose.[232] Fischer[233] also found that some glycosides of thiouracil had similar properties. Investigations by Ballou and Link,[234] particularly, have extended knowledge of this class of substance.

Kuhn and Löw[235] noted that picrocrocin, the bitter principle of saffron, decomposed in aqueous alkali with formation of D-glucose and safranal; examples of other glycosides derived from alcohols substituted in the β position by a negative group are as follows

R = CH$_2$CH$_2$COMe
R = CH$_2$CH$_2$SO$_2$OEt
R = CH$_2$CH$_2$NO$_2$

The sensitivity of phenyl glycosides to alkali has been recognized for over a century. As early as 1844 Bouchardet[236] described the alkaline decomposition of salicin, a D-glucoside occurring in the leaves and bark of willow (*Salix helix*). However, Tanret[237] was the first to recognize that the degradation of phenyl β-D-glucopyranoside yields 1,6-anhydro-β-D-glucopyranose (levoglucosan). Subsequent studies have elucidated some of the interesting steric factors that control the reaction,[238,239] and the transformation has been applied in a useful laboratory procedure for the preparation of 1,6-anhydro-β-D-hexopyranoses.[239,240]

Structural Influences

The glycosides of some substituted phenols, such as 2,4,6-tribromophenyl β-D-glucopyranoside[241] and nitrophenyl β-D-glucopyranosides,[242] are particularly labile toward alkali. Snyder and Link[243] compared the effect of *o*- and *p*-nitro substitution in phenyl glycopyranosides on the rate of base-catalyzed degradation and found that the *o*-nitrophenyl glycosides are decomposed faster, even though the acidities of *o*- and *p*-nitrophenol are practically identical. Rydon and his co-workers examined the alkaline hydrolysis of a series of substituted phenyl β-[207] and α-[208] D-glucosides and showed that the acid and alkaline hydrolysis of the β-D-glycosides were affected oppositely by the nature of the substituents on the phenyl group; alkaline hydrolysis was facilitated by electron-withdrawing substituents. In agreement with the results of earlier, less extensive studies,[238,240] it was found that the alkaline hydrolysis of aryl α-D-glucosides resembles that of the β anomers in being strongly aided by electron-withdrawing substituents in the phenyl group (see Table XVII). In the α series the effect is rather more marked than for the β anomers.

TABLE XVII

ALKALINE HYDROLYSIS OF SUBSTITUTED PHENYL α-D-GLUCOSIDES[208]
(0.002M in 3.9M SODIUM HYDROXIDE AT 70°)^a

Substituent:	None	*o*-Me	*m*-Me	*p*-Me	*m*-OMe	*m*-Cl	*p*-Cl	*m*-NO$_2$	*p*-NO$_2$
$10^6 \times$ k (sec^{-1}):	0.1	0.03	0.03	0.05	0.12	0.15	0.33	17.8	*ca* 60,000

The *p*-methoxy compound was hydrolyzed too slowly and the *p*-nitro compound too rapidly under the standard alkaline conditions to enable k_{alk} to be determined for these compounds. The very approximate value given for the *p*-nitro compound was obtained by extrapolation from experiments in which weak alkali was used.

^{a} Reproduced by permission of the Editor of the Chemical Society and the senior author of the paper.

Dyfverman and Lindberg[244] found a close correlation between the acidity of the phenol and the alkali lability of a series of phenyl β-D-glucopyranosides

* *References start on p. 346.*

and demonstrated a linear relationship between the values of log k for the substituted phenyl β-D-glucosides and Hammett's σ values for the substituent.

TABLE XVIII

ALKALINE DEGRADATION OF PHENYL GLYCOSIDES[238, 239] AS RELATED TO *cis* OR *trans* CONFIGURATION AT C-1–C-2[a]

Phenyl glycoside	Product	Yield (%)	Time (hours)	KOH (molarity)
trans-1,2				
β-D-Xylopyranoside	Tar	—	3	1.3
β-D-Glucopyranoside	1,6-Anhydride	88	9	1.3
β-D-Galactopyranoside	1,6-Anhydride	91	9	1.3
α-D-Mannopyranoside	Large rotational change	—	336	1.3
2,3-Di-*O*-methyl-β-D-glucopyranoside	No reaction	—	48	2.6
2,3,4,6-Tetra-*O*-methyl-β-D-glucopyranoside	No reaction	—	336	1.3
cis-1,2				
α-D-Xylopyranoside	Recovered	—	48	1.3
α-D-Glucopyranoside	Recovered	—	336	1.3
α-D-Galactopyranoside	1,6-Anhydride	85	2688	2.6
β-D-Mannopyranoside	1,6-Anhydride	57	120	2.6

[a] Published by permission of the Editor of *Advan. Carbohyd. Chem.*

The results listed in Table XVIII show the relationship between the anomeric configuration of phenyl D-hexosides and their susceptibility to conversion into 1,6-anhydro-β-D-hexopyranoses by alkali.

It has not been possible to rationalize all the results on the basis of a single mechanism; three possibilities have been considered.

1. The most general, which applies consistently for the conversion of substituted phenyl β-D-glucopyranosides into the 1,6-anhydride, is a "double-inversion" process. This mechanism involves a 1,2-epoxide intermediate resulting from neighboring-group displacement by a *trans*-located hydroxyl group at C-2, which is attacked by the group at C-6 to give the 1,6-anhydride. In such a reaction the rate would depend on the concentration of both hydroxyl ion and glycoside, as has been found.[244]

There is indirect evidence that a 1,2-epoxide may be an intermediate. For example, when 2,4,6-tribromophenyl β-D-glucopyranoside was treated with sodium hydroxide in boiling methanol, both 1,6-anhydro-β-D-glucopyranose (75%) and methyl β-D-glucopyranoside (21%) were obtained, their proportions presumably being a measure of competitive attack by the solvent and the terminal hydroxymethyl group on the reactive epoxide intermediate.[231] Phenyl 2,3-di-O-methyl- and 2,3,4,6-tetra-O-methyl-β-D-glucopyranoside, in which the hydroxyl group at C-2 is protected, undergo no reaction with alkali.

Gasman and Johnson[245] investigated the kinetics of the base-catalyzed cleavage of p-nitrophenyl β-D-galactopyranoside and p-nitrophenyl α-D-mannopyranoside and compared the data with results for the corresponding 2-methyl ethers. It was concluded that the p-nitrophenyl glycosides react by neighboring-group participation of the C-2 oxyanion, whereas the reactions of the 2-O-methylglycoside proceed, at least in methanol–sodium methoxide, by bimolecular, nucleophilic, aromatic substitution.

2. Those degradations of aryl glycosides that occur in spite of unfavorable steric relationships within the molecule may proceed either by a "single-inversion" mechanism, as has been suggested to explain the reaction with alkali of phenyl α-D-galactopyranoside, or by (3) "ionic dissociation," as proposed for the degradation of phenyl β-D-mannopyranoside.

The mechanism of alkaline hydrolysis of p-nitrophenyl α-D-glucopyranoside has been investigated in detail.[245a] The liberation of p-nitrophenol does not proceed by simple cleavage to give D-glucose and p-nitrophenoxide. The first step involves O-1 to O-2 migration of the p-nitrophenyl group to give 2-O-p-nitrophenyl-D-glucose, and this step is followed by further migration of the O-aryl group to give 3-O-p-nitrophenyl-D-glucose. Concurrently these ether derivatives are isomerized to the D-manno analogues. The liberation of p-nitrophenoxide anion takes place from the 3-ethers by the β-elimination type of pathway (see Chapter 4), so that the sugar residue is released as a saccharinic acid. In contrast, the alkaline hydrolysis of p-nitrophenyl β-D-glucopyranoside appears to involve two competing processes, one leading to 1,6-anhydro-β-D-glucopyranose (presumably by the "double-inversion" pathway) and the other involving the type of sequential aryl-group migration observed with the α-D anomer.[245b]

D. HYDROGENOLYSIS

Freudenberg and associates[246] found that benzyl ethers of sugars are cleaved by hydrogenolysis with sodium amalgam and that catalytic hydrogenolysis can be effected in acetic acid in the presence of platinum metals. On palladium catalyst, hydrogen splits off the benzyl group of benzyl β-D-glycosides in the presence of hydrogen ions at room temperature and atmospheric pressure to afford toluene and the sugar.[247] In some cases aromatic glycosides are reduced to cyclohexyl glycosides; salicin, however, is reduced only to o-tolyl β-D-glucopyranoside.[247]

* References start on p. 346.

Some aromatic glycosides are cleaved by hydrogen on colloidal platinum, and it appears that the reaction of aromatic glycosides with hydrogen in the presence of platinum[247] can follow two routes—either cleavage of the glycosidic linkage or reduction of the aryl residue to afford a (substituted) cyclohexyl β-D-glycoside, which does not undergo subsequent hydrogenolysis.

Under appropriate conditions, more profound changes can be effected with glycosides. For example, hydrogen at 250 atmospheres pressure in the presence of a catalyst of copper–chromium oxide at 240° converts methyl β-L-arabinopyranoside into a mixture of products, consisting for the greater part of *cis*- and *trans*-3,4-dihydroxytetrahydropyran.[248] The use of milder conditions leads only to extensive isomerization, as shown by the formation of the methyl glycosides of D-ribose, D-xylose, D-lyxose, and L-lyxose from the L-arabinoside.[249] Treatment of methyl α-D-mannopyranoside with hydrogen and a copper–chromite catalyst at 180° afforded a mixture of isomeric glycosides, notably the talopyranoside, which was obtained likewise from other hexopyranosides. The β anomers are more stable under these conditions, especially the D-glucoside and D-mannoside, which were recovered substantially unchanged. Reactions at higher temperatures with methyl α-D-glucopyranoside led to a mixture of stereoisomeric 1,5-anhydro-4-deoxy-D-hexitols.[250] Pentosides were converted more readily into tetrahydropyrandiols.

E. CHLORINOLYSIS

Aqueous acidic solutions of chlorine have been found to convert methyl aldopyranosides into aldoses, aldonic acids, smaller proportions of "oxosugars," and substances derived by further oxidation at nonacetal sites[251,252] (see also Chapter 24). Formaldehyde (22%) was identified as a product from the reaction of methyl tetra-*O*-acetyl-β-D-glucopyranoside with chlorine in carbon tetrachloride.[253] A low yield of ethyl β-D-glucopyranoside was obtained when the reaction mixture was treated with a silver salt in ethanol followed by deacetylation, a result that suggests that an acetylated D-glucosyl chloride is an intermediate in the reaction and that the glycosidic bond is cleaved between the oxygen atom and C-1 of the pyranoid ring. The latter conclusion is at variance with an earlier proposal[254] that the glycosidic linkage was cleaved by chlorine between the oxygen and aglycon carbon atom. The scheme in Fig. 9 shows the changes that have been suggested[253] to account for the products formed when a glycoside is treated with chlorine. The carbonyl compound (66) (formaldehyde from a methyl glycoside) arises from the aglycon by dehydrohalogenation of a hypochlorite ester (65). Attack at C-1 of the pyranoid ring by chloride ion gives the glycosyl chloride (67). Chlorine in dry acetic acid converts methyl α-D-glucopyranoside into a mixture of products from which, after deacetylation, D-glucose (52%), unreacted methyl α-D-glucopyranoside

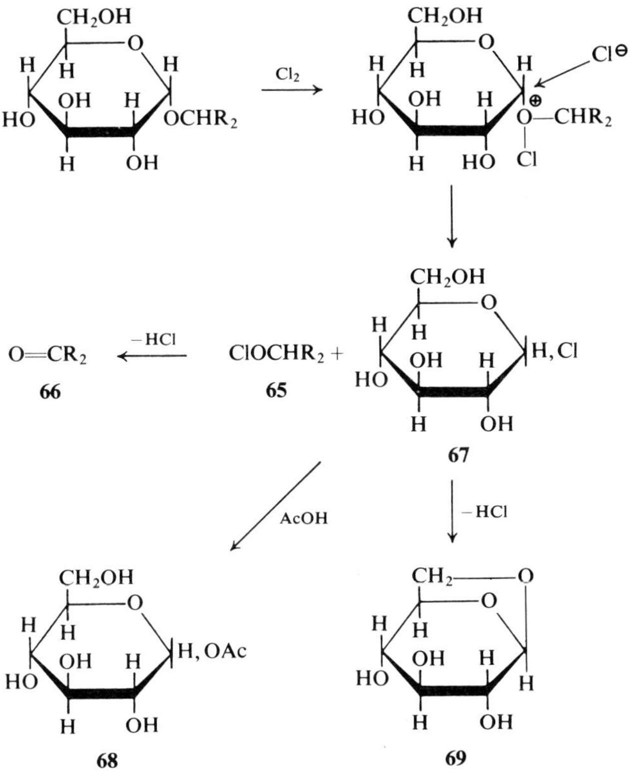

FIG. 9. Chlorinolysis of glycosides.

(22%), and a mixture (26%) of 1,6-anhydro-β-D-glucopyranose (69) and un-known material were obtained. Formation of these products can be rational-ized in terms of the proposed D-glucosyl chloride intermediate (67) undergoing solvolysis in acetic acid to give 68 and dehydrohalogenation to give the 1,6-anhydride (69). This mechanism also accounts for the major products of treat-ment of glycosides with chlorine in an aqueous medium. The intermediate (67) would be solvolyzed to an aldose, which would be converted by halogen oxida-tion into the aldonolactone.

F. PHOTOLYSIS AND RADIOLYSIS

The glycosidic bond is susceptible to cleavage by light, and breakage has been observed[255,256] during irradiation of aqueous solutions of a series of aryl glycosides with monochromatic light of 254 nm (see also Chapter 26). In this spectral region the aglycon is the absorbing center in the glycoside.

* *References start on p. 346.*

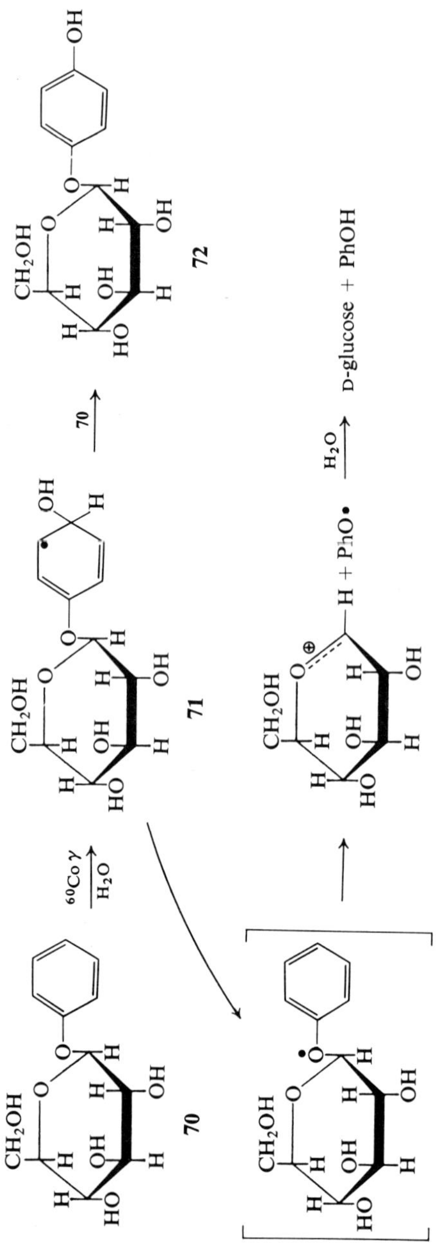
CH2OH
OH
60Co γ
H2O
70
71
72
H + PhO•
D-glucose + PhOH

The products of the main photochemical reaction with benzyl β-D-fructopyranoside are D-fructose and benzyl alcohol. When the aglycon is changed by replacing benzyl β- by benzyl α-, or phenyl α- and β-, or 2-phenylethyl α- and β-, there is no correlation between the ease of photochemical cleavage of the glycosidic link and its stability toward acid hydrolysis. Very little of the main photolysis is caused by photons directly absorbed at the glycosidic link (the reactive center), as shown when the benzyl group is replaced by the non-absorbing methyl group, as in methyl β-D-fructopyranoside. It appears that, after absorption by the aglycon, the energy of the absorbed photon must be transferred intramolecularly to the oxygen bridge. Although the effect is small, the efficiency of the transfer for the aryl glycosides is in the order benzyl > 2-phenylethyl > phenyl.

Irradiation of aryl D-glucosides in the solid state with γ-rays leads to glycosidic cleavage to give the phenol and D-glucose; in aqueous solution hydroxylation of the aglycon also occurs.[256a] The reaction with phenyl β-D-glucopyranoside (70) appears to proceed by way of the hydroxycyclohexadienyl radical (71), which undergoes further reaction either by a first-order process giving D-glucose by way of a D-glucosyl cation, or by a bimolecular process leading to the hydroxylated glycoside 72; an earlier suggestion[256b] that the glycosidic cleavage involved interaction of a solvated electron with the glycoside was not supported (see also Vol. IB, Chap. 26).

VI. 1-THIOGLYCOSIDES[257]

When the linkage between an alkyl, aryl, or aralkyl group and C-1 of an aldose in a cyclic form is effected by a sulfur atom, the compounds are known as 1-thioglycosides.

Complex glycosides of 1-thio-D-glucose occur naturally and most commonly in the mustard-oil glycosides of the *Cruciferae*, *Capparidaceae*, and *Resedaceae* plant families[257c] (see also Chapters 18 and 32).

A. PREPARATION

Methods for the preparation of 1-thioglycosides are to a considerable extent analogous to those used for *O*-glycosides.

1. *Direct Acid-Catalyzed Glycosidation*

The facile conversion of most sugars into their dithioacetals by the action of thiols in a strong acid makes the procedure unsatisfactory as a general method for the preparation of 1-thioglycosides. Examination of several sugars led to

* *References start on p. 346.*

the conclusion that, under mercaptalation conditions, formation of the dithioacetal is a fast, kinetically controlled reaction, and is followed by a slower change which at equilibrium gives a distribution of products, including free aldoses, thioglycosides, and dithioacetal, according to their thermodynamic stabilities in the system.[257d] Conditions have been described where, in individual instances, the 1-thioglycosides may be isolated, and in the cases investigated they have been found to be pyranosides. The configuration of the sugar influences the reaction, and derivatives of D-mannose appear to be particularly liable to form 1-thioglycosides. For example, treatment of D-mannose with ethanethiol and concentrated hydrochloric acid for 16 hours at room temperature gave a 31 % yield of ethyl 1-thio-α- and β-D-mannopyranosides (isolated as the tetraacetates)[258] with no dithioacetal detected, whereas if the reaction time were only 5 minutes the diethyl dithioacetal (63 % yield) was obtained.[259] In contrast, the ethanethiolysis of D-xylose (in ethanethiol–hydrogen chloride–N,N-dimethylformamide) was shown by radioactive-tracer methods to proceed to the dithioacetal by way of 1-thiofuranosides; at equilibrium not more than 6% of 1-thiopyranosides was present.[259a]

Although the mercaptolysis of simple O-glycosides has not been studied extensively, it is known that in some cases it leads to 1-thioglycosides. Treatment of methyl α- and β-D-mannopyranoside with ethanethiol and concentrated hydrochloric acid gives a low yield of ethyl 1-thio-β-D-mannopyranoside;[258] the β-D anomer gives, in addition, a small proportion of ethyl 1-thio-α-D-mannoside.

The Helferich method was applied to the preparation of thioglycosides by Hurd and Bonner,[260] who obtained phenyl 1-thio-β-D-glucopyranoside from thiophenol and penta-O-acetyl-β-D-glucopyranose with p-toluenesulfonic acid as catalyst. Lemieux[261] converted this acetate into ethyl 2,3,4,6-tetra-O-acetyl-1-thio-β-D-glucopyranoside (71 %) by zinc chloride-catalyzed ethanethiolysis, and this constitutes a good preparative method for this 1-thioglycoside. Under the same conditions α-D-glucopyranose pentaacetate remained unchanged: longer reaction times resulted in formation of a small amount of the β-D anomer. The D-galactose derivatives react similarly.

The reaction of sugar derivatives with ethanethiol and an acid catalyst, especially when a desiccant is added, can cause introduction of ethylthio groups at sites elsewhere in the molecule besides C-1 (see Vol. IB, Chap. 18).

2. From Dithioacetals

When an aqueous solution of one mole of D-glucose diethyl dithioacetal is treated at room temperature with one mole of mercuric chloride with addition of base, an ethyl 1-thio-D-glucoside is formed.[35] Other dialkyl 1-thio-D-glucosides are obtained similarly.[36] Pacsu and co-workers,[39,40] who modified the method by using yellow mercuric oxide to neutralize the hydrochloric acid as it forms, considered that the ethyl 1-thio-D-glucoside was an α-furanoside, and this conclusion was soon confirmed.[43] The 1-thio-α-D-glucofuranosides can be converted into the 1-thio-D-glucopyranosides by heating with 0.01M hydrochloric acid.[40] Although this route is a valuable general procedure for preparing 1-thio-D-glucosides, it is not applicable for all 1-thioaldoses. The partial demercaptalation of the diethyl dithioacetals of D-mannose, D-xylose, D-lyxose, and D-ribose has been studied chromatographically;[262] it was concluded that syntheses of 1-thioglycosides derived from D-mannose and D-xylose are impracticable by this route, since the reaction, even with a deficiency (0.5 mole) of mercuric chloride, proceeds readily to the free sugar. The D-lyxose derivative was found to form both a 1-thioglycoside and the free sugar, but ethyl and n-propyl 1-thio-α-D-ribofuranosides could be isolated in about 70% yield.

3. From Glycosyl Halides

The oldest method for preparing a thioglycoside, corresponding to the Michael synthesis, is the S$_N$2 reaction[101] of alkali salts of thiophenols or alkanethiols with O-acetylglycosyl halides. In this way phenyl 1-thio-β-D-glucopyranoside was prepared from tetra-O-acetyl-α-D-glucopyranosyl bromide and thiophenol in the presence of sodium hydroxide. The reaction has been used as a general method for synthesis of acetylated alkyl and aryl 1-thioglycosides,[36,263–267] the high nucleophilicity of sulfur facilitating the reaction, particularly with aryl thiols. Because the 1-thioglycosides possess the same ring structure as the glycosyl halide precursor, the reaction can be used to make both 1-thioglyco-pyranosides and -furanosides. Ethyl 1-thio-α-D-arabinofuranoside has been synthesized by attack of the ethanethiolate ion on 3,5-di-O-benzoyl-α,β-D-arabinofuranosyl chloride.[268]

4. Other Methods

A practical synthesis of acetylated 1-thio-β-D-glucopyranosides, especially valuable when appropriate thiols are not available, involves S-alkylation of 2,3,4,6-tetra-O-acetyl-1-thio-β-D-glucopyranose with alkyl bromides or iodides.[269,270] No change in anomeric configuration occurs during the reaction. An indirect, and usually efficient, method for preparing specifically aryl 1-thioglycosides involves reaction of diazonium salts with thiols, followed by thermal decomposition of the resulting diazo compound.[271] Alkyl 1-thio-β-D-galactopyranosides have been obtained from the corresponding 2-(β-D-galactopyranosyl)-2-thiopseudourea hydrobromide by heating with a solution of Na$_2$S$_2$O$_5$ followed by alkylation with various alkylating agents.[272]

* References start on p. 346.

Brigl's anhydride (3,4,6-tri-*O*-acetyl-1,2-anhydro-α-D-glucopyranose) reacts with ethanethiol in the presence of zinc chloride to afford ethyl 3,4,6-tri-*O*-acetyl-1-thio-β-D-glucopyranoside.[273]

B. REACTIONS

In comparing the reactions of glycosides and 1-thioglycosides it is worthwhile keeping in mind the following three features about sulfur:[257d]

1. It is less basic than oxygen and so has less affinity for protons, but its nucleophilicity is greater than that of oxygen.

2. It is more readily polarized than oxygen and also is able to expand its valence shell by use of its *d* orbitals, and in consequence can stabilize an adjacent carbanion by resonance.

3. In contrast to oxygen it readily undergoes oxidation to higher valence states.

1. *Hydrolysis*

1-Thioglycosides are much more resistant to acid hydrolysis than the analogous glycosides.[36, 40, 274] Comparative figures are shown in Table XIX.

The reversal of reactivity of alkyl and aryl 1-thio-D-glucosides, as compared with the analogous glycosides, is noteworthy. From Table XIX it will be seen

TABLE XIX[a]

THE RATES OF ACID-CATALYZED HYDROLYSIS (2*M* HCl) OF
D-GLUCOPYRANOSIDES AND 1-THIO-D-GLUCOPYRANOSIDES[275]

β-D-*Glucopyranoside*	10^5k (*sec*$^{-1}$) *at 70°*	E (*kcal. mole*$^{-1}$) (±0.6)	ΔS^+(*cal. deg.*$^{-1}$) (±1.7)
Methyl	7.10	34.3	18.3
Ethyl	7.07	—	—
Ethyl 1-thio	2.13	32.8	12.4
Phenyl	31.6	31.0	12.6
Phenyl 1-thio	0.088	33.7	10.8

[a] Reproduced by permission of the Editor of the Chemical Society and the senior author of the paper.

that the phenyl:ethyl ratio of rates for the 1-thio-D-glucosides is 1:25, whereas for the oxygen analogs it is 4:1.

The kinetic results[275] suggest that the hydrolysis of 1-thioglucosides proceeds by way of a cyclic carbonium ion formed by a rate-determining heterolysis of the protonated species. The low basicity of sulfur would certainly diminish the concentration of the conjugate acid in the case of the 1-thioglycosides.

1-Thioglycosides are unaffected by mild alkali, but react with hot, stronger alkali in the same manner as glycosides. The formation of 1,6-anhydro-β-D-glucopyranose from phenyl 1-thio-β-D-glucopyranoside indicates that the phenylthio group is eliminated on treatment with alkali.[276]

2. *Replacement Reactions at C-1*

The reaction of silver or mercury salts of carboxylic acids with 1-thioglycosides affords, under appropriate conditions, a route to 1-*O*-acylaldoses.[277] Thus, ethyl 1-thio-β-D-glucopyranoside reacts with silver benzoate in boiling acetonitrile to give a mixture which, on acetylation, yields tetra-*O*-acetyl-1-*O*-benzoyl-β-D-glucopyranose together with 1,3,4,6-tetra-*O*-acetyl-2-*O*-benzoyl-α-D-glucopyranose (probably formed by migration of the benzoyl group in the initially formed 1-*O*-benzoyl-α-D-glucopyranose).[278]

The behavior of 1-thioglycosides with bromine in acetic acid differs from that of the glycoside analogs. Whereas phenyl 2,3,4,6-tetra-*O*-acetyl-β-D-glucopyranoside initially undergoes bromination at the *para* position of the phenyl group,[279] the product from like treatment of phenyl 2,3,4,6-tetra-*O*-acetyl-1-thio-β-D-glucopyranoside is penta-*O*-acetyl-α-D-glucopyranose.[280] This reaction appears to be quite general for 1-thioglycosides having alkyl or aryl aglycons.

A mechanism, supported by kinetic results, has been proposed[280] that involves electrophilic bromination of the sulfur atom with subsequent conversion of the bromosulfonium ion into tetra-*O*-acetyl-α-D-glucopyranosyl bromide. This bromide then reacts with acetic acid in the presence of the bromine to give the observed product. Glycopyranosyl sulfones, which do not have an electron pair available for formation of a bromosulfonium ion, do not react under these conditions.

In an inert solvent such as carbon tetrachloride, high yields of the glycosyl halide are obtainable, and the reaction has been extended in this way to provide a convenient synthesis of glycosyl halides by treating acylated 1-thioglycosides with bromine in ether.[281] This route is a valuable one for the otherwise rather inaccessible aldofuranosyl halides.

3. *Oxidation*

The oxidation of 1-thioglycosides to convert them into sulfones has been studied in detail.[282] Acetylated 1-thioglycosides can be oxidized with a slight excess of aqueous potassium permanganate in acetic acid. Hydrogen peroxide

---

* *References start on p. 346.*

in acetic acid is also effective, but it causes some deacetylation. Peroxypropionic acid has been used also as oxidant. Presumably the oxidation takes place in two stages with a sulfoxide as intermediate, although oxidation of a 1-thioglycoside with one equivalent of permanganate gave a mixture of sulfone and unreacted material.[282] It is probable that the sulfoxide, once formed, is oxidized at a much higher rate to the sulfone.

The behavior of 1-thioglycosides on oxidation with periodate is complex, and the final uptake of the oxidant is always greater than that required for the Malaprade type of oxidation. The proportion of oxidant consumed may vary widely with concentration and experimental conditions.[283,284] Apparently, glycol cleavage is accompanied by oxidation at the sulfur atom, but the precise nature of the changes involved is not clear.

4. Reduction

Reductive desulfurization by means of Raney nickel with absorbed hydrogen of acetylated 1-thioglycopyranosides leads to products of the 1,5-anhydro type.[285] When a 1-thioaldofuranoside is reductively desulfurized, the product is the corresponding 1,4-anhydroalditol.[286]

REFERENCES

1. J. Conchie, G. A. Levvy, and C. A. Marsh, Advan. Carbohyd. Chem., 12, 157 (1957); R. J. Ferrier, Fortschr. Chem. Forsch., 14, 389 (1970).
2. E. Fischer, Ber., 26, 2400 (1893); ibid., 28, 1145 (1895).
3. See W. G. Overend and M. Stacey, Advan. Carbohyd. Chem., 8, 45 (1953).
4. W. N. Haworth, E. L. Hirst, and J. I. Webb, J. Chem. Soc., 651 (1930).
5. C. B. Purves and C. S. Hudson, J. Amer. Chem. Soc., 56, 708 (1934).
6. E. M. Montgomery and C. S. Hudson, J. Amer. Chem. Soc., 59, 992 (1937); R. K. Ness, H. W. Diehl, and H. G. Fletcher, Jr., ibid., 76, 763 (1954).
7. E. Fischer, Ber., 47, 1980 (1914).
8. J. W. Green, Advan. Carbohyd. Chem., 21, 95 (1966).
9. P. W. Austin, F. E. Hardy, J. G. Buchanan, and J. Baddiley, J. Chem. Soc., 5350 (1963).
10. I. Augestad and E. Berner, Acta Chem. Scand., 8, 251 (1954).
11. W. N. Haworth and C. R. Porter, J. Chem. Soc., 2796 (1929).
12. W. N. Haworth, C. R. Porter, and A. C. Waine, J. Chem. Soc., 2254 (1932).
13. F. Smith, J. Chem. Soc., 584 (1944).
14. D. D. Phillips, J. Amer. Chem. Soc., 76, 3598 (1954).
15. E. M. Osman, K. C. Hobbs, and W. E. Walston, J. Amer. Chem. Soc., 73, 2726 (1951).
16. J. E. Cadotte, F. Smith, and D. Spriestersbach, J. Amer. Chem. Soc., 74, 1501 (1952).
17. P. A. Levene, A. L. Raymond, and R. T. Dillon, J. Biol. Chem., 95, 699 (1932).
18. E. L. Eliel, "Stereochemistry of Carbon Compounds," McGraw-Hill, New York, 1962, p. 198.
19. J. A. Mills, Advan. Carbohyd. Chem., 10, 23 (1955).
20. C. T. Bishop and F. P. Cooper, Can. J. Chem., 40, 224 (1962).
21. C. T. Bishop and F. P. Cooper, Can. J. Chem., 41, 2743 (1963).
22. D. F. Mowery, Jr., J. Org. Chem., 26, 3484 (1961).

23. D. F. Mowery, Jr., and G. F. Ferrante, *J. Amer. Chem. Soc.*, **76**, 4103 (1954).
24. D. F. Mowery, Jr., *J. Amer. Chem. Soc.*, **77**, 1667 (1955).
25. D. D. Heard and R. Barker, *J. Org. Chem.*, **33**, 740 (1968).
26. R. J. Ferrier and L. R. Hatton, *Carbohyd. Res.*, **6**, 75 (1968).
27. B. Capon, G. W. Loveday, and W. G. Overend, *Chem. Ind.* (*London*), 1537 (1962).
28. B. Capon and D. Thacker, *J. Amer. Chem. Soc.*, **87**, 4199 (1965).
29. M. L. Wolfrom and S. W. Waisbrot, *J. Amer. Chem. Soc.*, **61**, 1408 (1939); see also H. A. Campbell and K. P. Link, *J. Biol. Chem.*, **122**, 635 (1938).
30. D. Thacker, Ph.D. Thesis, University of London (1965).
31. See L. L. Schaleger and F. A. Long, *Advan. Phys. Org. Chem.*, **1**, 27 (1963).
32. M. M. Kreevoy and R. W. Taft, *J. Amer. Chem. Soc.*, **77**, 5590 (1955).
33. J. C. McGowan, *J. Appl. Chem.*, **10**, 312 (1960).
34. B. Capon, *Quart. Rev. Chem. Soc.*, **18**, 105 (1965) and references therein.
35. W. Schneider and J. Sepp, *Ber.*, **49**, 2054 (1916).
36. W. Schneider, J. Sepp, and O. Stiehler, *Ber.*, **51**, 220 (1918).
37. E. Pacsu, *Ber.*, **58**, 509 (1925).
38. E. Pacsu and N. Ticharich, *Ber.*, **62**, 3008 (1929).
39. J. W. Green and E. Pacsu, *J. Amer. Chem. Soc.*, **59**, 1205, 2569 (1937); *ibid.*, **60**, 2056, 2288 (1938); *ibid.*, **61**, 1451, 1930 (1939).
40. E. Pacsu and E. J. Wilson, Jr., *J. Amer. Chem. Soc.*, **61**, 1450, 1930 (1939).
41. M. L. Wolfrom, L. J. Tanghe, R. W. George, and S. W. Waisbrot, *J. Amer. Chem. Soc.*, **60**, 132 (1938).
42. M. L. Wolfrom and S. W. Waisbrot, *J. Amer. Chem. Soc.*, **60**, 854 (1938).
43. M. L. Wolfrom, S. W. Waisbrot, D. I. Weisblat, and A. Thompson, *J. Amer. Chem. Soc.*, **66**, 2063 (1944).
44. M. A. Jermyn, *Aust. J. Chem.*, **8**, 403 (1955).
45. B. Helferich and E. Schmitz-Hillebrecht, *Ber.*, **66**, 378 (1933).
46. K. Shishido, *J. Chem. Soc. Ind., Jap.*, **39** (Suppl.), 217 (1936); *Chem. Abstr.*, **30**, 7118 (1936).
47. E. M. Montgomery, N. K. Richtmyer, and C. S. Hudson, *J. Amer. Chem. Soc.*, **64**, 690 (1942).
48. B. N. Stepanenko and G. K. Kryukova, *Dokl. Akad. Nauk SSSR*, **89**, 885 (1953); *Chem. Abstr.*, **48**, 5137 (1954).
49. T. H. Bembry and G. Powell, *J. Amer. Chem. Soc.*, **64**, 2419 (1942).
50. B. Helferich, S. Demant, J. Goerdeler, and R. Bosse, *Hoppe-Seyler's Z. Physiol. Chem.*, **283**, 179 (1948).
51. M. A. Jermyn, *Aust. J. Chem.*, **7**, 202 (1954).
52. R. U. Lemieux and W. P. Shyluk, *Can. J. Chem.*, **31**, 528 (1953).
53. T. Kariyone, M. Takahashi, K. Takaishi, and H. Isaka, *J. Pharm. Soc., Jap.*, **73**, 850 (1953).
54. C. D. Hurd and W. A. Bonner, *J. Org. Chem.*, **11**, 50 (1946).
55. K. W. Rosenmund and E. Güssow, *Arch. Pharm.* (*Weinheim*), **287**, 38 (1954).
56. H. Bretschneider and K. Beran, *Monatsh. Chem.*, **80**, 262 (1949).
57. W. E. Trevelyan, *Carbohyd. Res.*, **2**, 418 (1966).
58. H. Börjeson, P. Jerkeman, and B. Lindberg, *Acta Chem. Scand.*, **17**, 1705 (1963).
59. P. Jerkeman and B. Lindberg, *Acta Chem. Scand.*, **17**, 1709 (1963).
60. H. Feier and O. Westphal, *Ber.*, **89**, 589 (1956).
61. A. H. Fielding and L. Hough, *Carbohyd. Res.*, **1**, 327 (1965).
62. G. Zemplén, *Ber.*, **62**, 985 (1929).
63. B. Helferich and F. Eckstein, *Ber.*, **93**, 2467 (1960).

64. B. Helferich and L. Forsthoff, *Ber.*, **94**, 158 (1961).
65. B. Helferich, W. Piel, and F. Eckstein, *Ber.*, **94**, 491 (1961).
66. B. Helferich and M. Feldhoff, *Ber.*, **94**, 499 (1961).
67. E. K. Gladding and C. B. Purves, *J. Amer. Chem. Soc.*, **66**, 76 (1944).
68. G. Fleury and L. Brissaud, *Compt. Rend.*, **222**, 1051 (1946).
69. L. Fishbein, *J. Amer. Chem. Soc.*, **79**, 2959 (1957).
70. L. J. Morris, *Chem. Ind. (London)*, 1291 (1958).
71. M. L. Wolfrom and I. C. Gillam, *J. Org. Chem.*, **26**, 3564 (1961).
72. Yu A. Zhdanov, G. A. Korol'chenko, G. N. Dorofeenko, and G. I. Zhungietu, *Dokl. Akad Nauk SSSR*, **154**, 861 (1964).
73. A. Michael, *Am. Chem. J.*, **1**, 305 (1879); *Compt. Rend.*, **89**, 355 (1879).
74. W. Koenigs and E. Knorr, *Ber.*, **34**, 957 (1901).
75. E. Fischer and E. F. Armstrong, *Ber.*, **34**, 2885 (1901); *ibid.*, **35**, 833 (1902).
76. C. Mannich, *Ann.*, **394**, 223 (1912).
77. J. H. Fisher, W. L. Hawkins, and H. Hibbert, *J. Amer. Chem. Soc.*, **62**, 1412 (1940).
78. E. Glaser and W. Wulwek, *Biochem. Z.*, **145**, 514 (1924).
78a. A. N. de Belder, E. J. Bourne, and J. B. Pridham, *J. Chem. Soc.*, 4464 (1961).
79. P. W. Austin, F. E. Hardy, J. G. Buchanan, and J. Baddiley, *J. Chem. Soc.*, 2128 (1964).
80. P. W. Austin, F. E. Hardy, J. G. Buchanan, and J. Baddiley, *J. Chem. Soc.*, 1419 (1965).
81. J. W. H. Oldham, *J. Amer. Chem. Soc.*, **56**, 1360 (1934).
82. H. G. Fletcher, Jr., R. K. Ness, and C. S. Hudson, *J. Amer. Chem. Soc.*, **73**, 3698 (1951).
83. D. D. Reynolds and W. L. Evans, *J. Amer. Chem. Soc.*, **60**, 2559 (1938).
84. B. Helferich and J. Goerdeler, *Ber.*, **73**, 532 (1940).
85. C. M. McCloskey, R. E. Pyle, and G. H. Coleman, *J. Amer. Chem. Soc.*, **66**, 349 (1944).
86. B. Helferich, E. Bohn, and S. Winkler, *Ber.*, **63**, 989 (1930).
87. G. Zemplén and Z. Csürös, *Ber.*, **64**, 993 (1931); G. Zemplén, *ibid.*, **74A**, 75 (1941).
88. E. Fischer and L. von Mechel, *Ber.*, **49**, 2813 (1916).
89. E. Fischer and K. Raske, *Ber.*, **43**, 1750 (1910).
90. B. Helferich, A. Doppstadt, and A. Gottschlich, *Naturwissenschaften*, **40**, 441 (1953).
91. H. H. Schlubach and G. A. Schröter, *Ber.*, **61**, 1216 (1928).
92. H. H. Schlubach and K. Meisenheimer, *Ber.*, **67**, 429 (1934).
93. F. Micheel and F. Suckfüll, *Ber.*, **66**, 1957 (1933).
94. C. P. J. Glaudemans and H. G. Fletcher, Jr., *J. Amer. Chem. Soc.*, **87**, 2456 (1965).
95. F. H. Newth and G. O. Phillips, *J. Chem. Soc.*, 2896, 2900, 2904 (1953).
96. E. P. Painter, *J. Amer. Chem. Soc.*, **75**, 1137 (1953).
97. R. U. Lemieux and G. Huber, *Can. J. Chem.*, **33**, 128 (1955).
98. L. J. Haynes and F. H. Newth, *Advan. Carbohyd. Chem.*, **10**, 207 (1955).
99. G. L. Mattok and G. O. Phillips, *J. Chem. Soc.*, 1836 (1956); *ibid.*, 268 (1957).
100. R. U. Lemieux and A. R. Morgan, *J. Amer. Chem. Soc.*, **85**, 1889 (1963).
101. B. Capon, P. M. Collins, A. A. Levy, and W. G. Overend, *J. Chem. Soc.*, 3242 (1964).
102. L. R. Schroeter, J. W. Green, and D. C. Johnson, *J. Chem. Soc.*, **B**, 447 (1966).
102a. R. S. Tipson, *J. Biol. Chem.*, **130**, 55 (1939).
103. H. L. Frush and H. S. Isbell, *J. Res. Nat. Bur. Stand.*, **27**, 413 (1941); H. S. Isbell, *Ann. Rev. Biochem.*, **9**, 65 (1940); E. Pacsu, *Advan. Carbohyd. Chem.*, **1**, 77 (1945).
104. R. U. Lemieux, *Advan. Carbohyd. Chem.*, **9**, 1 (1954); cf. R. D. Guthrie and J. F. McCarthy, *ibid.*, **22**, 11 (1967).
105. G. L. Mattok and G. O. Phillips, *J. Chem. Soc.*, 130 (1958).
106. P. Brigl, *Hoppe-Seyler's Z. Physiol. Chem.*, **122**, 245 (1922).
107. W. J. Hickinbottom, *J. Chem. Soc.*, 1676 (1929).

108. W. F. Goebel, F. H. Babers, and O. T. Avery, *J. Exp. Med.*, **55**, 761 (1932); B. Helferich and W. M. Müller, *Ber.*, **104**, 671 (1971).
109. R. U. Lemieux, C. Brice, and G. Huber, *Can. J. Chem.*, **33**, 134 (1955).
110. R. Barker and H. G. Fletcher, Jr., *J. Org. Chem.*, **26**, 4605 (1961).
111. H. S. Isbell and H. L. Frush, *J. Res. Nat. Bur. Stand.*, **43**, 161 (1949).
112. H. R. Goldschmid and A. S. Perlin, *Can. J. Chem.*, **39**, 2025 (1961).
113. C-S. Giam, H. R. Goldschmid, and A. S. Perlin, *Can. J. Chem.*, **41**, 3074 (1963).
114. N. K. Kochetkov, A. J. Khorlin, and A. F. Bochkov, *Tetrahedron Lett.*, 289 (1964); *Tetrahedron*, **23**, 693 (1967).
115. N. K. Kochetkov, A. J. Khorlin, A. F. Bochkov, and I. G. Yazlovetsky, *Carbohyd. Res.*, **2**, 84 (1966); N. K. Kochetkov, A. F., Bochkov, and I. G. Yazlovetsky, *ibid.*, **9**, 49 (1969).
115a. N. K. Kochetkov, A. F. Bochkov, T. A. Sokolovskaya, and V. N. Snyatkova, *Carbohyd. Res.*, **16**, 17 (1971).
116. N. E. Franks and R. Montgomery, *Carbohyd. Res.*, **3**, 511 (1967); **6**, 286 (1968).
117. J. K. Dale, *J. Amer. Chem. Soc.*, **46**, 1046 (1924).
118. A. S. Perlin, *Can. J. Chem.*, **41**, 555 (1963).
119. E. Pacsu, *Advan. Carbohyd. Chem.*, **1**, 77 (1945).
120. C. B. Purves and C. S. Hudson, *J. Amer. Chem. Soc.*, **59**, 1170 (1937).
121. T. G. Bonner, E. J. Bourne, and S. McNally, *J. Chem. Soc.*, 761 (1962).
122. J. B. Lee and M. M. El Sawi, *Tetrahedron*, **6**, 91 (1959).
123. C. S. Hudson, *Org. Syn. Coll. Vol.*, **I**, 371 (1947).
124. L. Maquenne, *Bull. Soc. Chim. Fr.*, **33**, 469 (1905).
125. H. S. Isbell and H. L. Frush, *J. Res. Nat. Bur. Stand.*, **24**, 125 (1940).
126. R. Kuhn and H. H. Baer, *Ber.*, **86**, 724 (1953).
127. C. S. Hudson and D. H. Brauns, *J. Amer. Chem. Soc.*, **38**, 1216 (1916).
128. P. Brigl, *Hoppe-Seyler's Z. Physiol. Chem.*, **122**, 245 (1922).
129. R. U. Lemieux and G. Huber, *J. Amer. Chem. Soc.*, **75**, 4118 (1953).
130. M. L. Wolfrom, A. Thompson, and R. B. Ward, *J. Amer. Chem. Soc.*, **81**, 4623 (1959).
131. H. S. Hill, *J. Amer. Chem. Soc.*, **50**, 2725 (1928); W. H. Watanabe and L. E. Conlon, *ibid.*, **79**, 2825 (1957).
132. See R. J. Ferrier, *Advan. Carbohyd. Chem.*, **20**, 67 (1965); *Advan. Carbohyd. Chem. Biochem.*, **24**, 199 (1969), for lists of examples.
133. K. Wallenfels and J. Lehmann, *Ann.*, **635**, 166 (1960).
134. R. J. Ferrier, *J. Chem. Soc.*, 5443 (1964).
135. R. J. Ferrier, W. G. Overend, and A. E. Ryan, *J. Chem. Soc.*, 3667 (1962).
136. E. Fischer, M. Bergmann, and H. Schotte, *Ber.*, **53**, 509 (1920).
137. R. U. Lemieux and B. Fraser-Reid, *Can. J. Chem.*, **42**, 532 (1964).
138. See R. U. Lemieux and B. Fraser-Reid, *Can. J. Chem.*, **42**, 539, 547 (1964); *ibid.*, **43**, 1460 (1965); R. U. Lemieux and S. Levine, *ibid.*, **42**, 1473 (1964).
139. G. R. Inglis, J. C. P. Schwarz, and L. McLaren, *J. Chem. Soc.*, 1014 (1962).
140. P. T. Manolopoulos, M. Mednick, and N. N. Lichtin, *J. Amer. Chem. Soc.*, **84**, 2203 (1962).
141. E. L. Hirst and C. B. Purves, *J. Chem. Soc.*, 1352 (1923).
142. E. L. Hirst, *J. Chem. Soc.*, 350 (1926).
143. W. N. Haworth, E. L. Hirst, and E. J. Miller, *J. Chem. Soc.*, 2436 (1927).
144. F. Micheel and F. Suckfüll, *Ann.*, **507**, 138 (1933).
145. W. N. Haworth, E. L. Hirst, and A. Learner, *J. Chem. Soc.*, 1040, 2432 (1927); J. Avery, W. N. Haworth, and E. L. Hirst, *ibid.*, 2308 (1927).
146. W. Brock Neely, *Advan. Carbohyd. Chem.*, **12**, 13 (1957).

147. H. Spedding, *Advan. Carbohyd. Chem.*, **19**, 23 (1964).
148. R. S. Tipson and H. S. Isbell, *J. Res. Nat. Bur. Stand.*, **64A**, 239, 405 (1960); *ibid.*, **65A**, 249 (1961); H. S. Isbell, F. A. Smith, E. C. Creitz, H. L. Frush, J. D. Moyer, and J. E. Stewart, *ibid.*, **59**, 41 (1957).
149. R. U. Lemieux and D. R. Lineback, *Ann. Rev. Biochem.*, **32**, 155 (1963); R. J. Ferrier and N. R. Williams, *Chem. Ind. (London)*, 1697 (1964); L. D. Hall, *Advan. Carbohyd. Chem.*, **19**, 51 (1964).
150. P. L. Durette and D Horton, *Advan. Carbohyd. Chem. Biochem.*, **26**, 49 (1971).
151. P. L. Durette and D. Horton, *Carbohyd. Res.*, **18**, 289, 403 (1971).
151a. B. Capon and D. Thacker, *Proc. Chem. Soc.*, 369 (1964).
152. G. A. Jeffrey and R. D. Rosenstein, *Advan. Carbohyd. Chem.*, **19**, 23 (1964); G. Strahs, *Advan. Carbohyd. Chem. Biochem.*, **25**, 53 (1970).
153. C. J. Brown, *Acta Crystallogr.*, **13**, 1049 (1960).
154. S. Furberg and H. Hammer, *Acta Chem. Scand.*, **15**, 1190 (1961).
155. B. Sheldrick and J. H. Robertson, *Acta Crystallogr.*, **16**, A54 (1963).
156. C. R. Noller and W. C. Rockwell, *J. Amer. Chem. Soc.*, **60**, 2076 (1938).
157. W. W. Pigman and N. K. Richtmyer, *J. Amer. Chem. Soc.*, **64**, 369 (1942).
158. I. G. Farbenind., A. G., Fr. Patent 838863 (1939); *Chem. Abstr.*, **33**, 6996 (1939).
159. R. T. Williams, "Detoxification Mechanisms," Chapman & Hall, London, 1959, p. 284.
160. B. Capon, W. G. Overend, and M. Sobell, *J. Chem. Soc.*, 5172 (1961).
161. I. Listowsky, G. Avigad, and S. England, *J. Amer. Chem. Soc.*, **87**, 1765 (1965).
162. O. Hassel and B. Ottar, *Acta Chem. Scand.*, **1**, 929 (1947).
163. J. T. Edward, *Chem. Ind. (London)*, 1102 (1955); S. Wolfe, A. Rauk, L. M. Tel, and I. G. Csizmadia, *J. Chem. Soc. (B)*, 136 (1971).
164. R. U. Lemieux and W. P. Shyluk, *Can. J. Chem.*, **33**, 120 (1955).
165. R. U. Lemieux, R. K. Kullnig, H. J. Bernstein, and W. G. Schneider, *J. Amer. Chem. Soc.*, **80**, 6098 (1958).
166. E. Pacsu, *Ber.*, **61**, 1508 (1928): *J. Amer. Chem. Soc.*, **52**, 2563, 2568, 2571 (1930).
167. E. Pacsu, *Ber.*, **61**, 137 (1928).
168. See B. Lindberg, *Acta Chem. Scand.*, **4**, 49 (1950).
169. R. E. Reeves and L. W. Mazzeno, *J. Amer. Chem. Soc.*, **76**, 2219 (1954).
170. B. Lindberg, *Ark. Kemi, Mineral. Geol. Ser.* **B18** [No. 9], 1 (1944).
171. B. Lindberg, *Acta Chem. Scand.*, **2**, 426, 534 (1948).
172. B. Capon, *Chem. Commun.*, 21 (1967).
173. J. Swiderski and A. Temeriusz, *Carbohyd. Res.*, **3**, 225 (1966); A. Temeriusz, *Rocz. Chem.*, **40**, 825 (1966).
174. R. J. Ferrier, L. R. Hatton, and W. G. Overend, *Carbohyd. Res.*, **8**, 56 (1968).
175. B. Lindberg, *Acta Chem. Scand.*, **3**, 1350 (1949); see E. P. Painter, *J. Amer. Chem. Soc.*, **75**, 1137 (1953).
176. J. Janson and B. Lindberg, *Acta Chem. Scand.*, **14**, 877 (1960).
177. E. M. Montgomery, R. M. Hann, and C. S. Hudson, *J. Amer. Chem. Soc.*, **59**, 1124 (1937).
178. S. Winstein and K. C. Schreiber, *J. Amer. Chem. Soc.*, **74**, 2165 (1952); S. Winstein, and R. Heck, *ibid.*, **74**, 5584 (1952).
179. W. N. Haworth, L. N. Owen, and F. Smith, *J. Chem. Soc.*, 88 (1941); W. N. Haworth, J. Jackson, and F. Smith, *J. Chem. Soc.*, 620 (1940).
180. A. B. Foster, W. G. Overend, and G. Vaughan, *J. Chem. Soc.*, 3625 (1954).
181. J. N. BeMiller, *Advan. Carbohyd. Chem.* **22**, 25 (1967); J. Szejtli, *Staerke*, 19, 145 (1967).
182. W. Alberda van Ekenstein, *Rec. Trav. Chim.*, **13**, 183 (1894).

183. E. F. Armstrong and W. H. Glover, *Proc. Roy. Soc. (London)*, **80B**, 312 (1908).
184. E. A. Moelwyn-Hughes, *Trans. Faraday Soc.*, **24**, 309, 321 (1928); **25**, 81, 503 (1929); cf. J. Szejtli, *Staerke*, **19**, 173 (1967); *Acta Chim. Acad. Sci. Hung.*, **56**, 175 (1968).
185. C. N. Riiber and N. A. Sørensen, *Kgl. Norske Videnskab. Selskab, Skrifter*, No. 1 (1938): *Chem. Abstr.*, **33**, 4962 (1939).
186. H. S. Isbell and H. L. Frush, *J. Res. Nat. Bur. Stand.*, **24**, 125 (1940).
187. L. J. Heidt and C. B. Purves, *J. Amer. Chem. Soc.*, **66**, 1385 (1944).
188. W. N. Haworth, *Ber.*, **65A**, 43 (1932).
189. C. B. Purves and C. S. Hudson, *J. Amer. Chem. Soc.*, **59**, 1170 (1937).
190. C. A. Bunton, T. A. Lewis, D. R. Llewellyn, and C. A. Vernon, *J. Chem. Soc.*, 4419 (1955).
191. C. Armour, C. A. Bunton, S. Patai, L. H. Selman, and C. A. Vernon, *J. Chem. Soc.*, 412 (1961).
192. A. J. A'Court, B. Capon, and W. G. Overend, unpublished results, (1963).
193. J. N. Brönsted and W. F. K. Wynne-Jones, *Trans. Faraday Soc.*, **25**, 59 (1929).
194. See M. M. Kreevoy and R. W. Taft, *J. Amer. Chem. Soc.*, **77**, 3146 (1955).
195. W. G. Overend, C. W. Rees, and J. S. Sequeira, *J. Chem. Soc.*, 3429 (1962).
196. See J. C. Hornel and J. A. V. Butler, *J. Chem. Soc.*, 1361 (1936); W. J. C. Orr and J. A. V. Butler, *ibid.*, 330 (1937); M. Kilpatrick, *J. Amer. Chem. Soc.*, **85**, 1036 (1963).
197. See L. P. Hammett, "Physical Organic Chemistry," McGraw-Hill, New York, 1940, p. 273.
198. D. McIntyre and F. A. Long, *J. Amer. Chem. Soc.*, **76**, 3240 (1954).
199. See J. F. Bunnett, *J. Amer. Chem. Soc.*, **83**, 4978 (1961).
200. T. E. Timell, *Can. J. Chem.*, **42**, 1456 (1964).
201. R. J. Withey and E. Whalley, *Can. J. Chem.*, **41**, 1849 (1963).
201a. A. Thompson, M. L. Wolfrom, and E. J. Quinn, *J. Amer. Chem. Soc.*, **75**, 3003 (1953).
202. B. E. C. Banks, Y. Meinwald, A. J. Rhind-Tutt, I. Sheft, and C. A. Vernon, *J. Chem. Soc.*, 3240 (1961).
203. F. Shafizadeh, *Advan. Carbohyd. Chem.*, **13**, 9 (1958): F. Shafizadeh and A. Thompson, *J. Org. Chem.*, **21**, 1059 (1956).
204. A. B. Foster and W. G. Overend, *Chem. Ind. (London)*, 566 (1955).
205. A. J. Rhind-Tutt, Ph.D. Thesis, University of London (1957).
206. B. Capon and W. G. Overend, *Advan. Carbohyd. Chem.*, **15**, 11 (1960).
207. R. L. Nath and H. N. Rydon, *Biochem. J.*, **57**, 1 (1954).
208. A. N. Hall, S. Hollingshead, and H. N. Rydon, *J. Chem. Soc.*, 4290 (1961).
209. B. N. Stepanenko and O. G. Serdyak, *Dokl. Akad. Nauk SSSR*, **139**, 1132 (1961); *Chem. Abstr.*, **56**, 1516 (1962); compare J. N. BeMiller and E. R. Doyle, *Carbohyd. Res.*, **20**, 23 (1971).
210. See ref. 197, p. 186.
211. J. H. Fisher, W. L. Hawkins, and H. Hibbert, *J. Amer. Chem. Soc.*, **63**, 3031 (1941).
212. T. E. Timell, *Chem. Ind. (London)*, 1208 (1963).
213. P. Mamalis and H. N. Rydon, *J. Chem. Soc.*, 1049 (1955).
214. J. A. Snyder and K. P. Link, *J. Amer. Chem. Soc.*, **74**, 1883 (1952).
215. B. Capon, *Tetrahedron Lett.*, **14**, 911 (1963).
216. K. K. De and T. E. Timell, *Carbohyd. Res.*, **4**, 72 (1967).
217. M. S. Feather and J. F. Harris, *J. Org. Chem.*, **30**, 153 (1965).
218. R. D. Marshall, *Nature*, **199**, 998 (1963).
219. G. N. Richards, *Chem. Ind. (London)*, 228 (1955).
220. T. E. Timell, W. Enterman, F. Spencer, and E. J. Soltes, *Can. J. Chem.*, **43**, 2296 (1965).
221. L. M. Semke, N. S. Thompson, and D. G. Williams, *J. Org. Chem.*, **29**, 1041 (1964); J. Nakano and B. G. Rånby, *Svensk Papperstidn.*, **2**, 29 (1962).

222. E. Tomita, *Yakugaku Zasshi*, **87**, 485 (1967); E. Tomita, Y. Hirota, and Y. Nitta, *ibid.*, **87**, 479 (1967); M. D. Saunders and T. E. Timell, *Carbohyd. Res.*, **5**, 453 (1967); **6**, 12 (1968).
223. T. E. Timell, *Chem. Ind. (London)*, 503 (1964).
224. J. A. Mills, *Advan. Carbohyd. Chem.*, **10**, 1 (1955).
225. I. Augestad and E. Berner, *Acta Chem. Scand.*, **10**, 911 (1956).
225a. B. Capon and D. Thacker, *J. Chem. Soc.*, **B**, 185 (1967).
226. C. E. Ballou, *Advan. Carbohyd. Chem.*, **9**, 59 (1954).
227. J. Janson and B. Lindberg, *Acta Chem. Scand.*, **14**, 2051 (1960).
228. E. Dryselius, B. Lindberg, and O. Theander, *Acta Chem. Scand.*, **12**, 340 (1958).
229. J. Janson and B. Lindberg, *Acta Chem. Scand.*, **13**, 138 (1959).
230. B. Lindberg, *Svensk Papperstidn.*, **59**, 531 (1956).
231. L. H. Koehler and C. S. Hudson, *J. Amer. Chem. Soc.*, **72**, 981 (1950).
232. E. Fischer and B. Helferich, *Ber.*, **47**, 210 (1914).
233. E. Fischer, *Ber.*, **47**, 1377 (1914).
234. C. E. Ballou and K. P. Link, *J. Amer. Chem. Soc.*, **71**, 3743 (1949); *ibid.*, **72**, 3147 (1950).
235. R. Kuhn and I. Löw, *Ber.*, **74**, 219 (1941).
236. G. Bouchardet, *Compt. Rend.*, **19**, 1174 (1844).
237. C. Tanret, *Compt. Rend.*, **119**, 158 (1894).
238. C. M. McCloskey and G. H. Coleman, *J. Org. Chem.*, **10**, 184 (1945).
239. E. M. Montgomery, N. K. Richtmyer, and C. S. Hudson, *J. Amer. Chem. Soc.*, **65**, 3 (1943).
240. E. M. Montgomery, N. K. Richtmyer, and C. S. Hudson, *J. Amer. Chem. Soc.*, **64**, 1483 (1942); *ibid.*, **65**, 1848 (1943).
241. E. Fischer and H. Strauss, *Ber.*, **45**, 2467 (1912); compare J. Staněk and E. Hamernikova, *Ann.*, **738**, 136 (1970).
242. E. Glaser and Ch. Thaler, *Arch. Pharm. (Weinheim)*, **264**, 228 (1926).
243. J. A. Snyder and K. P. Link, *J. Amer. Chem. Soc.*, **74**, 1883 (1952).
244. A. Dyfverman and B. Lindberg, *Acta Chem. Scand.*, **4**, 878 (1950).
245. R. C. Gasman and D. C. Johnson, *J. Org. Chem.*, **31**, 1830 (1966).
245a. D. Horton and A. E. Luetzow, *Chem. Commun.*, 79 (1971).
245b. A. E. Luetzow, Ph.D. Dissertation, The Ohio State University, 1971.
246. K. Freudenberg, W. Dürr, and H. von Hochstetter, *Ber.*, **61**, 1735 (1928).
247. N. K. Richtmyer, *J. Amer. Chem. Soc.*, **56**, 1633 (1934).
248. H. F. Bauer and D. E. Stuetz, *J. Amer. Chem. Soc.*, **78**, 4097 (1956).
249. A. S. Perlin, E. von Rudloff, and A. P. Tulloch, *Can. J. Chem.*, **36**, 921 (1958).
250. P. A. J. Gorin, *Can. J. Chem.*, **38**, 641 (1960): T. Francis and A. S. Perlin, *ibid.*, **37**, 1229 (1959).
251. J. T. Henderson, *J. Amer. Chem. Soc.*, **79**, 5304 (1957).
252. O. Theander, *Svensk Papperstidn.*, **61**, 581 (1958).
253. R. L. Whistler, T. W. Mittag, and T. R. Ingle, *J. Amer. Chem. Soc.*, **87**, 4218 (1965).
254. N. N. Lichtin and M. H. Saxe, *J. Amer. Chem. Soc.*, **77**, 1875 (1955).
255. L. J. Heidt, *J. Amer. Chem. Soc.*, **61**, 2981 (1939).
256. G. Tanret, *Compt. Rend.*, **201**, 1057 (1935); **202**, 881 (1936).
256a. J. S. Moore and G. O. Phillips, *Carbohyd. Res.*, **16**, 79 (1971); G. O. Phillips, W. G. Filby, J. S. Moore, and G. V. Davies, *ibid.*, **16**, 89, 105 (1971).
256b. N. K. Kochetkov, L. I. Kudryashov, and M. A. Chlenov, *Zh. Obshch. Khim.*, **35**, 897, (1965); L. I. Kudryashov, T. Ya. Livertovskaya, S. V. Voznesenskaya, Yu. I. Kovalev, V. A. Sharpatyi, and N. K. Kochetkov, *ibid.*, **40**, 1133 (1970).

257. For reviews see (a) A. L. Raymond, *Advan. Carbohyd. Chem.*, **1**, 129 (1945); (b) G. Zinner, *Deut. Apotheker-Ztg.*, **98**, 335 (1958); (c) A. Kjær, in "Handbuch der Pflanzenphysiologie," (W. Ruhland, ed.), Vol. 9, Springer-Verlag, Berlin, 1958, p. 71; (d) D. Horton and D. H. Hutson, *Advan. Carbohyd. Chem.*, **18**, 123 (1963).

258. J. Fried and D. E. Walz, *J. Amer. Chem. Soc.*, **71**, 140 (1949).

259. P. A. Levene and G. M. Meyer, *J. Biol. Chem.*, **74**, 695 (1927).

259a. R. J. Ferrier, L. R. Hatton, and W. G. Overend, *Carbohyd. Res.*, **6**, 87 (1968).

260. C. D. Hurd and W. A. Bonner, *J. Org. Chem.*, **11**, 50 (1946).

261. R. U. Lemieux, *Can. J. Chem.*, **29**, 1079 (1951).

262. H. Zinner, A. Koine, and H. Nimz, *Ber.*, **93**, 2705 (1960).

263. E. Fischer and K. Delbrück, *Ber.*, **42**, 1476 (1909).

264. C. B. Purves, *J. Amer. Chem. Soc.*, **51**, 3619, 3627 (1929).

265. E. M. Montgomery, N. K. Richtmyer, and C. S. Hudson, *J. Org. Chem.*, **11**, 301 (1946).

266. B. Helferich, H. Grünewald, and F. Langenhoff, *Ber.*, **86**, 873 (1953).

267. B. Helferich and D. Türk, *Ber.*, **89**, 2215 (1956); compare G. Wagner and R. Metzner *Pharmazie*, **24**, 245 (1969).

268. E. J. Reist, P. A. Hart, L. Goodman, and B. R. Baker, *J. Amer. Chem. Soc.*, **81**, 5176 (1959).

269. W. Schneider, R. Gille, and K. Eisfeld, *Ber.*, **61**, 1244 (1928).

270. M. Černý and J. Pacák, *Chem. Listy*, **52**, 2090 (1958); *Collect. Czech. Chem. Commun.*, **24**, 2566 (1959).

271. M. Černý, D. Zachystalová, and J. Pacák, *Collect. Czech. Chem. Commun.*, **26**, 2206 (1961).

272. M. Černý, J. Staněk, and J. Pacák, *Monatsh. Chem.*, **94**, 290 (1963).

273. F. Weygand and H. Ziemann, *Ann.*, **657**, 179 (1962).

274. C. Wagner, E. Fickweiler, P. Nuhn and H. Pischel, *Z. Chem.*, **3**, 62 (1963).

275. C. Bamford, B. Capon, and W. G. Overend, *J. Chem. Soc.*, 5138 (1962); cf. M. D. Saunders and T. E. Timell, *Carbohyd. Res.*, **6**, 121 (1968).

276. E. M. Montgomery, N. K. Richtmyer, and C. S. Hudson, *J. Org. Chem.*, **10**, 194 (1945).

277. H. B. Wood, B. Coxon, H. W. Diehl, and H. G. Fletcher, Jr., *J. Org., Chem.*, **29**, 461 (1964).

278. C. Pedersen and H. G. Fletcher, Jr., *J. Amer. Chem. Soc.*, **82**, 3215 (1960).

279. C. D. Hurd and W. A. Bonner, *J. Amer. Chem. Soc.*, **67**, 1764 (1945).

280. W. A. Bonner, *J. Amer. Chem. Soc.*, **70**, 3491 (1948).

281. F. Weygand and H. Ziemann, *Ann.*, **657**, 179 (1962): F. Weygand, H. Ziemann, and H. J. Bestmann, *Ber.*, **91**, 2534 (1958).

282. W. A. Bonner and R. W. Drisko, *J. Amer. Chem. Soc.*, **70**, 2435 (1948).

283. M. L. Wolfrom and Z. Yosizawa, *J. Amer. Chem. Soc.*, **81**, 3474 (1959).

284. S. Okui, *Yakugaku Zasshi*, **75**, 1262 (1955).

285. H. G. Fletcher, Jr. and C. S. Hudson, *J. Amer. Chem. Soc.*, **69**, 921, 1672 (1947).

286. C. F. Huebner and K. P. Link, *J. Biol. Chem.*, **186**, 387 (1950).

10. ACYCLIC DERIVATIVES

MELVILLE L. WOLFROM[*]

I. INTRODUCTION

The early sugar chemists, such as Kiliani and Fischer, employed open-chain or acyclic formulas for the monosaccharides. It is a truism in organic chemistry that, when a molecule possesses two functions capable of reacting with each other, such a reaction will occur, intramolecularly, if space relations allow, and otherwise intermolecularly. Thus the formation of ring structures in these polyhydroxycarbonyl compounds follows from the fact that a carbonyl function will form a hemiacetal with a hydroxyl group. The experimental establishment of ring structures in the sugars caused the acyclic formulas to be replaced by the cyclic hemiacetal ring forms. Nevertheless, some sugar reactions result in the formation of acyclic derivatives, and the acyclic form is now an established type of isomerism in the sugars. Such structures may form in reactions

* Deceased. See *Advan. Carbohyd. Chem. Biochem.*, **26**, 1 (1971) for an account of the scientific work of the author, including a complete bibliography of his published work. The systematic elaboration of the chemistry of acyclic sugars was largely achieved in Professor Wolfrom's laboratory.

where the cyclizing hydroxyl groups become substituted, thus preventing formation of a cyclic hemiacetal. Whether such acyclic forms contribute to the dynamic tautomerism of the polyhydroxy carbonyl compounds in aqueous solution is a moot question, difficult to put to experimental proof.

II. DITHIOACETALS

A. FORMATION

Although Fischer was never able to obtain the acyclic acetals of the sugars, he did succeed in obtaining dithioacetals[1] of aldoses by the reaction of aldoses with ethanethiol in concentrated hydrochloric acid solution. He noted that

$$
\begin{array}{ccc}
\text{HC=O} & & \text{HC(SEt)}_2 \\
| & \xrightarrow[\text{H}^+]{\text{EtSH}} & | \\
\text{(CHOH)}_4 & & \text{(CHOH)}_4 \\
| & & | \\
\text{CH}_2\text{OH} & & \text{CH}_2\text{OH}
\end{array}
$$

these substances crystallized well from the reaction mixture and were easily isolated. In fact, the acetylated ethyl dithioacetal is the best derivative for the identification of D-glucose.[2] A few of the aldose dithioacetals, such as that of D-lyxose,[3] are water-soluble; the ketoses are too acid-sensitive for these conditions, and indirect methods are required to obtain them. With D-fructose, and other ketohexoses, the acyclic *keto*-D-fructose pentaacetate is available, and from this the dithioacetal can be formed, under mild conditions, with ethanethiol and zinc chloride.[4] Deacetylation then yields the ketose dithioacetal. Ethanethiol is the thiol most commonly employed in forming the sugar dithioacetals, but others have been used, especially α-toluenethiol (benzyl mercaptan), and Zinner and associates[5] have described others. The cyclic dithioacetals formed from 1,2-ethanedithiol were early recorded by Lawrence[6] from Fischer's laboratory; Lawrence also described the first dibenzyl dithioacetal (of D-glucose). Dithioacetals of the disaccharides maltose[7] and lactose[8] have been prepared, the former being characterized as the octaacetate.

Following Baumann,[9] Fischer designated the aldose dithioacetals as mercaptals, a term currently disapproved by nomenclature commissions. This name seems, nonetheless, to constitute the best basis for naming the reaction known as mercaptolysis. This term denotes the hydrolysis of a glycosidic linkage by strong acids in the presence of excess ethanethiol (ethyl mercaptan). Under these conditions, the reducing groups liberated are converted into dithioacetals. Like many others in the sugar series, this reaction was first applied to cellulose.[10] Mercaptolysis has been found especially useful in the structural elucidation of agar,[11] carrageenan,[12] streptomycin,[13, 14] and other antibiotics. If the mercaptolysis is performed at room temperature, rather than at the customary 0° to 12°, a 1-thiopyranoside may arise.[15]

B. REACTIONS

In his classical 1894 publication,[1] Fischer characterized, in test tube experiments, the main reactions of the sugar dithioacetals. These reactions have been elaborated and extended in modern times.

We need not be concerned herein with the hydroxyl-group reactions of the dithioacetals, since these reactions are typical of any series of hydroxyl functions. It need only be pointed out that the absence of a ring makes all of these hydroxyl groups available, and the dithioacetal group can be removed to regenerate a cyclic structure. Furthermore, the absence of a ring in the dithioacetals, as in the oxygen acetals derivable from them (see Section IV), makes the carbon chain more flexible and free of ring-conformational effects. Such a chain, extended, is depicted below. Baker and associates[16] have utilized this

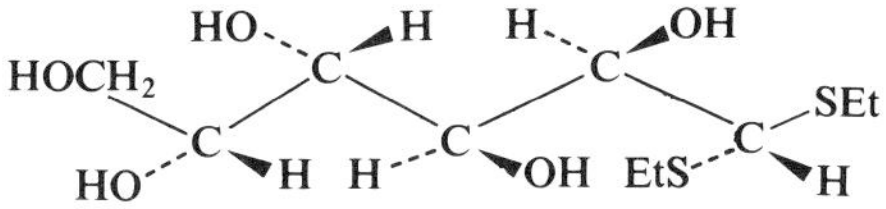

D-Galactose diethyl dithioacetal

property of opening and re-forming sugar rings to effect reactions not otherwise readily eventuated. Tritylation of D-galactose diethyl dithioacetal, acetylation, and "demercaptalation" (see Section V,A,1) led Micheel and co-workers[17] to septanoside structures.

After many vicissitudes, it was finally established that acetonation of D-glucose diethyl dithioacetal results in a mixture of di-isopropylidene acetals in which positions 2 and 4, respectively, are unsubstituted. Many cyclic acetals of aldose dithioacetals have been described by Zinner and associates.[18] Steric (or electronic) hindrance in the benzoylation of D-glucose diethyl dithioacetal leads to a tetrabenzoate with position 2 open.[19]

The most characteristic reaction of the dithioacetals is probably hydrolysis, readily effected by acids. Hydrolysis can also be caused, under mild conditions and in the presence of minimal amounts of water, by mercuric chloride and by bromine[1] (see Section V,A,1). A partial hydrolysis by mercuric chloride, leading to a 1-thiofuranoside, was established by Schneider and Sepp.[20] This interesting reaction was studied extensively by Pacsu and co-workers and has been reviewed by Green.[21] Mercuric chloride and yellow mercuric oxide were used in either aqueous[22] or alcoholic media.[23] Depending upon the sugar structure and the reaction conditions, the product formed may be a 1-thioaldo-

* *References start on p. 386.*

furanoside (D-glucose, D-galactose), an aldofuranoside (D-galactose, D-mannose, L-arabinose), an acetal (D-fructose, 6-deoxy-L-mannose), an aldopyranoside, the free sugar, or mixtures thereof. The method is useful for the synthesis of some aldofuranosides and 1-thioaldofuranosides. Acetal formation from a substituted dithioacetal is a significant reaction and will be discussed in Section IV.

Acetolysis of aldose dithioacetals leads to the production of the fully acetylated form of the aldose aldehydrol;[24] such a hepta-O-acetyl derivative of DL-galactose was obtained from agar by acetolysis.[24]

$$
\begin{array}{ccc}
\text{HC(SR)}_2 & & \text{HC(OAc)}_2 \\
| & \xrightarrow[\text{H}_2\text{SO}_4]{\text{Ac}_2\text{O–AcOH}} & | \\
\text{(CHOH)}_n & & \text{(CHOAc)}_n \\
| & & | \\
\text{CH}_2\text{OH} & & \text{CH}_2\text{OAc}
\end{array}
$$

The dithioacetals are stable toward such mild oxidizing agents as Fehling solution,[1] but toward agents of higher oxidation potential, such as organic peroxides, they form disulfones. The latter are degraded, by mild alkalies, to the next lower aldose.[25] The aldose disulfones are sensitive substances and can

$$
\begin{array}{ccccc}
\text{HC(SEt)}_2 & & \text{HC(SO}_2\text{Et)}_2 & & \text{C(SO}_2\text{Et)}_2 \\
| & \longrightarrow & | & \longrightarrow & \| \\
\text{HCOH} & & \text{HCOH} & & \text{HC} \quad \xrightarrow[\text{HOH}]{\text{NH}_4\text{OH}} \quad \text{HC=O} + \text{CH}_2(\text{SO}_2\text{Et})_2 \\
| & & | & & | \\
\end{array}
$$

undergo elimination reactions leading to unsaturated or anhydro derivatives.[26] Degradation of an aldose dithioacetal probably proceeds through the unsaturated derivative. Ketoses can be degraded by two carbon atoms.[27]

One alkylthio group of an acetylated dithioacetal can be replaced by halogen with phosphorus oxychloride and acetyl chloride,[28] acetyl bromide,[29] MeOCHCl$_2$ or MeOCHBr$_2$,[30] or bromine.[31]

$$
\begin{array}{ccc}
\text{HC(SR)}_2 & \longrightarrow & \text{HC}\!\!\begin{array}{c}\diagup \text{SR}\\ \diagdown \text{X}\end{array} \\
| & & | \\
\end{array}
$$

$$X = \text{Br or Cl}$$

The remarkable reductive desulfurization reaction discovered by Bougault et al.[32] was applied to the acetylated dithioacetals by Wolfrom and Karabinos[33] with resultant complete reduction of the carbonyl group to the hydrocarbon stage; there was thus added another route to the few available for attaining this objective. This method was the key reaction in the configurational correlation of D-glyceraldehyde to L-(+)-alanine.[34]

$$\underset{\substack{\text{H}\diamond\text{NHAc} \\ \text{AcO}-\text{H} \\ \text{H}-\text{OAc} \\ \text{H}\diamond\text{OAc} \\ \text{CH}_2\text{OAc}}}{\text{HC(SEt)}_2} \quad\xrightarrow[]{\text{[H]}}\quad \xrightarrow[]{\text{OH}^-}\quad \underset{\substack{\text{H}\diamond\text{NHAc} \\ \text{HO}-\text{H} \\ \text{H}-\text{OH} \\ \text{H}\diamond\text{OH} \\ \text{CH}_2\text{OH}}}{\text{CH}_3} \quad\xrightarrow[]{\text{Pb(OAc)}_4,}\quad \xrightarrow[]{\text{[O]}}\quad \underset{\substack{\text{AcHN}\diamond\text{H} \\ \text{CH}_3}}{\text{CO}_2\text{H}}$$

Conditions for effecting partial reduction of aldose dithioacetals in this reaction were established by Jones and Mitchell.[35]

$$\underset{|}{\text{HC(SEt)}_2} \quad\longrightarrow\quad \underset{|}{\text{CH}_2\text{SEt}}$$

Horton and Jewell[36] have irradiated methanolic solutions of D-galactose diethyl dithioacetal and obtained **1**; further irradiation produced **2** and minor amounts of the sulfoxide (**3**) and galactitol (**4**).

$$\underset{|}{\text{HC(SEt)}_2} \quad\xrightarrow[\text{MeOH}]{h\nu}\quad \underset{\substack{| \\ \textbf{1}}}{\text{CH}_2\text{SEt}} \quad\xrightarrow{h\nu}\quad \underset{\substack{| \\ \textbf{2}}}{\text{CH}_3} + \underset{\substack{| \\ \textbf{3}}}{\overset{\overset{\text{O}}{\uparrow}}{\text{CH}_2\text{SEt}}} + \underset{\substack{| \\ \textbf{4}}}{\text{CH}_2\text{OH}}$$

III. MONOTHIOACETALS

Monothioacetals were obtained from the acyclic halide (**5**) (see Section I,B above for synthesis) in the D-galactose[28] and D-glucose[29] series by a Koenigs–Knorr type of condensation with silver carbonate and an alcohol; the S-ethyl O-methyl monothioacetals of D-galactose and D-glucose (**6**, R = Me) were so obtained; likewise the diethyl monothioacetal (**6**, R = Et) of D-galactose.

$$\underset{\substack{| \\ (\text{CHOAc})_4 \\ | \\ \text{CH}_2\text{OAc} \\ \textbf{5}}}{\text{HC}\overset{\text{SEt}}{\underset{\text{X}}{\big<}}} \quad\xrightarrow[\text{Ag}_2\text{CO}_3]{\text{ROH}}\quad \underset{\substack{| \\ (\text{CHOAc})_4 \\ | \\ \text{CH}_2\text{OH}}}{\text{HC}\overset{\text{SEt}}{\underset{\text{OR}}{\big<}}} \quad\longrightarrow\quad \underset{\substack{| \\ (\text{CHOH})_4 \\ | \\ \text{CH}_2\text{OH} \\ \textbf{6}}}{\text{HC}\overset{\text{SEt}}{\underset{\text{OR}}{\big<}}}$$

X = Br or Cl

* References start on p. 386.

These substances are of interest as possible intermediates in the formation of 1-thiofuranosides or furanosides from dithioacetals, and their probable presence in such reactions has been studied and in part confirmed.[21, 37]

IV. ACETALS

Although glycosides can be considered as monocyclic acetals, we shall here reserve the term acetal for the acyclic structures. The first example in the sugar series was encountered by Hudson and co-workers[38] in a specific reaction series beginning with the acetolysis of the anomeric methyl D-arabino-pyranosides.

$$
\begin{array}{ccc}
& \text{OAc} & \text{Cl} \\
\boxed{\begin{array}{c} \text{HCOMe} \\ | \\ \text{HOCH} \\ | \end{array}} \longrightarrow &
\begin{array}{c} | \\ \text{HCOMe} \\ | \\ \text{AcOCH} \\ | \end{array} \longrightarrow &
\begin{array}{c} | \\ \text{HCOMe} \\ | \\ \text{AcOCH} \\ | \end{array} \longrightarrow
\end{array}
$$

$$
\begin{array}{cc}
\begin{array}{c} \text{HC(OMe)}_2 \\ | \\ \text{AcOCH} \\ | \end{array} \longrightarrow &
\begin{array}{c} \text{HC(OMe)}_2 \\ | \\ \text{HOCH} \\ | \end{array}
\end{array}
$$

A general synthesis of acetals involves the "demercaptalation" of the acetylated dithioacetals by mercuric chloride in alcoholic solution and in the presence of cadmium carbonate.[39] This reaction was first effected with

$$
\begin{array}{ccc}
\begin{array}{c} \text{HC(SR)}_2 \\ | \\ \text{CHOAc} \\ | \end{array}
\xrightarrow[\substack{\text{HgCl}_2 \\ \text{CdCO}_3}]{\text{ROH}} &
\begin{array}{c} \text{HC(OR)}_2 \\ | \\ \text{CHOAc} \\ | \end{array} \longrightarrow &
\begin{array}{c} \text{HC(OR)}_2 \\ | \\ \text{CHOH} \\ | \end{array}
\end{array}
$$

D-galactose and was later extended to D-glucose.[40] As mentioned previously (Section II,B), the dimethyl acetal of D-fructose was obtainable in good yield by demercaptalation of the unsubstituted dithioacetal, in methanolic solution at −80°, with mercuric chloride and yellow mercuric oxide.[41] The acetals are very sensitive to acids, by which they are hydrolyzed (or alcoholyzed). With D-glucose dimethyl acetal and methanolic hydrogen chloride, the final product isolated was methyl α-D-glucopyranoside.[42] The pentamethyl ethers of the dimethyl acetals of D-glucose, D-mannose, and D-galactose have been described by Levene and Meyer.[43]

3,6-Anhydro-D-galactose forms a dimethyl acetal[44] whose enantiomorph has been isolated from the methanolyzate of agar.[45] Acetals of glycolaldehyde and glyceraldehyde have long been known.

A remarkable finding was the observation of Breddy and Jones[46] that anhydrous D-xylose underwent reaction with methanolic hydrogen chloride and benzaldehyde to form a highly insoluble, crystalline di-*O*-benzylidene dimethyl acetal. This reaction could serve as a specific spot test and quantitative assay for the sugar.

Although Bishop and Cooper[45a] claimed to have proved the absence of the dimethyl acetal in the Fischer glycosidation reaction, more-sensitive radiochemical methods have demonstrated its presence.[45b, 45c] Kinetic data have been presented[45b] purporting to show that the acyclic acetal is not an obligatory intermediate, as Fischer[45d] had first suggested.

V. ALDEHYDO STRUCTURES

This section is concerned with: (1) the available methods of synthesis for obtaining aldehydo sugar derivatives by the use of appropriate blocking groups, and (2) the formation of such substances, occasionally encountered, in the reactions of cyclic aldoses. In Section VI the acyclic ketose structures are similarly considered. Hydroxyl protective groups utilized have been ethers, esters, and cyclic acetals.

A. FORMATION

1. *Demercaptalation*

The most general method for obtaining aldehydo sugar derivatives is by demercaptalation, by which is meant the hydrolysis of appropriately substituted dithioacetals. This was first effected with the methylated dithioacetals, and pentamethyl ethers of *aldehydo*-D-glucose, -D-galactose, and -D-mannose were obtained as distilled syrups.[43] Crystalline products became available in the acetate series by the use of mercuric chloride and cadmium carbonate in acetone solution.[47]

$$
\begin{array}{c}
HC(SEt)_2 \\
| \\
(CHOAc)_n \\
| \\
CH_2OAc
\end{array}
\quad
\xrightarrow[\text{CdCO}_3]{2HgCl_2}
\quad
\begin{array}{c}
HC{=}O \\
| \\
(CHOAc)_n \\
| \\
CH_2OAc
\end{array}
+ 2\ EtSHgCl + CdCl_2 + CO_2
$$

Although the above equation does not require any water for balancing, water is actually involved and is regenerated in equal amount from the carbonate. Very minimal amounts of water are required, and reagent-grade acetone may be used without any added water. The EtSHgCl formed is insoluble. The reaction requires a considerable excess of cadmium carbonate,[48] and it is inter-

* *References start on p. 386.*

esting that Schneider and Sepp,[49] who did not use such an excess, stated that the reaction failed. The cadmium salt appears to be a promoter,[50] and several other carbonates were not useful. The mercuric chloride–cadmium carbonate demercaptalation of acetylated dithioacetals has been extended to a number of aldoses, including maltose[51] and 2-acetamido-2-deoxy-D-glucose;[52] *aldehydo-*D-ribose tetrabenzoate has been reported.[53]

Brigl and Mühlschlegel[19a] demercaptalated benzoylated D-glucose diethyl dithioacetal by hydrolysis with formic acid, but this method has not been further utilized. Demercaptalation of acetylated dithioacetals by a controlled reaction with bromine[54, 55] is a useful and convenient method. Cyclic acetals can be employed for protecting the hydroxyl groups in the synthesis of acyclic sugar derivatives containing the free carbonyl group. These groups are stable to alkali. They can be used to substitute all hydroxyl groups in a pentose, but for a hexose an additional substituent must be introduced. Thus, 6-*O*-benzoyl-2,3,4,5-di-*O*-benzylidene-*aldehydo*-D-glucose has been prepared from the corresponding dithioacetal.[56] The 6-*O*-trityl group has also been so utilized.

2. *Other Methods*

After methods for obtaining the acylated aldonyl chlorides became available (see Section VI), the Rosemund[57] reduction was successfully used. *aldehydo-*D-Glucose pentaacetate[58] and tri-*O*-benzoyl-DL-threose[59] have been prepared in this manner; a dialdose can be formed.[60]

Acetylation of aldose oximes or semicarbazones may lead to acyclic acetates. When this occurs, the aldehydo form of the aldose acetate may be

$$
\begin{array}{ccc}
\text{HC=NHNHCONH}_2 & & \text{HC=O} \\
| & \xrightarrow{\ \text{HNO}_2\ } & | \\
\text{(CHOAc)}_n & & \text{(CHOAc)}_n \\
| & & | \\
\text{CH}_2\text{OAc} & & \text{CH}_2\text{OAc} \\
& & \uparrow\ \text{HNO}_2 \\
\text{HC=NOAc} & & \text{HC=NOH} \\
| & \xrightarrow{\ \text{(CO}_2\text{H)}_2\ } & | \\
\text{(CHOAc)}_n & & \text{(CHOAc)}_n \\
| & & | \\
\text{CH}_2\text{OAc} & & \text{CH}_2\text{OAc}
\end{array}
$$

obtained by the action of nitrous acid.[61] An interesting reaction of this general type was reported by Helferich and Schirp;[62] aldehydo forms of acetylated D-glucose, D-galactose, and D-arabinose were prepared.

$$\begin{array}{ccc} & & \overset{\overset{H}{|}}{\underset{\diagdown Ac}{C=O}} \\ HC=N-N & & \\ | & & \\ (CHOAc)_n & \xrightarrow{\;SeO_2\;} \\ | & & \\ CH_2OAc & & \end{array} \qquad \begin{array}{c} HC=O \\ | \\ (CHOAc)_n \\ | \\ CH_2OAc \end{array}$$

Glycol scission with metaperiodate ion or lead tetraacetate in partially substituted alditols or sugars may lead to aldehydo derivatives; some may be dialdose in nature. Thus, scission of di-O-isopropylidene-(−)-inositol with lead tetraacetate led to crystalline 1,2:3,4-di-O-isopropylidene-L-*manno*-hexodialdose.[63]

Ozonolyses of benzoates of suitable unsaturated polyhydric alcohols have produced aldehydo benzoates of glycolaldehyde and DL-glyceraldehyde.[64] Desulfurization of certain thiol esters with partially deactivated Raney nickel leads to aldehydo acetates.[65] The Raney nickel can be deactivated for this purpose by refluxing with acetone.

Racemic forms of the dimethyl acetals of the 2-deoxypentoses were synthesized from an acetylenic vinyl ether.[66]

$$HC{\equiv}CCH{=}CHOMe \xrightarrow[+\,H_2CO]{MeOK} HOCH_2C{\equiv}CCH_2CH(OMe)_2 \longrightarrow$$

$$HOCH_2CH{=}CHCH_2CH(OMe)_2 \longrightarrow HOCH_2(CHOH)_2CH_2CH(OMe)_2$$

It may be mentioned that in one case (D-*glycero*-D-*galacto*-heptose) direct acetylation of the sugar led to the formation of its aldehydo hexaacetate.[67] It can be predicted that this should occur with other higher sugars, probably because of the hindrance to ring closure in the long, extendable chain. It is established that 3,6-anhydro-2,4-di-O-methyl-D-glucose[68] (**7** and the D-galactose analog[44]) are aldehydo forms, because a structure in which a 3,6-anhydro ring and a pyranose ring are present together is in a state of molecular strain. 2,5-Anhydro-D-mannose (**8**) also exists as an aldehydo form because of strain in the bicyclic structure.[69, 70] In the case of 3,6-anhydro-

7   **8**

D-galactose, the dimethyl acetal is formed under the Fischer glycosidation conditions even though the hydroxyl group on C-5 is available for pyranose-ring formation.[44] Acid hydrolysis of agar[70a] produced the disaccharide

* *References start on p. 386.*

isoagarobiose (**8a**), which is an acetal of 3,6-anhydro-*aldehydo*-L-galactose with the C-5 and C-6 hydroxyl groups of a D-galactopyranose molecule. Compound **8a** was considered to be an artifact (acid reversion product).

8a

B. Aldehydrol Derivatives

HC⟨OH,OH	HC⟨OH,OR	HC⟨OH,SR	HC⟨OR,OR	HC⟨SR,SR	HC⟨SR,OR
9	**10**	**11**	**12**	**13**	**14**

HC⟨OAc,OAc	HC⟨OR,OAc	HC⟨SR,OAc	HC⟨OR,X	HC⟨Cl,Cl	HC⟨OAc,X
			X = Cl, Br		X = Br, Cl, I
15	**16**	**17**	**18**	**19**	**20**

HC⟨NHCR(=O),NHCR(=O)	HC⟨B,OR	HC⟨B,SR	HC⟨NHPh,NHPh	HC⟨OH,SO₂⁻K⁺	HC⟨NHR,SO₂⁻K⁺
	B = pyrimidine or purine base				
21	**22**	**23**	**24**	**25**	**26**

The above formulas show that eighteen types of crystalline derivatives have been obtained that may be considered as derived from the hydrated form of the aldehyde (aldehydrol). They are best named as a derivative of this aldehydrol. In the ketose series only types **12** and **13** are established. Types **11–14** are the acetals and thioacetals and have been already discussed. The more stable ones, **12–14, 21–23, 25,** and **26** are known with the hydroxyl stem unsubstituted. The others are known only with the stem hydroxyl groups protected, generally with the acetate function. These derivative types are not at all unique with the sugars but have been well established in general aldehyde chemistry. Possibly the nitrogen functions **21–24** may be unique.

Type **9** (compare chloral) has been established in the *aldehydo*-pentaacetates of D-galactose[71] and D-mannose.[72] As expected, they do not mutarotate in water.[72] Type **10** (hemiacetal) has been reported for the acetylated forms of D-galactose, 6-deoxy-L-galactose, D-mannose, maltose, and D-galacturonic acid, and possibly for D-glucose pentabenzoate.[19a] The predictable two forms have been isolated for the ethyl hemiacetals of methyl tetra-*O*-acetyl-*aldehydo*-D-galacturonate[73] and *aldehydo*-D-galactose pentaacetate.[74] Their mutarotation characteristics in chloroform are unusual (Fig. 1). The more stable form

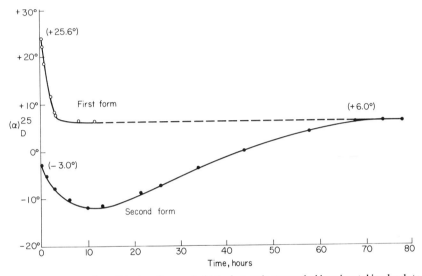

FIG. 1. Mutarotation of the two forms of *aldehydo*-D-galactose ethyl hemiacetal in absolute chloroform.

exhibits a slow mutarotation of the A $\rightleftharpoons$ B $\rightleftharpoons$ C type, whereby the intermediate B, very probably the aldehydo form, is so detectable. The methyl hemiacetal of *aldehydo*-D-galactose pentaacetate exhibits a typical aldose mutarotation curve in methanol (Fig. 2),[75] although no cyclic structures are present.

No suitable nomenclature has been proposed for the two isomeric forms epimeric on C-1, and their absolute configurations are unknown. Should the configurations be established, the isomers could be named on the Cahn–Ingold–Prelog system.

Types **11** and **17** have been characterized as intermediates in syntheses of **14**. Hexa-*O*-acetyl-1-*S*-ethyl-1-thio-D-glucose aldehydrol (type **17**) was prepared in two forms having $[\alpha]_D$ +12.5° and −1.8° (CHCl$_3$).[29] The single example for

* *References start on p. 386.*

19 was obtained by the action of phosphorus pentachloride on *aldehydo*-D-galactose pentaacetate.[76]

Examples of type **20** are known in several acetylated sugar structures, and some have been isolated in the predictable two forms. Like the cyclic, acetylated glycosyl halides, the optical rotations of these forms are quite large. Thus, the

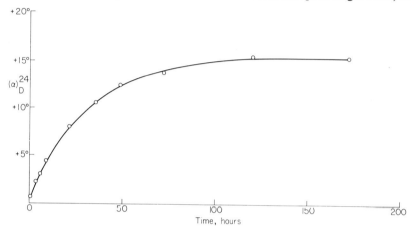

FIG. 2. Mutarotation of the methyl hemiacetal of *aldehydo*-D-galactose pentaacetate in methanol solution.

two forms of penta-*O*-acetyl-1-bromo-1-deoxy-D-arabinose aldehydrol show $[\alpha]_{22}^{D}$ +136° and −71° in chloroform.[77] The two forms of hexa-*O*-acetyl-1-chloro-1-deoxy-D-galactose aldehydrol are interconvertible in acetyl chloride containing zinc chloride to an equilibrium mixture containing ~75% of the dextrorotatory form and ~25% of the other.[78] The enantiomorph of penta-*O*-acetyl-1-bromo-1-deoxy-D-arabinose aldehydrol was obtained by Felton and Freudenberg[79] as a by-product (5 grams) in the direct treatment of L-arabinose (100 grams) with cold acetic anhydride and hydrogen bromide to form tetra-*O*-acetyl-L-arabinopyranosyl bromide. The derivatives of type **20** have not been found useful as intermediates in further reactions.

The known members of type **18** are useful for further synthetic reactions. They are examples of what has been designated α-halogen ethers in general aldehyde chemistry. Such halides are very reactive. The first example of this type of derivative was obtained by Hudson and co-workers[38] through an unusual reaction in the D-arabinose structure. Methyl β-D-arabinopyranoside was subjected to acetolysis with acetic acid–acetic anhydride–sulfuric acid, and there were isolated two forms of type **16** (R = Me) having rotations (D-line of sodium) in chloroform of +27° and +35°. The 1-acetoxy group of each of these isomers was replaced with chlorine by aluminum chloride to form the two isomeric halides, which were used to form the dimethyl acetal of D-arabinose by reaction with methanol and silver oxide followed by deacetylation.

The rotatory values of a number of isomeric pairs of types 16 and 20 have been recorded by Wolfrom and Brown,[77] and the Hudson isorotation rule calculations were applied to them.

Representatives of types 22 and 23, which may be designated acyclic nucleoside analogs, have been synthesized in the hope, so far unrealized, that they might be cyclized to furanose nucleosides. A typical synthesis follows (the sugar structure was D-galactose).[80]

HC(SEt)$_2$
|
(CHOAc)$_4$ $\xrightarrow{Br_2}$
|
CH$_2$OAc

HCBr(SEt)
|
(CHOAc)$_4$ $\xrightarrow[Ag_2CO_3]{MeOH}$
|
CH$_2$OAc

HC(OMe)(SEt)
|
(CHOAc)$_4$ $\xrightarrow{Br_2}$
|
CH$_2$OAc

HCBr(OMe)
|
(CHOAc)$_4$ +
|
CH$_2$OAc

$\longrightarrow$ $\longrightarrow$

CHOMe
|
(CHOH)$_4$
|
CH$_2$OH

Two forms
$[\alpha]_D$ +24°, −24° (H$_2$O)

A form of type 23 in the same sugar structure was obtained (R = Et and B = adenine),[81] and a thymine member of type 22 (R = Me and CH$_2$Ph) was also prepared.[82]

The fully acetylated members of type 15 may be obtained by acetylation of the aldehydo acetates; they are also encountered when cyclic sugar glycosides and polysaccharides are subjected to acetolysis. Hepta-O-acetyl-D-glucose aldehydrol[83] crystallizes especially well. It has previously been noted that these structures were obtained on the acetolysis of the aldose dithioacetals (Section III). Hepta-O-acetyl-D-glucose aldehydrol was isolated by Freudenberg and Soff[84] on acetolysis of methyl tetra-O-acetyl-D-glucosides (pyranose and furanose) in yields of 2 to 14%; the principal product was α-D-glucopyranose pentaacetate. Higher yields are obtainable with a perchloric acid catalyst, 80% of the heptaacetate being formed from benzyl tetra-O-acetyl-β-D-glucopyranoside.[85] Hexa-O-acetyl-D-arabinose aldehydrol is formed on acetolysis of methyl β-D-arabinopyranoside.[38] Whistler and associates[86] obtained the

* *References start on p. 386.*

corresponding (type **15**) acetates of D-galactose and D-xylose on acetolysis of the polysaccharide guaran, in yields of 3% and 10%, respectively.

Pirie[87] obtained racemic hepta-O-acetyl-DL-galactose on acetolysis of agar, and Micheel and co-workers[88] obtained the same substance by the zinc chloride-catalyzed acetylation of tetra-O-acetyl-6-deoxy-6-iodo-*aldehydo*-D-galactose. The galactose structure is unusual in that interchange of end groups leads to the enantiomorph. Apparently this may have occurred in these reactions, although it is true that agar contains both D- and L-galactose.

Type **24** is a very interesting structure and has been obtained in crystalline form with the 2,4:3,5-di-O-ethylidene-L-xylose stem.[89]

The diacetamido derivatives of the aldose sugars (type **21**) were encountered by Wohl[90] when he established his method for degradation of aldoses. They are thus easily the first acyclic aldehyde derivatives of the sugars to be obtained. They were not considered unusual in 1893, as at that time the ring structures were not established. Wohl obtained them on treatment of the acetylated nitriles with ammoniacal silver. These diacetamido compounds then required rather vigorous acid-hydrolytic conditions to obtain the desired lower-carbon sugar. Brigl and co-workers[19b] obtained 1,1-dibenzamido-1,1-dideoxy-D-glucose aldehydrol ("D-glucose dibenzamide") on ammonolysis of *aldehydo*-D-glucose pentabenzoate and 3,5,6-tri-O-benzoyl-D-gluco-furanose. Isbell and Frush[91] formed 1,1-diacetamido-1,1-dideoxy-L-arabinose aldehydrol by ammonolysis of *aldehydo*-L-arabinose tetraacetate, and they proposed a reaction mechanism, involving ortho ester-type intermediates, to explain its formation. Deulofeu and Deferrari[92] obtained a 21% yield of the 1,1-dibenzamido derivative of D-glucose by ammonolysis of α-D-gluco-pyranose pentabenzoate. Deulofeu, Deferrari, and associates[93] have studied the production of these aldose 1,1-diamido compounds by the ammonolysis of many other sugar structures. An interesting result was the finding by Hockett and Chandler[94] that N-acetyl-α-D-glucofuranosylamine was formed by the action of aqueous ammonia upon *aldehydo*-D-glucose pentaacetate and upon hexa-O-acetyl-D-*glycero*-D-*gulo*-heptononitrile.

Aldoses react rapidly and reversibly with bisulfites to form addition compounds[95,96] formulated as **25**, for which structure there is considerable evidence. These bisulfite compounds react[96] with amines (**26**). Such a bisulfite addition compound was obtained from *aldehydo*-D-glucose pentabenzoate.[19a]

VI. ACYCLIC ALDONIC ACID DERIVATIVES

The salts, esters, amides, and nitriles of the aldonic acids are acyclic. The esters were obtained by Nef and have a surprising stability. The aldonic acids form 1,5- and 1,4-lactones, the exocyclic carbonyl group causing these rings to have the reverse stability of that exhibited by the corresponding cyclic hemiacetals.

In this section attention is devoted to those aldonic acids in which all the stem hydroxyl groups have been substituted. The first example of this type of derivative was the penta-O-acetyl-D-gluconic acid of Major and Cook,[97] which was prepared by oxidation of *aldehydo*-D-glucose pentaacetate with buffered bromine and also by the zinc chloride-catalyzed acetylation of tetra-O-acetyl-D-gluconic acid monohydrate (obtained in turn by the acetylation of D-glucono-1,5-lactone). They were able to esterify this acid derivative with ethanol and hydrogen chloride. Methyl esters of these substances are readily obtained with diazomethane. These workers[98] also described the formation of penta-O-acetyl-D-gluconyl chloride by the action of phosphorus pentachloride on the acetylated acid; thionyl chloride can also be used for this purpose.

Gätzi and Reichstein[98a] oxidized 2,3:4,5-di-O-isopropylidene-*aldehydo*-D-arabinose with alkaline permanganate and obtained 2,3:4,5-di-O-isopropylidene-D-arabinonic acid. The alkali stability of the O-isopropylidene group is noteworthy.

A general method for synthesizing the acetylated acids is by deamination of the acetylated aldonamides with dinitrogen trioxide[99] or nitrosyl chloride.[100]

The amines on acetylation ordinarily do not undergo acetylation on the amide nitrogen; an exception was found with L-threonamide.[101, 102] Most of the aldonamides are unstable in aqueous solution wherein they form ammonium salts, perhaps through the formation of ring structures. The aldotetron-

amides are stable toward water. The corresponding postulated intermediate would in their case have a furanose structure.

A quite general method for obtaining the acetylated acids is by acetylation of their cadmium salts with acetic anhydride and hydrogen chloride.[103]

* *References start on p. 386.*

2-Acetamido-tetra-O-acetyl-2-deoxy-D-gluconic acid[104] and the pentanitrates of D-gluconic and D-galactonic acids[105] are known.

Penta-O-acetyl-D-gluconyl chloride was used to completely esterify the hydroxyl groups of several alditols and of methyl α-D-glucopyranoside.[106] An acyclic ortho ester was produced, as a minor product, in the following reaction, the sugar structure being D-galactose.[107]

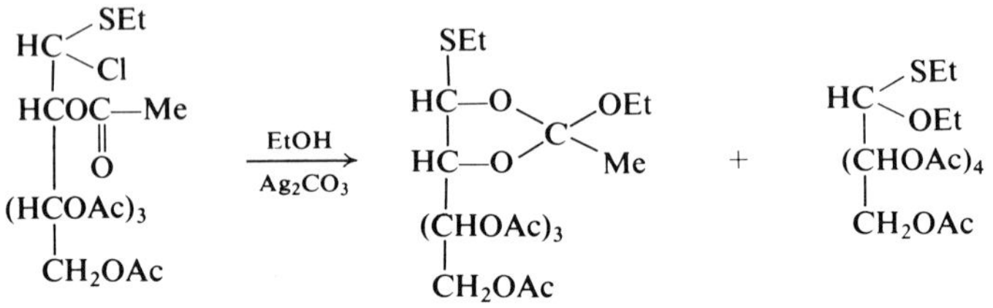

It has long been known that direct acetylation of galactaric acid gives a tetraacetate[108] that can be converted into tetra-O-acetylgalactaryl dichloride.[109] Methyl tetra-O-acetyl-D-*arabino*-hexulosonate (**27**) and penta-O-acetyl-D-*gulo*-heptulosonic acid were synthesized by Major and Cook.[98]

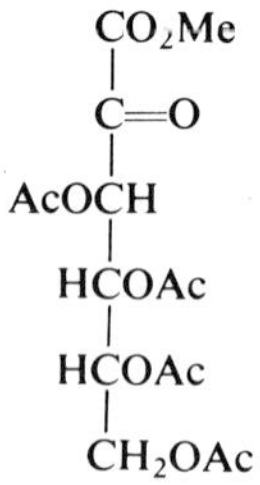

$$
\begin{array}{c}
CO_2Me \\
| \\
C{=}O \\
| \\
AcOCH \\
| \\
HCOAc \\
| \\
HCOAc \\
| \\
CH_2OAc
\end{array}
$$

27

VII. KETO STRUCTURES

A. FORMATION

Acyclic sugar derivatives containing the 2-keto group can be prepared through the diazomethyl ketone obtained in turn from a blocked acyclic aldonyl chloride (Section VI). Gätzi and Reichstein[110] obtained a diazomethyl ketone from di-O-ethylidene-L-xylonyl chloride, and Iwadare[111] obtained one from O-isopropylidene-D-glyceryl chloride, both preparations being syrups. A more general reaction, leading to crystalline products in most cases, was found in the acetate series.[112] The reaction sequence constitutes a higher-carbon sugar synthesis and leads to a synthesis of ketoses and of their keto acetates from the next lower aldonic acid.[113] By this synthesis many ketoses and their keto acetates were prepared; L-fructose was obtained in crystalline

$$\begin{array}{c}\text{C}\!=\!\text{O}\\|\\(\text{CHOH})_3\\|\\\text{CHO}\\|\\(\text{CHOH})_n\\|\\\text{CH}_2\text{OH}\end{array}\quad\xrightarrow{\text{NH}_3}\quad\begin{array}{c}\text{NH}_2\\|\\\text{C}\!=\!\text{O}\\|\\(\text{CHOH})_n\\|\\\text{CH}_2\text{OH}\end{array}\quad\longrightarrow\quad\begin{array}{c}\text{NH}_2\\|\\\text{C}\!=\!\text{O}\\|\\(\text{CHOAc})_n\\|\\\text{CH}_2\text{OAc}\end{array}\quad\xrightarrow{\text{N}_2\text{O}_3}\quad\begin{array}{c}\text{OH}\\|\\\text{C}\!=\!\text{O}\\|\\(\text{CHOAc})_n\\|\\\text{CH}_2\text{OAc}\end{array}\quad\longrightarrow$$

$$\begin{array}{c}\text{Cl}\\|\\\text{C}\!=\!\text{O}\\|\\(\text{CHOAc})_n\\|\\\text{CH}_2\text{OAc}\end{array}\quad\xrightarrow{\text{CH}_2\text{N}_2}\quad\begin{array}{c}\text{CHN}_2\\|\\\text{C}\!=\!\text{O}\\|\\(\text{CHOAc})_n\\|\\\text{CH}_2\text{OAc}\end{array}\quad\xrightarrow[\text{Cu}^{2+}]{\text{HOAc}}\quad\begin{array}{c}\text{CH}_2\text{OAc}\\|\\\text{C}\!=\!\text{O}\\|\\(\text{CHOAc})_n\\|\\\text{CH}_2\text{OAc}\end{array}\quad\xrightarrow{\text{Ba(OH)}_2}\quad\begin{array}{c}\text{CH}_2\text{OH}\\|\\\text{C}\!=\!\text{O}\\|\\(\text{CHOH})_n\\|\\\text{CH}_2\text{OH}\end{array}$$

form as its hemihydrate, and several new higher ketoses were produced, including a ketononose.

Oxidation of a partially benzylated alditol with an open C-2 hydroxyl group leads to the synthesis of a benzylated *keto*-sugar.[113a] The partially benzylated alditol is obtained by reduction of the corresponding partially benzylated aldose.

Acylation of the ketoses generally tends to produce mixtures of products in which the acyclic derivative is, in most cases, a major constituent. This was first found when the long-known D-fructose pentaacetate of Hudson and Brauns[114] was shown to be the acyclic *keto*-D-fructose pentaacetate.[115,116] Brigl and Schinle[117] isolated three crystalline benzoates by the direct benzoylation of D-fructose and furnished good evidence that they were:

β-D-Fructopyranose
tetrabenzoate

D-Fructofuranose
tetrabenzoate

$$\begin{array}{c}\text{CH}_2\text{OBz}\\|\\\text{C}\!=\!\text{O}\\|\\\text{BzOCH}\\|\\\text{HCOBz}\\|\\\text{HCOBz}\\|\\\text{CH}_2\text{OBz}\end{array}$$

keto-D-Fructose
pentabenzoate

* *References start on p.* 386.

One of the octaacetates of the disaccharide turanose has been proved to be a keto form.[115]

If a pyranose ring cannot form in the 1-arylaminoketose derivative obtained in the "Amadori rearrangement," the acyclic structure shown is present in the product.[118-120]

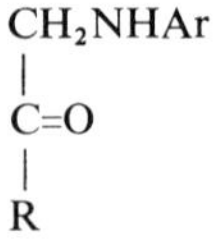

$$
\begin{array}{c}
CH_2NHAr \\
| \\
C=O \\
| \\
R
\end{array}
$$

B. Reactivity of Diazomethyl Ketones

The diazomethyl ketones of the sugar acetate series are stable, crystalline compounds with a light-yellow color. 1-Deoxy-1-diazo-*keto*-D-*galacto*-heptulose (**28**) was obtained on deacetylation of its pentaacetate.[121]

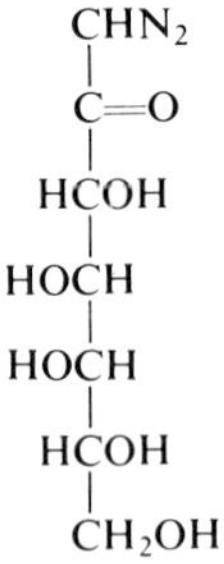

$$
\begin{array}{c}
CHN_2 \\
| \\
C=O \\
| \\
HCOH \\
| \\
HOCH \\
| \\
HOCH \\
| \\
HCOH \\
| \\
CH_2OH
\end{array}
$$

28

These diazomethyl ketones show the reactivity expected of such a function. They undergo reaction with acids, the rate being dependent upon the strength of the acid. Acetic acid reacts rather poorly, but the halogen acids react well and the 1-deoxy-1-halo derivatives of the ketoses are thus readily available. In the synthesis of diazomethyl ketoses an excess of diazomethane must be

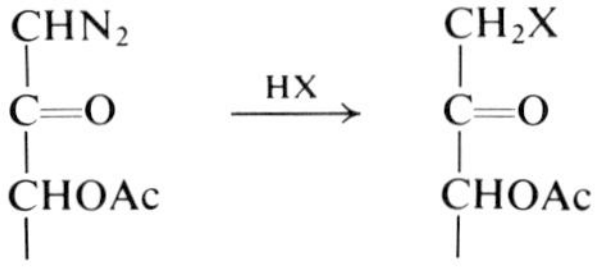

$$
\begin{array}{ccc}
CHN_2 & & CH_2X \\
| & & | \\
C=O & \xrightarrow{HX} & C=O \\
| & & | \\
CHOAc & & CHOAc \\
| & & |
\end{array}
$$

used to avoid the formation of 1-deoxy-1-halo derivatives. The 1-bromo compounds may be converted into the 1-acetates by the action of acetic anhydride and potassium acetate.[113]

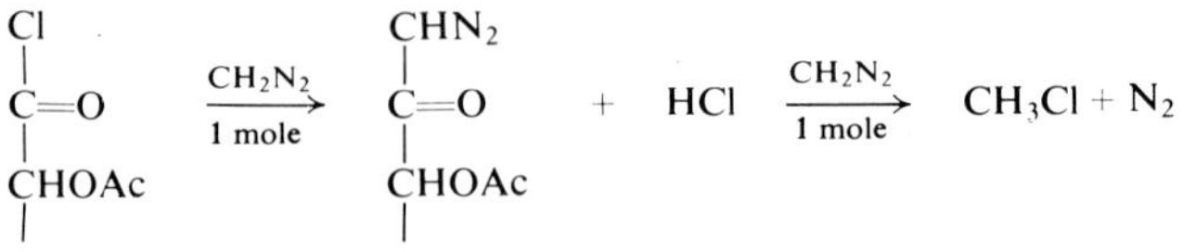

$$
\begin{array}{ccccccc}
Cl & & CHN_2 & & & & \\
| & & | & & & & \\
C=O & \xrightarrow[\text{1 mole}]{CH_2N_2} & C=O & + & HCl & \xrightarrow[\text{1 mole}]{CH_2N_2} & CH_3Cl + N_2 \\
| & & | & & & & \\
CHOAc & & CHOAc & & & & \\
| & & | & & & &
\end{array}
$$

The diazomethyl ketones undergo the Wolff rearrangement.[122] They can be reduced to the 1-deoxy derivatives by hydrogen iodide.[122] In the presence

$$
\begin{array}{c}
\mathrm{CHN_2} \\
| \\
\mathrm{C{=}O} \\
|
\end{array}
\quad
\xrightarrow[\mathrm{HOH}]{\mathrm{Ag_2O}}
\quad
\begin{array}{c}
\mathrm{CO_2H} \\
| \\
\mathrm{CH_2} \\
|
\end{array}
\quad + \quad \mathrm{N_2}
$$

of light, two atoms of iodine were introduced into penta-*O*-acetyl-1-deoxy-1-diazo-*keto*-D-*galacto*-heptulose.[123] It is possible to reduce the diazomethyl

$$
\begin{array}{c}
\mathrm{CHN_2} \\
| \\
\mathrm{C{=}O} \\
| \\
\mathrm{CHOAc} \\
|
\end{array}
\quad
\xrightarrow[h\nu]{\mathrm{I_2}}
\quad
\begin{array}{c}
\mathrm{CHI_2} \\
| \\
\mathrm{C{=}O} \\
| \\
\mathrm{CHOAc} \\
|
\end{array}
\quad + \quad \mathrm{N_2}
$$

ketones without loss of nitrogen.[124]

$$
\begin{array}{c}
\mathrm{HCN_2} \\
| \\
\mathrm{C{=}O} \\
|
\end{array}
\quad
\xrightarrow{\mathrm{HS^-}}
\quad
\begin{array}{c}
\mathrm{HC{=}NNH_2} \\
| \\
\mathrm{C{=}O} \\
|
\end{array}
$$

Penta-*O*-acetyl-1-deoxy-1-diazo-*keto*-D-*gluco*-heptulose (**29**) was converted in boiling benzene, in the presence of cupric oxide, into crystalline 1,2-bis-(penta-*O*-acetyl-D-gluconyl)ethylene (**31**), presumably by dimerization of a postulated carbene (divalent carbon) intermediate[125] (**30**).

By the same reaction with the D-galacto analog of **29**, a carbohydrate side-chain was attached to the β position of indole (**32**) and to the α position of thiophene (**33**); both compounds were crystalline.[125]

* References start on p. 386.

C. Acyclic Diketose Structures

Tetra-O-acetylgalactaryl dichloride (Section VI) was converted into the bis(diazomethyl) ketone (34), and this on reaction with acetic acid gave hexa-O-acetyl-*galacto*-2,7-octodiulose[113] (35) and, on reduction, tetra-O-acetyl-1,8-dideoxy-*galacto*-2,7-octodiulose[126] (36).

VIII. GLYCOSULOSES

The outstanding examples of acyclic glycosulose (aldose–ketose) derivatives are the sugar osazones (see Chapter 21). They are 1,2-bis(arylhydrazones) of aldosuloses and are now known to have the acyclic sugar chain and a chelated ring, as shown. This was established by chemical evidence[127] (lack of formazan

formation except under alkaline conditions sufficiently strong to disrupt the chelate ring) and by nuclear magnetic resonance spectroscopy.[128] The O-acetylosazones have been characterized.

New oxidizing agents[129] have allowed the insertion of carbonyl groups into protected glycosides of aldoses (see Vol. IB, Chapter 23). Related oxidation products may exist as acyclic keto forms. The structures of the polysaccharide oxidation products, termed "dialdehydes," while not glycosulose in nature, are not established with certainty. It is known that the D-glucan "dialdehydes" can be reduced stepwise. This has led to the suggestion that they may have a *p*-dioxane structure (37) in which the one aldehyde group may be hydrated (see Vol. IB, Chapter 25).

37

IX. GENERAL PROPERTIES OF ACYCLIC DERIVATIVES

A. Physical Properties†

1. *Ultraviolet and Infrared Absorption*

The ultraviolet absorption spectra of several aldehydo and keto acetates were measured in absolute chloroform solution by Lowry and associates.[130] They all showed strong bands for the characteristic carbonyl absorption at 290 nm. Bredereck and co-workers[131] found that keto acetates show strong absorption, near this wavelength, in methanol solution. Therefore, the carbonyl group of the keto acetates must undergo little, if any, hemiacetal formation in methanol solution. The ester-carbonyl absorption of the acyclic (aldehydo and keto) acetates masks the carbonyl-group absorption in the 1667- to 1818-cm^{-1} region[132] of their infrared spectra, but the 4,6-*O*-benzylidene derivatives of the acyclic 1-arylamino-1-deoxy-ketohexoses show this absorption (1715 to 1725 cm^{-1}).[118] Isbell and co-workers,[132] from infrared data, postulated hydrogen bonding of hydroxyl hydrogen atoms with the carbonyl oxygen atom of the 2-acetoxyl group in the aldehydrol and ethyl hemiacetal of *aldehydo*-D-galactose pentaacetate.

2. *Optical Rotatory Dispersion and Circular Dichroism*

Lowry and associates[130] measured the optical rotatory dispersion and circular dichroism (ellipticity) of several aldehydo and keto acetates, in absolute chloroform solution, in the region of 250 to 650 nm, where they all displayed

† See also Vol. IB, Chap. 27.

* *References start on p. 386.*

strong Cotton effects. These measurements were made with a Werner Kuhn photographic spectropolarimeter, and pure spectral lines from metal arcs were used as light sources. Quartz polarimeter tubes were employed. The optical rotatory dispersion of these derivatives included (1) a levorotation, due to the induced asymmetry of the aldehydic group, and (2) a dextrorotation (for the D series) with a characteristic frequency in the Schumann region, which was attributed to the fixed asymmetry of the $>$CHOAc groups. In the cases of tetra-O-acetyl-*aldehydo*-L-arabinose and penta-O-acetyl-*keto*-D-fructose, where the three asymmetric carbon atoms have the same relative configuration, but are of opposite sign, the Schumann terms canceled out and the whole of the observed rotation was due to the aldehydic group.

3. Nuclear Magnetic Resonance

A number of applications of this technique to the acyclic sugar derivatives have been reported.[132a] The signals for the aldehydic protons of 1,2:3,4-di-O-isopropylidene-α-D-*galacto*-hexodialdo-1,5-pyranose and 4-acetamido-4,5-dideoxy-2,3-O-isopropylidene-L-xylose appeared in chloroform-*d* at low field as singlets at τ 0.24 and 0.67, respectively.[133] The technique provides a convenient method for observing aldehyde–aldehydrol equilibria in aqueous media.[133c] The H-1 signal for tetra-O-benzoyl-L-arabinose diethyl dithioacetal appeared as a doublet, $J_{1,2}$ 6.9 Hz, at a higher field than the signals for all other hydrogen atoms on the chain.[134] Analysis of the n.m.r. spectra of acetylated diethyl dithioacetals of the pentoses and some 6-deoxyhexoses has provided information on the conformations of substituted acyclic sugar chains.[134a] *trans*-2,3-Deoxy-4,5:6,7-di-O-isopropylidene-*aldehydo*-L-*arabino*-hept-2-enose

$$\text{(possessing the grouping } -\overset{\text{H}}{\underset{\text{H}}{\text{C}}}{=}\overset{\text{H}}{\text{C}}-\text{C}{=}\text{O} \text{) showed the H-1 signal as a doublet,}$$

$J_{1,2}$ 7 Hz.[133a]

4. Mass Spectra[135]

The mass spectra of the sugar dithioacetals contain relatively intense peaks of molecular ion, so that the molecular weight of a monosaccharide can be directly determined by using the mass-spectral method.[136] This is an important advantage, as, in the monosaccharide series, the molecular weight can be determined only indirectly for almost all of the compounds except this type of derivative. It is possible to relate the characteristics of the mass spectra to the position of deoxy groups in these molecules and to distinguish between aldoses and ketoses. The dithioacetals have been used for allocating the acetamido group in acetamido sugars by the mass-spectral method.[137]

The fragmentation of aldehydo and keto derivatives begins with the ionization of the carbonyl group. In the mass spectrum of *aldehydo*-D-arabinose tetraacetate,[138] as a typical example, the molecular ion (M^{+}) is not observed, but peaks M + 1 and M + 43 are detected. The corresponding ions are due to ion–molecule collisions. The most intense peak of the spectrum is that at m/e 43 (CH_3CO^+).

B. CHEMICAL PROPERTIES

1. Introduction

In preceding sections, many of the chemical properties of the acyclic sugar derivatives have been treated, but emphasis was placed upon their formation by deliberate synthetic methods or as products in reactions of the cyclic sugars. Further reactions, some of which are common to both aldehydo and keto forms, are discussed in this section.

2. Color Tests

Like other aldehydes, the aldehydo derivatives give a Schiff test. Since most of these substances are not very soluble in water, a sensitive reagent[139] is to be recommended. Pacsu and co-workers[115, 140] have described a color test characteristic of a keto acetate and not given by an aldehydo acetate.

3. Oxidation–Reduction

The oxidation of aldehydo acetates to the acetylated aldonic acids has been noted in Section VI. Reduction of the carbonyl groups in both aldehydo and keto acetates is not easily done, probably because of steric hindrance from the acetate groups. Pacsu and Rich[115] effected the reduction[115] of keto-D-fructose pentaacetate and keto-turanose octaacetate, in ethanolic solution, with platinum and hydrogen. keto-L-Sorbose pentaacetate was reduced in ethereal suspension with the same catalyst.[140] As with ketoses in general, a mixture of two alditols was obtained in all of these cases.

aldehydo-D-Ribose tetraacetate could not be reduced with a platinum catalyst but was successfully reduced, in p-dioxane solution, with Raney nickel and hydrogen at room temperature, or slightly above, and with 40 lb in.$^{-2}$ pressure of hydrogen. The 2,3,4,5-tetra-O-acetyl-D-ribitol was crystalline.[141]

4. Reactions with Nitrogen Compounds

The aldehydo acetates react readily with hydroxylamine and semicarbazide to form oximes and semicarbazones. The keto acetates do not react so readily. Pacsu and co-workers[115, 140] failed to obtain such compounds with keto acetates. Wolfrom and co-workers[126] formed a crystalline oxime of keto-D-sorbose pentaacetate, and Bredereck and co-workers[142] finally found conditions

* References start on p. 386.

for obtaining a syrupy oxime of *keto*-D-fructose pentaacetate from which a crystalline hexaacetate was formed.

Schiff bases of 3,4,5,6-tetra-*O*-benzoyl-*aldehydo*-D-glucose[19b] and a 2-acetamido-tri-*O*-acetyl-2-deoxy-D-glucose[143] are known.

The unsubstituted oximes and semicarbazones of the reducing sugars are very probably in tautomeric equilibrium with ring and chain forms. They mutarotate and on acetylation provide a mixture of such ring–chain acetylated isomers (Section V,2).

aldehydo-D-Glucose oxime hexaacetate loses acetic acid on heating, and penta-*O*-acetyl-D-gluconitrile is formed. This is the reactive intermediate in the acetylation of D-glucose oxime to form the acetylated nitrile and was isolated from the reaction mixture when the reaction was interrupted before completion.[144] The acetylation of D-glucose oxime to form the acetylated nitrile is a step in the Wohl degradation. An isomeric cyclic hexaacetate was

also isolated from the reaction mixture.

Reaction of *aldehydo*-D-glucose pentaacetate with phenylhydrazine yielded a crystalline product,[145] also obtained by the acetylation of D-mannose phenylhydrazone,[145] which was shown by n.m.r. spectroscopy to be the phenylazo compound (**38**). The spectrum established the olefinic protons as geometrically *trans*.

38

The phenylhydrazone of D-galactose yields penta-*O*-acetyl-*aldehydo*-D-galactose phenylhydrazone on mild acetylation, this substance being identical with that produced directly from *aldehydo*-D-galactose pentaacetate and phenylhydrazine. The acetylated phenylhydrazone undergoes conversion into a phenylazo derivative on heating in ethanol solution.[145]

Two isomers of D-glucose phenylhydrazone are known, and all evidence would indicate that they are cyclic in structure.[146] On the other hand, it is probable that D-galactose phenylhydrazone is acyclic. The acyclic structures of the sugar osazones have been noted (Section VIII).

Pyridine derivatives, such as **39**, were obtained[147] in a series of steps beginning with the product formed from *aldehydo*-D-glucose pentaacetate and MeCO—CH=CHNH$_2$.

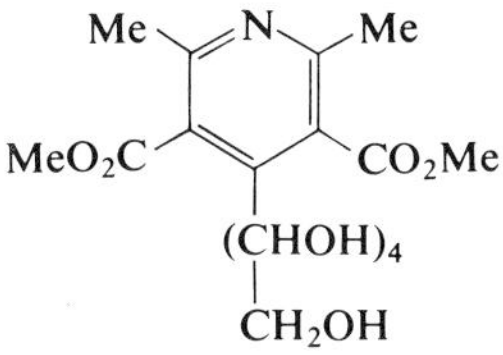

$$\text{MeO}_2\text{C} \quad \text{Me} \quad \text{N} \quad \text{Me} \quad \text{CO}_2\text{Me}$$

(CHOH)$_4$

CH$_2$OH

39

The reactions of diazomethane with aldehydo and keto acetates can be complex because diazomethane can form a methylene radical (carbene), which then becomes a reaction intermediate. The first step with an aldehydo acetate is the formation of a methyl ketone.

$$\text{HC}{=}\text{O} + \text{CH}_2\text{N}_2 \longrightarrow \overset{\text{CH}_3}{\underset{}{\text{C}}}{=}\text{O} + \text{N}_2$$

Further reaction can lead to an ethyl ketone.[148]

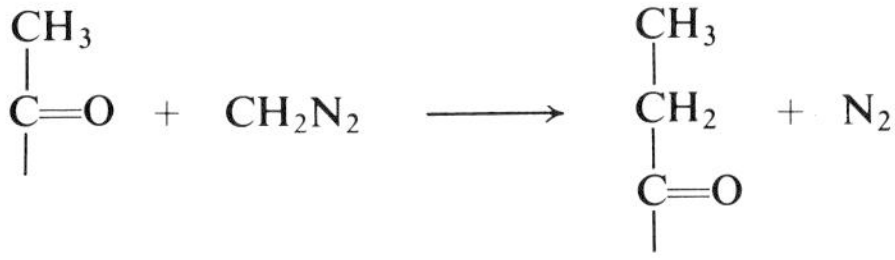

$$\overset{\text{CH}_3}{\underset{}{\text{C}}}{=}\text{O} + \text{CH}_2\text{N}_2 \longrightarrow \overset{\text{CH}_3}{\underset{\overset{}{\text{C}{=}\text{O}}}{\text{CH}_2}} + \text{N}_2$$

Keto acetates react with diazomethane to form epoxides, which may be opened with halogen acids.[149]

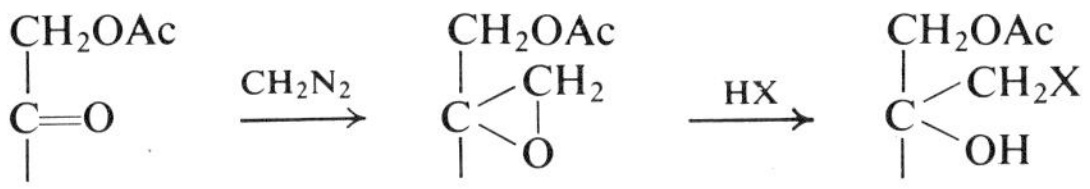

$$\underset{\text{C}{=}\text{O}}{\overset{\text{CH}_2\text{OAc}}{}} \xrightarrow{\text{CH}_2\text{N}_2} \underset{}{\overset{\text{CH}_2\text{OAc}}{\text{C}\langle\overset{\text{CH}_2}{\underset{\text{O}}{}}}} \xrightarrow{\text{HX}} \underset{}{\overset{\text{CH}_2\text{OAc}}{\text{C}\langle\overset{\text{CH}_2\text{X}}{\underset{\text{OH}}{}}}}$$

5. *Other Aldehydo Reactions*

Since aldehydes are more reactive than ketones, a number of reactions limited to aldehydo acetates will be treated herein. This type of investigation will assuredly continue.

a.   *Organometallics.* — 2,3:4,5-Di-*O*-isopropylidene-*aldehydo*-D-arabinose was brought into reaction with methylmagnesium iodide, and the two theoretically possible 6-deoxyalditols were obtained and related to substances of known configuration.[98a] This same aldehydo derivative, and its enantiomorph,

* *References start on p. 386.*

were brought into reaction with phenyl-, cyclohexyl-, and 1-naphthyl-magnesium halides to obtain, after hydrolysis, a series of 1-C-substituted pentitols.[150] In each case, however, only one of the two possible diastereoisomers was isolated. Bonner[151] later established the D-glucitol configuration for the 1-C-phenyl-D-pentitol obtained by English and Griswold[150] and isolated the second (D-mannitol) diastereoisomer from the reaction.

An aldehydo derivative (40), obtained by a glycol-scission reaction, underwent reaction with methylmagnesium iodide to form, exclusively, the L-*ido* isomer (42), probably through an intermediate (41) having a planar five-membered ring that would direct the addition to give the L-*ido* product.[153]

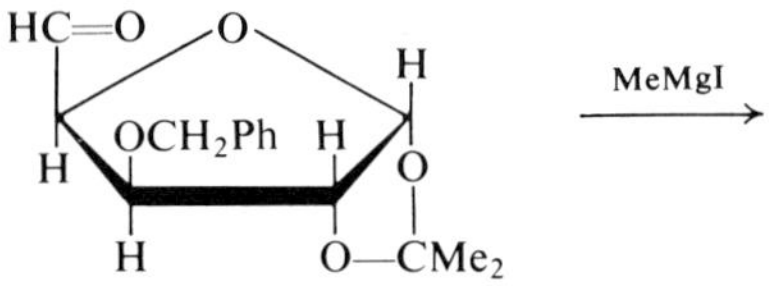

40

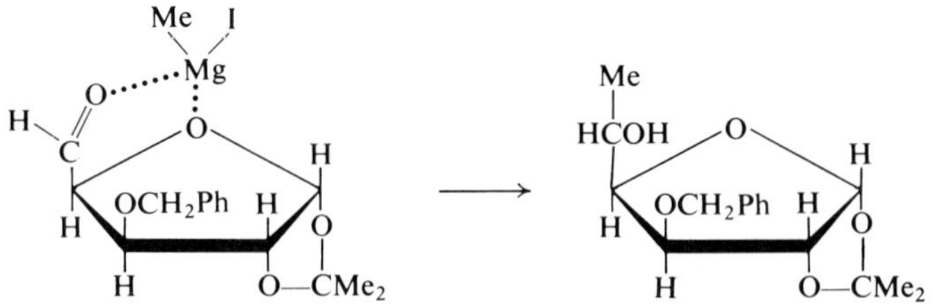

41   42

The L-idose derivative (42) was then converted into a methyl ketone (43) with chromium trioxide–pyridine, and this on treatment with methylmagnesium iodide formed the branched-chain sugar[152] 44.

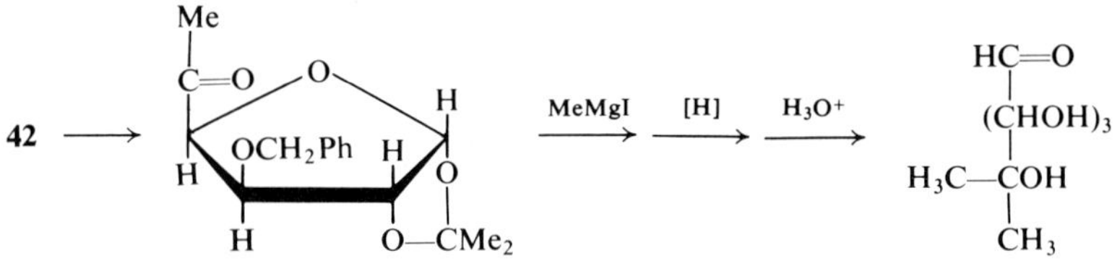

43   44

The reaction of di-O-isopropylidene-*aldehydo*-D-arabinose, -D-ribose, and -D-xylose with 2,4-benzyloxy-5-lithiouracil attached these aldose derivatives to C-5 of uracil by a carbon-to-carbon linkage. Removal of substituents from the first two *aldehydo*-pentose products yielded substances of structure 45, whereas the D-xylose product formed a 1,5-anhydro or "pseudonucleoside" structure (**46**). The configurations at C-1′ were established by optical rotatory dispersion data.[153]

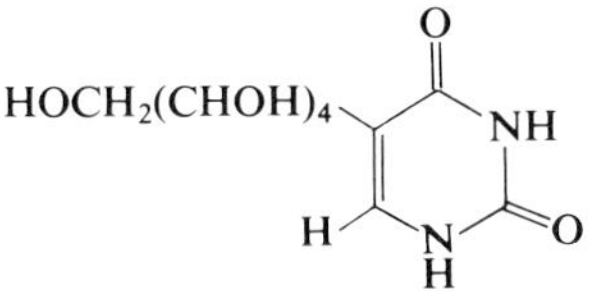

HOCH$_2$(CHOH)$_4$

45

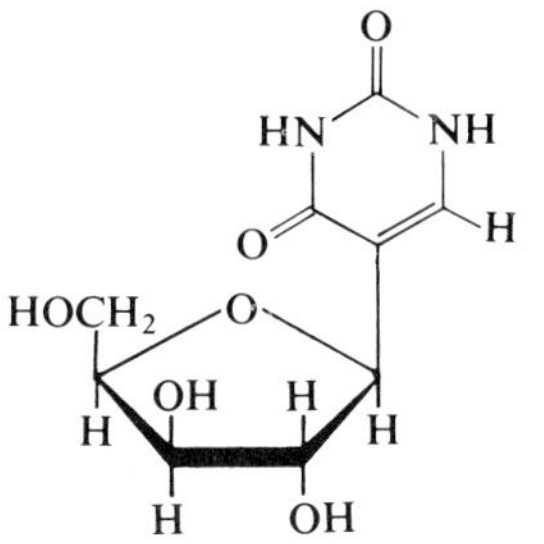

46

b. *Beta-Elimination.*—Brief treatment[154] of penta-*O*-methyl-*aldehydo*-D-glucose (**47**) with lime water yielded the enol ether (**48**) by a β-elimination mechanism (Isbell[155]). Treatment[154] with dilute acid gave **49**, probably by two successive β-eliminations.

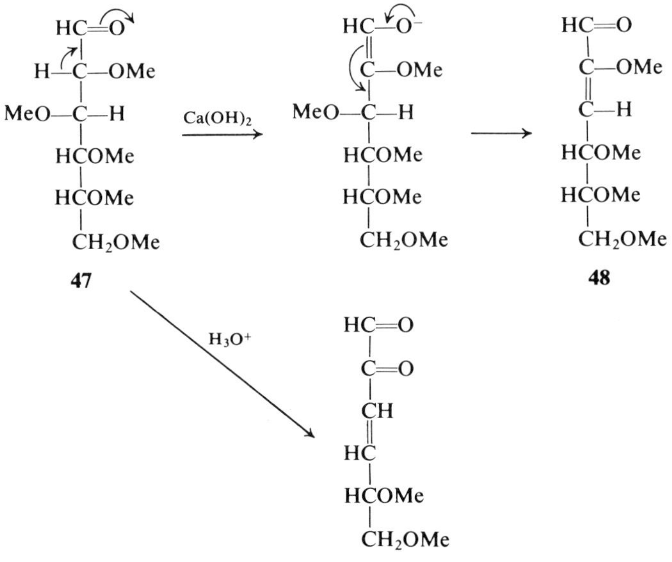

c. *Knoevenagel–Doebner Condensation.*[156]—Kochetkov and Dmitriev condensed 2,3:4,5-di-*O*-isopropylidene-*aldehydo*-L-arabinose,[98a, 150, 157] 2,3:4,5-di-*O*-isopropylidene-*aldehydo*-D-xylose, 2,4:3,5-di-*O*-benzylidene-*aldehydo*-D-ribose, and 2,3:4,5-di-*O*-isopropylidene-6-*O*-trityl-*aldehydo*-D-galactose with malonic acid, under the conditions of Knoevenagel–Doebner,[158] to obtain the

* *References start on p. 386.*

α,β-unsaturated acids. These could then be hydroxylated, hydrolyzed, and the lactones reduced to the aldose, producing, in overall reaction, a higher sugar by two atoms at a time.

$$R—CHO \rightarrow R—CH{=}CH—CO_2H \rightarrow R—CHOH—CHOH—CO_2H \rightarrow$$

$$R—CHOH—CHOH—CHO$$

The same type of reaction was extended to an acidic reaction with barbituric acid and several aldehydo acetates to give condensation products containing one or two sugar residues.[159] Condensation was also effected with 4-hydroxy-

$$R = HOCH_2(CHOH)_5{-}$$

coumarin.[160]

d. *Wittig Reaction.*—This reaction was first applied to 2,3-O-isopropylidene-D-glyceraldehyde to yield an α,β-unsaturated acid.[161] It was later extended to

$$R—CHO + Ph_3P{=}CHCO_2Et \rightarrow R—CH{=}CH—CO_2Et$$

several aldehydo acetates.[162] As above, these unsaturated esters could be converted into aldoses.

The Wittig reaction was utilized on the oxazoline (**50**) to yield eventually the N-benzoyl-C_{18}(and C_{20})-phytosphingosine[163] (**51**).

The Wittig reaction has been utilized in a synthesis of 3-deoxy-D-*manno*-octulosonic acid,[163a] and detailed applications of the reaction in the carbohydrate field have been reviewed.[163b]

50   **51**

e. *Nef Ethynylation.*—Horton and co-workers[133a, 164] ethynylated 2,3:4,5-di-*O*-isopropylidene-*aldehydo*-L-arabinose, reduced the product to the olefin, and separated, at this stage, the two isomers formed. The structure of each was established by ozonolysis to the hexose derivative. The acetylene derivative was transformed into the α,β-unsaturated aldehyde with bis(1,2-dimethylpropyl)borane followed by hydrogen peroxide, according to Brown and Zweifel.[165] The reaction has also been used as a route to uronic acids.[166]

R—CHO → R—CHOH—C≡CH → R—CHOH—CH=CH_2 → R—CHOH—CHO

R—CHOH—C≡CH → → R—C(H)=C(H)—C(H)=O

f. *Aldol Condensation.*—The dialdose derivative (**52**), possessing an aldehydo function, underwent self-condensation in lime water to yield the branched-chain trialdononose derivative (**53**), the 4-formyl group of which was considered, by Schaffer and Isbell, to be in a pyranose ring-form.[167] These authors favored the L-*xylo*-L-*ido* configuration, supposedly in a *B3* central ring.

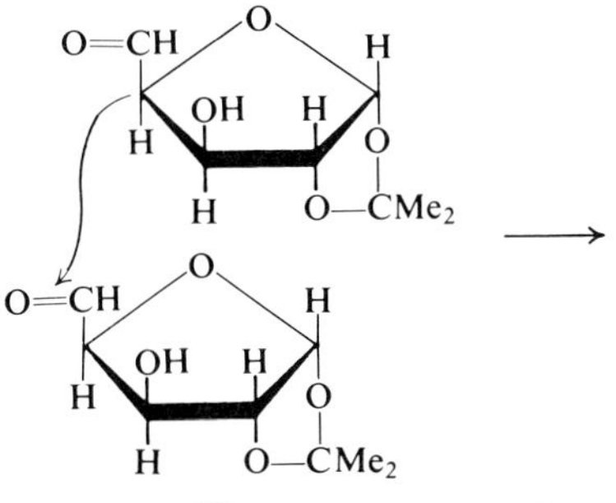

52

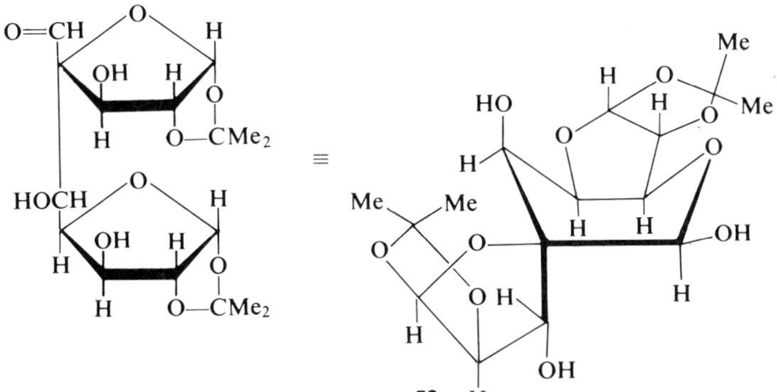

53

* *References start on p.* 386.

When **52** was brought into reaction with an excess of alkaline formaldehyde, the mixed aldol product (**53**) was not isolated, but its reduced form (**54**) was

obtained in a crossed-Cannizzaro type of reaction.[168] This procedure was applied to 2,3:4,5-di-O-isopropylidene-*aldehydo*-L-arabinose to obtain a derivative containing the structure **55**, which was then used as an intermediate in a synthesis of apiose.[169]

$$\text{HC} \overset{\text{CH}_2\text{OH}}{\underset{\text{R}}{\big|}} {}_{\text{CH}_2\text{OH}}$$

55

X ACYCLIC FORMS IN AQUEOUS SOLUTION[169a]

In preceding sections of this Chapter, the formation of small amounts of acyclic sugar derivatives has been shown to occur in many normal reactions of the cyclic structures. It can therefore be stated with assurance that the acyclic forms can be present as chemical reaction intermediates. Certainly it is logical to consider that a hemiacetal should be able to open. Whether such structures are present in aqueous solutions is another matter. Very probably no significant amounts are there under equilibrium conditions. The situation here is complicated by the possibility of hydration of the carbonyl group. The tautomeric equilibria of the reducing sugars are strongly pH-dependent, and in general it would appear that an increase in pH favors the formation of the acyclic forms. Assuredly, when an aqueous solution of an aldose is reduced to the alditol, the ring must open; otherwise an anhydroalditol would be formed.

The acyclic form of a glycose is the logical intermediate in the cyanohydrin reaction. Supporting evidence for such an intermediate is furnished by the initial hydrogen cyanide-binding capacity of sugars.[170] An increase in pH elevates the binding capacity for hydrogen cyanide to a maximum point, and it is of interest that a slight alkalinity is needed for the cyanohydrin reaction. For additional information on acyclic forms, see ref. 169a.

Periodate-oxidation results have been interpreted to indicate that aldohexoses react in their open-chain forms in water at pH 6.5 in the presence of a phosphate buffer.[171] 2,3-Di-O-methyl-D-ribose is stated to behave as an aldehydo sugar toward periodate.[172]

From electrophoretic studies on derivatives of D-glucose, the significant fact emerges that in alkaline borate solution all of the hydroxyl groups appear to be involved in complex-formation to a greater or lesser extent.[173]

In acidic, methanolic solutions of sugars, the acyclic carbonyl forms were detected, after permethylation, by gas–liquid chromatography. So determined were 0.2 to 2.1 % of aldehydo form for D-galactose, 1.6 to 3.4 % of keto form for D-fructose, and a maximum of only 0.1 % of aldehydo form for D-glucose.[174] This conforms to the greater tendency of the ketoses to form acyclic derivatives in such reactions as acetylation.

Considerable work has been done in attempts to measure acyclic forms in aqueous solutions of sugars by means of the sensitive polarograph of Heyrovský. Heyrovský and Smoleř[175] showed that the ketohexoses D-fructose and L-sorbose were electroreducible at −1.80 volts and that the wave height was temperature-dependent. Aldoses, however, showed only very small effects at room temperature. This behavior is the reverse of that found for simple, unsubstituted aliphatic, carbonyl compounds wherein the aldehydes show pH-dependent electroreducibility at 1.5 volts (pH 5) and the ketones are apparently not reducible under the usual conditions. The anomalous behavior of the sugars may be due to the greater tendency of ketoses to react in the acyclic form, or to a lesser hydration of the ketose carbonyl group. Cantor and Peniston[176] found conditions for obtaining measurable, though small, polarographic effects from aldoses in aqueous solution. These measurements were interpreted by Wiesner[177] as due to the rate of formation of the reducible tautomer, the aldehydo form, at the electrode. Overend and associates[178] considerably extended these measurements. At any event, all the data are in qualitative agreement in regard to variation with structure; they are also in good agreement with the relative order as determined by hydrogen cyanide-binding capacity. D-Ribose, 2-deoxy-D-(or L)-erythro-pentose, and 2-deoxy-D-ribo-hexose all show very high rates of ring–aldehydo transformations at the surface of the mercury drop. For all of the aldoses measured, these rates increase with pH.

Nuclear magnetic resonance spectroscopy can be used to measure aldehyde–aldehydrol equilibria of substituted aldehydo derivatives in water (see Section IX,A,3).[133c] The common aldoses in water show no detectable proportion of nonhydrated aldehydo form, but the presence of hydrated aldehydes in appreciable proportion cannot be discounted for certain configurations, because the signal of the C-1 proton would appear in the same spectral region as the signals of anomeric protons of the various cyclic tautomers.

* References start on p. 386.

REFERENCES

1. E. Fischer, *Ber.*, **27**, 673 (1894).
2. M. L. Wolfrom and J. V. Karabinos, *J. Amer. Chem. Soc.*, **67**, 500 (1945).
3. M. L. Wolfrom and F. B. Moody, *J. Amer. Chem. Soc.*, **62**, 3465 (1940).
4. M. L. Wolfrom and A. Thompson, *J. Amer. Chem. Soc.*, **56**, 880 (1934).
5. H. Zinner, *Ber.*, **83**, 275 (1950) and succeeding publications; compare H. Zinner and R. Heinatz, *J. Prakt. Chem.*, **312**, 561 (1970); H. Zinner, R. Kleeschätzky, and M. Schlutt, *Carbohyd. Res.*, **19**, 71 (1971).
6. W. T. Lawrence, *Ber.*, **29**, 547 (1896).
7. M. L. Wolfrom, M. R. Newlin, and E. E. Stahly, *J. Amer. Chem. Soc.*, **53**, 4379 (1931).
8. J. Staněk and J. Šáda, *Collect. Czech. Chem. Commun.*, **14**, 540 (1949).
9. E. Baumann, *Ber.*, **18**, 885 (1885).
10. M. L. Wolfrom and J. C. Sowden, *J. Amer. Chem. Soc.*, **60**, 3009 (1938).
11. C. Araki and S. Hirase, *Bull. Chem. Soc. Jap.*, **26**, 463 (1953).
12. A. N. O'Neill, *J. Amer. Chem. Soc.*, **77**, 2837 (1955).
13. F. A. Kuehl, Jr., E. H. Flynn, N. G. Brink, and K. Folkers, *J. Amer. Chem. Soc.*, **68**, 2096 (1946).
14. I. R. Hooper, L. H. Klemm, W. J. Polglase, and M. L. Wolfrom, *J. Amer. Chem. Soc.*, **68**, 2120 (1946); **69**, 1052 (1947).
15. J. Fried and D. E. Walz, *J. Amer. Chem. Soc.*, **71**, 140 (1949).
16. F. J. McEvoy, B. R. Baker, and M. J. Weiss, *J. Amer. Chem. Soc.*, **82**, 209 (1960).
17. F. Micheel and F. Suckfüll, *Ann.*, **502**, 85 (1933); *ibid.*, **507**, 138 (1933); *Ber.*, **66**, 1957 (1933); F. Micheel and W. Spruck, *Ber.*, **67**, 1665 (1934).
18. H. Zinner, G. Rembarz, H.-W. Linke, and G. Ulbricht, *Ber.*, **90**, 1761 (1957) and succeeding publications (see ref. 5).
19. (a) P. Brigl and H. Mühlschlegel, *Ber.*, **63**, 1551 (1930); (b) P. Brigl, H. Mühlschlegel, and R. Schinle, *ibid.*, **64**, 2921 (1931); (c) P. Brigl and R. Schinle, *ibid.*, **65**, 1890 (1932).
20. W. Schneider and J. Sepp, *Ber.*, **49**, 2054 (1916).
21. J. W. Green, *Advan. Carbohyd. Chem.*, **21**, 95 (1966).
22. E. Pacsu and E. J. Wilson, Jr., *J. Amer. Chem. Soc.*, **61**, 1450 (1939).
23. J. W. Green and E. Pacsu, *J. Amer. Chem. Soc.*, **59**, 1205 (1937).
24. N. W. Pirie, *Biochem. J.*, **30**, 374 (1936).
25. D. L. MacDonald and H. O. L. Fischer, *J. Amer. Chem. Soc.*, **74**, 2087 (1952); *Biochim. Biophys. Acta*, **12**, 203 (1953).
26. R. Barker and D. L. MacDonald, *J. Amer. Chem. Soc.*, **82**, 2297 (1960).
27. D. L. MacDonald and H. O. L. Fischer, *J. Amer. Chem. Soc.*, **77**, 4348 (1955).
28. M. L. Wolfrom and D. I. Weisblat, *J. Amer. Chem. Soc.*, **62**, 878 (1940).
29. M. L. Wolfrom, D. I. Weisblat, and A. R. Hanze, *J. Amer. Chem. Soc.*, **62**, 3246 (1940).
30. I. Farkaš, M. Menyhárt, R. Bognár, and H. Gross, *Ber.*, **98**, 1419 (1965).
31. F. Weygand, H. Ziemann, and H. J. Bestmann, *Ber.*, **91**, 2534 (1958).
32. J. Bougault, E. Cattelain, and P. Chabrier, *Bull. Soc. Chim. Fr.*, **6** [5], 34 (1939); *ibid.*, **7** [5], 780, 781 (1940).
33. M. L. Wolfrom and J. V. Karabinos, *J. Amer. Chem. Soc.*, **66**, 909 (1944).
34. M. L. Wolfrom, R. U. Lemieux, and S. M. Olin, *J. Amer. Chem. Soc.*, **71**, 2870 (1949).
35. J. K. N. Jones and D. L. Mitchell, *Can. J. Chem.*, **36**, 206 (1958).
36. D. Horton and J. S. Jewell, *J. Org. Chem.*, **31**, 509 (1966).
37. M. L. Wolfrom, D. I. Weisblat, and A. R. Hanze, *J. Amer. Chem. Soc.*, **66**, 2065 (1944).
38. E. M. Montgomery, R. M. Hann, and C. S. Hudson, *J. Amer. Chem. Soc.*, **59**, 1124 (1937).

39. M. L. Wolfrom, L. J. Tanghe, R. W. George, and S. W. Waisbrot, *J. Amer. Chem. Soc.*, **60**, 132 (1938); H. A. Campbell and K. P. Link, *J. Biol. Chem.*, **122**, 635 (1938).
40. M. L. Wolfrom and S. W. Waisbrot, *J. Amer. Chem. Soc.*, **60**, 854 (1938).
41. E. Pacsu, *J. Amer. Chem. Soc.*, **61**, 1071 (1939).
42. M. L. Wolfrom and S. W. Waisbrot, *J. Amer. Chem. Soc.*, **61**, 1408 (1939).
43. P. A. Levene and G. M. Meyer, *J. Biol. Chem.*, **69**, 175 (1926); *ibid.*, **74**, 695 (1927).
44. W. N. Haworth, J. Jackson, and F. Smith, *J. Chem. Soc.*, 620 (1940).
45. C. Araki, *J. Chem. Soc. Jap.*, **65**, 725 (1944); *Chem. Abstr.*, **41**, 3496 (1947).
45a. C. T. Bishop and F. P. Cooper, *Can. J. Chem.*, **40**, 224 (1962); *ibid.*, **41**, 2743 (1963).
45b. D. Heard and R. Barker, *J. Org. Chem.*, **33**, 740 (1968).
45c. R. J. Ferrier and L. R. Hatton, *Carbohyd. Res.*, **6**, 75 (1968).
45d. E. Fischer, *Ber.*, **28**, 1145 (1895).
46. L. J. Breddy and J. K. N. Jones, *J. Chem. Soc.*, 738 (1945); L. E. Wise and E. K. Ratliff, *Anal. Chem.*, **19**, 694 (1947).
47. M. L. Wolfrom, *J. Amer. Chem. Soc.*, **51**, 2188 (1929).
48. M. L. Wolfrom and C. C. Christman, *J. Amer. Chem. Soc.*, **58**, 39 (1936).
49. W. Schneider and J. Sepp, *Ber.*, **51**, 220 (1918).
50. B. Holmberg, *J. Prakt. Chem.*, **243**, 57 (1932).
51. M. L. Wolfrom and M. Konigsberg, *J. Amer. Chem. Soc.*, **62**, 1153 (1940).
52. M. W. Whitehouse, P. W. Kent, and C. A. Pasternak, *J. Chem. Soc.*, 2315 (1954).
53. R. K. Ness, H. W. Diehl, and H. G. Fletcher, Jr., *J. Amer. Chem. Soc.*, **76**, 763 (1954).
54. B. Gauthier and C. Vaniscotte, *Bull. Soc. Chim. Fr.*, 30 (1956).
55. F. Weygand, H. J. Bestmann, and H. Ziemann, *Ber.*, **91**, 1040 (1958).
56. M. L. Wolfrom and L. J. Tanghe, *J. Amer. Chem. Soc.*, **59**, 1497 (1937).
57. K. W. Rosenmund, *Ber.*, **51**, 585 (1918).
58. E. W. Cook and R. T. Major, *J. Amer. Chem. Soc.*, **58**, 2410 (1936).
59. W. W. Lake with J. W. E. Glattfeld, *J. Amer. Chem. Soc.*, **66**, 1091 (1944).
60. M. L. Wolfrom and E. Usdin, *J. Amer. Chem. Soc.*, **75**, 4318 (1953).
61. M. L. Wolfrom, L. W. Georges, and S. Soltzberg, *J. Amer. Chem. Soc.*, **56**, 1794 (1934).
62. B. Helferich and H. Schirp, *Ber.*, **86**, 547 (1953).
63. S. J. Angyal, C. G. Macdonald, and N. K. Matheson, *J. Chem. Soc.*, 3321 (1953).
64. C. D. Hurd and E. M. Filachione, *J. Amer. Chem. Soc.*, **61**, 1156 (1939).
65. M. L. Wolfrom and J. V. Karabinos, *J. Amer. Chem. Soc.*, **68**, 1455 (1946).
66. F. Weygand and H. Leube, Ger. Pat. 1,010,957 (1957); *Chem. Abstr.*, **54**, 3248 (1960).
67. E. Montgomery and C. S. Hudson, *J. Amer. Chem. Soc.*, **56**, 2463 (1934).
68. W. N. Haworth, L. N. Owen, and F. Smith, *J. Chem. Soc.*, 88 (1941).
69. B. C. Bera, A. B. Foster, and M. Stacey, *J. Chem. Soc.*, 4531 (1956).
70. S. Peat, *Advan. Carbohyd. Chem.*, **2**, 63 (1946); J. Defaye, *Advan. Carbohyd. Chem. Biochem.*, **25**, 181 (1970).
70a. C. Araki and K. Arai, *Bull. Chem. Soc. Jap.*, **40**, 1452 (1967).
71. M. L. Wolfrom, *J. Amer. Chem. Soc.*, **52**, 2464 (1930).
72. M. L. Wolfrom, M. Konigsberg, and F. B. Moody, *J. Amer. Chem. Soc.*, **62**, 2343 (1940).
73. R. J. Dimler and K. P. Link, *J. Amer. Chem. Soc.*, **62**, 1216 (1940).
74. M. L. Wolfrom and W. H. Decker, *Ann.*, **690**, 163 (1965).
75. M. L. Wolfrom and W. M. Morgan, *J. Amer. Chem. Soc.*, **54**, 3390 (1932).
76. M. L. Wolfrom and D. I. Weisblat, *J. Amer. Chem. Soc.*, **62**, 1149 (1940).
77. M. L. Wolfrom and R. L. Brown, *J. Amer. Chem. Soc.*, **65**, 951 (1943).
78. M. L. Wolfrom and R. L. Brown, *J. Amer. Chem. Soc.*, **63**, 1246 (1941).
79. G. E. Felton and W. Freudenberg, *J. Amer. Chem. Soc.*, **57**, 1637 (1935).

80. M. L. Wolfrom, A. B. Foster, P. McWain, W. von Bebenburg, and A. Thompson, *J. Org. Chem.*, **26**, 3095 (1961).

81. M. L. Wolfrom, P. McWain, and A. Thompson, *J. Org. Chem.*, **27**, 3549 (1962); compare M. L. Wolfrom, H. G. Garg, and D. Horton, *ibid.*, **30**, 1096 (1965).

82. M. L. Wolfrom, W. von Bebenburg, R. Pagnucco, and P. McWain, *J. Org. Chem.*, **30**, 2732 (1965); compare M. L. Wolfrom and P. J. Conigliaro, *Carbohyd. Res.*, **20**, 369 (1971); G. Giovanninetti, L. Nobile, M. Amorosa, and J. Defaye, *ibid.*, **21**, 320 (1972).

83. M. L. Wolfrom, *J. Amer. Chem. Soc.*, **57**, 2498 (1935).

84. K. Freudenberg and K. Soff, *Ber.*, **70**, 264 (1937).

85. H. Bredereck, A. Wagner, G. Hagelloch, and G. Faber, *Ber.*, **91**, 515 (1958).

86. R. L. Whistler, E. Heyne, and J. Bachrach, *J. Amer. Chem. Soc.*, **71**, 1476 (1949).

87. N. W. Pirie, *Biochem. J.*, **30**, 369 (1936).

88. F. Micheel, H. Ruhkopf, and F. Suckfüll, *Ber.*, **68**, 1523 (1935).

89. E. J. Bourne, W. M. Corbett, and M. Stacey, *J. Chem. Soc.*, 2810 (1952).

90. A. Wohl, *Ber.*, **26**, 730 (1893).

91. H. S. Isbell and H. L. Frush, *J. Amer. Chem. Soc.*, **71**, 1579 (1949).

92. V. Deulofeu and J. O. Deferrari, *J. Org. Chem.*, **17**, 1087 (1952).

93. J. O. Deferrari and V. Deulofeu, *J. Org. Chem.*, **17**, 1093 (1953) and succeeding publications; compare A. B. Zanlungo, J. O. Deferrari, and R. A. Cadenas, *Carbohyd. Res.*, **14**, 245 (1970).

94. R. C. Hockett and L. B. Chandler, *J. Amer. Chem. Soc.*, **66**, 957 (1944); A. S. Cerezo and V. Deulofeu, *Carbohyd. Res.*, **2**, 35 (1966).

95. J. B. S. Braverman, *J. Sci. Food Agr.*, **4**, 540 (1953).

96. D. L. Ingles, *Aust. J. Chem.*, **12**, 97 (1959).

97. R. T. Major and E. W. Cook, *J. Amer. Chem. Soc.*, **58**, 2474 (1936).

98. R. T. Major and E. W. Cook, *J. Amer. Chem. Soc.*, **58**, 2477 (1936).

98a. K. Gätzi and T. Reichstein, *Helv. Chim. Acta*, **21**, 914 (1938).

99. C. D. Hurd and J. C. Sowden, *J. Amer. Chem. Soc.*, **60**, 235 (1938).

100. M. L. Wolfrom, M. Konigsberg, and D. I. Weisblat, *J. Amer. Chem. Soc.*, **61**, 574 (1939).

101. M. L. Wolfrom and R. B. Bennett, *J. Org. Chem.*, **30**, 458 (1965).

102. M. L. Wolfrom, R. B. Bennett, and J. D. Crum, *J. Amer. Chem. Soc.*, **80**, 944 (1958).

103. K. Ladenburg, M. Tishler, J. W. Wellman, and R. D. Babson, *J. Amer. Chem. Soc.*, **66**, 1217 (1944).

104. M. L. Wolfrom and M. J. Cron, *J. Amer. Chem. Soc.*, **74**, 1715 (1952).

105. M. L. Wolfrom and A. Rosenthal, *J. Amer. Chem. Soc.*, **75**, 3662 (1953).

106. M. L. Wolfrom and P. W. Morgan, *J. Amer. Chem. Soc.*, **64**, 2026 (1942).

107. M. L. Wolfrom and D. I. Weisblat, *J. Amer. Chem. Soc.*, **66**, 805 (1944).

108. L. Maquenne, *Bull. Soc. Chim. Fr.*, **48** [2], 719 (1887); Z. H. Skraup, *Monatsh. Chem.*, **14**, 488 (1893).

109. O. Diels and F. Löflund, *Ber.*, **47**, 2351 (1941); J. Müller, *ibid.*, 2654.

110. K. Gätzi and T. Reichstein, *Helv. Chim. Acta*, **21**, 186 (1938).

111. K. Iwadare, *Bull. Chem. Soc. Jap.*, **14**, 131 (1939).

112. M. L. Wolfrom, D. I. Weisblat, W. H. Zophy, and S. W. Waisbrot, *J. Amer. Chem. Soc.*, **63**, 201 (1941).

113. M. L. Wolfrom, S. W. Waisbrot, and R. L. Brown, *J. Amer. Chem. Soc.*, **64**, 2329 (1942).

113a. Y. Rabinsohn and H. G. Fletcher, Jr., *J. Org. Chem.*, **32**, 3452 (1967).

114. C. S. Hudson and D. H. Brauns, *J. Amer. Chem. Soc.*, **37**, 2736 (1915).

115. E. Pacsu and F. V. Rich, *J. Amer. Chem. Soc.*, **55**, 3018 (1933).

116. M. L. Wolfrom and A. Thompson, *J. Amer. Chem. Soc.*, **56**, 880 (1934).

117. P. Brigl and R. Schinle, *Ber.*, **66**, 325 (1933); *ibid.*, **67**, 127 (1934).

118. F. Micheel and A. Frowein, *Ber.*, **90**, 1599 (1957).
119. R. Kuhn and G. Krüger, *Ann.*, **628**, 240 (1959).
120. T. van Es and R. L. Whistler, *J. Org. Chem.*, **29**, 1087 (1964).
121. M. L. Wolfrom, R. L. Brown, and E. F. Evans, *J. Amer. Chem. Soc.*, **65**, 1021 (1943).
122. M. L. Wolfrom, S. W. Waisbrot, and R. L. Brown, *J. Amer. Chem. Soc.*, **64**, 1701 (1942).
123. M. L. Wolfrom and R. L. Brown, *J. Amer. Chem. Soc.*, **65**, 1516 (1943).
124. M. L. Wolfrom and J. B. Miller, *J. Amer. Chem. Soc.*, **80**, 1678 (1958).
125. Y. A. Zhdanov, V. I. Kornilov, and G. V. Bogdanova, *Carbohyd. Res.*, **3**, 139 (1966).
126. M. L. Wolfrom, S. M. Olin, and E. F. Evans, *J. Amer. Chem. Soc.*, **66**, 204 (1944).
127. L. Mester, *J. Amer. Chem. Soc.*, **77**, 4301 (1955).
128. L. Mester, E. Moczar, and J. Parello, *Tetrahedron Lett.*, 3223 (1964); H. El Khadem, M. L. Wolfrom, and D. Horton, *J. Org. Chem.*, **30**, 838 (1965); L. Mester, H. El Khadem, and D. Horton, *J. Chem. Soc. (C)*, 2567 (1970).
129. R. F. Butterworth and S. Hanessian, *Synthesis*, 70 (1971).
130. H. Hudson, M. L. Wolfrom, and T. M. Lowry, *J. Chem. Soc.*, 1179 (1933); W. C. G. Baldwin, M. L. Wolfrom, and T. M. Lowry, *ibid.*, 696 (1935).
131. H. Bredereck, G. Höschele, and W. Huber, *Ber.*, **86**, 1271 (1953).
132. H. S. Isbell, F. A. Smith, E. C. Creitz, H. L. Frush, J. D. Moyer, and J. E. Stewart, *J. Res. Nat. Bur. Stand.*, **59**, 41 (1957).
132a. P. L. Durette and D. Horton, *Advan. Carbohyd. Chem. Biochem.*, **26**, 49 (1971).
133. (a) D. Horton, J. B. Hughes, and J. M. J. Tronchet, *Chem. Commun.*, 481 (1965); (b) A. E. El-Ashmawy and D. Horton, *Carbohyd. Res.*, **1**, 164 (1965); (c) D. Horton, M. Nakadate, and J. M. J. Tronchet, *ibid.*, **7**, 56 (1968); D. Horton and J. D. Wander, *ibid.*, **16**, 477 (1971).
134. J. L. Godman, D. Horton, and J. M. J. Tronchet, *Carbohyd. Res.*, **4**, 392 (1967).
134a. D. Horton and J. D. Wander, *Carbohyd. Res.*, **10**, 279 (1969); **15**, 271 (1970); compare J. Defaye, D. Gagnaire, D. Horton, and M. Muesser, *ibid.*, in press (1972).
135. N. K. Kochetkov and O. S. Chizhov, *Advan. Carbohyd. Chem.*, **21**, 84 (1966).
136. D. C. DeJongh, *J. Amer. Chem. Soc.*, **86**, 3149 (1964).
137. D. C. DeJongh and S. Hanessian, *J. Amer. Chem. Soc.*, **87**, 1408 (1965).
138. D. C. DeJongh, *J. Org. Chem.*, **30**, 453 (1965).
139. H. N. Alyea and H. L. J. Bäckström, *J. Amer. Chem. Soc.*, **51**, 97 (1929).
140. F. B. Cramer and E. Pacsu, *J. Amer. Chem. Soc.*, **59**, 1467 (1937).
141. H. H. Fox, *J. Org. Chem.*, **13**, 580 (1948); E. J. Reist, H. H. Hamlow, I. G. Junga, R. M. Silverstein, and B. R. Baker, *ibid.*, **25**, 1455 (1960).
142. H. Bredereck, G. Höschele, and T. Heinkel, *Ber.*, **87**, 531 (1954).
143. Y. Inouye, K. Onodera, and S. Kitaoka, *J. Agr. Chem. Soc. Jap.*, **29**, 139 (1955); *Chem. Abstr.*, **50**, 825 (1956).
144. M. L. Wolfrom and A. Thompson, *J. Amer. Chem. Soc.*, **53**, 622 (1931).
145. M. L. Wolfrom, A. Thompson, and D. R. Lineback, *J. Org. Chem.*, **27**, 2563 (1962).
146. M. L. Wolfrom and M. G. Blair, *J. Amer. Chem. Soc.*, **68**, 2110 (1946).
147. F. Micheel and W. Möller, *Ann.*, **670**, 63 (1963).
148. M. L. Wolfrom, D. I. Weisblat, E. F. Evans, and J. B. Miller, *J. Amer. Chem. Soc.*, **79**, 6454 (1957).
149. M. L. Wolfrom, J. B. Miller, D. I. Weisblat, and A. R. Hanze, *J. Amer. Chem. Soc.*, **79**, 6299 (1957).
150. J. English, Jr., and P. H. Griswold, Jr., *J. Amer. Chem. Soc.*, **67**, 2039 (1945); *ibid.*, **70**, 1390 (1948).
151. W. A. Bonner, *J. Amer. Chem. Soc.*, **73**, 3126 (1951).
152. M. L. Wolfrom and S. Hanessian, *J. Org. Chem.*, **27**, 1800, 2107 (1962).
153. W. Asbun and S. B. Binkley, *J. Org. Chem.*, **31**, 2215 (1966).

154. E. F. L. J. Anet, *Carbohyd. Res.*, **3**, 251 (1966).

155. H. S. Isbell, *J. Res. Nat. Bur. Stand.*, **32**, 45 (1944).

156. N. K. Kochetkov and B. A. Dmitriev, *Tetrahedron*, **21**, 803 (1965) and references cited therein.

157. N. K. Kochetkov and B. A. Dmitriev, *Chem. Ind.* (*London*), 2147 (1962).

158. E. Knoevenagel, *Ber.*, **29**, 172 (1896); **31**, 730 (1898); O. Doebner, *Ber.*, **33**, 2140 (1900).

159. Y. A. Zhdanov and G. V. Bogdanova, *Khim. Geterotsikl. Soedin.*, *Akad. Nauk Latv. SSR*, 56 (1966).

160. Y. A. Zhdanov, G. V. Bogdanova, and V. G. Zolotukhina, *Dokl. Akad. Nauk SSSR*, **157**, 917 (1964); *Chem. Abstr.*, **61**, 16137 (1964).

161. R. Kuhn and R. Brossmer, *Angew. Chem.*, **74**, 252 (1962).

162. N. K. Kochetkov and B. A. Dmitriev, *Chem. Ind.* (*London*), 864 (1963).

163. J. Gigg, R. Gigg, and C. D. Warren, *J. Chem. Soc.* (C), 1872 (1966).

163a. N. K. Kochetkov, B. A. Dmitriev, and L. V. Backinowsky, *Carbohyd. Res.*, **11**, 193 (1969).

163b. Y. A. Zhdanov, Y. E. Alexeev, and V. G. Alexeeva, *Advan. Carbohyd. Chem. Biochem.*, **27**, in press (1972).

164. D. Horton and J. M. J. Tronchet, *Carbohyd. Res.*, **2**, 315 (1966); D. Horton, J. B. Hughes, and J. K. Thomson, *J. Org. Chem.*, **33**, 728 (1968).

165. H. C. Brown and G. Zweifel, *J. Amer. Chem. Soc.* **83**, 3834 (1961).

166. D. Horton and F. O. Swanson, *Carbohyd. Res.*, **14**, 159 (1970).

167. R. Schaffer and H. S. Isbell, *J. Amer. Chem. Soc.*, **81**, 2178 (1959).

168. R. Schaffer, *J. Amer. Chem. Soc.*, **81**, 5452 (1959).

169. D. T. Williams and J. K. N. Jones, *Can. J. Chem.*, **42**, 69 (1964).

169a. W. Pigman and H. S. Isbell, *Advan. Carbohyd. Chem.*, **23**, 11 (1968); **24**, 13 (1969).

170. F. Lippich, *Biochem. Z.*, **248**, 280 (1932).

171. P. F. Fleury, J. E. Courtois, and A. Bieder, *Bull. Soc. Chim. Fr.*, **19** [5], 118 (1952).

172. G. R. Barker and D. C. C. Smith, *J. Chem. Soc.*, 1323 (1955).

173. A. B. Foster, *J. Chem. Soc.*, 982 (1953); A. B. Foster and M. Stacey, *ibid.*, 1778 (1955).

174. E. Bayer and R. Widder, *Ann.*, **686**, 181, 197 (1965).

175. J. Heyrovský and I. Smoleř, *Collect. Czech. Chem. Commun.*, **4**, 521 (1932).

176. S. M. Cantor and Q. P. Peniston, *J. Amer. Chem. Soc.*, **62**, 2113 (1940).

177. K. Wiesner, *Collect. Czech. Chem. Commun.*, **12**, 64 (1947).

178. W. G. Overend, A. R. Peacocke, and J. B. Smith, *J. Chem. Soc.*, 3487 (1961).

11. CYCLIC ACETAL DERIVATIVES OF SUGARS AND ALDITOLS

A. B. FOSTER

I. FORMATION OF CYCLIC ACETALS

In the presence of a mineral acid or Lewis acid catalyst (zinc chloride, copper sulfate), alcohols react reversibly with aldehydes and ketones to form acetals.

$$\underset{R'}{\overset{R}{>}}C{=}O \underset{+H_2O}{\overset{\underset{-H_2O}{R''OH}}{\rightleftarrows}} \left[\underset{R'}{\overset{R}{>}}C\underset{OR''}{\overset{OH}{<}} \right] \underset{+H_2O}{\overset{\underset{-H_2O}{R''OH}}{\rightleftarrows}} \underset{R'}{\overset{R}{>}}C\underset{OR''}{\overset{OR''}{<}}$$

The position of the equilibrium can be controlled; removal of water facilitates acetal formation, and acetals are hydrolyzed by dilute acid. If the alcohol and aldehyde groups are suitably located in the same molecule (as in a sugar), formation of a cyclic hemiacetal occurs spontaneously and extensively (see Chapter 4), but a catalyst is usually necessary to effect reaction with another alcohol molecule to give the full acetal (glycoside). Diols react with aldehydes and ketones to give cyclic acetals (termed O-alkylidene and O-arylidene derivatives) and, as with simple alcohols, the intermediate hemiacetal is not usually isolated. Numerous possibilities exist for cyclic acetal formation when polyhydric alcohols are involved. Since Wurtz's early work[1] on the acetaldehyde–ethylene glycol reaction and the demonstration by Meunier[2] that acetal formation was catalyzed by acid, a continuing study of the formation of carbohydrate cyclic acetals has revealed reaction patterns and has given value to cyclic acetals as protecting groups in syntheses.

Cyclic acetal formation, which is usually effected at room temperature with the aldehyde or ketone often serving as solvent for the reaction, may be represented as follows:

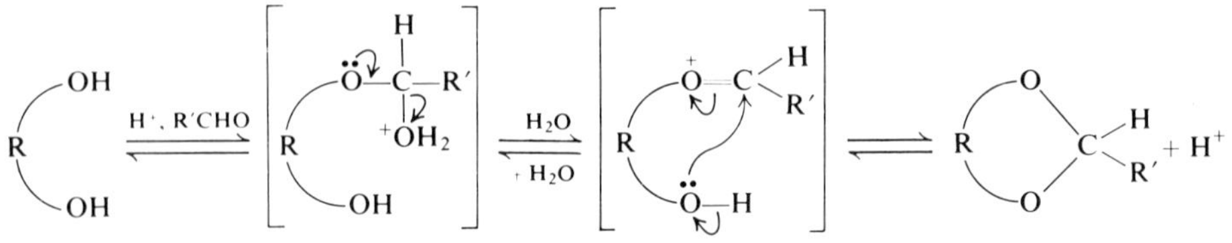

Rings of five (1,3-dioxolane), six (1,3-dioxane), and seven members (1,3-dioxepan) are usually formed. At equilibrium, ketones give rise mainly to 1,3-dioxolane derivatives but 1,3-dioxane structures are known, whereas aldehydes preferentially give 1,3-dioxane derivatives although 1,3-dioxolane compounds are quite common; for example, L-threitol gives 1,3:2,4-di-O-benzylidene and 1,2:3,4-di-O-isopropylidene derivatives. If R in the above equation is not symmetrical, and providing that the carbonyl compound is not formaldehyde or a symmetrical ketone, then the cyclic acetal contains a new asymmetric carbon atom and two isomers are possible. These are obtained when 1,3-dioxolane derivatives are produced, whereas only one isomer results on formation of a 1,3-dioxane ring. The most commonly used carbonyl compounds in carbohydrate chemistry are acetaldehyde, acetone, benzaldehyde, and formaldehyde.

Kinetic and thermodynamic stages occur in the formation of cyclic acetals. When water has been lost from the protonated hemiacetal (1), the resultant oxocarbonium ion (2) reacts rapidly with the nearest hydroxyl group to give the first (kinetic) product (3) of reaction. However, rearrangement may subsequently occur, and, if several hydroxyl groups are attached to R, migration may result, to give the thermodynamically most stable product, which will preponderate at equilibrium. If the acetalation reaction is monitored, it may be found that the cyclic acetals formed initially may be quite different from those present at equilibrium. Thus, benzaldehyde reacts[3] with 1,4-anhydro-erythritol (4) to give first the benzylidene acetal (5) having an endo-phenyl

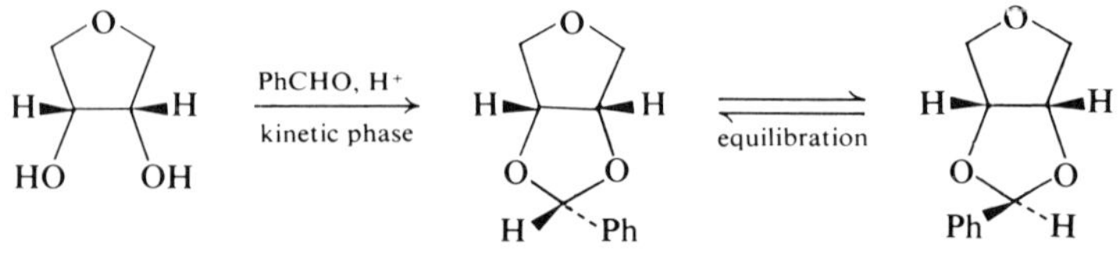

group. Subsequently, equilibration occurs and, at equilibrium, a near-equimolar mixture of the *endo-* (**5**) and *exo*-phenyl isomers (**6**) is formed. When
D-glucitol reacts[4] with an equimolar amount of *n*-butyraldehyde in *N* sulfuric
acid, the first isolable product is the 2,3-butylidene acetal (**7**), but the 2,4- (**8**)
and 3,4-butylidene (**9**) acetals are subsequently formed. Only since 1965 has the

kinetic phase of polyhydric alcohol–aldehyde reactions been studied; the vast
majority of carbohydrate cyclic acetal derivatives described in the literature
are equilibrium products.

The thermodynamic stability of the substituted 1,3-dioxolane derivatives

10 and **11** are comparable, and hence, at equilibrium, near-equimolar mixtures
are usually formed, with the *cis* isomer (**10**) slightly preponderating.[5] When
1,3-dioxane derivatives are formed, the products have structures of the type
12, with the group R equatorial. Isomers of the type **13** are markedly destabilized, and, although they have not been isolated after acid-catalyzed
acetalation, they may be obtained indirectly by treatment of suitably protected
sugar derivatives with an alkylidene halide (RCHBr$_2$) in the presence of a
base.[6] They are stable in basic media but, on exposure to traces of acid,
rearrange into the stable isomers (**12**).

* *References start on p.* 401.

II. CYCLIC ACETALS FROM ACYCLIC SUGAR DERIVATIVES

A. WITH ALDEHYDES

It is convenient to classify acetalation reactions into two groups involving acyclic (pentitols, hexitols, and the like) and cyclic sugar derivatives (such as anhydro sugars and glycosides). From an examination of a wide range of data on alditol–aldehyde reactions, Barker and Bourne[7] were able to extend and elaborate Hann and Hudson's empirical rules[8] for predicting reaction patterns. The modified rules are likely to be of general validity in those cases where the stability of cyclic acetals is such as to allow true equilibrium to be established after reasonably short reaction times. To simplify the presentation, the Greek letters α, β, and γ will be used to signify the relative position of the two hydroxyl groups engaged in cyclization, along the carbon chain of an alditol. The terms *erythro* and *threo* will indicate whether these groups are disposed *cis* or *trans* in the usual Fischer projection formula; *erythro* and *threo* will be required only when both alcohol groups are secondary.† The extended rules are: (1) the first preference is for a β-*erythro* ring; (2) the second, for a β ring; (3) the third, for an α, α-*threo*, β-*threo*, or γ-*threo* ring; (4) in *O*-methylenation a β-*threo* ring takes precedence over an α-*threo* or γ-*threo* ring; (5) in *O*-benzylidenation and *O*-ethylidenation an α-*threo* ring takes precedence over a β-*threo* or a γ-*threo* ring; (6) rules (4) and (5) may not apply when one or both of the carbon atoms carrying the hydroxyl groups concerned are already part of a ring system.

CH₂OH	CH₂OH
HCOH	HOCH
HOCH	HOCH
HCOH	HCOH
HCOH	HCOH
CH₂OH	CH₂OH
D-Glucitol	D-Mannitol

(1st, β-*erythro*) 2,4-*O*-benzylidene	(1st, β,β) 1,3:4,6-di-*O*-benzylidene
(2nd, β-*erythro*, β) 1,3:2,4-di-*O*-benzylidene	(2nd, β,β,γ-*threo*) 1,3:2,5:4,6-tri-*O*-benzylidene
(3rd, β-*erythro*,β,α) 1,3:2,4:5,6-tri-*O*-benzylidene	

† The symbols C and T in the original Barker and Bourne nomenclature have been replaced by *erythro* and *threo*, respectively.[9] Although the terms *cis* and *trans* have meaning when applied to pairs of hydroxyl groups in the Fischer projection formulae of acyclic compounds, they have no direct geometrical significance when applied to the actual molecules.

The rules do not mention an *α-erythro* ring, since compounds containing acetal rings of this type have been obtained only by indirect routes.[10]

Graded acidic hydrolysis of di- and triacetals is an important method in structural determination, and the many examples given by Barker and Bourne[7] were important in the formulation of the extended rules. Selective acetolysis with acetic anhydride–acetic acid–sulfuric acid mixtures is of particular value in methylene acetal chemistry. Methylene acetals involving a primary position are preferentially cleaved; thus, acetolysis[11] of 1,3:2,5:4,6-tri-*O*-methylene-D-mannitol followed by saponification gave the 2,5-methylene acetal. The stability sequence (*γ-threo* > *β*) is the reverse of that predicted by the above rules for acid hydrolysis.

The empirical rules have been rationalized in terms of the extents of deformation of alditols from a (supposedly favored) planar, zigzag conformation on cyclic acetal formation,[12] and also in terms of the relative thermodynamic stabilities of the cyclic acetals formed, according to the established tenets of conformational analysis.[13] Mills' approach is more general (it also covers cyclic sugar–aldehyde reactions), but it relates to the molecular situation at true equilibrium. Briefly, the thermodynamic stability sequence of acetal rings is 6 > 5 > 7. For six-membered rings, the most stable ring will have equatorial substituents and higher symmetry (that is, *β-erythro* is favored over *β*), whereas axial substituents at positions 4 and/or 6 (but not position 5) in the 1,3-dioxane ring will cause relative destabilization. Thus, for 1,3:2,4:5,6-tri-*O*-benzylidene-D-glucitol (**16**) the 2,4-ring (*β-erythro*) has three and the 1,3-ring (*β*) two equatorial substituents; each ring also has one axial substituent in the noncritical position. Any other combination of six-membered rings would be sterically less favorable. When the 1,3:2,4-positions are occupied, the sole remaining possibility is for a five-membered (*α*) ring. Formulas **14**, **15**, and **16** illustrate three ways in which 1,3:2,4:5,6-tri-*O*-benzylidene-D-glucitol may be represented involving, respectively, the traditional Fischer projection (**14**), the type of formula (**15**) widely used in alicyclic chemistry and introduced into the carbohydrate field by Mills,[13] and the conformational depiction (**16**).

14   15   16

α-*erythro*-Acetal rings are relatively unstable because of the nonbonded interactions associated with the vicinal *cis* substituents. Among the alditols, glycerol is exceptional in that condensation with aldehydes can be controlled[14] to give preponderantly five- or six-membered cyclic acetals at equilibrium.

B. WITH KETONES

The situation with regard to ketone–alditol reactions is not well documented. The preferred formation of five-membered rings (α, α-*threo*) is possibly due to the fact that six-membered rings must, of necessity, have an axial substituent at the acetal carbon atom and will therefore be destabilized. The reaction of acetone with alditols to give six-membered cyclic acetals is rare; 3-*O*-methyl-D-glucitol gives mainly the 1,2:5,6-di-*O*-isopropylidene derivative (two α-rings) together with ~10% of the 2,4:5,6 isomer (α, β-*erythro* rings).[15] The formation of acetals from acetone and D-glucitol has been studied in detail.[15a]

Variation of the catalyst in acetonation reactions may have a notable effect. In the presence of zinc chloride, D-mannitol affords the 1,2:5,6-diisopropylidene acetal,[16] whereas mineral acid catalyzes formation of the 1,2:3,4:5,6-triacetal.[17] Graded acidic hydrolysis of the latter compound yields 3,4-*O*-isopropylidene-D-mannitol. The pattern of acetonation of some glycosides is also influenced by the nature of the catalyst.[18] Alditols occurring naturally, in fruits[19] and berries,[20] have been characterized as cyclic acetal derivatives.

III. CYCLIC ACETALS FROM CYCLIC SUGAR DERIVATIVES

A. FORMATION

There may be differences in the patterns of reaction of aldehydes and ketones with a particular cyclic sugar derivative. Thus, benzylidenation[21] of D-galactose affords mainly the 4,6-benzylidene acetal with ~10% of the 1,2:3,4-diacetal, whereas on acetonation[22] 1,2:3,4-di-*O*-isopropylidene-D-galactose (18) is the preponderant product. The acetonation of reducing sugars elegantly illustrates the influence of steric effects on reaction pattern. Under comparable reaction conditions, the products from D-glucose, D-galactose, and D-mannose are, respectively, the 1,2:5,6- (17), 1,2:3,4- (18), and 2,3:5,6-diacetals (19). Thus, acetonation of reducing sugars can result in the preponderance of furanose or pyranose products; the reader is referred to Mills' review[13] for a detailed analysis of the steric situation. Until recently it appeared that acetonation of most reducing sugars gave, in each case, a single product. However, the application of analytical methods such as gas–liquid chromatography and mass spectrometry may reveal other isomers, as in the case of D-galactose, which affords small proportions of the 1,2:5,6-diisopropylidene acetal in addition to the 1,2:3,4-diacetal.[23] D-Fructose gives a mixture of

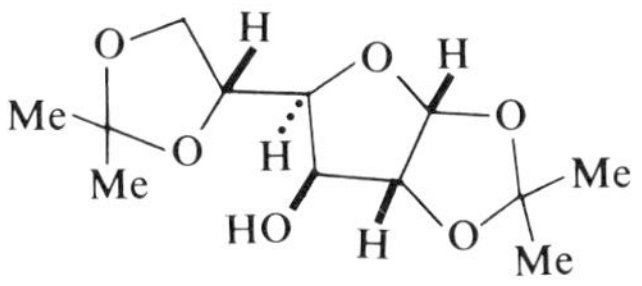

17

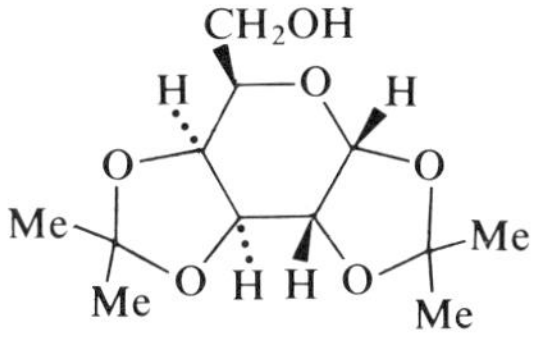

18

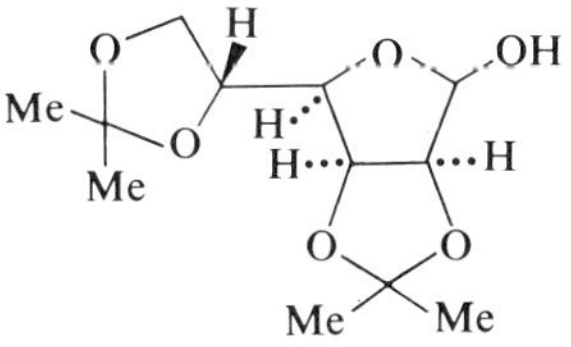

19

1,2:4,5- and 2,3:4,5-diisopropylidene acetals on acetonation.[24] Treatment of D-glucose with acetone containing 4% of sulfuric acid gives,[24a] in addition to the well known 1,2:5,6-diacetal, 1,2:3,5-di-*O*-isopropylidene-α-D-glucofuranose and 1,2:3,4- and 2,3:4,5-di-*O*-isopropylidene-D-glucoseptanose. Acetonation of D-allose gives mainly 2,3:5,6-di-*O*-isopropylidene-β-D-allofuranose.[24b]

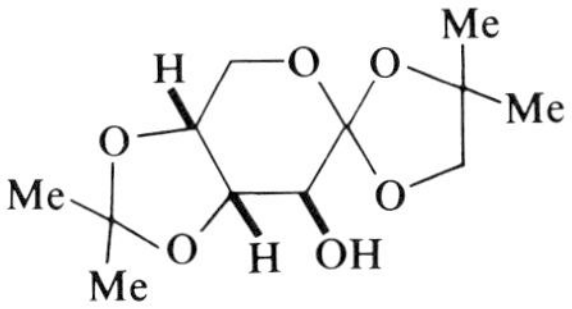

1,2:4,5-Di-*O*-isopropylidene-
β-D-fructopyranose

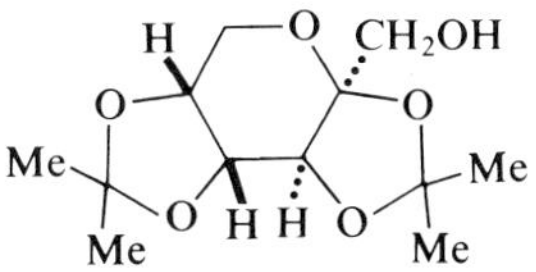

2,3:4,5-Di-*O*-isopropylidene-
β-D-fructopyranose

Ketones usually react with vicinal *cis*-hydroxyl groups attached to five- or six-membered cyclic sugar derivatives. For example, methyl α-D-arabinopyranoside affords the 2,3-isopropylidene acetal **20** (but compare ref. 18). On the other hand, aldehydes react to give five-, six-, and seven-membered cyclic acetals. Thus, benzylidenation of methyl α-D-galactopyranoside affords the 4,6-benzylidene acetal (**21**), whereas methyl α-D-mannopyranoside gives

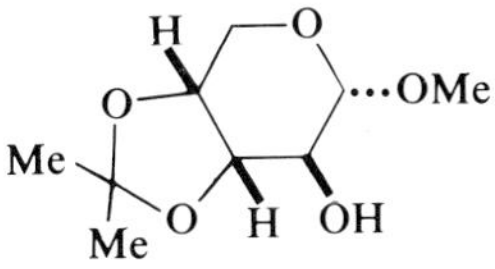

20

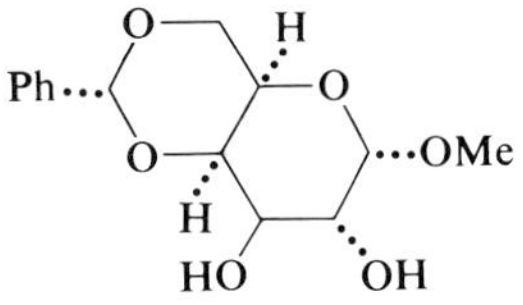

21

* *References start on p.* 401.

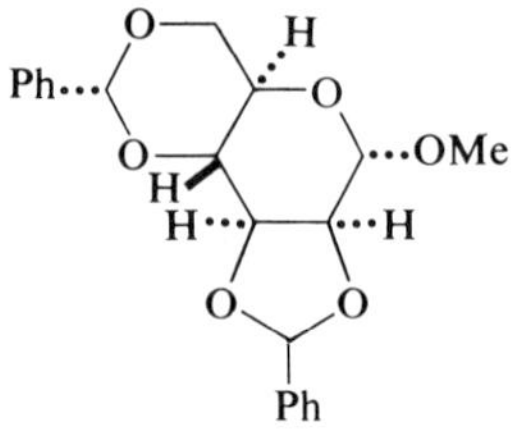

two isomeric 2,3:4,6-dibenzylidene acetals (**22**), differing only in the stereo-chemistry at the 2,3-acetal carbon atom. Ethylidenation[25] of methyl α-D-glucopyranoside affords methyl 4,6-*O*-ethylidene-2,3-*O*-oxidodiethylidene-α-D-glucopyranoside (**23**) if paraldehyde is used; the oxidodiethylidene ring is not formed if 1,1-diethoxyethane[26] is the reagent. The ring system in methyl 4,6-*O*-benzylidene-α-D-glucopyranoside is related to *trans*-decalin, whereas that in the galactose analog is *cis*-decalin in type.

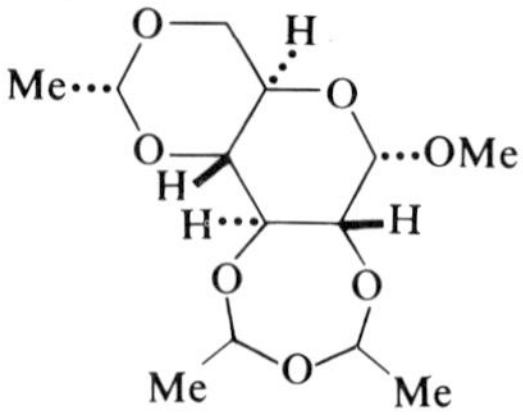

B. Applications in Synthesis

The cyclic acetal derivatives of carbohydrates are especially useful in synthesis, since they provide a convenient means of specific and partial blocking, and they have been involved in syntheses of deoxy, amino, methyl-ated, branched-chain, and other sugar derivatives.[27] Since cyclic acetals are stable under neutral and quite strongly basic conditions, they survive esterifica-tion and etherification reactions and the application of a variety of oxidants (including sodium periodate and lead tetraacetate) and reductants. For example, methyl 4-6-*O*-benzylidene-α-D-glucopyranoside is readily converted into a 2- or 2,3-di-*p*-toluenesulfonate and then by the action of base into 2,3-anhydro-D-mannose or D-allose derivatives, as shown in the above reaction scheme. The susceptibility of the anhydro ring to nucleophiles[28] offers a route to a wide variety of derivatives, as illustrated by the conversion into 3-amino-3-deoxy and 2-thio-D-altrose derivatives.

O-Isopropylidene derivatives are involved[29] in the commercial synthesis of vitamin C (see Chapter 23).

Cyclic acetals of carbohydrates are of great value in the synthesis of keto sugars[30] (see Chapter 23). For example, 1,2:5,6-di-*O*-isopropylidene-α-D-glucofuranose is converted into the 3-keto compound by phosphorus penta-oxide and methyl sulfoxide.[31] Such carbonyl compounds provide access to amino-, deoxy-, and branched-chain sugar derivatives (see Vol. IB, Chapters 16 and 17).

An isopropylidene acetal spanning vicinal *cis*-hydroxyl groups in a furanose ring is a particularly stable arrangement. Thus, 1,2:5,6-di-*O*-isopropylidene-

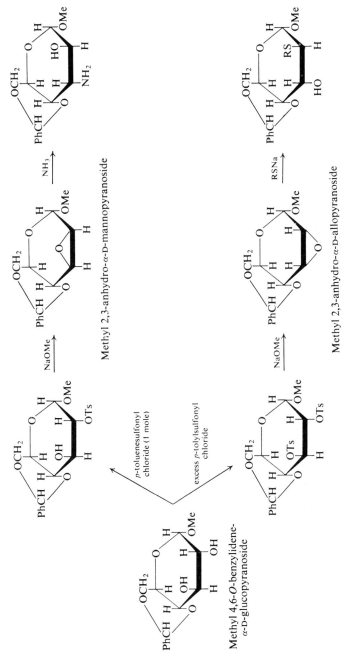

Methyl 2,3-anhydro-α-D-mannopyranoside

Methyl 2,3-anhydro-α-D-allopyranoside

Methyl 4,6-O-benzylidene-α-D-glucopyranoside

α-D-glucofuranose is readily converted by mild acid treatment into the 1,2-isopropylidene acetal. The exposed vicinal diol grouping can then be

cleaved with periodate. Treatment of the resulting aldehyde with $H^{14}CN$ followed by hydrolysis and reduction regenerates 1,2-*O*-isopropylidene-α-D-glucofuranose-6-^{14}C, from which D-glucose-6-^{14}C is obtained on acidic hydrolysis.[37] Derivatives of sugars labeled with ^{14}C (see Chapter 3) are of great value in studies of metabolism and mechanism. Cyclic acetal formation may be used to confer rigidity on a sugar molecule and to allow selective reaction to be accomplished more easily.[33] Thus, methyl 4,6-*O*-ethylidene-α-D-mannopyranoside has a relatively rigid *trans*-decalin type of ring system (24) and, with 1 mole of *p*-toluenesulfonyl chloride in pyridine, preferential sulfonylation of the sterically more accessible, equatorial hydroxyl group at C-3 occurs (25).

              24   25

The stability of cyclic acetals toward acid is controlled by the nature of the group attached to the acetal carbon atom. Benzylidene and ethylidene acetals are much more sensitive to acid than are methylene acetals, and when a CCl_3 or CF_3 group is present, marked resistance to acid may be conferred.[34] 1,3:4,6-Di-*O*-methylenegalactitol is stable to boiling $0.1N$ hydrochloric acid;[35] most benzylidene and ethylidene acetals would be hydrolyzed rapidly under these conditions. Benzylidene acetals are of particular value, since they can be removed under neutral conditions by hydrogenolysis[27, 36] or by treatment with *N*-bromosuccinimide.[37] Nuclear magnetic resonance spectroscopy has been used to assign structure and conformation to a variety of carbohydrate cyclic acetals.[38] Isopropylidene acetals are converted by triphenylmethyl fluoroborate into α-keto alcohols; this reaction promises to be useful in synthetic transformations of carbohydrates.[39]

TABLE I

CARBOHYDRATE ACETAL DERIVATIVES OF PARTICULAR VALUE IN SYNTHETIC WORK[27]

Derivative	Reference
1,2:5,6-Di-*O*-isopropylidene-α-D-glucofuranose	a
1,2:3,4-Di-*O*-isopropylidene-α-D-galactopyranose	b
2,3:5,6-Di-*O*-cyclohexylidene-D-mannofuranose	c
4,6-*O*-Benzylidene-D-glucopyranose	d
2,4-*O*-Benzylidene-D-glucitol	e
1,3:2,4-Di-*O*-ethylidene-D-glucitol	f
1,2:5,6-Di-*O*-isopropylidene-D-mannitol	g
cis-1,3-*O*-Benzylideneglycerol	h
Methyl 3,4-*O*-isopropylidene-α-L-arabinopyranoside	i
Methyl 2,3-*O*-isopropylidene-α-L-rhamnopyranoside	j
Methyl 3,4-*O*-isopropylidene-α-D-galactopyranoside	k
Methyl 4,6-*O*-benzylidene-α-D-glucopyranoside	l
Methyl 4,6-*O*-benzylidene-α-D-galactopyranoside	m
2,3-*O*-Isopropylidene-D-ribono-1,4-lactone	n

[a]E. Fischer, *Ber.*, **28**, 1145 (1895); R. S. Tipson, H. S. Isbell, and J. E. Stewart, *J. Res. Nat. Bur. Stand.*, **62**, 257 (1959). [b]P. A. Levene and G. M. Meyer, *J. Biol. Chem.*, **64**, 473 (1925). [c]R. D. Guthrie and J. Honeyman, *J. Chem. Soc.*, 853 (1959). [d]P. Brigl and H. Grüner, *Ber.*, **65**, 1428 (1932). [e]S. J. Angyal and J. V. Lawler, *J. Amer. Chem. Soc.*, **66**, 837 (1944). [f]R. C. Hockett and F. C. Schaefer, *J. Amer. Chem. Soc.*, **69**, 849 (1947). [g]E. Baer, J. M. Grosheintz, and H. O. L. Fischer, *J. Amer. Chem. Soc.*, **61**, 2607 (1939). [h]See ref. 14. [i]J. Honeyman, *J. Chem. Soc.*, 990 (1946). [j]P. A. Levene and J. Compton, *J. Biol. Chem.*, **111**, 325 (1935). [k]R. G. Ault, W. N. Haworth, and E. L. Hirst, *J. Chem. Soc.*, 1012 (1935). [l]W. Alberda van Ekenstein and J. J. Blanksma, *Rec. Trav. Chim.*, **25**, 153 (1906). [m]G. J. Robertson and R. A. Lamb, *J. Chem. Soc.*, 1321 (1934). [n]J. Baddiley, J. G. Buchanan, and F. E. Hardy, *J. Chem. Soc.*, 2180 (1961).

REFERENCES

1. A. Wurtz, *Compt. Rend.*, **53**, 378 (1861); *Ann.*, **120**, 328 (1861).
2. J. Meunier, *Compt. Rend.*, **107**, 910 (1888).
3. F. S. Al-Jeboury, N. Baggett, A. B. Foster, and J. M. Webber, *Chem. Commun.*, 222 (1966).

4. T. G. Bonner, E. J. Bourne, P. J. V. Cleare, and D. Lewis, *Chem. Ind.* (London), 1268 (1966); compare T. G. Bonner, E. J. Bourne, P. J. V. Cleare, R. F. J. Cole, and D. Lewis, *J. Chem. Soc.* (*B*), 957 (1971).

5. R. U. Lemieux, in "Molecular Rearrangements," P. de Mayo, Ed., Interscience, New York, 1964, p. 709.

6. N. Baggett, J. M. Duxbury, A. B. Foster, and J. M. Webber, *Carbohyd. Res.*, **1**, 22 (1965); compare N. Baggett, M. Mosihuzzaman, and J. M. Webber, *ibid.*, **11**, 263 (1969).

7. S. A. Barker and E. J. Bourne, *Advan. Carbohyd. Chem.*, **7**, 137 (1952).

8. R. M. Hann and C. S. Hudson, *J. Amer. Chem. Soc.*, **66**, 1909 (1944).

9. See footnote in K. W. Buck, A. B. Foster, B. H. Rees, J. M. Webber, and J. Lehmann, *Carbohyd. Res.*, **1**, 329 (1965).

10. N. Baggett, K. W. Buck, A. B. Foster, R. Jefferis, B. H. Rees, and J. M. Webber, *J. Chem. Soc.*, 3382 (1965).

11. A. T. Ness, R. M. Hann, and C. S. Hudson, *J. Amer. Chem. Soc.*, **66**, 665 (1944).

12. S. A. Barker, E. J. Bourne, and D. H. Whiffen, *J. Chem. Soc.*, 3865 (1952).

13. J. A. Mills, *Advan. Carbohyd. Chem.*, **10**, 1 (1955); see also J. F. Stoddart, "Stereochemistry of Carbohydrates," Wiley-Interscience, 1971, pp. 186–220.

14. N. Baggett, J. M. Duxbury, A. B. Foster, and J. M. Webber, *Carbohyd. Res.*, **2**, 216 (1966).

15. A. B. Foster, M. H. Randall, and J. M. Webber, *J. Chem. Soc.*, 3388 (1965).

15a. T. G. Bonner, E. J. Bourne, R. F. J. Cole, and D. Lewis, *Carbohyd. Res.*, **21**, 29 (1972).

16. E. Baer and H. O. L. Fischer, *J. Amer. Chem. Soc.*, **61**, 761 (1939).

17. L. F. Wiggins, *J. Chem. Soc.*, 13 (1946).

18. J. G. Buchanan and R. M. Saunders, *J. Chem. Soc.*, 1796 (1964).

19. Y. Asahina and H. Shinoda, *J. Pharm. Soc. Japan*, **50**, 1 (1930).

20. H. H. Strain, *J. Amer. Chem. Soc.*, **56**, 1756 (1934).

21. H. Zinner and W. Thielebeule, *Ber.*, **93**, 2791 (1960); J. Pacák and M. Černý, *Collect. Czech. Chem. Commun.*, **26**, 2212 (1961).

22. K. Freudenberg and R. M. Hixon, *Ber.*, **56**, 2119 (1923).

23. D. C. DeJongh and K. Biemann, *J. Amer. Chem. Soc.*, **86**, 67 (1964).

24. C. P. Barry and J. Honeyman, *Advan. Carbohyd. Chem.*, **7**, 68 (1952).

24a. J. D. Stevens, *Chem. Commun.*, 1140 (1969).

24b. M. Haga, M. Takano, and S. Tejima, *Carbohyd. Res.*, **14**, 237 (1970).

25. H. Appel and W. N. Haworth, *J. Chem. Soc.*, 793 (1938).

26. D. O'Meara and D. M. Shepherd, *J. Chem. Soc.*, 2778 (1955).

27. A. N. de Belder, *Advan. Carbohyd. Chem.*, **20**, 220 (1965); R. F. Brady, *Advan. Carbohyd. Chem. Biochem.*, **26**, 197 (1971).

28. S. Peat, *Advan. Carbohyd. Chem.*, **2**, 37 (1946); N. R. Williams, *Advan. Carbohyd. Chem. Biochem.*, **25**, 109 (1970).

29. T. Reichstein and A. Grüssner, *Helv. Chim. Acta*, **17**, 311 (1934).

30. J. S. Brimacombe, *Chem. Ind.* (London), 99 (1966); R. F. Butterworth and S. Hanessian, *Synthesis*, 70 (1971).

31. K. Onodera, S. Hirano, and N. Kashimura, *J. Amer. Chem. Soc.*, **87**, 4651 (1965).

32. J. C. Sowden, *J. Amer. Chem. Soc.*, **74**, 4377 (1952).

33. G. O. Aspinall and G. Zweifel, *J. Chem. Soc.*, 2271 (1957).

34. E. J. Bourne, A. J. Huggard, M. Stacey, and J. C. Tatlow, *J. Chem. Soc.*, 2716 (1960).

35. R. M. Hahn, W. T. Haskins, and C. S. Hudson, *J. Amer. Chem. Soc.*, **64**, 986 (1942).

36. H. B. Wood, H. W. Diehl, and H. G. Fletcher, *J. Amer. Chem. Soc.*, **79**, 1986 (1957).

37. S. Hanessian, *Carbohyd. Res.*, **2**, 86 (1966); S. Hanessian and N. R. Plessas, *J. Org. Chem.*, **34**, 1053 (1969).

38. N. Baggett, K. W. Buck, A. B. Foster, and J. M. Webber, *J. Chem. Soc.*, 3401 (1965).

39. P. D. Magnus, D. H. R. Barton, G. Smith, and D. Zurr, *Chem. Commun.*, 861 (1971).

12. ETHERS OF SUGARS

J. K. N. JONES AND G. W. HAY

External and internal ethers of sugars are recognized. The occurrence and chemistry of the former only will be discussed in this chapter. Many methyl ethers of sugars and of sugar derivatives are known.

I. TYPES OF SUGAR ETHERS AND THEIR OCCURRENCE

The great majority of the methylated sugars characterized have been prepared synthetically or have been isolated from the hydrolysis products of methylated di-, oligo-, or polysaccharides. Some of the more unusual and rare types, however, have been encountered as constituents of antibiotic substances (see Vol. IIA, Chap. 31) and of naturally occurring, methylated polymers (see Vol. IIA, Chap. 37). Most of these sugars have now been synthesized (see Chapter 31). Some of them contain tertiary hydroxyl groups, which are particularly difficult to methylate *in vitro*.

Methyl ethers of sugars and of polysaccharides are of considerable academic and industrial importance. Until recently the determination of structure of polysaccharides depended mainly on the separation and identification of methyl ethers resulting from the hydrolysis of methylated polysaccharides. Indeed this procedure is still of great utility, and separation of the methyl ethers, either alone or as their acetyl,[1] trimethylsilyl,[2] or dimethylsilyl[3] derivatives, by gas–liquid, thin-layer,[4] paper,[5] and column chromatography[6] has greatly facilitated the qualitative identification and quantitative determination of these derivatives. In some cases it is possible to identify the position of ether groups in a sugar by physical means, such as by nuclear magnetic resonance[7] or mass spectrometry,[7a] or by the use of electrophoresis in certain complexing buffer solutions,[8] or by oxidation[9] with lead tetraacetate or periodate. The use of gas–liquid chromatography in combination with mass spectrometry for the structural identification of partially methylated sugars as their alditol acetates or per-O-trimethylsilyl derivatives is a most valuable tool in structural work on polysaccharides.[10] It is still desirable, however, to synthesize the unknown · sugar for definitive characterization. Such syntheses may involve the intermediate preparation of several other ether derivatives of the sugars. For example, the O-benzyl group[11] may be used as a protecting group which is relatively stable to both acids and bases but can be removed by hydrogenolysis,[12] or the triphenylmethyl (trityl) group may be employed;[13] the latter may be removed by hydrogenolysis,[14] or by treatment with dilute acids.[15]

TABLE I

RATE OF REACTION OF SUGAR DERIVATIVES WITH CHLOROTRIPHENYLMETHANE[a]

Substance	Excess of chlorotriphenylmethane	k in hr^{-1} at 20°
1,2:3,4-Di-O-isopropylidene-α-D-galacto-pyranose	4-fold	0.014
	8-fold	0.036
2,3:4,6-Di-O-isopropylidene-L-sorbofuranose	4-fold	0.0052
	8-fold	0.0055
1,2:5,6-Di-O-isopropylidene-α-D-glucofuranose	4-fold	0.00012
	8-fold	0.00016

[a] R. C. Hockett, H. G. Fletcher, Jr., and J. B. Ames, *J. Amer. Chem. Soc.*, **63**, 2516 (1941).

However, it may be removed by alcoholysis,[16] which, at times, is a disadvantage. O-Tritylation usually proceeds much more rapidly with primary hydroxyl groups[17] than with secondary hydroxyl groups[18] and has been used to determine quantitatively the number of primary hydroxyl groups in a molecule.[19] This procedure is not always satisfactory, as in some cases equatorial secondary hydroxyl groups may be attacked quite rapidly. Axial groups react much more slowly. A pentose sugar that normally exists mainly in the pyranose form—for example, L-arabinose—may react in the furanose form, in which the more reactive primary hydroxyl group is present.[20] The O-allyl derivatives[21] of carbohydrates have also been used as protecting groups. They can be prepared in the usual way (see below), but in methyl sulfoxide solution and in the presence of potassium *tert*-butoxide they undergo isomerization to 1-propenyl ethers, which are readily removed by acids under very mild conditions.[22] Vinyl ether derivatives are known and have been used to make polymers containing sugar residues.[23] They have also been used in the synthesis of plasmalogens. In this procedure, cyclic α-bromoacetals are treated with sodium to yield 1-alkenyl ethers.[24] The extension of such syntheses to give ethynyl ethers[24a] affords access to a substituent of potentially great interest. Ethyl ethers of sugars have been prepared,[25] either directly or by reduction of the corresponding vinyl ether derivatives. 2-Hydroxyethyl ethers of sugars are also known and have been prepared from ethylene oxide or ethylene chlorohydrin and the appropriate sugars,[26] in order to identify the O-2-hydroxyethyl derivatives resulting from the hydrolysis of O-(2-hydroxyethyl)-cellulose[27] or other O-(2-hydroxyethyl)-ated polysaccharides.[28] Other ethers, such as carboxymethyl,[29] 2-cyanoethyl,[30] and 2-(diethylamino)ethyl[31] ethers are of interest mainly in connection with the commercial production of polysaccharide derivatives. Some of them, such as O-(carboxymethyl)cellulose (CM-cellulose), O-(2-diethylaminoethyl)cellulose (DEAE-cellulose), and O-(2-diethylaminoethyl)-Sephadex (DEAE-Sephadex), are important, as they can be used for chromatographic fractionation of charged polymers such as proteins and acidic (or potentially acidic) polysaccharides.[32] 2,4-Dinitrophenyl ethers of sugars have been prepared[33] but have not received much attention. Methylene iodide and benzylidene chloride react with sugars in the presence of base to produce cyclic derivatives that are not ethers but acetals[34] (see Chapter 11). Tetrahydropyranyl ethers may be used for the synthesis of partially methylated sugars,[35] but they suffer for the disadvantage that two diastereoisomers result when they are formed. However, the reagent 4-methoxy-5,6-dihydro-2*H*-pyran can be used to replace dihydropyran as it forms only one ether derivative on reaction with a alcohol, whereas dihydropyran forms two.[35a] Unusual ethers of methyl α-D-glucoside have been made by a base-catalyzed reaction with epoxy-octadecane,[35b] and ethers containing an oxirane or a thiirane ring have been introduced at C-3 of 1,2:5,6-di-O-isopropylidene-D-glucofuranose and were subsequently converted into glycol ethers.[35c] Paulsen *et al.*[35d] have reported the formation of 2,3,4-tri-O-acetyl-6-O-(dichloromethyl)-α-D-glucopyranosyl chloride as one product of the reaction of 2,3,4-tri-O-acetyl-1,6-anhydro-β-D-glucopyranose with an excess of dichloromethyl methyl ether in chloroform in the presence of a catalytic amount of anhydrous zinc chloride. An intermediate 6-formic ester was postulated. The reaction can also be effected with the

* *References start on p. 416.*

bromide analogs. Benzylidene acetals may be converted into benzyl ethers on reaction with aluminum hydride, a procedure that has been exploited to make various methyl ethers of D-mannose.[35e] An unusual method of obtaining methyl ethers is via the (methylthio)methyl ethers,[35f] obtained as byproducts in the oxidation of sugars by the methyl sulfoxide–N,N'-dicyclohexylcarbodiimide reagent.[35g] These (methylthio)methyl ethers are reduced by Raney nickel to a mixture of the alcohol and its methyl ether. Thioetherification has been achieved by the action of ethanethiol and boron trifluoride on α-L-arabinopyranose tetraacetate, whereupon a 5-S-ethyl-5-thio derivative of L-arabinose diethyl dithioacetal results.[36] The Procean dyestuffs react with hydroxyl groups in polysaccharides to give stable ether derivatives.

II. THE REMOVAL OF ETHER GROUPS

Sugars can be regenerated from their benzyl and trityl ethers by hydrogenolysis[12, 14] or by addition of sodium to the substance dissolved in ethanol or liquid ammonia. Closely related derivatives are the benzyloxycarbonic esters (see Chapter 6), which are also susceptible to hydrogenolysis.[37] The O-trityl group can be removed readily by treatment of the compound with hydrogen chloride or hydrogen bromide in acetic acid. It is much more difficult to recover a sugar in good yield from its methyl ether. In some instances O-demethylation by hydrogen bromide has been used,[38] the product then being subjected to chromatography to identify the parent sugar. O-Demethylation by hydrogen iodide is used to determine quantitatively[39] the methoxyl content of a sugar. Boron trichloride is a much more satisfactory reagent and in some cases permits recovery of the parent sugar in fair yield.[40] Boron tribromide in dichloromethane has also been used to demethylate sugars. O-Demethylation may also be achieved with diborane and iodine.[41a] However, if the ether group is to be removed at a later stage to regenerate the sugar in good yield, the allyl ether is a more suitable derivative. The potential of this ether group has as yet been little exploited.[21] Demethylation at C-2 is also achieved in formation of a phenylosazone, usually with much degradation. Methyl ether groups can be removed from sugars by oxidizing them first to formic esters with chromium trioxide in acetic acid, with subsequent hydrolysis of the ester. For example, 1,2:5,6-di-O-isopropylidene-3-O-methyl-D-glucose is oxidized to the corresponding 3-formate.[41b]

III. METHODS FOR SYNTHESIS OF PARTIALLY METHYLATED SUGARS

Methyl ethers of the aldoses, ketoses, uronic acids, and aminodeoxy sugars are of great importance, as they are key compounds used in the determination of polysaccharide structure. For this reason much effort has been made to synthesize many partially methylated sugars. These methylations may be performed in a random fashion, the products being separated by fractionation, with subsequent identification, or they may be prepared from derivatives of known structure in which hydroxyl groups are protected by such groups as acetals, ortho esters, benzeneboronates,[42] and benzyl ethers.[42a] In some cases the O-acetyl or O-benzoyl[43] group has been used for the purpose of blocking, but as there is a tendency for these groups to migrate[43a] under the influence of base the resultant methyl ether may not be identified unambiguously. The benzyloxythiocarbonyl group[44] does not wander under mildly alkaline conditions and may, therefore, be used as a protecting group, as may the nitrate ester[45] (see Chapter 6). Carbamate derivatives (from isocyanates) have also been used as protecting groups.[46]

A. METHYLATION PROCEDURES

Barker et al.[47] showed that partial methylation of methyl β-D-glucopyranoside with methyl iodide and thallium hydroxide led to the formation of the highly crystalline 2,4,6-trimethyl ether. Similarly, methyl β-D-xylopyranoside, on partial methylation, either by the thallium hydroxide–methyl iodide procedure[48] or by the Haworth technique (methyl sulfate–aqueous sodium hydroxide), yielded mainly the 2,4-dimethyl ether.[49] More satisfactory syntheses are those of 2- and 3-O-methyl-D-mannose,[50] obtained by preferential esterification of the equatorial C-3 hydroxyl group of methyl 4,6-O-ethylidene-α-D-mannoside. Similarly, 1,6-anhydro-β-D-mannopyranose is esterified preferentially at the C-2 hydroxyl group (equatorial); this product thus affords a route for synthesis of 3,4-di-O-methyl-D-mannose. Partial methylation of sugar diethyl dithioacetals also leads to the isolation, sometimes in good yield, of the 2-methyl ether.[51, 51a]

Opening of an epoxide ring by sodium methoxide can, in principle, yield two isomers, but sometimes one isomer preponderates greatly. This procedure has been used to prepare mono- and dimethyl ethers that are difficult to prepare by the usual procedures. Methyl 2,3-anhydro-4,6-O-benzylidene-α-D-alloside is opened by sodium methoxide to yield mainly the 2-O-methyl-D-altroside derivative.[52] The stereoselectivity of this process may be attributed to the obligatory diaxial opening (Fürst–Plattner rule) of the locked pyranose ring system.[53] A second example of the synthesis of methyl ether that is difficult to obtain by other procedures is the opening of the epoxide ring, after methylation, of methyl 2,3-anhydro-β-D-ribopyranoside, (1) with sodium hydroxide, which leads to practically exclusive formation of 4-O-methyl-β-D-xylopyrano-

* References start on p. 416.

side or (2) with sodium methoxide, which gives methyl 3,4-di-O-methyl-β-D-xylopyranoside.[54] The results of opening of epoxide rings (see this volume, Chap. 13) are not always as clear-cut as this. In some cases, if there is a hydroxyl group suitably situated, the epoxide ring may migrate[55] or a 2,6-, 3,6-, 2,5-, or 5,6-anhydro ring may result.[56]

Methylation of sugar acetals has been used frequently for synthesis of methyl ethers of sugars.[57] For example, methylation of 1,2:5,6-di-O-isopropylidene-α-D-glucofuranose yields the 3-methyl ether,[58] and 1,2-O-isopropylidene-α-D-glucofuranose gives the 3,5,6-trimethyl ether.[59] The position of the methoxyl groups in these sugars was used, in fact, to determine the structure of the O-isopropylidene derivatives.[60] As a modification of the first reaction, the 3-benzyl ether may be formed. This ether is relatively stable to acids, and with methanolic hydrogen chloride it yields mainly methyl 3-O-benzyl-α-D-glucopyranoside, which may then be methylated to the 2,4,6-trimethyl ether.[61] The observations that acetone usually condenses with adjacent *cis*-hydroxyl groups to form a five-membered ring, and that benzaldehyde reacts to form a six-membered ring, have permitted the synthesis of many methyl ethers of sugars.[57] However, it is always advisable to check the structure of the methyl ethers, because acetone sometimes condenses with the formation of a six-membered ring,[62] and benzaldehyde may yield a five-membered ring[63] (see this volume, Chap. 11). Migration of the O-isopropylidene group has also been observed.[64] Formaldehyde condenses to yield O-methylene derivatives, which may then be methylated. In this fashion 6-O-methyl-D-glucose and 6-O-methyl-D-galactose are readily prepared.[65]

A disadvantage which may be encountered in the methylation of a *p*-tolyl-sulfonyl or methylsulfonyl derivative is the possible formation of an anhydro ring[66] by participation of a suitably located, adjacent *trans*-hydroxyl group under the influence of the base (silver oxide) used in the methylation reaction. A second possibility which should also be considered is the migration of anhydro rings during methylation reactions, again under the influence of base,[55] and a third is the direct replacement of a primary methylsulfonyloxy group by methoxyl when the reaction is performed in N,N-dimethylformamide.[66] It has been shown[66a] that methyl 2,3-di-O-methyl-6-O-methylsulfonyl-β-D-galactopyranoside gave methyl 2,3,6-tri-O-methyl-β-D-galactopyranoside as well as methyl 3,6-anhydro-2-O-methyl-β-D-galactopyranoside, the latter by methyl-group participation, on solvolysis in 50% aqueous methanol. Esters of carboxylic acids are not frequently used as protecting groups, for the reasons already stated.[66b] A notable exception to this is the ortho ester grouping. Attempts to prepare a methyl glycoside, from a poly-O-acetylglycosyl bromide having the halogen atom *trans* to an acetoxy group at C-2, may give an ortho ester.[67] Ortho esters are comparatively stable to base and are not affected when acetyl groups are removed by base. The synthesis of 3,4,6-tri-O-methyl-D-mannose[68] and of 3,4-di-*O*-methyl-L-rhamnose follows from this observation.[69]

B. USE OF ACYCLIC DERIVATIVES

Acyclic derivatives of the sugars are sometimes used in the synthesis of methyl ethers, as the hydroxyl groups normally engaged in ring formation can remain free.[70,71] For example, 2,3:5,6-di-*O*-isopropylidene-D-glucose dimethyl acetal and the corresponding D-galactose derivative have a hydroxyl group free at C-4, which permits the synthesis of the otherwise difficultly accessible 4-methyl ethers of these sugars.

It should be noted that some methyl ethers of aldoses are reduced by sodium borohydride more readily than are others. For example, the 3-methyl ethers of the hexoses and 4-*O*-methylhexuloses are reduced[72] much more slowly than are those that have no substituent at C-3 and C-4.

C. METHYLATION REAGENTS

1. *Diazomethane*

This reagent methylates acidic and enolic[73] hydroxyl groups rapidly and methylates other more reactive hydroxyl groups, such as primary hydroxyl groups, slowly.[74] Polysaccharides cannot be methylated completely by this reagent even when catalysts such as boron trifluoride are added.†

Acetylated sugars that possess one hydroxyl group free can be methylated with diazomethane in the presence of boron trifluoride etherate,[74a] and it has been shown[74b] that diazomethane in the presence of stannous chloride, methanol, and *N,N*-dimethylformamide will methylate sugars. For example, methyl α-D-glucopyranoside yields the 3-methyl ether, and diazoethane gives 92% of the 3-ethyl ether (see also ref. 74c).

2. *Purdie's Method and Modifications*

The first methylation procedure was devised by Purdie[75] and was very successful for compounds that are soluble in methyl iodide. The solution is usually heated under reflux with freshly prepared, dry silver(I) oxide in the presence of a drying agent such as calcium sulfate. This procedure has been

† In one experiment the authors[75a] treated ash-free cherry gum with methanolic and ethereal diazomethane over a period of three years. At the end of this time the methoxyl content of the product was ~28%, and on hydrolysis it gave fully methylated arabinofuranose and partially and unmethylated sugars. No solvent was found that would dissolve the partially methylated polysaccharide.

* *References start on p. 416.*

modified over the years, and methylation is now effected commonly in N,N-dimethylformamide or methyl sulfoxide solution. These two dipolar, aprotic solvents facilitate the reaction by stabilizing the transition state, and they do not form hydrogen bonds as does methanol, for example. The sulfoxide has also been used as a diluent and solvent. Barium hydroxide and barium oxide,[76] barium oxide alone,[77] sodium,[78] and sodium hydride[79] have all been used as substitutes for silver oxide. The disadvantage in using N,N-dimethylformamide and methyl sulfoxide is their high boiling point and the tendency of N,N-dimethylformamide to dissolve silver salts. Tetrahydrofuran[80] and 1,2-dimethoxyethane have also been used as solvents.

A significant advance in the methylation of polysaccharides was achieved by Hakomori,[80a] who used sodium methylsulfinyl carbanion in methyl sulfoxide solution to convert the polysaccharide into its sodium salt under anhydrous conditions and then made a methylated polysaccharide by addition of methyl iodide to the solution. This procedure was based on an observation of Corey and Chakovsky[80b] and is very useful for the methylation of milligram quantities of material. Claims have been made that the technique results in a lowering of molecular weight of the polymer.[80c] Coupled with gas–liquid chromatography–mass spectrometric techniques for identification of partially methylated sugars the Hakomori procedure provides a rapid method for methylation linkage-analysis applicable with polysaccharides on the milligram scale.[10] This procedure has been applied with particular effectiveness in elucidating the structures of *Salmonella* and other bacterial lipopolysaccharides.[10, 80d]

3. *Haworth Methylation Procedure*

Haworth[81] used methyl sulfate and 30% aqueous sodium hydroxide to methylate sugars. The procedure is particularly successful for methylating nonreducing mono- and oligosaccharides[82] and water-soluble polysaccharides or acetylated polysaccharides.[83] Reducing sugars are decomposed rapidly by the sodium hydroxide solution, but they can be methylated in fair yield if the rate of addition of the reagents is such that the solution remains at or near pH 7 until all the reducing sugar has been converted into glycosides. A disadvantage is that several glycosides and a mixture of methylated sugars may result.

4. *Methyl Chloride Procedure*[83a]

The procedure is used industrially to methylate cellulose. Cellulose is steeped in concentrated sodium hydroxide, squeezed dry, and then heated under pressure in closed tanks in the presence of methyl chloride. Salts are washed out of the methylated polymer at the end of the reaction. Methyl chloride is rarely used in the laboratory to methylate sugars.

5. *Thallium Hydroxide*

Concentrated, warm thallium hydroxide solution, made from a solution of thallium formate by passage down a column of anion-exchange resin, is added to an aqueous solution of a polysaccharide or nonreducing sugar derivative. The yellow, insoluble complex which forms is collected, dried *in vacuo*, powdered, and heated in suspension in methyl iodide.[47] The disadvantages are that thallium salts are cumulative poisons and that some materials are degraded during formation of the complex One operation rarely results in the production of a fully methylated derivative.

6. *Muskat Procedure*[84]

Liquid ammonia is used as a solvent in this procedure. The whole experiment is conducted in an enclosed apparatus. The sugar derivative (nonreducing), dissolved in liquid ammonia, is treated with the calculated quantity of sodium metal, and methyl iodide is then added to the solution. This procedure has been developed as a method for micro methylation.[85]

7. *Use of Sugar Sulfonate Derivatives*

A *p*-tolylsulfonyloxy[86] or (methylsulfonyl)oxy[87] group present on the primary alcohol of a suitably substituted sugar may be replaced by a methoxyl group by heating the compound in methanolic sodium methoxide solution. In some cases a secondary, axial, (methylsulfonyl)oxy group may be replaced by methoxyl, with inversion.[87a]

8. *Use of Deoxy-nitro Sugars*

Treatment of a 1-nitro-1-alkene with sodium methoxide yields the 2-methoxy-1-nitro derivative. For example, 1-*C*-nitro-D-*arabino*-3,4,5,6-tetra-acetoxy-1-nitro-1-hexene, on treatment with sodium methoxide, yields a mixture of products from which 2-*O*-methyl-D-glucose and 2-*O*-methyl-D-mannose can be isolated by way of the Nef reaction[88] (see Chapter 3).

9. *Epimerization*

Reducing sugars that are methylated are epimerized in basic solution. This reaction may facilitate the identification of an unknown methylated sugar.[38]

10. *Use of Periodate and Lead Tetraacetate.*

These reagents split vicinal glycol groups and thus can be used to make methylated sugars containing fewer carbon atoms than the starting material

* *References start on p. 416.*

(see Vol. IB, Chap. 25). For example, 3,4,6-tri-O-methyl-D-glucose is oxidized to 2,3,5-tri-O-methyl-D-arabinose. In a more complicated case, 1,2:5,6-di-O-isopropylidene-3-O-methyl-α-D-glucofuranose after partial hydrolysis affords the 1,2-O-isopropylidene derivative which, after periodate oxidation, reduction with sodium borohydride, and hydrolysis, yields 3-O-methyl-D-xylose. Other methods of descent of the series, such as the Weerman reaction,[89,90] the Ruff degradation,[91] or the Wohl procedure,[92] may also be used.

D. METHYLATION OF ACIDIC AND BASIC SUGARS

Complications sometimes occur in the methylation of acidic and basic sugars. For example, if 5,6-O-isopropylidene-L-ascorbic acid is methylated with diazomethane, a 2,3-dimethyl ether results.[93] However, if the methylation is performed on the sodium salt with methyl sulfate, some C-methyl sugar is produced.[94] Similarly, if an attempt is made to prepare a O-benzyl derivative with benzyl chloride and alkali, some C-benzyl derivatives are obtained; the relative proportions vary according to the solvent employed.[95] Uronic acids, when methylated with silver oxide and methyl iodide, may undergo oxidation. Elimination reactions may also occur. For example, Smith[96] found that D-glucurono-6,3-lactone yielded a 2-pyrone derivative. It has been observed that methylation of glycuronans, even in the absence of oxygen, causes breakdown of the polymer molecule, possibly because of β-elimination reactions involving groups on C-4 and C-5. The resultant unsaturated compound is then very susceptible to further degradation.

The methylation of aminodeoxy sugars (compare ref. 76) is complicated by the tendency of the amino group to undergo N-methylation and subsequent quaternization. The products then undergo elimination reactions. For this reason, N-acyl, N-benzenesulfonyl,[97] N-2,4-dinitrophenyl,[98] or N-salicylidene derivatives of the amino sugars and of their derivatives are commonly used as intermediates when the synthesis of partially etherified derivatives is required, as, for example, in the synthesis of N-acetylmuramic acid [2-acetamido-3-O-(D-1-carboxyethyl)-2-deoxy-α-D-glucose] and its derivatives.[99]

N-Methylation occurs when acetamidodeoxy sugars are methylated by the Kuhn method.[76] The 3-, 4-, and 6-mono-, 3,4-, 3,6-, and 4,6-di-, and 3,4,6-trimethyl ethers of 2-deoxy-2-methylamino-D-glucose and the galacto analogs have been prepared as reference compounds for structural work, and the gas–liquid chromatographic retention times of their O-trimethylsilyl derivatives have been recorded.[99a] The corresponding methyl ethers of 2-amino-2-deoxy-D-glucose and -D-galactose have been prepared, mostly by Jeanloz.[99b]

IV. ETHERS OF SUGARS OF LOW MOLECULAR WEIGHT

The foregoing types of reactions, involving both the selective reactivity of sugars with the etherifying reagents,[17-20] and also the purposeful introduction of determinative protecting groups, have been applied broadly and successfully for the synthesis of ethers of the monosaccharides of lower molecular weight. As will be noted later, from time to time some ether derivatives have been obtained by unique or less common methods.

A. ETHERS OF TRIOSES AND ALDONIC ACIDS

The preparation of partially etherified sugars demands a variety of synthetic approaches. Thus, methylation of the 1-benzyl ether of the dimeric cyclic acetal of glyceraldehyde was employed to obtain 3-O-methylglyceraldehyde,[100] whereas 2-O-methylglyceraldehyde was synthesized by the condensation of formaldehyde with 2-methoxyethanal,[101] and 2-O-benzylglyceraldehyde was prepared by oxidative glycol cleavage of suitable hexitol[102] and pentitol[103] ethers.

The 2-O-[103] and 3-O-benzylglyceric acids[104] have been obtained by oxidation of the etherified aldehyde precursors. This sequence of etherification and oxidation has been utilized in the preparation of a number of ether derivatives of L-[51a, 105, 106] and D-arabinonic[107] and D-xylonic acids[49, 108] and, in somewhat modified form, in the preparation of 2-deoxy-4-O-methyl-D-erythronic acid.[109] However, the 5-benzyl[109] and 5-methyl ethers of D-ribono-1,4-lactone were prepared by etherification of the 2,3-isopropylidene acetal of the lactone.[110] 3-O-Benzyl-D-erythronic acid[111] and the 2-,[112] 3-, 2,3-di-, and 2,3,4-tri-O-methyl-L-threonic acids[113] also were prepared by direct etherification of suitable derivatives. As lactones can be reduced[114] to aldoses, these reactions provide routes to the tetrose ethers. The oxidation of ethers of alditols[115] can give rise to ketose derivatives. 3,4-O-Isopropylidene-1,5-di-O-trityl-D-*erythro*-pentulose[116] and diethers of 1,3-dihydroxypropanone[115, 117-119] have been synthesized in this manner.

B. ETHERS OF TETROSES

2,4-Di-O-methyl-D- and -L-erythrose and -threose,[120] 3-O-methyl-D-*glycero*-tetrulose,[121] 4-O-trityl-D-erythrose,[122] and 2-[89] and 4-O-methyl-D-threose[121] have all been synthesized by conventional (albeit often ingenious) synthetic routes. The acid-stability of formic esters resulting from glycol cleavage has been exploited as a means of generating protecting groups *in situ*[123]

* *References start on p. 416.*

for the synthesis of 2-O-methyl-L-threose,[124] 2-O-methyl-D-arabinose,[125] and 4-O-methyl-D-erythrose.[126, 127] The use of olefinic compounds[23] such as 2-butene or its derivatives, as starting compounds for the synthesis of partially etherified tetritols[128] and other carbohydrates,[23] is an interesting example of the application of common olefinic reactions that may be of potential value in carbohydrate chemistry.

C. Ethers of Pentoses

1. *Arabinose*

The chemistry[129] and synthesis of methyl,[130] benzyl,[11] and trityl[13] ethers of pentoses have been reviewed extensively. Of the possible methyl ethers of arabinose, only the 4,5-di-, and 3,4,5-tri-O-methyl-D, and the 4,5-di-, 2,4,5-, and 3,4,5-tri-O-methyl-L derivatives remain unsynthesized.

Oxidative glycol scission has been of much value in the synthesis of partially substituted ethers of D-[89, 131, 132] and L-arabinose.[51a] Thus, controlled oxidation of 3-O-methyl-D-glucose with lead tetraacetate affords 2-O-methyl-D-arabinose,[125] a reaction which may involve a ring contraction during the oxidation.[133]

The syntheses of 3-O-methyl-D-[133] and -L-arabinose[105] have been known for several years. However, studies by Williams and Jones[86, 134] have contributed a new route to these and a number of other methyl ethers of L-arabinose. Of particular importance is the observation of a direct displacement of a primary p-tolylsulfonyloxy group by methoxide,[86, 134] which has greatly simplified the previous routes to 5-[135] and 3,5-di-O-methyl-L-arabinose.[105, 120] This same investigation has evolved a new synthesis[86, 134] of 4-O-methyl-L-arabinose[106] and has partially revised the syntheses of 2,3-[17] and 3,4-di-O-methyl-L-arabinose.[136]

Some previously unsynthesized ethers of D- and L-arabinose have been reported more recently. The formation of the tripotassium alcoholate of benzyl β-D-arabinopyranoside was effected by treatment with potassium *tert*-butoxide.[137] Benzyl chloride reacted smoothly with the alcoholate to give the 2,3,4-tribenzyl ether, but when methoxymethyl chloride was used a more complex reaction ensued which afforded benzyl 2-O-methoxymethyl-3,4-O-methylene-β-D-arabinopyranoside.[137] Other workers have reported the synthesis of 1,4-anhydro-2,3-di-O-methyl-L-arabinopyranose[138] and 2,3-di-O-benzyl-5-O-p-nitrobenzoyl-α-D-arabinopyranose.[139]

With the exceptions noted above, the synthesis of other di-[140] and tri-O-methyl-D-,[89, 90, 107] tri-O-methyl-L-,[134, 141] and the 2,3,5-tri-O-benzyl-D-arabinose[142, 143] derivatives are well documented.

Except for a very few derivatives of L-lyxose[144] and L-ribose,[144, 145] known ethers of the remaining pentoses are restricted to the D series.

2. Xylose

The modern literature describes the synthesis of the following D-xylose ethers: 2-,[146, 146a] 3-,[89, 146a–149] 4-,[54, 150] and 5-;[130, 149] 2,3-[130, 151] 2,4-,[49, 148, 152] 2,5-,[130] 3,4-,[54, 153] and 3,5-di-;[108, 153] 2,3,4-,[49, 154] and 2,3,5-trimethyl;[130] 5-trityl;[108, 155] 2,3-[155] and 3,5-dibenzyl;[146a, 156] 3,5-diallyl;[157] and tetrakis(trimethylsilyl).[158]

3. Ribose

A small number of mono-,[159–163] di-,[159, 160, 164, 165] and tri-[116, 130, 143, 160, 166] methyl,[116, 159, 164–166] benzyl,[143] and trityl[160, 162] ethers of D-ribose have been prepared. Of interest in this regard is the use of an orthobenzoyl group to protect three hydroxyl groups simultaneously, thus providing a facile synthesis of 3-O-methyl-D-ribose.[167] The preparation of some methyl ethers of 2-deoxy-L-erythro-pentose;[145] O-methyl,[109] 5-O-benzyl-,[168] and 5-O-trityl-2-deoxy-D-erythro-pentose;[169] and methyl 4-O-benzyl-2,3-dideoxy-β-L-glycero-hex-2-enopyranoside[144] have been published.

4. Lyxose

The 2,3,4-[170] and 2,3,5-trimethyl[171] ethers of D-lyxose have been known for many years. More recently, syntheses of 4-O-,[172] and 2-O-[172a] and 2,3-di-O-methyl-D-lyxose[173] and of ethyl 3-O-methyl-D-lyxoside[163] have been reported. The first ether of L-lyxose to be prepared—1-O-benzoyl-2,3,4-tri-O-benzyl-4-O-methyl-aldehydo-L-lyxose—was formed from 2,3,5-tri-O-benzyl-4-O-p-tolylsulfonyl-D-ribose dimethyl acetal by an apparent 1 → 4-migration of a methoxyl group on treatment with tert-butylammonium benzoate in N-methylpyrrolidin-2-one.[144] The synthesis of L-lyxose[174] and the projected synthesis of its allyl ethers[22] give promise of the imminent extension of knowledge of the ether derivatives of this sugar. 4-O-Methyl-L-lyxose has been prepared from 2,3,5-tri-O-benzyl-4-O-tosyl-D-ribose dimethyl acetal by a displacement reaction that involves methoxyl-group participation and migration.[175] 2-O-Methyl-L-lyxose is a component of everninomycin B and D.[176]

5. Other Sugars

A list of methyl ethers of monosaccharides is given in "Methods in Carbohydrate Chemistry".[177] Mass-spectral data have been tabulated for numerous trimethylsilyl ethers of sugars, including partially methylated sugars,[178] and similar data have been recorded for methylated alditols of oligosaccharides.[179]

* References start on p. 416.

REFERENCES

1. C. T. Bishop, *Methods Biochem. Anal.*, **10**, 1 (1962); *Advan. Carbohyd. Chem.*, **19**, 95 (1964); G. O. Aspinall, *J. Chem. Soc.*, 1676 (1963).
2. R. Bentley, C. C. Sweeley, M. Makita, and W. W. Wells, *J. Amer. Chem. Soc.*, **85**, 2497 (1963); compare A. E. Pierce, "Silylation of Organic Compounds," Pierce Chemical Co., 1968; G. G. S. Dutton, *Advan. Carbohyd. Chem. Biochem.*, **27**, in press (1972).
3. W. R. Supina, *Gas Chromatogr. News Letter* No. 2, April (1966).
4. G. W. Hay, B. A. Lewis, and F. Smith, *J. Chromatogr.*, **11**, 479 (1963).
5. L. Hough, *Methods Biochem. Anal.*, **1**, 205 (1954); G. N. Kowkabany, *Advan. Carbohyd. Chem.*, **9**, 303 (1954).
6. W. W. Binkley, *Advan. Carbohyd. Chem.*, **10**, 55 (1955).
7. W. Hofheinz, H. Grisebach, and H. Friebolin, *Tetrahedron*, **18**, 1265 (1962); A. C. Richardson, *J. Chem. Soc.*, 2758 (1962); E. G. Gros, I. Mastronardi, and A. R. Frasca, *Carbohyd. Res.*, **16**, 232 (1971); E. B. Rathbone, A. M. Stephen, and K. G. R. Pachler, *ibid.*, **21**, 73, 83 (1972).
7a. N. K. Kochetkov, O. S. Chizhov, and B. M. Zolotarev, *Carbohyd. Res.*, **2**, 89 (1966).
8. H. Weigel, *Advan. Carbohyd. Chem.*, **18**, 61 (1963); A. B. Foster, J. Lehmann, and M. Stacey, *J. Chem. Soc.*, 4649 (1961).
9. A. S. Perlin, *Advan. Carbohyd. Chem.*, **14**, 43 (1959); J. M. Bobbitt, *Advan. Carbohyd. Chem.*, **11**, 35 (1956).
10. H. Björndal, C. G. Hellerqvist, B. Lindberg, and S. Svensson, *Angew. Chem.*, **16**, 643 (1970); H. Björndal, B. Lindberg, Å. Pilotti, and S. Svensson, *Carbohyd. Res.*, **15**, 339 (1970).
11. C. M. McCloskey, *Advan. Carbohyd. Chem.*, **12**, 137 (1957); M. E. Tate and C. T. Bishop, *Can. J. Chem.*, **41**, 1801 (1963).
12. K. Freudenberg, W. Dürr, and H. von Hochstetter, *Ber.*, **61**, 1735 (1928).
13. B. Helferich, *Advan. Carbohyd. Chem.*, **3**, 79 (1948); compare M. Smith, D. H. Rammler, I. H. Goldberg and H. G. Khorana, *J. Amer. Chem. Soc.*, **84**, 430 (1962); H. Schiller, G. Weimann, B. Lerch, and H. G. Khorana, *ibid.*, **85**, 3823 (1963).
14. P. E. Verkade, F. D. Tollenaar, and T. A. P. Posthumus, *Rec. Trav. Chim.*, **61**, 373 (1942).
15. R. Kuhn, H. Rudy, and F. Weygand, *Ber.*, **69**, 1546 (1936); M. L. Wolfrom, W. J. Burke, and S. W. Waisbrot, *J. Amer. Chem. Soc.*, **61**, 1827 (1939).
16. Ref. 13, p. 81.
17. F. Smith, *J. Chem. Soc.*, 753 (1939).
18. G. J. Halliburton and R. J. McIlroy, *J. Chem. Soc.*, 299 (1949).
19. E. L. Hirst and J. K. N. Jones, *J. Chem. Soc.*, 1065 (1947).
20. K. Zeile and W. Kruckenberg, *Ber.*, **75**, 1127 (1942).
21. J. Gigg and R. Gigg, *J. Chem. Soc.* (*C*), 82 (1966); R. R. McLaughlin and D. B. Mutton, *Can. J. Chem.*, **33**, 646 (1955).
22. J. Cunningham, R. Gigg, and C. D. Warren, *Tetrahedron Lett.*, 1191 (1964).
23. J. C. Sowden, M. S. Oftedahl, and A. Kirkland, *J. Org. Chem.*, **27**, 1791 (1962).
23a. R. L. Whistler, P. Panzer, and J. T. Goatley, *J. Org. Chem.*, **27**, 2961 (1962); E. Schwab, V. Stannett, and J. J. Hermans, *Tappi*, **44**, 251 (1961); S. P. Rowland, V. O. Cirino, and A. L. Bullock, *Can. J. Chem.*, **44**, 1051 (1966); Y. Shinohara, *Kobunshi Kagaku*, **17**, 197 (1960); *Chem. Abstr.*, **55**, 18117 (1961); W. A. Black, E. T. Dewar, and D. Rutherford, *Chem. Ind.* (*London*), 1624 (1962).
24. C. Pantadosi, A. F. Hirsch, C. L. Yarbo, and C. E. Anderson, *J. Org. Chem.*, **28**, 2425 (1963).

24a. A. E. Favorskiǐ and M. F. Shostakovakiǐ, *Zh. Obshch. Khim.*, **13**, 1 (1943); *Chem. Abstr.*, **38**, 330 (1944).
25. A. R. Padgett and E. F. Degering, *J. Org. Chem.*, **1**, 336 (1936).
26. J. T. Marvel, S. K. Sen, F. T. Uenaka, J. W. Berry, and A. J. Deutschman, Jr., *Carbohyd. Res.*, **6**, 18 (1968).
27. H. H. Brownell and C. B. Purves, *Can. J. Chem.*, **35**, 677 (1957); C. W. Tasker and C. B. Purves, *J. Amer. Chem. Soc.*, **71**, 1023 (1949).
28. M. Wiedersheim, *Arch. Int. Pharmocodyn. Ther.*, **111**, 353 (1957); *Chem. Abstr.*, **52**, 1448 (1958); Henkel and Co., Fr. Patent 1,422,766, December 24, 1965.
29. J. V. Karabinos and M. Hindert, *Advan. Carbohyd. Chem.*, **9**, 289 (1954); R. M. Reinhardt and J. D. Reid, *Textile Res. J.*, **27**, 59 (1957).
30. S. A. Barker, J. S. Brimacombe, M. R. Harnden, and J. A. Jarvis, *J. Chem. Soc.*, 3403 (1963).
31. M. Kimura, *Bull. Nat. Inst. Ind. Health (Kawasaki, Japan)*, **2**, 48 (1959); *Chem. Abstr.*, **54**, 21755 (1960).
32. N. Toccaceli and F. Della Berta, *Produits Pharm.*, **16**, 212 (1961); *Chem. Abstr.*, **55**, 18014 (1961).
33. M. L. Wolfrom, B. O. Juliano, M. S. Toy, and A. Chaney, *J. Amer. Chem. Soc.*, **81**, 1446 (1959).
34. N. Baggett, J. M. Duxbury, A. B. Foster, and J. M. Webber *Chem. Ind. (London)*, 1832 (1964); cf. S. A. Barker, J. S. Brimacombe, J. A. Jarvis, and J. M. Williams, *J. Chem. Soc.*, 3158 (1962).
35. J. W. Mench and M. S. Lounsbery, U.S. Patent 2,639,280, May 19, 1953; *Chem. Abstr.*, **47**, 7777 (1953); L. A. Cohen and J. A. Steele, *J. Org. Chem.*, **31**, 2333 (1966).
35a. C. B. Reese, R. Saffhill, and J. E. Sultson, *Tetrahedron*, **26**, 1023 (1970).
35b. P. Micheel and P. Schiller, *Ber.*, **101**, 3721 (1968).
35c. R. E. Wing, W. M. Doane, and C. E. Rist, *Carbohyd. Res.*, **12**, 285, 347 (1970).
35d. R. Bognar, I. Farkas, M. Menyhart, H. Gross, and H. Paulsen, *Carbohyd. Res.*, **6**, 404 (1968).
35e. S. S. Bhattacharjee and P. A. J. Gorin, *Can. J. Chem.*, **47**, 1195 (1969).
35f. J. S. Jewell and W. A. Szarek, *Tetrahedron Lett.*, 43 (1969); E. H. Williams, W. A. Szarek, and J. K. N. Jones, *Can. J. Chem.*, **47**, 4467 (1969).
35g. K. E. Pfitzner and J. G. Moffat, *J. Amer. Chem. Soc.*, **87**, 5661, 5670 (1965).
36. P. Brigl, H. Mühlschlegel, and R. Schinle, *Ber.*, **64**, 2921 (1931); M. L. Wolfrom and T. E. Whiteley, *J. Org. Chem.*, **27**, 2109 (1962); compare J. Harness and N. A. Hughes, *Chem. Commun.*, 811 (1971).
37. M. R. Salmon and G. Powell, *J. Amer. Chem. Soc.*, **61**, 3507 (1939).
38. L. Hough, J. K. N. Jones, and W. H. Wadman, *J. Chem. Soc.*, 1703 (1950); L. Hough and R. S. Theobald, *Methods Carbohyd. Chem.*, **2**, 203 (1963).
39. G. Ingraham, *Analyst*, **69**, 269 (1944).
40. T. G. Bonner and E. J. Bourne, *Methods Carbohyd. Chem.*, **2**, 206 (1963); but see M. A. Bukhari, A. B. Foster, and J. M. Webber, *Carbohyd. Res.*, **1**, 474 (1966).
41. T. G. Bonner, E. J. Bourne, and S. McNally, *J. Chem. Soc.*, 2929 (1960).
41a. L. H. Long and G. F. Freeguard, *Nature*, **207**, 403 (1965).
41b. S. J. Angyal and K. James, *Carbohyd. Res.*, **12**, 147 (1970).
42. R. J. Ferrier, D. Prasad, and A. Rudowski, *J. Chem. Soc.*, 859 (1965).
42a. S. Tejima, R. K. Ness, R. F. Kaufman, and H. G. Fletcher, Jr., *Carbohyd. Res.*, **7**, 485 (1968); S. Koto, T. Tsumura, Y. Kato, and S. Umezawa, *Bull. Chem. Soc. Japan*, **41**, 2765 (1968); E. Zissis and H. G. Fletcher, Jr., *Carbohyd. Res.*, **12**, 361 (1970).
43. F. Brown, L. Hough, and J. K. N. Jones, *J. Chem. Soc.*, 1125 (1950).

43a. Cf. R. L. Whistler and S. J. Kazeniac, *J. Amer. Chem. Soc.*, **76**, 3044, 5812 (1954); W. A. Bonner, *J. Org. Chem.*, **24**, 1388 (1959).

44. J. J. Willard, J. S. Brimacombe, and R. P. Brueton, *Can. J. Chem.*, **42**, 2560 (1964).

45. J. Honeyman and J. W. W. Morgan, *Advan. Carbohyd. Chem.*, **12**, 117 (1955); A. L. Fink and G. W. Hay, *Can. J. Chem.*, **47**, 845 (1969).

46. M. L. Wolfrom and D. E. Pletcher, *J. Amer. Chem. Soc.*, **62**, 1151 (1940); M. R. Salmon, and G. Powell, *ibid.*, **61**, 3507 (1939).

47. C. C. Barker, E. L. Hirst, and J. K. N. Jones, *J. Chem. Soc.*, 1695 (1938); W. N. Haworth and W. G. Sedgewick, *ibid.*, 2573 (1926).

48. C. C. Barker, E. L. Hirst, and J. K. N. Jones, *J. Chem. Soc.*, 783 (1946).

49. O. Wintersteiner and A. Klingsberg, *J. Amer. Chem. Soc.*, **71**, 939 (1949).

50. G. O. Aspinall and G. Zweifel, *J. Chem. Soc.*, 2271 (1957); A. S. Perlin, *Can. J. Chem.*, **42**, 1365 (1964); V. L. N. Murty and I. R. Siddiqui, *Carbohyd. Res.*, **11**, 273 (1969); N. Handa and R. Montgomery, *ibid.*, **11**, 467 (1969); M. H. Randall, *ibid.*, **11**, 173 (1969).

51. G. G. S. Dutton and K. Yates, *Can. J. Chem.*, **36**, 550 (1958).

51a. G. G. S. Dutton and Y. Tanaka, *Can. J. Chem.*, **39**, 1797 (1961).

52. G. J. Robertson and C. F. Griffith, *J. Chem. Soc.*, 1193 (1935).

53. Cf. A. J. Dick and J. K. N. Jones, *Can. J. Chem.*, **44**, 79 (1966).

54. L. Hough and J. K. N. Jones, *J. Chem. Soc.*, 4349 (1952).

55. F. H. Newth, *Quart. Rev. (London)*, **13**, 30 (1959); J. G. Buchanan and E. M. Oakes, *Tetrahedron Lett.*, 2013 (1964).

56. Cf. S. Peat, *Advan. Carbohyd. Chem.*, **2**, 52–56 (1946); N. R. Williams, *Advan. Carbohyd. Chem. Biochem.*, **25**, 109 (1970).

57. Cf. A. N. de Belder, *Advan. Carbohyd. Chem.*, **20**, 265–301 (1965).

58. J. C. Irvine and J. P. Scott, *J. Chem. Soc.* 564 (1913).

59. C. G. Anderson, W. Charlton, and W. N. Haworth, *J. Chem. Soc.*, 1329 (1929).

60. W. N. Haworth, in "The Constitution of the Sugars," Edward Arnold, London, 1929, Chapter 7.

61. K. Freudenberg and E. Plankenhorn, *Ann.*, **536**, 257 (1938).

62. J. K. N. Jones, *Can. J. Chem.*, **34**, 840 (1956).

63. N. Baggett, K. W. Buck, A. B. Foster, B. H. Rees, and J. M. Webber, *J. Chem. Soc. (C)*, 212 (1966).

64. R. M. Hann, W. D. Maclay and C. S. Hudson, *J. Amer. Chem. Soc.*, **61**, 2432 (1939).

65. L. Hough, J. K. N. Jones, and N. S. Magson, *J. Chem. Soc.*, 1525 (1952).

66. R. C. Chalk, D. H. Ball, and L. Long, Jr., *J. Org. Chem.*, **31**, 1509 (1966).

66a. J. S. Brimacombe and O. A. Ching, *Chem. Commun.*, 781 (1968); O. A. Ching, *Bol. Soc. Quim. Peru*, **36**, 60 (1970); compare E. Allred and S. Winstein, *J. Amer. Chem. Soc.*, **89**, 3991 (1967).

66b. J. S. Brimacombe, F. Hunedy, and A. Husain, *Carbohyd. Res.*, **10**, 144 (1969).

67. E. Fischer, M. Bergmann, and A. Rabe, *Ber.*, **53**, 2362 (1920).

68. H. G. Bott, W. N. Haworth, and E. L. Hirst, *J. Chem. Soc.*, 1395 (1930).

69. W. N. Haworth, E. L. Hirst, and E. J. Miller, *J. Chem. Soc.*, 2469 (1929).

70. E. J. C. Curtis and J. K. N. Jones, *Can. J. Chem.*, **38**, 890 (1960).

71. E. J. C. Curtis and J. K. N. Jones, *Can. J. Chem.*, **43**, 2508 (1965).

72. P. D. Bragg and L. Hough, *J. Chem. Soc.*, 4347 (1957).

73. T. Reichstein and A. Grüssner, *Helv. Chim. Acta*, **17**, 311 (1934).

74. L. Hough and R. S. Theobald, *Methods Carbohyd. Chem.*, **2**, 162 (1963); L. Schmid, *Ber.*, **58**, 1963 (1925); L. Hough and J. K. N. Jones, *Chem. Ind. (London)*, 380 (1952); R. Kuhn and H. H. Baer, *Ber.*, **86**, 724 (1953).

74a. J. O. Deferrari, E. G. Gros, and I. O. Mastronardi, *Carbohyd. Res.*, **4**, 432 (1967); compare A. Liptak, *Acta Chim.* (Budapest), **66**, 315 (1970); E. G. Gros and E. M. Gruneiro, *Carbohyd. Res.*, **14**, 409 (1970).

74b. M. Aritoni and T. Kawasaki, *Chem. Pharm. Bull.* (*Japan*), **18**, 677 (1970).

74c. E. G. Gros and I. O. Mastronardi, *Carbohyd. Res.*, **10**, 318 (1969).

75. T. Purdie and J. C. Irvine, *J. Chem. Soc.*, **83**, 104 (1903).

75a. L. Hough and J. K. N. Jones, unpublished results (1953).

76. R. Kuhn, H. H. Baer, and A. Seeliger, *Ann.*, **611**, 236 (1958).

77. H. C. Srivastava, S. W. Harshe, and P. P. Singh, *Tetrahedron Lett.*, 1869 (1963).

78. N. D. Scott, J. F. Walker, and V. L. Hansley, *J. Amer. Chem. Soc.*, **58**, 2442 (1936); V. E. Diner, F. Sweet, and R. K. Brown, *Can. J. Chem.*, **44**, 1591 (1966).

79. B. Flaherty, W. G. Overend, and N. R. Williams, *J. Chem. Soc.* (*C*), 399 (1966).

80. E. L. Falconer and G. A. Adams, *Can. J. Chem.*, **34**, 338 (1956).

80a. S. Hakomori, *J. Biochem.* (*Tokyo*), **55**, 205 (1964); compare P. A. Sanford and H. E. Conrad, *Biochemistry*, **5**, 1508 (1966); G. Keilich, P. Salminen, and E. Husemann, *Makromol. Chem.*, **141**, 117 (1971).

80b. E. J. Corey and M. Chaykovsky, *J. Amer. Chem. Soc.* **84**, 866 (1962).

80c. D. M. W. Anderson and G. M. Cree, *Carbohyd. Res.*, **2**, 162 (1966); D. Martin, A. Weise, and H. J. Niclas, *Angew. Chem. Intern. Ed. Engl.*, **6**, 326 (1967); D. M. W. Anderson, I. C. M. Dea, P. A. Maggs, and A. C. Munro, *Carbohyd. Res.*, **5**, 489 (1967).

80d. O. Lüderitz, *Angew. Chem. Intern. Ed. Engl.*, **9**, 649 (1970).

81. W. N. Haworth, *J. Chem. Soc.*, **107**, 13 (1915); W. S. Denham and H. Woodhouse, *ibid.*, **103**, 1735 (1913).

82. W. N. Haworth and G. C. Leitch, *J. Chem. Soc.*, **113**, 188 (1918).

83. W. N. Haworth and E. L. Hirst, *J. Chem. Soc.*, **119**, 193 (1921).

83a. L. Lillienfeld, Brit. Patent 12,854, May 31 (1912): *Chem. Abstr.*, **7**, 3839 (1913); H. Dreyfus, Fr. Patent 462,274, Nov. 18 (1912); *Chem. Abstr.*, **8**, 3859 (1914); G. K. Greminger, Jr. and A. B. Savage, *in* "Industrial Gums" (R. L. Whistler, ed.), Academic Press, New York, 1959, p. 565.

84. T. I. Muskat, *J. Amer. Chem. Soc.*, **56**, 693, 2449 (1934).

85. H. S. Isbell, H. L. Frush, B. H. Bruckner, C. Wampler, and G. N. Kowkabany, *Anal. Chem.*, **29**, 1523 (1957).

86. S. C. Williams and J. K. N. Jones, *Can. J. Chem.*, **43**, 3440 (1965).

87. A. K. Mitra, D. H. Ball, and L. Long, Jr., *J. Org. Chem.*, **27**, 160 (1962).

87a. P. W. Kent, D. W. A. Farmer, and N. F. Taylor, *Proc. Chem. Soc.*, 187 (1959).

88. J. C. Sowden, M. L. Oftedahl, and A. Kirkland, *J. Org. Chem.*, **27**, 1791 (1962).

89. R. U. Lemieux and J. D. T. Cipera, *Can. J. Chem.*, **34**, 906 (1956); G. W. Huffman, Bertha A. Lewis, F. Smith, and D. R. Spriestersbach, *J. Amer. Chem. Soc.*, **77**, 4346 (1955).

90. R. A. Weerman, *Rec. Trav. Chim.*, **37**, 16 (1917); W. N. Haworth, S. Peat, and J. Whetstone, *J. Chem. Soc.*, 1975 (1938).

91. O. Ruff, *Ber.*, **31**, 1573 (1898); *ibid.*, **34**, 1362 (1901).

92. A. Wohl, *Ber.*, **26**, 730 (1893).

93. W. N. Haworth, E. L. Hirst, and F. Smith, *J. Chem. Soc.*, 1556 (1934); cf. F. Smith, *ibid.*, 510 (1944).

94. K. G. A. Jackson and J. K. N. Jones, unpublished results.

95. K. G. A. Jackson and J. K. N. Jones, *Can. J. Chem.*, **43**, 450 (1965).

96. L. N. Owen, S. Peat, and W. J. G. Jones, *J. Chem. Soc.*, 341 (1941); F. Smith, *Chem. Ind.* (*London*), **57**, 450 (1938); W. N. Haworth, D. Heslop, E. Salt, and F. Smith, *J. Chem. Soc.*, 217 (1944).

97. M. L. Wolfrom and R. Wurmb, *J. Org. Chem.*, **30**, 3058 (1965).
98. D. K. Stearns, R. G. Naves, and R. W. Jeanloz, *J. Org. Chem.*, **26**, 901 (1961).
99. T. Osawa and R. W. Jeanloz, *Carbohyd. Res.* **1**, 181 (1965).
99a. P. A. J. Gorin and A. J. Finlayson, *Carbohyd. Res.*, **18**, 269 (1971); P. A. J. Gorin, *ibid.*, **18**, 281 (1971).
99b. R. W. Jeanloz, *Advan. Carohyd. Chem.*, **13**, 189 (1958).
100. H. O. L. Fischer and E. Baer, *Ber.*, **65**, 337 (1932).
101. E. I. duPont de Nemours and Co., Brit. Patent, 540,418, Oct. 16, 1941; *Chem. Abstr.*, **36**, 4131 (1942); D. J. Loder, U.S. Patent, 2,286,037, June 9, 1942; *Chem. Abstr.*, **36**, 7030 (1942); W. F. Gresham, U.S. Patent, 2,286,034, June 9, 1942; *Chem. Abstr.*, **36**, 7030 (1942).
102. W. T. Haskins, R. M. Hann, and C. S. Hudson, *J. Amer. Chem. Soc.*, **64**, 132 (1942); C. E. Ballou and H. O. L. Fischer, *ibid.*, **77**, 3329 (1955).
103. F. Wold, *J. Org. Chem.*, **26**, 197 (1961).
104. C. E. Ballou and H. O. L. Fischer, *J. Amer. Chem. Soc.*, **76**, 3188 (1954).
105. E. L. Hirst, J. K. N. Jones, and E. Williams, *J. Chem. Soc.*, 1062 (1947).
106. I. R. Siddiqui and C. T. Bishop, *Can. J. Chem.*, **40**, 233 (1962).
107. P. W. Kent and M. W. Whitehouse, *J. Chem. Soc.*, 2501 (1953).
108. R. A. Laidlaw, *J. Chem. Soc.*, 2941 (1952).
109. J. Kenner and G. N. Richards, *J. Chem. Soc.*, 2916 (1956); J. Baddiley, J. G. Buchanan, and F. E. Hardy, *J. Chem. Soc.*, 2180 (1961).
110. L. Hough, J. K. N. Jones, and D. L. Mitchell, *Can. J. Chem.*, **36**, 1720 (1958).
111. R. Barker and F. Wolfe, *J. Org. Chem.*, **28**, 1847 (1963).
112. K. Gätzi and T. Reichstein, *Helv. Chim. Acta*, **20**, 1298 (1937).
113. K. Gätzi and T. Reichstein, *Helv. Chim. Acta*, **21**, 195 (1938).
114. R. K. Hulyalkar, *Can. J. Chem.*, **44**, 1594 (1966).
115. J. Kenner and G. N. Richards, *J. Chem. Soc.*, 2240 (1953).
116. D. H. Rammler and D. L. MacDonald, *Arch. Biochem. Biophys.*, **78**, 359 (1958).
117. J. R. Geigy, A.-G., Brit. Patent 546,286, July 6, 1942; *Chem. Abstr.*, **37**, 3104 (1943).
118. K. Heyns and W. Stein, *Ann.*, **558**, 194 (1947).
119. Y. H. Abouzeid and W. H. Linnell, *J. Pharm. Pharmacol.*, **1**, 235 (1949).
120. G. G. S. Dutton and K. N. Slessor, *Can. J. Chem.*, **42**, 614 (1964).
121. G. N. Richards, *J. Chem. Soc.*, 3222 (1957).
122. C. E. Ballou, H. O. L. Fischer, and D. L. MacDonald, *J. Amer. Chem. Soc.*, **77**, 2658, 5967 (1955).
123. A. S. Perlin and C. Brice, *Can. J. Chem.*, **34**, 541 (1956).
124. M. Cantley, L. Hough, and A. O. Pittet, *Chem. Ind. (London)*, 1253 (1959).
125. W. Mackie and A. S. Perlin, *Can. J. Chem.*, **43**, 2645 (1965).
126. A. J. Charlson and A. S. Perlin, *Can. J. Chem.*, **34**, 1200 (1956).
127. P. A. J. Gorin and J. F. T. Spencer, *Can. J. Chem.*, **43**, 2978 (1965).
128. W. F. Beech, *J. Chem. Soc.*, 2483 (1951); H. Pasedach, U.S. Patent 2,694,736, Nov. 16, 1954; *Chem. Abstr.*, **50**, 4217 (1956); B. W. Horrom and H. E. Zaugg, *J. Amer. Chem. Soc.*, **79**, 1754 (1957); J. Kiss and F. Sirokman, *Helv. Chim. Acta*, **43**, 334 (1960).
129. J. K. N. Jones and F. Smith, *Advan. Carbohyd. Chem.*, **4**, 243 (1949); R. L. Whistler, *ibid.*, **5**, 269 (1950); R. W. Jeanloz and H. G. Fletcher, Jr., *ibid.*, **6**, 135 (1951); F. Smith and R. Montgomery, *Amer. Chem. Soc. Monogr.*, No. 141, pp. 133–240, 1959; J. Staněk, M. Černý, J. Kocourek, and J. Pacák, "The Monosaccharides," Academic Press, New York, 1963.
130. R. A. Laidlaw and E. G. V. Percival, *Advan. Carbohyd. Chem.*, **7**, 1 (1952); G. G. Maher, *ibid.*, **10**, 257 (1955).

131. J. C. P. Schwartz and M. MacDougall, *J. Chem. Soc.*, 3065 (1965); I. J. Goldstein, H. Sorger-Domenigg, and F. Smith, *J. Amer. Chem. Soc.*, **81**, 444 (1959).
132. H. G. Fletcher, Jr. and H. W. Diehl, *J. Org. Chem.*, **30**, 2321 (1965).
133. A. S. Perlin, *Can. J. Chem.*, **42**, 2365 (1964); E. E. Percival and R. Zobrist, *J. Chem. Soc.*, 564 (1953).
134. S. C. Williams, M.Sc. Thesis, Queen's University, Kingston, March, 1966.
135. G. G. S. Dutton, Y. Tanaka, and K. Yates, *Can. J. Chem.*, **37**, 1955 (1959); I. R. Siddiqui, C. T. Bishop, and G. A. Adams, *ibid.*, **39**, 1595 (1961).
136. P. Andrews, D. H. Ball, and J. K. N. Jones, *J. Chem. Soc.*, 4090 (1953).
137. V. Bruckner, J. Császár, and K. Kovacs, *Rev. Chim. Acad. Rep. Populaire Roumaine*, **7**, 715 (1962); *Chem. Abstr.*, **61**, 5740 (1964).
138. J. Kops and C. Schuerch, *J. Org. Chem.*, **30**, 3951 (1965).
139. C. P. J. Glaudemans and H. G. Fletcher, Jr., *J. Amer. Chem. Soc.*, **87**, 4636 (1965).
140. J. P. Verheijden and P. J. Stoffyn, *Bull. Soc. Chim. Belges*, **68**, 699 (1959).
141. J. K. N. Jones, *J. Chem. Soc.*, 1055 (1947); J. K. N. Jones and G. H. S. Thomas, *Can. J. Chem.*, **39**, 192 (1961).
142. R. Barker and H. G. Fletcher, Jr., *J. Org. Chem.*, **26**, 4605 (1961).
143. S. Tejima and H. G. Fletcher, Jr., *J. Org. Chem.*, **28**, 2999 (1963).
144. N. A. Hughes and P. R. H. Speakman, *Chem. Commun.*, 199 (1965); N. F. Taylor and G. M. Riggs, *J. Chem. Soc.*, 5600 (1963).
145. R. E. Deriaz, W. G. Overend, M. Stacey, and L. F. Wiggins, *J. Chem. Soc.*, 2836 (1949).
146. W. D. S. Bowering and T. E. Timell, *Can. J. Chem.*, **36**, 283 (1958); compare P. Kováč and M. Petrícová, *Carbohyd. Res.*, **16**, 492 (1971).
146a. G. G. S. Dutton and Y. Tanaka, *Can. J. Chem.*, **40**, 1899 (1962).
147. E. J. C. Curtis and J. K. N. Jones, *Can. J. Chem.*, **38**, 1305 (1960).
148. R. J. Ferrier, D. Prasad, A. Rudowski, and I. Sangster, *J. Chem. Soc.*, 3330 (1964).
149. D. A. Applegarth, G. G. S. Dutton, and Y. Tanaka, *Can. J. Chem.*, **40**, 2177 (1962).
150. P. J. Garegg, *Acta Chem. Scand.*, **14**, 957 (1960).
151. E. G. Meek, *J. Chem. Soc.*, 219 (1956); S. K. Chanda, E. E. Percival, and E. G. V. Percival, *ibid.*, 260 (1952); P. Kováč and M. Petríková, *Carbohyd. Res.*, **19**, 249 (1971).
152. E. L. Hirst, E. G. V. Percival, and C. B. Wylam, *J. Chem. Soc.*, 189 (1954).
153. E. E. Percival and R. Zobrist, *J. Chem. Soc.*, 4306 (1952).
154. R. Kuhn, I. Löw, and H. Trischmann, *Ber.*, **90**, 203 (1957).
155. D. V. Myhre and F. Smith, *J. Org. Chem.*, **26**, 4609 (1961).
156. M. E. Tate and C. T. Bishop, *Can. J. Chem.*, **41**, 1801 (1963).
157. V. F. Kazimirova and K. V. Levitskaya, *Zh. Obshch. Khim.* **30**, 723 (1960); *Chem. Abstr.*, **55**, 395 (1961).
158. R. J. Ferrier and M. F. Singleton, *Tetrahedron*, **18**, 1143 (1962); S. L. Liu and C-H. Ho, *J. Chinese Chem. Soc.* (*Taiwan*), **6**, 137 (1960); *Chem. Abstr.*, **55**, 3445 (1961).
159. G. R. Barker, T. M. Noone, D. C. C. Smith, and J. W. Spoors, *J. Chem. Soc.*, 1327 (1955).
160. H. Bredereck and W. Greiner, *Ber.*, **86**, 717 (1953).
161. G. R. Barker and J. W. Spoors, *J. Chem. Soc.*, 1192 (1956); E. B. Rauch and D. Lipkin, *J. Org. Chem.*, **27**, 403 (1962); H. B. Wood, Jr., H. W. Diehl, and H. G. Fletcher, Jr., *J. Amer. Chem. Soc.*, **78**, 4715 (1956).
162. H. Zinner, *Ber.*, **86**, 496 (1953).
163. G. R. Barker and D. C. C. Smith, *J. Chem. Soc.*, 1323 (1955).
164. G. R. Barker and M. V. Lock, *J. Chem. Soc.*, 23 (1950).
165. G. R. Barker and J. W. Spoors, *J. Chem. Soc.*, 2656 (1956).
166. G. R. Barker, *J. Chem. Soc.*, 2035 (1948).

167. H. G. Fletcher, Jr. and R. K. Ness, *J. Amer. Chem. Soc.*, **77**, 5337 (1955).
168. G. W. Kenner, C. W. Taylor, and A. R. Todd, *J. Chem. Soc.*, 1620 (1949).
169. H. Zinner, H. Nimz, and H. Venner, *Ber.*, **91**, 638 (1958).
170. E. L. Hirst and J. A. B. Smith, *J. Chem. Soc.*, 3147 (1928).
171. P. A. Levene, *J. Biol. Chem.*, **133**, 767 (1940).
172. J. Piotrovsky, J. P. Verheijden, and P. J. Stoffyn, *Bull. Soc. Chim. Belges*, **73**, 969 (1964).
172a. J. S. Brimacombe, A. M. Mofti, and A. K. Al-Radhi, *J. Chem. Soc.* (*C*), 1363 (1971).
173. J. P. Verheijden and P. J. Stoffyn, *Tetrahedron*, **1**, 253 (1957).
174. R. K. Hulyalkar and M. B. Perry, *Can. J. Chem.*, **43**, 3241 (1965).
175. N. A. Hughes and P. R. H. Speakman, *J. Chem. Soc.* (*C*), 1182 (1967).
176. J. S. Brimacombe and A. M. Mofti, *Chem. Commun.*, 241 (1971).
177. J. N. BeMiller, *Methods Carbohyd. Chem.*, **5**, 298–357 (1965).
178. G. Petersson, *Arch. Mass. Spectral Data*, **1**, 624–680 (1970); compare S. Korody and S. H. Pines, *Tetrahedron*, **26**, 4527 (1970).
179. J. Kärkkäinen, *Carbohyd. Res.*, **14**, 27 (1970); **17**, 1, 11 (1971).

13. GLYCOSANS AND ANHYDRO SUGARS

R. D. Guthrie

I. INTRODUCTION

This Chapter is concerned principally with those sugar derivatives in which a molecule of water has been eliminated intramolecularly, and to a minor extent with those compounds in which two molecules of water have been split out intermolecularly between two sugar molecules.

These compounds can be broken down into two distinct classes: firstly, those involving elimination of water from two hydroxyl groups, one of which is the hemiacetal hydroxyl group, and secondly, those involving elimination of water between two hydroxyl groups, neither of which is the hemiacetal hydroxyl group. The former class of compounds may be named glycosans, the latter anhydro sugars. In the Rules of Carbohydrate Nomenclature, both classes are named as anhydro sugars (see Vol. IIB, Chapter 46, Rule 33).

The topic of this chapter has been reviewed,[1] and stereochemical aspects have been discussed.[2] The following treatment is selective rather than exhaustive.[2a]

It is important to stress the difference between glycosans and anhydro sugars, the former being internal acetals, the latter being internal ethers. Glycosans are opened readily under acidic conditions, whatever the ring size. The three-membered anhydro sugars of the ethylene oxide type (oxiranes; colloquially, but strictly incorrectly, called epoxides†) are readily opened, but the larger-ring anhydro sugars resemble the simple aliphatic ethers and are stable to the majority of reagents.

The mass spectra of glycosans and anhydro sugars have been studied.[2b] Each class of compound showed one or more characteristic peaks (see Vol. IB, Chapter 27).

II. GLYCOSANS

A. 1,2-ANHYDROPYRANOSES

1

These compounds are internal 1,2:1,5 acetals (**1**) if derived from pyranose sugars. The most familiar example in this class is 1,2-anhydro-α-D-glucopyranose 3,4,6-triacetate (**3**)—"Brigl's anhydride"—which can be prepared[3] from β-D-glucopyranose pentaacetate (**2**) by the sequence shown. The anhydride (**3**) is extremely reactive and opens, as expected, exclusively at the anomeric center.

† See IUPAC Rules of Organic Nomenclature, Rules B.24 and B.52.

Under mild conditions, the products have the β-D-*gluco* configuration.[3-5] However, under forcing conditions, for example[4] with phenol for 20 hours at 80° to 100°, α-D-glucosides are formed. This reaction has been put to use in the synthesis of α-D-linked disaccharides,[6-9] in particular of sucrose.[10,11] Lemieux and Huber[10] have explained the formation of α-D-glucosides under these conditions by participation of the C-6 acetoxy group, which directs the incoming alcohol group to the α position, as shown in Scheme I. Opening of Brigl's

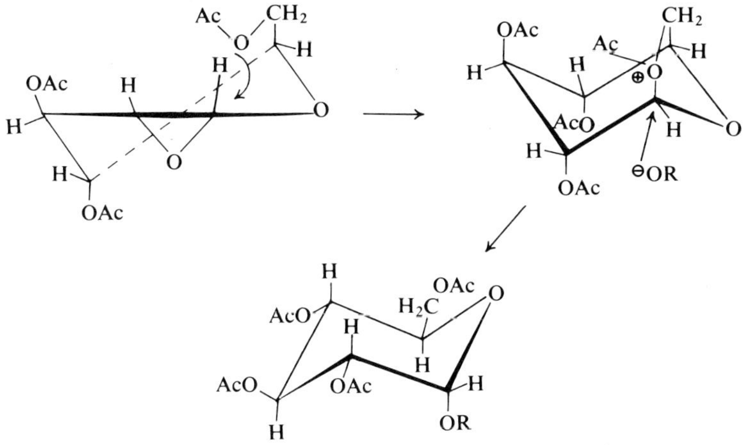

SCHEME I

anhydride, with titanium tetrachloride in chloroform containing a trace of alcohol, gave 3,4,6-tri-*O*-acetyl-α-D-glucopyranosyl chloride.[12]

A supposed synthesis of the unsubstituted anhydride is claimed to result from the pyrolysis of α-D-glucose at 150° under diminished pressure.[13,14] No satisfactory structural proof for this product has yet been put forward, although on treatment with hydrogen sulfide under pressure in *N,N*-dimethylformamide it gives 1-thio-D-glucose[15] (see Vol. IB, Chapter 18).

The unsubstituted 1,2-anhydro-α-D-hexopyranose has been suggested as an intermediate in the formation of 1,6-anhydropyranoses from aryl β-D-aldohexopyranosides (see Section II,D, p. 427).

Polymerization of Brigl's anhydride at 118° gave a syrup from which kojiobiose [2-*O*-(α-D-glucopyranosyl)-D-glucose] has been isolated.[16]

B. 1,3-ANHYDROPYRANOSES

The only compound known in this class appears to be a 1,3:1,5-pyranose internal acetal, namely 1,3-anhydro-β-D-galactopyranose (4). This was formed, together with the 1,6-anhydropyranose (see Section II,D below) from D-galactose by pyrolysis.[17] Structure 4 was assigned because the compound did not

* *References start on p. 471.*

reduce periodate and because it gave D-galactose (isolated as its phenylosazone) after acid hydrolysis.

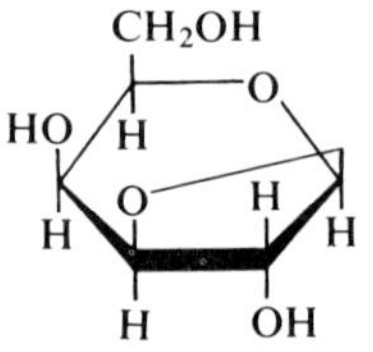

4

C. 1,4-ANHYDROPYRANOSES (1,5-ANHYDROFURANOSES)

5 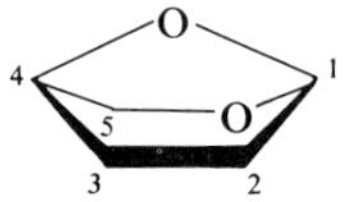 6

These glycosans, which have the general formula **5**, are 1,4:1,5 internal acetals and can be regarded as either pyranose or furanose derivatives; the pyranose ring must be in a boat form as in **6**. Treatment of 2,3,6-tri-*O*-methyl-4-*O*-*p*-tolylsulfonyl-D-glucose (**7**) with alkali gave a product initially believed to be 1,4-anhydro-2,3,6-tri-*O*-methyl-D-idose,[18] later shown[19] to be 1,4-anhydro-2,3,6-tri-*O*-methyl-β-D-galactopyranose (**8**). Similarly, 2,3-di-*O*-methyl-4-*O*-*p*-tolylsulfonyl-D-xylose (contaminated with some 5-*O*-*p*-tolylsulfonyl furanose derivative) gave 1,4-anhydro-2,3-*O*-methyl-L-arabinopyranose (contaminated with some 1,4-anhydro-2,3-di-*O*-methyl-D-xylopyranose).[19] The action of alkali on 1-*O*-acetyl-2,3,6-tri-*O*-methyl-5-*O*-*p*-tolylsulfonyl-α-D-glucofuranose (**9**) gave an anhydride,[20] presumably

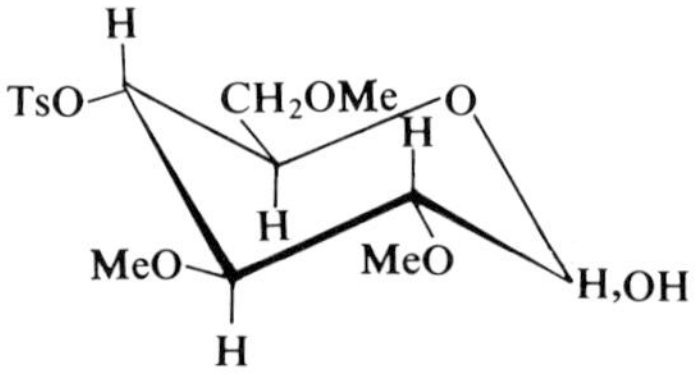

7 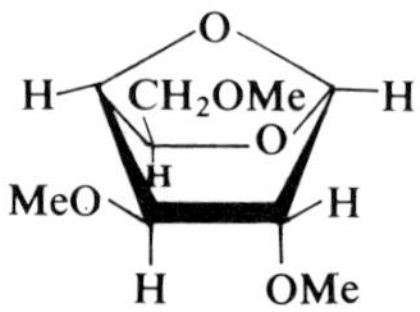 8

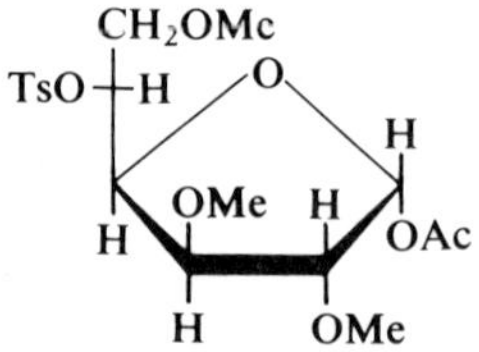

9 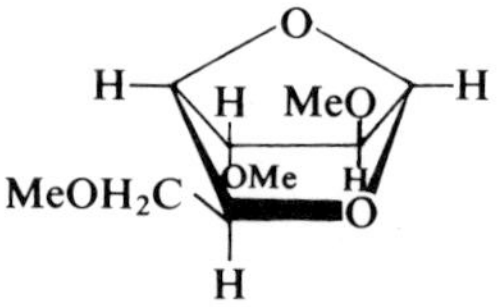 10

1,4-anhydro-2,3,6-tri-*O*-methyl-α-L-idopyranose (**10**). Other known glycosans of this type include 1,4-anhydro-β-D-ribopyranose[21] and its 2,3-benzylidene acetal,[21] 1,4-anhydro-2,3,6-tri-*O*-methyl-α-D-glucopyranose,[22] and 1,4-anhydro-α-L-rhamnopyranose 2,3-diacetate.[23]

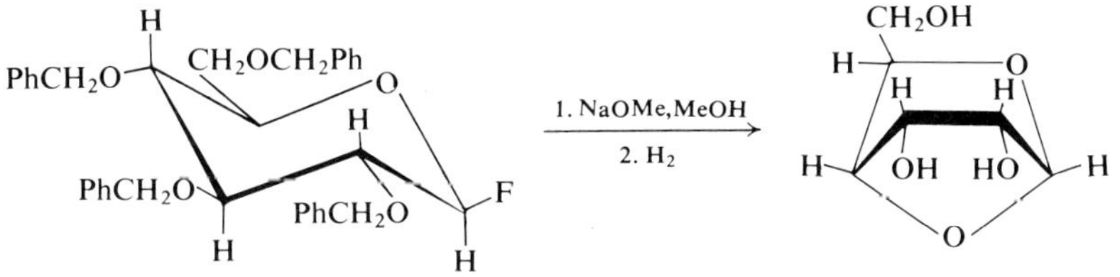

10a   10b

Crystalline 1,4-anhydro-β-D-glucopyranose (**10b**) was prepared[24] by treating 2,3,6-tri-*O*-benzyl-β-D-glucopyranosyl fluoride (**10a**) with methanolic sodium methoxide, with subsequent debenzylation. The product readily underwent hydrolysis by aqueous acid to give D-glucose.

D. 1,6-ANHYDROPYRANOSES[24a]

This class of glycosan is the most commonly encountered, and 1,6-anhydro-β-D-glucopyranose (**11**) has been the most studied. As can be seen from

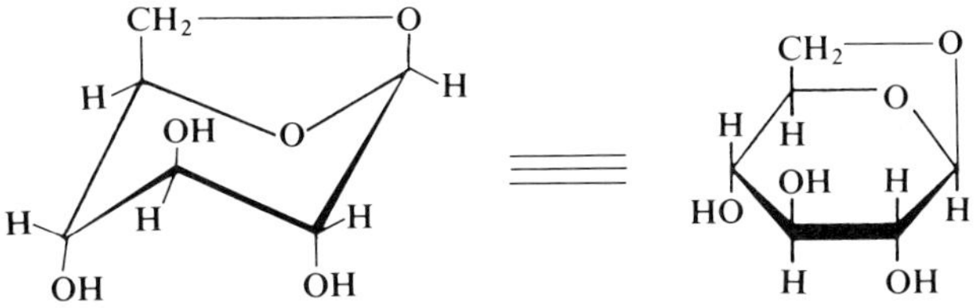

11

formula **11**, the molecule in a chair form must exist in the *1C* (D) conformation (*C1* in the L series) to allow the anhydro bridge to be formed. This conformation is supported by the work of Reeves[25] and is entirely confirmed by subsequent n.m.r. spectral work (see Vol. IB, Chapter 27).

The 1,6-anhydroaldohexopyranoses, having a rigid ring system, are amenable to correlation[25a] of their specific rotations by the concept of "optical superposition." A more general approach, based on the algebraic summation of a series of empirical terms for conformational elements of asymmetry, essentially by the Whiffen–Brewster approach (see Vol. IB, Chapter 27), has also been made.[26] The procedure gives calculated rotations in excellent agreement with experimental data for the eight 1,6-anhydro-β-D-aldohexopyranoses (and their

* *References start on p. 471.*

triacetates), including the D-*talo* isomer, the last example of the series to have been synthesized.[27] Extension to deoxy analogs and also to the 2,7-anhydro-heptulopyranoses likewise permits accurate prediction of specific rotations.[26]

Trivial names that have been used for 1,6-anhydro-β-D-glucopyranose (**11**) are levoglucosan and "β-glucosan." Several different methods are available for the preparation of 1,6-anhydrohexopyranoses, and these will be dealt with in turn.

Pyrolysis of cellulose,[28] starch,[28, 29] β-D-glucosides,[28] and β-D-glucose[29] all gave levoglucosan (**11**), as did treatment of cellulose with superheated steam at 24 to 30 mmHg.[30] Pyrolysis of agar gave 1,6-anhydro-β-D-galactopyranose,[31] also obtained by pyrolysis of β-D-galactose.[32] A mixture of levoglucosan and 1,6-anhydro-β-D-galactopyranose was obtained by pyrolysis of lactose.[31, 33] Pyrolysis of ivory-nut mannan gave 1,6-anhydro-β-D-mannopyranose.[34, 35] Some of these pyrolyses also gave rise to small proportions of 1,6-anhydro-β-D-aldohexofuranoses (see Section II,E, p. 431).

Levoglucosan was first prepared by Tanret[36] by the action of hot aqueous barium hydroxide on a number of naturally occurring β-D-glucopyranosides. This reaction has been studied extensively.[37] It was shown that, although phenyl β-D-glucopyranoside gave levoglucosan with hot alkali, the α-D-glycoside did not and was quite stable.[38] Phenyl β-D-mannopyranoside[39] and the corresponding D-galactoside[38] also gave rise to 1,6-anhydrides. Micheel and Micheel[40] postulated a 1,3-anhydride as an intermediate, but this suggestion was disproved when 1,3-anhydro-β-D-galactopyranose (**4**) was shown to be stable to alkali.[41] Evidence that C-2 was involved in the reaction of the aryl β-D-glucopyranoside was provided when it was shown that phenyl 2-*O*-methyl-β-D-glucopyranoside,[42] and the corresponding 2,3-dimethyl and 2,3,4-trimethyl ethers, were all stable to alkali, but that the 3-methyl ether gave the corresponding 1,6-anhydride.[42, 43] The mechanism accepted for the reaction (Scheme II), was finally confirmed when it was shown that 3,4,6-tri-*O*-acetyl-1,2-anhydro-α-D-glucopyranose (**3**) gave levoglucosan with alkali.[42, 44]

Details for routine preparation of 1,6-anhydro-β-D-glucopyranose and 1,6-anhydro-β-D-galactopyranose have been given, based on the above reactions.[45, 46]

Treatment of phenyl, *p*-chlorophenyl, *p*-tolyl, or *p*-nitrophenyl 2-deoxy-α-D-*arabino*-hexopyranosides with hot 0.05 *M* aqueous potassium hydroxide gave 1,6-anhydro-2-deoxy-β-D-*arabino*-hexopyranoside.[47] Treatment of the phenyl β-D-glycosides of lactose, cellobiose, and D-*glycero*-β-D-*gulo*-heptose gave the corresponding 1,6-anhydrides.[48]

1-*O*-Mesitoyl-β-D-glucopyranose (**12**) and its tetraacetate yielded levoglucosan when treated with alkali,[49, 50] whereas the fully acetylated α-D anomer[49, 50] or the 1-*O*-benzoyl analog[49] of **12** did not. The explanation advanced for this difference was that, in the mesitoate (**12**), steric hindrance

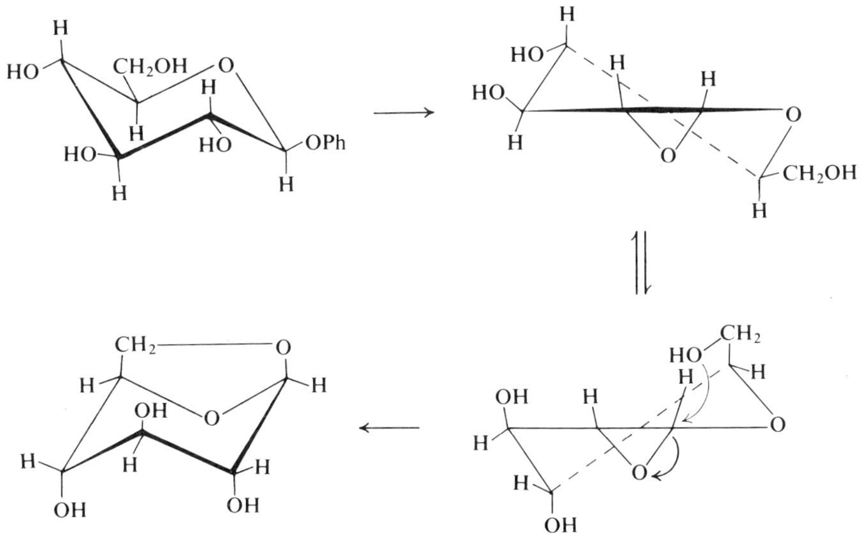

SCHEME II

prevented attack at the carbonyl group, and so the C-1–O linkage was split preferentially.

1,6-Anhydro-β-D-glucopyranose and the D-galactose analog have also been prepared by the action of alkali on the appropriate (2,3,4,6-tetra-O-acetyl-β-D-aldohexopyranosyl)trimethylammonium bromides.[51,52] The mechanism is

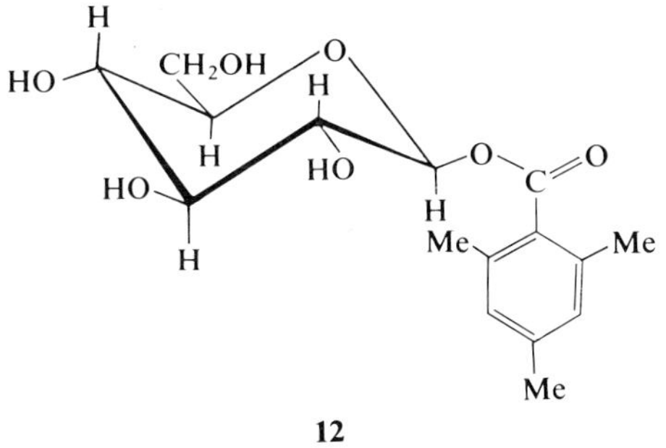

12

probably similar to that described above for the aryl β-D-glycosides. Treatment of tetra-O-acetyl-α-D-glucopyranosyl bromide with alkali gave levoglucosan,[53] as did methyl 2,3,4-tri-O-acetyl-α-D-glucoside 6-nitrate[54] and 2,3,4,6-tetra-O-acetyl-α-D-glucosyl nitrate.[54] Micheel et al. have shown that hexosyl fluorides and azides react with alkali to give the corresponding 1,6-anhydrides.[55-58]

* References start on p. 471.

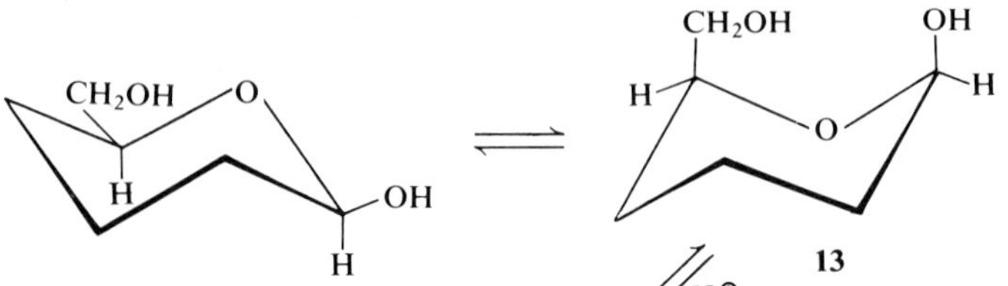

13

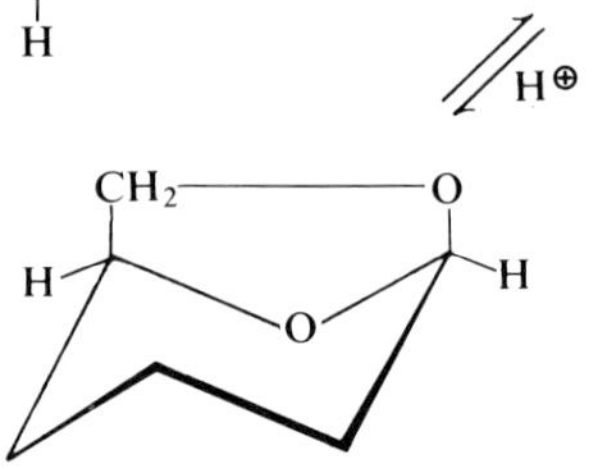

14

In dilute aqueous acid, an equilibrium is established[59-68] between the free aldohexose and its 1,6-anhydride (**14**). Formation of the anhydro sugar can take place only from the aldohexose in the $1C$ (D) [or $C1$ (L)] conformation (**13**), and the proportion of 1,6-anhydro sugar formed can be correlated with the ease with which this conformation is formed[25] (see also Chapter 5). Table I

TABLE I

PERCENTAGE OF 1,6-ANHYDROALDOSE FORMED FROM
ALDOHEXOSES

D or L Aldohexose	Percent of 1,6-anhydride at equilibrium	Reference
Idose	75	61
Altrose	57	62
Gulose	43	62
Talose	12	68
Allose	14	63
Galactose[a]	0.7	68
Mannose	0.6	64
Glucose	2.3	65
	0.36	66
	0.2	67

[a] Plus 0.9% of 1,6-anhydrogalactofuranose.[68]

shows the percentage of anhydro sugar formed in equilibrium with each hexose. As would be expected, an axial group at C-3 in **14** is a very destabilizing factor because of 1,3-interactions with O-1 and C-6; compare, for example, the *altro* and *manno* derivatives, which in **14** have the 3-hydroxyl group equatorial and axial, respectively.[69] Similar formation of 1,6-anhydropyranoses has been shown to occur with aldoheptoses;[70, 71] in one case, some 1,7-anhydroheptose was also formed.[71]

Levoglucosan has been polymerized thermally[72–74] to yield branched-chain polysaccharides having molecular weights[75] in the range 2 to 5 × 10⁴. Polymerization in the presence of zinc chloride gave substances containing 2 to 8 monomer residues.[76–78] Attempted polymerization of trisubstituted levo-glucosans gave only starting material or products of decomposition, which led to the suggestion that polymerization takes place by way of a 1,2-epoxide ring.[79] Polymerization of 1,6-anhydro-β-D-galactopyranose also gave branched-chain polysaccharides of high molecular weight.[80] A detailed thermal analysis of levoglucosan has been reported.[80a]

15

A large number of completely and partially substituted derivatives of 1,6-anhydroaldohexopyranoses have been prepared, including various ethers[81–87] and esters.[88–90] Oxidation of 1,6-anhydro-β-D-galactopyranose (**15**) with oxygen in the presence of platinum gave the 3-oxo derivative (preferential oxidation of the axial hydroxyl group at C-3)[91] (see Vol. IB, Chapter 23, for further details of oxidation products). Periodate-oxidized levoglucosan has been treated with nitromethane and the product reduced to give 3-amino-3-deoxy derivatives of D-gulose, D-altrose, and D-idose[92] (see Vol. IB, Chapter 16). A number of 2,3- and 3,4-epoxides of 1,6-anhydroaldohexopyranoses are known and are referred to elsewhere in this chapter.

Treatment of tri-O-acetyl-1,6-anhydro-β-D-glucopyranose with titanium tetrachloride gave 2,3,4,6-tetra-O-acetyl-α-D-glucosyl chloride.[12] A detailed p.m.r. study has been made of 1,6-anhydroaldohexoses and their derivatives[93] (see Vol. IB, Chapter 27).

E. 1,6-ANHYDROFURANOSES

This class of compound, whose basic skeleton is shown in formula **16**, has been reviewed by Dimler.[94] 1,6-Anhydro-β-D-glucofuranose has been found

* *References start on p.* 471.

in small proportion among the products from the pyrolysis of starch[95] and cellulose,[96] and from treatment of D-glucose with dilute sulfuric acid.[66] 1,6-Anhydro-α-D-galactofuranose has been prepared in low yield from α-D-galactose by pyrolysis,[32] or by treatment with dilute hydrochloric acid.[68] Pyrolysis of D-mannose gives 1,6-anhydro-β-D-mannofuranose in low yield; the

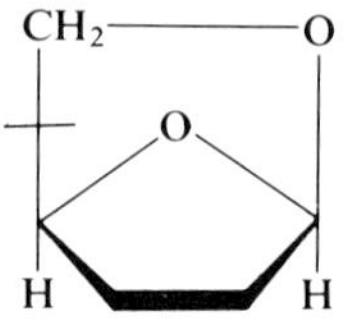

16

latter was converted by inversion at C-5 via a 5-ketone derivative into the α-L-*gulo* isomer.[94a]

Both 1,6-anhydro-β-D-glucofuranose and its α-D-*galacto* analog are unusual in that they contain a vicinal diol group that is not attacked by periodate. Glycol cleavage is prevented because the two hydroxyl groups are held rather rigidly, with a dihedral angle of about 120 degrees[32,95] (see Vol. IB, Chapter 25).

F. Miscellaneous Aldosans

Detritylation of 1,2,3-tri-O-acetyl-5-O-trityl-D-ribofuranose with hydrogen bromide in acetic acid gave di-D-ribofuranose 1,5′:1′,5-dianhydride.[97,98]

Fusion of 5,6-anhydro-1,2-O-isopropylidene-D-glucofuranose with 1,2-O-isopropylidene-α-D-glucofuranose or with 1,2:3,4-di-O-isopropylidene-α-D-galactopyranose gave 6,6′-anhydrides in both cases.[99]

The glycosan (**17**) is a component of an antibiotic isolated from *Streptomyces aureofaciens*.[100]

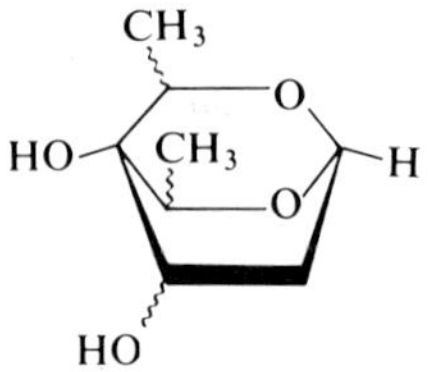

17

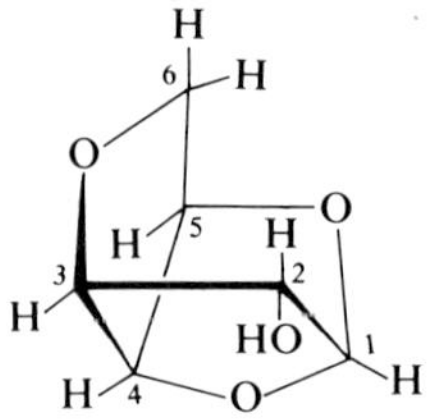

18

The dianhydro sugar, 1,5:3,6-dianhydro-α-D-glucofuranose (**18**), has been prepared in low yield from the gasification of wood[101] and from the vacuum pyrolysis of amylose or 3,6-anhydro-D-glucose.[102] The D-*manno* isomer has also been isolated from the products of vacuum pyrolysis of D-mannose.[102] The p.m.r. spectra of the *gluco* and *manno* isomers have been studied in detail.[102]

G. KETOSE DERIVATIVES

With the exception of the diketose dianhydrides, most of the known anhydro ketose derivatives belong to the classes of glycosan discussed above for aldoses.

An anhydro ketose of the ethylene oxide type has been described, namely 2,3-anhydro-D-fructofuranose 1,4,6-trinitrate, obtained by nitration of D-fructose with dinitrogen pentaoxide–sodium fluoride.[103] 2,3-Anhydro-D-*altro*-heptulose has been postulated as an intermediate in the reaction of phenyl D-*altro*-heptuloside with sodium methoxide.[104]

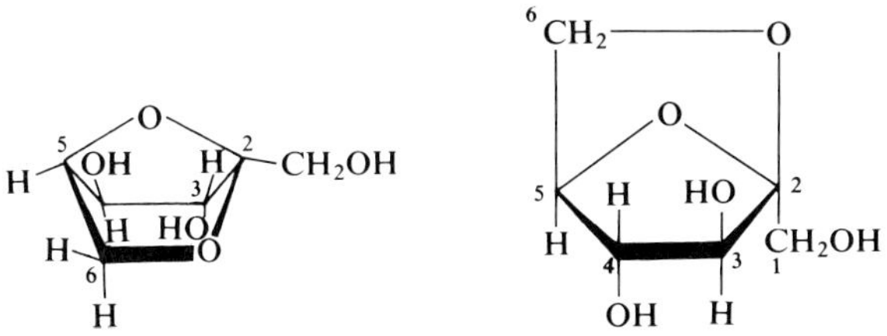

19

2,5-Anhydro-D-fructopyranose (alternatively named 2,6-anhydro-D-fructofuranose) (**19**) has been prepared by the action of base on D-fructosyl fluoride[105] and has been found among the products resulting[106] from hydrogenolysis of sucrose at 180°.

Treatment of heptuloses with dilute aqueous acids sets up an equilibrium between them and their 2,7-anhydrides (compare 1,6-anhydroaldohexoses, Section II,D, p. 430). This reaction has been studied for D-*altro*-heptulose (sedoheptulose),[107-109] L-*gulo*-heptulose,[110] and D-*ido*-heptulose.[111]

A number of diketose dianhydrides are known that are formed by intermolecular loss of two molecules of water from between two ketose units.[112] The D-fructose derivatives have been made by the action of strong acid (such as fuming nitric acid) on D-fructose[113-117] or on inulin.[118-121] Six di-D-fructose dianhydrides have been isolated; they are shown as follows.

Diheterolevulosan I[113, 115, 117]

Di-D-fructopyranose 1,2′:2,1′-dianhydride[114, 125]

Diheterolevulosan II[115, 117]

D-Fructofuranose D-fructopyranose 1,2′:2,1′-dianhydride[114, 115]

---

* *References start on p.* 471.

Diheterolevulosan III[116]

    Di-D-fructopyranose 1,2':2,3'-dianhydride or anomer of diheterolevulosan II[116]

Di-D-fructose anhydride I[117, 118, 120-122]

    Di-D-fructofuranose 1,2':2,1'-dianhydride[120]

Di-D-fructose anhydride II[123]

    Di-D-fructofuranose 1,2':2,4'-dianhydride[123, 124]

Di-D-fructose anhydride III[126]

    Di-D-fructofuranose 1,2':2,3'-dianhydride or anomer of di-D-fructose anhydride II[126]

The action of hydrochloric acid on L-sorbose gave two di-L-sorbose dianhydrides, one of which was shown to be the dipyranose 1,2':2,1'-dianhydride.[127]

III. ANHYDRO SUGARS

A. INTRODUCTION

Anhydro sugars are here defined as compounds formed by elimination of one molecule of water from between two hydroxyl groups within the same molecule; neither of the hydroxyl groups is at the anomeric center. These compounds are internal ethers and therefore, with the exception of the ethylene oxide type, are stable and will persist through most reactions that a compound may undergo. The types most commonly encountered are the oxirane (epoxide or ethylene oxide type),[1b] which is extremely useful in synthesis, and the five-membered tetrahydrofuran type, such as the 3,6- and 2,5-anhydro sugars.[1c] Nearly all the possible types of anhydro sugar are known.

The general principles involved in synthesis of anhydro sugars will be outlined, and the various classes of anhydro sugars will then be considered in turn.

B. SYNTHESIS OF ANHYDRO SUGARS

One fundamental step is involved in most syntheses of anhydro sugars. This is the displacement of a leaving group (Y) by intramolecular nucleophilic attack of an internal anion produced by removal of the proton from a hydroxyl group by base, as shown in Scheme III. The leaving group, Y, is generally a sulfonic ester such as a p-tolylsulfonyloxy or a methylsulfonyloxy group, but it may also be a sulfuric or nitric ester or a halide ion, such as bromide or iodide. A similar type of leaving group is generated by diazotization of an amino sugar, where the leaving group has been supposed to be N_2^+ as shown in Scheme IV (but see ref. 1b).

These internal nucleophilic attacks are essentially irreversible and are all of the SN2 type, and so the stereochemical features necessary for formation of an

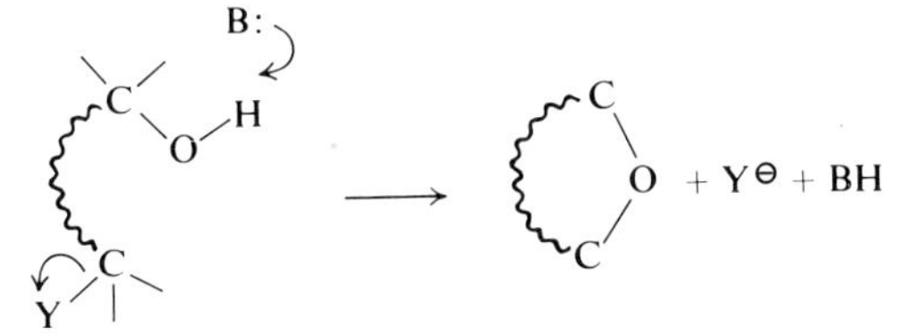

SCHEME III

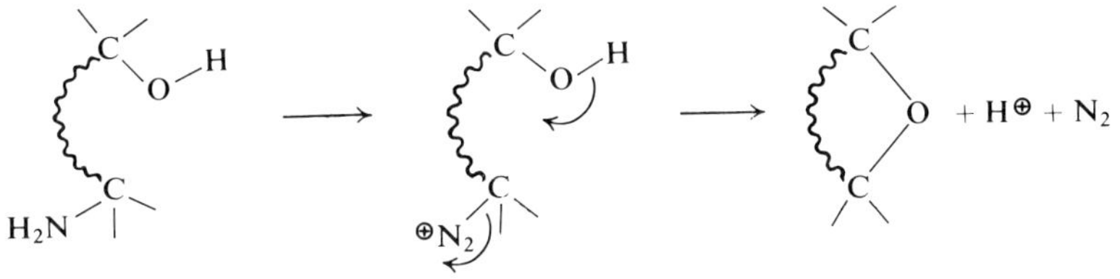

SCHEME IV

anhydro sugar in a cyclic system are defined. The attacking hydroxyl group must be in a position to attack from the back of the leaving group. The important result of this rearside attack is that inversion occurs at the carbon atom bearing the leaving group.

The general situations encountered in sugar systems are set out below; Y represents any good leaving group of the types discussed above. For formation of an epoxide on a pyranose ring the interacting groups must be *trans* and essentially antiparallel (diaxial) as in **20a** and not gauche (diequatorial) as in **20b**. In the *cis* equatorial–axial system, in neither possible conformation (**21a** or **21b**) can the hydroxyl group attack from the back of Y. Similarly, for furanose systems, formation of an epoxide is possible from a *trans* precursor (**22**) but not from a *cis* precursor (**23**). Epoxide formation can obviously occur easily in an acyclic chain, as in **24**. Examples leading to 3,6-anhydro-furanoses and -pyranoses are also shown. Formation of such anhydro rings can occur only where the stereochemistry permits it, as in examples **25–27**. For pyranose derivatives, the molecule must be able to adopt a conformation such that the two groups involved are essentially *syn*-diaxial; formation of an anhydro ring is not possible with the systems **28** and **29**. For similar reasons, the furanoid examples **30** and **31** do not give 3,6-anhydrides. In situations where there is the possibility of forming either a three- or a five-membered ring, as in **32**, the three-membered ring is usually formed. These general principles will be illustrated in the succeeding sections.

* *References start on p. 471.*

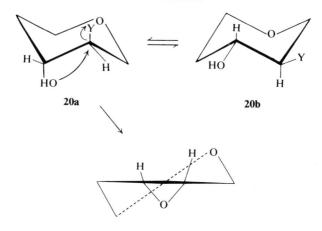

20a   20b

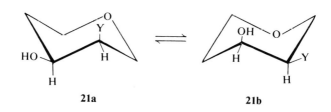

21a   21b

22   23

24

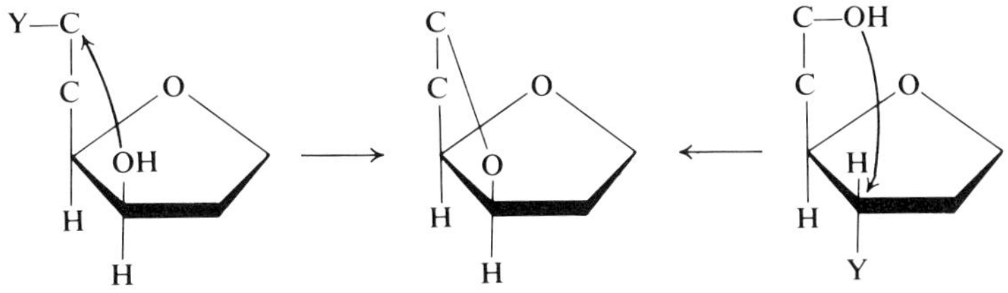

25   **26**

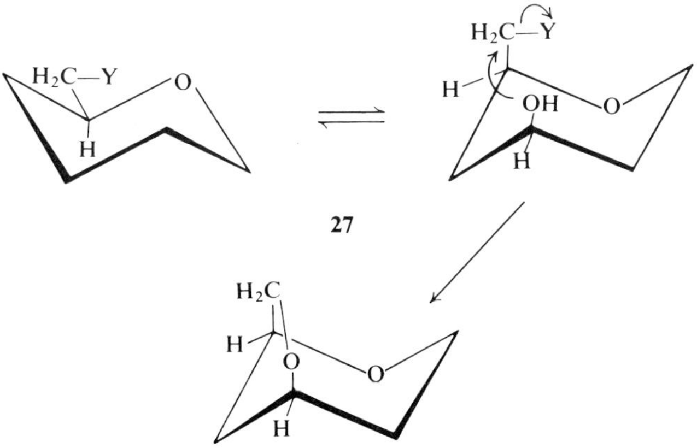

27

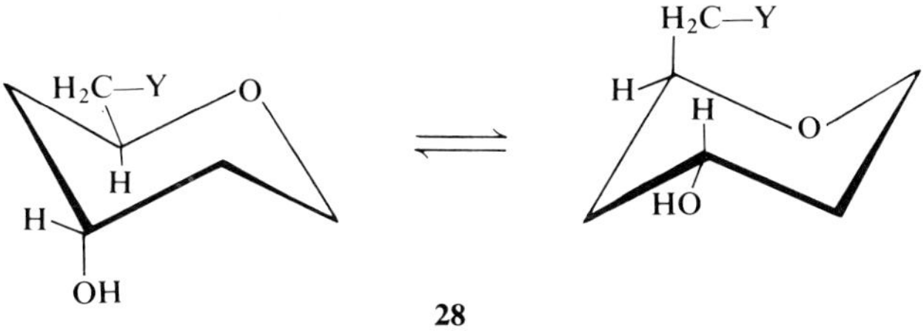

28

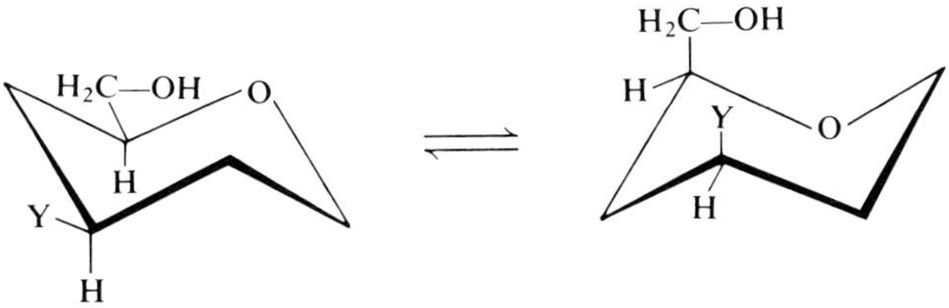

29

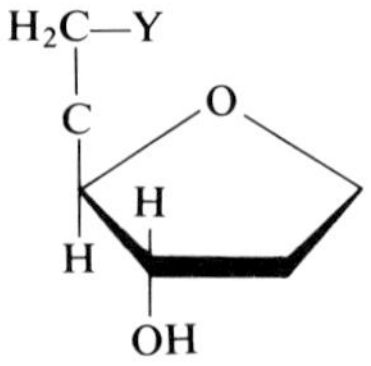

30 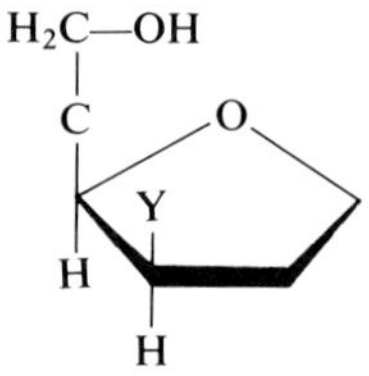   **31** 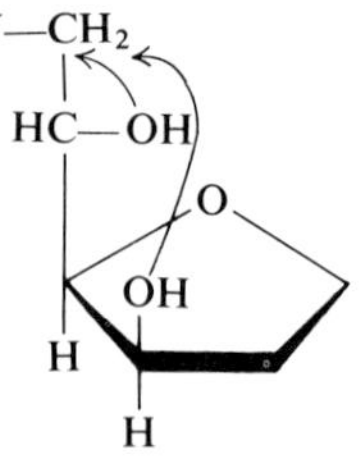   **32**

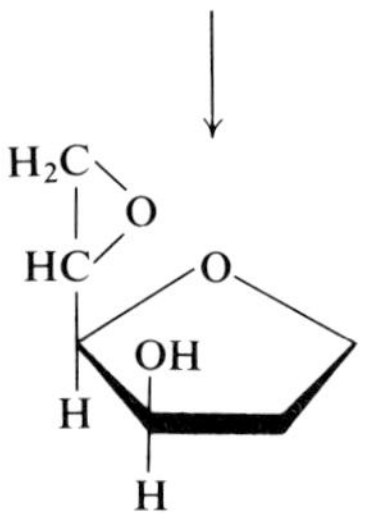

C. EPOXIDES[16, 128]

1. *Synthesis from Halogen Compounds*

6-Bromo-6-deoxy-1,2-*O*-isopropylidene-α-D-glucofuranose reacts with sodium ethoxide to give[129] the 5,6-anhydride (**33**). Treatment of methyl 3,4,6-tri-*O*-acetyl-2-bromo-2-deoxy-α-D-glucoside[130] or the 2-chloro analog[131]

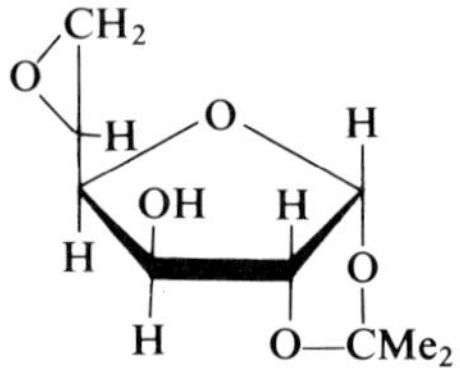

33

with ammonia gave a 3-amino-3-deoxy-D-altroside derivative, presumably formed by way of an intermediate 2,3-epoxide.

2. *Synthesis from Amino Compounds*

Deamination of 6-amino-6-deoxy-1,2-*O*-isopropylidene-α-D-glucofura-nose[132] with nitrous acid gave the epoxide (**33**). Deamination of methyl 3-amino-4,6-*O*-benzylidene-3-deoxy-α-D-altroside[133] (**34**) and of 4-amino-1,6-anhydro-β-D-glucopyranose[132] (**36**) gave methyl 2,3-anhydro-4,6-*O*-benzyl-idene-α-D-mannoside (**35**) and 1,6:3,4-dianhydro-β-D-galactoside (**37**), respectively. In both cases the interacting groups were axial in the favored

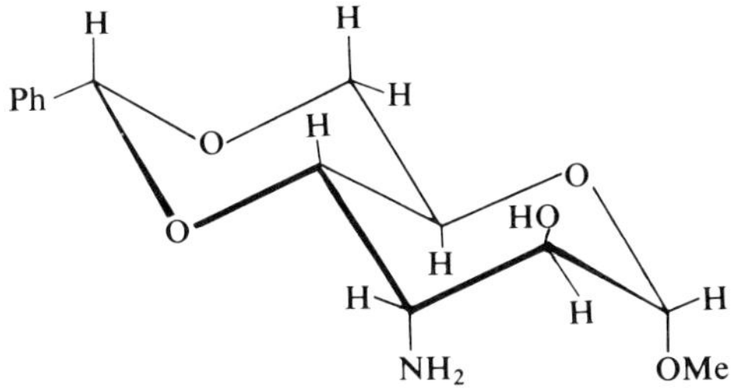

29

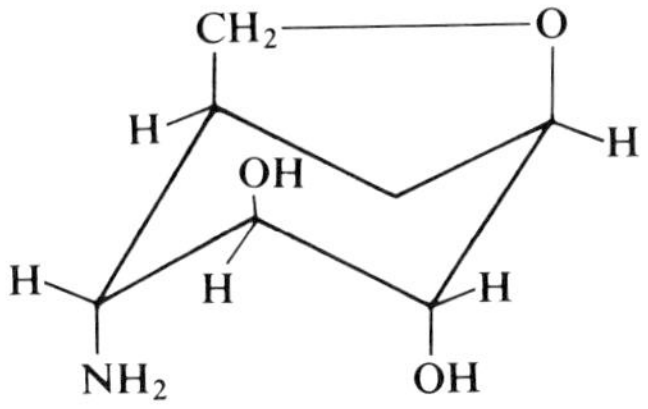

36

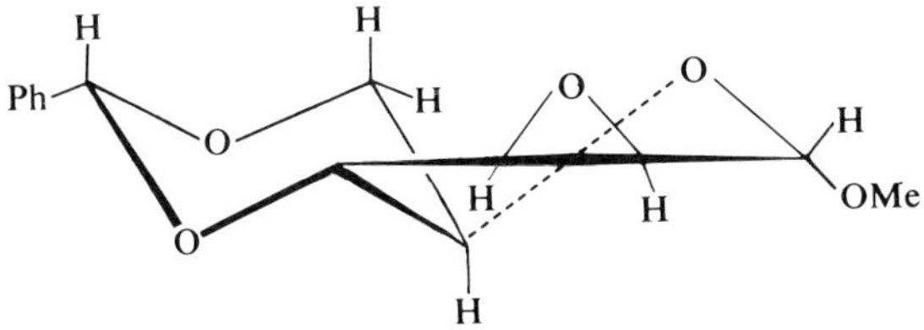

35

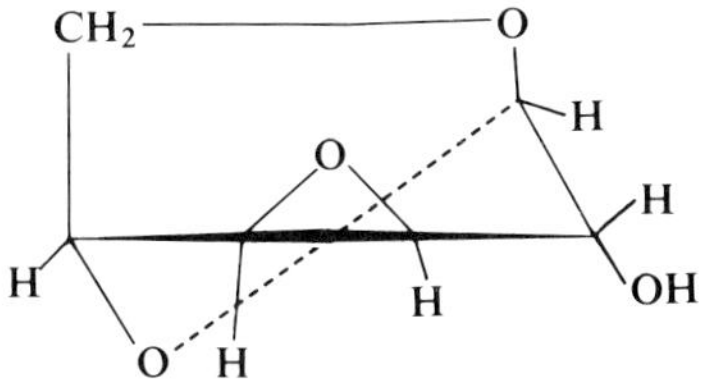

37

conformation. A 4,5-epoxide having the D-*galacto* configuration has been proposed as one of the initial products in the deamination of methyl 4-amino-4-deoxy-α-D-glucopyranoside by nitrous acid.[133a]

3. *Synthesis from Monosulfonates*

The acyclic epoxide (**33**) has been synthesized by treatment of 1,2-*O*-isopropylidene-6-*O*-*p*-tolylsulfonyl-α-D-glucofuranose with alkali.[134]

The action of alkali on methyl 2,3,6-tri-*O*-acetyl-4-*O*-*p*-tolylsulfonyl-β-D-glucoside (**38**) gave an anhydro sugar[135, 136] shown to be methyl 3,4-anhydro-β-D-galactoside[136] (**39**). Müller *et al.*[137] later studied the methylsulfonyl analog of **38** and its C-4 epimer (**40**) and showed that the only effect of alkali on **40** was to remove the acetyl groups, whereas the *gluco* compound again gave the epoxide (**39**). This was the first demonstration that *trans* arrangement of the TsO–C–C–OH group (as in **38**) is necessary for formation of an epoxide.

* *References start on p.* 471.

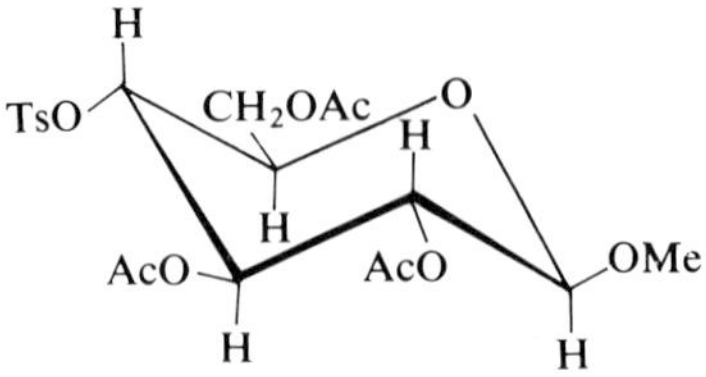

38

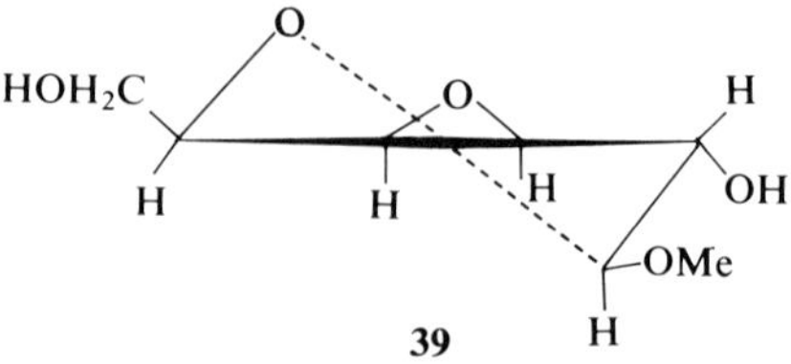

39

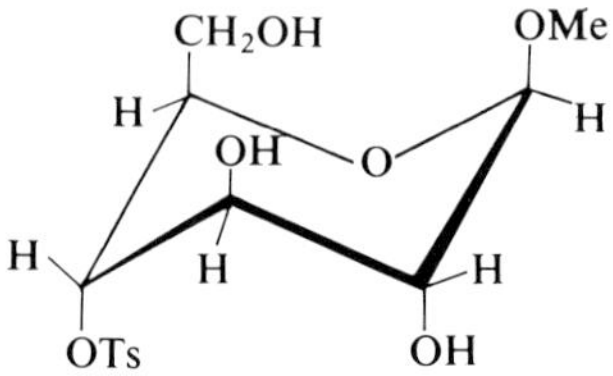

38a

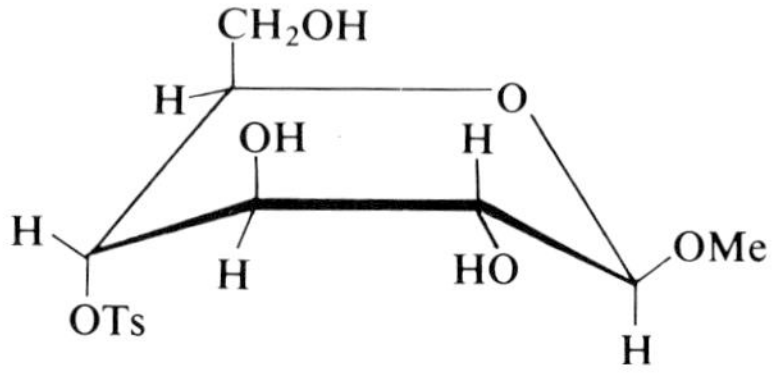

38b

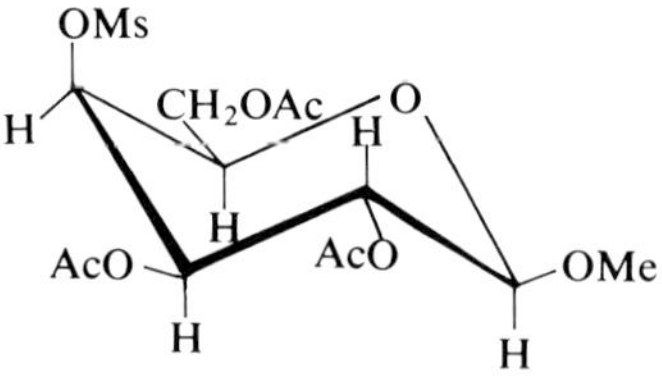

40

Compound **38** has the reacting group *trans*, and even though the favored conformation does not have these groups antiparallel, the necessary geometry can be reached by way of some other conformation, such as **38a** or **38b**. The *galacto* derivative (**40**) cannot have both groups at C-3 and C-4 axial, whatever conformation it adopts. The first reported synthesis of a 2,3-epoxide of a sugar appears to be that of methyl 2,3-anhydro-β-D-mannoside from methyl 3,4,6-tri-O-acetyl-2-O-p-tolylsulfonyl-β-D-glucoside.[138]

Some other examples of epoxide formation in flexible systems are methyl 2,3-anhydro-β-L-ribopyranoside from methyl 2-O-p-tolylsulfonyl-β-L-arabino-pyranoside;[139] methyl 3,4-anhydro-β-L-ribopyranoside† from methyl 4-O-p-tolylsulfonyl-α-D-lyxoside;[140] and methyl 2,3-anhydro-6-deoxy-α-D-mannoside from methyl 6-deoxy-2-O-p-tolylsulfonyl-α-D-glucoside.[141]

For compounds of rigid conformation, such as aldohexopyranoses having a *trans*-4,6-O-benzylidene group or a 1,6-anhydro ring, the possibilities of conformational change are more limited. Because of this factor, study of such systems has been useful in defining fine stereochemical features in the formation of epoxides. Robertson and Griffith[142] showed that methyl 2-O-benzoyl-4,6-O-benzylidene-3-O-p-tolylsulfonyl-α-D-glucoside (**41**) gave methyl

† Incorrectly described as β-D-*lyxo* in the original paper.[140]

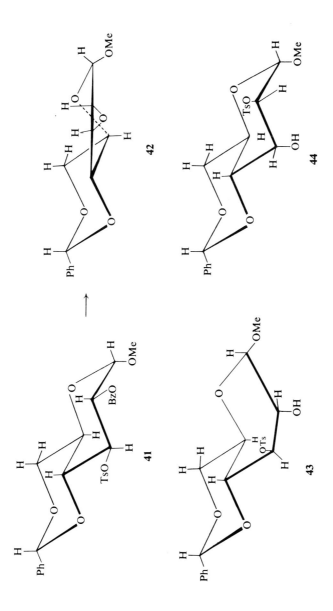

2,3-anhydro-4,6-*O*-benzylidene-α-D-alloside (**42**) on treatment with alkali and that methyl 3-*O*-benzoyl-4,6-*O*-benzylidene-2-*O*-*p*-tolylsulfonyl-α-D-glucoside gave the anhydro-D-mannoside (**35**) on similar treatment,[142] as did the corresponding 3-hydroxy compound.[143] The *C1* (D) (2,3-diequatorial) conformation is locked by the 4,6-substituent and cannot change to the alternative *1C* (D) chair conformation. A boat-like conformation (**43**) (the *O*-benzoyl group is saponified prior to formation of the epoxide) of the pyranose ring is, however, possible. A similar conformational change presumably occurs with the corresponding 3-*O*-benzoyl-2-*O*-*p*-tolylsulfonyl-D-glucoside. This idea of a conformational change is supported by the fact that methyl 4,6-*O*-benzylidene-2-*O*-*p*-tolylsulfonyl-α-D-altroside (**44**), which has the 3-OH–2-OTs system in a *trans*-diaxial arrangement in its favored conformation, forms the anhydro-

45, R = H
46, R = OMe

45a, R = H
46a, R = OMe

D-alloside (**42**) very readily in the presence of a trace of alkali, or even on chromatography on alumina.[144] Newth[145] has shown that 1,5-anhydro-4,6-*O*-benzylidene-2-*O*-*p*-tolylsulfonyl-D-glucitol (**45**) forms an epoxide more easily than does methyl 4,6-*O*-benzylidene-2-*O*-*p*-tolylsulfonyl-α-D-glucoside (**46**). This difference has been attributed to a smaller passing interaction on **45** passing into **45a** than in the D-glucoside (**46**) changing into **46a**, namely H/OTs against OMe/OTs.

Similar examples are available in the 4,6-*O*-benzylidene-D-galactopyranose series, although, since the molecule is flexible because of the *cis* fusion of the rings, it is not known whether epoxide formation proceeds by way of a boat conformation as above, or through the alternative chair conformation. Both methyl 2,3-anhydro-4,6-*O*-benzylidene-α-[146, 147] and β-D-taloside,[146] and the corresponding β-D-guloside,[61] have been prepared from the appropriate mono-*p*-toluenesulfonate.

The stereochemical features leading to formation of epoxides have also been studied by using derivatives of 1,6-anhydro-β-aldohexopyranoses.[24a] As expected, mild alkaline treatment of 1,6-anhydro-4-*O*-*p*-tolylsulfonyl-β-D-mannose[148] (**47**) and 1,6-anhydro-2-*O*-(methylsulfonyl)-β-D-galactose[149] (**48**)

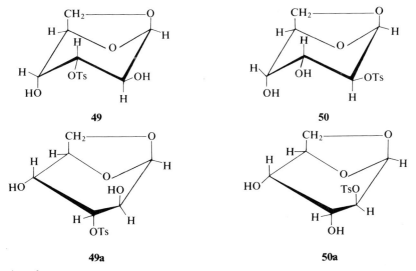

gave the anticipated epoxides: both **47** and **48** have the reacting groups diaxial already in the favored conformation of the molecules.

Although 1,6-anhydro-3-*O*-*p*-tolylsulfonyl-β-D-altropyranose (**49**) could be converted into an epoxide, the 2-*O*-*p*-tolylsulfonyl analog (**50**) was resistant to alkali, and no epoxide was formed.[150] (Note that both compounds have the same geometry at C-2 and C-3.) For formation of an epoxide, **49** must attain the conformation **49a** in which there are destabilizing interactions of the 2-OH group with O-1 and C-5–C-6; also there will be a 4-OH/3-OTs passing interaction on going from **49** to **49a**. These interactions are evidently not sufficient to prevent epoxide formation.   However, in the change from **50** to **50***, although there the 3-OH/4-OH passing interaction is no apparent limitation, the conformation **50a** has the very destabilizing interaction of the 2-OTs group with O-1 and C-5–C-6, and this factor presumably prevents formation of an epoxide. Furthermore, 1,6-anhydro-3,4-di-*O*-*p*-tolylsulfonyl-β-D-altropyranose did not give an epoxide with alkali.[150] In this example the interactions in the necessary boat conformation are approximately the same as those in **49a**, but to reach this conformation a 3-OTs/4-OTs passing interaction is involved. This interaction appears to be the factor preventing formation of an epoxide. These three compounds provide an excellent example of the fine balance of steric effects that can prevent or allow epoxide formation.

Furanose epoxides can be prepared in a similar way from monosulfonates— for example, methyl 2,3-anhydro-α-D-lyxofuranoside from methyl 2-*O*-*p*-tolylsulfonyl-α-D-xylofuranoside.[151] Ketose epoxides have also been prepared by this method.[152, 153]

4. *Synthesis from Disulfonates*

Epoxides can be prepared from *trans*-vicinal disulfonates, as well as from monosulfonates; one of the sulfonyl groups evidently undergoes S–O cleavage by base. It has been postulated[154] that the removal of one sulfonyl group will

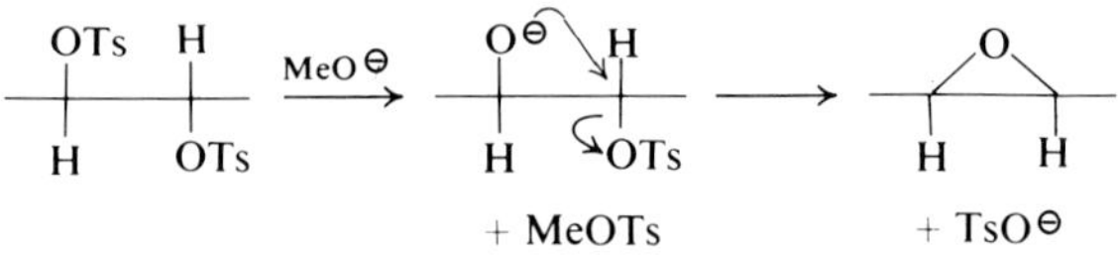

SCHEME V

be assisted by the inductive effect of the other and that the first to be removed will be the more accessible one; this is shown diagrammatically in Scheme V. The best-known example in this class of reaction is the formation of only the anhydro-D-alloside (**42**) from methyl 4,6-*O*-benzylidene-2,3-di-*O*-*p*-tolylsulfonyl-α-D-glucoside.[142] This means that S–O cleavage occurs preferentially at the 2-*p*-tolylsulfonyloxy group. This proposition is supported by the isolation of some methyl 4,6-*O*-benzylidene-3-*O*-*p*-tolylsulfonyl-α-D-glucoside on brief treatment of the di-*p*-toluenesulfonate with alkali.[155] Also, methyl 4,6-di-*O*-methyl-2,3-di-*O*-*p*-tolylsulfonyl-α-D-glucoside gave methyl 2,3-anhydro-4,6-di-*O*-methyl-α-D alloside on treatment with base.[155a] Similarly, methyl

4,6-*O*-benzylidene-2,3-di-*O*-*p*-tolylsulfonyl-α-D-altroside gave only the *manno* epoxide;[156] again O–S cleavage occurs at the 2-*p*-tolylsulfonyloxy group. In the D-galactose series the situation is less simple. Methyl 4,6-*O*-benzylidene-2,3-di-*O*-*p*-tolylsulfonyl-β-D-galactoside gave only the corresponding *talo* epoxide,[61] whereas the α-D anomer gave a mixture of the corresponding *gulo* and *talo* epoxides.[147] More work is needed for a fuller understanding of these reactions of disulfonates. The formation of epoxides from a series of 4-azido-4-deoxy-2,3-di-*O*-(methylsulfonyl)pentosides has been discussed in terms of steric and electronic factors,[157] and the whole subject of reactions of sugar sulfonates has been reviewed in detail.[157a]

5. Synthesis from Sulfuric and Nitric Esters

Treatment of the barium salt of 1,2-*O*-isopropylidene-3-*O*-methyl-α-D-glucofuranose 6-sulfate with methoxide gave[158] the 5,6-epoxide (3-methyl ether of compound **33**).

A nitric ester can undergo cleavage in two ways—either by C–O fission (compare sulfonates) or by O–N fission (compare carboxylates).[159] The situation is complex, and again further work is necessary before any theory can be solidly presented. The results showed[155, 160] that, for derivatives of 4,6-acetals of methyl α-D-glucoside, the 3-nitrate, 2,3-dinitrate, 3-nitrate 2-*p*-toluenesulfonate, and 2-nitrate 3-*p*-toluenesulfonate all gave the *allo* epoxide (**42**). Thus in all cases the C-2 group provided the anion for attack on C-3.

6. Synthesis from Unsaturated Sugars

An example of this type of synthesis is illustrated by the formation of the spiro epoxide, methyl 2,3,4-tri-*O*-acetyl-5,6-anhydro-α-D-glucopyranoside (**52**), by treatment of methyl 2,3,4-tri-*O*-acetyl-α-D-*xylo*-hex-5-enopyranoside (**51**) with *p*-nitroperoxybenzoic acid.[161] Spiro epoxides can also be obtained from carbonyl derivatives of sugars by the action of diazomethane.[162] (See also Vol. IB, Chap. 17.)

51   **52**

7. Ring-Opening Reactions of Epoxides

Epoxides are synthetically the most useful class of anhydro sugars and their chemistry has been the subject of several reviews.[1, 128, 163]

* *References start on p. 471.*

Three types of reaction can be distinguished—namely, opening of an epoxide ring on a flexible ring system, on a rigid system, or where one end of the ring is on a primary carbon atom.

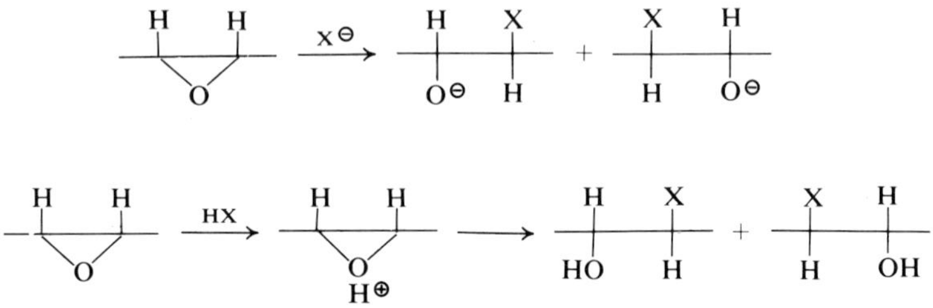

SCHEME VI

The general statement of ring opening with nucleophiles or acid reagents is shown in the two processes illustrated in Scheme VI, the second reaction being the faster one. Opening of epoxides that have one primary carbon atom occurs stereospecifically at that atom, to give one product only (see examples later). For ring systems, the conformation of the pyranose or furanose ring to which the epoxide ring is attached must be considered.

a. Pyranose epoxides—The opening of epoxide rings on pyranoid systems has been much discussed,[1, 128, 163–169] and a clear pattern has emerged. Cyclohexene oxide has been shown[170] to exist in the half-chair conformations, 53 and

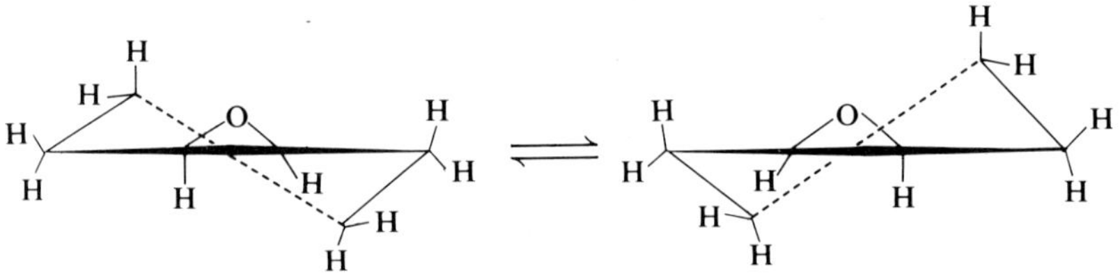

53   54

54. For pyranose epoxides that are flexible, possible attack on both forms of the epoxide must be considered; for rigid systems, one epoxide conformation will normally preponderate.

The Fürst–Plattner rule of diaxial opening of steroid epoxides was used by Mills[164] to rationalize the reactions of pyranose epoxides. The geometry of axial opening of epoxides (Scheme VII) is obviously a less hindered process than equatorial opening (Scheme VIII), since the former goes via a coplanar transition state. Thus for epoxides held in a rigid conformation by a trans-

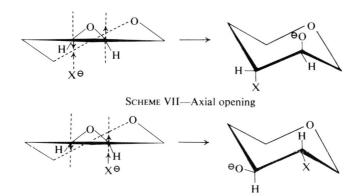

Scheme VII—Axial opening

Scheme VIII—Equatorial opening

4,6-acetal group or by a 1,6-anhydro bridge, one principal product would be expected and is in fact found (for examples, see below). For flexible epoxides, assuming axial attack, two products would be expected, one arising from each of the possible half-chair conformations (Scheme IX).

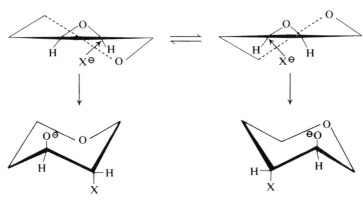

Scheme IX

In many cases, an initial diaxial product is actually more stable in the diequatorial form and will change to this form if the conformation is not locked by a bridging group; this factor accounts for the early so-called "exceptions" to the Fürst–Plattner rule.

The locked-ring, pyranose epoxides that have been studied most are methyl 2,3-anhydro-4,6-O-benzylidene-α-D-alloside (**42**) and mannoside (**35**). The opening of these epoxides with nucleophiles gives diaxial products, accompanied in some cases by 5 to 10% of the diequatorial products. Exceptions to

* *References start on p. 471.*

this generalization are openings by acidic reagents in which the acetal blocking group is removed before the epoxide is opened, openings by Grignard reagents, and openings by hydrogenation in the presence of nickel. The wide range of nucleophiles used to open the epoxides **42** and **35** are shown below.

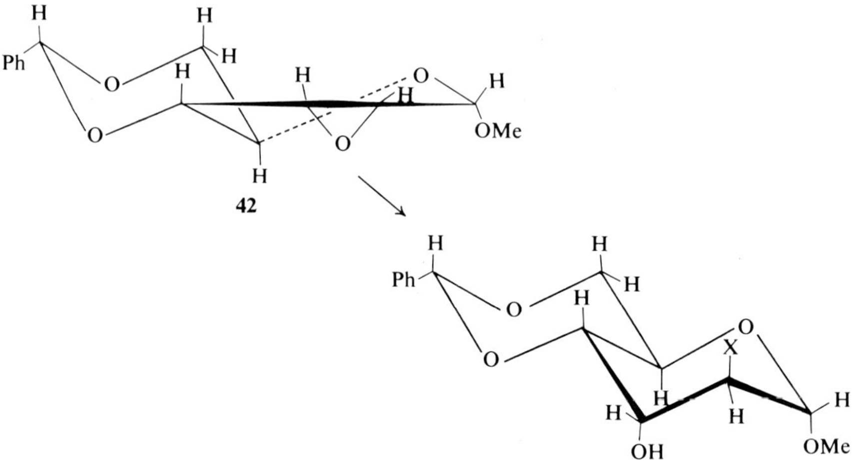

42

X:

NH_2[171,173]	$PhCH_2S$[175]
$OP(OCH_2Ph)_2$[172]	MeS[176]
$\parallel$	N_3[177]
O	H[178]
$NH_2 \cdot NH$[156]	OH[142]
MeO[142]	
CH_2CO_2Et [from $^\ominus CH(CO_2Et)_2$][174]	

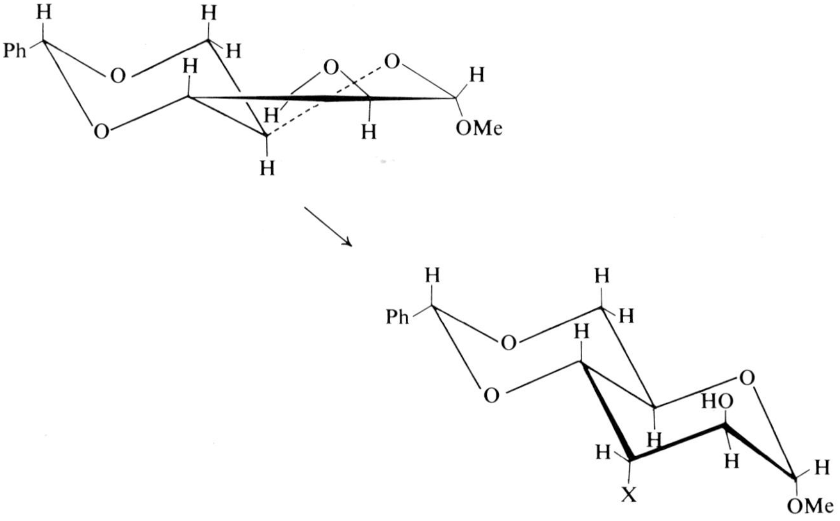

X:	NH$_2$[173,179]	SCN[183] (from NH$_4$SCN)
	SCH$_2$Ph[175,181]	N$_3$[177]
	OH[142]	PhNHNH[184]
	OMe[142,180]	H[178]
	SMe[182,186]	Me (from MeLi)[185]

Opening of the analogous methyl 2,3-anhydro-4,6-O-benzylidene-D-talo-sides and D-gulosides, which have two *cis*-fused six-membered rings, and therefore nominally have a flexible epoxide system, gives predominantly products from axial opening, assuming that the compounds have the "O-inside" conformation—for example, **55**. Nucleophiles used on these systems have been MeO$^-$,[61,146,147] HO$^-$,[146] NH$_3$,[187] and N$_3$$^-$.[188]

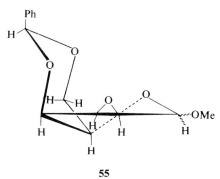

55

Epoxides of 1,6-anhydroaldopyranoses are also conformationally rigid. In the few cases studied, diaxial opening occurs predominantly;[149,189] for example, the epoxide **56** is converted by methoxide into **57**. An exception to this generalization is the opening of 1,6:2,3-dianhydro-4-deoxy-D-*lyxo*-hexo-pyranose, which gives a mixture of D-*gluco* and D-*altro* products when treated

with potassium hydroxide. It has been suggested[189] that this is because of a competition between steric and electronic factors. Under acidic conditions of ring-opening and in the presence of participating neighboring-groups the simple principle of *trans*-axial ring-opening is not applicable; examples of

* *References start on p.* 471.

net *cis*-opening and *trans*-diequatorial opening have been demonstrated with
2-*O*-acyl-1,6:3,4-dianhydro-β-D-galactopyranose;[189a] these reactions occur
via an intermediate 2,3-acyloxonium ion.[189b]

The pyranose epoxides having flexible rings studied have had the epoxide
ring in both possible positions, namely 2,3 or 3,4. In the latter series, the
epoxide group of methyl 3,4-anhydro-β-D-galactoside (58) has been opened
with a variety of reagents, (H$_2$S,[190] MeSH,[191], LiAlH$_4$,[192] and MeO$^-$[136]). In
every case the 3-substituted D-*gulo* product is obtained. However, treatment of
methyl 3,4-anhydro-β-L-riboside (59)† with hydrogen bromide[140] or various
amines[193] caused opening exclusively at C-4 to give methyl 4-substituted 4-
deoxy-α-D-lyxosides. The difference between these two 3,4-epoxides can be

58a   58b

59a   59b

rationalized on steric grounds. Conformation **58b** can be expected to be more
stable than **58a**, and **59a** more stable than **59b**. Also, assuming diaxial attack,
it can be shown for both compounds that attack at C-4 in **58a** or at C-3 in **59b**
would lead to transition states for each molecule in which there are two pairs
of incipient 1,3 interactions, presumably a large destabilizing factor. Hence **58**
opens from conformation **58b** at C-3 and **59** at C-4 by way of **59a**. In the opening of **58** with hydrogen sulfide[190] the product is 3,3′-thiobis(methyl 3-deoxy-β-
D-gulopyranoside), since the 3-thio-D-*gulo* derivative formed initially is a
better nucleophile than hydrogen sulfide.

† This epoxide is erroneously described as β-L-*lyxo* in one paper[140] and as α-L-*ribo* in the
other.[193]

A wider variety of flexible-ring 2,3-anhydroaldopyranosides have been opened with various nucleophiles. A few generalizations can be made. Opening of the epoxide group of various 2,3-anhydro-β-D-ribosides having the general formula **60** (or its enantiomorph) occurs predominantly at C-3 in all

60

cases so far studied. Groups introduced have been for R = H, R' = Me, SMe,[194] Br,[195] Cl,[196] H,[196] NH$_2$,[197] Ph,[185] Me$_2$C=CH,[185] PhC≡C;[185] for R = CH$_2$Ph, R' = Me, F;[198] and for R = H, R' = CH$_2$Ph, F.[199] Variously substituted 2,3-anhydrohexopyranosides have been studied similarly, but insufficient examples have been reported to permit generalizations. The nucleophiles used have been those described above and also dimethylamine; the latter provides a method of introducing the dimethylamino group in a synthesis of mycaminose.[200]

An example in which hydrogen bonding apparently controls the fission of flexible epoxides has been noted[201] with methyl 2,3-anhydro-6-deoxy-α-L-taloside (**61**) and its 4-methyl ether (**62**). The methyl ether (**62**) reacted with methoxide ion to give the product expected from the stable half-chair conformation, whereas the nonetherified analog (**61**) gave predominantly the alternative product. It can be supposed that hydrogen bonding in **61** between the hydroxyl group and the oxygen atoms of the epoxide and pyranose rings stabilizes the "unstable" half-chair conformation. A similar explanation can be used to explain results obtained with methyl 3,4-anhydro-6-deoxy-α-L-taloside[201, 201a] and its 2-methyl ether,[201] and with the glycosides of 2,3-anhydro-β-D-ribopyranoside or its L isomer (see foregoing).

The opening of epoxide rings by acidic reagents, in systems initially held rigid by a *trans*-4,6-O-benzylidene group, gives products as obtained from the corresponding debenzylidenated epoxide. Presumably the acetal group is removed before the epoxide ring is opened. Thus, methyl 2,3-anhydro-4,6-O-benzylidene-α-D-alloside reacts with hydrogen halides to give mainly the methyl 3-deoxy-3-halo-α-D-glucoside.[202, 203] The course of the reaction of the corresponding D-mannoside with hydrogen chloride has been described differently by different workers.[202, 204]

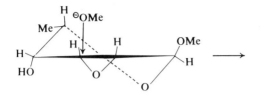

61

62

The influence of a neighboring *trans*-acetoxy group on epoxide ring opening has been studied.[205] Compound **63**, when treated with 80% acetic acid, gave a mixture of the D-*ido* and D-*manno* products. Under the same conditions the

63, R = Tr } R′ = CH$_2$Ph
64, R = Ac }

65, R′ = CH$_2$Ph

6-acetoxy analog (**64**) gave exclusively the D-*ido* derivative, with only a trace of the D-*manno* product; the reaction was postulated as proceeding through the intermediate (**65**). Other examples of participation by a neighboring, acetoxy group have been studied.[198a, 206]

The reaction of sugar epoxides with Grignard reagents and with dialkyl and diaryl magnesium compounds has been studied in detail by Richards and his co-workers.[207] No consistent pattern emerges from the reactions so far examined, which have been summarized diagrammatically by Newth.[128]

Reduction of sugar epoxides by hydrogen in the presence of a catalyst of Raney nickel[178, 182, 208, 209] does not follow the usual rules of epoxide ring opening, presumably because of the non-ionic nature of the reaction.

Treatment of methyl 2,3-anhydro-4,6-*O*-benzylidene-α-D-alloside and -mannoside with thiourea gave the 2,3-epithio-D-mannoside and -D-alloside, respectively.[210]

A detailed proton magnetic resonance study has been made on a number of 2,3-pyranose epoxides.[211, 211a]

b. 2,3-Anhydrofuranose Derivatives.—Compounds having both possible relative configurations—namely, 2,3-anhydro-*lyxo* and 2,3-anhydro-*ribo*—almost all undergo scission at C-3 by a variety of nucleophiles. Derivatives studied have been the methyl glycosides,[151, 212-218] their 5-methyl ethers,[151, 216] and also nucleoside derivatives.[219-221] The groups introduced have been OMe,[151] OH,[219] NH$_2$,[213-216, 220, 222] SCH$_2$Ph,[217, 218] SEt,[212, 221] SAc,[212] F,[198] and SH (plus the corresponding sulfide).[212] The favored opening at C-3 has been attributed to the greater inductive effect of C-1 over C-4, making the C-3–O bond the easier one to break.[222] Predominant opening at C-2 occurred in the reaction of sodium α-toluenethioxide on methyl 2,3-anhydro-β-D-lyxoside, where the ratio of C-2 to C-3 products was 2:3; no explanation was offered for this result.[218] The α-anomer reacts, as expected,[217] at C-3.

* *References start on p. 471.*

Treatment of methyl 2,3-anhydro-α-D-ribofuranoside (66) with aqueous potassium thiocyanate gave the thietane derivative (66a), the formation of which was explained as shown[223] (Scheme X). A similar sequence occurred

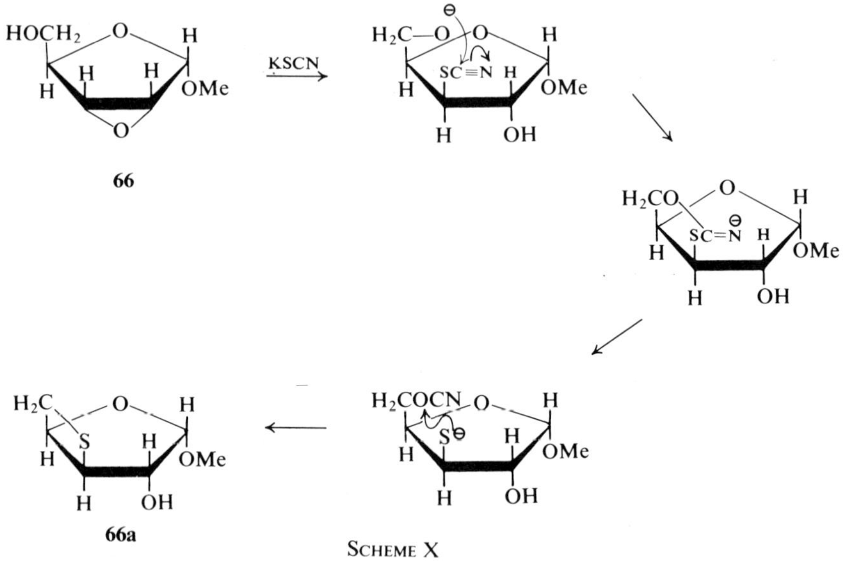

SCHEME X

with the β-D anomer, and the corresponding thietane was formed in higher yield. It was suggested that steric reasons favored attack at C-3 in the β-D anomer.

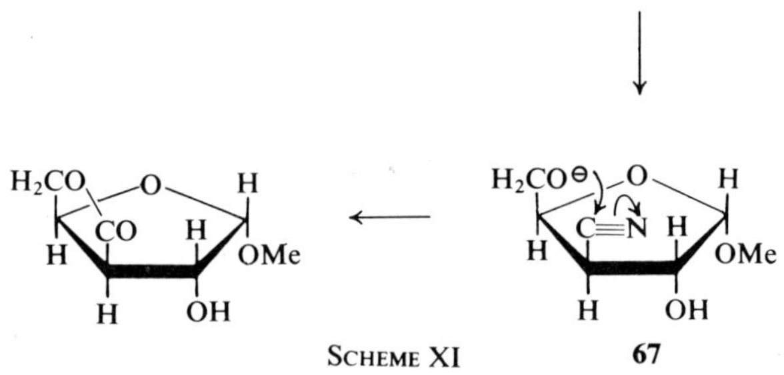

SCHEME XI

Methyl 2,3-anhydro-α-D-ribofuranoside (66) reacts with aqueous potassium cyanide at 100° to give the lactone (67), presumed[223a] to be formed as in Scheme XI.

c. 5,6-Anhydroaldohexofuranose Derivatives.—The two compounds most studied in this class are 5,6-anhydro-1,2-O-isopropylidene-α-D-glucofuranose (33) and the L-*ido* analog (68). A wide range of nucleophiles all react with these compounds, by the expected attack at C-6. The groups introduced have

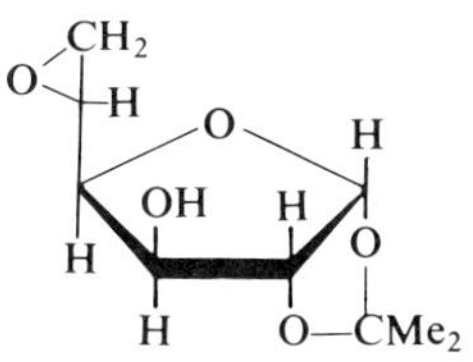

33 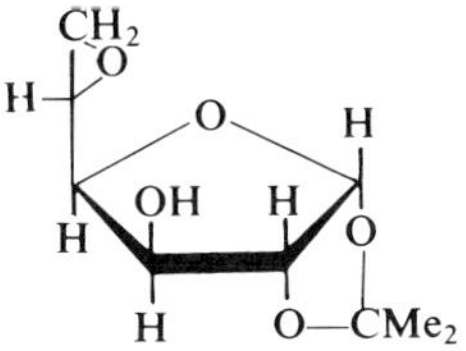   68 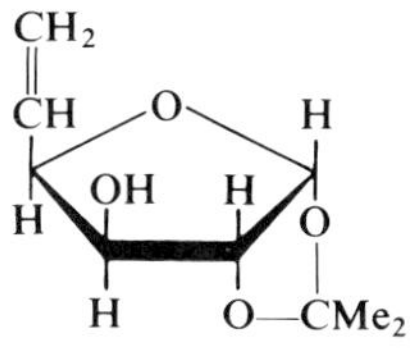   69

included SH,[224] Br,[222] I,[224] PhS,[224] BzS,[225] NH$_2$,[226] N(CH$_2$CH$_2$OH)$_2$,[226] NPh$_2$,[227] NHPh,[228] N(alkyl)$_2$,[229] phthalimido,[230] OAr,[231,232] OR,[233] Ph,[234] OP(O)(OH)$_2$ (as the barium salt[235]), and a wide variety of amino acid derivatives.[236,237] Attack of methyllithium in ether on 33 or 68 gave,[234] in each case, the 5,6-alkene (69).

The anhydro-glucoside (33) reacts with carbon disulfide in alkali to give the 5,6-D-*gluco*-trithiocarbonate,[238] treatment of 33 with thiourea gives the L-*ido*-5,6-epithio derivative;[239] the L-*ido* epoxide (68) similarly gave the D-*gluco*-epithio compound.[239]

Reduction of either 33 or 68 with hydrogen in the presence of Raney nickel gives mainly the 6-deoxy derivatives.[240–242] It has been reported that the 5-deoxy sugars are produced in significant amounts only if the nickel is de-ionized exhaustively before use.[242]

The anhydro-D-glucose derivative (33) has been reported[243] to react with boron trifluoride in methyl sulfoxide to give the 6-aldehyde derivative (70).

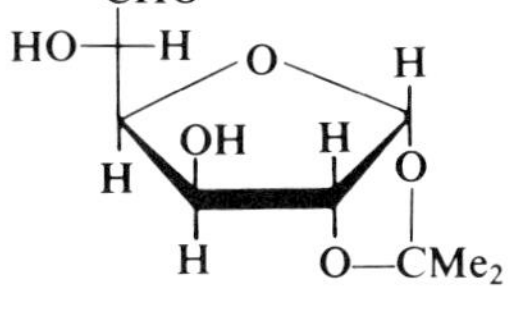

70

D. Anhydro Sugars of the Tetrahydrofuran Type[1c]

1. *3,6-Anhydroaldohexopyranoses*

These compounds, if locked in the pyranose ring as, for example, in a glyco-pyranoside, must exist in the *1C* (D) or *C1* (L) conformation so that the groups on C-3 and C-6 are both axial, as in **71**. 3,6-Anhydro derivatives are possible

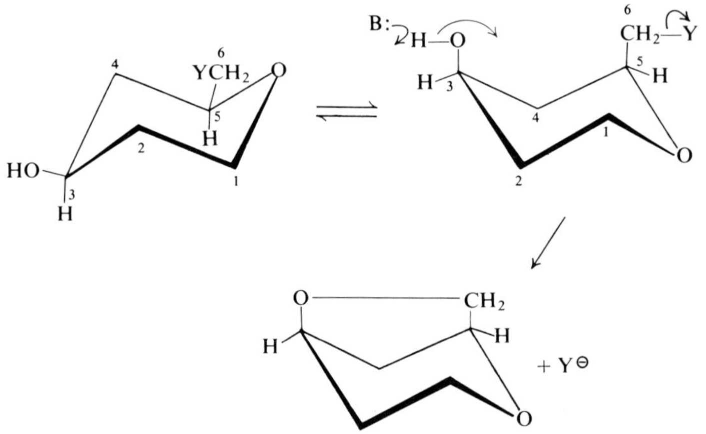

71

Scheme XII

only in the *gluco, manno, galacto,* and *talo* series. The route for synthesis in all cases has been the alkaline hydrolysis of a 6-substituted glycoside (Scheme XII) or a 2,3,4-triester thereof. Methyl 3,6-anhydro-α-D-glucopyranoside has been synthesized from the appropriate 6-bromo,[244] 6-iodo,[245] 6-*O*-*p*-tolylsul-fonyl,[246] and 6-sulfate[247] derivatives, and the β anomer from the 6-bromo derivative[248] and the 6-sulfate.[247] The same type of precursors[247, 249] were used in the synthesis of methyl 3,6-anhydro-α-D-mannopyranoside. Methyl 3,6-anhydro-α-D-galactopyranoside and its derivatives have been prepared by way of the appropriate 6-sulfate,[247] 6-bromide,[250] or 6-*p*-toluenesulfonate.[251–253] The β anomer has been prepared from the 6-sulfate[247] and 6-*p*-toluenesulfo-nate[252] and also by treatment of 2,3,4-tri-*O*-acetyl-6-*O*-*p*-tolylsulfonyl-α-D-galactosyl bromide with silver carbonate in aqueous methanol.[254]

Most of the work on the 3,6-anhydrohexopyranosides has been on the study of their acid-catalyzed rearrangement or hydrolysis, and a number of 3,6-anhydro-2-deoxyhexopyranosides were synthesized to discover[255] the effect of the hydroxyl group at C-2. As would be expected, the 3,6-anhydro ring is not broken under acid conditions that cleave the glycoside, and the structure adopted by the resulting tetrahydrofuran derivative is of interest. Early work showed that aqueous acid converted the methyl 3,6-anhydro-D-galactopyranosides into products that reduced Schiff's reagent[250] (indicating a free aldehyde group), and methanolic acids gave 3,6-anhydro-*aldehydo*-D-galactose dimethyl acetal.[252] Treatment of the 2,4-dimethyl ether of the anhydro-D-galactopyranoside with aqueous acid again gave a Schiff-positive product. A trace of acid converted methyl 3,6-anhydro-2,4-di-O-methyl-α-D-galactopyranoside into the β-D anomer. Methanolic hydrogen chloride converted both anomers into the dimethyl acetal, which on treatment with gaseous hydrogen chloride gave the β-D pyranoside.[252]

However, in the D-*gluco* series, treatment of either methyl 3,6-anhydro-α- or -β-D-glucopyranoside rapidly gave a mixture of the α- and β-D-furanosides, even with ethereal hydrogen chloride.[256] The 2,4-dimethyl ether was converted by methanolic hydrogen chloride into 3,6-anhydro-2,4-di-O-methyl-D-*aldehydo*-glucose dimethyl acetal.[256] Methyl 3,6-anhydro-α-D-manno-pyranoside gave the α-furanoside on treatment with methanolic hydrogen chloride.[249]

Thus, the 3,6-anhydrohexopyranosides, representing a strained ring system, tend to rearrange to relieve the strain to give furanosides if possible, as in the *gluco* and *manno* series. Where formation of a furanoside is impossible, as in the *galacto* series, the *aldehydo* form is produced.

Peat[1] has suggested that the rearrangement in the D-*gluco* derivative (**72**) proceeds by the sequence shown, leading to the *cis*-fused, five-membered ring system (**74**) via the carbonium ion (**73**). Clearly in the C-4 epimer (D-*galacto*) the 4-OH group is not sterically able to attack C-1 in the ion (**75**), and formation of a furanoside is prohibited. The subsequent reaction merely leads to the acetal, or, if water is present, to the *aldehydo* sugar.

The preceding reactions were further investigated by Foster and Overend by using 2-deoxy derivatives. The foregoing reaction paths were substantiated and were considered conformationally.[255] From rate measurements it was suggested that the methyl 3,6-anhydroglucosides experience more strain than the corresponding mannosides. This strain can be visualized by comparing formula **72**, which shows a 1,3 interaction between the C-2 and the C-4 hydroxyl groups, with its C-2 epimer, the *manno* derivative, in which this interaction is removed (because the 2-hydroxyl group is equatorial). The action of acid on phenyl 3,6-anhydro-α-D-glucopyranoside and -galactopyranoside was also investigated,[256a] and it was shown that for the D-*gluco* derivative there was no pyranose → furanose conversion, and for the *galacto* derivative there was no anomerization. These results substantiated Peat's mechanism,[1] since the phenyl group stabilizes the C-1–O bond and diminishes the tendency for a carbonium ion to form.

2. 3,6-Anhydroaldohexofuranoses (See also Section III,D,1)

Compounds having this ring system can be obtained from 1,2-*O*-isopropylidene-α-D-glucofuranose derivatives. Treatment of the 6-*p*-toluenesulfonate[257] (**76**) or the corresponding 6-sulfate[258] with base gives 3,6-anhydro-1,2-*O*-isopropylidene-α-D-glucofuranose (**77**). Mild alkaline treatment of 1,2-*O*-isopropylidene-5,6-di-*O*-*p*-tolylsulfonyl-α-D-glucofuranose gives the 5-*p*-toluenesulfonate of **77** (and not the 5,6-epoxide), from which the 5-substituent

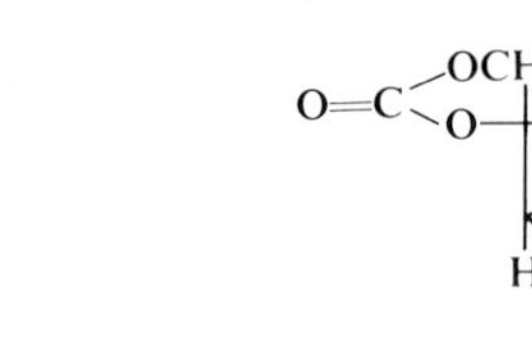

76

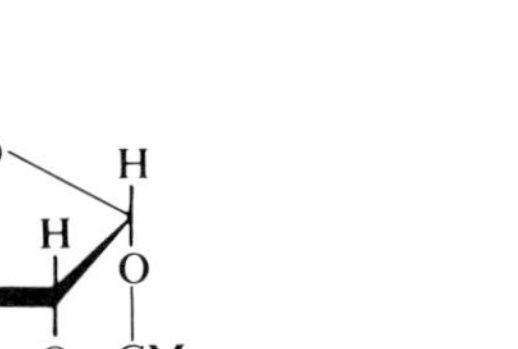

77

78

group can be removed with strong alkali.[259] The anhydro sugar (**77**) has been oxidized with permanganate to give 1,2-*O*-isopropylidene-D-xyluronic acid.[260]

Compound **77** has also been prepared by treatment of 1,2-*O*-isopropylidene-α-D-glucofuranose 5,6-carbonate (**78**) with a weak base. The reaction was shown not to proceed via the 5,6-epoxide.[261]

3. 2,5-Anhydro Sugars

The chemistry of 2,5-anhydro sugars has been reviewed.[1c,262]

The best-known compound in this class is 2,5-anhydro-D-mannose ("chitose"),[263,264] (**79**) prepared by deamination of 2-amino-2-deoxy-D-glucose ("chitosamine") with nitrous acid. Much work was devoted to confirming the structure of "chitose"[265-268] and of the related acids, "chitonic acid" (from oxidation of "chitose") and "chitaric acid" (from deamination of 2-amino-2-deoxy-D-gluconic acid).[269] These acids were shown to be epimers; the former has the D-*manno* and the latter the D-*gluco* configuration.[267,270] A mechanism

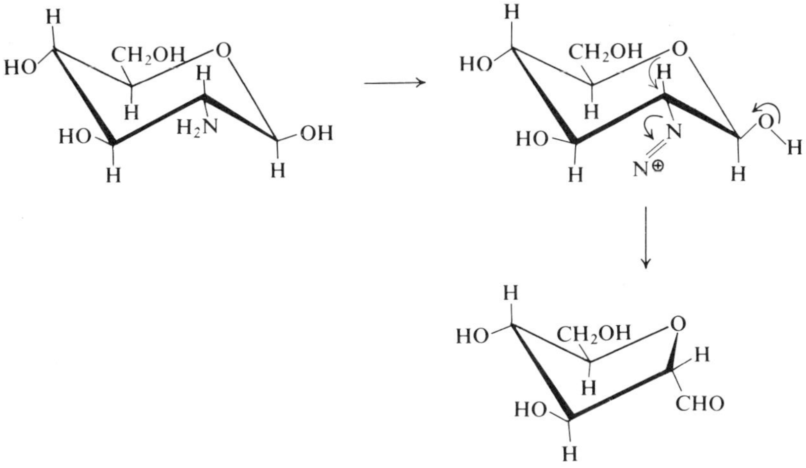

79

SCHEME XIII

for the inversion at C-2 in the deamination of 2-amino-2-deoxy-D-glucose has been proposed[1a] (Scheme XIII); the retention of configuration at C-2 in 2-amino-2-deoxy-D-gluconic acid has been interpreted by Foster[271] as involving participation of the carboxyl group (Scheme XIV).

Similarly, deamination of 2-amino-2-deoxy-D-galactose with nitrous acid gave 2,5-anhydro-D-talose.[272-274]

Brominolysis of methyl 3,4,6-tri-*O*-acetyl-2-deoxy-2-iodo-β-D-glucoside (**80**) gave 3,4,6-tri-*O*-acetyl-2,5-anhydro-D-mannose monomethyl hemiacetal

* *References start on p. 471.*

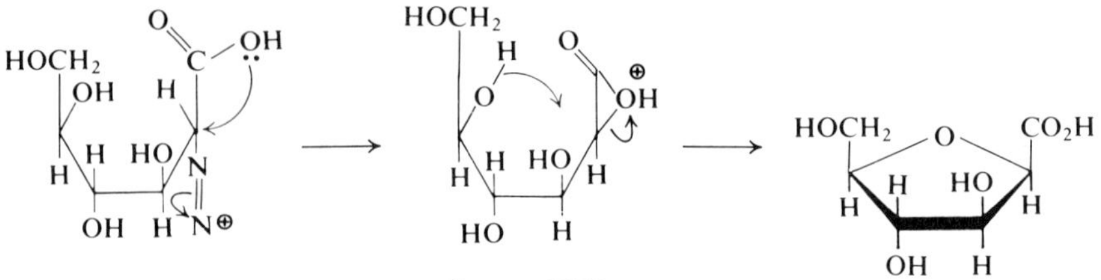

SCHEME XIV

acetate (81), which on treatment with methanolic hydrogen chloride gave the dimethyl acetal (82) of 2,5-anhydro-D-mannose.[275] A closely analogous reaction is the solvolysis in water of methyl 2-O-(p-nitrophenylsulfonyl)-α-D-glucopyranoside, which leads[275a] to 2,5-anhydro-D-mannose (79).

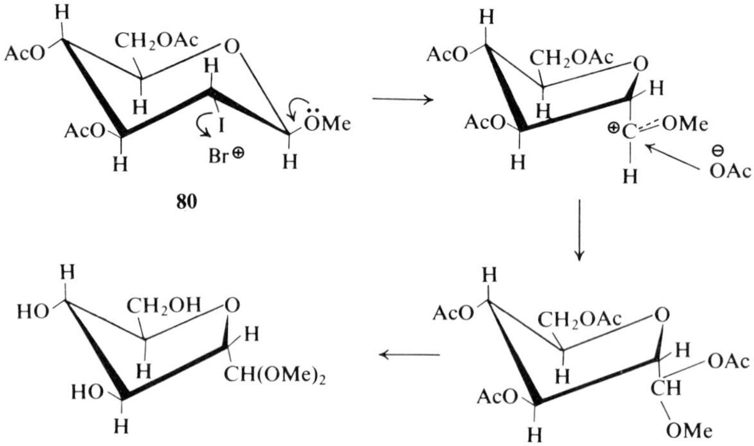

Attempted hydrolysis of 5,6-anhydro-1,2-O-isopropylidene-α-D-glucofuranose (33) with very dilute acid gave a product identified as 2,5-anhydro-L-idose (83). The mechanism proposed for this reaction involves attack at the secondary carbon atom of a terminal epoxide.[276]

Treatment of dialkyl dithioacetals of L-lyxose[277] and D-ribose[278] with p-toluenesulfonyl chloride and pyridine gave the corresponding 2,5-anhydro-pentose dithioacetals, presumably via the 5-p-toluenesulfonates. The D-xylose analogs react similarly, but in the D-arabinose series the reaction leads to stable 5-p-toluenesulfonates. These differences in reactivity have been interpreted in terms of steric and conformational factors involved in closure of the anhydro ring.[278a] A similar stereochemical dependence was observed in the conversion by acid-catalyzed methanolysis of 1,2-O-isopropylidene-5-O-p-tolylsulfonylpentofuranoses into the corresponding 2,5-anhydropentose dimethyl acetals.[278b]

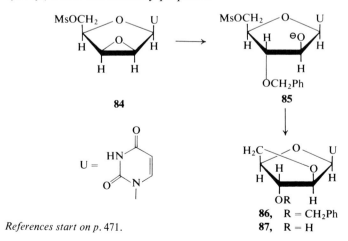

33

83

Methyl or ethyl 5-*O*-p-tolylsulfonyl-α-L-arabinofuranoside reacts with alkali to give, in low yield, methyl or ethyl 2,5-anhydro-α-L-arabinofuranoside. The products are hydrolyzed, even by water alone, to the free 2,5-anhydro sugar, which appears to exist in the *aldehydo* form, since it gives a positive Schiff's test. With methanolic hydrogen chloride the free anhydro sugar gave the dimethyl acetal.[279]

Treatment of 1-[2,3-anhydro-5-*O*-(methylsulfonyl)-β-D-lyxosyl]uracil (**84**) with sodium benzyloxide gave 1-(2,5-anhydro-3-*O*-benzyl-β-D-arabinosyl)-uracil (**86**), presumably by way of the intermediate (**85**). Reductive debenzylation gave the 3-hydroxy derivative (**87**), also prepared by treatment of 1-[5-*O*-(methylsulfonyl)-β-D-arabinosyl]uracil with alkali.[280] 1-(2,5-Anhydro-β-D-lyxosyl)uracil was similarly prepared.[280]

84

85

U =

86, R = CH₂Ph
87, R = H

E. MISCELLANEOUS ANHYDRO SUGARS

1. 2,6-Anhydropyranosides

Methyl 2,3,4-tri-O-benzoyl-6-O-p-tolylsulfonyl-α-D-altroside (**88**) reacts with alkali to give methyl 2,6-anhydro-α-D-altroside (**90**); the same compound was obtained from the 6-bromo and 6-iodo analogs[281] of **88**. The reaction

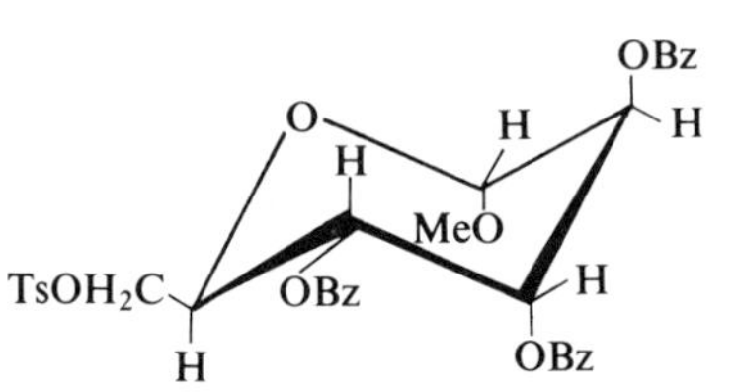

88

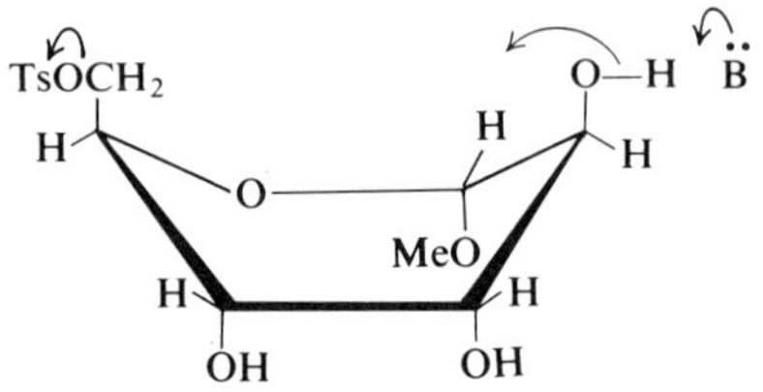

89

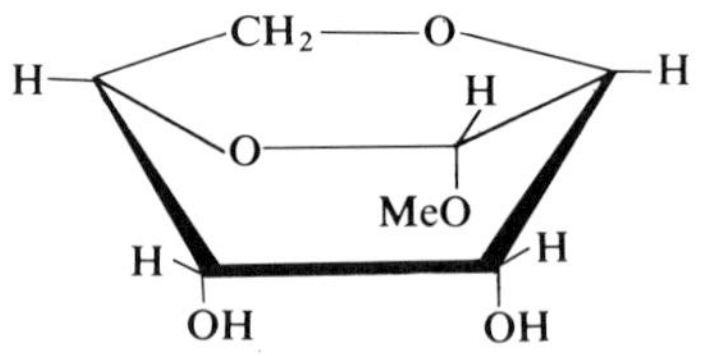

90

presumably proceeds by way of the boat conformation (**89**), which has the interacting groups in the correct steric orientations, and the product has a bicyclo[2.2.2] ring system. The 2-thio analog of **90** has also been prepared, and by oxidation has been converted into a mixture of two sulfoxides that are stereoisomeric at the sulfur atom.[281a]

Methyl 3,4-di-O-methyl-2-O-(methylsulfonyl)-α-D-glucoside is converted by sodium in ether into methyl 2,6-anhydro-3,4-di-O-methyl-α-D-mannoside.[282]

2. 3,5-Anhydrofuranoses

3,5-Anhydro-1,2-O-isopropylidene-α-D-xylofuranose (**92**), a derivative having a bicyclo[3.2.0] system containing fused oxetane and oxolane rings, was prepared[283] by alkaline treatment of 1,2-O-isopropylidene-5-O-p-tolylsulfonyl-α-D-xylofuranose (**91**). Opening of the oxetane ring in **92** could be effected at C-5, to give, for example, the 5-methyl ether[283] and various 5-thio derivatives.[283a]

The same type of fused-ring system present in **92** and **95** is present in the 4,6-anhydroketohexofuranose system, and examples of this type have been prepared[283b] from L-sorbose.

2'-Deoxy-3',5'-di-O-(methylsulfonyl)thymidine (**93**) has been converted[284] into 3',5'-anhydrothymidine (**95**) by way of the 2,3'-anhydro-5'-O-(methylsulfonyl) derivative (**94**). 3',5'-Anhydro derivatives of 1-(β-D-lyxosyl)- and 1-(β-D-xylosyl)uracil have also been synthesized.[280]

IV. REARRANGEMENTS OF ANHYDRO SUGARS

Reactions discussed in this section involve migration of an anhydro ring within a molecule, or conversion of one type of anhydro ring into another.

A. CONVERSION OF EPOXIDES INTO 3,6-ANHYDRO DERIVATIVES

Methyl 2,5,6-tri-O-acetyl-3-O-p-tolylsulfonyl-β-D-glucofuranoside (**96**) is converted rapidly by alkali into methyl 3,6-anhydro-β-D-glucofuranoside (**99**). Deacetylation presumably occurred first to give methyl 3-O-p-tolylsulfonyl-β-D-glucofuranoside (**97**); this compound cannot form a 3,6-anhydride

directly because of steric reasons. It was proposed, therefore, that the 2,3-anhydro alloside (**98**) is initially formed from **97**, and rapid attack by the 6-OH group on the anhydro ring ensues.[257] The same workers[257] also prepared the 3,6-anhydride (**99**) by alkaline treatment of methyl 2,3-anhydro-5,6-di-O-benzoyl-β-D-allofuranoside.

Treatment of methyl 2,3-anhydro-α-D-allopyranoside[285] (**100**) or its 4-methyl ether[286] with alkali gave methyl 3,6-anhydro-α-D-glucopyranoside or its 4-methyl ether. This reaction has been rationalized conformationally.[285, 286] The epoxide (**100**) can exist in either of two conformations, (**100a**) or (**100b**), but only in the latter is the geometry suitable for attack by the 6-OH group on the epoxide ring; diaxial opening occurs to give the 3,6-anhydro derivative (**101**).

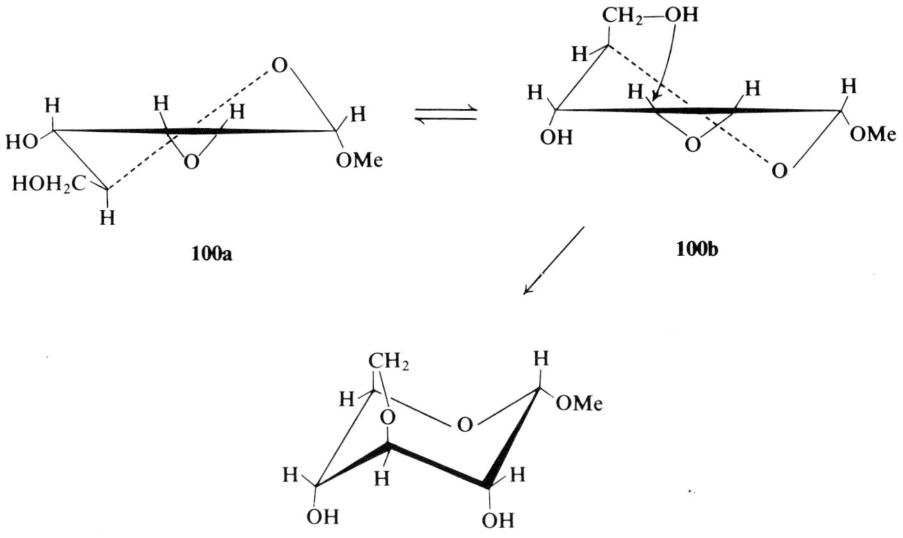

100a **100b**

101

Methyl 3-*O*-*p*-tolylsulfonyl-β-D-glucopyranoside (**102**) is converted by alkali into a mixture[287] of methyl 2,3-anhydro-β-D-alloside (**103**), methyl 3,4-anhydro-β-D-alloside (**104**), and methyl 3,6-anhydro-β-D-glucoside (**105**). The last compound cannot be formed (as in the furanose example discussed earlier) directly from the 3-*p*-toluenesulfonate **102**; but clearly it can arise from the 2,3-anhydro-D-alloside, as in the α anomer (**100**). That compound **105** cannot

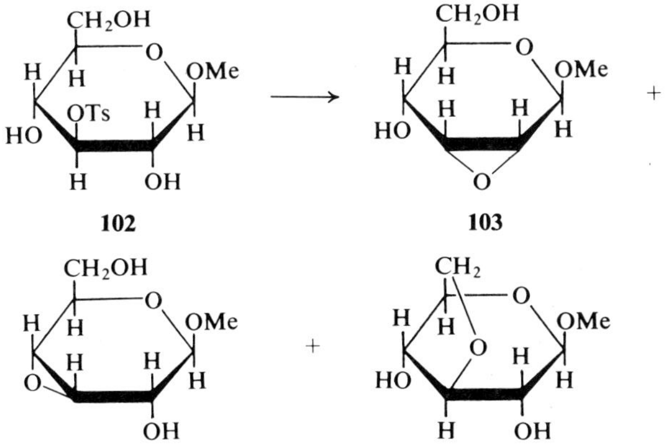

102 **103**

104 **105**

* *References start on p.* 471.

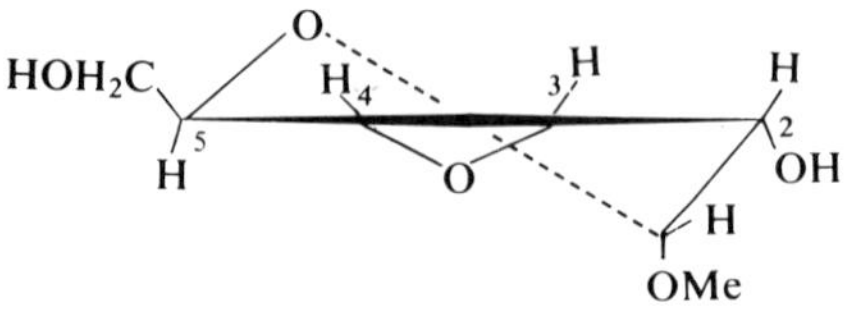

HOH$_2$C   H$_4$ $_3$H   H
H   O   OH
OMe

104a

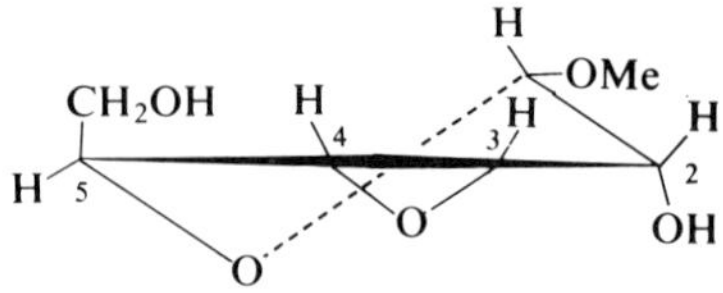

H   OMe
CH$_2$OH   H$_4$ $_3$H   H
H$_5$ O   OH

104b

CH$_2$OH
H   H   H
O   H   H
H

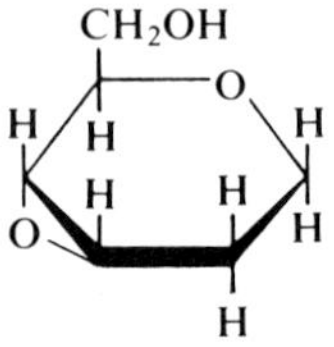

106

arise from the 3,4-anhydride was demonstrated by the fact that the reaction of 1,5:3,4-dianhydro-2-deoxy-D-*ribo*-hexitol (**106**) with alkali did not lead to a 3,6-anhydride.[285] This explanation is reasonable, for in neither possible conformation of the anhydro-D-alloside (**104a** or **104b**) is the 6-OH group favorably placed to attack the epoxide ring at C-3. The preponderance of the 2,3-anhydro-D-alloside (**103**) over the 3,4-anhydro-D-alloside (**104**) has been discussed in terms of passing interactions involved in the approach to the appropriate intermediate conformations.[128]

When a number of methyl 2,3-anhydroaldohexopyranosides or their 4,6-benzylidene acetals were hydrolyzed by acid, 3,6-anhydroaldohexoses were the main products. By varying the stereochemistry of the systems it was shown that hydrolysis of the glycoside preceded the attack of the 6-OH group on the epoxide ring.[288] For example, methyl 2,3-anhydro-α-D-guloside (**107**) and

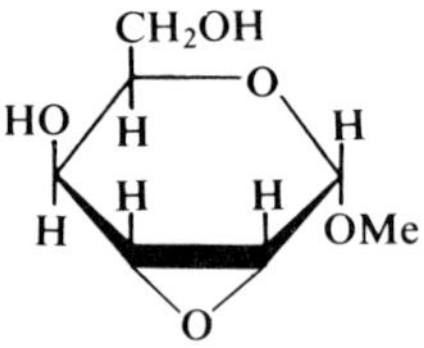

CH$_2$OH
HO   H   H
H   H   H   OMe
O

107

CH$_2$OH
HO   H   H
H   O   OMe
H   H

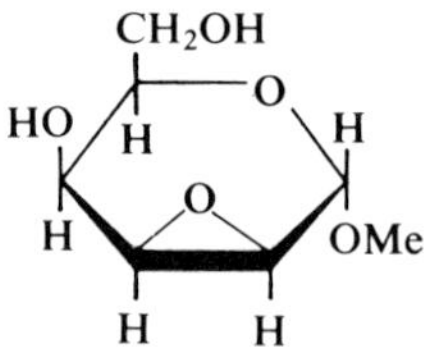

108

-taloside (**108**) both gave the appropriate 3,6-anhydroaldohexose on treatment with acid; clearly, whatever conformation compound **108** adopts, the 6-OH group cannot attack the epoxide ring until hydrolysis of the glycosidic bond has taken place.

Treatment of 5,6-anhydro-1,2-O-isopropylidene-α-D-glucofuranose (**33**) with alkali gave a product initially believed to be L-idose,[134] but later it was shown to be 3,6-anhydro-1,2-O-isopropylidene-α-D-glucofuranose,[289] resulting from attack of the anion from the 3-OH group on the epoxide ring (see also

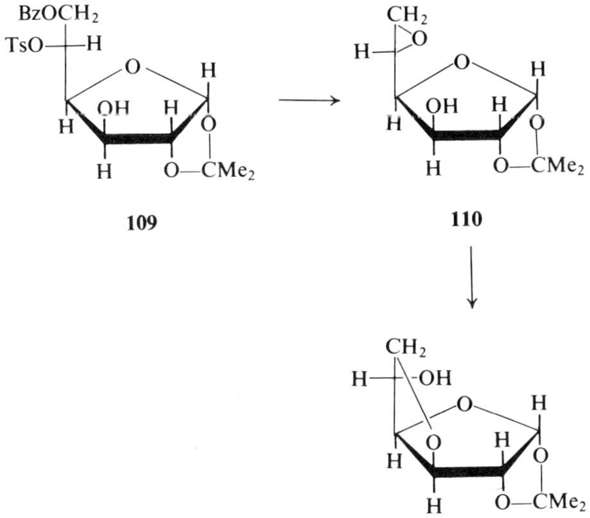

109   110

111

Section IV,B below). When 6-O-benzoyl-1,2-O-isopropylidene-5-O-p-tolylsulfonyl-α-D-glucofuranose (**109**) was treated with methanolic ammonia, 3,6-anhydro-1,2-O-isopropylidene-α-L-idofuranose (**111**) was isolated,[290] presumably formed by way of the 5,6-anhydro-L-*ido* derivative (**110**) (but see below).

B. INTERCONVERSION OF EPOXIDE AND OXETANE SYSTEMS

The action of aqueous alkali on 5,6-anhydro-1,2-O-isopropylidene-α-L-idofuranose (**110**) has been studied in some detail.[291] In addition to 1,2-O-isopropylidene-α-L-idofuranose,[240] there was isolated by chromatography 3,6-anhydro-1,2-O-isopropylidene-α-L-idofuranose (**111**) together with a new anhydride shown to be 3,5-anhydro-1,2-O-isopropylidene-α-D-glucofuranose (**112**). Subjecting the latter oxetane (**112**) to the same reaction conditions as the epoxide (**110**) gave the same mixture of products. It appears, therefore, that these oxides are interconvertible under alkaline conditions and that the more-reactive, oxirane ring undergoes irreversible ring scission to the 3,6-anhydro

* *References start on p.* 471.

derivative. Reinvestigation of the action[289] of aqueous alkali on 5,6-anhydro-1,2-O-isopropylidene-α-D-glucofuranose (33) showed that an oxetane, 3,5-anhydro-1,2-O-isopropylidene-α-L-idofuranose (113), is produced in low yield. Again the oxetane (113) gave the same products as the oxirane (33) with aqueous alkali.

112   113

Aqueous alkali converts methyl 2,3-anhydro-β-D-ribofuranoside (114) into methyl 3,5-anhydro-β-D-xylofuranoside (115).[223a] This reaction was shown to be reversible, with the equilibrium in favor of the oxetane. Treatment of the oxetane (115) with methanolic sodium methoxide leads mainly to methyl

114   115   116

5-O-methyl-β-D-xylofuranoside (116),[223a] indicating preponderant attack of methoxide ion at C-5. Only a small proportion of products formed by opening of the epoxide ring was found, and this result was attributed to steric hindrance to attack on this ring. The α-D anomer of 114 gave no 5-methyl ether with methoxide ion, but only products from epoxide opening, presumably because the three-membered ring is more accessible than in the anomer.

C. Epoxide Migrations[128]

These migrations, much studied in recent years, led to confusion in early work because investigators were unaware of their occurrence. Various examples of 2,3 ⇌ 3,4 shifts on pyranoid rings have been studied. The first such migration to be postulated appears to be the conversion of methyl 2,3-anhydro-β-D-mannoside into methyl 3,4-anhydro-β-D-altroside, since both compounds resulted after treatment of methyl 2-O-p-tolylsulfonyl-β-D-glucoside with alkali.[292] For migration to occur, there must be a hydroxyl group adjacent and *trans* to the epoxide ring. This is shown schematically in Scheme XV, and more informatively in Scheme XVI, where it can be seen that for the migration to be reversed, a change in conformation must occur. Therefore, in predicting the position of equilibrium in such a system, it is necessary to consider the non-bonded interactions in all the species involved. In restricted systems, such as

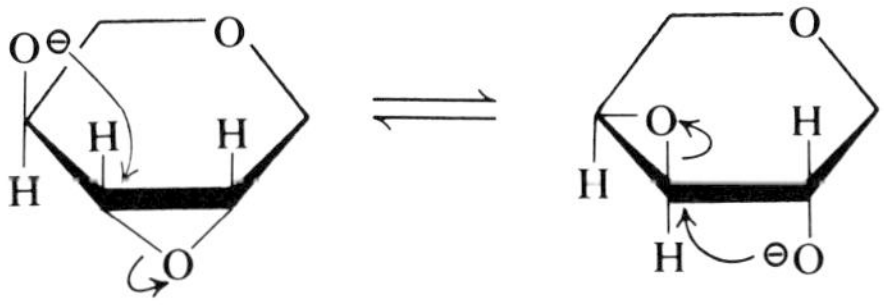

SCHEME XV

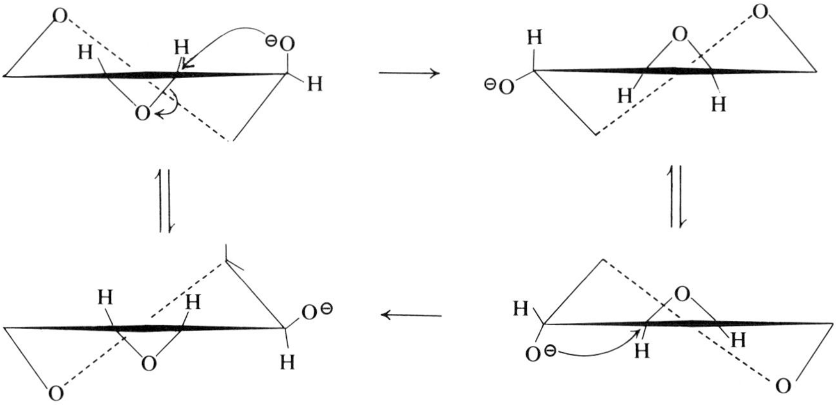

SCHEME XVI

1,6-anhydro-β-D-aldohexopyranose derivatives, this conformational mobility is removed, and only one conformation of each epoxide is possible. For such systems it has been postulated[150, 154] that, provided the ring conformations are the same, that compound having the hydroxyl group equatorial will be the more stable. In this way was explained[150] the formation of 1,6:3,4-dianhydro-β-D-altropyranose (118) and not 1,6:2,3-dianhydro-β-D-mannopyranose (117) from 1,6-anhydro-3-O-p-tolylsulfonyl-β-D-altropyranose (49); compound 117 rearranges to the more stable 118. The reconversion of the anhydro altroside (118) into the anhydro mannoside (117) would be extremely difficult. This equilibrium has been examined by independent investigators, and the *altro* isomer was again found to preponderate.[293] It has also been shown[294] that the equilibrium between 1,6:2,3-dianhydro-β-D-gulopyranose and 1,6:3,4-dian-

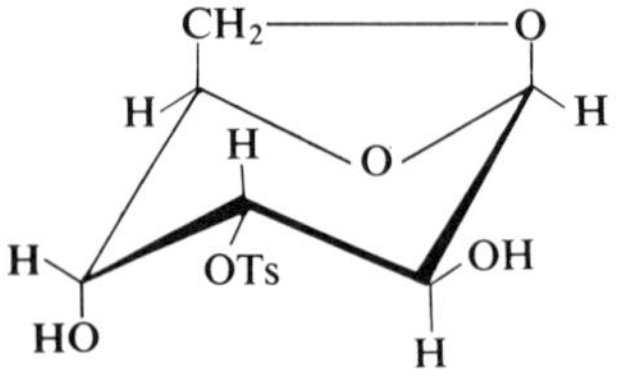

49

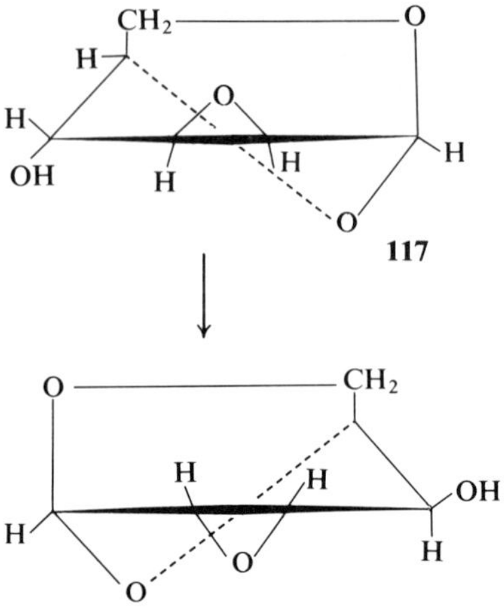

117

118

hydro-β-D-galactopyranose lies on the side of the *gulo* isomer (equatorial hydroxyl group).

The situation in systems of flexible epoxides is obviously more complex because there are four conformations to consider. The examples studied so far do not allow clear-cut predictions to be made. In the equilibrium between the 2,3-anhydro-D-gulosides (**119**) and the 3,4-anhydro-D-galactosides (**120**), the *gulo* isomers (OH axial) were slightly favored.[295, 296] However, in the

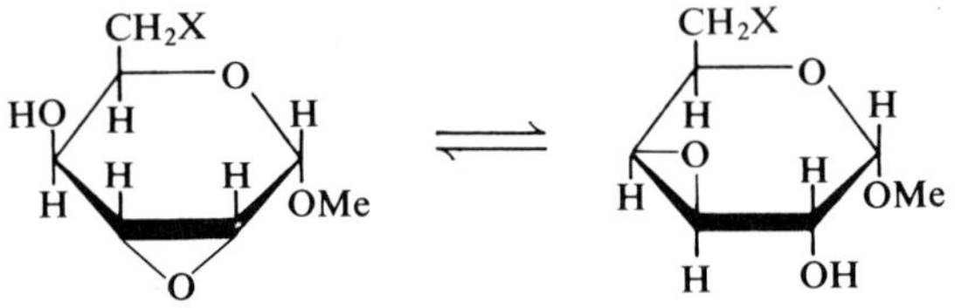

119   120

X = OTr,[277] OCH$_2$Ph,[277] or H[278]

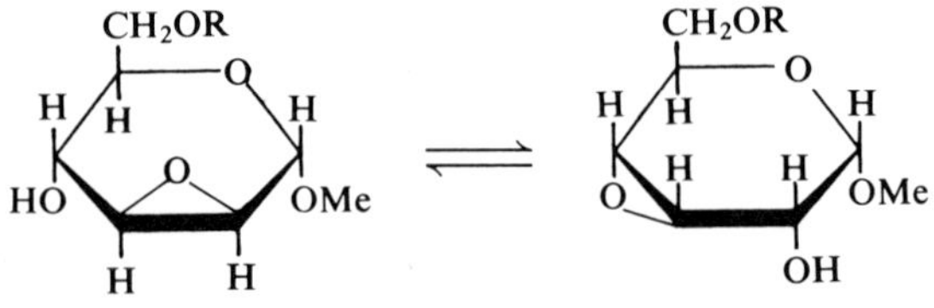

121   122

R = H or Tr

equilibrium between the 2,3-anhydro-D-mannosides (121) and the 3,4-anhydro-D-altrosides (122), the *manno* isomers (OH equatorial) preponderated.[211a, 297] Early work[298, 299] on ring-opening reactions of methyl 3,4-anhydro-α-D-galactoside led to apparently anomalous results. It was not appreciated at the time that the starting material could in fact be a mixture of the foregoing 3,4-anhydro-D-galactoside and methyl 2,3-anhydro-α-D-gulopyranoside, the latter being formed by epoxide migration. This possibility was noted by Buchanan,[30, 301] who re-examined the reaction sequences and confirmed that the *gulo* epoxide was present in the starting material. The observed products were thus derived from two epoxides and not one.

REFERENCES

1. (a) S. Peat, *Advan. Carbohyd. Chem.*, **2**, 37 (1946); (b) N. R. Williams, *Advan. Carbohyd. Chem. Biochem.*, **25**, 109 (1970); (c) J. Defaye, *ibid.*, **25**, 181 (1970).
2. J. A. Mills, *Advan. Carbohyd. Chem.*, **10**, 1 (1955).
2a. See "Specialist Periodical Reports, Carbohydrate Chemistry" (R. D. Guthrie, Senior Reporter), Chemical Society, London, **1**, 38 (1968); **2**, 34 (1969); **3**, 32 (1970); **4**, 22 (1971) for detailed annual surveys of the literature on anhydro sugars.
2b. K. Heyns and H. Scharmann, *Carbohyd. Res.*, **1**, 371 (1966); compare A. Rosenthal, *Carbohyd. Res.*, **8**, 61 (1968); O. S. Chizhov, B. M. Zolotarev, A. I. Usov, M. A. Rechter, and N. K. Kochetkov, *ibid.*, **16**, 29 (1971).
3. P. Brigl, *Hoppe-Seyler's Z. Physiol. Chem.*, **116**, 1 (1921); **122**, 245 (1922).
4. W. J. Hickinbottom, *J. Chem. Soc.*, 3140 (1928).
5. E. Hardegger and J. de Pascual, *Helv. Chim. Acta*, **31**, 281 (1948).
6. W. N. Haworth and W. J. Hickinbottom, *J. Chem. Soc.*, 2847 (1931).
7. R. U. Lemieux, *Can. J. Chem.*, **31**, 949 (1953).
8. R. U. Lemieux and H. F. Bauer, *Can. J. Chem.*, **32**, 340 (1954).
9. L. J. Sargent, J. G. Buchanan, and J. Baddiley, *J. Chem. Soc.*, 2184 (1962).
10. R. U. Lemieux and G. Huber, *J. Amer. Chem. Soc.*, **75**, 4118 (1953); **78**, 4117 (1956).
11. H. Tsuchida and M. Komoto, *Agric. Biol. Chem.* (Tokyo), **29**, 239 (1965).
12. Z. Csürös, G. Deák, and M. Haraszthy, *Acta Chim. Acad. Sci. Hung.*, **21**, 181, 193 (1959); Z. Csürös, G. Deák, I. Gyurkovics, M. Haraszthy-Papp, and E. Zara-Kaczián, *ibid.*, **67**, 93 (1971).
13. A. Pictet and P. Castan, *Helv. Chim. Acta*, **3**, 645 (1920).
14. M. Cramer and E. H. Cox, *Helv. Chim. Acta*, **5**, 884 (1922).
15. V. Prey and F. Grundschober, *Monatsh. Chem.*, **91**, 358 (1960).
16. S. Haq and W. J. Whelan, *Nature*, **178**, 1222 (1956).
17. R. M. Hann and C. S. Hudson, *J. Amer. Chem. Soc.*, **63**, 2241 (1941).
18. K. Hess and F. Neumann, *Ber.*, **68**, 1360 (1935).
19. J. Kops and C. Schuerch, *J. Org. Chem.*, **30**, 3951 (1965).
20. K. Hess and K. E. Heumann, *Ber.*, **72**, 137 (1939).
21. E. Vis and H. G. Fletcher, Jr., *J. Org. Chem.*, **23**, 1393 (1958); E. Vis and H. G. Fletcher, Jr., *J. Amer. Chem. Soc.*, **79**, 1182 (1957).
22. K. Freudenberg and E. Braun, *Ann.*, **460**, 288 (1928); *Ber.*, **66**, 780 (1933); *Ber.*, **68**, 1988 (1935).

23. F. Micheel and H. Micheel, *Ber.*, **63**, 2862 (1930); compare J. S. Brimacombe and L. C. N. Tucker, *Carbohyd. Res.*, **5**, 36 (1967).
24. F. Micheel and U. Kreuzer, *Ann.*, **722**, 228 (1969).
24a. For a review, see M. Černý and J. Staněk, *Fortschr. Chem. Forsch.*, **14**, 389 (1970).
25. R. E. Reeves, *J. Amer. Chem. Soc.*, **71**, 2116 (1949); **72**, 1499 (1950).
25a. M. Černý, J. Pacák, and J. Staněk, *Chem. Ind.* (London), 1559 (1966).
26. D. Horton and J. D. Wander, *J. Org. Chem.*, **32**, 3780 (1967); D. Horton and J. D. Wander, *Carbohyd. Res.*, **14**, 83 (1970).
27. D. Horton and J. S. Jewell, *Carbohyd. Res.*, **3**, 255 (1966–1967); **5**, 149 (1967).
28. A. Pictet and J. Sarasin, *Helv. Chim. Acta*, **1**, 87 (1918); *ibid.*, **2**, 698 (1919).
29. P. Karrer, *Helv. Chim. Acta*, **3**, 258 (1920); P. Karrer and J. O. Rosenberg, *ibid.*, **5**, 575 (1922).
30. Ya. V. Ephstein, O. P. Golova, and L. I. Durynina, *Bull. Acad. Sci. U.S.S.R.*, 1089 (1959).
31. R. M. Hann and C. S. Hudson, *J. Amer. Chem. Soc.*, **63**, 1484 (1941).
32. B. H. Alexander, R. J. Dimler, and C. L. Mehltretter, *J. Amer. Chem. Soc.*, **73**, 4658 (1951).
33. R. M. Hann and C. S. Hudson, *J. Amer. Chem. Soc.*, **64**, 2435 (1942).
34. G. Zemplén, A. Gerecs, and T. Valatin, *Ber.*, **73**, 575 (1940).
35. A. E. Knauf, R. M. Hann, and C. S. Hudson, *J. Amer. Chem. Soc.*, **63**, 1447 (1941).
36. C. Tanret, *Compt. Rend.*, **119**, 158 (1894); *Bull. Soc. Chim.*, France, [3], **211**, 949 (1894).
37. C. E. Ballou, *Advan. Carbohyd. Chem.*, **9**, 59 (1954); D. Shapiro, *J. Org. Chem.*, **35**, 1464 (1970).
38. E. M. Montgomery, N. K. Richtmyer, and C. S. Hudson, *J. Amer. Chem. Soc.*, **65**, 3 (1943).
39. E. M. Montgomery, N. K. Richtmyer, and C. S. Hudson, *J. Amer. Chem. Soc.*, **64**, 1483 (1942).
40. F. Micheel and H. Micheel, *Ber.*, **65**, 258 (1932).
41. E. M. Montgomery, N. K. Richtmyer, and C. S. Hudson, *J. Org. Chem.*, **10**, 194 (1945).
42. M. P. Bardolph and G. H. Coleman, *J. Org. Chem.*, **15**, 169 (1950).
43. C. M. McCloskey and G. H. Coleman, *J. Org. Chem.*, **10**, 184 (1945).
44. A. Dyfverman and B. Lindberg, *Acta Chem. Scand.*, **4**, 878 (1950).
45. G. H. Coleman, C. M. McCloskey, and R. Kirby, *Ind. Eng. Chem.*, **36**, 1040 (1944).
46. A. Fernez and P. J. Stoffyn, *Tetrahedron*, **6**, 139 (1959); see also P. Seib, *ibid.*, 2252 (1969).
47. R. J. Ferrier, W. G. Overend, and A. E. Ryan, *J. Chem. Soc.*, 3484 (1965).
48. E. M. Montgomery, N. K. Richtmyer, and C. S. Hudson, *J. Amer. Chem. Soc.*, **65**, 1848 (1943).
49. F. Micheel and G. Baum, *Chem. Ber.*, **88**, 2020 (1955).
50. H. B. Wood, Jr. and H. G. Fletcher, Jr., *J. Amer. Chem. Soc.*, **78**, 207 (1956).
51. P. Karrer and A. P. Smirnoff, *Helv. Chim. Acta*, **4**, 817 (1921).
52. F. Micheel, *Ber.*, **62**, 687 (1920).
53. G. Zemplén, R. Bognár, and G. Pongor, *Acta Chim. Acad. Sci. Hung.*, **19**, 285 (1959).
54. E. K. Gladding and C. B. Purves, *J. Amer. Chem. Soc.*, **66**, 76, 153 (1944).
55. F. Micheel and A. Klemer, *Chem. Ber.*, **85**, 187 (1952).
56. F. Micheel, A. Klemer, and G. Baum, *Chem. Ber.*, **88**, 475 (1955).
57. F. Micheel, A. Klemer, and R. Flitsch, *Chem. Ber.*, **91**, 194, 663 (1958).
58. F. Micheel and G. Baum, *Chem. Ber.*, **88**, 479 (1955).
59. N. K. Richtmyer and C. S. Hudson, *J. Amer. Chem. Soc.*, **57**, 1716 (1935); **61**, 214 (1939); **63**, 961 (1941).
60. E. L. Jackson and C. S. Hudson, *J. Amer. Chem. Soc.*, **62**, 958 (1940).

61. E. Sorkin and T. Reichstein, *Helv. Chim. Acta*, **28**, 1 (1945).
62. L. C. Stewart and N. K. Richtymer, *J. Amer. Chem. Soc.*, **77**, 1021 (1955).
63. J. W. Pratt and N. K. Richtmyer, *J. Amer. Chem. Soc.*, **77**, 1906 (1955).
64. E. Zissis, L. C. Stewart, and N. K. Richtmyer, *J. Amer. Chem. Soc.*, **79**, 2593 (1957).
65. L. D. Ough and R. G. Rohwer, *J. Agr. Food Chem.*, **4**, 267 (1956).
66. S. Peat, W. J. Whelan, T. E. Edwards, and P. Owen, *J. Chem. Soc.*, **586** (1958).
67. A. Thompson, K. Anno, M. L. Wolfrom, and M. Inatome, *J. Amer. Chem. Soc.*, **76**, 1309 (1954).
68. N. K. Richtmyer, *Arch. Biochem. Biophys.*, **78**, 376 (1958).
69. J. W. Pratt and N. K. Richtmyer, *J. Amer. Chem. Soc.*, **79**, 2597 (1957).
70. J. W. Pratt, N. K. Richtmyer, and C. S. Hudson, *J. Amer. Chem. Soc.*, **75**, 4503 (1953).
71. L. C. Stewart and N. K. Richtmyer, *J. Amer. Chem. Soc.*, **77**, 424 (1955).
72. A. Pictet, *Helv. Chim. Acta*, **1**, 226 (1918); compare Y. Houminer and S. Patai, *J. Polym. Sci.*, *Part A*-1, **7**, 3005 (1969); B. Lazdina, R. Pernikis, and J. Surna, *Latv. PSR Zinat.*, *Akad. Vestis, Kim. Ser.*, 602 (1970).
73. J. C. Irvine and J. W. H. Oldham, *J. Chem. Soc.*, 2903 (1925).
74. M. L. Wolfrom, A. Thompson, and R. B. Ward, *J. Amer. Chem. Soc.*, **81**, 4623 (1959).
75. J. S. Caravalho, W. Prins, and C. Schuerch, *J. Amer. Chem. Soc.*, **81**, 4054 (1959).
76. A. Pictet and J. H. Ross, *Helv. Chim. Acta*, **5**, 876 (1922); *Compt. Rend.*, **174**, 1113 (1922).
77. A. Pictet and J. Pictet, *Helv. Chim. Acta*, **4**, 788 (1921).
78. H. Pringsheim and K. Schmalz, *Ber.*, **55**, 3001 (1922).
79. A. J. Mian, E. J. Quinn, and C. Schuerch, *J. Org. Chem.*, **27**, 1895 (1962).
80. A. Bhattacharya and C. Schuerch, *J. Org. Chem.*, **26**, 3101 (1961).
80a. F. Shafizadeh, C. W. Philpot, and N. Ostojic, *Carbohyd. Res.*, **16**, 279 (1971).
81. J. C. Irvine and J. W. H. Oldham, *J. Chem. Soc.*, 2729 (1925).
82. G. Zemplén, Z. Csürös, and S. J. Angyal, *Ber.*, **70**, 1848 (1937).
83. L. F. Wiggins, *J. Chem. Soc.*, 1590 (1949).
84. N. Baggett, P. J. Stoffyn, and R. W. Jeanloz, *J. Org. Chem.*, **28**, 1041 (1963).
85. N. Baggett and R. W. Jeanloz, *J. Org. Chem.*, **28**, 1845 (1963).
86. N. M. Merlis, Z. V. Volodina, and O. P. Golova, *J. Gen. Chem.*, *U.S.S.R.*, **34**, 3870 (1964); P. A. Seib and P. C. Wollwage, *J. Chem. Soc.* (*C*), 3143 (1971).
87. J. C. Irvine and J. W. H. Oldham, *J. Chem. Soc.*, 1744 (1921).
88. R. W. Jeanloz, A. M. C. Rapin, and S.-I. Hakomori, *J. Org. Chem.*, **26**, 3939 (1961).
89. R. W. Jeanloz and A. M. C. Rapin, *J. Org. Chem.*, **28**, 2978 (1963).
90. M. Černý, V. Gut, and J. Pacák, *Coll. Czech. Chem. Comm.*, **26**, 2542 (1961).
91. K. Heyns, J. Weyer, and H. Paulsen; *Chem. Ber.*, **98**, 327 (1965).
92. A. C. Richardson and H. O. L. Fischer, *J. Amer. Chem. Soc.*, **83**, 1132 (1961).
93. L. D. Hall and L. Hough, *Proc. Chem. Soc.*, 382 (1962); K. Heyns and J. Weyer, *Ann.*, **718**, 224 (1968).
94. R. J. Dimler, *Advan. Carbohyd. Chem.*, **7**, 37 (1952).
94a. K. Heyns, P. Koell, and H. Paulsen, *Ber.*, **104**, 830 (1971).
95. R. J. Dimler, H. A. Davis, and G. E. Hilbert, *J. Amer. Chem. Soc.*, **68**, 1377 (1946).
96. O. P. Golova, N. M. Merlis, and Z. V. Volodina, *J. Gen. Chem.*, *U.S.S.R.*, 978 (1959).
97. H. Bredereck, M. Köthnig, and E. Berger, *Ber.*, **73**, 956 (1940).
98. R. W. Jeanloz, G. B. Barker, and M. V. Lock, *Nature*, **167**, 42 (1951).
99. R. L. Whistler and A. Frowein, *J. Org. Chem.*, **26**, 3946 (1961).
100. J. S. Webb, R. W. Broschard, D. B. Cosulich, J. H. Mowat, and J. E. Lancaster, *J. Amer. Chem. Soc.*, **84**, 3183 (1962).

101. D. Tishchenko and N. Nosova, *Zhur. Obshch. Khim.*, **18**, 1193 (1948); *Chem. Abstr.*, **43**, 1726 (1949).
102. G. R. Bedford and D. Gardiner, *Chem. Comm.*, 287 (1965).
103. M. Sarel-Imber and J. Leibowitz, *J. Org. Chem.*, **24**, 141, 1897 (1959).
104. E. Zissis and N. K. Richtmyer, *J. Org. Chem.*, **30**, 462 (1965).
105. F. Micheel and E. A. Kleinheidt, *Chem. Ber.*, **98**, 1669 (1965).
106. H. R. Goldschmid and A. S. Perlin, *Can. J. Chem.*, **38**, 2178 (1960).
107. F. B. LaForge and C. S. Hudson, *J. Biol. Chem.*, **30**, 61 (1917).
108. J. W. Pratt, N. K. Richtmyer, and C. S. Hudson, *J. Amer. Chem. Soc.*, **74**, 2200 (1952).
109. N. K. Richtmyer and J. W. Pratt, *J. Amer. Chem. Soc.*, **78**, 4717 (1956).
110. L. C. Stewart, N. K. Richtmyer, and C. S. Hudson, *J. Amer. Chem. Soc.*, **74**, 2206 (1952).
111. J. W. Pratt, N. K. Richtmyer, and C. S. Hudson, *J. Amer. Chem. Soc.*, **74**, 2210 (1952).
112. E. J. McDonald, *Advan. Carbohyd. Chem.*, **2**, 253 (1946).
113. A. Pictet and J. Chavan, *Helv. Chim. Acta*, **9**, 809 (1926).
114. M. L. Wolfrom and M. G. Blair, *J. Amer. Chem. Soc.*, **70**, 2406 (1948).
115. M. L. Wolfrom, W. W. Binkley, W. L. Shilling, and H. W. Hilton, *J. Amer. Chem. Soc.*, **73**, 3553 (1951).
116. M. L. Wolfrom, H. W. Hilton, and W. W. Binkley, *J. Amer. Chem. Soc.*, **74**, 2867 (1952).
117. A. H. Shamgar and J. Leibowitz, *J. Org. Chem.*, **25**, 430 (1960).
118. J. C. Irvine and J. W. Stevenson, *J. Amer. Chem. Soc.*, **51**, 2197 (1929).
119. R. F. Jackson and S. M. Goergen, *J. Res. Nat. Bur., Stand.*, **3**, 27 (1929); **5**, 733 (1930).
120. W. N. Haworth and H. R. L. Streight, *Helv. Chim. Acta*, **15**, 693 (1932).
121. E. W. Bodycote, W. N. Haworth, and C. S. Woolvin, *J. Chem. Soc.*, 2389 (1932).
122. H. H. Schlubach and H. Knoop, *Ann.*, **504**, 19 (1933); **511**, 140 (1934).
123. E. J. McDonald and R. F. Jackson, *J. Res. Nat. Bur. Stand.*, **35**, 497 (1945).
124. E. J. McDonald and A. L. Turcotte, *J. Res. Nat. Bur. Stand.*, **38**, 423 (1947).
125. H. H. Schlubach and C. Behre, *Ann.*, **508**, 16 (1934).
126. R. F. Jackson and E. J. McDonald, *J. Res. Nat. Bur. Stand.*, **6**, 709 (1931); **24**, 181 (1940).
127. M. L. Wolfrom and H. W. Hilton, *J. Amer. Chem. Soc.*, **74**, 5334 (1952).
128. F. H. Newth, *Quart. Rev.* (London), **13**, 30 (1959).
129. K. Freudenberg, H. Toepfer, and C. C. Anderson, *Ber.*, **61**, 1750 (1928).
130. E. Fischer, M. Bergmann, and H. Schotte, *Ber.*, **53**, 509 (1920); compare M. L. Wolfrom, Y.-L. Hung, and D. Horton, *J. Org. Chem.*, **30**, 3394 (1965).
131. P. A. Levene and G. M. Meyer, *J. Biol. Chem.*, **55**, 221 (1922).
132. V. G. Bashford and L. F. Wiggins, *Nature*, **165**, 566 (1950).
133. L. F. Wiggins, *Nature*, **157**, 300 (1946).
133a. N. M. K. Ng Ying Kin, J. M. Williams, and A. Horsington, *Chem. Commun.*, 971 (1969).
134. H. Ohle and L. v. Vargha, *Ber.*, **62**, 2435 (1929).
135. B. Helferich and A. Müller, *Ber.*, **63**, 2142 (1930).
136. A. Müller, *Ber.*, **67**, 421 (1934); **68**, 1094 (1935).
137. S. Müller, M. Moricz, and G. Verner, *Ber.*, **72**, 745 (1939).
138. W. N. Haworth, E. L. Hirst, and L. Panizzon, *J. Chem. Soc.*, 154 (1934); W. H. G. Lake and S. Peat, *ibid.*, 1417 (1938).
139. J. Honeyman, *J. Chem. Soc.*, 990 (1946).
140. P. W. Kent and P. F. V. Ward, *J. Chem. Soc.*, 416 (1953).
141. J. Jarý, K. Čapek, and J. Kovár, *Coll. Czech. Chem. Commun.*, **29**, 930 (1964).
142. G. J. Robertson and C. F. Griffith, *J. Chem. Soc.*, 1193 (1935).
143. H. R. Bolliger and D. A. Prins, *Helv. Chim. Acta*, **28**, 465 (1945).
144. K. S. Ennor, J. Honeyman, C. J. G. Shaw, and T. C. Stening, *J. Chem. Soc.*, 2921 (1958).

145. F. H. Newth, *J. Chem. Soc.*, 2717 (1959).
146. L. F. Wiggins, *J. Chem. Soc.*, 522 (1944).
147. M. Gyr and T. Reichtstein, *Helv. Chim. Acta*, **28**, 226 (1945).
148. R. M. Hann and C. S. Hudson, *J. Amer. Chem. Soc.*, **64**, 925 (1942).
149. S. P. James, F. Smith, M. Stacey, and L. F. Wiggins, *J. Chem. Soc.*, 625 (1946).
150. F. H. Newth, *J. Chem. Soc.*, 441 (1956).
151. E. E. Percival and R. Zobrist, *J. Chem. Soc.*, 564 (1953).
152. M. Ohle and F. Just, *Ber.*, **68**, 601 (1935).
153. H. Ohle and C. A. Schultz, *Ber.*, **71**, 2302 (1938).
154. S. J. Angyal and P. T. Gilham, *J. Chem. Soc.*, 3691 (1957).
155. J. Honeyman and J. W. W. Morgan, *J. Chem. Soc.*, 3660 (1955).
155a. D. S. Mathers and G. J. Robertson, *J. Chem. Soc.*, 1076 (1933).
156. G. J. Robertson and W. Whitehead, *J. Chem. Soc.*, 319 (1940).
157. A. J. Dick and J. K. N. Jones, *Can. J. Chem.*, **44**, 79 (1966).
157a. R. S. Tipson, *Advan. Carbohyd. Chem.*, **8**, 107 (1953); D. H. Ball and F. W. Parrish, *ibid.*, **23**, 233 (1968); **24**, 139 (1969).
158. R. B. Duff and E. G. V. Percival, *J. Chem. Soc.*, 1675 (1947).
159. J. Honeyman and J. W. W. Morgan, *Advan. Carbohyd. Chem.*, **12**, 117 (1957).
160. E. G. Ansell and J. Honeyman, *J. Chem. Soc.*, 2778 (1952).
161. J. Defaye, *Compt. Rend.*, **255**, 2794 (1962).
162. D. Horton and J. S. Jewell, *Carbohyd. Res.*, **2**, 251 (1966).
163. R. E. Parker and N. S. Isaacs, *Chem. Rev.*, **59**, 737 (1959).
164. J. A. Mills [cited by F. H. Newth and R. F. Homer, *J. Chem. Soc.*, 989 (1953)].
165. F. H. Newth, *Chem. Ind.* (London), 1257 (1953); W. G. Overend and G. Vaughan, *Chem. Ind.* (London) 995 (1955).
166. R. C. Cookson, *Chem. Ind.* (London), 223, 1512 (1954).
167. S. J. Angyal, *Chem. Ind.* (London), 1230 (1954).
168. G. Huber and O. Schier, *Helv. Chim. Acta*, **43**, 129 (1960).
169. A. K. Bose, O. K. R. Chaudhuri, and A. K. Bhattacharyya, *Chem. Ind.* (London), 869 (1953).
170. B. Ottar, *Acta Chem. Scand.*, **1**, 283 (1947).
171. S. Peat and L. F. Wiggins, *J. Chem. Soc.*, 1810 (1938).
172. W. E. Harvey, J. J. Michalski, and A. R. Todd, *J. Chem. Soc.*, 2271 (1951).
173. W. H. Myers and G. J. Robertson, *J. Amer. Chem. Soc.*, **65**, 8 (1943).
174. N. K. Kochetkov, L. I. Kudryashov, and A. P. Khyagina, *J. Gen. Chem. U.S.S.R. (Engl. Transl.)*, **32**, 402 (1962).
175. N. C. Jamieson and R. K. Brown, *Can. J. Chem.*, **39**, 1765 (1961).
176. R. W. Jeanloz, D. A. Prins, and T. Reichstein, *Helv. Chim. Acta*, **29**, 371 (1946).
177. R. D. Guthrie and D. Murphy, *J. Chem. Soc.*, 5288 (1963).
178. D. A. Prins, *J. Amer. Chem. Soc.*, **70**, 3955 (1948).
179. L. F. Wiggins, *J. Chem. Soc.*, 18 (1947).
180. C. A. Grob and D. A. Prins, *Helv. Chim. Acta*, **28**, 840 (1945).
181. L. Goodman and J. E. Christensen, *J. Org. Chem.*, **28**, 158 (1963).
182. H. R. Bolliger and D. A. Prins, *Helv. Chim. Acta*, **29**, 1061 (1946).
183. J. C. Christensen and L. Goodman, *J. Amer. Chem. Soc.*, **82**, 4738 (1960).
184. G. J. F. Chittenden and R. D. Guthrie, *J. Chem. Soc.*, C, 695 (1966).
185. A. A. J. Feast, W. G. Overend, and N. R. Williams, *J. Chem. Soc.*, 7378 (1965); but see also M. Sharma and R. K. Brown, *Can. J. Chem.*, **46**, 757 (1968).
186. R. W. Jeanloz, D. A. Prins, and T. Reichstein, *Experientia*, **1**, 336 (1945).
187. J. G. Buchanan and K. J. Miller, *J. Chem. Soc.*, 3392 (1960).

188. S. Hanessian and T. H. Haskell, *J. Org. Chem.*, **30**, 1080 (1965).
189. M. Černý and J. Pacák, *Coll. Czech. Chem. Commun.*, **27**, 94 (1962); compare A. D. Barford, A. B. Foster, J. H. Westwood, L. D. Hall, and R. N. Johnson, *Carbohyd. Res.*, **19**, 49 (1971).
189a. M. Prystaš, H. Gustafsson, and F. Šorm, *Collect. Czech. Chem. Commun.*, **36**, 1487 (1971).
189b. H. Paulsen, *Advan. Carbohyd. Chem. Biochem.*, **26**, 127 (1971).
190. M. Dahlgard, *J. Org. Chem.*, **30**, 4352 (1965).
191. M. Dahlgard, B. H. Chastain, and R. L. Han, *J. Org. Chem.*, **27**, 932 (1962).
192. M. Dahlgard, B. H. Chastain, and R. L. Han, *J. Org. Chem.*, **27**, 929 (1962).
193. W. G. Overend, A. C. White, and N. R. Williams, *Chem. Ind.* (London), 1840 (1963).
194. S. Mukherjee and A. R. Todd, *J. Chem. Soc.*, 969 (1947).
195. P. W. Kent, M. Stacey, and L. F. Wiggins, *J. Chem. Soc.*, 1232 (1949).
196. R. Allerton and W. G. Overend, *J. Chem. Soc.*, 1480 (1951).
197. B. R. Baker and R. E. Schaub, *J. Org. Chem.*, **19**, 646 (1954).
198. N. F. Taylor, R. F. Childs, and R. V. Brunt, *Chem. Ind.* (London), 928 (1964); J. A. Wright and N. F. Taylor, *Carbohyd. Res.*, **3**, 333 (1967); **6**, 347 (1968); compare A. B. Foster and J. H. Westwood, *Angew. Chem. Intern. Ed. Engl.*, **11**, in press (1972).
199. S. Cohen, D. Levy, and E. D. Bergmann, *Chem. Ind.* (London), 1802 (1964).
200. A. B. Foster, T. D. Inch, J. Lehmann, M. Stacey, and J. M. Webber, *J. Chem. Soc.*, 2116 (1962).
201. G. Charalambous and E. Percival, *J. Chem. Soc.*, 2443 (1954).
201a. J. Jarý, K. Čapek, and J. Kovář, *Coll. Czech. Chem. Commun.*, **28**, 2171 (1963).
202. S. Mukherjee and H. C. Srivastava, *Proc. Indian Acad. Sci.*, **35A**, 178 (1952).
203. F. H. Newth, W. G. Overend, and L. F. Wiggins, *J. Chem. Soc.*, 10 (1947).
204. F. H. Newth and R. F. Homer, *J. Chem. Soc.*, 989 (1953).
205. J. G. Buchanan and R. M. Saunders, *J. Chem. Soc.*, 1791 (1964).
206. J. G. Buchanan and R. M. Saunders, *J. Chem. Soc.*, 1796 (1964).
207. G. N. Richards, L. F. Wiggins, and W. S. Wise, *J. Chem. Soc.*, 496 (1956) and references therein.
208. D. A. Prins, *Helv. Chim. Acta*, **29**, 1 (1946).
209. E. J. Hedgley, W. G. Overend, and R. A. C. Rennie, *J. Chem. Soc.*, 4701 (1963).
210. R. D. Guthrie, *Chem. Ind.* (London), 2121 (1962); R. D. Guthrie and D. Murphy, *J. Chem. Soc.*, 6666 (1965).
211. D. H. Buss, L. Hough, L. D. Hall, and J. F. Manville, *Tetrahedron*, **21**, 69 (1965).
211a. J. G. Buchanan, R. Fletcher, and W. A. Thomas, *J. Chem. Soc.* (*B*), 377 (1969).
212. C. D. Anderson, L. Goodman, and B. R. Baker, *J. Amer. Chem. Soc.*, **81**, 898 (1959).
213. C. D. Anderson, L. Goodman, and B. R. Baker, *J. Amer. Chem. Soc.*, **80**, 5247 (1958).
214. B. R. Baker, R. E. Schaub, J. P. Joseph, and J. H. Williams, *J. Amer. Chem. Soc.*, **76**, 4044 (1954).
215. B. R. Baker, R. E. Schaub, and J. H. Williams, *J. Amer. Chem. Soc.*, **77**, 7 (1955).
216. J. M. Anderson and E. Percival, *J. Chem. Soc.*, 819 (1956).
217. J. E. Christensen and L. Goodman, *J. Org. Chem.*, **28**, 2995 (1963).
218. G. Casini and L. Goodman, *J. Amer. Chem. Soc.*, **85**, 235 (1963); **86**, 1427 (1964).
219. W. W. Lee, A. Benitez, L. Goodman, and B. R. Baker, *J. Amer. Chem. Soc.*, **82**, 2648 (1960).
220. B. R. Baker and R. E. Schaub, *J. Amer. Chem. Soc.*, **77**, 5900 (1955).
221. J. Davoll, B. Lythgoe, and S. Trippett, *J. Chem. Soc.*, 2230 (1951).
222. R. E. Schaub and M. J. Weiss, *J. Amer. Chem. Soc.*, **80**, 4683 (1958).
223. L. Goodman, *J. Amer. Chem. Soc.*, **86**, 4167 (1964).

223a. P. W. Austin, J. G. Buchanan, and E. M. Oakes, *Chem. Commun.*, 374 (1965).
224. H. Ohle, W. Mertens, M. Andrée, and E. Euler, *Ber.*, **68**, 2176 (1935).
225. J. Kocourek, *Carbohyd. Res.*, **3**, 502 (1967).
226. S. N. Danilov and I. S. Lishanskiĭ, *Zh. Obshch. Khim.*, **21**, 366 (1951); *Chem. Abstr.*, **45**, 7529 (1951); V. I. Kovalenko, *Sb. Stateĭ Obshch. Khim.*, *Akad. Nauk S.S.S.R.*, **1**, 482 (1953); *Chem. Abstr.*, **49**, 881 (1955).
227. H. Ohle, H. Friedberg, and G. Haeseler, *Ber.*, **69**, 2311 (1936).
228. H. Ohle and M. Andrée, *Ber.*, **71**, 27 (1938).
229. H. Ohle, E. Euler, and W. Malerczyk, *Ber.*, **69**, 1636 (1936).
230. H. Ohle and E. Euler, *Ber.*, **69**, 1022 (1936).
231. J. Kocourek and V. Jiříček, *Coll. Czech. Chem. Commun.*, **22**, 806 (1957).
232. H. Ohle, E. Euler, and R. Voullième, *Ber.*, **71**, 2250 (1938).
233. H. Ohle and K. Tessmar, *Ber.*, **71**, 1843 (1938).
234. J. English, Jr. and M. F. Levy, *J. Amer. Chem. Soc.*, **78**, 2846 (1956).
235. G. P. Lampson and H. A. Lardy, *J. Biol. Chem.*, **181**, 693 (1949).
236. B. Helferich and R. Mittag, *Ber.*, **71**, 1585 (1938).
237. M. Kh. Gluzman and V. I. Kovalenko, *Sb. Stateĭ Obshch. Khim.*, *Akad. Nauk S.S.S.R.*, **1**, 473 (1953); *Chem. Abstr.*, **49**, 880 (1955), and references therein.
238. G. P. McSweeney and L. F. Wiggins, *Nature*, **168**, 874 (1951).
239. L. D. Hall, L. Hough, and R. A. Pritchard, *J. Chem. Soc.*, 1537 (1961).
240. A. S. Meyer and T. Reichstein, *Helv. Chim. Acta*, **29**, 152 (1946).
241. F. Blindenbacher and T. Reichstein, *Helv. Chim. Acta*, **31**, 1669 (1948).
242. E. J. Hedgley, O. Méréz, W. G. Overend, and R. Rennie, *Chem. Ind.* (London), 938 (1960).
243. G. Henseke and G. Hanisch, *Angew. Chem. Intern. Ed.*, **2**, 324 (1963).
244. B. Helferich, W. Klein, and W. Schafer, *Ber.*, **59**, 79 (1926).
245. W. G. Overend and S. A. Brooks, *Chem. Ind.* (London), 471 (1960).
246. W. T. Haskins, R. M. Hann, and C. S. Hudson, *J. Amer. Chem. Soc.*, **68**, 628 (1946).
247. R. B. Duff and E. G. V. Percival, *J. Chem. Soc.*, 830 (1941).
248. E. Fischer and K. Zach, *Ber.*, **45**, 456, 2068 (1912).
249. F. Valentin, *Coll. Czech. Chem. Commun.*, **6**, 354 (1934).
250. F. Valentin, *Coll. Czech. Chem. Commun.*, **4**, 364 (1932).
251. H. Ohle and H. Thiel, *Ber.*, **66**, 525 (1933).
252. W. N. Haworth, J. Jackson, and F. Smith, *J. Chem. Soc.*, 620 (1940).
253. P. A. Rao and F. Smith, *J. Chem. Soc.*, 229 (1944).
254. I. A. Forbes and E. G. V. Percival, *J. Chem. Soc.*, 1844 (1939).
255. A. B. Foster, W. G. Overend, M. Stacey, and G. Vaughan, *J. Chem. Soc.*, 3367 (1954).
256. W. N. Haworth, L. N. Owen, and F. Smith, *J. Chem. Soc.*, 88 (1941).
256a. A. B. Foster, W. G. Overend, and G. Vaughan, *J. Chem. Soc.*, 3625 (1954).
257. H. Ohle and H. Wilke, *Ber.*, **71**, 2316 (1938).
258. E. G. V. Percival, *J. Chem. Soc.*, 119 (1945).
259. H. Ohle, L. v. Vargha, and H. Erlbach, *Ber.*, **61**, 1211 (1928).
260. H. Ohle and H. Erlbach, *Ber.*, **62**, 2758 (1929).
261. L. D. Hall and L. Hough, *J. Chem. Soc.*, 5301 (1963).
262. J. Defaye and S. D. Gero, *Bull. Soc. Chim. Biol.*, **47**, 1765 (1965).
263. F. Tiemann, *Ber.*, **17**, 241 (1884).
264. E. Fischer and F. Tiemann, *Ber.*, **27**, 138 (1894).
265. F. Shafizadeh, *Advan. Carbohyd. Chem.*, **13**, 1 (1958).
266. P. A. Levene and F. B. LaForge, *J. Biol. Chem.*, **21**, 345, 351 (1915).
267. P. A. Levene, *Biochem. Z.*, **124**, 37 (1921); P. A. Levene, *J. Biol. Chem.*, **59**, 135 (1924).

268. C. Neuberg, H. Wolff and W. Neimann, *Ber.*, **35**, 4009 (1902).
269. D. Horton, in "The Amino Sugars," Vol. IA (R. W. Jeanloz, Ed.), Academic Press, Inc., New York, N.Y., 1969, pp. 128–132.
270. P. A. Levene and E. P. Clark, *J. Biol. Chem.*, **46**, 19 (1921).
271. A. B. Foster, *Chem. Ind.* (London), 627 (1955).
272. P. A. Levene, *J. Biol. Chem.*, **31**, 609 (1917).
273. P. A. Levene, *J. Biol. Chem.*, **36**, 89 (1918).
274. J. Defaye, *J. Bull. Soc. Chim. France*, 999 (1964).
275. R. U. Lemieux and B. Fraser-Reid, *Can. J. Chem.*, **42**, 547 (1964).
275a. P. W. Austin, J. G. Buchanan, and R. M. Saunders, *J. Chem. Soc.*, *C*, 372 (1967).
276. C. A. Dekker and T. Hashizume, *Arch. Biochem. Biophys.*, **78**, 348 (1958); compare J. Defaye and V. Ratovelomanana, *Carbohyd. Res.*, **17**, 57 (1971).
277. J. Defaye, *Bull. Soc. Chim. France*, 2686 (1964).
278. H. Zinner, H. Brandhoff, H. Schmandke, H. Kristen, and R. Haun, *Chem. Ber.*, **92**, 3151 (1959).
278a. J. Defaye and D. Horton, *Carbohyd. Res.*, **14**, 128 (1970).
278b. J. Defaye, D. Horton, and M. Muesser, *Carbohyd. Res.*, **20**, 305 (1971).
279. M. Cifonelli, J. A. Cifonelli, R. Montgomery, and F. Smith, *J. Amer. Chem. Soc.*, **77**, 121 (1955).
280. I. L. Doerr, J. F. Codington, and J. J. Fox, *J. Org. Chem.*, **30**, 467, 476 (1965).
281. D. A. Rosenfeld, N. K. Richtmyer, and C. S. Hudson, *J. Amer. Chem. Soc.*, **70**, 2201 (1948).
281a. A. B. Foster, J. M. Duxbury, T. M. Inch, and J. M. Webber, *Chem. Commun.*, 881 (1967).
282. E. D. M. Eades, D. H. Ball, and L. Long, Jnr., *J. Org. Chem.*, **30**, 3949 (1965).
283. P. A. Levene and A. L. Raymond, *J. Biol. Chem.*, **102**, 331 (1933).
283a. R. L. Whistler, T. J. Lutenegger, and R. M. Rowell, *J. Org. Chem.*, **33**, 396 (1968).
283b. L. Hough and B. A. Otter, *Carbohyd. Res.*, **4**, 126 (1967).
284. J. P. Horowitz, J. Chua, J. A. Urbanski, and M. Noel, *J. Org. Chem.*, **28**, 942 (1963).
285. A. B. Foster, M. Stacey, and S. V. Vardheim, *Acta Chem. Scand.*, **12**, 1819 (1958).
286. P. Chang and Chan-Ming Hu, *Hua Hsueh Pao*, **29**, 166 (1963) (in Chinese); *Sci. Sinica*, **13**, 441 (1964) (in English).
287. S. Peat and L. F. Wiggins, *J. Chem. Soc.*, 1088 (1938).
288. J. G. Buchanan and J. Conn, *J. Chem. Soc.*, 201 (1965).
289. E. Seebeck, A. Meyer, and T. Reichstein, *Helv. Chim. Acta*, **27**, 1142 (1944).
290. H. Ohle and R. Lichtenstein, *Ber.*, **63**, 2905 (1930).
291. J. G. Buchanan and E. M. Oakes, *Carbohyd. Res.*, **1**, 242 (1965).
292. W. H. G. Lake and S. Peat, *J. Chem. Soc.*, 1069 (1939).
293. M. Černý, J. Pacák, and J. Staněk, *Coll. Czech. Chem. Commun.*, **30**, 1151 (1965).
294. M. Černý, I. Burban and J. Pacák, *Coll. Czech. Chem., Commun.*, **28**, 1569 (1963).
295. J. G. Buchanan and R. Fletcher, *J. Chem. Soc.*, 6316 (1965).
296. J. Jarý and K. Čapek, *Coll. Czech. Chem. Commun.*, **31**, 315 (1966).
297. J. G. Buchanan and J. C. P. Schwarz, *J. Chem. Soc.*, 4770 (1962); compare J. G. Buchanan and R. Fletcher, *J. Chem. Soc. (C)*, 1926 (1966).
298. J. W. H. Oldham and G. J. Robertson, *J. Chem. Soc.*, 685 (1935).
299. V. Y. Labaton and F. H. Newth, *J. Chem. Soc.*, 992 (1953).
300. J. G. Buchanan, *J. Chem. Soc.*, 995 (1958).
301. J. G. Buchanan, *J. Chem. Soc.*, 2511 (1958).

14. ALDITOLS AND DERIVATIVES

J. S. BRIMACOMBE AND J. M. WEBBER

I. INTRODUCTION

Alditols are acyclic, polyhydric alcohols that are formally derived from aldoses or ketoses by reduction of the carbonyl group. The alditols may be subdivided, on the basis of the number of carbon atoms in the normal chain, into tetritols, pentitols, hexitols, and so on.[1]

As a group, the alditols are crystalline substances that range in taste from faintly sweet to very sweet. Members of the tetritol to heptitol classes have long been known to occur in Nature, where their distribution is, in general, limited to plants belonging to both the higher and lower orders. In studies of the avocado, the first isolation of a naturally occurring octitol has been reported. Alditols occur naturally in both the free and the combined forms; glycosides having an alditol as the aglycon occur in plants, and an acetic acid ester of mannitol has been found in algae. Cell-wall teichoic acids of certain Gram-positive bacteria are polymers of ribitol phosphate to which other constituents are attached.

Alditols, particularly ethylene glycol, glycerol,[1a] D-glucitol, and D-mannitol, have widespread commercial applications, often as a result of their hygroscopic properties. Partial esters of alditols, particularly those from long-chain fatty acids, may have surface-active properties that make them of industrial importance as emulsifying agents, but the ester products are usually mixtures, since the conditions of commercial esterification lead to simultaneous formation of anhydro derivatives. Nitric acid esters of alditols are useful explosives and pharmaceuticals. The acetal derivatives of alditols have been extensively prepared and examined, but have not yet found a practical application. Terminal dimethanesulfonic esters of alditols are of interest as potential antitumor agents.

II. CONFIGURATIONS, OCCURRENCE, AND PREPARATION

A. Tetritols

All of the theoretically possible tetritols are known.

$$
\begin{array}{cc}
CH_2OH & CH_2OH \\
| & | \\
HCOH & HOCH \\
| & | \\
HCOH & HCOH \\
| & | \\
CH_2OH & CH_2OH \\
\mathbf{1} & \mathbf{2} \\
\text{Erythritol} & \text{D-Threitol}
\end{array}
$$

1. **Erythritol**[2] (**1**), m.p. 120°; tetraacetate,[3] m.p. 89–90°; dibenzylidene acetal,[2] m.p. 202°.

This alditol has been isolated from roots of *Primula officinalis*.[4] It also occurs naturally in certain algae,[5] lichens,[6] and grasses,[7] and is produced by the fermentative action of certain fungi.[8,9]

Erythritol has been synthesized[3] by *trans*-hydroxylation of 1,4-diacetoxy-*trans*-2-butene, and also from epichlorohydrin by the following sequence of reactions:[10]

$$
\begin{array}{c}
\text{CH}_2 \\
| \quad \rangle\text{O} \\
\text{CH} \\
| \\
\text{CH}_2\text{Cl}
\end{array}
\xrightarrow{\text{HCN}}
\begin{array}{c}
\text{CN} \\
| \\
\text{CH}_2 \\
| \\
\text{CHOH} \\
| \\
\text{CH}_2\text{Cl}
\end{array}
\xrightarrow{\text{PCl}_5}
\begin{array}{c}
\text{CN} \\
| \\
\text{CH}_2 \\
| \\
\text{CHCl} \\
| \\
\text{CH}_2\text{Cl}
\end{array}
\xrightarrow[\text{NaOH}]{\text{aq.}}
\text{(lactone)}
\xrightarrow[\text{2. Na/Hg}]{\text{1. Ba(MnO}_4)_2}
\begin{array}{c}
\text{CH}_2\text{OH} \\
| \\
\text{HCOH} \\
| \\
\text{HCOH} \\
| \\
\text{CH}_2\text{OH}
\end{array}
$$

More-convenient syntheses of erythritol may be effected by reduction of dialkyl erythrarates (*meso*-tartrates)[11] or by the hydrogenolysis of periodate-oxidized starch. Conditions for the latter reaction have been extensively investigated,[12] and by the use of a one-stage, reductive hydrolysis with Raney nickel in the presence of carbon, erythritol and ethylene glycol have been obtained in yields of 75 to 95%. Erythritol has also been prepared in high yield[13] from 4,6-*O*-ethylidene-D-glucose by sequential treatment with sodium periodate, sodium borohydride, and aqueous acid; a similar sequence may be applied to 4,6-*O*-benzylidene-D-glucose.[14]

2. D-**Threitol**[15] (2), m.p. 88°, [α]$_D$ +4.3° (water); tetrabenzoate,[11] m.p. 97°.
This alditol is rare in Nature but is found[16] in the edible fungus *Armellaria mellea* as a large proportion of the dry weight. It was synthesized[15] from D-xylose by using the Wohl degradation, followed by reduction with sodium amalgam.

L-**Threitol** has not been reported to occur naturally, but it has been isolated from plants that had been fed on L-sorbose.[17] It has been synthesized by bacterial oxidation of erythritol, followed by reduction of the resulting L-*glycero*-tetrulose with sodium amalgam.[18] L-Threitol has been prepared in high yield by acid hydrolysis of 1,3-*O*-benzylidene-L-threitol obtained by sequential treatment of 1,3-*O*-benzylidene-L-arabinitol with sodium periodate and sodium borohydride.[19]

DL-**Threitol**,[2] m.p. 72°; tetraacetate,[3] m.p. 54–55°; dibenzylidene acetal,[20] m.p. 221–223°.
This alditol has been synthesized by *trans*-hydroxylation of 1,4-dihydroxy-*cis*-2-butene,[20] or its diacetate,[3] as follows:

$$
\begin{array}{c}
\text{HOCH}_2\text{CH} \\
\| \\
\text{HOCH}_2\text{CH}
\end{array}
\xrightarrow{\text{PhCO}_3\text{H}}
\begin{array}{c}
\text{CH}_2\text{OH} \\
| \\
\text{CH} \\
\text{O} \langle \quad | \\
\text{CH} \\
| \\
\text{CH}_2\text{OH}
\end{array}
\xrightarrow{\text{H}_2\text{SO}_4}
\begin{array}{c}
\text{CH}_2\text{OH} \\
| \\
\text{HOCH} \\
| \\
\text{HCOH} \\
| \\
\text{CH}_2\text{OH}
\end{array}
\; + \;
\begin{array}{c}
\text{CH}_2\text{OH} \\
| \\
\text{HCOH} \\
| \\
\text{HOCH} \\
| \\
\text{CH}_2\text{OH}
\end{array}
$$

DL-Threitol has also been synthesized from 3,4-epoxy-1-butene.[21]

* *References start on p. 513.*

D-, L-, and DL-Threitols may all be conveniently synthesized by reduction of the appropriate dialkyl threarates (tartrates)[11] (or their 2,3-acetals[22]) with lithium borohydride or lithium aluminum hydride.

B. PENTITOLS

All of the theoretically possible pentitols are known. As with the tetritols, only the D configurations of the optically active pentitols will be depicted.

CH₂OH	CH₂OH	CH₂OH

$$
\begin{array}{ccc}
\text{CH}_2\text{OH} & \text{CH}_2\text{OH} & \text{CH}_2\text{OH} \\
\text{HOCH} & \text{HCOH} & \text{HCOH} \\
\text{HCOH} & \text{HCOH} & \text{HOCH} \\
\text{HCOH} & \text{HCOH} & \text{HCOH} \\
\text{CH}_2\text{OH} & \text{CH}_2\text{OH} & \text{CH}_2\text{OH} \\
\mathbf{3} & \mathbf{4} & \mathbf{5} \\
\text{D-Arabinitol} & \text{Ribitol} & \text{Xylitol} \\
\text{(D-Lyxitol)} & &
\end{array}
$$

1. D-**Arabinitol** (D-lyxitol)[23] (**3**), m.p. 102°, $[\alpha]_D$ +7.82° (borax); penta-acetate, m.p. 76°.

This alditol has been found in many lichens[24,25] both in the free form and as the D-galactoside, umbilicin[24] (see Vol. IIA, Chap. 32). It has also been found in the field mushroom, *Fistulina hepatica*, to the extent of 9.5% of the dry weight.[26] D-Arabinitol is produced by the fermentative action of various microorganisms,[9,27] and, under favorable conditions, D-glucose has been converted into the pentitol in yields as high as 35 to 40%. Amongst higher plants, D-arabinitol has been found in avocado seeds[27a] and in Pichi tops[27b] (the dried herbiage of *Fabiana imbricata*).

Synthetically, D-arabinitol was originally obtained by reduction of D-arabinose[23] or D-lyxose[28] by means of sodium amalgam, but this reduction can be effected more conveniently with sodium borohydride.

L-Arabinitol,[29] $[\alpha]_D$ −5.4° (borax), has not been shown to occur naturally, although it has been detected in the urine of patients suffering from pentosuria.[30] The alditol has been prepared by reduction of L-arabinose[31] and by employing the Cannizzaro reaction with L-arabinose in the presence of nickel.[32]

DL-Arabinitol,[33] m.p. 105°; pentaacetate, m.p. 97–98°. This alditol has not been found in Nature but was synthesized (together with ribitol) from acrolein by Lespieau.[34] Raphael has described[33] a more-convenient synthesis, based on the reaction of epichlorohydrin with sodium acetylide, in which the ribitol and DL-arabinitol products are readily separated.

CH₂
| \
CH   O + NaC≡CH —NH₃→
| /
CH₂Cl

$$
\begin{array}{c}
CH_2OH \\
| \\
CH \\
|| \\
CH \\
| \\
C \\
||| \\
CH
\end{array}
\quad \xrightarrow{HCO_3H} \quad
\begin{array}{c}
CH_2OH \\
| \\
CHOH \\
| \\
CHOH \\
| \\
C \\
||| \\
CH
\end{array}
$$

$$
\begin{array}{c}
CH_2OAc \\
| \\
HCOAc \\
| \\
HCOAc \\
| \\
HCOH \\
| \\
CH_2Br
\end{array}
\quad + \quad
\begin{array}{c}
CH_2OAc \\
| \\
HCOAc \\
| \\
HCOAc \\
| \\
HOCH \\
| \\
CH_2Br
\end{array}
\quad \xleftarrow[\text{in water}]{N\text{-bromosuccinimide}} \quad
\begin{array}{c}
CH_2OAc \\
| \\
CHOAc \\
| \\
CHOAc \\
| \\
CH \\
|| \\
CH_2
\end{array}
$$

(soluble in ether)   (insoluble in ether)

Ac₂O
AcOH
KOAc

Ribitol pentaacetate   DL-Arabinitol pentaacetate

2. Ribitol (adonitol)[35] (**4**), m.p. 102°; dibenzylidene acetal, m.p. 164–165°.

This alditol occurs naturally in the plants *Adonis vernalis*[36] and *Bupleurum falactum* root (the Chinese drug, Chei-Hou).[37] In the bound form, it is a constituent of riboflavin (vitamin B₂), of the capsular polysaccharides[38,39] of *Pneumococcus* Type VI and Type XXXIV, and of ribitol teichoic acids. The ribitol teichoic acids,[40] which occur in the cell walls of *Bacillus subtilis*, *Lactobacillus arabinosus*, and *Staphylococcus aureus*, are composed of chains of ribitol residues joined through phosphate diester groups at C-1 and C-5. The ribitol residues bear glycosyl and D-alanine ester substituents (see Chapter 41).

Synthetic ribitol has been prepared by the reduction of L-ribose with sodium amalgam,[35] and by the reduction of D-ribose with sodium borohydride or by catalytic hydrogenation. As mentioned previously, ribitol was obtained by Lespieau[34] and by Raphael[33] in their syntheses of DL-arabinitol from noncarbohydrate precursors.

3. Xylitol (**5**), m.p. 61–61.5° (metastable modification),[41] 93–94.5° (stable modification);[42] tetraacetate, m.p. 62–63°.

This alditol has been isolated from the field mushroom, *Psalliota campestris*, to the extent of 0.01 % of the dry weight,[16] and also from roots of *Primula*

* *References start on p. 513.*

officinalis.[4] D-Xylose is of widespread occurrence in pentosans that accumulate as agricultural waste products (such as, cottonseed hulls and corncobs), and methods have been described[43] for its conversion into xylitol by acid hydrolysis of these materials, followed by catalytic hydrogenation.

In the laboratory, D-xylose has been reduced to xylitol by using sodium amalgam,[44] catalytic hydrogenation,[41, 42] or sodium borohydride.

Xylitol has been reported[45] to have a sweetening strength equal to that of sucrose in 10% solution.

C. HEXITOLS

Ten stereoisomeric hexitols are possible, and all are known. D-Glucitol may equally well be regarded as L-gulitol, but the former name is the recognized one by the principle of alphabetic precedence. The hexitol is most conveniently prepared by reduction of D-glucose. When the molecule is substituted at one of positions 4, 5, or 6, it is correctly named as a derivative of L-gulitol. D-Altritol is the same as D-talitol. Allitol and galactitol are *meso* forms, but the prefixes D- and L- are necessary when the molecules are rendered optically active by substitution. Sorbitol and dulcitol, the old names for D-glucitol and galactitol, respectively, have been superseded by the systematic names.

CH$_2$OH	CH$_2$OH	CH$_2$OH
HCOH	HCOH	HCOH
HCOH	HOCH	HOCH
HCOH	HOCH	HCOH
HCOH	HCOH	HCOH
CH$_2$OH	CH$_2$OH	CH$_2$OH
6	**7**	**8**
Allitol	Galactitol	D-Glucitol
(*meso*)	(*meso*)	

CH$_2$OH	CH$_2$OH	CH$_2$OH
HOCH	HOCH	HOCH
HCOH	HOCH	HCOH
HOCH	HCOH	HCOH
HCOH	HCOH	HCOH
CH$_2$OH	CH$_2$OH	CH$_2$OH
9	**10**	**11**
D-Iditol	D-Mannitol	D-Altritol

The physical properties of the hexitols and their most-accessible derivatives, the hexaacetates, are given in Table I; the properties of the D-enantiomorph only are recorded.

TABLE I
PHYSICAL PROPERTIES OF THE HEXITOLS[46]

Hexitol	M.p. (degrees)	$[\alpha]_D$ (in H_2O) (degrees)	Hexaacetate	
			M.p. (degrees)	$[\alpha]_D$ in $CHCl_3$ (degrees)
Allitol	150–151	meso	61	meso
D-Altritol	87–88	+3.2		
DL-Altritol	95–96	DL	85–86[c]	DL
Galactitol	188.5	meso	168–169	meso
D-Glucitol				
Stable form	97	−19[a]	99	+10.0
Labile form	92			
DL-Glucitol	136–138	DL	117–119	DL
D-Iditol	73.5	+3.5	121–122	+25.3
DL-Iditol			165–166	DL
D-Mannitol	166	−2.1[b]	126	+25.0
DL-Mannitol	170	DL	107–109[c]	DL

[a] $[\alpha]_D$ +108° in acidified ammonium molybdate.   [b] $[\alpha]_D$ +141° in acidified ammonium molybdate.   [c] Reported in reference 47.

1. **Allitol** was unknown in Nature until 1959, when it was found[48] in the leaves and branches of *Itea ilicifolia* and *I. virginica*.

A total synthesis of allitol was accomplished[49] prior to its unequivocal synthesis by the catalytic reduction of D-allose.[50] Hydroxylation of the diastereoisomeric forms of divinylglycol (CH_2=CH—CHOH—CHOH—CH=CH_2), with silver chlorate–osmium tetraoxide, yielded a mixture of allitol and DL-mannitol.[49] Similar treatment of the isomer having exclusively *cis*-hydroxyl groups gave[51] allitol together with a trace of galactitol. Hydroxylation of 1,2,5,6-tetrahydroxy-3-hexene (**14**, see below) also provides a route for total synthesis of allitol.[52]

2. **Galactitol** (dulcitol) (**7**) was originally isolated, in 1837, from *Melampyrum nemorosum*,[53] the old name for galactitol being melampyrum or melampyrite, but it has since been reported in plants ranging from red seaweeds and a

* References start on p. 513.

pentose-fermenting yeast (*Torula utilis*) to the mannas of higher plant life. Madagascar manna appears to be relatively pure galactitol,[54] and the manna of *Gymnosporia deflexa*[55] is another rich source.

Galactitol has been synthesized by reduction of D-galactose and by the Cannizzaro process of Delépine and Horeau.[32] Two total syntheses have been reported. The first was accomplished[52] from 1,6-dichloro-2,5-dihydroxy-3-hexyne (12), formed by condensation of two molecules of chloroacetaldehyde with a bifunctional Grignard reagent prepared from acetylene. Compound 12 was transformed into the diepoxide (13) and thence into 1,2,5,6-tetrahydroxy-3-hexene (14) on scission of the epoxide rings and partial hydrogenation. *cis*-Hydroxylation of hexenetetrol (14), with osmium

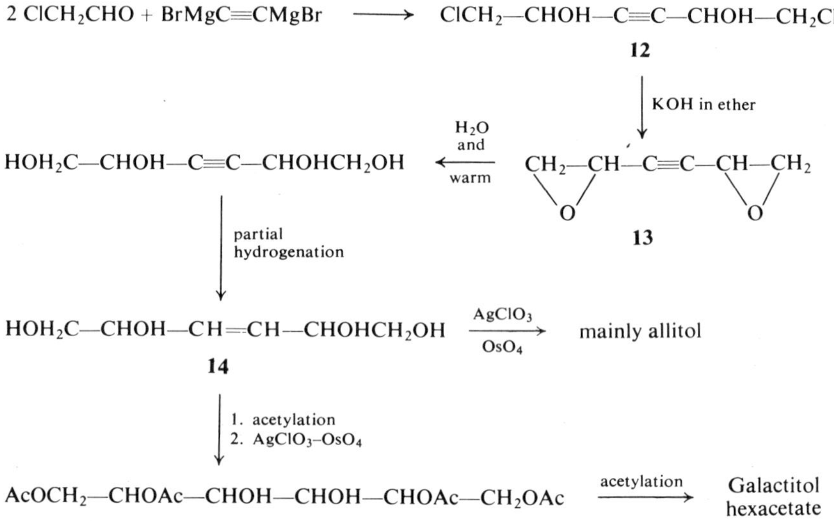

$$2 \; ClCH_2CHO + BrMgC\equiv CMgBr \longrightarrow ClCH_2-CHOH-C\equiv C-CHOH-CH_2Cl$$

12

KOH in ether

$$HOH_2C-CHOH-C\equiv C-CHOHCH_2OH \xleftarrow[\text{warm}]{\substack{H_2O \\ \text{and}}} CH_2-CH-C\equiv C-CH-CH_2$$

with epoxide O rings and 13

partial hydrogenation

$$HOH_2C-CHOH-CH=CH-CHOHCH_2OH \xrightarrow[\text{OsO}_4]{\text{AgClO}_3} \text{mainly allitol}$$

14

1. acetylation
2. AgClO₃–OsO₄

$$AcOCH_2-CHOAc-CHOH-CHOH-CHOAc-CH_2OAc \xrightarrow{\text{acetylation}} \substack{\text{Galactitol} \\ \text{hexacetate}}$$

tetraoxide–silver chlorate, gave principally allitol, whereas hydroxylation of the derived tetraacetate, followed by acetylation, afforded galactitol hexaacetate.

Alternatively, galactitol has been obtained by hydroxylation of 3,4-dihydroxyhexa-1,5-diene[56] with peroxyformic acid and subsequent acid treatment; DL-iditol was also formed and characterized as the hexaacetate.

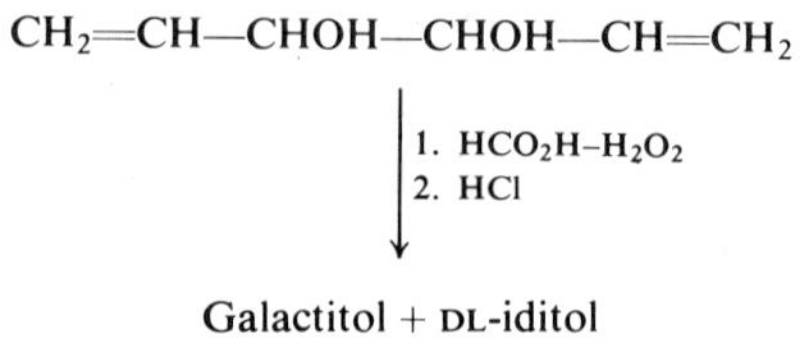

$$CH_2=CH-CHOH-CHOH-CH=CH_2$$

1. HCO₂H–H₂O₂
2. HCl

Galactitol + DL-iditol

3. D-**Glucitol** (L-gulitol, sorbitol, "sorbite") (**8**) was discovered in 1872 by Boussingault[57] in the fresh juice of the berries of the mountain ash (*Sorbus aucuparia* L.) where it may be oxidized, on keeping, to L-sorbose by bacteria such as *Acetobacter xylinum*. Pelouze[58] had reported the isolation of L-sorbose from the fermented juice of sorbus berries some twenty years earlier. D-Glucitol is now known to be of widespread occurrence among plants ranging from algae to the higher orders, especially in fruit and berries. The red seaweed *Bostrychia scorpoides*, for example, is composed of approximately 14% of D-glucitol.[59] The fruits of the plant family Rosaceae, which includes apples, pears, cherries, and apricots, contain appreciable amounts of the hexitol.[60] Isotopic labeling experiments have demonstrated a high turnover of D-glucitol in plum[61] and apple leaves.[62]

Considerable effort has been devoted to the preparation of D-glucitol from commercial sources. Reduction of D-glucose or D-glucono-1,4-lactone by pressure hydrogenation has been widely applied[63] on an industrial scale, and strenuous efforts have been made to achieve quantitative conversions. Nickel catalysts, to which various promoters have been added, are widely used in these hydrogenations.[64] More recently, ruthenium catalysts supported on carbon, sometimes with palladium added, have proved extremely efficient.[65] Hydrolytic hydrogenation of cellulose over a 0.5% ruthenium–carbon catalyst is claimed[66] to give D-glucitol in high yield. Continuous pressure hydrogenation seems to have supplanted the electrolytic process[67] for the manufacture of D-glucitol.

L-Glucitol (D-gulitol) is unknown in Nature. It has been prepared synthetically by reduction of D-gulose either with sodium amalgam or by catalytic hydrogenation,[29, 68] and from D-sorbose by using sodium amalgam.[69]

DL-Glucitol (DL-gulitol) has been made by mixing equimolar proportions of the two enantiomorphs. It has also been isolated,[68] in low yield, from a commercial "sorbitol" prepared by the electrolytic reduction of D-glucose under alkaline conditions.

4. D-**Iditol** (**9**) is best prepared by reduction of D-idose[70] or D-sorbose,[69] the latter reaction also yielding L-glucitol.

L-Iditol ("sorbierite") is of limited occurrence in Nature, but it has been isolated from the mother liquors of the juice of the mountain ash berry (*Sorbus aucuparia*) after the removal of D-glucitol by fermentation.[71] Chemical syntheses are available by the reduction of L-idonolactone[72] and L-sorbose.[47] In the latter instance, the D-glucitol also formed is conveniently removed as the relatively insoluble pyridine complex,[73] and the residual L-iditol can be crystallized from methanol. Although reduction of L-sorbose appears to yield approximately equal amounts of the two hexitols, reduction of the peracetyl-

* *References start on p. 513.*

ated ketose apparently favors the formation of L-iditol derivatives. Thus, a 60% yield of L-iditol hexaacetate resulted from the reduction of the acetylated ketose in ether over a platinum catalyst and acetylation of the hydrogenated product.[74] The proportion of L-iditol derivatives is increased even more dramatically, to approximately 90%, when the reduction is carried out in ethanol with a Raney nickel catalyst and a slight overpressure of hydrogen.[75] Catalytic isomerization of both D-glucitol and D-mannitol yields a complex mixture of hexitols from which L-iditol can be isolated.[47]

5. D-**Mannitol** (**10**) was the first crystalline alditol to be discovered[76] in Nature, and it is frequently found in exudates of plants. This factor, together with the high tendency to crystallize and moderate solubility, were most likely responsible for the early isolation of D-mannitol from the manna of the flowering or manna ash, *Fraxinus ornus*.[76] It is also present in high concentration in the exudates of olive and plane trees,[77] and, before more convenient sources became available, D-mannitol was obtained commercially in Sicily from the sap of *Fraxinus rotundifolis*.

Marine algae appear to be the greatest potential source of D-mannitol, but the fronds of only a few varieties, such as *Laminaria claustoni*, are suitable for commercial exploitation.[78] D-Mannitol is found in all brown algae, where it is invariably the major product of photosynthesis.[79] There have been reports[80] of the occurrence of D-mannitol in combined form in algae.

D-Mannitol has been isolated from onions,[81] grasses,[82] and Jerusalem artichokes,[83] and most varieties of mushroom were shown[84] to contain a small proportion of D-mannitol.

D-Mannitol has been synthesized by several methods. Its manufacture has been based on the epimerization of D-glucose under alkaline conditions and subsequent electrolytic or catalytic reduction to a mixture composed mainly of D-mannitol and D-glucitol.[67] Depending on the alkalinity, over 20% of D-glucose can be converted into D-mannitol in this way. D-Mannitol was obtained[47] in approximately 22% yield from the catalytic isomerization of D-glucitol at 170° with hydrogen under pressure and a nickel–Kieselguhr catalyst.

Catalytic or sodium borohydride reduction of D-mannose or the reduction of D-fructose or invert sugar offers the best synthesis of D-mannitol in the laboratory; catalytic reduction of D-mannono-1,5-lactone has also been used.[85] Certain microorganisms are known to produce D-mannitol, in high yield, from either D-glucose, D-fructose, or sucrose.[86] The hexitol is one of the products obtained[87] when *Aspergillus niger* is grown in the presence of acetate as the sole carbon source.

L-Mannitol is unknown in Nature but has been prepared by the reduction of L-mannose with sodium amalgam[88] or by the high-pressure hydrogenation of L-mannono-1,5-lactone, over a platinum catalyst containing a little iron.[89]

Total syntheses of DL-mannitol (α-acritol) have been described. It was originally obtained by reduction of DL-fructose (α-acrose) with sodium amalgam, a reaction forming part of Emil Fischer's monumental researches[90] into the synthesis of sugars from noncarbohydrate precursors. As indicated earlier, DL-mannitol is one of the products resulting from hydroxylation of divinylglycol.[49]

A total synthesis of DL-mannitol has been achieved[91] commencing from acetoacetic ester. Oxidation of the ester with iodine and saponification of the product gave 2,5-hexanedione, which afforded hexane-2,5-diol on reduction. The diol was converted into the dibromide and thence into a hexadiene on dehydrobromination. Bromination gave a tetrabromohexane that was transformed, by a similar sequence of reactions, into a hexabromide. This gave DL-mannitol on treatment with alcoholic potassium hydroxide.

6. **D-Altritol** (D-talitol) (**11**) is known only from synthesis; it may be prepared by reduction of D-talose[92] or D-talonolactone[93] with sodium amalgam, or, preferably, by the catalytic reduction of D-altrose.[94] L-Altritol (L-talitol) is likewise prepared from L-altrose[95] and, incidentally, was the last of the hexitols to be synthesized.

New routes to D-altritol have been developed from D-mannitol derivatives. Oxidation of 3-O-benzoyl-1,2:5,6-di-O-isopropylidene-D-mannitol with the chromium trioxide–pyridine complex afforded[96] 4-O-benzoyl-1,2:5,6-di-O-isopropylidene-D-*arabino*-3-hexulose, which was concomitantly de-esterified and reduced, with lithium aluminum hydride, to a mixture of 1,2:5,6-di-O-isopropylidene-D-mannitol and 1,2:5,6-di-O-isopropylidene-D-altritol (**18**). The altritol acetal was separated by chromatography on Magnesol–Celite and yielded crystalline D-altritol on hydrolysis. It is known[97] that 1,2:5,6-di-O-isopropylidene-3-O-(methylsulfonyl)-D-mannitol is readily oxidized to 1,2:5,6-di-O-isopropylidene-4-O-(methylsulfonyl)-D-*arabino*-3-hexulose, with N,N'-dicyclohexylcarbodiimide and phosphoric acid in methyl sulfoxide, and this reaction could also provide a route to D-altritol.

The sulfonic ester group of 4-O-benzoyl-1,2:5,6-di-O-isopropylidene-3-O-(methylsulfonyl)-D-mannitol (**15**) is solvolyzed,[98] with sodium acetate in wet N,N-dimethylformamide, to products which, on debenzoylation, afforded 1,2:5,6-di-O-isopropylidene-D-altritol (**18**). The solvolysis is anchimerically assisted by the neighboring benzoyloxy group, and the products (**17**) result from the attack of water on the intermediate (**16**).

DL-Altritol was originally obtained[95] by recrystallizing the enantiomorphs in equimolar proportion. It was subsequently isolated from the product of

$$Me_2C \big<{}^{OCH_2}_{OCH} \quad MeOCH \quad HCOBz \quad HCO \big<{}_{H_2CO}^{}CMe_2 \quad \mathbf{15}$$

$$\xrightarrow[\text{HCONMe}_2\text{–H}_2\text{O}]{\text{NaOAc}}$$

$$Me_2C \big<{}^{OCH_2}_{OCH} \quad HCO \!\!\!\underset{HCO}{\overset{}{\cdots}}\!\! \overset{+}{:}CPh \quad HCO \big<{}_{H_2CO}^{}CMe_2 \quad \mathbf{16}$$

$$\xrightarrow{\text{H}_2\text{O}}$$

$$Me_2C \big<{}^{OCH_2}_{OCH} \quad \left.{}^{HCO}_{HCO}\right\} {}^{H}_{Bz} \quad HCO \big<{}_{H_2CO}^{}CMe_2 \quad \mathbf{17}$$

$$\xrightarrow{\text{NaOMe}}$$

$$Me_2C \big<{}^{OCH_2}_{OCH} \quad HCOH \quad HCOH \quad HCO \big<{}_{H_2CO}^{}CMe_2 \quad \mathbf{18}$$

catalytic isomerization of D-mannitol, D-glucitol, and galactitol, and was characterized as the previously unreported hexaacetate.[47]

D. Heptitols

This group contains four identical pairs of alditols and therefore only twelve different isomers, all of which have been synthesized (see Table II). The configurations of the known optically active (D series) and *meso* heptitols are illustrated in this section.

Three main systems of nomenclature have been used for the higher-carbon sugars (for a summary, see ref. 99), but the system based entirely on the configuration of the asymmetric carbon atoms actually present has now been generally adopted. This system is defined in Rules 22 and 23 of the Rules for Nomenclature of Carbohydrates[100] and is discussed further in Chapter 1.

Two heptitols have long been known to occur in Nature, and both have well-known trivial names, volemitol and perseitol. Volemitol (D-*glycero*-D-*talo*-heptitol) (**28**), originally found in the mushroom *Lactarius volemus*,[101] has also been isolated from the roots of various *Primulae*[4, 102] and from certain lichens[103] and algae.[104] In the alga *Pelvetia canaliculata*, volemitol occurs glycosidically bound to D-glucose, as well as in the free form. Perseitol (D-*glycero*-D-*galacto*-heptitol) (**26**) was first isolated[105] from the avocado, where it exists together with D-*manno*-heptulose. It has now been shown that the best source of perseitol is the seed of the avocado,[106] but the crystalline heptitol has also been obtained from leaves[107] and wound exudates[108] of the avocado tree. Subsequent reports have described the isolation of perseitol from Pichi (*Fabiana imbricata*) tops,[27b] of volemitol from avocado seeds,[27a] and of both of these heptitols from *Sedum* plants.[108a]

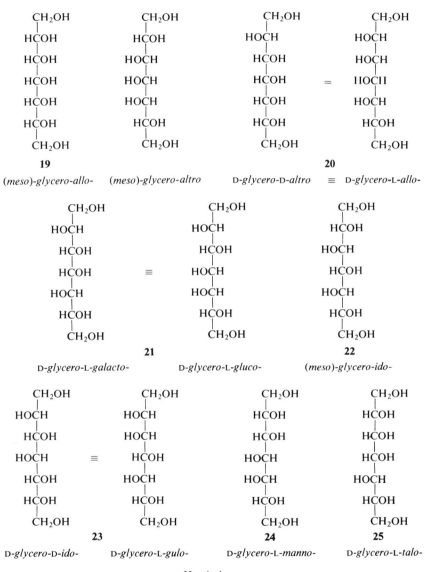

19

(*meso*)-*glycero-allo-*   (*meso*)-*glycero-altro*   D-*glycero*-D-*altro*   ≡   D-*glycero*-L-*allo-*

20

21

D-*glycero*-L-*galacto-*   D-*glycero*-L-*gluco-*   (*meso*)-*glycero-ido-*

22

23

D-*glycero*-D-*ido-*   D-*glycero*-L-*gulo-*   D-*glycero*-L-*manno-*   D-*glycero*-L-*talo-*

24   25

Heptitols

The natural occurrence of a heptulose together with a related heptitol appears to be a general feature,[108b] since sedoheptulose (D-*altro*-heptulose) is accompanied by (1) volemitol in the root stock of *Primula elatior* (L) Hill,[109] and (2) D-*glycero*-D-*gluco*-heptitol ("β-sedoheptitol") (**27**) in the leaves and stem of *Sedum spectabile*.[110] D-*glycero*-D-*gluco*-Heptitol has also been isolated[4] from the roots of *Primula officinalis* Jacq., where it is accompanied by volemitol as well as sedoheptulose. In this context, the illustrated transformations that may be effected amongst the natural heptuloses and heptitols[111] are noteworthy.

The heptitols have been synthesized by reduction of the appropriate aldoses, ketoses, or lactones; references to particular syntheses are given in Table II. The reduction of L-*gluco*-heptulose (**29**) with sodium amalgam is of interest,

TABLE II

HEPTITOLS AND HEPTITOL ACETATES[99]

			Heptaacetate	
Heptitol	M.p. (degrees)	$[\alpha]_D$ (degrees) (in H_2O)	M.p. (degrees)	$[\alpha]_D$ (degrees) (in $CHCl_3$)
(*meso*)-glycero-allo	144.5–146	—		
(*meso*)-glycero-altro[a]	syrup	—	70	—
D-*glycero*-D-altro- ⎫ D-*glycero*-L-allo- ⎭	125–128	−0.3 +53.2[b]		
D-*glycero*-D-galacto- (perseitol)	187–188	−1.04 +4.9[c]	119	−13.4
D-*glycero*-L-galacto- ⎫ D-*glycero*-L-gluco- ⎭	141–142	+2.4	118	+11.4
D-*glycero*-D-gluco- ("β-sedoheptitol")	128–129	−0.75 +4.3[c]		
(*meso*)-glycero-gulo-	127–128	—	113–115	—
(*meso*)-glycero-ido-	110–112	—	175–176	—
D-*glycero*-D-ido- ⎫ D-*glycero*-L-gulo- ⎭	128–129	+1.20	181–182[d]	+25.1[d]
D-*glycero*-D-manno- ⎫ D-*glycero*-D-talo- ⎭	152–153	+2.15	63–64	+36.1
D-*glycero*-L-manno-	187–188	+1.1 −4.35[c]	119	+13.4
D-*glycero*-L-talo-	127–129	+0.95 −4.6[c]		

[a] See reference 99a.   [b] In 5% ammonium molybdate solution.   [c] In saturated borax solution.   [d] Heptabenzoate.

```
CH2OH              CH2OH              CH2OH              CH2OH
 |                  |                  |                  |
C=O                HCOH               HCOH               C=O
 |                  |        bacterial |        reduction |
HOCH      ≡        HOCH     oxidation HOCH     ─────────▶ HOCH
 |                  |       ─────────▶  |                  |
HCOH               HOCH               HOCH               HOCH
 |                  |                  |                  |
HCOH               HCOH               HCOH               HCOH
 |                  |                  |                  |
HOCH               HCOH               HCOH               HCOH
 |                  |                  |                  |
CH2OH              CH2OH              CH2OH              CH2OH

                                        26                D-manno-
L-Perseulose                  D-glycero-D-galacto-        Heptulose
                                    Heptitol
                                    (Perseitol)
```

```
                                                 bacterial  ‖ reduction
                                                 oxidation  ‖

CH2OH              CH2OH              CH2OH              CH2OH
 |                  |                  |                  |
C=O                HOCH               HOCH               HOCH
 |                  |        reduction |                  |
HOCH      ≡        HOCH     ─────────▶ HOCH      ≡       HOCH
 |                  |       bacterial   |                  |
HCOH               HCOH     oxidation  HCOH               HCOH
 |                  |                  |                  |
HCOH               HCOH               HCOH               HCOH
 |                  |                  |                  |
HCOH               C=O                HCOH               HCOH
 |                  |                  |                  |
CH2OH              CH2OH              CH2OH              CH2OH

                  Sedoheptulose      D-glycero-D-talo-Heptitol   28
                                         (Volemitol)      D-glycero-D-manno-
                                                              Heptitol
```

```
CH2OH
 |
HCOH
 |        reduction
HOCH     ─────────▶
 |
HCOH
 |
HCOH
 |
HCOH
 |
CH2OH

  27
D-glycero-D-
gluco-Heptitol
```

since it gave an alcohol "α-glucoheptulitol," which differed from both (*meso*)-*glycero-gulo*-heptitol (**30**), and L-*glycero*-L-*ido*-heptitol (**31**), the expected products of the reduction.[112, 113] An enantiomorphous product was similarly prepared from D-*gluco*-heptulose,[113] whereas catalytic hydrogenation with Raney nickel gave only the expected heptitols.[114] Subsequent investigations[115] suggest that "α-glucoheptulitol" comprises a mixture of **30** and an unidentified, optically active impurity; the problem awaits reexamination.

```
    CH₂OH              CH₂OH
     |                  |
     C=O              HOCH
     |                  |
   HOCH               HOCH
     |                  |
   HCOH               HCOH
     |                  |
   HOCH               HOCH
     |                  |
   HOCH               HOCH
     |                  |
    CH₂OH              CH₂OH
     29                 30
```

L-*gluco*-Heptulose   (*meso*)-*glycero-gulo*-Heptitol

```
    CH₂OH
     |
   HCOH
     |
   HOCH
     |
   HCOH
     |
   HOCH
     |
   HOCH
     |
    CH₂OH
     31
```

L-*glycero*-L-*ido*-Heptitol

The heptitols can be differentiated by gas–liquid chromatography of their heptaacetates.[99a]

E. OCTITOLS

Only one naturally occurring octitol has been reported. This is D-*erythro*-D-*galacto*-octitol (**36**), which was isolated[110] from the avocado (Calavo, Fuerte variety), where it occurs alongside the closely related D-*glycero*-D-*manno*-octulose.[116, 117]

The physical constants of the known octitols and their acetates, together with the methods used for their preparation, are listed in Table III. The configurations of these octitols are shown in the following formulas.

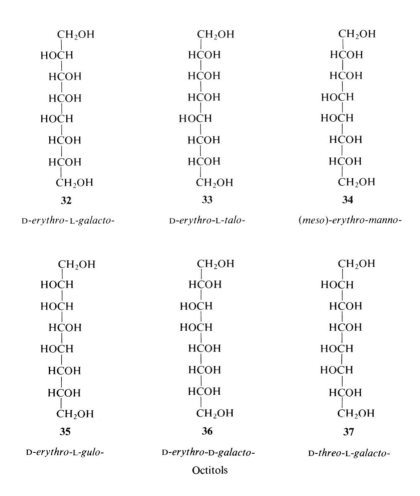

	32	33	34
	D-*erythro*-L-*galacto*-	D-*erythro*-L-*talo*-	(*meso*)-*erythro*-*manno*-

	35	36	37
	D-*erythro*-L-*gulo*-	D-*erythro*-D-*galacto*-	D-*threo*-L-*galacto*-

Octitols

---

* *References start on p. 513.*

TABLE III

OCTITOLS AND OCTITOL ACETATES

Octitol	M.p. (degrees)	$[\alpha]_D$ (degrees) (in H_2O)	Octaacetate			
			M.p. (degrees)	$[\alpha]_D$ (degrees) (in $CHCl_3$)	Method of preparation[a]	References
D-erythro-L-galacto-	153	+2.4	88–89	+20.7	a, b	118, 119
D-erythro-L-talo-	161–162	−0.8	101–102	+17.4	a	118
meso-erythro-manno-	262		166		a	120
D-erythro-L-gulo-	164.5–165	171[b]	110–111	+47	b	119
D-erythro-D-galacto-[c] (monohydrate)	170–171	−12[b] +62[d]	99–100	+1	a	110
D-threo-L-galacto-	230	0.0 −0.5[e]	141	+40.4	a	121

[a] a, By reduction of aldose prepared by the cyanohydrin method; b, by reduction of D-gluco-L-glycero-3-octulose. [b] In 5% ammonium molybdate solution. [c] L-Enantiomorph (monohydrate), m.p. 168–169°, [α]_D −70° (acidified molybdate). [d] In acidified molybdate solution. [e] In saturated borax solution.

F. NONITOLS AND DECITOLS

No naturally occurring member of this group of alditols has yet been reported.

D-*arabino*-L-*galacto*-Nonitol (38) has been synthesized[122] from D-*erythro*-L-*manno*-octose by application of the cyanohydrin method. Reduction with potassium borohydride of D-*erythro*-L-*galacto*-nonulose (isolated from the avocado, or synthesized by application of the 2-nitroethanol method to D-*glycero*-D-*gulo*-heptose or the diazomethane method to sodium D-*erythro*-L-*galacto*-octonate) gave two crystalline nonitols, which were identified[123] as D-*arabino*-D-*manno*-nonitol (39), m.p. 192–193°, and D-*arabino*-D-*gluco*-nonitol (40), m.p. 180–181°, on the basis of the structures of the nonuloses formed when the nonitols were oxidized with *Acetobacter suboxydans*.

The D-*arabino*-D-*manno*-nonitol was shown to be identical with the "α,α,α-D-glucononitol," m.p. 192–194° (m.p. 198°, ref. 124), $[\alpha]_D$ +1.5° (water), obtained by reduction of the most accessible nonose resulting from application of the cyanohydrin synthesis to D-glucose.[124, 125] As predicted by Hudson,[126] the amorphous nonose may therefore be assigned the D-*arabino*-D-*manno* configuration. Extension of the cyanohydrin synthesis to this nonose gave[125] a decitol ("α,α,α,α-D-glucodecitol"), m.p. 222°, $[\alpha]_D$ +1.2° (water); decaacetate, m.p. 149–150°, $[\alpha]_D$ +16° (chloroform). The confirmation of the nonose

CH₂OH	CH₂OH	CH₂OH	CH₂OH
HOCH	HOCH	HCOH	HCOH
HCOH	HOCH	HOCH	HOCH
HCOH	HCOH	HCOH	HOCH
HOCH	HCOH	HCOH	HCOH
HOCH	HOCH	HOCH	HCOH
HCOH	HCOH	HCOH	HOCH
HCOH	HCOH	HCOH	HCOH
CH₂OH	CH₂OH	CH₂OH	HCOH
			CH₂OH
38	**39**	**40**	**41**
D-*arabino*-L-galacto-Nonitol	D-*arabino*-D-manno-Nonitol	D-*arabino*-D-gluco-Nonitol	D-*gluco*-D-galacto-Decitol

structure lends weight to Hudson's further prediction[126] that the decitol is D-*gluco*-D-*galacto*-decitol (41).

* *References start on p. 513.*

An unidentified dodecitol has been isolated as a by-product during the alkaline electroreduction of D-glucose; it apparently resulted from a reductive coupling of two hexose molecules.[127]

III. SYNTHESIS

A. MISCELLANEOUS METHODS

Alditols are conveniently prepared by reduction of the appropriate aldoses or ketoses; an aldose gives one product, whereas a ketose gives two products that are epimeric at C-2. In the laboratory, the reduction of glycoses was originally effected with sodium amalgam, but these reductions are now effected more conveniently (1) by catalytic hydrogenation[128] or (2) with sodium (or potassium) borohydride.[129] The commercially preferred methods of reduction are pressure hydrogenation and electrolytic reduction (see under D-glucitol). Sufficient hydrogen is adsorbed on Raney nickel to effect direct reduction of glycoses without use of added hydrogen.[130] Alditols can be prepared by direct reduction of aldonolactones with sodium (or lithium) borohydride or with lithium aluminum hydride,[129, 131] and the latter reagent also reduces aldonic acids to alditols. Thiol esters of aldonic acids are converted into alditols with Raney nickel.[132] Preparations of alditols-*1-t* by the reduction of aldoses and aldonolactones with lithium borohydride-*t* have been described; similar treatment of ketoses gives epimeric pairs of 2-labeled alditols[133, 134] (see Chapter 3).

The cyanohydrin synthesis was the first method to be used for the preparation of higher-carbon sugars, and the corresponding alditols may then be prepared by direct reduction of the aldonolactone intermediates of these syntheses or by reduction of the aldoses. Other methods that may be used to prepare higher-carbon sugars are based on the reaction of aldonyl chlorides with diazomethane or of aldoses with nitroalkanes. The use of these methods in synthesis of higher-carbon sugars has been reviewed,[99] and the methods are discussed, together with other methods of increasing the length of the carbon chain, in Chapter 3.

Aminoalditols, derived from nitroalditols and glycosylamines (Chapter 20), may be converted into alditols by treatment with nitrous acid, but this reaction may be accompanied by the formation of anhydro rings (Chapter 13).

Methods are available for the synthesis of various deoxy alditols having three to six carbon atoms, by using acetylene as the starting material.[135] Typical intermediates in such syntheses are 1-propyn-3-ol and 1-butyn-3-ol obtained by the reaction of formaldehyde or acetaldehyde with acetylene in the presence of copper acetylide catalyst. Other methods of total synthesis of alditols are

$$HC{\equiv}CH \xrightarrow{\text{HCHO}} HOH_2C{-}C{\equiv}CH \xrightarrow{\text{HCHO}} HOH_2C{-}C{\equiv}C{-}CH_2OH$$

discussed in the sections devoted to tetritols, pentitols, and hexitols. These total syntheses are of interest because of their departure from the methods of classical carbohydrate chemistry.

B. CATALYTIC ISOMERIZATION OF ALDITOLS

Catalytic isomerization has been used[47] to provide a synthetic route to certain hexitols. When concentrated, aqueous solutions of either D-glucitol or an equimolar mixture of D-mannitol and L-iditol at pH 8.8 are treated with hydrogen under pressure at 170°, in the presence of a nickel–Kieselguhr catalyst, a quasi-equilibrium, 2 D-glucitol ⇌ D-mannitol + L-iditol, is established. The approximate equilibrium concentrations of the hexitols are 41.4 ± 2.5 wt. % D-glucitol, 31.5 ± 2.4 wt. % D-mannitol, and 26.5 ± 2.3 wt. % L-iditol.

The products isolated in reasonable purity from the catalytic isomerizations of D-glucitol, D-mannitol, L-iditol, and galactitol are summarized in Table IV. The hexitols were obtained from the product mixtures either by fractional crystallization or as the hexaacetates. It should be stressed that Table IV indicates the percentage of products actually isolated and not their "equilibrium" concentrations.

The isomerizations are effected under slightly alkaline conditions, but are inhibited in slightly acid solutions, indicating that the reaction proceeds by the formation of a small, but catalytically significant, amount of the related aldohexose. With D-glucitol, D-mannitol, and L-iditol, isomerization was principally confined to the ends of the molecule. The scheme below illustrates the mechanism proposed[47] for the isomerization at one end of the D-glucitol molecule. The initial dehydrogenation step may also occur at C-6 of D-glucitol to yield a mixture of L-gulose, L-idose, and L-sorbose, but there appeared to be a slight preference for this step to take place at C-1. The products arise by hydrogenation of the equilibrium mixture of hexoses. From this mechanism, it may be predicted that the primary products of isomerization of the above hexitols will have the configuration at C-2 and C-5 inverted. A small amount of secondary isomerization of primary products occurs, and, with galactitol, 2,3-enediols appear to be significant intermediates of isomerization.

Hydrogenolysis of cotton-meal xylitol with nickel–Kieselguhr has been reported[136] to give erythritol (1 to 5%), propane-1,2-diol (22%), and ethylene glycol (26%).

* *References start on p. 513.*

TABLE IV

PRODUCTS FROM ISOMERIZATION OF HEXITOLS

Hexitol	Percent of product isolated from isomerization						
	Allitol	Galactitol	D-Glucitol	L-Iditol	D-Mannitol	DL-Altritol
D-Glucitol	—	2.0	27.0	3.0	22.4	4.0
D-Mannitol	—	1.0	50.0	11.0	11.5	2.0
L-Iditol	—	—	34.0	35.0	—	—
Galactitol	2.4	25.2	10.3[a]	—	0.4[a]	10.2

[a] Product is DL-racemate.

CH₂OH ... (structures)

```
   CH₂OH              CHO                CHOH               CH₂OH
    |                  ‖                  ‖                  |
   HCOH               HCOH               COH                C=O
    |         -H₂       |       alkali     |                  |
   HOCH     ⇌          HOCH    ⟶          HOCH    ⇌          HOCH
    |         +H₂       |                  |                  |
   HCOH               HCOH               HCOH               HCOH
    |                  |                  |                  |
   HCOH               HCOH               HCOH               HCOH
    |                  |                  |                  |
   CH₂OH              CH₂OH              CH₂OH              CH₂OH
  D-Glucitol         D-Glucose        1,2-Enediol         D-Fructose
```

```
   CH₂OH              CHO
    |                  |
   HOCH               HOCH
    |                  |
   HOCH      +H₂       HOCH
    |        ⟶          |
   HCOH      -H₂       HCOH
    |                  |
   HCOH               HCOH
    |                  |
   CH₂OH              CH₂OH
  D-Mannitol        D-Mannose
```

IV. PROOFS OF STRUCTURE AND CONFIGURATION

The structures and configurations of the stereoisomeric alditols have generally followed from the proof of structure and configuration of the parent sugar, determined as described in Chapter 1. In certain cases, however, the configuration of the parent sugar and that of the derived alditol have been established simultaneously by conversion of the sugar into the alditol. This approach is of particular value when the alditol, resulting from the reduction of a sugar of partially known configuration, is found to be optically inactive (*meso* structure).

Fischer[124] used this method in establishing the structure of the aldoses obtained by means of the cyanohydrin synthesis, and, since this synthesis produces derivatives that are epimeric at C-2, the configuration of both epimers may be deduced. As an example, the determination of the structures of the heptoses and heptitols derived from D-allose may be considered. Application of the cyanohydrin method to D-allose gives two heptoses,[137] and the alditol obtained by reduction of one of these was found to be optically inactive and therefore has a *meso* configuration. Of the two possible structures, only

* References start on p. 513.

glycero-allo-heptitol (**19**) has a *meso* configuration and must therefore be derived from D-*glycero*-D-*allo*-heptose. The alditol derived from the second aldose must be epimeric with **19** at C-2 and is therefore D-*glycero*-D-*altro*-heptitol (**20**).

Because of their ease of crystallization in most cases, the alditols are superior to the aldaric acids for this type of structural proof. However, the low values of their specific rotations sometimes cause difficulties that may lead to an erroneous assignment of a *meso* configuration to a substance that is optically active. For example, "β-sedoheptitol" obtained by reduction of sedoheptulose was first thought to be optically inactive,[138] but the assignment[139] of the D-*altro*-heptulose structure to sedoheptulose meant that the derived "β-sedoheptitol" could not have the *meso* structure that had been assigned as a result of the apparent lack of optical activity. A careful reexamination[140] confirmed this view and showed that "β-sedoheptitol" is D-*glycero*-D-*gluco*-heptitol (**27**), having $[\alpha]_D$ −0.75° (water), +4.3° (borax).

In the presence of saturated borax, alditols show enhanced optical activity, although in some cases the enhancement is insufficiently large to give a dependable value. For example, D-*threo*-L-*galacto*-octitol (**37**) has $[\alpha]_D$ 0.0° (water) and −0.5° (borax).[121] Ammonium molybdate solution may also be used to increase the specific rotation of alditols. The values obtained over a wide range of concentrations are nearly constant when acidified molybdate[26,141,142] is used; some values[26,141] for D-mannitol are given in Table V.

TABLE V

SPECIFIC ROTATIONS OF D-MANNITOL IN AMMONIUM MOLYBDATE SOLUTION[a]

D-*Mannitol* (mmoles)	MoO_3 (mmoles)	$Ratio \dfrac{MoO_3}{\text{D-mannitol}}$	$[\alpha]^{20}_{5461}$ (degrees)
0.747	9.6	12.85	+168.8
0.861	4.8	5.57	+169.1
1.689	4.8	2.84	+168.8
2.298	4.8	2.10	+167.8
2.479	4.8	1.94	+164.4
2.716	4.8	1.76	+148.8

[a] Weighed amounts of D-mannitol were added to a solution of 5 ml of 0.5M H_2SO_4 and 5 or 10 ml of 0.1M ammonium paramolybdate and diluted to 50 ml.

These values are very large for an alditol such as D-mannitol, whose $[\alpha]_D$ value in water is only −2.1°. Similar results have been obtained with other alditols.

An alternative approach is to prepare the alditol peracetates, which have pronounced rotations in chloroform when the configurations are not *meso*.[126]

A related method for establishing the configuration of a synthetic alditol may be applied when the same alditol is obtained from two different aldoses. Thus, application of the cyanohydrin synthesis to D-mannose and L-galactose yields four heptoses, which, on reduction, give three different alditols, one of which (D-*glycero*-D-*galacto*-heptitol, **26**) is produced from both heptoses. Hudson[126] has discussed in detail the way in which this observation enables the configurations of the four heptoses and three heptitols to be deduced. Similar methods and arguments have been applied to certain of the octitols and nonitols.[122]

The configurations of two nonitols were established[123] by identification of the nonuloses formed by oxidation of the nonitols with *Acetobacter suboxydans*.

The conformations of the stereoisomeric pentitols and hexitols in the solid state have been determined[142a] by X-ray crystallography; it was established that the extended-chain conformation is the favored form only when this conformation does not give rise to a 1,3-interaction between two oxygen atoms (see Chapter 5 for further details).

V. REACTIONS

A. ESTERIFICATION

The esterification methods used for sugars (Chapter 6) are applicable to the alditols. However, fully esterified derivatives are generally unobtainable by direct esterification with organic acids because of the formation of internal anhydrides (Chapter 13). Partial esters, especially diesters, may be obtained by the use of an amount of acid chloride that is insufficient for complete esterification.

Various inorganic esters, including nitric, sulfuric, and phosphoric esters and halohydrins, were discussed in a previous edition of this book,[143] and reference may also be made to Chapter 8.

Of particular interest are the complexes formed by the alditols with various inorganic polybasic acids such as boric, molybdic, and tungstic acids. It is believed that these complexes are true esters formed with one or more molecular proportions of the alditol. The complexes are usually known only in solution and have been observed through such effects as increased conductivity and acidity of solution, increased rotation of optically active alditols (see Table V), and marked changes in volume. Two crystalline monoborates of D-mannitol are known,[144,145] and crystalline sodium borate complexes of acyclic 1,3-diols have also been described.[146] Treatment of D-mannitol with metaboric acid in anhydrous acetone afforded[147] a crystalline dimetaboric

* *References start on p. 513.*

ester that gave D-mannitol 1,6-dibenzoate on benzoylation and hydrolysis of the boric ester groupings. Crystalline esters of alditols completely esterified with benzeneboronic acid have been reported.[148] Cyclic butaneboronic esters of alditols are useful, volatile derivatives that can be used for gas–liquid chromatographic resolution of alditols, and the gross structures of the product esters can be determined by mass spectrometry.[148a]

Although the observation that alditols form complexes with inorganic acids is by no means new, the application of this property to provide structural information is of more recent origin and arose from the development of paper electrophoresis.

Paper Electrophoresis.—Basically, the migration of a polyhydroxy compound in an electric field depends on (1) the ionization of hydroxyl groups, as occurs in solutions in aqueous sodium hydroxide, or (2) the formation of anionic or cationic complexes with an inorganic complexing agent. In the latter instances, the structure of the inorganic complexing agent and the polyhydroxy compound, will, together, determine the stability of the complex formed, and this may be correlated with the electrophoretic mobility. The paper electrophoresis of carbohydrates has been comprehensively reviewed.[149, 150]

Paper electrophoresis of polyhydroxy compounds is also discussed in Vol. IB, Chap. 28, and the examples given below are confined to those complexing reagents that have been studied in some detail. The data in Tables VI and VII demonstrate the value of paper electrophoresis (1) as a means of rapid identification and (2) in the structural investigation of alditols and their derivatives.

Böeseken's observation[151] on the conductivity of certain polyhydroxy compounds in borate solution was the forerunner of paper electrophoresis of neutral carbohydrates in this electrolyte.[149] The esters and complexes between boric acid and borate ions, respectively, and polyhydroxy compounds can be formulated as follows:

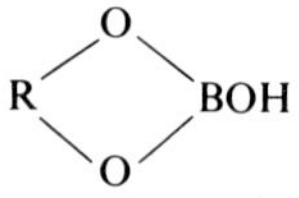

42

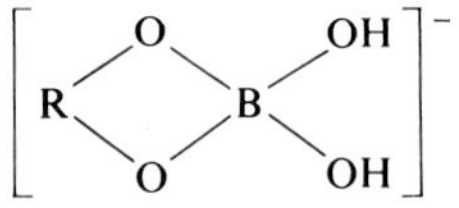

43

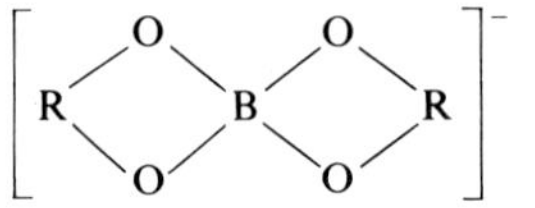

44

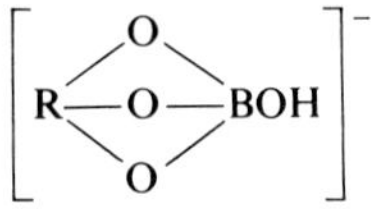

45

The ionic species **43**, **44**, and **45** migrate during electrophoresis. It has been suggested[150] that the borate ion is able to complex with those compounds in which the oxygen atoms of at least two hydroxyl groups are separated by, or can approach each other to, a distance of approximately 2.4 Å. In the planar, zigzag conformations of acyclic *erythro*- and *threo*-1,2-diols, the O–O distance is too great for complexing. However, in the eclipsed conformations, the O–O distance is suitable for complexing, and the greater electrophoretic mobility of the acyclic *threo*-1,2-diol, as compared with that of the *erythro*-diol, has been attributed to differences of nonbonded interactions in the essentially planar complex.[149, 152] Similar conformational principles have been applied to the complexes of 1,3-diols.[146]

Oxidation of the D-glucitol–borate complex with a limited amount of periodate afforded[153] L-xylose as a major product (compare p. 509) and indicated that the hydroxyl groups on C-2 and C-4 are involved in complex formation. This technique has been applied in other cases to establish the sites of the hydroxyl groups which participate in complex formation.

Lindberg and Swan[154] have investigated the electrophoresis of polyhydroxy compounds in sodium germanate solution of pH 10.7, where anionic complexes of structure **46** or **47** are probably formed. It has been calculated that complexes will form in germanate if the oxygen atoms can approach to a distance of approximately 2.64 Å.

The results of electrophoresis in tungstate[155, 156] (pH 5) and molybdate[156, 157] (pH 5) solutions indicate that the ditungstate ion, $W_2O_7^{2-}$, and the dimolybdate ion, $Mo_2O_7^{2-}$, are the complexing species. A close parallel has

46   **47**

been established between changes in the optical rotations of substituted D-glucitols in molybdate and tungstate and their electrophoretic behavior in these electrolytes. The following observations have been made: (1) 3-Substituted D-glucitols do not form a complex; (2) 2-substituted D-glucitols and 1-substituted L-gulitols form complexes containing the substituted hexitol and metal atom in the ratio of 1:2; and (3) 3-substituted L-gulitols form complexes in which this ratio is 1:1, but in this case a much weaker complex is formed with tungstate than with molybdate. It is significant that no acyclic compound containing fewer than four hydroxyl groups migrates during electrophoresis

* *References start on p. 513.*

in either electrolyte, and the mobility is diminished or lost if the hydroxyl groups are not located on contiguous carbon atoms.

Oxidation with periodate of several complexes which contained an alditol and the molybdenum (tungsten) atom in the ratio of 1:2 have demonstrated that a vicinal tetritol system is involved in complex formation. Structures **48** and **49** are postulated[155] for such complexes in which the hydroxyl groups on C-2 and C-3 are in *erythro* and *threo* arrangement, respectively. Complexes are formed with the hydroxyl groups on C-1 to C-4 of allitol and D-mannitol and with those on C-2 to C-5 of D-altritol and galactitol. In these arrangements, the interaction between R_2 and R_3 of the complexes formed is the smallest.

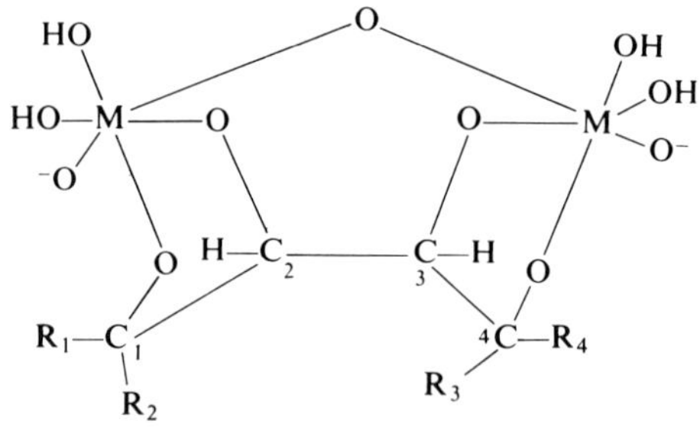

48

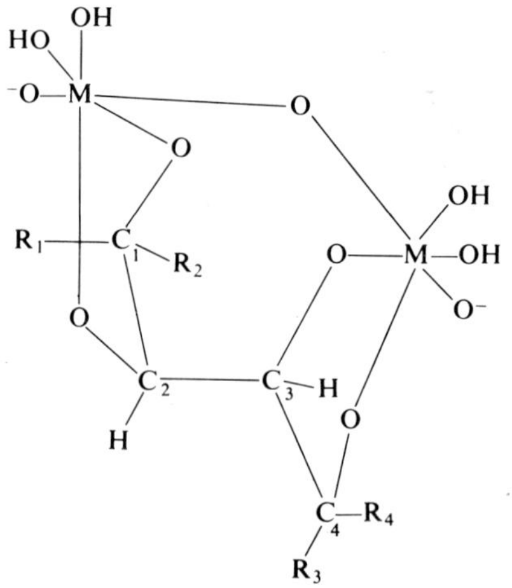

49

Electrophoretic mobility of polyhydroxy compounds in sodium hydroxide solution is almost certainly due to ionization of a hydroxyl group, rather than complexing.[158] The behavior of pentitols and hexitols in sodium hydroxide (see Table VII) shows a good correlation between the mobility and the number of *erythro*-1,2-diol systems.

Most alditols form complexes to a greater or lesser extent with copper(II) ions,[158a] and thus migrate at different rates toward the cathode on electrophoresis in a solution of copper(II) acetate or basic copper(II) acetate. Aldoses show negligible mobility in this system. The resolution of mixtures of alditols can be achieved on a column of cation-exchange resin in the Cu^{2+} form.[158a]

TABLE VI

REFERENCE COMPOUNDS

Electrolyte (E)	Reference compound (R)	Mobility $(10^5\ cm^2 V^{-1}\ sec^{-1})$	Nonmigrating marker[a]
Borate (B)	D-glucose (G)	14.8 (pH 9.2)	a, b
		12.2 (pH 10)	
Sulfonated benzeneboronic acid (PhB)	D-mannitol (M)	9.4 (pH 6.5)	c
Germanate (Ge)	D-glucose (G)	6.2–8.1 (pH 10.7; 40°)	a
Stannate (Sn)	D-glucitol (S)	14.3 (pH 11.5)	c
Arsenite (As)	D-ribose (Ri)	5.9 (pH 9.6)	a, b
Molybdate (Mo)	D-glucitol (S)	17.0 (pH 5)	c, d
Tungstate (W)	D-glucitol (S)	17.0 (pH 5)	c, d
Sodium hydroxide (Na)	D-ribose (Ri)	9.6 (0.1M)	b, e
Basic lead acetate (Pb)	D-ribose (Ri)	8.1 (pH 6.8)	a, b

[a] Nonmigrating markers: a, 2,3,4,6-tetra-O-methyl-D-glucose; b, caffeine; c, 5-(hydroxymethyl)-2-furaldehyde; d, glycerol; e, 1,4-dideoxy-L-threitol (L-*threo*-2,3-butanediol).

Mills[159] has provided evidence for the existence in dilute aqueous solutions of complexes of neutral polyhydroxy compounds with cations of the alkali and alkaline earth metals. The migration of an acyclic polyhydroxy compound in these solutions can be related to the number of *threo*-1,2-diol groupings in the molecule. A similar relationship holds for the anionic complexes formed in stannate[150] and arsenite[158] solutions and for the cationic complex presumed to be present in basic lead acetate solution.[158] Little is known of the nature of the complexes formed in these cases. Some alditols form crystalline esters with arsenious oxide.[160]

* References start on p. 513.

TABLE VII

Relative Mobilities ($10^2 \times M_R$)[a] of Alditols and Some Derivatives[150]

Compound	B	PhB	Ge	Sn	Pb	As	Mo	W	Na
1,4-Dideoxyerythritol (cis-2,3-butanediol)	13	0		2	0	6	0	0	0
1,4-Dideoxythreitol (trans-2,3-butanediol)	51	30		4	0	33	0	0	0
Glycerol	49	0	40	23	3	24	0	0	0
Erythritol	75	10	100	57	3	53	100	90	3
L-Threitol	75	30		62	11	96	50	24	3
D-Arabinitol	87			95	14	124	110	104	7
1-deoxy-				58			103	109	
L-Arabinitol		60	180						
D-Lyxitol, 1-deoxy-				78			95	65	
L-Lyxitol, 2-O-methyl-		50	50						
Ribitol	85	30	120	72	4	76	110	103	10
Xylitol	79	90	170	100	25	155	110	104	3
1-deoxy-D-				88			96	82	
Allitol	90			88	9	92	94	97	23
D-Altritol	89			95	17	138	99	97	16
1-deoxy-				80			98	100	
Galactitol	97	100	210	99	32	145	100	100	8
1-deoxy-L-				87			100	103	
D-Glucitol	83	130	190	100	47	161	100	100	11
1-deoxy-		90		89			98	98	
2-O-methyl-		120	150						
3-O-methyl-		10	80	30		36	0	0	
L-Gulitol									
1-deoxy-				94			94	98	
1-O-methyl-		60					93	89	
3-O-methyl-		130	140	85			47	9	
L-Iditol	81				57	173			7
D-Mannitol	91	100	190	93	23	130	100	100	12
1-deoxy-				94			100	100	
1,2-di-O-methyl-				66			100	95	
2-O-methyl-				88			100	98	
D-Talitol, 1-deoxy-							104	94	
Heptitols									
(meso)-glycero-allo-	95				11	100			54
D-glycero-D-altro-	92				27	144			44
D-glycero-D-galacto-	98				51	140			11
D-glycero-D-gluco-	88				53	171			30
D-glycero-L-gluco-	95				59	176			17
(meso)-glycero-gulo-	85				72	160			27
D-glycero-D-ido-	85				71	168			20
(meso)-glycero-ido-	78				79	182			17
D-glycero-D-talo-	93				34	140			24

[a] M_R = true distance of migration of substance/true distance of migration of reference compound.

B. PERIODATE OXIDATION

Acyclic *threo*-diols are usually oxidized[161] more rapidly than the *erythro*-diols, and relative configurations have been assigned[162] on this basis. With a limited quantity of periodate, D-mannitol, galactitol, and D-glucitol are likewise oxidized preferentially at a *threo*-diol group;[163] D-glucitol, for example, yields glyceraldehyde and D-erythrose. Similar oxidation of erythritol occurs more readily at a terminal —CHOH·CH$_2$OH group than at a —CHOH·CHOH— group, but the reverse has been found for hexitols.[164]

A detailed investigation[153] has revealed the sequence of periodate oxidation of all the diol pairs in D-glucitol as 3,4(α-*threo*) > 2,3(α-*threo*) > 4,5(α-*erythro*) > 5,6(α) > 1,2(α). Oxidation with periodate is thought to involve formation of a five-membered cyclic ester intermediate.[161] By considering the nonbonded interactions in the intermediates for the three diol systems, with the inclusion of an entropy term for the α-diol, a rationalization[153] was made for the observed susceptibility to oxidation.

C. OXIDATION

Strong oxidizing agents, such as nitric acid, convert alditols into aldaric acids (see Chapter 23). Identification of galactitol has been achieved by this procedure, which gives the very insoluble galactaric (mucic) acid; the same product is given by the galactoses and galacturonic acids.

Milder agents oxidize alditols to aldoses or ketoses. For example, aldoses are formed by treatment of alditols with hydrogen peroxide in the presence of ferrous salts,[165] whereas mixtures of aldoses and ketoses are formed with halogens[166] and by electrolytic oxidation at platinum electrodes in the absence of an electrolyte.[167] Oxidation at carbon electrodes with sodium bromide as electrolyte gives principally ketoses.[168] Formation of ketoses may also be effected with mercuric acetate, which, with ribitol, gave a mixture of *erythro*-3-pentulose and DL-*erythro*-pentulose.[169]

A method that has recently been re-investigated[170] is oxidation in the presence of a platinum-on-carbon catalyst. With this system, alditols give a mixture of products from which it is sometimes possible to isolate a particular sugar. For example, L-gulose has been isolated in 20% yield by the oxidation of D-glucitol.[171] The oxidative process may continue to give acidic products.

By using protected alditol derivatives, it is possible to achieve selective oxidation of a particular hydroxyl group and, if necessary, to make use of more powerful reagents. For example, 4-O-benzoyl-1,2:5,6-di-O-isopropylidene-D-*arabino*-3-hexulose has been prepared[96] by oxidation of 3-O-benzoyl-1,2:5,6-di-O-isopropylidene-D-mannitol with chromic acid–pyridine. Similar oxidations have also been effected by use of the Pfitzner–Moffatt reagent[97] or methyl sulfoxide–phosphorus pentaoxide.[172]

Another important method for the synthesis of ketoses involves the oxidation of alditols by certain bacteria, the most useful of which is *Acetobacter suboxydans*. This bacterium effects oxidation of a secondary hydroxyl group only when it is adjacent to a primary hydroxyl group and also forms part of an *erythro*-diol group.[173, 174] In applying this specificity rule to terminal, deoxy derivatives of alditols, the methyl group is considered as a hydrogen atom, but the specificity is less marked. For example, 6-deoxy-L-galactitol is reported to give 6-deoxy-L-*lyxo*-hexulose in addition to the expected 1-deoxy-D-*xylo*-3-hexulose.[175] Oxidation of the appropriate alditols with *Acetobacter suboxydans* has been used to prepare L-sorbose (for use in the synthesis of vitamin C),[176] L-*galacto*-heptulose,[174] L-*glycero*-L-*gluco*-octulose,[173] D-*glycero*-L-*gluco*-octulose,[176a] and D-*erythro*-L-*gluco*-nonulose.[123]

For further information on these methods for the synthesis of ketoses, see Chapter 3.

D. Reduction

The reduction of alditols to secondary alkyl halides by treatment with hydriodic acid is mainly of historical interest.[177] Under catalytic conditions of reduction, alditols may be degraded to lower alcohols. Thus, D-glucitol gives, among other products, glycerol and propane-1,2-diol,[178] and xylitol gives a small proportion of erythritol as well as ethylene glycol and propane-1,2-diol.[136] As discussed previously, isomerization of alditols may result from high-pressure hydrogenation.

An alternative mode of reduction involves conversion of alditols into deoxy derivatives, and these are discussed below.

E. Etherification and Acetalation

The procedures for preparing alditol ethers are the same as those described for other carbohydrates in Chapter 12. Alditols readily form anhydro derivatives and cyclic acetals, and these are discussed in Chapters 13 and 11, respectively.

VI. QUALITATIVE AND QUANTITATIVE DETERMINATION

The alditols are best separated from each other, or from closely related compounds, by means of the various chromatographic techniques. The paper-chromatographic[179] and paper-electrophoretic[150] behaviors of the alditols have been reviewed (see also foregoing Section VA), and fractionation may be achieved by ion-exchange chromatography of the free alditols[180] or their borate complexes.[181] Alditols are readily separated by gas–liquid chromatography of their peracetates[99a, 182] or trimethylsilyl ethers,[183] and these techniques have the advantage over other analytical methods that precise, quantitative data may be obtained.

Periodate oxidation may be of use for the quantitative determination of alditols. By using macro or semimicro techniques (see Vol. IB, Chap. 25), it is possible to determine both the amounts of oxidant consumed and the amounts of formaldehyde and formic acid produced. The ratios of these figures may provide information on the structure of the alditol, and, if the structure is known, any one of these values may be used for the quantitative determination of the alditol present. In the oxidation of partially substituted alditols, it is desirable that the above analyses should be confirmed by isolation and identification of the fragments that are resistant to oxidation.

Particular alditols have been identified by specific reactions. For example, galactitol yields the insoluble galactaric (mucic) acid, and D-glucitol forms a relatively insoluble complex with pyridine.[73] It appears that the only other alditol that forms an analogous complex is 2-deoxy-D-*arabino*-hexitol.[184] The pyridine complex has been used to isolate D-glucitol from complex mixtures,[73] and the complex decomposes at ordinary humidities to give crystalline D-glucitol.

VII. MONODEOXY DERIVATIVES

In general, the deoxyalditols are known only as synthetic products, although 1-deoxy-D-*glycero*-D-*talo*-heptitol (1-deoxyvolemitol) has been isolated from the lichen, *Siphyla ceratites*.[185]

When the appropriate monodeoxyaldose or monodeoxyaldonic acid is known, the alditol may be readily prepared by the methods outlined in Section II. The availability of monodeoxyaldoses has been reviewed.[186]

Direct syntheses of 1(or ω)-deoxyalditols may be effected by reductive desulfurization of aldose dithioacetals with Raney nickel[187] or by catalytic reduction of a terminal deoxyiodo derivative.[188] The latter derivatives are prepared from terminal sulfonates by treatment with sodium iodide. Direct reduction of primary sulfonates with lithium aluminum hydride also yields 1(or ω)-deoxyalditols,[189] although in certain cases formation of the terminal alcohol may be a competing reaction.[190] It has been noted that use of ethane-sulfonates may minimize the extent of the latter reaction.[190]

* *References start on p. 513.*

Another method used to prepare 1-deoxyalditols is the reduction with sodium borohydride of the appropriate aldose *p*-tolylsulfonylhydrazone.[191] A similar reduction of D-*erythro*-pentulose *p*-tolylsulfonylhydrazone gives 2-deoxy-D-*erythro*-pentitol.[192]

Terminal deoxyalditols have also been prepared by irradiation with ultraviolet light of 1-*S*-ethyl-1-thioalditols,[192a] aldose dithioacetals,[192a] and 6-deoxy-6-iodohexose derivatives.[192b]

Compilations of physical constants for some of the monodeoxyalditols have been published.[193]

VIII. BIOCHEMISTRY

The widespread occurrence of the alditols in Nature, particularly in lower forms of life, points to their physiological importance; the biochemistry of the alditols has been authoritatively reviewed.[194] A number of physiological functions can be attributed to, or inferred for, the alditols, namely: (1) structural, as in the ribitol phosphate polymers in bacterial cell walls; (2) energy storage, as in plants containing large amounts of mannitol polymers; (3) coenzyme regulation, as in bacteria that excrete large amounts of alditols formed by reduction of sugars in culture media; and (4) sugar interconversion. In the latter instance, it has shown that the conversions of D-glucitol and D-fructose into liver glycogen follow[195] closely similar routes, and that D-glucitol is transformed directly into D-fructose in an enzymic reaction[196] catalyzed by nicotinamide adenine dinucleotide.

It is now generally recognized that D-glucitol is a normal metabolite in mammals, namely as a precursor of D-fructose. Hers[197] has suggested that the biosynthesis of D-fructose in seminal plasma follows the route D-glucose → D-glucitol → D-fructose rather than the route D-glucose → D-glucose 6-phosphate → D-fructose 6-phosphate → D-fructose, suggested[198] earlier.

The pentitols are of rather more limited occurrence compared to the hexitols and, until comparatively recently, did not appear to have any physiological importance. This situation has altered since the discovery of xylitol as a metabolic intermediate in mammalian liver,[199] DL-arabinitol in human urine,[30, 194] and ribitol as a constituent of bacterial cell walls.[40]

Of the tetritols, erythritol has been isolated from the urine of fasting males[200] and from the placenta and other fetal tissues of cattle,[201] where it acts as a growth stimulant for *Brucella abortus*, an organism that grows preferentially in fetal tissues; threitol and other alditols are inactive in this respect. L-Threitol, labeled in the terminal carbon atoms, yielded $^{14}CO_2$ and D-glucose-^{14}C in rat-liver slices, and it is probable that the conversion occurred by way of L-*glycero*-tetrulose.[202] Certain terminal disulfonic esters of the alditols, such as 1,6-di-O-methylsulfonyl-L-threitol,[203] possess some antitumor activity, but it has been pointed out[204] that conversion of the disulfonates into diepoxides may take place under physiological conditions.

REFERENCES

1. For general reviews on alditols, see: D. H. Lewis and D. C. Smith, *New Phytologist*, **66**, 143, 185 (1967); V. Plouvier, *Bull. Soc. Chim. Biol.*, **45**, 1079 (1963); F. R. Benson, "Polyhydric Alcohols," in Kirk-Othmer Encyclopedia of Chemical Technology, Vol. I (A to Aluminum), Interscience Publishers, 1962, pp. 569–588.

1a. Although ethylene glycol and glycerol (diol and triol, respectively) may be classified as alditols, they are not considered here, since they have been adequately reviewed elsewhere (see G. O. Curme, Jr., and F. Johnston, *Amer. Chem. Soc. Monogr.* No. 114, 1952; C. S. Miner and N. N. Dalton, *Amer. Chem. Soc. Monogr.* No. 117, 1953; I. Mellan, "Polyhydric Alcohols," Spartan Books, Washington D.C., 1962).

2. L. Maquenne and G. Bertrand, *Compt. Rend.*, **132**, 1565 (1901).

3. R. A. Raphael, *J. Chem. Soc.*, 401 (1952).

4. R. Begbie and N. K. Richtmyer, *Carbohyd. Res.*, **2**, 272 (1966).

5. M. Bamberger and A. Landsiedl, *Monatsh. Chem.*, **21**, 571 (1900); J. Tischer, *Hoppe-Seyler's Z. Physiol. Chem.*, **243**, 103 (1936).

6. O. Hesse, *Ann.*, **117**, 297 (1861); *J. Prakt. Chem.*, **92** [2], 425 (1915); A. Goris and P. Ronceray, *Chem. Zentr.*, **78**, I, 111 (1907).

7. A. W. Hofmann, *Ber.*, **7**, 508 (1874).

8. G. J. Hajny, J. H. Smith, and J. C. Garver, *Appl. Microbiol.*, **12**, 240 (1964).

9. H. Onishi and T. Saito, *Agr. Biol. Chem.* (*Tokyo*), **26**, 804 (1962).

10. R. Lespieau, *Bull. Soc. Chim. Fr.*, **1** [4], 1112 (1907).

11. P. W. Kent, K. R. Wood, and V. A. Welch, *J. Chem. Soc.*, 2493 (1964).

12. F. H. Otey, J. W. Sloan, C. A. Wilham, and C. L. Mehltretter, *Ind. Eng. Chem.*, **53**, 267 (1961); F. H. Otey, C. A. Wilham, and C. L. Mehltretter, U.S. Patent No. 3,046,312, (1962); *Chem. Abstr.*, **58**, 1348 (1963).

13. R. Barker and D. L. McDonald, *J. Amer. Chem. Soc.*, **82**, 2301 (1960); S. A. Barker, A. B. Foster, A. H. Haines, J. Lehmann, J. M. Webber, and G. Zweifel, *J. Chem. Soc.*, 4161 (1963).

14. N. Baggett, K. W. Buck, A. B. Foster, B. H. Rees, and J. M. Webber, *J. Chem Soc.*, (C), 212 (1966).

15. L. Maquenne, *Compt. Rend.*, **130**, 1402 (1900).

16. K. Kratzl, H. Silbernagel, and K. H. Bässler, *Naturwissenschaften*, **50**, 154 (1963); *Monatsh. Chem.* **94**, 106 (1963).

17. E. A. McComb and V. V. Rendig, *Arch. Biochem. Biophys.*, **103**, 84 (1963).

18. G. Bertrand, *Compt. Rend.*, **130**, 1472 (1900).

19. A. B. Foster, A. H. Olavesen, and J. M. Webber, *J. Chem. Soc.*, 5095 (1961); A. S. Perlin, *Methods Carbohyd. Chem.*, **1**, 68 (1962).

20. J. L. Bose, A. B. Foster, and R. W. Stephens, *J. Chem. Soc.*, 3314 (1959).

21. H. Pariselle, *Compt. Rend.*, **150**, 1343 (1910).

22. See, for example, K. Abildgaard, Belg. Patent No. 622,353 (1962); *Chem. Abtsr.*, **59**, 9798 (1963).

23. O. Ruff, *Ber.*, **31**, 1576 (1898); *ibid.*, **32**, 554 (1899).

24. B. Lindberg, A. Misiorny, and C. A. Wachtmeister, *Acta Chem. Scand.*, **7**, 591 (1953).
25. G. P. Bruner, G. E. Gream, and N. V. Riggs, *Aust. J. Chem.*, **13**, 277 (1960); K. Aghora-murthy, K. G. Sarma, and T. R. Seshadri, *Tetrahedron*, **12**, 173 (1961).
26. M. Frèrejacque, *Compt. Rend.*, **208**, 1123 (1939).
27. G. J. Hajny, *Appl. Microbiol.*, **12**, 87 (1964); F. B. Anderson and G. Harris, *J. Gen. Microbiol.*, **33**, 137 (1963).
27a. N. K. Richtmyer, *Carbohyd. Res.*, **12**, 135 (1970).
27b. N. K. Richtmyer, *Carbohyd. Res.*, **12**, 233 (1970).
28. O. Ruff and G. Ollendorf, *Ber.*, **33**, 1798 (1900).
29. E. Fischer and R. Stahel, *Ber.*, **24**, 538 (1891).
30. O. Touster and S. O. Harwell, *J. Biol. Chem.*, **230**, 1031 (1958).
31. H. Kiliani, *Ber.*, **20**, 1234 (1887).
32. M. Delépine and A. Horeau, *Bull. Soc. Chim. Fr.*, **4** [5], 1524 (1937).
33. R. A. Raphael, *J. Chem. Soc.*, S44 (1949).
34. R. Lespieau, *Compt. Rend.*, **203**, 145 (1936); *Bull. Soc. Chim. Fr.*, **5** [5], 1638 (1938).
35. E. Fischer, *Ber.*, **26**, 633 (1893).
36. W. V. Podwykssozki, *Arch. Pharm. (Weinheim)*, **227**, 141 (1889); E. Merck, *ibid.*, **231**, 129 (1893).
37. F. Wessely and S. Wang, *Monatsh. Chem.*, **72**, 168 (1938).
38. P. A. Rebers and M. Heidelberger, *J. Amer. Chem. Soc.*, **81**, 2415 (1959); *ibid.*, **83**, 3056 (1961).
39. W. K. Roberts, J. G. Buchanan, and J. Baddiley, *Biochem. J.*, **88**, 1 (1963).
40. A. R. Archibald and J. Baddiley, *Advan. Carbohyd. Chem.*, **21**, 354 (1966).
41. M. L. Wolfrom and E. J. Kohn, *J. Amer. Chem. Soc.*, **64**, 1739 (1942).
42. J. F. Carson, S. W. Waisbrot, and F. T. Jones, *J. Amer. Chem. Soc.*, **65**, 1777 (1943).
43. S. F. Dronov and K. A. Vasil'eva, *Gidrolizn. i Lesokhim. Prom.*, **15**, 21 (1962); *Chem. Abstr.*, **57**, 1129 (1962).
44. E. Fischer, *Ber.*, **27**, 2487 (1894).
45. J. Gutschmidt and G. Ordynsky, *Deut. Lebensm. Rundschau*, **57**, 321 (1961); *Chem. Abstr.*, **56**, 10610 (1962).
46. R. Lohmar and R. M. Goepp, Jr., *Advan. Carbohyd. Chem.*, **4**, 211 (1949).
47. L. Wright and L. Hartmann, *J. Org. Chem.*, **26**, 1588 (1961).
48. V. Plouvier, *Compt. Rend.*, **249**, 2828 (1959).
49. R. Lespieau and J. Wiemann, *Compt. Rend.*, **194**, 1946 (1932); *Bull. Soc. Chim. Fr.*, **53** [4], 1107 (1933).
50. M. Steiger and T. Reichstein, *Helv. Chim. Acta*, **19**, 184 (1936).
51. J. Wiemann, *Ann. Chim. (Paris)*, **5**, [11], 267 (1936).
52. R. Lespieau and J. Wiemann, *Compt. Rend.*, **198**, 183 (1934); R. Lespieau, *Bull. Soc. Chim. Fr.*, **1** [5], 1374 (1934).
53. F. L. Hünefeld, *J. Prakt. Chem.*, **7**, 233 (1836); **9**, 47 (1836).
54. G. Bouchardat, *Ann. Chim. Phys.*, **27** [4], 68 (1872).
55. J. R. Furlong and L. E. Campbell, *Proc. Chem. Soc. (London)*, **29**, 128 (1913).
56. J. Wiemann and J. Gardan, *Bull. Soc. Chim. Fr.*, **25** [5], 433 (1958).
57. J. Boussingault, *Compt. Rend.*, **74**, 939 (1872).
58. J. Pelouze, *Ann. Chim. Phys.*, **35** [3], 222 (1852).
59. P. Hass and T. G. Hill, *Biochem. J.*, **26**, 987 (1932).
60. C. Vincent and Delachanal, *Compt. Rend.*, **109**, 676 (1889).
61. J. D. Anderson, P. Andrews, and L. Hough, *Biochem. J.*, **81**, 149 (1961).
62. A. Hutchinson, C. D. Taper, and G. H. N. Towers, *Can. J. Biochem. Physiol.*, **37**, 901 (1959).

63. R. M. Goepp, Jr., M. T. Saunders, and S. Soltzberg, in "Encyclopedia of Chemical Technology," R. E. Kirk and D. F. Othmer, Eds., Wiley (Interscience), New York, 1947, Vol. 1, p. 321; L. G. Britton, *Ind. Chemist*, **39**, 233 (1963).
64. M. A. Phillips, *Brit. Chem. Eng.*, **8**, 767 (1963).
65. G. G. Boyers, U. S. Patent No. 2,868,847 (1959); Brit. Patent No. 867,689 (1961).
66. A. A. Balandin, N. A. Vasyunina, S. V. Chepigo, and G. S. Barysheva, *Dokl. Akad. Nauk SSSR*, **128**, 941 (1959).
67. R. L. Taylor, *Chem. Met. Eng.*, **44**, 588 (1937); M. J. Creighton, *Trans. Electrochem. Soc.*, **75**, 389 (1939); *Can. Chem. Process Ind.*, **26**, 690 (1942).
68. M. L. Wolfrom, B. W. Lew, R. A. Hales, and R. M. Goepp, Jr., *J. Amer. Chem. Soc.*, **68**, 2342 (1946).
69. C. A. Lobry de Bruyn and W. Alberda van Ekenstein, *Rec. Trav. Chim.*, **19**, 1 (1900).
70. E. Fischer and I. W. Fay, *Ber.*, **28**, 1975 (1895).
71. G. Bertrand, *Ann. Chim. Phys.*, 3 [8], 181 (1904); *Bull. Soc. Chim. Fr.*, **33** [3], 166, 264 (1905).
72. G. Bertrand and A. Lazenberg, *Bull. Soc. Chim. Fr.*, **35** [3], 1073 (1906).
73. H. H. Strain, *J. Amer. Chem. Soc.*, **56**, 1756 (1934).
74. F. B. Cramer and E. Pacsu, *J. Amer. Chem. Soc.*, **59**, 1467 (1937).
75. Y. Khouvine and G. Arragon, *Bull. Soc. Chim. Fr.*, **5** [5], 1404 (1938).
76. M. Proust, *Ann. Chim. Phys.*, **57**, [1] 144 (1806).
77. E. Jandrier, *Compt. Rend.*, **117**, 498 (1893).
78. K. C. Reid, *Proc. 2nd Int. Seaweed Symp.*, Trondheim, p. 105, 1955.
79. R. G. S. Bidwell, *Can. J. Bot.*, **36**, 337 (1958).
80. B. Lindberg, *Acta Chem. Scand.*, **7**, 1119, 1218 (1953); B. Lindberg and J. McPherson, *ibid.*, **8**, 1547 (1954); S. Peat, W. J. Whelan, and H. G. Lawley, *Chem. Ind. (London)*, 35 (1955).
81. C. R. Raha, *Sci. Cult. (Calcutta)*, **22**, 637 (1957).
82. V. D. Harwood, *J. Sci. Food Agr.*, **5**, 453 (1954).
83. H. H. Schlubach, German Patent No. 871,736 (1953).
84. P. Paris, M. Durand, and J. L. Bonnet, *Ann. Pharm. Fr.*, **15**, 677 (1957).
85. J. W. E. Glattfeld and G. W. Schimff, *J. Amer. Chem. Soc.*, **57**, 2204 (1935).
86. J. H. Birkinshaw, J. H. V. Charles, A. Hetherington, and H. Raistrick, *Trans. Roy. Soc. (London)* Ser. **B220**, 153 (1931); V. Bolcato and G. Pasquini, *Ind. Saccar. Ital.*, **32**, 408 (1939).
87. S. A. Barker, A. Gomez-Sanchez, and M. Stacey, *J. Chem. Soc.*, 2583 (1958).
88. E. Fischer, *Ber.*, **23**, 370 (1890).
89. E. Baer and H. O. L. Fischer, *J. Amer. Chem. Soc.*, **61**, 761 (1939).
90. See summary by E. Fischer, *Ber.*, **23**, 2114 (1890); K. Freudenberg, *Advan. Carbohyd. Chem.*, **21**, 1 (1966).
91. E. Pace, *Arch. Farmacol. Sper.*, **42**, 167 (1926).
92. G. Bertrand and P. Bruneau, *Compt. Rend.*, **146**, 482 (1908); *Bull. Soc. Chim. Fr.*, 3 [4], 495 (1908).
93. E. Fischer, *Ber.*, **27**, 1524 (1894).
94. R. M. Hann, W. T. Haskins, and C. S. Hudson, *J. Amer. Chem. Soc.*, **69**, 624 (1947).
95. F. L. Humoller, M. L. Wolfrom, B. W. Lew, and R. M. Goepp, Jr., *J. Amer. Chem. Soc.*, **67**, 1226 (1945).
96. J. M. Sugihara and G. U. Yuen, *J. Amer. Chem. Soc.*, **79**, 5780 (1957).
97. B. R. Baker and D. H. Buss, *J. Org. Chem.*, **30**, 2304 (1965).
98. B. R. Baker and A. H. Haines, *J. Org. Chem.*, **28**, 438 (1963).
99. J. M. Webber, *Advan. Carbohyd. Chem.*, **17**, 15 (1962).

99a. R. Young and G. A. Adams, *Can. J. Chem.*, **44**, 32 (1966).

100. Rules of Carbohydrate Nomenclature, *J. Org. Chem.*, **28**, 281 (1963); see Vol. IIB, Chap. 46.

101. E. Bourquelot, *J. Pharm. Chim.*, **2** [6], 285 (1895); V. Ettel, *Collect. Czech. Chem. Commun.*, **1**, 288 (1929).

102. J. Bougault and G. Allard, *Compt. Rend.*, **135**, 796 (1902).

103. Y. Asahina and M. Kagitani, *Ber.*, **67**, 804 (1934); B. Lindberg, *Acta Chem. Scand.*, **9**, 917 (1955).

104. B. Lindberg and J. Paju, *Acta Chem. Scand.*, **8**, 817 (1954); B. Lindberg, *ibid.*, **9**, 1097 (1955).

105. J. B. Avequin, *J. Chim. Med. Pharm. Toxicol.*, **7** [1], 467 (1831); L. Maquenne, *Compt. Rend.*, **107**, 583 (1888); R. M. Hann and C. S. Hudson, *J. Amer. Chem. Soc.*, **61**, 336 (1939); E. M. Montgomery and C. S. Hudson, *ibid.*, **61**, 1654 (1939).

106. N. K. Richtmyer, *Methods Carbohyd. Chem.*, **2**, 90 (1963).

107. A. Nordal and A. A. Benson, *J. Amer. Chem. Soc.*, **76**, 5054 (1954).

108. J. K. N. Jones and R. A. Wall, *Nature*, **189**, 746 (1961).

108a. N. K. Richtmyer, *Carbohyd. Res.*, **12**, 139 (1970).

108b. See also, refs. 27a and 108a.

109. A. Nordal and D. Öiseth, *Acta Chem. Scand.*, **5**, 1289 (1951).

110. A. J. Charlson and N. K. Richtmyer, *J. Amer. Chem. Soc.*, **82**, 3428 (1960).

111. C. Long, *Sci. Progr.*, **41**, 282 (1953).

112. G. Bertrand and G. Nitzberg, *Compt. Rend.*, **186**, 1172, 1773 (1928); Y. Khouvine and G. Nitzberg, *ibid.*, **196**, 218 (1933).

113. Y. Khouvine, *Compt. Rend.*, **198**, 985 (1934).

114. Y. Khouvine, *Compt. Rend.*, **204**, 983 (1939).

115. F. L. Humoller, S. J. Kuman, and F. H. Snyder, *J. Amer. Chem. Soc.*, **61**, 3370 (1939).

116. H. H. Sephton and N. K. Richtmyer, *J. Org. Chem.*, **28**, 1691 (1963).

117. A. J. Charlson and N. K. Richtmyer, *J. Amer. Chem. Soc.*, **81**, 1512 (1959).

118. R. M. Hann, A. T. Merrill, and C. S. Hudson, *J. Amer. Chem. Soc.*, **66**, 1912 (1944).

119. R. Schaffer and A. Cohen, *J. Org. Chem.*, **28**, 1929 (1963).

120. R. M. Hann, W. D. Maclay, A. E. Knauf, and C. S. Hudson, *J. Amer. Chem. Soc.*, **61**, 1268 (1939).

121. W. D. Maclay, R. M. Hann, and C. S. Hudson, *J. Amer. Chem. Soc.*, **60**, 1035 (1938).

122. E. Zissis, D. R. Strobach, and N. K. Richtmyer, *J. Org. Chem.*, **30**, 79 (1965); compare J. C. Sowden and D. R. Strobach, *J. Amer. Chem. Soc.*, **82**, 956 (1960).

123. H. H. Sephton and N. K. Richtmyer, *Carbohyd. Res.*, **2**, 289 (1966).

124. E. Fischer, *Ann.*, **270**, 64 (1892).

125. L.-H. Philippe, *Ann. Chim. Phys.*, **26** [8], 289 (1912).

126. C. S. Hudson, *Advan. Carbohyd. Chem.*, **1**, 1 (1945).

127. M. L. Wolfrom, W. W. Binkley, C. C. Spencer, and B. W. Lew, *J. Amer. Chem. Soc.*, **73**, 3357 (1951).

128. L. Hough and R. S. Theobald, *Methods Carbohyd. Chem.*, **1**, 94 (1962).

129. M. L. Wolfrom and H. B. Wood, Jr., *J. Amer. Chem. Soc.*, **73**, 2933 (1951); M. L. Wolfrom and K. Anno, *ibid.*, **74**, 5583 (1952).

130. J. V. Karabinos and A. T. Ballun, *J. Amer. Chem. Soc.*, **75**, 4501 (1953); also see M. L. Wolfrom and J. N. Schumacher, *ibid.*, **77**, 3318 (1955).

131. R. K. Ness, H. G. Fletcher, Jr., and C. S. Hudson, *J. Amer. Chem. Soc.*, **73**, 4759 (1951).

132. O. Jeger, J. Norymberski, S. Szpilfogel, and V. Prelog, *Helv. Chim. Acta*, **29**, 684 (1946).

133. H. L. Frush, H. S. Isbell, and A. J. Fatiadi, *J. Res. Nat. Bur. Stand.*, **64A**, 433 (1960).

134. H. S. Isbell and H. L. Frush, *Methods Carbohyd. Chem.*, **1**, 417 (1962).

135. J. W. Reppe, "Acetylene Chemistry," Meyer, New York, 1949.
136. N. A. Vasyunina, A. A. Balandin, V. I. Karzhev, B. Ya. Rabinovich, S. V. Chepigo, E. S. Grigoryan, and R. L. Slutskin, *Khim. Prom.*, 82 (1962); *Chem. Abstr.*, **58**, 3304 (1963).
137. J. W. Pratt and N. K. Richtmyer, *J. Amer. Chem. Soc.*, **77**, 6326 (1955).
138. F. B. LaForge, *J. Biol. Chem.*, **42**, 375 (1920).
139. V. Ettel, *Collect. Czech. Chem. Commun.*, **4**, 513 (1932).
140. A. T. Merrill, W. T. Haskins, R. M. Hann, and C. S. Hudson, *J. Amer. Chem. Soc.*, **69**, 70 (1947).
141. M. Frèrejacque, *Compt. Rend.*, **200**, 1410 (1935).
142. N. K. Richtmyer and C. S. Hudson, *J. Amer. Chem. Soc.*, **73**, 2249 (1951).
142a. G. A. Jeffrey and H. S. Kim, *Carbohyd. Res.*, **14**, 207 (1970).
143. R. L. Lohmar, in "The Carbohydrates," W. Pigman, Ed., Academic Press, New York, 1957.
144. J. J. Fox and A. J. H. Gauge, *J. Chem. Soc.*, **99**, 1075 (1911).
145. W. H. Holst, paper presented before the Division of Sugar Technol., Amer. Chem. Soc., April, 1939.
146. J. Dale, *J. Chem. Soc.*, 922 (1961).
147. P. Brigl and H. Grüner, *Ann.*, **495**, 70, 72 (1932).
148. H. G. Kuivila, A. H. Keough, and E. J. Soboczenski, *J. Org. Chem.*, **19**, 780 (1954); J. M. Sugihara and C. M. Bowman, *J. Amer. Chem. Soc.*, **80**, 2443 (1958).
148a. F. Eisenberg, Jr., *Carbohyd. Res.*, **19**, 135 (1971).
149. A. B. Foster, *Advan. Carbohyd. Chem.*, **12**, 81 (1957).
150. H. Weigel, *Advan. Carbohyd. Chem.*, **18**, 61 (1963).
151. See J. Böeseken, *Advan. Carbohyd. Chem.*, **4**, 189 (1949).
152. D. H. R. Barton and R. C. Cookson, *Quart. Rev. (London)*, **10**, 44 (1956).
153. D. H. Hutson and H. Weigel, *J. Chem. Soc.*, 1546 (1961).
154. B. Lindberg and B. Swan, *Acta Chem. Scand.*, **14**, 1043 (1960).
155. H. J. F. Angus, E. J. Bourne, and H. Weigel, *J. Chem. Soc.*, 21 (1965).
156. H. J. F. Angus and H. Weigel, *J. Chem. Soc.*, 3994 (1964).
157. E. J. Bourne, D. H. Hutson, and H. Weigel, *J. Chem. Soc.*, 35 (1961).
158. J. L. Frahn and J. A. Mills, *Aust. J. Chem.*, **12**, 65 (1959).
158a. E. J. Bourne, F. Searle, and H. Weigel, *Carbohyd. Res.*, **16**, 185 (1971).
159. J. A. Mills, *Biochem. Biophys. Res. Commun.*, **6**, 418 (1961/2).
160. M. L. Wolfrom and M. J. Holm, *J. Org. Chem.*, **26**, 273 (1961).
161. See B. Sklarz, *Quart. Rev. (London)*, **21**, 3 (1967).
162. P. Zuman, J. Sicher, J. Krupicka, and M. Svoboda, *Collect. Czech. Chem. Commun.*, **23**, 1237 (1958).
163. J. C. P. Schwarz, *J. Chem. Soc.*, 276 (1957).
164. J. E. Courtois and M. Guernet, *Bull. Soc. Chim. Fr.*, 1388 (1957).
165. H. J. H. Fenton and H. Jackson, *J. Chem. Soc.*, **75**, 1 (1899).
166. J. W. Green, *Advan. Carbohyd. Chem.*, **3**, 129 (1948); R. Bognár and L. Somogyi, *Acta Chim. Acad. Sci. Hung.*, **14**, 407 (1958).
167. C. Neuberg, *Biochem. Z.*, **17**, 270 (1909).
168. J. E. Hunter, *Iowa State Coll. J. Sci.*, **15**, 78 (1940).
169. R. J. Stoodley, *Can. J. Chem.*, **39**, 2593 (1961).
170. K. Heyns and H. Paulsen, *Advan. Carbohyd. Chem.*, **17**, 169 (1962).
171. K. Heyns and M. Beck, *Ber.*, **91**, 1720 (1958).
172. J. S. Brimacombe, J. G. H. Bryan, A. Husain, M. Stacey, and M. S. Tolley, *Carbohyd. Res.*, **3**, 318 (1967).

173. R. M. Hann, E. B. Tilden, and C. S. Hudson, *J. Amer. Chem. Soc.*, **60**, 1201 (1938).
174. E. B. Tilden, *J. Bacteriol.*, **37**, 629 (1939).
175. D. J. Wilham and J. K. N. Jones, *Can. J. Chem.*, **45**, 741 (1967).
176. P. A. Wells, J. J. Stubbs, L. B. Lockwood, and E. T. Roe, *Ind. Eng. Chem.*, **29**, 1385 (1937).
176a. N. K. Richtmyer, *Carbohyd. Res.*, **17**, 401 (1971).
177. E. Erlenmeyer and J. A. Wanklyn, *Ann.*, **135**, 129 (1865).
178. C. W. Lenth and R. N. DuPuis, *Ind. Eng. Chem.*, **37**, 152 (1945).
179. G. N. Kowkabany, *Advan. Carbohyd. Chem.*, **9**, 303 (1954); also see E. J. Bourne, E. M. Lees, and H. Weigel, *J. Chromatogr.*, **11**, 253 (1963).
180. J. K. N. Jones and R. A. Wall, *Can. J. Chem.*, **38**, 2290 (1960).
181. L. P. Zill, J. X. Khym, and G. M. Cheniae, *J. Amer. Chem. Soc.*, **75**, 1339 (1953).
182. C. T. Bishop, *Advan. Carbohyd. Chem.*, **19**, 95 (1964); J. S. Sawardeker, J. H. Sloneker, and A. Jeanes, *Anal. Chem.*, **37**, 1602 (1965).
183. C. C. Sweeley, R. Bentley, M. Makita, and W. W. Wells, *J. Amer. Chem. Soc.*, **85**, 2497 (1963); B. Smith and O. Carlsson, *Acta Chem. Scand.*, **17**, 455 (1963).
184. M. L. Wolfrom, M. Konigsberg, F. B. Moody, and R. M. Goepp, Jr., *J. Amer. Chem. Soc.*, **68**, 122 (1946).
185. B. Lindberg and H. Meier, *Acta Chem. Scand.*, **16**, 543 (1962).
186. S. Hanessian, *Advan. Carbohyd. Chem.*, **21**, 143 (1966).
187. J. V. Karabinos, *Methods Carbohyd. Chem.*, **2**, 77 (1963).
188. R. S. Tipson, *Advan. Carbohyd. Chem.*, **8**, 107 (1953).
189. O. T. Schmidt, *Methods Carbohyd. Chem.*, **1**, 198 (1963).
190. R. Lukes and J. Jary, *Chem. Listy*, **51**, 920 (1957).
191. A. N. de Belder and H. Weigel, *Chem. Ind. (London)*, 1689 (1964).
192. S. David and P. Jaymond, *Bull. Soc. Chim. Fr.*, **26** [5], 157 (1959).
192a. D. Horton and J. S. Jewell, *J. Org. Chem.*, **31**, 509 (1966).
192b. W. W. Binkley and R. W. Binkley, *Carbohyd. Res.*, **11**, 1 (1969).
193. W. W. Pigman and R. M. Goepp, Jr., "Chemistry of The Carbohydrates," Academic Press, New York, 1948, p. 257; J. Staněk, M. Černý, J. Kocourek, and J. Pacák, "The Monosaccharides," Academic Press, New York, 1963, p. 643.
194. O. Touster and D. R. D. Shaw, *Physiol. Rev.*, **42**, 181 (1962).
195. H. G. Hers, *J. Biol. Chem.*, **214**, 373 (1955).
196. R. L. Blakley, *Biochem. J.*, **49**, 257 (1951).
197. H. G. Hers, *Biochim. Biophys. Acta*, **37**, 127 (1960).
198. T. Mann and C. Lutwak-Mann, *Biochem. J.*, **48**, xvi (1951).
199. S. Hollmann and O. Touster, *J. Amer. Chem. Soc.*, **78**, 3544 (1956); O. Touster, V. H. Reynolds, and R. M. Hutcheson, *J. Biol. Chem.*, **221**, 697 (1956); S. Hollmann and O. Touster, *ibid.*, **225**, 87 (1957); D. B. McCormick and O. Touster, *ibid.*, **229**, 451 (1957).
200. O. Touster, S. O. Hecht, and W. M. Todd, *J. Biol. Chem.*, **235**, 951 (1960).
201. H. Smith, A. E. Williams, J. H. Pearce, J. Keppie, P. W. Harris-Smith, R. B. Fitz-George, and K. Witt, *Nature*, **193**, 47 (1962); J. Keppie, A. E. Williams, K. Witt, and H. Smith, *Brit. J. Exp. Pathol.*, **46**, 104 (1965); J. D. Anderson and H. Smith, *J. Gen. Microbiol.*, **38**, 109 (1965).
202. D. R. Batt, F. Dickens, and D. H. Williamson, *Biochem. J.*, **77**, 281 (1960).
203. R. Jones, W. B. Kessler, H. E. Lessmer, and L. Rane, *Cancer Chemother. Rept.*, **10**, 99 (1960); F. R. White, *ibid.*, **24**, 95 (1962); P. W. Feit, *J. Med. Chem.*, **7**, 14 (1964); see also L. Vargha and T. Horváth, *Acta Unio Intern. Contra Cancrum*, **20**, 76 (1964).
204. W. Davis and W. C. J. Ross, *Biochem. Pharmacol.*, **12**, 915 (1963); see also M. J. Tisdale, *Carbohyd. Res.*, **19**, 117 (1971).

15.   THE CYCLITOLS

LAURENS ANDERSON

---

* *References start on p. 573.*

I. INTRODUCTION

In addition to the monosaccharides and their derivatives, which may have either heterocyclic or open-chain structures, the simple carbohydrates include a substantial number of carbocyclic substances. Primarily these constitute the polyhydroxycyclohexanes and -hexenes and several series of derivatives related to the hydroxy compounds by processes (actual or formal) of O-substitution, C-substitution, or replacement of hydroxyl groups by other groups. However, polyhydroxycycloalkanes and derivatives having rings other than six-membered are now known. The term *cyclitol* is used to designate all of these compounds.

While the cyclitols were recognized for many years as a small group of natural products, their status has changed as a result of extensive investigations, since 1940, on their synthesis, chemical reactions, and biochemistry. These aspects, and the use of cyclitols as model compounds for the sugars, are summarized in the sections that follow. During this same period over a dozen new cyclitols were discovered in Nature, and the task of determining the constitutions (structure and configuration) of these and the longer-known, natural cyclitols was essentially completed. The work on constitution is important, but it is mentioned only briefly here, since it was accomplished largely by classical methods, which will be of diminishing importance in the future.

For a complete account of the constitutional aspects, and for a fuller treatment of the other topics, the reader is referred to the excellent monograph of Posternak.[1] This author, pre-eminent in the development of cyclitol chemistry, has described all of the significant chemical and biochemical work in the field up to shortly before the respective dates of publication of the two editions. Other useful references are the previous editions of this book, reviews which have appeared in *Advances in Carbohydrate Chemistry*,[2, 3] and a listing, complete when it was published, of the known natural occurrences of cyclitols in plants.[4]

A. NOMENCLATURE

The naming and numbering of the cyclitols poses some difficult problems, for which several different solutions have been advocated and used, with the result that the literature on the subject is often confusing. This situation should, however, be alleviated by the adoption, by the IUPAC/IUB Commission on Biochemical Nomenclature and the IUPAC Commission on the Nomenclature of Organic Chemistry, of a set of rules proposed by a Joint Cyclitol Nomenclature Subcommittee.[4a] The nomenclature embodied in the 1967 IUPAC/IUB Tentative Rules for Cyclitols, as they may be called, is used in this chapter. Explanatory notes are provided at several points, and when names specified by the rules differ markedly from those in current usage, the latter are mentioned parenthetically.

II. THE INOSITOLS (CYCLOHEXANEHEXOLS) AND THEIR *O*-SUBSTITUTION PRODUCTS

Over 100 years ago Scherer isolated an optically inactive isomer of cyclohexanehexol and christened the compound *inosit*, after the Greek ἴς, ἰνός (variously translated as sinew, fiber, or muscle).[5] The name was applied to the other isomers as these were discovered or synthesized, so that it is now a generic term (with the suffix *ol* added in English and French). The original isomer came to be called *meso-* or *i*-inositol, and more recently, in the English language literature at least, *myo*-inositol. As shown theoretically by Bouveault in 1894, there are nine inositols in all—seven meso forms and one DL pair.[6] These have stereoformulas **1–9** and are distinguished from each other by the configurational prefixes shown with the formulas (1967 Rules).

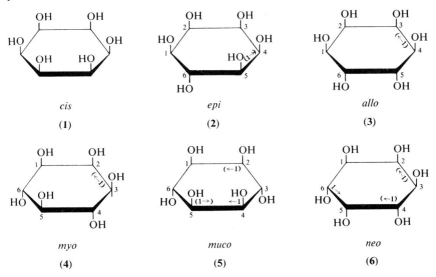

cis	*epi*	*allo*
(1)	**(2)**	**(3)**

myo	*muco*	*neo*
(4)	**(5)**	**(6)**

* References start on p. 573.

D-*chiro* (+)   L-*chiro* (−)   *scyllo*

(7)   (8)   (9)

The prefixes for isomers **1–3**, **5**, and **6** are those established by usage. For isomer **4** the Rules specify *myo*, reversing the decision of an earlier IUPAC commission to retain *meso*. Isomer **9** is called *scyllo*-inositol instead of scyllitol, and the prefix *chiro* (from chirality) is introduced to designate the relative configuration (**7/8**) found in the active inositols.

Cyclitols named as inositols include not only the parent compounds **1–9** and their *O*- and *C*-substitution products, but also derivatives in which up to three hydroxyl groups are replaced by other univalent substituents.

The numbering of the ring positions (or hydroxyl groups) in the inositols is necessarily based on configurational factors. In all systems, the hydroxyl groups are considered as two sets, one set projecting above the plane of the ring, the other below. The principle most commonly employed, and formalized in the 1967 Rules, is then to assign lowest possible numbers to one of the sets—the more numerous one if there is such. Application of this principle leads to two or more numberings for all the inositols; hence an additional stipulation is made: Meso inositols are normally numbered so that C-1 has the L-configuration (see below). This gives clockwise numbering, as shown on formulas **1–6** and **9**, when the 1-OH group projects upward, counterclockwise when it projects downward. Note that: (1) the six positions in unmodified *cis*-inositol (**1**) are indistinguishable, and the same is true for *scyllo*-inositol (**9**); (2) in both *muco*- (**5**) and *neo*- (**6**) inositol there are still two, identical possibilities; the second is shown by the symbol 1→ without parentheses. In the *chiro*-inositols (**7** and **8**) the two sets of hydroxyl groups are superposable by rotation of the ring. Either set may be given lowest numbers, and when the choice is made the direction of numbering is thereby established.

In substituted inositols the groups other than unmodified hydroxyl groups are given the lowest possible numbers compatible with a fixed numbering based on the configuration of the parent inositol. For substituted *chiro*-inositols and meso derivatives of the meso inositols the principles stated above are sufficient. In the case of *chiral* ("asymmetric") derivatives of meso inositols, only one enantiomorph of each DL pair is numbered by the "clockwise when 1-OH (or R-1) projects upward" rule. For the other, one uses *alternative, mirror-image numbering*, shown by the symbol (1→), proceeding counterclockwise when 1-OH (or R-1) projects upward. This gives corresponding, substituted positions identical numbers in both enantiomorphs. Position in the alphabet, or degree of complexity, may be invoked, when necessary, to establish precedence in polysubstituted derivatives.

In their methods for designating the absolute configurations of the chiral cyclitols, the various nomenclature systems differ more than in any other feature. The solution adopted in the IUPAC/IUB 1967 Rules is based on suggestions made by McCasland: a Fischer projection of the molecular formula is made, with C-1 at the top as shown, and the configuration about the *lowest numbered* chiral center is then examined. If the substituent at the center projects to the right, the compound is D; if to the left, it is L. To distinguish the new convention (parallel to that used with amino acids) from the sugar convention, which has been applied to cyclitols by the present author, it is recommended that in most cases the positional numeral of the reference center be affixed to the chirality symbol (that is, 1D, 1L, or sometimes 2D, 3L, etc.).

L– *chiro*

The inositols are beautifully crystalline, high-melting substances, markedly stable to heat, acids, and alkalis. *neo*-Inositol and *scyllo*-inositol have surprisingly low solubilities in water (0.1 % and about 1 %, respectively, in the cold). Among organic solvents only the very polar ones such as formamide, *N,N*-dimethylformamide, and methyl sulfoxide dissolve inositols to an appreciable extent.

A. THE MAJOR NATURALLY OCCURRING INOSITOLS
(*myo*, D- AND L-*chiro*, *scyllo*)

The most widely distributed isomer is *myo*-inositol (**4**). It appears to be present, both free and combined, in the tissues of nearly all living species.[1] Efforts to detect it have failed only with a few bacteria. In animals and microorganisms the combined *myo*-inositol is mostly in the form of phospholipids, and these also account for part of the *myo*-inositol in higher plants. In addition, plants may contain substantial portions of *myo*-inositol as phytate (see phosphates, below) or as methyl ethers. *myo*-Inositol is available commercially as a by-product of the corn wet-milling industry. It is prepared from phytate salts isolated from steep liquor.

The next most abundant inositols are the optically active isomers, D-(+)- and L-(−)-*chiro*-inositol (**7** and **8**), previously designated as *d*- and *l*-, *dextro*- and *levo*-, D- and L-, or (+)- and (−)-inositol. These occur in higher plants, predominantly as the methyl ethers D-(+)-pinitol and L-(−)-quebrachitol, respectively (see Section II,C, p. 525).† D-*chiro*-Inositol is a structural component of the antibiotic kasugamycin,[7] but aside from this, the *chiro*-inositols and their derivatives have not been found in microorganisms or animals. D- and L-*chiro*-

---

† L-*chiro*-Inositol in unsubstituted form is characteristically found in plants of the family Compositae.

* *References start on p. 573.*

Inositol are normally prepared by the O-demethylation of their ethers in re-fluxing hydriodic acid, but a cheaper reagent could no doubt be used if the reaction had to be carried out on a large scale. D-*chiro*-Inositol has been detected in a species of *Chlorella* (green algae).[7a]

scyllo-Inositol (9, also called scyllitol) has been isolated from plants[4] (red algae, several scattered species of angiosperms) and from animals (notably plagiostomous fish).[4] It has been detected in several species of insects,[7b] and in mammalian urines.[1] Mammalian tissues that have been subjected to careful analysis have 2 to 10% as much *scyllo*-inositol as *myo*-inositol.[7c, 7d] Urinary *scyllo*-inositol may be partly of dietary origin.[8] *scyllo*-Inositol is most readily obtained by the reduction of the appropriate inosose (cyclitol ketone, 68), as described in Section VI.

The natural cyclitols that are available in quantity (*myo*-inositol, D-pinitol, L-quebrachitol) are important as starting materials for the synthesis of other compounds of the group. Although the total synthesis of nearly any desired cyclitol is now possible (see Section IV), procedures that utilize preformed cyclitols are often more practical. D-Pinitol and L-quebrachitol and their parent inositols are particularly useful, for these compounds can be employed to prepare optically active end products without recourse to resolution or a biological step.

The *structure* of the inositols was inferred in the nineteenth century from: (1) the elemental analyses of *myo*-, *scyllo*-, and D- and L-*chiro*-inositol; (2) the fact that they gave hexaacetates and hexabenzoates; (3) the conversion of ino-sitols into benzene derivatives by certain vigorous procedures; and (4) the formation of hexaoxygenated, cyclic oxidation products (tetrahydroxyquinone, etc.; see Section II,F) on treatment with nitric acid.[1] This latter observation was particularly important because it indicated that each ring-carbon atom carries a hydroxyl group.

Oxidative cleavages to give mixtures containing hexaric acids were the prin-cipal means for determining the configurations of the common inositols, which was not accomplished until the period 1936–1942. Careful oxidation with permanganate served to establish formula 7 for (+)-inositol (D-*chiro*) and 8 for (−)-inositol (L-*chiro*).[9] The unequivocal assignment of formula 4 to *myo*-inositol depended on a study of the oxidation of the inosose[10] (68), and on the discovery of a procedure[11] (acetonation–acetylation–deacetonation) for block-ing the inositol so that lead tetraacetate cleavage could be effected specifically at C-1–C-2.

B. The Other Inositols (*epi, muco, allo, neo, cis*)

Of the remaining inositols, two have been detected in Nature—*neo* in soil phytate and *muco* as a methyl ether in many plant sources. These inositols were earlier prepared by synthesis, however, and this is still the only practical means of obtaining *neo*-inositol. The other three (*epi, allo, cis*) are so far strictly synthetic products.

epi-Inositol (**2**), like the *scyllo* isomer, can be made by the reduction of an inosose (**67**). The configurations of both these inositols are established by the relationship of the inosose precursors to *myo*-inositol (see Section VI).

The original syntheses of *muco*- (**5**) and *allo*-inositol (**3**) were accomplished by the hydroxylation of derivatives of a natural 5-cyclohexene-1,2,3,4-tetrol (conduritol, Table III).[12] Because conduritol is scarce, *muco*- and *allo*-inositol were for a long time rather rare compounds. Practical syntheses of both are now available, however. *muco*-Inositol can be made from *myo*-inositol by way of sulfonic ester and epoxide intermediates. Similarly *allo*-inositol, and also *neo*-inositol (**6**), can be prepared from the *p*-toluenesulfonic esters of the *chiro*-inositols. Further details are given in Section V. The epimerization reactions of the inositol acetates (Section X) also provide practical routes to *muco*- and *neo*-inositol.

The last of the inositols to become known, and the most difficult to obtain, is *cis*-inositol (**1**). It was first made by the hydrogenation of benzenehexol (Section IV,A), but an improved synthesis, from *epi*-inositol, is now available.[12a]

C. METHYL ETHERS OF INOSITOLS

The cyclitols of plants, as already mentioned, include a number of methyl ethers of inositols.[1] Five of these—D-(+)-pinitol, L-(−)-quebrachitol, L-(+)-bornesitol, dambonitol, and sequoyitol—have been known for many years, and subsequent to its original discovery each was from time to time detected in additional sources. A systematic investigation of the distribution of these compounds has been carried out by Plouvier, who in the course of examining hundreds of species of gymnosperms and dicotyledons for alditols discovered the additional methyl ethers L-(−)-pinitol, D-(−)-bornesitol, D-(+)-ononitol, and D-(−)-liriodendritol, as well as two previously unknown lower cyclitols.[4]

It appears from Plouvier's work that, to some extent, individual cyclitols are characteristic of certain plant families: D-pinitol and L-quebrachitol, for instance, are seldom found in the same family. But there are exceptions (Apocyanaceae, Euphorbiaceae) and indeed one case (mistletoe) where both are found in the same plant. On the whole, it must be said that the natural occurrence of cyclitols other than *myo*-inositol is somewhat irregular. Often a "characteristic" cyclitol appears in only a few of the species of a family, and, more strikingly, the several families in which a particular cyclitol—for example, quebrachitol—is found are not necessarily closely related. Little is known about the occurrence of cyclitols other than *myo*-inositol in monocotyledons, lower vascular plants, and bryophytes.

Table I summarizes our present information about the naturally occurring methyl ethers of inositols. It will be noted that the list includes four of the six possible mono-*O*-methyl-*chiro*-inositols,† four of the six possible mono-*O*-methyl-*myo*-inositols, and two di-*O*-methyl-*myo*-inositols.‡

Very thorough examination of a number of plant species by Kindl, Hoffmann-Ostenhof, and collaborators has shown that most contain a greater variety of cyclitols than previously realized. The data collected by these authors have been tabulated.[12b]

The extensive use of D-pinitol and L-quebrachitol in synthetic work is due not so much to their wide distribution (Table I) as to the fact that each is accumulated in remarkably high concentration by one particular plant species. The sugar pine (*Pinus lambertiana* Dougl.) is the usual source of D-pinitol, which accounts for 4% (average) of the dry weight of the heartwood.[13] Similarly, the rubber tree (*Hevea brasiliensis*) provides L-quebrachitol, which constitutes 1 to 1.5% (12 to 18% of the solids) of the aqueous phase (serum) of the latex.[14] Since each of these plants is the basis of a major industry—sugar pine is an important lumber tree in California and Oregon—both D-pinitol and L-quebrachitol could be produced in large tonnages.

Once the configurations of the parent inositols were known, the problem of determining the stereoformulas of the methyl ethers became one of finding the positions of the methyl groups. For D-pinitol and L-quebrachitol this was accomplished by acetonation, which indicated the number of pairs of *cis*-hydroxyl groups, and by degradations of the isopropylidene acetals.[15–17] The formula given to pinitol was confirmed by an unequivocal synthesis of the L-isomer.[17a] Formulas for the *myo*-inositol monomethyl ethers were then elucidated through definitive syntheses in which D-pinitol and L-quebrachitol were epimerized at specific positions by oxidation and subsequent reduction (see Section VI).[18, 19]

Most of the naturally occurring mono- and dimethyl ethers of inositols, and several other compounds of this class, have been synthesized. Regrettably, only a few of the syntheses are suitable for preparative work. Good results have been obtained by displacing *p*-tolylsulfonyloxy groups from derivatives of the *O*-methyl-*chiro*-inositols, and D-ononitol is now accessible by this route.[19a] Procedures that would make the other mono-*O*-methyl-*myo*-inositols more readily available for biological and chemical investigation would be especially helpful. A few of the many possible tri-, tetra-, and penta-*O*-methyl-*myo*-inositols are known, some of them as by-products of the methylation

---

† Each of the enantiomorphic *chiro*-inositols can be monosubstituted in three ways. Hydroxyl groups at positions 4, 5, and 6 are superposable, respectively, on hydroxyl groups at positions 3, 2, and 1 of the same enantiomorph. See p. 522.

‡ "Asteritol," an alleged methyl ether of an inositol isolated from methanol-preserved specimens of a starfish, has been identified as methyl α-D-glucopyranoside.[1]

TABLE I

THE NATURALLY OCCURRING METHYL ETHERS OF THE INOSITOLS

Trivial name (systematic name)	Formula	Melting point, degrees [α]ᴅ, degrees (water)	Distributionᵃ
(1ᴅ-1-O-Methyl-chiro-inositol)		207–208 +61	Afrormosia elataᵇ (Leguminosae)
D-(+)-Pinitol (1ᴅ-3-O-Methyl-chiro-inositol)		186–187 +65	Gymnosperms (9 families)ᶜ Dicotyledons (15 families)
L-(−)-Quebrachitol (1ʟ-2-O-Methyl-chiro-inositol)		190–194 −81	Dicotyledons (11 families)
L-(−)-Pinitol (1ʟ-3-O-Methyl-chiro-inositol)		186 −65	Artemisia dracunculus (Compositae)

TABLE I

The Naturally Occurring Methyl Ethers of the Inositols

Trivial name (systematic name)	Formula	Melting point, degrees $[\alpha]_D$, degrees (water)	Distribution[a]
(1D-1-O-Methyl-muco-inositol)[c]		Amorphous −53 (EtOH)	Conifers (exc. Pinaceae), Taxopsida, Cistaceae[c]
L-(+)-Bornesitol (1L-1-O-Methyl-myo-inositol)		201–202 +32	Apocyanaceae (7 genera), Sarcocephallus diderrichii (Rubiaceae)
D-(−)-Bornesitol (1D-1-O-Methyl-myo-inositol)		203–204 −32	Dicotyledons (6 families) Caryota urens (a palm)
D-(+)-Ononitol (1D-4-O-Methyl-myo-inositol)		172 +6.6	Dicotyledons (2 families)

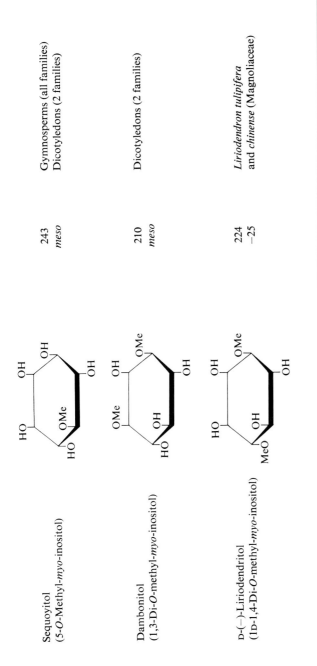

Sequoyitol
(5-O-Methyl-myo-inositol)

243
meso

Gymnosperms (all families)
Dicotyledons (2 families)

Dambonitol
(1,3-Di-O-methyl-myo-inositol)

210
meso

Dicotyledons (2 families)

D-(−)-Liriodendritol
(1D-1,4-Di-O-methyl-myo-inositol)

224
−25

Liriodendron tulipifera
and chinense (Magnoliaceae)

[a] Data from Plouvier,[4] and Kindl and Hoffmann-Ostenhof,[12b] except as noted otherwise.
[b] C. D. Foxall and J. W. W. Morgan, J. Chem. Soc., 5573 (1963).
[c] P. Dittrich, M. Gietl, and O. Kandler, Phytochemistry, in press.

analysis of mycobacterial lipids (see below). The hexamethyl ether is readily prepared by treating myo-inositol with methyl sulfate, barium oxide, and barium hydroxide in N,N-dimethylformamide–methyl sulfoxide.[20]

D. QUANTITATIVE DETERMINATION OF CYCLITOLS

Because of the biological importance of myo-inositol, much effort has been spent on its quantitative determination. Among the methods developed for this purpose are microbiological and specific enzymic assays, and periodate oxidation and gas chromatographic procedures. In the microbiological assays the yeasts Saccharomyces cerevisiae and S. carlsbergensis have been the most frequently used test organisms, but other yeasts and a mutant Neurospora have also served.[1] These assays are largely but not absolutely specific for myo-inositol, since few of the other cyclitols have substantial growth-factor activity in microorganisms. Although the microbiological methods are tedious and less than satisfactorily reliable, they have, with some assist from periodate oxidation techniques, served for the compilation of data on the myo-inositol content of tissues from a great variety of organisms. Some typical figures are given in Table II.

TABLE II
myo-INOSITOL CONTENT OF REPRESENTATIVE BIOLOGICAL MATERIALS[a]
(FRESH-WEIGHT BASIS)

Tissue	myo-Inositol, per cent	Tissue	myo-Inositol, per cent
Serratia marcescens (bacterium)	0.028	Rat skeletal muscle	0.013[c]
Brewers yeast	0.028[b]	Rat brain	0.056[c]
Tetrahymena geleii (protozoön)	0.043	Rat kidney	0.088[c]
Cockroach	0.036	Rat seminal vesicle	0.550
Lima beans	0.170	Rabbit thyroid	0.229[c]
Carrots	0.048	Human plasma	0.0008
Lettuce	0.055	Human lens	0.500

[a] Data from A. M. Woods, J. Taylor, M. J. Hofer, G. A. Johnson, R. L. Lane, and J. R. McMahan, Univ. Texas Publ. No. 4237, 84 (1942), and various other sources (see Posternak[1]).
[b] Dry-weight basis. [c] Free inositol—no hydrolysis or autolysis of the sample prior to assay.

Quantitative analyses for cyclitols other than myo-inositol have not been widely attempted, but the bases for suitable procedures are now well established. Thus, animal tissues have been assayed for scyllo-inositol and 2,4,6/3,5-pentahydroxycyclohexanone, as well as myo-inositol, by gas–liquid chromatography.[7c] However, most of the available data on the proportions of various cyclitols in plant material are in terms of isolated, crystalline substance. The figures range from a few thousandths to a few hundredths of a per cent of the fresh weight of the tissue; occasionally proportions of 1 % or more are obtained.[1]

Periodate oxidation should in principle be applicable to the determination of any individual cyclitol. For success in practice it is necessary to provide for the removal of interfering substances and to establish conditions for the rapid and reproducible oxidation of the cyclitol to be measured (see Section II,F). Spectrophotometric methods may then be used for analyses on a micro scale. In one modern procedure, periodate consumption is measured by the disappearance of absorption at 260 nm, permitting the estimation of as little as 2 μg of myo-inositol.[21] Increased sensitivity is achieved by working at 222 nm, or by converting the excess periodate to iodine and reading the latter at 352 nm.[21a]

For a more general solution to analytical problems involving cyclitols, the most promising technique is gas–liquid chromatography (see Chapter 45). The trimethylsilyl ethers of a variety of cyclitols have been chromatographed successfully on polyester, silicone, and Carbowax columns, and it appears that nearly any desired separation can be accomplished by the proper choice of liquid phase and operating conditions.[22, 23] The acetates have been used for the gas-chromatographic analysis of inositols and their methyl ethers,[24] and data have been reported for the trifluoroacetates of the inositols and a few inosamines.[24a] Specific procedures have been published for the quantitation of myo-inositol,[23, 25] and the components of the mixture myo-inositol–sequoyitol–pinitol,[25a] as trimethylsilyl ethers. The ready isolation and crystallization of the O-trimethylsilyl derivatives of some common cyclitols suggests that these ethers may be interesting compounds in their own right.[25b]

For the purification of tissue extracts prior to derivatization and chromatography, the best general method would seem to be the alkali treatment used in conjunction with periodate analysis.[21] All reducing sugars are converted to acids (as barium salts) by heating the extract with barium hydroxide. After precipitation of the barium ions (not always necessary[25a]), the sugar acids and other ionic materials are removed with ion-exchange resins. Extracts thus purified will contain any open-chain alditols present in the original sample, permitting these to be determined along with the cyclitols.

E. Inositol Phosphates, Lipids, and Glycosides

The major inositol phosphate, from a quantitative viewpoint, is phytic acid, or myo-inositol hexaphosphate.[1, 26] Salts of this acid, accompanied by smaller amounts of related inositol polyphosphate salts, are found in plants and in soil.

* *References start on p. 573.*

The term "phytate" refers to salts of phytic acid per se, and also to these naturally occurring mixtures. Phytates are present generally in plant tissues, but they accumulate particularly in seeds until, at maturity, 65 to 90% of the phosphate of the seed is in this form. Calcium magnesium phytate is obtained by the extraction of plant material, and is usually supposed to be the native form of phytate. Association with other ions in the tissue is not excluded, however. Lower phosphates of *myo*-inositol (fewer than six phosphate groups) have been obtained from plants, and detected chromatographically in soil phytate. In addition, soil phytates contain small amounts of D-*chiro*-,† *scyllo*-, and *neo*-inositol polyphosphates.[27, 28] The origin of these compounds is an intriguing question. The only reported occurrence of phytate in animal tissues is in avian and reptilian erythrocytes. According to a recent study, the principal component of chicken erythrocyte phytate is *myo*-inositol 1,3,4,5,6-pentaphosphate.[28a]

In the past there have been doubts whether the six phosphate residues of phytic acid are in the form of simple monoester groups. However, electrometric titrations of carefully purified preparations[29] reveal the twelve acidic hydrogen atoms expected for the hexa-orthophosphate structure **10**. This structure has been confirmed by an X-ray crystallographic study of sodium phytate,[29a] which, in the crystal, has the conformation having five phosphate groups axially disposed and one equatorial. The ^{31}P n.m.r. spectrum of sodium phytate has been determined.[29]

A procedure for the synthesis of phytic acid by heating *myo*-inositol with polyphosphoric acid was published long ago by S. Posternak.[29b] Attempts to repeat this procedure have given products of questionable identity, but a modification of the method has been used successfully to prepare the hexaphosphates of *myo*-, *scyllo*-, *neo*-, and D-*chiro*-inositol.[29c]

Mixtures of the lower phosphates of *myo*-inositol are produced by the controlled chemical or enzymic (phytase) hydrolysis of phytate. Several of the individual components have been isolated and characterized. Small amounts of *myo*-inositol monophosphate have been isolated from various biological materials, and all the position-isomeric *myo*-inositol monophosphates have been prepared (the 4-phosphate only as the racemate) by the phosphorylation of appropriately blocked precursors.[1] A few monophosphates of other inositols have been made in the same way,[1] or by a variant of the "benzeneglycol" synthesis[29d] (see Section IV). The treatment of inositol phosphates with hot acid induces migration of the phosphate groups, easily to neighboring *cis*-hydroxyl groups, more slowly to *trans*.

The synthesis of individual inositol monophosphates is straightforward, given the availability of specifically blocked precursors (see Section V for

---

† Not DL-*chiro*, as previously assumed. (D. J. Cosgrove, personal communication.)

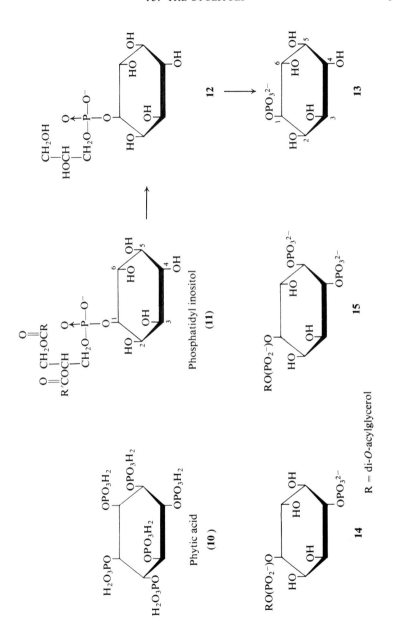

Phosphatidyl inositol
(11)

Phytic acid
(10)

R = di-O-acylglycerol

optically active *myo* derivatives). Isomerically pure, multiply but incompletely phosphorylated inositols cannot be made from such precursors with the usual phosphorylating agents, but successful preparations have been accomplished with *N*-benzoylphosphoramidic acid.[29e]

Isolations of *myo*-inositol and *myo*-inositol phosphates from hydrolyzates of phospholipids led to the recognition, around 1930, of the inositol lipids or phosphoinositides[1] (see Chapter 44). The simplest and most widely distributed of these is phosphatidyl inositol, which typically constitutes a small portion (usually less than 10%) of the total phospholipid from animal and plant tissues. In structure, phosphatidyl inositol (**11**) is a close analog of the longer known glycerophosphatides (lecithin, phosphatidyl ethanolamine, etc.).

Substantial amounts of more highly phosphorylated inositide were found in brain, and the original preparation was termed diphosphoinositide by its discoverer.[30] Subsequently, brain phosphoinositide was shown to be a mixture of phosphatidyl inositol, phosphatidyl inositol monophosphate (**14**, "diphosphoinositide"), and phosphatidyl inositol diphosphate (**15**, "triphosphoinositide").[31] Trace amounts of the latter two lipids accompany phosphatidyl inositol in other animal organs. More recently the phosphatidylinositols have been detected in various microorganisms.

The problem of characterizing these lipids was essentially one of determining the points of attachment of the phosphate groups to the inositol. Stepwise degradation of phosphatidyl inositol gave 1D-*myo*-inositol 1-phosphate (**13**), via an inositol glycerol phosphate[32] which must have the structure **12**. The absolute configuration of **13** was established by the synthesis of its enantiomorph from galactinol (see below).[33] Degradation of the brain inositides gave *myo*-inositol di- and triphosphates, which were characterized by successive periodate oxidation, reduction, and dephosphorylation to alditols.[34,35] Syntheses of phosphatidylinositol have been reported.[35a-35c]

Phosphoinositides having sugar residues attached to the inositol have been reported in a number of instances, and the materials from *Mycobacteria* and from seeds (phytoglycolipids) have been investigated in some detail. Some of the other lipids alleged to be of this type are probably association complexes of simple lipids with oligo- or polysaccharides.[36] In the characterization of the glycosylated phosphoinositides, vigorous alkaline hydrolysis is used to liberate "oligosaccharides" containing *myo*-inositol. These can then be studied by the usual methods of carbohydrate chemistry.

The phosphoinositides of the *Mycobacteria* (*M. tuberculosis*, *M. phlei*) are phosphatidyl inositols bearing from one to five D-mannose residues[37] (see Chapter 44). All have a single D-mannose residue linked at C-2 of the *myo*-inositol (see formula **11**). The higher members of the series carry in addition a single D-mannose residue or a D-mannose oligosaccharide at C-6. Phytoglycolipids[36,38,39] are derivatives of phytosphingosine or dehydrophytosphingosine (Vol. IIB, p. 733), and they also constitute families of related compounds. In the best studied type, the carbohydrate moieties form a series beginning with the "trisaccharide" O-2-amino-2-deoxy-α-D-glucopyranosyl-(1 → 4)-O-(α-D-glucopyranosyluronic acid)-(1 → 4D)-myo-inositol.† This is elaborated by the addition of an α-D-mannopyranosyl residue at position 2 of the myo-inositol, then addition to the 2-amino-2-deoxy-D-glucose of a D-galactose residue, or an oligosaccharide grouping containing galactose, arabinose, and sometimes fucose.[40, 41]

An interesting feature of the mannosyl phosphatidyl inositols of *Myco-bacteria* is that some of them have three or four fatty acid residues.[41a] Obviously, not more than two of these can be attached to the glycerol moiety. An indication as to the placement of the others is given by the structure of related lipids found in *Propionibacteria*. These lipids are 2-O-α-D-mannopyranosyl-myo-inositols carrying fatty acyl groups at position 6 of the D-mannose moiety and a 1 position of the inositol.[41b]

The best studied simple glycoside of an inositol is galactinol, 1L-1-O-α-D-galactopyranosyl-myo-inositol.[42] The occurrence of this widely distributed galactoside parallels that of the galactosylsucroses raffinose and stachyose, with which it has a biochemical relationship (see Section II,G).[42a] An additional compound of this family, O-α-D-galactopyranosyl-(1 → 6)-O-α-D-galacto-pyranosyl-(1 → 1L)-myo-inositol, has been isolated from the seeds of the vetch *Vicia sativa*.[42b] A β-D-glucopyranosyl-myo-inositol was found among the oligo-saccharides of potato tubers after two months' storage,[43] and other glycosyl-myo-inositols have been obtained as degradation products of lipids. Attempts to synthesize glycosylinositols by chemical or enzymic means have given generally disappointing results, but this situation will change as better methods for synthesis of oligosaccharides are devised. Successful applications of the Koenigs–Knorr reaction to inositol acceptors include the preparation of 1D-3-O-β-D-glucopyranosyl-4-O-methyl-chiro-inositol in good yield, from tetra-O-acetyl-α-D-glucopyranosyl bromide and di-O-isopropylidene-D-pinitol (formula **55**, H instead of Ts),[44] and the nearly quantitative condensation of tetra-O-benzoyl-α-D-mannopyranosyl bromide with 3-bromo-3-deoxy-1,2:4,5-di-O-isopropylidene-muco-inositol.[45]

A very interesting discovery is that the seeds of corn contain indoleacetic acid (the plant growth factor) in the form of esters with myo-inositol and an arabinosyl-myo-inositol.[46] In the two simple esters which were isolated, the indoleacetyl group occupies positions 1 and 2, respectively, of the inositol. One of these isomers may be an artifact.[46]

---

† (1 → 4D) indicates linkage to O-4 of myo-inositol, numbered normally as stipulated above. With this numbering O-4 is to the right in a Fischer projection.

* *References start on p. 573.*

F. Oxidation with Periodate and Nitric Acid

The behavior of inositols with periodate is generally regarded as anomalous because, at room temperature and below, a greater-than-expected amount of reagent is consumed, and some carbon dioxide is produced.[47] However, normal oxidation (six moles of periodate consumed and six moles of formic acid produced per mole of inositol) is achieved in procedures for the quantitative determination of *myo*-inositol. These procedures employ elevated temperatures (38 to 65°) and dilute solutions, sometimes buffered a little above pH 4.[21, 21a] It is believed that, in any case, the initial cleavage product is a hexodialdose (for example, **16** from D-*chiro*-inositol), arising by preferential attack at a pair of *cis*-hydroxyl groups. In most instances, the dialdose is rapidly oxidized in the open-chain form. Thus, under the usual conditions a *portion* of the dialdose is

cleaved between C-3 and C-4 to give hydroxymalonaldehyde (tartronaldehyde), with the remainder presumably being converted into first glyoxal, and then formic acid. The eventual "overoxidation" of the hydroxymalonaldehyde accounts for the anomalous stoichiometry.[48–49a] Occasionally, as in the case of dambonitol (1,3-di-*O*-methyl-*myo*-inositol), the open-chain form of the initial dialdose is not readily attacked by periodate. The dialdose then goes over to a cyclic hemiacetal form, and on oxidation yields a formate ester that undergoes slow hydrolysis, as is often observed in the periodate oxidation of sugars[3] (see Chapter 25). Inositols (except *myo*) having *cis*-1,2,3-triol groupings that can assume the *a–e–a* conformation appear to form stable complexes with periodate at pH 7, in common with other cyclic *cis*-1,2,3-triols. Evidence for formation of a complex is the fact that oxidation at pH 7 levels off when the periodate consumption reaches 2 to 4 moles per mole of inositol.[49b] In $0.1 N$ acid at 0°, inositols yield some glyoxylic acid (~0.75 moles per mole of inositol), which is not further oxidized under these conditions.[49c] The stoichiometry is the same as for complete conversion into formic acid.

The oxidation of inositols (but not lower cyclitols) with nitric acid, followed by isolation under alkaline conditions, gives benzenehexol (19) and an interesting series of quinonoid compounds (20, 21, 23) which can be derived from it by stepwise oxidation.[50] The fact that some of these compounds form highly colored salts with Ca^{2+} and other metal ions is the basis of the classical Scherer test for inositols.[1] Tetrahydroxyquinone (20) and rhodizonic acid (21) can be made on a preparative scale by oxidation of *myo*-inositol,[51,52] but benzene-hexol (19) and croconic acid (23) are best obtained from 20 in separate steps.[53] Two additional compounds of the series, triquinoyl (22) and leuconic acid (24), are prepared by the action of nitric acid on tetrahydroxyquinone and croconic acid, respectively.[53]

The *initial* residue from the treatment of inositols with nitric acid is colorless, whereas tetrahydroxyquinone is a deeply colored substance. This and much other evidence suggests that compounds 19–24 are secondary products, arising during workup by processes of enolization, air oxidation, and benzilic acid rearrangement. A precursor (or precursors) of the secondary products is

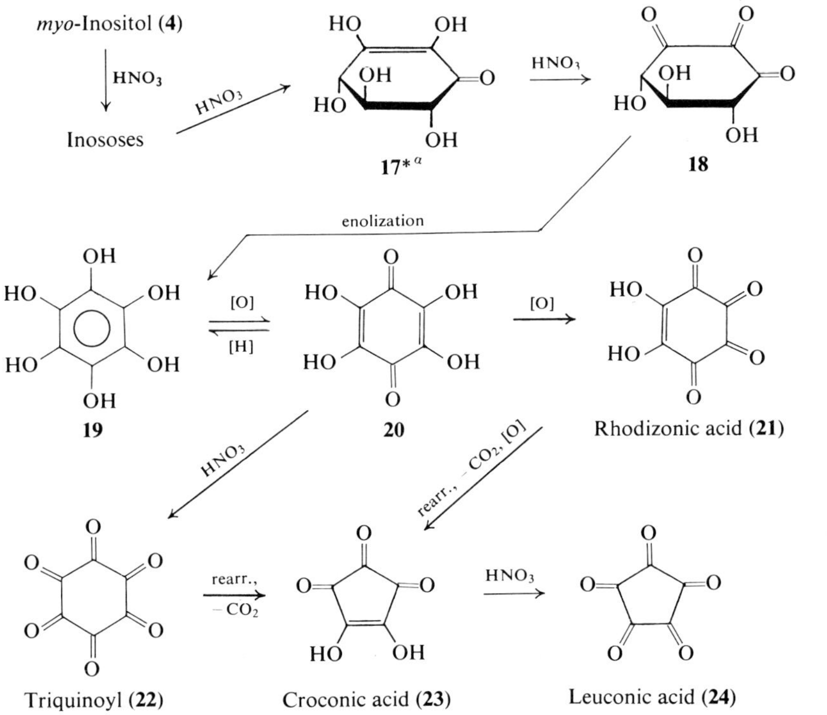

myo-Inositol (4)

Inososes

17*[a]

18

enolization

19

20

Rhodizonic acid (21)

Triquinoyl (22)

Croconic acid (23)

Leuconic acid (24)

[a] When formula numbers carry an asterisk (*) the compound is a DL mixture, of which one enantiomorph, chosen arbitrarily, is shown.

* References start on p. 573.

formed during the primary, nitric acid oxidation, the details of which are only partly elucidated. The sequence inositol → inososes → **17** → **18** is suggested by the fact that an inosose (**67**) or the enediolic acid **17** can be obtained in low yields when the oxidation is carefully controlled.[54, 55] Further treatment of **17** converts it to **18**, which on acetylation in the presence of a basic catalyst readily gives hexaacetoxybenzene.[55] The triketone **18** may thus be an important primary oxidation product.

G. BIOCHEMISTRY

Since the identification of *myo*-inositol as one of the "bios" complex of yeast growth factors,[56] there have been numerous studies on the role of this inositol in the nutrition of living organisms. A definite requirement for *myo*-inositol is shown by a few yeasts, and by a number of plant and animal tissues when they are grown in artificial culture. In many other cases growth is not dependent on *myo*-inositol, but is increased when it is added to the nutrient medium or diet.[1, 57]

There were early claims[58] that a deficiency of *myo*-inositol depressed the growth of higher animals, and caused an alopecia (loss of hair), but these claims have not been fully substantiated. Stimulation of the growth of higher animals by *myo*-inositol has indeed been observed,[1] but the effect in most cases appears to have been indirect, for the basal diets used have usually been deficient in other components, or unbalanced in some way. In the most careful experiments with diets low in *myo*-inositol but adequate in all other respects, no retardation of growth or other deficiency symptoms were seen.[59–61]

Although many organisms do not require an external source of *myo*-inositol, it is clear that this compound is an essential item in the life economy of almost all cells. One of its major functions is to serve as a precursor of more complex molecules, such as phosphoinositides and phytates. The possibility of other functions is often suggested, and cannot be discounted, but so far no biological response to *myo*-inositol has been demonstrated to result from the action of the free cyclitol at the molecular level.

Intensive research on the various bound forms of *myo*-inositol is slowly furnishing insight into the physiological roles of these substances. The phosphatidyl inositols **11**, **14**, and **15**, along with other phospholipids, are components of cellular-membrane structures. They may be required for membranes to function normally in the transport of ions,[61a] amino acids,[61b] and other metabolites. Phytate clearly serves as a storage form of phosphate, and the possibility is being investigated that a portion of this phosphate has a sufficiently high $\Delta G°$ of hydrolysis to act directly as an energy source in metabolism.[61c] Accumulating evidence also assigns a biochemical niche to galactinol. This glycoside appears to be the galactosyl donor in the synthesis of oligosaccharides of the raffinose series (raffinose, stachyose, verbascose).[61d,61e] The possible involvement of other glycosylinositols in synthesis of complex saccharides is of obvious interest.

Considerable progress has been made, primarily during the 1960's, in elucidating the metabolic reactions of the inositols. A major achievement is the confirmation of the hypothesis, suggested 80 years ago by Maquenne, that the biosynthesis of *myo*-inositol involves the cyclization of D-glucose with the chain intact. This was shown in experiments with yeast,[62,63] higher plant,[64,65] and mammalian systems,[66] by using D-glucose (**25**) labeled as shown in Scheme A. 1L-*myo*-Inositol 1-phosphate (**27**) (erroneously called D-inositol 1-phosphate in one paper) is an intermediate; it is formed from D-glucose 6-phosphate

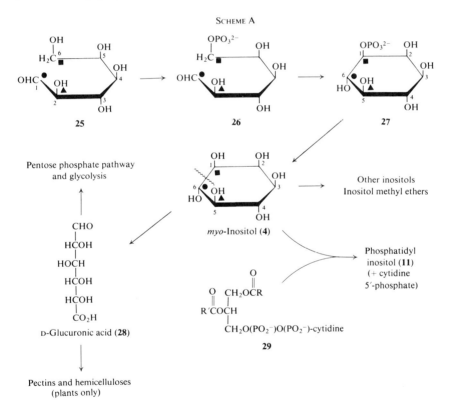

SCHEME A

(**26**) and hydrolyzed by a phosphatase.[67,68] The cyclization reaction probably includes oxidation and reduction steps, since the purified enzyme system requires nicotinamide adenine dinucleotide (NAD^+) as a cofactor.[68a]

D-Glucose 6-phosphate labeled in various ways with deuterium or tritium has been used in several studies of the enzymic cyclization. The results support the hypothesis that enzyme-bound D-*xylo*-hexos-5-ulose ("5-ketoglucose") 6-phosphate or its enol is the cyclizing species.[69, 69a] The non-enzymic, base-catalyzed cyclization of D-*xylo*-hexos-5-ulose to inososes has been demonstrated.[69b]

Tracer experiments carried out by Hoffmann-Ostenhof and co-workers have established the strong presumption that *myo*-inositol is the precursor of the other inositols and the inositol methyl ethers found in plants.[65] Direct methylation of *myo*-inositol, catalyzed by position-specific enzymes, appears to be the route of biosynthesis of the methyl ethers sequoyitol, D-ononitol, and D- and L-bornesitol (Table I).[12b, 69c] The dimethyl ether dambonitol arises, in one plant, by further methylation of D-bornesitol,[12b] but L-bornesitol may turn out to be the intermediate in other species. The epimerization of the *myo*- to the *scyllo*- and D- and L-*chiro*-configurations is accomplished by dehydrogenation to intermediate inososes, and then rehydrogenation of these inososes in the opposite steric sense. The formation of *scyllo*-inositol by this process has been demonstrated in animals[7b, 8] and in an actinomycete.[69d] In some higher plants, but not in *Chlorella*,[7a] conversion into the D-*chiro*-series is effected on sequoyitol (D-pinitol is the product) rather than on *myo*-inositol itself.[12b, 65] *myo*-Inositol does not seem to be an intermediate in the biosynthesis of either the lower cyclitols (Section VII),[12b] or the *C*-methylinositols (Section IX).[7a]

In some bacteria the degradation of *myo*-inositol is initiated by the dehydrogenation of a carbinol group, and in one species (*Acetobacter suboxydans*, see Section VI) action is apparently limited to this step. Higher organisms—inositol-degrading yeasts, higher plants, and mammals—are equipped with an inositol oxygenase,[12b, 70, 70a] which effects a cleavage between C-1 and C-6 according to the equation

$$myo\text{-Inositol} + O_2 \rightarrow D\text{-glucuronate} + H_2O + H^+$$

Half of the oxygen consumed is incorporated into the carboxyl group of the D-glucuronate (**28**). This compound is then converted by well-established reactions into D-*threo*-pentulose ("D-xylulose") and then, via the pentose phosphate sequence, to the intermediates of glycolysis.[71–73] In this way, *myo*-inositol is channeled into the mainstream of carbohydrate metabolism. In plants, part of the D-glucuronate is converted into D-galacturonic acid, D-xylose, and L-arabinose residues in the cell-wall polysaccharides (pectins, hemicelluloses)[72] (see Chapters 37 and 39). *myo*-Inositol thus becomes an intermediate in one pathway for the synthesis of these substances from D-glucose.

Studies with cell-free enzyme systems have indicated that *myo*-inositol is incorporated into phosphatidyl inositol by reaction of the free cyclitol with a cytidine 5′-(diglyceride pyrophosphate) (29).[21, 74] Diphosphoinositide (14) and triphosphoinositide (15) appear to be formed, in brain, by the stepwise phosphorylation of phosphatidyl inositol.[75, 76] Similarly, *Mycobacteria* possess enzymes for the synthesis of their complex inositides by the mannosylation and further acylation of phosphatidyl inositol.[76a]

The biosynthesis of phytic acid, not yet fully understood, may involve successive phosphorylations of a *myo*-inositol residue that is part of a larger molecule. Cleavage to phytate and the "carrier" substance would be the last step. The simple lower phosphates of *myo*-inositol do not seem to be intermediates (Asada *et al.*[76b]). The preformed, free inositol is incorporated into phytate by several systems (Asada *et al.*,[76b] Loewus[76b]), but it may not be an obligatory intermediate.[12b]

The proceedings of a symposium held in the autumn of 1968 provide discussions of the then current status of many aspects of inositol biochemistry.[76b]

III. CYCLITOLS AND CONFORMATIONAL ANALYSIS

From the general tenets of conformational analysis the following deductions can be made about the conformations of cyclitols having six-membered rings:[77] (1) these cyclitols should exist as chair forms in most situations; (2) when a cyclitol [for example, *myo*- (30), *epi*- (69), and the *chiro*- (71) inositols] has more axial substituents in one of the two chair forms than the other, the form with the fewer axial substituents should preponderate; (3) when there are equal numbers of axial and equatorial hydroxyl groups there may be only one conformer [chair–chair conversion gives superposable conformations— for example, *muco*-inositol (61) and *cis*-inositol], or there may be two nonsuperposable conformers; (4) in the latter case the conformers may be enantiomorphic (*allo*-inositol) or nonequivalent (1,2/3,4-cyclohexanetetrol); (5) conformations other than the chair (half-chair, skew form) may be favored when constraints such as double bonds or fused rings are present in the structure.

The configurations of cyclitols that cannot be named as inositol derivatives are indicated by the Maquenne system, in which the positional numerals of all the substituents are arranged as a fraction. The numerals for one set of substituents constitute the numerator of the fraction, and those for the other set the denominator. Typically the set containing the lowest number is used as the numerator, regardless of the orientation of the formula.

The foregoing analysis led to the rationalization of much of the previously known chemistry of the cyclitols, and it has been the basis of many successful predictions. Some early direct evidence that the conformations of cyclitols are as stated in statements 1 to 5 came from dipole moment data and infrared spectra,[3] and p.m.r. spectroscopy has since provided ample verification. X-ray analysis has shown that *myo*-inositol has the expected conformation (30) in the crystal.[78] Similarly it was found[78a] that *epi*-inositol adopts a relatively undistorted chair conformation, but that the *syn*-diaxial oxygen atoms are displaced outward to 2.96 Å from their "ideal" distance of 2.50 Å.

The mass spectra of structurally isomeric, unsubstituted cyclitols differ in characteristic ways. Stereoisomers can be distinguished, in some cases, by differences in the intensities of key peaks.[78b] Such differences are marked in the spectra of the trimethylsilyl ethers of the inositols.[78c]

30   31   32

A. NONBONDED INTERACTION ENERGIES

Important contributions to the conformational analysis of sugars have resulted from studies on cyclitols, particularly from the work of Angyal and McHugh.[79,80] In a study of borate complexing, these authors found that cyclitols having *cis*-hydroxyl groups in a 1,3,5-relationship were in simple equilibrium with their 1:1 complexes. More complicated equilibria were found with other types of cyclitols, indicative of both 1:1 and 2:1 (borate–cyclitol) complexes. It was postulated that cyclitols of the first group react with borate in their less-favored 1,3,5-triaxial forms (for example, 31 for *myo*-inositol) to give tridentate cage structures such as 32.* Data on the free-energy changes associated with complex formation (from measurements of the equilibrium constants) were then used to calculate the energies of the nonbonded interactions of hydroxyl groups, methyl groups, and hydrogen atoms in carbohydrate molecules (see Chapter 5). The values for $\Delta G°$ (overall reaction) ranged from -0.95 kcal mole^{-1} for 1,3,5/2,4-cyclohexanepentol (65) to -8.15 kcal mole^{-1} for *cis*-inositol (which is 1,3,5-triaxial in the single possible chair conformation).

---

* *scyllo*-Inositol, which in its less favored chair form would have two separate sets of triaxial (1,3,5) hydroxy groups, forms an isolable complex containing borate and cyclitol in the ratio 2:1. The doubly tridentate structure proposed for this compound is supported by the p.m.r. spectrum[80a] (only one peak, a broadened singlet at τ 5.61).

Three types of nonbonded interaction were considered to be significant in the unsubstituted cyclitols: OH having neighboring, gauche OH or OR groups $(O_1:O_2)$; OH having a diaxially opposed H atom $(O_a:H_a)$; and OH having diaxially opposed OH group $(O_a:O_a)$. For each cyclitol one could then write an equation of the type

$$\Delta G° = \Delta G_B° + (G_C - G_F)$$

where $\Delta G_B°$ is the standard free-energy change of the esterification part of the process, and G_C and G_F are, respectively, the sums of the nonbonded interaction energies in the complex and in the free cyclitol in its favored conformation. The equation for *myo*-inositol, for example, is

$$\Delta G_{myo}° = \Delta G_B° + [(2O_1:O_2 + 2O_a:H_a + O_a:O_a) - (6O_1:O_2 + 2O_a:H_a)]$$

$$= \Delta G_B° - 4O_1:O_2 + O_a:O_a = -1.90 \text{ kcal mole}^{-1}$$

A least-squares method was used to determine values for $O_1:O_2$, $O_a:H_a$, and $O_a:O_a$ which best fit the set of simultaneous equations written for the six inositols and cyclohexanepentols used. The same procedures, applied to a set of *C*-methyl cyclitols, gave values (less reliable) for the H:Me and OH:Me interactions. The internal consistency of all the values suggests their general validity and supports the structure proposed for the complexes.

B. Use of Proton Magnetic Resonance Spectroscopy

Proton magnetic resonance spectroscopy has found extensive use in the cyclitol field. The conformation, and hence the configuration, of a cyclitol can often be inferred from the p.m.r. spectrum, preferably by evaluating some of the vicinal coupling constants of the ring protons (those attached to the carbon atoms of the ring). Since these tend to form complex spin systems, the use of 100 MHz or higher frequencies is helpful, and spin decoupling is often necessary.[81] Also useful is the introduction of electronegative substituents at some positions. This spreads the chemical shifts of the ring protons over a wider range, giving spectra more amenable to "first-order" analysis. An example is the spectrum of the selectively methylated and *O*-acetylated deoxystreptamine shown in Fig. 1; a study of this spectrum confirmed the all-equatorial disposition of the functional groups, and hence the all-*trans* configuration of the parent aminocyclitol (see p. 563).

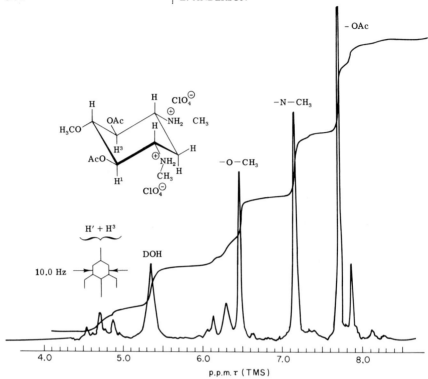

FIG. 1. The p.m.r. spectrum of 1,3-di-O-acetyl-4,6-di-N-methyl-2-O-methyldeoxystrept-amine bis(hydrogen perchlorate) in deuterium oxide; compare R. U. Lemieux and R. J. Cushley, *Can. J. Chem.*, **41**, 858 (1963). Positions numbered according to systematic nomenclature, which differs from that of the original paper.

It is also useful to examine the spectra of fully acetylated derivatives in chloroform solution. The chemical shifts in this solvent of the acetyl-methyl protons of a large number of acetylated carbohydrates have been recorded. In the spectra of fully acetylated cyclitols the lines for axial acetyl groups of a given type (*O*- or *N*-) fall in regions quite distinct from those characteristic of equatorial groups of the same type (Fig. 2). The positions of acetyl-methyl lines are perturbed when the compound is not fully acetylated, and when aromatic or certain other types of substituent are present, but in the absence of such perturbations are good indicators of the conformation of the molecule.[81a] Slight downfield shifts of the characteristic ranges are seen in mixtures of $CDCl_3$ and CD_3OD, which have greater solvent power than $CDCl_3$ alone.[81b] Improved separation of the ranges is observed in CD_3SOCD_3.[81a]

The ^{13}C n.m.r. spectra of a series of inositols and inositol methyl ethers have been reported.[81c] Detailed p.m.r. studies have been made on a series of derivatives of (−)-quinic acid.[81d, 81e]

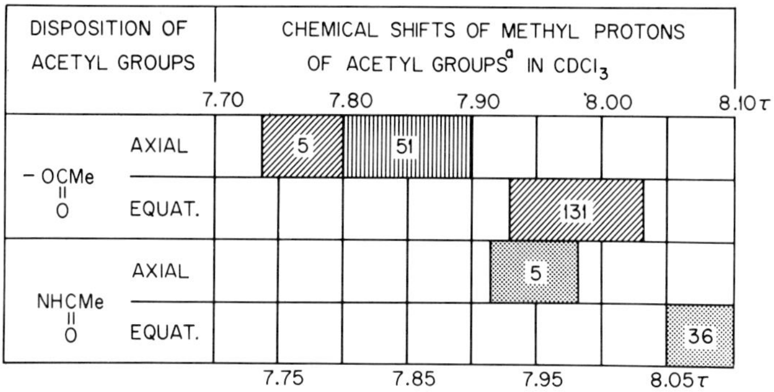

DISPOSITION OF ACETYL GROUPS		CHEMICAL SHIFTS OF METHYL PROTONS OF ACETYL GROUPS[a] IN CDCl$_3$				
		7.70	7.80	7.90	8.00	8.10 τ
− OCMe $\overset{\text{II}}{\text{O}}$	AXIAL		5 51			
	EQUAT.				131	
NHCMe $\overset{\text{II}}{\text{O}}$	AXIAL				5	
	EQUAT.					36
		7.75	7.85	7.95	8.05 τ	

FIG. 2. Relationship between chemical shifts of the methyl resonances of O- and N-acetyl groups and their conformational disposition. Adapted from F. W. Lichtenthaler and P. Emig, ref. 81a.

[a] The number in each shaded area is the number of individual acetyl-methyl signals found in that range.

C. OPTICAL ROTATION

The rotations of optically active cyclitols agree well, in most cases, with values calculated by Whiffen or Brewster's methods (see Chapter 7). Since the calculations are based on conformational factors, this agreement strongly supports the generally accepted conformational assignments.[81] In the characterization of new cyclitols, rotational calculations may be used to check a proposed configuration, or to choose between alternatives when other types of evidence have reduced the number of the possibilities to a very few. It is usually essential to be able to assume a favored conformation for each configuration considered.

IV. METHODS FOR THE TOTAL SYNTHESIS OF CYCLITOLS

A. SYNTHESIS FROM SMALLER MOLECULES

A synthesis of cyclitols from smaller molecules is provided by the hydrogenation of tetrahydroxybenzoquinone or benzenehexol, since the latter compounds can be made by the self-condensation of glyoxal[53,82] or carbon monoxide.[83] Some of the earlier work on this hydrogenation could not be repeated, but a careful modern study has delineated its nature.[84] Because the

* *References start on p. 573.*

reaction is accompanied by much hydrogenolysis the products range from cyclohexanetriols to inositols. The exact composition of the mixture varies according to the catalyst and conditions used: With a precipitated palladium catalyst *myo*-inositol was the main product (17%), whereas with palladium on charcoal *cis*-inositol was predominant. None of the mixtures was completely resolved, but by chromatography and fractional crystallization a dozen cyclitols in all were isolated, including every inositol except *muco*-inositol.

The condensation of ^{14}CO in the presence of metallic potassium has been used in making uniformly labeled *myo*-inositol via benzenehexol.[85]

exo + endo

33   34   35

The construction of a cyclitol ring by the Diels–Alder reaction has been successfully accomplished in several instances. The use of vinylene carbonate (34) as the dienophile provides two potential hydroxyl groups, and additional oxygen functions can be introduced via the diene [for example, furan (33),[86, 86a] or 1,4-diacetoxybutadiene[87]].

B. The "Benzeneglycol" Synthesis

For the only truly general total synthesis of cyclitols, devised by Nakajima and his collaborators, the starting materials are *cis*- and *trans*-3,5-cyclohexadiene-1,2-diol, referred to as the "benzeneglycols."[88, 89] These glycols (37, 38) are made as shown in Scheme B from 3,6/4,5-tetrachloro-1-cyclohexene (36), itself one of several products of the iodine-inhibited photochlorination of benzene. The hydroxylation of the "benzeneglycols" gives 5-cyclohexene-1,2,3,4-tetrols ("conduritols"),† and is accomplished via the sequences illustrated (Scheme B) for the *cis*-glycol.[89, 90] In the *trans*-hydroxylation, oxygen from peroxybenzoic acid adds to a double bond cis to a neighboring, allylic hydroxyl group, to give an epoxide intermediate which is subsequently hydrolyzed. *cis*-Hydroxylation is usually effected with permanganate; addition occurs preferentially trans to a neighboring, allylic acetoxy group.

In a variation of the synthesis the diacetates of the glycols (for example, 40) are carefully converted into epoxides with peroxybenzoic acid, and the epoxides are treated with methanolic ammonia.[91] In all cases the ammonia attacks the

† See p. 560 for comment on this nomenclature.

SCHEME B

oxirane ring at the position next to the double bond, yielding 6-amino-4-cyclohexene-1,2,3-triols ("conduramines"†—for example, **44** and **45**).

Nakajima and his co-workers have shown, in a series of elegant papers, that the cyclohexenetetrols and aminocyclohexenetriols can be further converted into inositols,[92] inositol alkyl ethers,[93] cyclohexanepentols,[94] deoxyhalo-inositols,[94] and aminodeoxyinositols (inosamines and inosadiamines, see Section VIII). These syntheses have greatly enriched the chemistry of the cyclitols even though, because of the many steps involved and incomplete stereospecificity at some points, they will probably not often be employed for preparative work.

† See p. 560 for comment on this nomenclature.

* References start on p. 573.

C. The Nitroinositol Synthesis

The condensation of nitromethane with aldehydes, an important general reaction of carbohydrates (Chapter 3), has been used to make cyclitols from sugars.[95] In the most thoroughly studied procedure the mixture (47) of 6-deoxy-6-nitro-D-glucose and -L-idose formed by the action of nitromethane, and then acid, on 1,2-*O*-isopropylidene-5-*aldehydo*-α-D-*xylo*-pentodialdo-1,4-furanose (46) is cyclized in basic solution.[96, 97] If barium hydroxide is used (precipitation of barium *aci*-nitro salts) one obtains a mixture of deoxynitro-inositols having the *muco* (48), *myo* (49), and *scyllo* (50) configurations, but with sodium hydroxide only the *myo* and *scyllo* isomers. Apparently deoxy-nitro-*muco*-inositol is formed first and is then isomerized to the more stable *myo* and *scyllo* forms.[97a] Only the latter are present at equilibrium, with *scyllo*

preponderating. The same set of products results if unsubstituted *xylo*-pentodialdose is treated with nitromethane and base,[98] or if methyl 6-deoxy-6-nitrohexopyranosides are treated with mild base.[99] In the latter case, the opening of the sugar ring presumably involves an elimination. The reduction of deoxynitroinositols affords inosamines, and these can be deaminated (Section VIII) to inositols.

Work on the nitromethane synthesis through early 1967 has been comprehensively reviewed by Lichtenthaler.[99a]

V. REACTIVE CYCLITOL DERIVATIVES

A. SELECTIVE BLOCKING

The selective blocking of some of the hydroxyl groups in a carbohydrate molecule is contingent on these groups being more reactive than others toward blocking agents. In the cyclitols such selectivity is almost entirely dependent on conformational factors. Formation of cyclic acetals (see Scheme C for examples, Section X for discussion) is the primary blocking reaction most frequently used in this series. The hydroxyl groups not affected by this step may then be subjected to a desired reaction (for example, **51** → **52** or **53**), or they may be protected by acetylation, benzylation, etc. A treatment with mild acid then accomplishes hydrolysis of the cyclic acetal function, unmasking the hydroxy groups involved in acetal formation, as in the preparation of **58**. Some of the free hydroxyl groups of partially protected cyclitols may be disposed axially, and others equatorially, and in some cases (for example, **58**) the equatorial groups can be substituted selectively. Preference for equatorial hydroxyl groups is shown in reactions (acetylation, sulfonylation, methylation, benzylation, etc.) where the number of ligands to the attacking atom of the reagent is increased in the transition state.[100] Reactions that proceed via carbonium ions (for example, pyranylation) are less sensitive to steric hindrance. In some cyclitol derivatives, for instance in 1,2,3,4-tetra-O-benzyl-*chiro*-inositol, the free axial hydroxyl groups appear to be no more hindered than the equatorial ones.[101]

The acetolysis of 1,4,5,6-tetra- and 1,3,4,5,6-penta-O-benzyl-*myo*-inositol is remarkably selective. The groups at positions 4, 5, and 6 are quickly cleaved off to give 1-mono- and 1,3-di-O-benzyl-*myo*-inositol, respectively, in good yield.[101a]

The resolution of racemic, partially blocked *myo*-inositols is one possible route to the optically active intermediates needed in biochemically oriented syntheses. The enantiomorphic forms of 1,4,5,6-tetra-O-benzyl- and 1,2,4,5,6-penta-O-benzyl-*myo*-inositol have been separated via their diastereoisomeric 3,4,6-tri-O-acetyl-D-mannose 1,2-orthoacetate derivatives.[101b, 101c]

B. SULFONIC ESTERS

The mono- or disulfonyl (mostly *p*-tolylsulfonyl) derivatives readily obtainable from partially blocked cyclitols are useful intermediates. Monosulfonates having a *trans*-hydroxyl group vicinal to the sulfonic ester function (for example, **52**), and appropriately blocked vicinal *trans*-disulfonates (for example,

* *References start on p. 573.*

SCHEME C

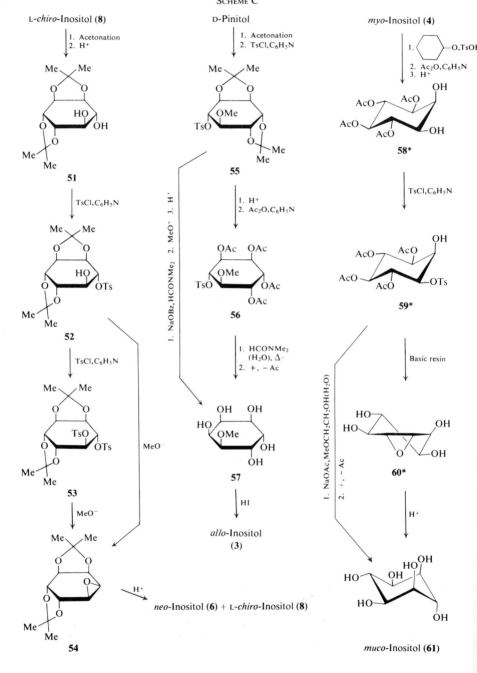

L-*chiro*-Inositol (**8**)

D-Pinitol

myo-Inositol (**4**)

51

55

58*

52

56

59*

53

57

60*

allo-Inositol
(**3**)

54

neo-Inositol (**6**) + L-*chiro*-Inositol (**8**)

muco-Inositol (**61**)

53), like their analogs in the sugar series, are converted into epoxides by treatment with bases. In the transformation of the disulfonates the first sulfonyl group is presumably attacked at the sulfur atom, with resultant S–O cleavage.

The nucleophilic displacement reactions of the cyclitol sulfonates parallel those of the secondary sulfonates of pyranoid sugar derivatives. The conditions required for reaction, and the neighboring-group effects observed, are similar in the two series. So far the reagent most used with cyclitols has been azide ion, which gives intermediates reducible to inosamines (see Section VIII). Displacement by a neighboring acetoxy group and direct displacement by benzoate ion have also been explored, in the *chiro*-inositol series.[19a] These reactions proceed smoothly, for the most part, in boiling *N,N*-dimethylformamide. Thus the di-*O*-isopropylidene-*O*-*p*-tolylsulfonyl derivative (**55**) of D-pinitol can be converted into 1L-1-*O*-methyl-*allo*-inositol (**57**) via the tetraacetate **56**, or directly by treatment with sodium benzoate followed by deblocking (see Scheme C). These reactions provide the best presently available synthesis of *allo*-inositol, with the direct displacement giving the better yield.

An excellent synthesis of *muco*-inositol (**61**) depends on the neighboring-group displacement of a sulfonate ion from the readily available 1,4,5,6-tetra-*O*-acetyl-3-*O*-*p*-tolylsulfonyl-*myo*-inositol (**59**), or the corresponding methyl-sulfonyl derivative.[101d] In this case, presumably, an initially formed acetoxonium ion of the *chiro* configuration undergoes epimerization (Section IX) before hydrolysis to a mixture of *muco*-inositol tetraacetates. Inversion of the configuration at position 1 of the *chiro*-inositols, to give optically active, partially blocked *myo*-inositol derivatives, is still another potentially important application of the sulfonate displacement reaction.[101e]

The *p*-toluenesulfonates of the inositols undergo solvolysis in boiling acetic acid if they are free or largely free of other substituents.[102] In some cases the sulfonic ester group is replaced mainly with stereochemical inversion, in others with preponderant retention of configuration. This ambiguity limits the usefulness of the reaction.

Iodide ion, under vigorous conditions, causes the elimination of vicinal sulfonyloxy groups from cyclitols with the formation of cyclohexenetetrols in fair yields. Improved yields would probably result if zinc dust were included in the reaction mixture, as recently described for an open-chain disulfonate.[102a] The reaction is facilitated if, instead of *p*-tolylsulfonyl, the more electron-withdrawing *p*-nitrophenylsulfonyl group is used.[103]

C. Anhydroinositols

The "1,2"-anhydroinositols, derived from sulfonic esters or formed by the action of peroxybenzoic acid on cyclohexenetetrols, can be transformed in a

* *References start on p. 573.*

number of ways. If blocking groups, carried over from sulfonic ester precursors, are present, it may be useful to retain them during subsequent operations. However, the oxirane rings of anhydroinositols are sufficiently stable to permit the removal of some blocking groups, such as the isopropylidene groups of the diacetal (54) of 1,2-anhydro-*allo*-inositol.[104]

In the reactions of anhydroinositols with nucleophiles the preponderant product is usually the one expected from the diaxial ring opening of the favored half-chair form of the cyclitol.[77] Thus, for example, 1,2-anhydro-*chiro*-inositol (60) yields mainly *muco*-inositol and only a little *myo*-inositol on acid hydrolysis.[105] The two half-chair forms of 1,2-anhydro-*allo*-inositol each have two axial or pseudoaxial hydroxyl groups, and are apparently favored about equally; acid hydrolysis gives about as much *chiro*-inositol as *neo*-inositol.[106]

An interesting special case is the reaction of anhydroinositols with sodium borohydride in methanol and ethanol to give the alkyl ethers corresponding to diequatorial ring opening.[93] It was suggested that here the active nucleophiles are the bulky tetraalkoxyboron anions, $(RO)_4B^-$, and that these are unable to attack the oxirane ring in the normal manner. Another possibility might be that the anhydroinositols are converted into complex borates whose structures guide the reaction in a seemingly anomalous course. A preponderance of anomalous ring-opening is also observed, in several cases, when anhydroinositols are heated in moist N,N-dimethylformamide containing sodium benzoate.[19a]

A "1,4"-anhydroinositol (as its benzylidene acetal) is, surprisingly, the product of the reaction of *myo*-inositol with benzaldehyde in the presence of *p*-toluenesulfonic acid.[107] The adducts (35) formed from furan and vinylene carbonate, and various of their transformation products, are also 1,4-anhydrocyclitols.[86, 86a] On vigorous acid hydrolysis the anhydro ring is opened with inversion at the point of opening.

D. DEOXYHALOINOSITOLS

The modern studies of McCasland and colleagues have clarified the reactions that occur when inositol acetates are heated with HBr (or HCl)—acetic acid in sealed tubes, or when the free inositols are treated with acetyl bromide or chloride.[108-110] Since *myo*- and *scyllo*-inositol, as well as the *chiro*-inositols and their methyl ethers, all give the same mixture of two monobromo and two dibromo compounds, epimerization (Section X) is involved. The opening of an acetoxonium intermediate (62) by bromide ion would explain the configurations of the two monobromodeoxyinositols (pentaacetates, formulas 63 and 64),[81, 105] and an extension of the overall process could lead from DL-1-bromo-1-deoxy-*chiro*-inositol pentaacetate (63) to the dibromo compounds, which

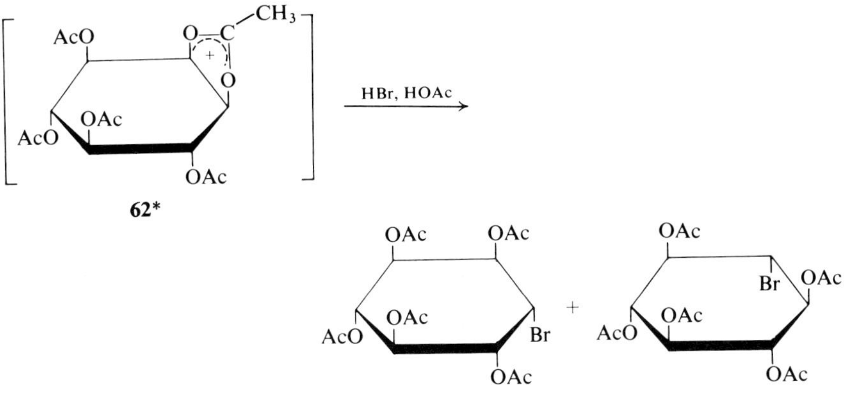

62*

HBr, HOAc
⟶

63* + **64**

are DL-2,5-dibromo-2,5-dideoxy-*chiro*-inositol and DL-1,5-dibromo-1,5-dideoxy-*chiro*-inositol.[81]

The action of concentrated aqueous hydrohalogen acids (except hydrogen fluoride) on anhydroinositols at room temperature affords a more definitive route to deoxyhaloinositols.[81] A large number of chloro-, bromo-, and iodo-cyclitols of known configuration have been made in this way. A different approach will evidently be required for the synthesis of fluorocyclitols.

Deoxyhaloinositols readily undergo hydrogenolysis with loss of the halogen, and with bases they give the products expected on the assumption that anhydro intermediates are formed. The acetates of the bromo compounds are smoothly converted into cyclohexenetetrol acetates by the action of zinc dust–acetic acid.[81]

E. THIOINOSITOLS

The replacement of one hydroxyl group of a cyclitol by —SH or —SR should be easily accomplished by treating 1,2-anhydro compounds with appropriate thiol derivatives, and this has been done in one case.[111] With potassium methylxanthate the epoxide oxygen of anhydroinositols is either replaced by a *trans*-trithiocarbonate (—S—CS—S—) function, or an elimination reaction takes place.[81]

VI. CYCLITOL KETONES (INOSOSES)

Ketones of the cyclitol series played an important part in studies on the configurations of cyclitols and are useful intermediates for cyclitol synthesis.

* *References start on p. 573.*

As already noted (p. 540), they also serve the latter function in Nature. In keeping with this biosynthetic role, small proportions of the common inosose **68** have been found in animal tissues,[7e] and in *Streptomyces griseus*.[69d] Being polyhydroxyketones, these compounds are formally sugars, whence the term "inosose" by which they are commonly known.

Under the IUPAC 1967 Rules these compounds are named preferably as polyhydroxy-cyclohexanones, with position 1 assigned to the carbonyl group, as in standard organic nomenclature. The numbering of the hydroxyl positions follows the principles used for the inositols (p. 522), except that the lowest possible number for such positions is 2 (or 3). The relative configurations of the hydroxyl groups are indicated by the Maquenne system (p. 541), and assignment to D- and L-series is made as for the inositols. "Inosose" is retained as a useful generic term, but the now-prevalent practice of naming specific ketones by attaching prefixes and numerals to the word inosose is discouraged. Some of these names are given in parentheses, however, for correlation with the existing literature.

With few exceptions, inososes have been obtained by the oxidation of inositols or lower cyclitols. The methods that have been used are: (1) carefully controlled oxidation with nitric acid; (2) biological oxidation with *Acetobacter suboxydans*, which possesses a dehydrogenase system for cyclitols similar to the one with which it oxidizes open chain alditols to ketoses; and (3) catalytic oxidation, with air or oxygen and platinum catalysts. Newer oxidation procedures employing ruthenium tetraoxide[112, 113] or methyl sulfoxide in various combinations[114] will no doubt be of considerable use for the oxidation of blocked cyclitols. For example 1,2:3,4-di-*O*-isopropylidene-6-*O*-methyl-*epi*-inositol has been converted into an inosose by the Pfitzner–Moffatt procedure, in which methyl sulfoxide is the oxidant.[114a]

The nitric acid method is employed only for the preparation,[115] from *myo*-inositol, of DL-2,3,4,6/5-pentahydroxycyclohexanone ["(±)-*epi*-inosose-2," formula **67**], the first of the inososes to be characterized.[54] The low yield (10%) in this preparation has discouraged further study of the method and stimulated work on the alternatives, biological and catalytic oxidation, both of which have wide applicability. Both of these procedures are specific for the oxidation of axial hydroxyl groups in saturated cyclitols, but they differ in several details. Bacterial oxidation,[1, 116, 117] at least with *A. suboxydans* ATCC 621, is subject to additional specificity rules, can produce an optically active inosose from a meso precursor, and gives diketones or higher oxidation products in some cases. Catalytic oxidation[118, 119] stops at the monoketone stage, and may be specific for one of two differently situated axial hydroxyl groups, although the factors governing the specificity are not understood. Some of these considerations are illustrated in Scheme D, where it may be noted that the application of either method to *myo*-inositol gives 2,4,6/3,5-pentahydroxycyclohexanone ("*myo*-inosose-2," formula **68**), the commonest of the inososes.[120, 121] Catalytic oxidation of the 5-cyclohexene-1,2,3,4-tetrols takes place at one of the positions adjacent to the double bond.[122]

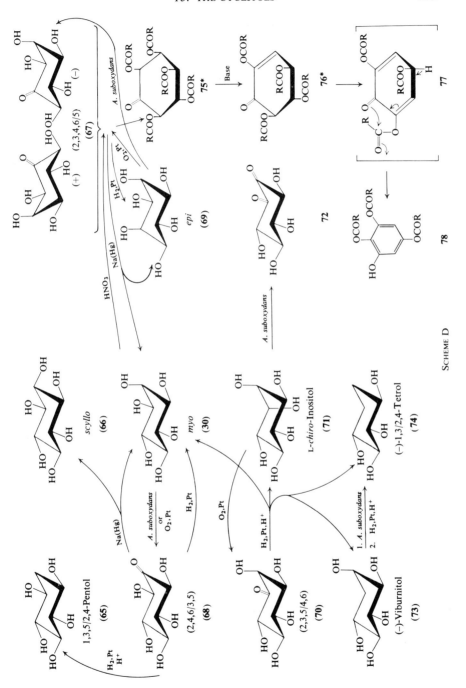

SCHEME D

The isolation of inososes from oxidation mixtures is often accomplished via the phenylhydrazones, which readily form in the cold. The inososes are regenerated by treating the phenylhydrazones with benzaldehyde[115] or by heating them with a cation-exchange resin of the sulfonic acid type.[123] An alternative method of treating oxidation mixtures would be to pass them through ion-exchange columns in the bisulfite form, which allow unoxidized starting materials to pass but retain truly ketonic carbohydrates* for subsequent elution by acetone or other appropriate reagent.[124]

The phenylhydrazone of the common inosose (68) is converted into the osazone only with difficulty.[125] A few other cyclitol osazones have been obtained directly from 1,2-diketones.[1, 125a] A 1,3-bis(phenylhydrazone) exhibiting an interesting phenylazo–phenylhydrazono tautomerism is obtained[125b] from the 1,2,3-trione 18.

The reduction of an inosose generates two epimeric products, one having the new hydroxy group axial and one having it equatorial. In catalytic hydrogenations the axial epimer usually preponderates, but sodium amalgam normally gives a substantial portion of the equatorial epimer, and so does sodium borohydride in some cases.[1] An oxidation–reduction cycle can thus often be used to invert the configuration of one —CHOH— group in a cyclitol, as in the syntheses of epi-inositol[126] (69) and scyllo-inositol[127] (66) from myo-inositol (Scheme D). The reduction of the pentaacetate of the inosose 68 with sodium borohydride gives almost exclusively the product having the scyllo configuration.[128]

In aqueous mineral acid over a platinum catalyst inososes undergo a facile hydrogenolysis by which the carbonyl group is converted into a methylene group.[1] The reaction is a general one for this group of compounds, but it does not always proceed exclusively to the deoxyinositol. Some inososes give a mixture of products including inositols (simple reduction), cyclohexanepentols (hydrogenolysis of the carbonyl groups), cyclohexanetetrols, and even triols (loss of hydroxyl groups) (see Scheme D).[119, 129]

The inososes are powerful reducing agents in alkaline solution, presumably by virtue of their conversion into enediolate anions. The pentaacetates and pentabenzoates exhibit lowered melting points in soft-glass capillaries as a result of extremely facile, base-catalyzed β-elimination.[1] The initial products are α,β-unsaturated ketones (76),[128, 130] and these are probably intermediates in the transformation—effected by pyridine or sodium acetate under normal acetylating conditions—of inosose derivatives into derivatives of 1,2,3,5-benzenetetrol.[1] A reasonable mechanism[130, 131] for this transformation is shown in Scheme D.

---

* Those having a keto group that cannot participate in ring formation.

VII. THE LOWER CYCLITOLS

Cyclitols having six-membered rings but fewer than six hydroxyl or substituted hydroxyl groups, and those with ring-size less than six, are treated under this heading. The "higher cyclitols" would, correspondingly, be those alicyclic polyols having a ring size greater than six, but there are as yet very few examples. Only six of the lower cyclitols are definitely known from biological sources (Table III), but extensive synthetic work has been done on all the groups of cyclitols of this class.

The 1967 Rules provide that the basic method for naming cyclitols other than inositols shall be standard organic nomenclature (IUPAC). The lower cyclitols are accordingly named as cycloalkanepolyols, with lowest possible numbers given to the positions carrying hydroxyl groups. Decisions between the alternatives which then remain, and assignment to D and L series, are made as for the inositols (p. 522). The relative configurations of the hydroxyl groups are indicated by the Maquenne system (p. 541).

A. THE CYCLOHEXANEPENTOLS

The first known cyclohexanepentol was a dextrorotatory cyclitol obtained from the acorns of *Quercus* species (oaks), hence the name (+)-quercitol. The structure and configuration were elucidated, as in the case of the inositols, by conversion into aromatic substances, and by oxidative cleavage to aldaric acids.[1] Since the cyclohexanepentols correspond, in symmetry properties, to the open-chain heptitols, there are ten possible diastereoisomeric forms—four meso and six DL-pairs. The availability of (+)- and (−)-quercitol by isolation, and the development of syntheses for all the other isomers, some of them in optically active forms, made possible the recent statement that the cyclohexanepentols constitute "perhaps the largest *all-known* family of diastereoisomers in organic chemistry. . . ."[81] The same author later reported a synthesis of the quercitols.[132]

"Quercitol" has been used as a generic term for cyclohexanepentol, and it has been suggested that the individual pentols all be named as quercitols, with a configurational prefix to designate each diastereoisomer. With the adoption of the 1967 Rules these usages are discontinued, to avoid undue proliferation of trivial nomenclature.

The conversion of inososes into cyclohexanepentols by hydrogenolysis was described above. Equally useful precursors are the anhydroinositols, which may either be hydrogenolyzed directly or first converted into deoxyhaloinositols.[81] Since both procedures appear to give the same products, direct hydrogenolysis seems preferable.

Like the inositols, the cyclohexanepentols form isopropylidene acetals, and they are converted into ketones by *Acetobacter* or by catalytic oxidation.

TABLE III

The Naturally Occurring Lower Cyclitols

Trivial name (systematic name)	Formula	Melting point, degrees [α]$_D$, degrees (water)	Distribution[a]
(+)-Quercitol (1L-1,3,4/2,5-Cyclohexanepentol)		232–237 +26	Dicotyledons (10 families) *Chamaerops humilis* (a palm)
(−)-Quercitol (1D-1,3,4/2,5-Cyclohexanepentol)		237 −27	*Eucalyptus populnea* (Myrtaceae)
(−)-Viburnitol (1L-1,2,4/3,5-Cyclohexanepentol)		180–181 −50[b]	Dicotyledons (5 families)

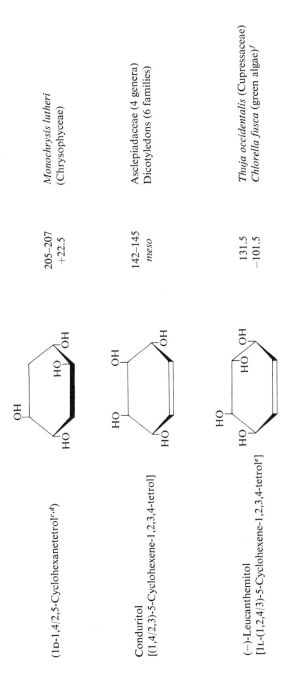

(1D-1,4/2,5-Cyclohexanetetrol[c,d])

205–207
+22.5

Monochrysis lutheri
(Chrysophyceae)

Conduritol
[[(1,4/2,3)-5-Cyclohexene-1,2,3,4-tetrol]

142–145
meso

Asclepiadaceae (4 genera)
Dicotyledons (6 families)

(−)-Leucanthemitol
[1L-(1,2,4/3)-5-Cyclohexene-1,2,3,4-tetrol[e]]

131.5
−101.5

Thuja occidentalis (Cupressaceae)
Chlorella fusca (green algae)[f]

[a] Data from Plouvier,[4] and Kindl and Hoffmann-Ostenhof,[12b] except as noted otherwise.   [b] For the monohydrate.
[c] J. D. Ramanathan, J. S. Craigie, J. McLachlan, D. G. Smith, and A. G. McInnes, *Tetrahedron Lett.*, 1527 (1966). Probably also in *Porphyridium* (red algae), compare J. S. Craigie, J. McLachlan, and R. D. Tocher, *Can. J. Botany*, **46**, 605 (1968).
[d] The occurrence of "betitol," which was described many years ago as a natural cyclohexanetetrol, has never been confirmed.
[e] Configuration: see ref. 1, also H. Kindl and O. Hoffmann-Ostenhof, *Phytochemistry*, **5**, 1091 (1966).
[f] Reference 7a.

Malonaldehyde is formed during their oxidation by periodate, with, usually, consequent overoxidation and high consumption of reagent.[133] Stoichiometric conversion into malonaldehyde can be accomplished, however, permitting the determination of the pentols by the thiobarbituric acid method.[49a] Under the conditions used, oxidation is slow, apparently owing to the generation of formate esters via the cyclic hemiacetals of the initially formed dialdoses (see Section II,F above).

B. THE CYCLOHEXENETETROLS

The 5-cyclohexene-1,2,3,4-tetrols, of which there are six diastereoisomers (two meso forms and four DL-pairs), are the only unsaturated cyclitols that have been studied extensively.[1]

"Conduritol" has been used as a generic term for 5-cyclohexene-1,2,3,4-tetrol, and the designations "conduritol-A," "-B," etc., appear in the literature. However, these usages, and the term "conduramine" for the 6-amino-4-cyclohexene-1,2,3-triols are not sanctioned by the 1967 Rules. "Conduritol" is retained here solely as the trivial name for the 1,4/2,3-enetetrol.

The role of these compounds as intermediates in the preparation of other cyclitols was pointed out in the discussion of the "benzeneglycol" synthesis (Section IV,B). Conduritol (meso) and all of the racemic isomers can be made by this method. Several of the enetetrols have also been prepared from sulfonic esters or bromodeoxy derivatives of inositols by elimination reactions,[81] and these methods, or Diels–Alder syntheses, are usually to be preferred when they are applicable. The Diels–Alder condensation of vinylene carbonate (34) and *trans-trans*-1,4-diacetoxybutadiene yields the 1,2,3,4/0-isomer.[87] A convenient synthesis of the 1,3/2,4-isomer has been reported.[133a]

L-Leucanthemitol (Table III), fairly widely distributed in the plant kingdom, is apparently the biological precursor of the less-common conduritol. *myo*-Inositol did not serve as a precursor for L-leucanthemitol in experiments with detached leaves of *Chrysanthemum leucanthemum*.[133b]

C. THE CYCLOHEXANETETROLS AND TRIOLS

When the number of hydroxyl groups attached to the cyclohexane ring is reduced to four, position isomerism as well as stereoisomerism becomes possible. There are thus three families of cyclohexanetetrols: 1,2,3,4 (the largest family—two meso forms and four DL-pairs), 1,2,4,5, and 1,2,3,5.

Nearly all of the possible cyclohexanetetrols have been synthesized, primarily by the application of hydroxylation reactions (see Section IV,B) to unsaturated precursors. Stepwise hydroxylations of 1,3-cyclohexadiene gave the saturated 1,2,3,4-tetrols via *cis*- and *trans*-3-cyclohexene-1,2-diol.[134] Osmium tetroxide–

silver chlorate, as well as permanganate, was effective for the *cis* hydroxylation of these diols. 1,4-Cyclohexadiene similarly served as the precursor of the 1,2,4,5-tetrols.[81] The skillful use of *Acetobacter* oxidation enabled Posternak and his students to resolve the racemic 1,2,3,4-tetrols and determine the absolute configurations of the enantiomorphic forms.[134,135] (+)- and (−)-1,2/3,4-Cyclohexanetetrol are available by reduction of the corresponding enetetrols,[103] and two of the active 1,2,3,5-cyclohexanetetrols can be made from quinic acid (p. 569).[81] The hydroxylation of 1,3-cyclohexadiene has been restudied.[135a]

The cyclohexanetriols may be regarded as cyclitols, and certain of them are of interest in this context. They are made by hydroxylations of cyclohexenols, or by the hydrogenation of benzenetriols.[1] 1,3,5/0-Cyclohexanetriol (α-phloroglucitol) appears to form a tridentate complex with borate (see Section III),[80] and with oxyacids of phosphorus it gives triesters to which a similar cage (adamantane) structure has been attributed.[136] A detailed p.m.r. study has been carried out on the cyclohexanetriols.[137]

D. CYCLOPENTANE CYCLITOLS

The cyclitol series has recently been expanded to include the polyhydroxycyclopentanes by the synthetic work of Sable and Posternak and their collaborators.[138] Application of the hydroxylation reactions used for the lower C_6 cyclitols enabled these investigators to make five of the six possible diastereoisomeric 1,2,3,4-cyclopentanetetrols, and a true inositol homolog, 1,2,4/3,5-cyclopentanepentol. Further extension of the work has led to amino derivatives (Section VIII). No naturally occurring cyclopentanepolyols have been found as yet.

The precursors of the C_5 cyclitols are *cis*- and *trans*-3-cyclopentene-1,2-diol (**79**), *cis*- and *trans*-4-cyclopentene-1,3-diol (**80**), (1,3/2)-4-cyclopentene-1,2,3-triol (**81**), and their acetates, all derived from cyclopentadiene (Scheme E). The epoxidation of the free alcohols with peroxybenzoate is governed, as in the C_6 compounds, by the *cis*-directing influence of the allylic hydroxyl group. Both the permanganate hydroxylation and the peroxybenzoate epoxidation of the acetates take place on the less-hindered side of the ring, which is perhaps in many cases defined more sharply than in the C_6 series.[139,140] The conformational factors which dominate the chemistry of the C_6 epoxides are absent in the lower homologs, and as a result any one of several other effects may regulate the direction of ring opening in the C_5 epoxides.[140,141] Thus, in acid-catalyzed reactions of the anhydrotetrol **82**, the entering nucleophile (water, or Br⁻) attacks the less-hindered carbon atom (steric effect). The behavior of the isomeric epoxide **84** is thought to be explained by the inductive effect of the hydroxyl group adjacent to one of the carbon atoms of the oxirane ring. This

---

* *References start on p. 573.*

SCHEME E

carbon atom is destabilized with respect to formation of an incipient carbonium ion; hence the ring is opened at the other carbon atom, next to the methylene group. A still different factor operates in the diacetate (**83**) of **82**; participation (apparently) by the *trans*-acetoxy group causes opening at the hindered position.

The operation of the steric and inductive effects postulated in the foregoing examples has been substantiated by extensive further work on the synthesis of the cyclopentanetetrols[141a] and -triols[141b] via epoxides and bromohydrins. The hydroxylation of cyclopentadiene has been restudied,[135a] and all of the possible cyclopentanetetrols and -pentols have now been made.[141c]

The configurations of the cyclopentanetetrols were deduced by cross-comparison of the products from different precursors, and by examination of the p.m.r. spectra of the tetrols and related compounds.[142,143] These spectra provide information that should be useful in studies of furanoid sugar derivatives.

Considerable insight on the conformations of cyclopentane cyclitol derivatives in nonpolar media is afforded by their infrared spectra, which disclose the presence of internal hydrogen bonds and permit an estimate of their length.[141b]

VIII. AMINOCYCLITOLS (INOSAMINES)

Cyclitols bearing amino or substituted amino groups are components of many antibiotics (Chapter 31). The first example encountered was streptidine, obtained by hydrolysis of the streptomycins. Streptidine was shown by classical means to be 1,3-dideoxy-1,3-diguanidino-*scyllo*-inositol (**85**)[144-146] and the closely related bluensidine (from bluensomycin) turned out to be 1-*O*-carbamoyl-3-deoxy-3-guanidino-*scyllo*-inositol.[147] These compounds are readily hydrolyzed to streptamine (1,3-diamino-1,3-dideoxy-*scyllo*-inositol, **87**) and aminodeoxy-*scyllo*-inositol, respectively. The most frequently encountered natural inosamine is deoxystreptamine, which has primary amino groups; it is (1,3/2,4,6)-4,6-diaminocyclohexane-1,2,3-triol[148] (**91**). Other inosamines from antibiotics are 2-amino-2-deoxy-*neo*-inositol ("*neo*-inosamine-2") and the methylamino compounds actinamine [1,3-dideoxy-1,3-bis(methylamino)-*myo*-inositol] and hyosamine (deoxy-*N*-methylstreptamine).[148-150]

This entire group may be referred to as "inosamines," although this name was originally applied to the monoaminomonodeoxyinositols. The term "inosadiamine" is sometimes used for the diamino compounds. These terms are used here only in the generic sense, however. Under the 1967 Rules individual aminocyclitols are named preferably as aminodeoxyinositols when this nomenclature is appropriate (cyclohexane ring, six univalent substituents of which at least three are hydroxyl or substituted hydroxyl groups), otherwise as aminocycloalkanepolyols. In the latter case, the procedures for numbering and configurational designation are the same as for the lower cyclitols (p. 557). The *hydroxyl groups*, being cited in the suffix, have numbering precedence.

The discovery of streptidine stimulated work on the synthesis of aminocyclitols, and interest was sustained by the growth of the list of natural inosamines. All of these except bluensidine, and many purely synthetic inosamines, have been obtained by application of the following methods:

1. Reduction of the phenylhydrazones and oximes of inososes.[3,151] Catalytic hydrogenation normally gives the epimer having the amino group axial. Good yields of equatorial epimers have been obtained by reduction of the oximes with sodium amalgam.

2. Reduction of deoxynitroinositols obtained by the cyclization reaction with nitromethane (Section IV,C).[97,99a,152] In an extension of this method, vicinal triamines are formed by aminating the deoxynitrocyclitol prior to reduction.[152a,152b] When dialdehydes derived from acetamido- or diacetamidocyclopentane cyclitols are used, the nitroinositol synthesis gives 1,3-inosadiamines and 1,3,5-inosatriamines, respectively.[81b,152c]

3. Action of ammonia on anhydroinositols and deoxyhaloinositols.[153,154] With the halo compounds, anhydroinositols are formed, presumably, as intermediates. The configurations of the products are determined by the rule of diaxial ring opening. Probably related mechanistically is the reaction of 1,3-di-*O*-*p*-tolylsulfonyl-*myo*-inositol with hydrazine and 1,2-dimethylhydrazine, which affords an elegant synthesis of 1,3-diamino-1,3-dideoxy-*myo*-inositol[154a] and its *N,N'*-dimethylderivative (actinamine)[154b] via bridged 1,3-dideoxy-1,3-hydrazo intermediates.

* *References start on p. 573.*

4. Displacement of sulfonyloxy groups by azide.[155] If there are neighboring *cis*-acetyl groups, the sulfonyloxy group may be displaced directly, with inversion of configuration. *trans*-Acetyl groups participate, and the (presumed) acetoxonium ion intermediate can be attacked at either carbon atom. This method, not always high yielding, has been extensively exploited[101d, 155a] by Suami, who has extended it to the bromodeoxyinositols.[155b]

5. Hydrogenation of aromatic compounds. Triaminophloroglucinol gives 1,3,5-triamino-1,3,5-trideoxy-*cis*-inositol in 65% yield.[156] The hydrogenation of rhodizonic acid 1,4-diimide has also been reported, but the product(s) has not been fully characterized.[157] 4,6-Dinitropyrogallol gives a mixture of 4,6-diaminocyclohexane-1,2,3-triols (the 1,2,3,4,6/0 isomer can be made to preponderate[157a]) and diamines having fewer hydroxyl groups.[157b]

6. From acetylated aminocyclohexenetriols (**86**). These intermediates, derived from the "benzeneglycols" (Section IV,B), have been converted into a whole array of aminocyclitols by Nakajima and co-workers, as shown in Scheme F.[91, 158–160] A key reaction in the scheme is the epoxidation of the acetamidotriols with peroxybenzoate. The course of the reaction is governed by the *cis*-directing effect of the acetamido group.

7. Combinations of the foregoing.[160, 161] In making inosadiamines one amino group may be generated by one method, the second by another.

The key position of streptamine has made its synthesis the goal of many efforts, and methods 1, 2, and 4 have been used successfully.[162–164] The most recent, and most direct, preparation is by the reduction of the 1,3-bisphenylhydrazone of 4,6/5-trihydroxycyclohexane-1,2,3-trione (**18**).[165] Streptamine is readily converted into streptidine by standard methods. The biogenesis of streptidine starts with *myo*-inositol and appears to involve two nearly identical reaction sequences, each of five steps. Each sequence accomplishes the replacement of a hydroxyl group by a guanidino group. The steps are: dehydrogenation to an inosose, transamination to form an inosamine, phosphorylation, transamidination, and dephosphorylation.[166, 167, 167a]

The benzeneglycol synthesis has been elaborated to give not only streptamine,[167b] but also deoxystreptamine[151] and (−)-hyosamine.[167c] Syntheses of deoxystreptamine from 2,4-dinitropyrogallol[157a] and 1,3-diamino-1,3-dideoxy-*myo*-inositol[167d] have also been reported.

Most of the methods listed above could be used for the preparation of inosamines in quantity if the desired precursors were readily available. So far, only a few inosose derivatives, some aromatic compounds, and a limited number of cyclitol sulfonates have satisfied this criterion.

The characterization of new, synthetic inosamines is nowadays relatively easy, since the method of preparation generally limits the possible configurations to two or three. The p.m.r. spectrum of the acetate may be used to decide between these, as discussed in Section III,A.

SCHEME F

Several aminocyclitols of the cyclopentane series have been prepared by Hasegawa and Sable, employing methods closely related to methods 3 and 6 listed above. Aminocyclopentenediols, homologs of the intermediates used by Nakajima (method 6), were made by allylic bromination of the acetates of the 3-cyclopentene-1,2-diols (79), and displacement of the bromine with ammonia.[140] Azide was used, instead of ammonia, in the introduction of a nitrogen atom via anhydro compounds.[139] The carbon atom of the oxirane ring attacked by azide ion was not, in some cases, the same as the one attacked in acid-catalyzed ring opening. This was interpreted as indicative of the more purely SN2 nature of the reaction with azide.

* References start on p. 573.

The inosamines themselves have no antibiotic activity, but they might yield new antibiotics by coupling with sugars. A few efforts have been made along these lines, with only modest success in terms of new compounds. The problem is essentially one in oligosaccharide synthesis, and substantial progress will be possible only when improved methods for such syntheses become available. The total syntheses of the kanamycins reported from the laboratories of Nakajima[167e] and S. Umezawa[167f] are outstanding achievements that promise to be the basis of much further work.[167g]

The deamination of inosamines with nitrous acid gives mixtures of products, as might be expected.[168] However, O-acetylated inosamines react somewhat more smoothly and in many cases, but not all, the replacement of the amino group by a hydroxyl group occurs with preponderant inversion of configuration.[168a] The deamination reaction has been used to prepare *myo*-inositol-2-^{14}C, of key importance in metabolic studies, from 1-deoxy-1-nitro-*scyllo*-inositol-1-[14]C (**50**) via 1-amino-1-deoxy-*scyllo*-inositol-1-^{14}C pentaacetate.[169, 170]

IX. CYCLITOLS HAVING SIDE CHAINS

A. With C-Methyl and Substituted C-Methyl Groups

Nature has provided two examples of C-methylinositols, namely mytilitol (**92**), found so far in three species of invertebrate animals and in two red algae, and (−)-laminitol (**93**), from several red and brown algae.[4] Both mytilitol and laminitol have been detected in *Chlorella fusca* (green algae).[7a] The investigations of Posternak and co-workers on these compounds have led to the synthesis of a variety of other C-substituted cyclitols.

The addition of methylmagnesium iodide to the ketonic carbonyl group of an acetylated inosose gives the axial C-methylinositol (mytilitol) in the case of **94**, but the equatorial product from **96**, probably because of hindrance by the axial 3-acetoxy group.[171, 172] Both **94** and **96**, or the free inososes, react with diazomethane to form spiro epoxides in which the methylene group is equatorial (for example, **95**).[171, 172] By hydrogenolysis the spiro epoxides are converted into C-methylinositols, and by hydrolysis they give C-hydroxymethylinositols with retention of configuration at the spiro carbon atom. The chloromethyl, bromomethyl, and a series of aminomethyl derivatives have been prepared by the action of the appropriate reagents on the epoxide **95**.[1] Hydriodic acid transformed **95** into the very useful C-methylene compound, presumably via an iodomethyl intermediate.[173]

myo-Inositol apparently is not a biological precursor of the C-methylinositols, nor is methionine the source of their methyl groups.[7a, 173a]

C-Alkyldeoxynitroinositols are obtained when higher nitroalkanes are used in the nitroinositol synthesis, and these compounds can be reduced to the

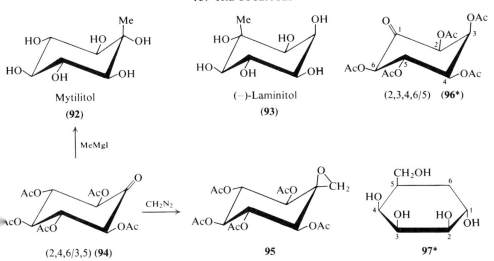

Mytilitol (92)   (−)-Laminitol (93)   (2,3,4,6/5) (96*)

(2,4,6/3,5) (94)   95   97*

corresponding inosamines—for example, 1-amino-1-deoxy-1-C-methyl-*scyllo*-inositol ("mytilamine").[174]

A different type of C-alkylcyclitol is the "pseudohexose" **97** [(1/2,3,4,5)-5-(hydroxymethyl)cyclohexane-1,2,3,4-tetrol], made by a variation of a Diels–Alder reaction used for the synthesis of shikimic acid.[175] The 1,4,5/2,3 and 1,2/3,4,5 isomers of **97** have also been reported.[175a] A five-membered C-alkyl-cyclitol, (1,2/3,5)-3-amino-5-(hydroxymethyl)cyclopentane-1,2-diol, has been prepared as an intermediate in the synthesis of the potentially very interesting "pseudoribose" "nucleosides."[175b]

Branched-chain inosamines and cyclohexanetriols have been isolated by hydrogenolysis and subsequent hydrolysis of the validamycin antibiotics, and the structures of the products were determined by n.m.r. spectroscopy.[175c] From validamycin A there was obtained 1L-(1,3,4/2,6)-4-amino-6-(hydroxy-methyl)cyclohexane-1,2,3-triol (validamine), and a corresponding 1,2,3,5-tetrol (hydroxyvalidamine) was obtained from validamycin B. Both valid-amycins also gave (1,3,4/2)-4-(hydroxymethyl) cyclohexane-1,2,3-triol (prob-ably the 1D-isomer), and the corresponding 4-methyl derivative.

B. CYCLITOL CARBOXYLIC ACIDS

Two acids of the cyclitol series have long been known as plant constituents. Quinic acid (**98**) is a common and abundant plant acid.[1] Shikimic acid (**102**), once thought to be a rare substance, is now also known to be widely distri-

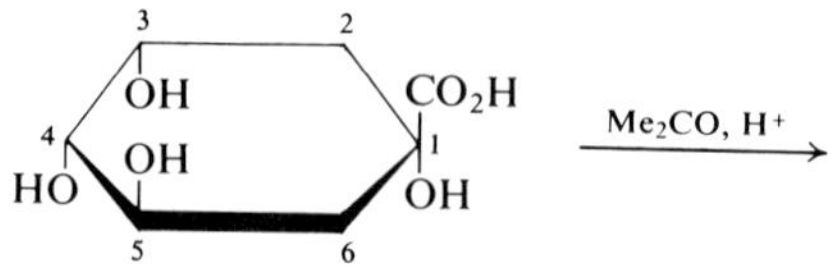

L-(−)-Quinic acid [a]

(98)

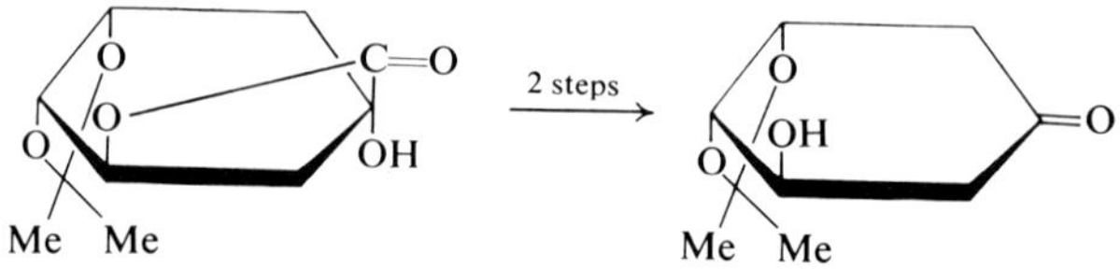

99

100

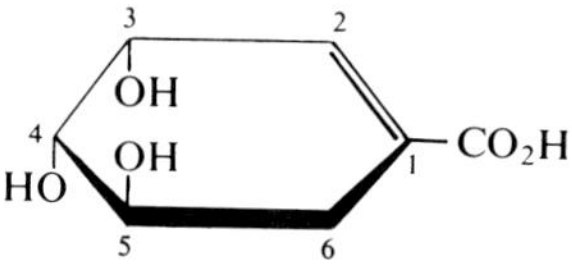

L-(−)-Shikimic acid [b]

(102)

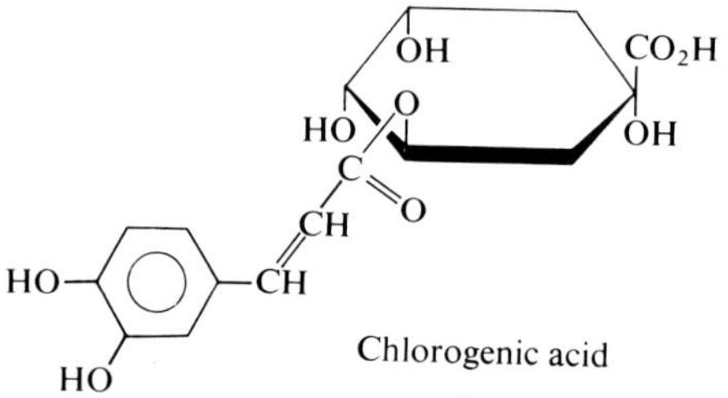

Chlorogenic acid

(101)

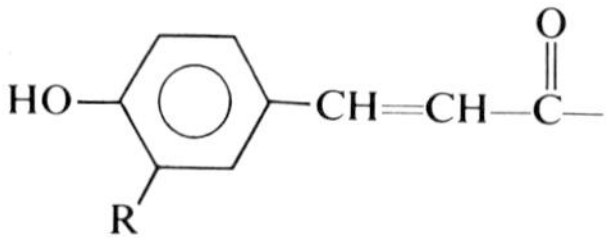

R = H, p-coumaryl

R = OH, caffeoyl

R = OMe, feruloyl

[a] Assignment to the L-series based on the projection of the OH group on C-1.
[b] Assignment to the L-series based on the configuration at C-3.

buted.[176] Shikimic acid is an intermediate in the biosynthesis of aromatic compounds, at least in bacteria, and quinic acid is a side product of the process.[177] Much of the quinic acid in plants is in the form of depsides, of which the best known is chlorogenic acid, the 5-caffeoyl derivative (101). Other caffeic esters of quinic acid have been characterized, and p-coumaric and ferulic esters as well.[1, 178] Evidence indicates that the related depsides of shikimic acid also occur in plants.[176]

The established convention for quinic and shikimic acids has been to number them in the sense opposite that shown here. This gave the number 5 to the positions designated 3 in 98 and 102, but such numbering violates the elementary conventions of systematic nomenclature. It is to be hoped that the older convention will be dropped.

A variety of syntheses of quinic and shikimic acids have been devised, most of them involving Diels–Alder condensations.[1,176,178a] Quinic acid can be converted, via the *O*-isopropylidene derivative (99) of its lactone, quinide, into the ketone (100), which is a source of 1,2,3,5-cyclohexanetetrols and related compounds. The prolonged heating of quinic acid gives a mixture of lactones, some formed by configurational inversions.[1] Ring-opening reactions of a diepoxide derived from quinic acid have been described.[179]

An improved method for the synthesis of shikimic acid from 1,3-butadiene and propiolic acid has been reported. After conversion to the 3,4-cyclohexyl-idene acetal, synthetic DL-shikimic acid can be resolved *via* the α-phenylethyl-amine salts.[179a] A synthesis from D-arabinose has been reported.[179b]

Other cyclitol acids include the reduction products of gallic acid, and the inositolcarboxylic acids obtained by oxidation of the *C*-(hydroxymethyl)-inositols, or the related *C*-methylenecyclohexanepentols.[1]

X. CYCLITOL REACTIONS OF GENERAL SIGNIFICANCE IN CARBOHYDRATE CHEMISTRY

Although the cyclitols are functionally alditols, their ring structures relate them closely to the sugars. But since the cyclitols are free of the structural complications—ring oxygen, anomeric center, and (for the most part) side chains—characteristic of sugars, their reactions are dominated by conformational effects. They are thus good model compounds for study of the conformational factors in reactions common to all carbohydrates.

A. FORMATION OF CYCLIC ACETALS

The isopropylidene acetals, which have been the most extensively studied, are in recent practice made by treating cyclitols with the dimethyl or diethyl acetal of acetone and an acid catalyst.[180,181] Acetal rings formed in cyclitol derivatives are always five-membered (involving adjacent hydroxyl groups); six-membered (1,3) acetals have not been observed in this series.

Inositols having two pairs of vicinal *cis*-hydroxyls readily form diacetals. If at this stage the acetal functions are contiguous (1,2:3,4),† and the remaining two hydroxy groups are *trans*, these latter hydroxy groups are nevertheless readily bridged by a third acetal function. This is true at least in the *chiro*-inositols and *epi*-inositol.[16] When only one *cis*-acetal group is possible, addi-

---

* *References start on p. 573.*

† Numbering in the general, structural sense, not according to the conventions for any specific inositol.

tional, *trans*-fused acetal rings are not formed so readily, but these can be generated under forcing conditions. Thus *myo*-inositol when refluxed with cyclohexanone and *p*-toluenesulfonic acid gives the three possible position-isomeric diacetals, and eventually some of the tri-*O*-cyclohexylidene derivative.[182, 183] A still greater degree of acetalation is achieved by employing cyclohexanone dimethyl acetal in *N,N*-dimethylformamide at 70°, with continuous removal of the by-product methanol.[183a] Even *scyllo*-inositol (**9**, all-trans) is converted into the triacetal when subjected to this procedure.

Infrafred spectral studies[184] of the internal hydrogen-bonding in cyclitol acetals in dilute solution in carbon tetrachloride indicate that the compounds having two *cis*-acetal functions have a twist (skew) conformation (for example, di-*O*-isopropylidene-D-inositol, **103**). The ready closure of a third, *trans*-acetal ring (for example, **103** → **104**) in these compounds is thus easily understood:

103   104

the two *trans*-hydroxyl groups have a much smaller dihedral angle in the twist form than in the original chair form of the cyclitol. On the other hand, cyclitol diacetals having one *cis*- and one *trans*-acetal ring, in either the 1,2:3,4 or the 1,2:4,5 arrangement, appear to retain the chair conformation, somewhat distorted.

B. Acyl Migration

Comparisons of the rates of migration of acetyl groups between amino and neighboring hydroxyl functions ($N \to O$ in acid, $O \to N$ in base) were used in early work on the characterization of inosamines.[3] Epimers having amino groups cis to the adjacent hydroxyl groups showed rates of migration more rapid than the trans compounds, as predicted. Acyl migration proceeds through cyclic intermediates, and one might expect these intermediates to form more readily between cis groups.

Acyl migration between trans groups nevertheless occurs rather easily in the inosamines, and also in the partially acetylated *myo*-inositols. The 1,3,4,5,6-derivative (**105**), for example, is readily isomerized in the presence of moist silver oxide or in aqueous pyridine to a mixture of the four possible *myo*-inositol pentaacetates (**105–108**).[185, 186] Only when dry, neutral silver oxide and a non-polar solvent were used was isomerization restricted to the cis migration

1,3,4,5,6 (**105**)   1,2,4,5,6 (**106***)

1,2,3,4,5 (**107***)   1,2,3,4,6 (**108**)

105 → 106.[185] Migration even occurs in anhydrous pyridine, although only at temperatures of 100° and above.[187]

Conformational theory suggests that the nonbonded interactions of an axial hydroxyl group are not greatly increased by acetylation. If this is the case the isomers **105**, **106**, **107**, and **108** should all have approximately equal energy contents, and at equilibrium should be present in the ratio 1:2:2:1. (Formulas **106** and **107** represent two individual, enantiomorphic components each.) The proportions obtained in aqueous pyridine deviate from this, but there is no indication that the axial 2-position is a less favorable location for the acetyl group than the average equatorial position.[186]

C. EPIMERIZATION

The epimerization of acylated cyclitols has been studied in refluxing acetic acid containing a little strong acid, and in liquid hydrogen fluoride.[188, 189] In both media, inversions take place at the middle carbon atoms of *trans-cis* sequences—that is, carbons bearing an acyloxy group flanked by one trans and one cis neighbor. It is supposed that these reactions involve the initial formation of a cyclic acyloxonium ion,[190] of which **109** is one possible type, and that conversion into an epimeric acyloxonium ion occurs as shown. This latter step is reversible. The isolated products, always a mixture, are formed by the eventual hydrolysis of the various isomeric acyloxonium ions.

The formation of acyloxonium ions from some cyclopentane cyclitols and related compounds has been verified by the isolation of several acyloxonium

109   **110**   **111**

salts in crystalline form. The isomerization of the ions in solution can be followed by p.m.r. spectroscopy.[189a, 190]

The inositols that epimerize may be grouped in two interconverting systems which, at equilibrium in acetic acid–strong acid, have the compositions: *myo*:DL-*chiro*:*muco* = 54:41:5 and *neo*:*allo*:*epi* = 58:21:15.[188] The $\Delta G°$ values associated with these ratios were used to calculate nonbonded interaction energies (AcO:H, AcO:AcO; see Section III), and the results agreed well with those derived in other ways. Inositol methyl ethers and lower cyclitols are, of course, epimerized in acetic acid–strong acid, but side reactions, such as demethylation in the case of the ethers, and other processes intrude. The epimerization of *epi*-inositol in this medium is probably the simplest route to *neo*-inositol, which can be isolated readily from the reaction mixture.

In liquid hydrogen fluoride the same products arise as in acetic acid–strong acid, except that methyl ethers are not cleaved.[189] There is little quantitative information on the composition of the mixtures formed in hydrogen fluoride, but *myo*-inositol hexaacetate can be converted into *muco*-inositol (after hydrolysis) in up to 70% yield in this medium.[181,189] The preponderance of the usually unfavorable *muco* configuration indicates that the equilibrium species, if any, in this case are not simple acetates. Perhaps cyclic orthoester derivatives are involved. It is not certain, however, that equilibrium is achieved in liquid hydrogen fluoride, especially if water is present.

D. Epoxide Migration

By epoxide migration is meant the base-catalyzed conversion of an epoxycarbinol structure such as **112** into an isomeric structure, **113**, in which the

112   **113**

oxirane ring is trans to the original one and displaced one position along the carbon skeleton.[191] The transformation of several anhydroinositols, in the presence of bases, into isomeric anhydroinositols has shown that epoxide migration does occur in carbohydrates, and in fact quite readily.[104] Presumably, the mechanism is rearward attack on the oxirane ring by the neighboring, *trans*-hydroxyl group. The reaction is of course reversible, and with cyclitols the composition of the system at equilibrium appears to depend on simple conformational considerations: the isomer having the fewer axial groups preponderates.

REFERENCES

1. T. Posternak, "The Cyclitols" (Fr. ed.), Hermann, Paris, 1962; Engl. ed. (updated), Hermann, Paris, and Holden-Day, San Francisco, California, 1965.
2. H. G. Fletcher, Jr., *Advan. Carbohyd. Chem.*, **3**, 45 (1948).
3. S. J. Angyal and L. Anderson, *Advan. Carbohyd. Chem.*, **14**, 135 (1959).
4. V. Plouvier, *Bull. Soc. Chim. Biol.*, **45**, 1079 (1963); also *in* "Chemical Plant Taxonomy" (T. Swain, Ed.), p. 313. Academic Press, New York, 1963.
4a. *IUPAC Information Bull.*, **32**, 51 (1968); *Eur. J. Biochem.*, **5**, 1 (1968); *Biochim. Biophys. Acta*, **165**, 1 (1968); *J. Biol. Chem.*, **243**, 5809 (1968); *Biochem. J.*, **112**, 17 (1969); *Hoppe-Seyler's Z. Physiol. Chem.*, **350**, 523 (1969) (German language version).
5. J. Scherer, *Ann.*, **73**, 322 (1850).
6. L. Bouveault, *Bull. Soc. Chim. Fr.*, **11** [3], 144 (1894).
7. Y. Suhara, K. Maeda, and H. Umezawa, *Tetrahedron Lett.*, 1239 (1966).
7a. G. Wöber and O. Hoffmann-Ostenhof, *Monatsh. Chem.*, **100**, 369 (1969).
7b. D. J. Candy, *Biochem. J.*, **103**, 666 (1967).
7c. W. R. Sherman, M. A. Stewart, M. M. Kurien, and S. L. Goodwin, *Biochim. Biophys. Acta*, **158**, 197 (1968).
7d. R. F. Seamark, M. E. Tate, and T. C. Smeaton, *J. Biol. Chem.*, **243**, 2424 (1968).
8. T. Posternak, W. H. Schopfer, B. Kaufmann-Boetsch, and S. Edwards, *Helv. Chim. Acta*, **46**, 2676 (1963).
9. T. Posternak, *Helv. Chim. Acta*, **19**, 1007 (1936).
10. T. Posternak, *Helv. Chim. Acta*, **25**, 746 (1942).
11. G. Dangschat, *Naturwissenschaften*, **30**, 146 (1942).
12. G. Dangschat and H. O. L. Fischer, *Naturwissenschaften*, **27**, 756 (1939).
12a. S. J. Angyal and R. J. Hickman, *Carbohyd. Res.*, **20**, 97 (1971).
12b. H. Kindl and O. Hoffmann-Ostenhof, *Fortschr. Chem. Org. Naturstoffe*, **24**, 149 (1966).
13. A. B. Anderson, *Ind. Eng. Chem.*, **45**, 593 (1953).
14. J. van Alphen, *Ind. Eng. Chem.*, **43**, 141 (1951).
15. A. B. Anderson, D. L. MacDonald, and H. O. L. Fischer, *J. Amer. Chem. Soc.*, **74**, 1479 (1952).
16. S. J. Angyal and C. G. Macdonald, *J. Chem. Soc.*, 686 (1952).
17. T. Posternak, *Helv. Chim. Acta*, **35**, 50 (1952).
17a. S. J. Angyal, C. G. Macdonald, and N. K. Matheson, *J. Chem. Soc.*, 3321 (1953).
18. L. Anderson, E. S. DeLuca, A. Bieder, and G. G. Post, *J. Amer. Chem. Soc.*, **79**, 1171 (1957).
19. G. G. Post and L. Anderson, *J. Amer. Chem. Soc.*, **84**, 478 (1962).
19a. S. J. Angyal and T. S. Stewart, *Aust. J. Chem.*, **20**, 2117 (1967).
20. R. Kuhn and H. Trischmann, *Ber.*, **96**, 284 (1963).

21. B. W. Agranoff, R. M. Bradley, and R. O. Brady, *J. Biol. Chem.*, **233**, 1077 (1958).

21a. M. K. Gaitonde and M. Griffiths, *Anal. Biochem.*, **15**, 532 (1966).

22. Y. C. Lee and C. E. Ballou, *J. Chromatogr.*, **18**, 147 (1965).

23. W. W. Wells, T. A. Pittman, and H. J. Wells, *Anal. Biochem.*, **10**, 450 (1965).

24. Z. S. Krzeminski and S. J. Angyal, *J. Chem. Soc.*, 3251 (1962).

24a. T. Ueno, N. Kurihara, and M. Nakajima, *Agr. Biol. Chem. (Tokyo)*, **31**, 1189 (1967).

25. R. N. Roberts, J. A. Johnston, and B. W. Fuhr, *Anal. Biochem.*, **10**, 282 (1965).

25a. N. V. Riggs and F. M. Strong, *Anal. Biochem.*, **19**, 351 (1967).

25b. F. Loewus, *Carbohyd. Res.*, **3**, 130 (1966).

26. D. J. Cosgrove, *Rev. Pure Appl. Chem.*, **16**, 209 (1966).

27. D. J. Cosgrove, *Nature*, **194**, 1265 (1962).

28. D. J. Cosgrove and M. E. Tate, *Nature*, **200**, 568 (1963).

28a. L. F. Johnson and M. E. Tate, *Can. J. Chem.*, **47**, 63 (1969).

29. R. Barré, J. E. Courtois, and G. Wormser, *Bull. Soc. Chim. Biol.*, **36**, 455 (1954).

29a. G. E. Blank, J. Pletcher, and M. Sax, *Biochem. Biophys. Res. Commun.*, **44**, 319 (1971).

29b. S. Posternak, *Helv. Chim. Acta*, **4**, 150 (1921).

29c. D. J. Cosgrove, *J. Sci. Food Agr.*, **17**, 550 (1966).

29d. N. Kurihara, H. Shibata, H. Saeki, and M. Nakajima, *Ann.*, **701**, 225 (1967).

29e. S. J. Angyal and A. F. Russell, *Aust. J. Chem.*, **22**, 391 (1969).

30. J. Folch, *J. Biol. Chem.*, **177**, 497, 505 (1949).

31. H. Brockerhoff and C. E. Ballou, *J. Biol. Chem.*, **236**, 1907 (1961).

32. D. M. Brown, B. F. C. Clark, G. E. Hall, and R. Letters, *Proc. Chem. Soc.*, 212 (1960).

33. C. E. Ballou and L. I. Pizer, *J. Amer. Chem. Soc.*, **82**, 3333 (1960).

34. C. Grado and C. E. Ballou, *J. Biol. Chem.*, **236**, 54 (1961).

35. R. V. Tomlinson and C. E. Ballou, *J. Biol. Chem.*, **236**, 1902 (1961).

35a. S. P. Kozlova, N. B. Tarusova, and N. A. Preobrazhenskii, *J. Gen. Chem. USSR*, **39**, 2403 (1969).

35b. B. A. Klyashchitskii, E. G. Zhelvakova, V. I. Shvets, R. P. Evstigneeva, and N. A. Preobrazhenskii, *Tetrahedron Lett.*, 587 (1970).

35c. P. A. Gent, R. Gigg, and C. D. Warren, *Tetrahedron Lett.*, 2575 (1970).

36. H. E. Carter, D. S. Galanos, H. S. Hendrickson, B. Jann, T. Nakayama, Y. Nakazawa, and B. Nichols, *J. Amer. Oil Chem. Soc.*, **39**, 107 (1962).

37. Y. C. Lee and C. E. Ballou, *Biochemistry*, **4**, 1395 (1965).

38. H. E. Carter, W. D. Celmer, D. S. Galanos, R. H. Gigg, W. E. M. Lands, J. H. Law, K. L. Mueller, T. Nakayama, H. H. Tomizawa, and E. Weber, *J. Amer. Oil Chem. Soc.*, **35**, 335 (1958).

39. H. E. Carter, R. H. Gigg, J. H. Law, T. Nakayama, and E. Weber, *J. Biol. Chem.*, **233**, 1309 (1958).

40. H. E. Carter, B. E. Betts, and D. R. Strobach, *Biochemistry*, **3**, 1103 (1964).

41. H. E. Carter, D. R. Strobach, and J. N. Hawthorne, *Biochemistry*, **8**, 383 (1969).

41a. M. C. Pangborn and J. A. McKinney, *J. Lipid Res.*, **7**, 627 (1966).

41b. N. Shaw and F. Dinglinger, *Biochem. J.*, **112**, 769 (1969).

42. E. A. Kabat, D. L. MacDonald, C. E. Ballou, and H. O. L. Fischer, *J. Amer. Chem. Soc.*, **75**, 4507 (1953).

42a. M. Senser and O. Kandler, *Phytochemistry*, **6**, 1533 (1967).

42b. F. Petek, E. Villarroya, and J.-E. Courtois, *Compt. Rend. D*, **263**, 195 (1966).

43. B. Urbas, *Can. J. Chem.*, **46**, 49 (1968).

44. K. A. Caldwell, S. P. Raman, and L. Anderson, *Nature*, **199**, 373 (1963).

45. H. Shibata, D. Nishimura, N. Kurihara, and M. Nakajima, *Agr. Biol. Chem.* (Tokyo), **32**, 1002 (1968).

46. P. B. Nicholls, *Planta*, **72**, 258 (1967).

47. P. Fleury and L. LeDizet, *Bull. Soc. Chim. Biol.*, **37**, 1099 (1955).

48. J. C. P. Schwarz, *Chem. Ind.* (London), 1388 (1955).

49. S. J. Angyal and J. E. Klavins, *Aust. J. Chem.*, **14**, 577 (1961).

49a. P. Szabó, *in* "Advances in Chemistry Series" (R. F. Gould, Ed.), American Chemical Society, Washington, D.C., 1968, No. 74, p. 94.

49b. G. R. Barker, *J. Chem. Soc.*, 624 (1960).

49c. S. R. Sarfati and P. Szabó, *Carbohyd. Res.*, **11**, 571 (1969).

50. O. Gelormini and N. E. Artz, *J. Amer. Chem. Soc.*, **52**, 2483 (1930).

51. F. A. Hoglan and E. Bartow, *J. Amer. Chem. Soc.*, **62**, 2397 (1940).

52. P. W. Preisler and L. Berger, *J. Amer. Chem. Soc.*, **64**, 67 (1942).

53. A. J. Fatiadi, H. S. Isbell, and W. F. Sager, *J. Res. Nat. Bur. Stand.*, **A67**, 153 (1963).

54. T. Posternak, *Helv. Chim. Acta*, **19**, 1333 (1936).

55. A. J. Fatiadi and H. S. Isbell, *J. Res. Nat. Bur. Stand.*, **A68**, 287 (1964).

56. E. V. Eastcott, *J. Phys. Chem.*, **32**, 1094 (1928).

57. L. Anderson and K. E. Wolter, *Ann. Rev. Plant Physiol.*, **17**, 209 (1966).

58. D. W. Woolley, *J. Biol. Chem.*, **139**, 29 (1941).

59. M. H. McCormick, P. N. Harris, and C. A. Anderson, *J. Nutr.*, **52**, 337 (1954).

60. M. E. Reid, *Proc. Soc. Exp. Biol. Med.*, **85**, 547 (1954).

61. L. Anderson, R. H. Coots, and J. W. Halliday, *J. Nutr.*, **64**, 167 (1958).

61a. M. R. Hokin and L. E. Hokin, *in* "Metabolism and Physiological Significance of Lipids" (R. M. C. Dawson and D. N. Rhodes, Eds.), Wiley, New York, 1964, p. 423.

61b. K. Lembach and F. C. Charalampous, *J. Biol. Chem.*, **242**, 2606 (1967).

61c. R. K. Morton and J. K. Raison, *Nature*, **200**, 429 (1963).

61d. W. Tanner and O. Kandler, *Eur. J. Biochem.*, **4**, 233 (1968).

61e. L. Lehle, W. Tanner, and O. Kandler, *Hoppe Seyler's Z. Physiol. Chem.*, **351**, 1494 (1970).

62. I.-W. Chen and F. C. Charalampous, *J. Biol. Chem.*, **239**, 1905 (1964).

63. H. Kindl, J. Biedl-Neubacher, and O. Hoffmann-Ostenhof, *Biochem. Z.*, **341**, 157 (1965).

64. F. A. Loewus, *Phytochemistry*, **2**, 109 (1963).

65. H. Kindl, R. Scholda, and O. Hoffmann-Ostenhof, *Angew. Chem.*, **78**, 198 (1966); *Int. Ed. Engl.*, **5**, 165 (1966).

66. F. Eisenberg, Jr., A. H. Bolden, and F. A. Loewus, *Biochem. Biophys. Res. Commun.*, **14**, 419 (1964).

67. F. Eisenberg, Jr., *J. Biol. Chem.*, **242**, 1375 (1967).

68. I.-W. Chen and F. C. Charalampous, *J. Biol. Chem.*, **241**, 2194 (1966).

68a. I.-W. Chen and F. C. Charalampous, *J. Biol. Chem.*, **240**, 3507 (1965).

69. J. E. G. Barnett and D. L. Corina, *Biochem. J.*, **108**, 125 (1968).

69a. W. R. Sherman, M. A. Stewart, and M. Zinbo, *J. Biol. Chem.*, **244**, 5703 (1969).

69b. D. E. Kiely and H. G. Fletcher, Jr., *J. Org. Chem.*, **34**, 1386 (1969).

69c. I. Wagner, H. Hofmann, and O. Hoffmann-Ostenhof, *Hoppe-Seyler's Z. Physiol. Chem.*, **350**, 1460, 1465 (1969).

69d. W. H. Horner and I. H. Thaker, *Biochim. Biophys. Acta*, **165**, 306 (1968).

70. F. C. Charalampous, *J. Biol. Chem.*, **235**, 1286 (1960).

70a. K. M. Gruhner and O. Hoffmann-Ostenhof, *Hoppe-Seyler's Z. Physiol. Chem.*, **347**, 278 (1966).

71. S. Hollmann and O. Touster, "Nonglycolytic Pathways of Metabolism of Glucose," Academic Press, New York, 1964, p. 90.

72. F. Loewus, ref. 76b, p. 577.

73. P. Dworsky and O. Hoffmann-Ostenhof, *Biochem. Z.*, **343**, 394 (1965).
74. H. Paulus and E. P. Kennedy, *J. Biol. Chem.*, **235**, 1303 (1960).
75. H. Brockerhoff and C. E. Ballou, *J. Biol. Chem.*, **237**, 49 (1962).
76. C.Prottey, J. G. Salway, and J. N. Hawthorne, *Biochim. Biophys. Acta*, **164**, 238 (1968).
76a. P. Brennan and C. E. Ballou, *J. Biol. Chem.*, **243**, 2975 (1968).
76b. F. Eisenberg, Jr., Ed., "Cyclitols and Phosphoinositides: Chemistry, Metabolism, and Function," *Ann. N. Y. Acad. Sci.*, **165**, 509–819 (1969).
77. S. J. Angyal, *Quart. Rev.* (London), **11**, 212 (1957).
78. I. N. Rabinowitz and J. Kraut, *Acta Crystallogr.*, **17**, 159 (1964).
78a. G. A. Jeffrey and H. S. Kim, *Carbohyd. Res.*, **15**, 310 (1970).
78b. A. Buchs, E. Charollais, and T. Posternak, *Helv. Chim. Acta*, **51**, 695 (1968).
78c. W. R. Sherman, N. C. Eilers, and S. L. Goodwin, *Org. Mass Spectrom.*, **3**, 829 (1970).
79. S. J. Angyal and D. J. McHugh, *Chem. Ind.* (London), 1147 (1956).
80. S. J. Angyal and D. J. McHugh, *J. Chem. Soc.*, 1423 (1957).
80a. T. Posternak, E. A. C. Lucken, and A. Szente, *Helv. Chim. Acta*, **50**, 326 (1967).
81. G. E. McCasland, *Advan. Carbohyd. Chem.*, **20**, 11 (1965).
81a. F. W. Lichtenthaler and P. Emig, *Carbohyd. Res.*, **7**, 121 (1968).
81b. A. Hasegawa and H. Z. Sable, *J. Org. Chem.*, **33**, 1604 (1968).
81c. D. E. Dorman, S. J. Angyal, and J. D. Roberts, *J. Amer. Chem. Soc.*, **92**, 1351 (1970).
81d. E. Haslam and M. J. Turner, *J. Chem. Soc.* (C), 1496 (1971).
81e. J. Corse and R. E. Lundin, *J. Org. Chem.*, **35**, 1904 (1970).
82. A. J. Fatiadi and W. F. Sager, *Org. Syn.*, **42**, 66, 90 (1962).
83. R. Nietzki and T. Benckiser, *Ber.*, **18**, 1833 (1885).
84. S. J. Angyal and D. J. McHugh, *J. Chem. Soc.*, 3682 (1957).
85. F. Weygand and E. Schulze, *Z. Naturforsch.*, **B11**, 370 (1956).
86. Yu. K. Yur'ev and N. S. Zefirov, *Zh. Obshch. Khim.*, **31**, 685 (1961) (Engl. transl., p. 629); *Chem. Abstr.*, **55**, 24573 (1961).
86a. S. Sarel and H. Kowarsky, *Bull. Res. Council Israel*, **A9**, 72 (1960).
87. R. Criegee and P. Becher, *Ber.*, **90**, 2516 (1957).
88. M. Nakajima, I. Tomida, A. Hashizume, and S. Takei, *Ber.*, **89**, 2224 (1956).
89. M. Nakajima, I. Tomida, and S. Takei, *Ber.*, **92**, 163 (1959).
90. M. Nakajima, I. Tomida, and S. Takei, *Ber.*, **90**, 246 (1957).
91. M. Nakajima, A. Hasegawa, and N. Kurihara, *Ber.*, **95**, 2708 (1962).
92. M. Nakajima, I. Tomida, N. Kurihara, and S. Takei, *Ber.*, **92**, 173 (1959).
93. M. Nakajima, N. Kurihara, and T. Ogino, *Ber.*, **96**, 619 (1963).
94. M. Nakajima and N. Kurihara, *Ber.*, **94**, 515 (1961).
95. F. W. Lichtenthaler, *Angew. Chem.*, **76**, 84 (1964); *Int. Ed. Engl.*, **3**, 211 (1964).
96. J. M. Grosheintz and H. O. L. Fischer, *J. Amer. Chem. Soc.*, **70**, 1479 (1948).
97. F. W. Lichtenthaler, *Ber.*, **94**, 3071 (1961).
97a. H. H. Baer and W. Rank, *Can. J. Chem.*, **43**, 3462 (1965).
98. F. W. Lichtenthaler, *Angew. Chem.*, **75**, 93 (1963); *Int. Ed. Engl.*, **1**, 662 (1962).
99. H. H. Baer and W. Rank, *Can. J. Chem.*, **43**, 3330 (1965).
99a. F. W. Lichtenthaler, *Newer Methods Preparative Org. Chem.*, **4**, 155 (1968).
100. S. J. Angyal and M. E. Tate, *J. Chem. Soc.*, 6949 (1965).
101. S. J. Angyal and T. S. Stewart, *Aust. J. Chem.*, **19**, 1683 (1966).
101a. S. J. Angyal, M. H. Randall, and M. E. Tate, *J. Chem. Soc.*, 919 (1967).
101b. B. A. Klyashchitskii, G. D. Strakhova, V. I. Shvets, S. D. Sokolov, and N. A. Preobrazhenskii, *J. Gen. Chem. USSR*, **40**, 214 (1970).
101c. B. A. Klyashchitskii, V. V. Pimenova, V. I. Shvets, S. D. Sokolov, and N. A. Preobrazhenskii, *J. Gen. Chem. USSR*, **39**, 2311 (1969).
101d. T. Suami, F. W. Lichtenthaler, and S. Ogawa, *Bull. Chem. Soc. Jap.*, **40**, 1488 (1967).

101e. D. Mercier, J. E. G. Barnett, and S. D. Géro, *Tetrahedron*, **25**, 5681 (1969).

102. S. J. Angyal, V. J. Bender, P. T. Gilham, R. M. Hoskinson, and M. E. Pitman, *Aust. J. Chem.*, **20**, 2109 (1967).

102a. R. S. Tipson and A. Cohen, *Carbohyd. Res.*, **1**, 338 (1965).

103. S. J. Angyal and P. T. Gilham, *J. Chem. Soc.*, 375 (1958).

104. S. J. Angyal and P. T. Gilham, *J. Chem. Soc.*, 3691 (1957).

105. S. J. Angyal, V. Bender, and J. H. Curtin, *J. Chem. Soc.*, *C*, 798 (1966).

106. S. J. Angyal and N. K. Matheson, *J. Amer. Chem. Soc.*, **77**, 4343 (1955).

107. S. J. Angyal and R. M. Hoskinson, *J. Chem. Soc.*, 2043 (1963).

108. G. E. McCasland and E. C. Horswill, *J. Amer. Chem. Soc.*, **75**, 4020 (1953).

109. G. E. McCasland and E. C. Horswill, *J. Amer. Chem. Soc.*, **76**, 2373 (1954).

110. G. E. McCasland and J. M. Reeves, *J. Amer. Chem. Soc.*, **77**, 1812 (1955).

111. G. E. McCasland, S. Furuta, and A. Furst, *J. Org. Chem.*, **29**, 724 (1964).

112. P. J. Beynon, P. M. Collins, D. Gardiner, and W. G. Overend, *Carbohyd. Res.*, **6**, 431 (1968).

113. B. T. Lawton, W. A. Szarek, and J. K. N. Jones, *Carbohyd. Res.*, **10**, 456 (1969).

114. W. W. Epstein and F. W. Sweat, *Chem. Rev.*, **67**, 247 (1967).

114a. H. Fukami, H.-S. Koh, T. Sakata, and M. Nakajima, *Tetrahedron Lett.*, 4771 (1967).

115. T. Posternak, *Biochem. Prepn.*, **2**, 57 (1952).

116. B. Magasanik, R. E. Franzl, and E. Chargaff, *J. Amer. Chem. Soc.*, **74**, 2618 (1952).

117. L. Anderson, R. Takeda, S. J. Angyal, and D. J. McHugh, *Arch. Biochem. Biophys.*, **78**, 518 (1958).

118. K. Heyns and H. Paulsen, *Angew. Chem.*, **69**, 600 (1957).

119. G. G. Post and L. Anderson, *J. Amer. Chem. Soc.*, **84**, 471 (1962).

120. G. E. McCasland, *Methods Carbohyd. Chem.*, **1**, 291 (1962).

121. T. Posternak, *Methods Carbohyd. Chem.*, **1**, 294 (1962).

122. K. Heyns, H. Gottschalk, and H. Paulsen, *Ber.*, **95**, 2660 (1962).

123. A. J. Fatiadi, *Carbohyd. Res.*, **1**, 489 (1966).

124. K. Heyns, J. Lenz, and H. Paulsen, *Ber.*, **95**, 2964 (1962).

125. H. E. Carter, C. Belinskey, R. K. Clark, Jr., E. H. Flynn, B. Lytle, G. E. McCasland, and M. Robbins, *J. Biol. Chem.*, **174**, 415 (1948).

125a. A. J. Fatiadi, *Carbohyd. Res.*, **8**, 135 (1968).

125b. H. S. Isbell and A. J. Fatiadi, *Carbohyd. Res.*, **2**, 204 (1966).

126. T. Posternak, *Helv. Chim. Acta*, **29**, 1991 (1946).

127. T. Posternak, *Helv. Chim. Acta*, **24**, 1045 (1941).

128. N. Z. Stanacev and M. Kates, *J. Org. Chem.*, **26**, 912 (1961).

129. P. A. J. Gorin, *Can. J. Chem.*, **42**, 1748 (1964).

130. T. Posternak and J. Deshusses, *Helv. Chim. Acta*, **44**, 2088 (1961).

131. H. S. Isbell, *Ann. Rev. Biochem.*, **12**, 205 (1943).

132. G. E. McCasland, M. O. Naumann, and L. J. Durham, *J. Org. Chem.*, **33**, 4220 (1968).

133. P. Fleury, J. Courtois, W. C. Hammam, and L. LeDizet, *Bull. Soc. Chim. Fr.*, 1307 (1955).

133a. T. L. Nagabhushan, *Can. J. Chem.*, **48**, 383 (1970).

133b. H. Kindl, ref. 76b, p. 615.

134. T. Posternak and H. Friedli, *Helv. Chim. Acta*, **36**, 251 (1953).

135. T. Posternak and D. Reymond, *Helv. Chim. Acta*, **38**, 195 (1955).

135a. H. Z. Sable, K. A. Powell, H. Katchian, C. B. Niewoehner, and S. B. Kadlec, *Tetrahedron*, **26**, 1509 (1970).

136. H. Stetler and K. H. Steinacker, *Ber.*, **85**, 451 (1952).

137. G. E. McCasland, M. O. Naumann, and L. J. Durham, *J. Org. Chem.*, **31**, 3079 (1966).

578   L. ANDERSON

138. H. Z. Sable, T. Anderson, B. Tolbert, and T. Posternak, *Helv. Chim. Acta*, **46**, 1157 (1963).
139. A. Hasegawa and H. Z. Sable, *J. Org. Chem.*, **31**, 4149 (1966).
140. A. Hasegawa and H. Z. Sable, *J. Org. Chem.*, **31**, 4154 (1966).
141. A. Hasegawa and H. Z. Sable, *J. Org. Chem.*, **31**, 4161 (1966).
141a. B. Tolbert, R. Steyn, J. A. Franks, Jr., and H. Z. Sable, *Carbohyd. Res.*, **5**, 62 (1967).
141b. R. Steyn and H. Z. Sable, *Tetrahedron*, **25**, 3579 (1969).
141c. F. G. Cocu and T. Posternak, *Helv. Chim. Acta*, **54**, 1676 (1971).
142. H. Z. Sable, W. M. Ritchey, and J. E. Nordlander, *Carbohyd. Res.*, **1**, 10 (1965).
143. H. Z. Sable, W. M. Ritchey, and J. E. Nordlander, *J. Org. Chem.*, **31**, 3771 (1966).
144. R. U. Lemieux and M. L. Wolfrom, *Advan. Carbohyd. Chem.*, **3**, 337 (1948).
145. O. Wintersteiner and A. Klingsberg, *J. Amer. Chem. Soc.*, **73**, 2917 (1951).
146. H. Straube-Rieke, H. A. Lardy, and L. Anderson, *J. Amer. Chem. Soc.*, **75**, 694 (1953).
147. B. Bannister and A. D. Argoudelis, *J. Amer. Chem. Soc.*, **85**, 119 (1963).
148. J. D. Dutcher, *Advan. Carbohyd. Chem.*, **18**, 259 (1963).
149. A. L. Johnson, R. H. Gourlay, D. S. Tarbell, and R. L. Autrey, *J. Org. Chem.*, **28**, 300 (1963).
150. P. F. Wiley, M. V. Sigal, Jr., and O. Weaver, *J. Org. Chem.*, **27**, 2793 (1962).
151. M. Nakajima, A. Hasegawa, and N. Kurihara, *Ann.*, **689**, 235 (1965).
152. F. W. Lichtenthaler and H. O. L. Fischer, *J. Amer. Chem. Soc.*, **83**, 2005 (1961).
152a. F. W. Lichtenthaler, P. Voss, and N. Majer, *Angew. Chem.*, **81**, 221 (1969); *Intern. Ed. Engl.*, **8**, 211 (1969); compare T. Nakagawa, T. Sakakibara, and F. W. Lichtenthaler, *Bull. Chem. Soc. Jap.*, **43**, 3861 (1970).
152b. H. H. Baer and M. C. T. Wang, *Can. J. Chem.*, **46**, 2793 (1968).
152c. A. Hasegawa and H. Z. Sable, *Tetrahedron*, **25**, 3567 (1969).
153. G. R. Allen, Jr., *J. Amer. Chem. Soc.*, **79**, 1167 (1957).
154. M. L. Wolfrom, J. Radell, R. M. Husband, and G. E. McCasland, *J. Amer. Chem. Soc.*, **79**, 160 (1957).
154a. T. Suami, S. Ogawa, S. Naito, and H. Sano, *J. Org. Chem.*, **33**, 2831 (1968).
154b. T. Suami, S. Ogawa, and H. Sano, *Bull. Chem. Soc. Jap.*, **43**, 1843 (1970).
155. T. Suami, F. W. Lichtenthaler, and S. Ogawa, *Bull. Chem. Soc. Jap.*, **39**, 170 (1966) and papers there cited.
155a. T. Suami and K. Yabe, *Bull. Chem. Soc. Jap.*, **39**, 1931 (1966).
155b. T. Suami, S. Ogawa, and M. Uchida, *Bull. Chem. Soc. Jap.*, **43**, 3577 (1970)
156. F. W. Lichtenthaler and H. Leinert, *Ber.*, **99**, 903 (1966).
157. G. Quadbeck and E. Röhm, *Ber.*, **89**, 1645 (1956).
157a. D. Dijkstra, *Rec. Trav. Chim.*, **87**, 161 (1968).
157b. H. H. Baer and R. J. Yu, *Tetrahedron Lett.*, 807 (1967).
158. M. Nakajima, A. Hasegawa, and F. W. Lichtenthaler, *Ann.*, **669**, 75 (1963).
159. M. Nakajima, A. Hasegawa, and F. W. Lichtenthaler, *Ann.*, **680**, 21 (1964).
160. M. Nakajima, A. Hasegawa, and T. Kurokawa, *Ann.*, **689**, 229 (1965).
161. M. Nakajima, N. Kurihara, A. Hasegawa, and T. Kurokawa, *Ann.*, **689**, 243 (1965).
162. K. Heyns and H. Paulsen, *Ber.*, **89**, 1152 (1956).
163. M. L. Wolfrom, S. M. Olin, and W. J. Polglase, *J. Amer. Chem. Soc.*, **72**, 1724 (1950).
164. T. Suami and S. Ogawa, *Bull. Chem. Soc. Jap.*, **40**, 1925 (1967); S. Ogawa, T. Abe, H. Sano, and T. Suami, *ibid.*, **40**, 2405 (1967).
165. F. W. Lichtenthaler, H. Leinert, and T. Suami, *Ber.*, **100**, 2383 (1967).
166. W. H. Horner and G. A. Russ, *Biochim. Biophys. Acta*, **192**, 352 (1969).
167. J. B. Walker and M. S. Walker, *Biochemistry*, **6**, 3821 (1967).
167a. J. B. Walker and M. S. Walker, *Biochemistry*, **8**, 763 (1969).

167b. N. Kurihara, T. Kurokawa, and M. Nakajima, *Agr. Biol. Chem. (Tokyo)*, **31**, 1166 (1967).

167c. N. Kurihara, K. Hayashi, and M. Nakajima, *Agr. Biol. Chem. (Tokyo)*, **33**, 256 (1969).

167d. T. Suami, F. W. Lichtenthaler, S. Ogawa, Y. Nakashima, and H. Sano, *Bull. Chem. Soc. Jap.*, **41**, 1014 (1968).

167e. A. Hasegawa, N. Kurihara, D. Nishimura, and M. Nakajima, *Agr. Biol. Chem. (Tokyo)*, **32**, 1123, 1130 (1968).

167f. S. Umezawa, K. Tatsuta, and S. Koto, *Bull. Chem. Soc. Jap.*, **42**, 533 (1969), and accompanying papers.

167g. T. Ito, E. Akita, T. Tsuruoka, and T. Niida, *Agr. Biol. Chem. (Tokyo)*, **34**, 980 (1970).

168. T. Posternak, *Helv. Chim. Acta*, **33**, 1597 (1950).

168a. S. J. Angyal and J. S. Murdoch, *Aust. J. Chem.*, **22**, 2417 (1969).

169. T. Posternak, W. H. Schopfer, and R. Huguenin, *Helv. Chim. Acta*, **40**, 1875 (1957).

170. G. I. Drummond, J. N. Aronson, and L. Anderson, *J. Org. Chem.*, **26**, 1601 (1961).

171. T. Posternak, *Helv. Chim. Acta*, **27**, 457 (1944).

172. T. Posternak and J.-G. Falbriard, *Helv. Chim. Acta*, **43**, 2142 (1960).

173. T. Posternak and D. Reymond, *Helv. Chim. Acta*, **36**, 1370 (1953).

173a. G. Wöber and O. Hoffmann-Ostenhof, *Eur. J. Biochem.*, **17**, 393 (1970).

174. F. W. Lichtenthaler and H. K. Yahya, *Carbohyd. Res.*, **5**, 485 (1967).

175. G. E. McCasland, S. Furuta, and L. J. Durham, *J. Org. Chem.*, **31**, 1516 (1966).

175a. G. E. McCasland, S. Furuta, and L. J. Durham, *J. Org. Chem.*, **33**, 2835, 2841 (1968).

175b. Y. F. Shealy and J. D. Clayton, *J. Amer. Chem. Soc.*, **91**, 3075 (1969).

175c. S. Horii, T. Iwasa, E. Mizuta, and Y. Kameda, *J. Antibiot.* **24**, 59 (1971).

176. B. A. Bohm, *Chem. Rev.*, **65**, 435 (1965).

177. D. B. Sprinson, *Advan. Carbohyd. Chem.*, **15**, 235 (1960).

178. M. L. Scarpati and M. Guiso, *Tetrahedron Lett.*, 2851 (1964).

178a. J. Wolinsky, R. Novak, and R. Vasileff, *J. Org. Chem.*, **29**, 3596 (1964).

179. D. Mercier, J. Lebaul, J. Cléophax, and S. D. Géro, *Carbohyd. Res.*, **20**, 299 (1971).

179a. R. Grewe and S. Kersten, *Ber.*, **100**, 2546 (1967).

179b. H. J. Bestmann and H. A. Heid, *Angew. Chem.*, **83**, 329 (1971).

180. S. J. Angyal and R. M. Hoskinson, *J. Chem. Soc.*, 2985 (1962).

181. L. Anderson, K. H. Brackmann, and D. B. Finkelstein, unpublished data, 1963.

182. S. J. Angyal, M. E. Tate, and S. D. Gero, *J. Chem. Soc.*, 4116 (1961).

183. S. J. Angyal, G. C. Irving, D. Rutherford, and M. E. Tate, *J. Chem. Soc.*, 6662 (1965).

183a. F. H. Bisset, M. E. Evans, and F. W. Parrish, *Carbohyd. Res.*, **5**, 184 (1967).

184. S. J. Angyal and R. M. Hoskinson, *J. Chem. Soc.*, 2991 (1962).

185. S. J. Angyal and G. J. H. Melrose, *J. Chem. Soc.*, 6501 (1965).

186. S. J. Angyal and G. J. H. Melrose, *J. Chem. Soc.*, 6494 (1965).

187. S. J. Angyal, P. T. Gilham, and G. J. H. Melrose, *J. Chem. Soc.*, 5252 (1965).

188. S. J. Angyal, P. A. J. Gorin, and M. E. Pitman, *J. Chem. Soc.*, 1807 (1965).

189. E. J. Hedgley and H. G. Fletcher, Jr., *J. Amer. Chem. Soc.*, **84**, 3726 (1962).

189a. H. Paulsen and H. Behre, *Ber.*, **104**, 1281, 1299 (1971); H. Paulsen, H. Behre, and C.-P. Herold, *Fortschr. Chem. Forsch.*, **14**, 472 (1970).

190. H. Paulsen, *Advan. Carbohyd. Chem. Biochem.*, **26**, 127 (1971).

191. F. H. Newth, *Quart. Rev. (London)*, **13**, 30 (1959).

AUTHOR INDEX

Numbers in parentheses are reference numbers and indicate that an author's work is referred to, although his name is not cited in the text. Numbers in italics show the page on which the complete reference is listed.

SUBJECT INDEX

A

Acetalation
  of acyclic sugar derivatives, 394–396
  of alditols, 510
  of cyclic sugar derivatives, 396–398
  of cyclitols, 569
Acetals
  cyclic, of cyclitols, 569
    of sugars and alditols, 391–402
    syntheses with, 398–401
  dithio-, demercaptalation of, 360, 361
    formation of sugar, 356
    reactions of sugar, 357–359
    1-thioglycosides from, 342
  glycoside synthesis from, 287–291
  hydrolysis of, 317
  monothio-, preparation of, 359
  preparation of sugar, 360
  thio-, glycoside synthesis from, 291
Acetic acid, trifluoro-, sugar esters, 225
Acetolysis, of aldose dithioacetals, 358
Acetoxonium ion, in preparation of sugar
    acetates, 221
Acetylation, and anomerization of sugars,
    220–222
Acetyl group
  analysis for, 223
  migration of, in sugar acetates, 228
Acids, effect on mutarotation, 165–186
Acrosazone, α- and β-, history, 114
Acrose, history, 114
Actinamine, structure of, 563
Acyl groups
  migration of, 228
    in acylated glycosyl halides, 245
    in cyclitols, 570
Acyloxonium-ion rearrangements, of sugar
    benzoates, 224
Adonitol, see Ribitol
Agar
  mercaptolysis and structure of, 356
  pyrolysis product from, 428

Aglycon, definition, 280
Aldaric acids, stereoisomers, number of,
    13
Aldehyde–aldehydrol equilibria, and nuclear
    magnetic resonance spectra in aqueous
    solution, 376, 385
Aldehydrols
  equilibria with aldehydes, 376, 385
  structure, mutarotation and nomencla-
    ture, 364–368
Aldgamycin E, structure of component
    aldgarose, 230
Aldgarose, structure of, 230
Alditols
  acetalation and etherification of, 510
  biochemistry of, 512
  catalytic isomerization of, 499
  conformation of, 213
  cyclic acetals, 391–402
  deoxy, preparation of, 511
  and derivatives, 479–518
  determination of, qualitative and quanti-
    tative, 510
  esterification of, 503
  nomenclature and configuration of higher,
    52
  oxidation of, 151–154, 509
  paper electrophoresis of, 504–508
  reduction of, 510
  stereoisomers, number of, 13
  structure and configuration of, proof of,
    501–503
  syntheses of, 498
  tritium-labeled, 156
Aldofuranoses, homomorphology of,
    51
Aldofuranosides, methyl, acidic hydrolysis,
    mechanism of, 320
Aldoheptoses, occurrence, preparation and
    synonyms, 74, 99–101
Aldohexofuranoses, 3,6-anhydro-, synthe-
    sis of, 458

621

H